# Integrated Network Management V

# IFIP – The International Federation for Information Processing

IFIP was founded in 1960 under the auspices of UNESCO, following the First World Computer Congress held in Paris the previous year. An umbrella organization for societies working in information processing, IFIP's aim is two-fold: to support information processing within its member countries and to encourage technology transfer to developing nations. As its mission statement clearly states,

> IFIP's mission is to be the leading, truly international, apolitical organization which encourages and assists in the development, exploitation and application of information technology for the benefit of all people.

IFIP is a non-profitmaking organization, run almost solely by 2500 volunteers. It operates through a number of technical committees, which organize events and publications. IFIP's events range from an international congress to local seminars, but the most important are:

- the IFIP World Computer Congress, held every second year;
- open conferences;
- working conferences.

The flagship event is the IFIP World Computer Congress, at which both invited and contributed papers are presented. Contributed papers are rigorously refereed and the rejection rate is high.

As with the Congress, participation in the open conferences is open to all and papers may be invited or submitted. Again, submitted papers are stringently refereed.

The working conferences are structured differently. They are usually run by a working group and attendance is small and by invitation only. Their purpose is to create an atmosphere conducive to innovation and development. Refereeing is less rigorous and papers are subjected to extensive group discussion.

Publications arising from IFIP events vary. The papers presented at the IFIP World Computer Congress and at open conferences are published as conference proceedings, while the results of the working conferences are often published as collections of selected and edited papers.

Any national society whose primary activity is in information may apply to become a full member of IFIP, although full membership is restricted to one society per country. Full members are entitled to vote at the annual General Assembly, National societies preferring a less committed involvement may apply for associate or corresponding membership. Associate members enjoy the same benefits as full members, but without voting rights. Corresponding members are not represented in IFIP bodies. Affiliated membership is open to non-national societies, and individual and honorary membership schemes are also offered.

# Integrated Network Management V

## Integrated management in a virtual world

**Proceedings of the Fifth IFIP/IEEE International Symposium on Integrated Network Management San Diego, California, U.S.A., May 12-16, 1997**

Sponsored by IFIP TC6 WG6 on Network Management, co-sponsored by the IEEE Communications Society CNOM

Edited by

**Aurel A. Lazar**
*Department of Electrical Engineering and*
*Center for Telecommunications Research*
*Columbia University*
*New York, U.S.A.*

**Roberto Saracco**
*CSELT*
*Torino, Italy*

and

**Rolf Stadler**
*Department of Electrical Engineering and*
*Center for Telecommunications Research*
*Columbia University*
*New York, U.S.A.*

Published by Chapman & Hall on behalf of the
International Federation for Information Processing (IFIP)

 **CHAPMAN & HALL**

London · Weinheim · New York · Tokyo · Melbourne · Madras

**Published by Chapman & Hall, 2–6 Boundary Row, London SE1 8HN, UK**

Chapman & Hall, 2–6 Boundary Row, London SE1 8HN, UK

Chapman & Hall GmbH, Pappelallee 3, 69469 Weinheim, Germany

Chapman & Hall USA, 115 Fifth Avenue, New York, NY 10003, USA

Chapman & Hall Japan, ITP-Japan, Kyowa Building, 3F, 2-2-1 Hirakawacho, Chiyoda-ku, Tokyo 102, Japan

Chapman & Hall Australia, 102 Dodds Street, South Melbourne, Victoria 3205, Australia

Chapman & Hall India, R. Seshadri, 32 Second Main Road, CIT East, Madras 600 035, India

First edition 1997

© 1997 IFIP

Printed in Great Britain by TJ International, Padstow, Cornwall

ISBN  0 412 80960 5

A catalogue record for this book is available from the British Library

∞ Printed on permanent acid-free text paper, manufactured in accordance with ANSI/NISO Z39.48-1992 and ANSI/NISO Z39.48-1984 (Permanence of Paper).

# CONTENTS

# Preface

Welcome to IM'97! We hope you had the opportunity to attend the Conference in beautiful San Diego. If that was the case, you will want to get back to these proceedings for further readings and reflections. You'll find e-mail addresses of the main author of each paper, and you are surely encouraged to get in touch for further discussions. You can also take advantage of the CNOM (Committee on Network Operation and Management) web site where a virtual discussion agora has been set up for IM'97 (URL: http://www.cselt.stet.it/CNOMWWW/IM97.html). At this site you will find a brief summary of discussions that took place in the various panels, and slides that accompanied some of the presentations--all courtesy of the participants. If you have not been to the Conference, leafing through these proceedings may give you food for thought. Hopefully, you will also be joining the virtual world on the web for discussions with authors and others who were at the Conference.

At IM'97 the two worlds of computer networks and telecommunications systems came together, each proposing a view to management that stems from their own paradigms. Each world made clear the need for end-to-end management and, therefore, each one stepped into the other's field. We feel that there is no winner but a mutual enrichment. The time is ripe for integration and it is likely that the next Conference will bear its fruit.

The technical papers presented in this volume were selected from among 138 submissions through a rigorous review process. Each paper was reviewed by at least 3 referees and carefully evaluated by the program committee to ensure the highest quality. The contents of the proceedings include the 56 selected submissions, and abstracts from the keynote addresses presented by leading visionaries of integrated systems management, short descriptions of 6 panels involving some of the best technical experts in the field, and abstracts of papers presented as posters. The table of contents is organized following the conference framework (tracks/sessions). Three main tracks, including sessions, have been identified as follows:

❒ *Track I*
   ○ New Approaches to Customer Network Management
   ○ Web and Java Approaches
   ○ TMN - Today and Tomorrow
   ○ Standards, Protocols and Interoperability
   ○ Management Paradigms
❒ *Track II*
   ○ Broadband Network Management
   ○ Service Management
   ○ Management and Control of ATM Networks
   ○ Multimedia Services, Applications, Policies
   ○ Testing the Management Information Base

❐ *Track III*

    ○ CORBA-based Management

    ○ Network Monitoring Policies

    ○ Fault Management - I

    ○ Information Models

    ○ Fault Management - II

    ○ Intelligent Agents

❐ *Keynote and Panel Sessions*

    ○ Keynote Addresses

    ○ Panel Sessions

❐ *Posters*

When we read the proceedings, new challenges become apparent and new technologies appear to be full of promise. Managing new services on many different types of networks with many different players are among the challenges. Along with them is the need for seamless interoperability and independence between services and networks, between market strategies, niche markets and global markets, and between one-point management and independent management. These challenges are discussed in the sessions on Broadband Network Management, Service Management, Fault Management, Management and Control of ATM Networks, Network Monitoring Policies, Standards, Protocols and Interoperability, Information Models, Multimedia Services, Applications, and Policies, and Testing the Management Information Base. New technologies are promising different ways of approaching management issues, thereby leading to novel management requirements. These are discussed in the sessions on New Approaches to Customer Network Management, Web and Java Approaches, Corba-based Management, TMN Today and Tomorrow, Management Paradigms and Intelligent Agents.

The work included in this volume represents the collective contributions of authors, dedicated reviewers and a committed program committee. We wish to extend our gratitude to the authors of the technical papers and posters, without whom this symposium would not have been possible, the efforts of truly tireless reviewers, and the members of the Program Committee for their help with paper solicitation and review.

Special thanks go to Andreas Eggenberger for helping with the design and implementation of the paper handling software. Last but not least, we thank Constantin Adam, Cristina Aurrecoechea, Jit Biswas, Marco Borla, Cristina Calderaro, and Andreas Eggenberger for their help with the handling of paper submissions and many other tasks.

Aurel A. Lazar, Roberto Saracco and Rolf Stadler
January 30, 1997

# Symposium Committees

## ORGANIZING COMMITTEE MEMBERS

| | |
|---|---|
| Dan Stokesberry | U.S.A., General Chair |
| Seraphin B. Calo | IBM T.J. Watson Research Center, U.S.A., Vice-Chair, Acting Chair |
| Aurel A. Lazar | Columbia University, U.S.A., Program Co-Chair |
| Roberto Saracco | CSELT, Italy, Program Co-Chair |
| Rolf Stadler | Columbia University, U.S.A., Program Committee |
| Yvonne Hildebrand | Copper Mountain Networks, U.S.A., Local Arrangements Chair |
| Joseph Diamand | General Instruments, U.S.A., Vendor Program Chair |
| Varoozh Harikian | IBM Int'l Technical Support Organization, U.S.A., Tutorials Chair |
| Joseph L. Hellerstein | IBM T.J. Watson Research Center, U.S.A., Secretary |
| Jim Herman | Northeast Consulting, U.S.A., Special Technical Events Chair |
| Branislav Meandzija | Meta Communications, Inc., U.S.A., Advisory Board Chair |
| Anne-Marie Lambert | Parlance Corporation (subsidiary of BBN), U.S.A., Publicity and Marketing |
| Kenneth J. Lutz | Bellcore, U.S.A., IEEE/CNOM Coordinator |
| Tom Stevenson | IEEE ComSoc, U.S.A., IEEE ComSoc Coordinator |
| Binay Sugla | AT&T Research, U.S.A., Treasurer |

## ADVISORY BOARD MEMBERS

| | |
|---|---|
| Lawrence Bernstein | Price Waterhouse and Network Programs, Inc., U.S.A. |
| Seraphin B. Calo | IBM T.J. Watson Research Center, U.S.A. |
| Branislav Meandzija | Meta Communications, Inc., U.S.A., Chair |
| Wolfgang Zimmer | GMD FIRST, Germany |
| Douglas N. Zuckerman | Bellcore, U.S.A. |

# PROGRAM COMMITTEE MEMBERS

*Members of the Organizing Committee and the Advisory Board are also members of the Program Committee.*

| | |
|---|---|
| Nikos G. Anerousis | AT&T Labs, U.S.A. |
| Joseph Betser | The Aerospace Corporation, U.S.A. |
| Jit Biswas | Institute of Systems Science, Singapore |
| Walter Buga | Bell Labs, Lucent Technologies, U.S.A. |
| Laura Cerchio | CSELT, Italy |
| William Donnelly | Broadcom Eireann Research, Ireland |
| Janusz Filipiak | University of Cracow, Poland |
| Kurt Geihs | University of Frankfurt, Germany |
| German Goldszmidt | IBM T.J. Watson Research Center, U.S.A. |
| Rodney M. Goodman | California Inst. of Technology, U.S.A. |
| Gita Gopal | HP Labs, U.S.A. |
| Shri Goyal | GTE Laboratories, U.S.A. |
| Sigmund Handelman | IBM T.J. Watson Research Center, U.S.A. |
| Satoshi Hasegawa | NEC Corporation, Japan |
| Heinz-Gerd Hegering | University of Munich, Germany |
| James W. Hong | Pohang University of Science and Technology, Korea |
| Gautam Kar | IBM T.J. Watson Research Center, U.S.A. |
| Shaygan Kheradpir | GTE Laboratories, U.S.A. |
| Wolfgang Kleinoeder | IBM Zurich Research Laboratory, Switzerland |
| Kenichi Kitami | NTT, Japan |
| Subrata Mazumdar | Bell Laboratories, U.S.A. |
| Keith McCloghrie | Cisco, U.S.A. |
| Paul S. Min | Washington University, U.S.A. |
| George Mouradian | AT&T Labs, U.S.A. |
| Shoichiro Nakai | NEC Corporation, Japan |
| Giovanni Pacifici | IBM T.J.Watson Research Center, U.S.A. |
| Jong-Tae Park | Kyungpook National University, Korea |
| George Pavlou | University College London, U.K. |
| Pradeep Ray | University of Technology, Australia |
| Yves Raynaud | Universite Paul Sabatier, France |
| Jan Roos | University of Pretoria, South Africa |
| Izhak Rubin | University of California at Los Angeles, U.S.A. |
| Veli Sahin | NEC America, U.S.A. |
| Stelios Sartzetakis | ICS-FORTH, Greece |
| Adarshpal S. Sethi | University of Delaware, U.S.A. |
| Morris Sloman | Imperial College London, U.K. |

## PROGRAM COMMITTEE MEMBERS (Continued)

| | |
|---|---|
| Liba Svobodova | IBM Zurich Research Laboratory, Switzerland |
| Mark Sylor | Digital Equipment Corporation, U.S.A. |
| Paolo Tiribelli | Sodalia, Italy |
| Mitsuru Tsuchida | Mitsubishi, Japan |
| Fabienne Vincent-Franc | Centre National des Etudes Spatiales, France |
| Carlos Becker Westphall | Federal University of SC, Brazil |
| Yechiam Yemini | Columbia University, U.S.A. |

# List of Reviewers

A. Abdulmalak
A. Adas
N. Anerousis
M.S.M. Annoni Notare
C. Aurrecoechea
B. Baer
E. Bagnasco
C. Becker Westphall
M. Bert
J. Betser
A. Biliris
R. Bisio
J. Biswas
D.J. Bobko
S. Borioni
P.G. Bosco
S. Brady
R. Brockett
W. Buga
S.B. Calo
D. Caswell
T.A. Cauble
L. Cerchio
M.C. Chan
T. Chandra
G. Chen
W. Choe
C.W. Choi
S. Chutani
W.J. Davis
A.S.M. De Franceschi
L. Deri
J. Dilley
W. Donnelly
G. Dreo Rodosek
F. Duda
B. Dvais
R. Farsi
L. Feldkhun
H. Fossa
B. Fraley
P. Garg
C. Gbaguidi
K. Geihs

J.M. Goett
G. Goldszmidt
R.M. Goodman
G. Gopal
S. Goyal
E. Grasso
D. Griffin
H. Gruender
S. Gugliermetti
S. Handelman
V. Harikian
S. Hasegawa
H.-G. Hegering
S. Heilbronner
K. Heiler
G. Hjalmtysson
Y. Hoffner
G. Jakobson
D. Jordaan
S. Kaetker
G. Kaleeswaran
C. Kalmanek
P. Kalyanasundaram
G. Kar
G. Karayannopoulos
G. Karjoth
M. Karsten
A. Keller
P. Kelly
K. Kitami
B. Kitson
W. Kleinoeder
D.S. Klett
R. Koch
Q. Kong
W. Korfhage
L.F. Kormann
M. Kosarchyn
A.A. Lazar
D. Lewis
K.-S. Lim
G. Lo Russo
C. Louison
W.P. Lu

E. Lupu
K.J. Lutz
F. Malabocchia
J.B. Mangueira Sobral
A. Mann
M. Mansouri-Samani
F. Marconcini
M. Marini
J. Martinka
S. Mazumdar
K. McCloghrie
B. Meandzija
P.S. Min
P. Mishra
K. Moore
M. Mountzia
G. Mouradian
S. Nakai
B. Neumair
R. Ordower
G. Pacifici
G. Pavlou
E. Perry
R. Pillai
P. Porras
E. Pring
A. Puder
P. Putter
S. Ramanathan
G. Ratta
P. Ray
Y. Raynaud
B.G. Riso
J. Roos
C. Rosado-Sosa
I. Rubin
R. Saracco
B. Sarikaya
S. Sartzetakis
T. Saydam
A. Schade
S. Schonberger
S. Schwerdtner
A. Sethi

## LIST OF REVIEWERS (Continued)

H.V. Shah
S.K. Sharma
C. Sherwin
C. Shim
J. Siwko
M. Sloman
R. Stadler
P. Steenekamp
N. Stoffel
B. Sugla

C. Sundaramurthy
L. Svobodova
M. Sylor
P. Tiribelli
T. Tiropanis
M. Tsuchida
J. Turek
A. Valderruten Vidal
F. Vincent-Franc

J. White
W. Whitt
R. Wies
Y. Yemini
S.C. Young
M. Zapf
D. Zhu
W. Zimmer
D.N. Zuckerman

# Introduction

# Integrated management in a virtual world

*Seraphin B. CALO,* IBM Research Division, U.S.A.

## 1. The Spirit Continues

During the two years since our last International Symposium on Integrated Network Management, ISINM '95 in Santa Barbara, there has been an increasing recognition of the critical importance of network and systems management technologies. Business needs and global competition have led to a greater reliance on information processing systems, while the size and complexity of such systems has continued to grow. At the same time the Internet phenomenon has produced new opportunities for large scale electronic services. Effective and efficient management technologies are clearly necessary for the successful realization of a worldwide information infrastructure.

This trend towards increasing interconnection and electronic access, however, is producing large, heterogeneous, distributed systems whose size, scope, and organizational boundaries pose challenging management problems. Enterprise management solutions must make greater advances in scalability, automation, and integration in order to remain cost effective in complex, distributed, mission critical systems environments. Advances in network and systems management are thus at the center of this dynamic world of evolving electronic services; and, the demand for seamless integration of computer applications and communications services into network, systems and technology infrastructures that are robust, flexible and cost-effective continues to increase.

The fifth international symposium on Integrated Network Management, IM '97, covers many of these topics. Its principal theme, however, is Integrated Management in a Virtual World, emphasizing the pivotal role that integrated network management plays in worldwide information networks and distributed systems that cross geographical and political boundaries. Indeed, these networks extend beyond physical boundaries to support virtual corporations, virtual LANs, inter-enterprise inter-networking, real and virtual service management, outsourcing and electronic commerce.

The 1997 symposium offers a world-class program of high-quality technical sessions presented by recognized leaders in their field. They address the relevant current issues, future trends, and emerging technologies involved in formulating overall management solutions across all types of networks, enterprise communication systems, distributed computing systems and applications. The technical presentations show an excellent mix of topics, organizations and international contributions.

## 2. History of the IFIP/IEEE Symposium on Integrated Network Management

Known by the acronym "ISINM" until this year's change to "IM", which is much easier to say, this biennial symposium series has earned a worldwide reputation as one of the network management industry's leading events.

Since 1989, the symposium has provided a technical forum for the research, standards, development, systems integrator, vendor and user communities. Each ISINM program and its related theme has reflected the historic events in integrated network management, and has indeed helped shape them.

- 1989: Improving Global Communication Through Network Management When the first ISINM was held in Boston in 1989, the need for comprehensive network management capabilities was apparent after major disasters had occurred in the telecommunications industries in the years before. Standards for enabling integrated network management across multiple vendor networking resources were under development in international and regional arenas. While some thought that the development of these standards was the most difficult obstacle to integrated management solutions, many realized a few years later that standards were just the beginning of a long journey. Integrated network management emerged as one of the most complex and hard to solve problems in heterogeneous communications.

- 1991: Worldwide Advances in Integrated Network Management Two years later at the second ISINM in Washington, D.C., the need for enterprise-oriented management across data and telecommunications applications and distributed systems became increasingly apparent. The principal problems related to providing coherent, integrated network management solutions across standards-based, multi-vendor components. Multi-vendor demonstrations in North America, Europe and Japan seemed to indicate that the time had come when users could competitively procure network management products in any of several countries and be confident that they would inter-operate with comparable products in other parts of the world. That wasn't so.

- 1993: Strategies For The Nineties We have learned that we are not at the end of the road; we are not even in the middle. We are only at the beginning, and will remain there probably for the greater part of the nineties. Worldwide coordinated strategies are needed to evolve integrated network management in the best way. The beginning of the nineties was characterized by big political, ecological and technical changes in all areas worldwide. The exponential growth of inter-networking in general and new multimedia applications based on broadband and mobile network technology will remain the driving forces in the communications area. However, the element of uncertainty plays a dominant role in all environments. Down-sizing, while up-sizing in volume and time requires flexibility to change. A paradigm shift took place during these phases: network management systems used for crisis situations in the past evolved to powerful tools for the day-to-day management of systems, services, applications and, of course networks.

-1995: Rightsizing in the Nineties During this sometimes turbulent period of rightsizing in all areas, the need for management systems is greater than ever before. Management is a fundamental part of a reliable information infrastructure. It assures the correct, efficient and mission-directed behavior of the hardware, software, procedures and people that use and provide all the information services. Effective management of the information infrastructure is becoming as essential as marketing and selling products. In addition, it helps to raise customer satisfaction. In order to get the maximum benefit from management technologies, a number of challenges must be faced: management functions need to be considered an integral part of the total enterprise, and appropriately incorporated in newly re-engineered business processes; their value needs to be assessed in terms of cost avoidance and customer satisfaction; appropriate privacy and security protection has to be provided; and , management software must be extensible, meet high performance requirements, and be highly reliable. Integrated network management belongs to the enabling technologies of a worldwide information infrastructure. The global trends towards interconnection and electronic access continue.

The goal remains affordable and instant access to any information, independent of the geographical location of clients and servers worldwide. Integrated network management solutions are intrinsic to achieving that goal. This brings us to 1997 and 'Integrated Management in a Virtual World.'

## 3. Integrated Management in a Virtual World

Our theme reflects the increasing interest in overall management applied across all types of networks, distributed computing systems and applications in an integrated fashion. It challenges us to consider comprehensive management solutions in a world increasingly filled with virtual corporations, virtual LANs, inter-enterprise networking, real and virtual service management, outsourcing and electronic commerce.

The evolution and increased deployment of computer communications and information systems has produced several major trends in management technologies. Manageability has become a recognized systems requirement. The explosion in the number and types of users, and the concomitant need to support users of all skill levels means that component systems must evolve to be as self-managing as possible. Standards, protocols, and techniques are being developed that allow information about computing resources to be provided by the resources themselves. These capabilities form the basis for increased automation and remote management of systems components, easing the management burden on end users and system administrators alike.

In order to provide self-manageability, all elements of a resource (hardware, software, communications capabilities and information requirements) need to be more tightly integrated. This Task Oriented Integration is most evident in the functionality being provided in PC and workstation operating systems and middleware components.

Due to size, scope, and organizational boundaries, services will be provided by multiple independently controlled systems, so that the movement towards Distributed Control will continue. In this context, Policy-directed management will become important as a way to define overall goals, constraints, and requirements, and allow the delegation of responsibility to local systems to execute them appropriately in their environments.

While many of these trends can be seen within enterprises themselves, there will be an increasing Inter-Enterprise Orientation. As electronic commerce expands, and utilizes common network interconnections, the need for reliable, secure, inter-enterprise processes with predictable performance characteristics must be met.

The future of network computing will be driven by the desire of people to connect to other people and enterprises around the world, leveraging powerful new technologies that can transcend distance and time, lowering boundaries between markets, cultures and individuals. The Internet is the most prominent representation of global networking, with predictions that by the year 2000 more than 200 million people will be active users. The technologies underlying this phenomenon are already being deployed in Intranets within enterprises and Extranets between associated enterprises, for business and commercial uses. The implications for network and systems management are just beginning to be understood.

The sixteen paper sessions and six panel sessions that make up the IM '97 technical program thus cover a wide range of topics. There are presentations on basic technology issues, the impact of standards, system architecture, services, the application of internet technology, and new management approaches.

We believe that the papers in this collection present a comprehensive view of the current state of management technology. They should form an excellent basis for anyone interested in understanding the important and dynamic fields of inquiry that they cover.

## 4. Future Events

As the management world continues evolving, this ongoing series of international symposia will continue to foster and promote cooperation among individuals of diverse and complementary backgrounds, and to encourage international information exchange on all aspects of network and distributed systems management.

To broaden the scope of these symposia, the International Federation for Information Processing (IFIP) Working Group (WG) 6.6 on Network Management for Communication Networks, as the main organizer of IM events, has been successfully collaborating with the Institute of Electrical and Electronics Engineers (IEEE) Communications Society's (COMSOC) Committee on Network Operations and Management (CNOM). IM and the Network Operations and Management Symposium (NOMS) are the premier technical conferences in the area of network and systems management, operations and control. IM is held in odd-numbered years, and NOMS is held in even-numbered years. CNOM and IFIP WG 6.6 have been working together as a team to develop both these symposia.

NOMS '98 will take place in New Orleans, Louisiana, USA, February 14-18, 1998. The next International Symposium on Integrated Network Management (IM '99) will be held in the Spring of 1999, on the East Coast of North America.

Starting in 1990, IFIP WG 6.6 together with IEEE CNOM has also been organizing the International Workshops on Distributed Systems: Operations and Management (DSOM) which take place in October of every year and alternate in location internationally. DSOM '97 will be held in Sydney, Australia, October 21-23, 1997 and will be jointly hosted by the University of Technology, Sydney (UTS), and the University of Western Sydney, Nepean (UWSN), Australia.

For more information on future IM, NOMS, and DSOM events, and other related activities please get in touch with us.


## 5. Acknowledgments


IM '97 is the result of a great coordinated effort of a number of volunteers and organizations. First of all, we would like to thank our main sponsors, IFIP Working Group 6.6 on Network Management and the IEEE Communications Society Committee on Network Operations and Management (CNOM) for their support. Our thanks also to GMD-FIRST, AT&T Bell Laboratories, IBM and the various other organizations that make it possible for our volunteers to dedicate their time.

Following the very huge success of ISINM '95, a small corps of dedicated individuals took it upon themselves to see that the tradition would continue with an even better event in 1997. We owe a debt of gratitude to Branislav Meandzija, Wolfgang Zimmer, and Doug Zuckerman who worked so hard in the beginning to form the vision for an IM '97 that would most effectively meet the needs of the network management community.

A very special acknowledgment to Dan Stokesberry, who served as the General Chair of the IM'97 Organizing Committee throughout much of its existence. We were all saddened by his unexpected death last December. His contributions were many, and the success of IM'97 is a tribute to his leadership.

The organizing committee of IM '97 has been the main force behind the symposium. We would like to thank (in alphabetical order): Joe Diamand, Varoozh Harikian, Joe Hellerstein, Jim Herman, Yvonne Hildebrand, Anne-Marie Lambert, Aurel Lazar, Branislav Meandzija, Roberto Saracco, Tom Stevenson, and Binay Sugla for their unflagging effort throughout the long months of bringing it all together.

The program committee under the able leadership of Aurel Lazar and Roberto Saracco, and with the tireless assistance of Rolf Stadler, has once again defined the standard for conferences and proceedings in network management. Its creative work, represented in this book, clearly delineates the main problem areas of integrated network management and displays the most promising solutions available. Our deepest thanks go to Salah Aidarous, Nikos G. Aneroussis, Jit Biswas, Walter Buga, Laura Cerchio, German Goldszmidt, Sigmund Handelman, James Won-Ki Hong, Pramod Kalyanasundaram, Wolfgang Kleinoeder, Subrata Mazumdar, Shoichiro Nakai, Jong-Tae Park, George Pavlou, Pradeep Ray, Gabi Dreo Rodosek, Stelios Sartzetakis, Morris Sloman, Liba Svobodova, Yechiam Yemini and the members of the Organizing Committee who participated at the well attended Program Committee meeting in L'Aquila, all other members of the program committee, and all the additional reviewers who created the outstanding program.

Finally, we would like to thank Clark DesSoye for producing our main symposium brochures such as the advance and final programs, and the IM Web pages; Steve Adler for his enthusiastic pursuit of vendor patrons; and, last but certainly not least, all vendor patrons for their key role in the vendor program and showcase.

# TRACK I

## PART ONE

# New Approaches to Customer Network Management

## Chair: Ken Lutz, Bellcore

# 1

# Multi-level reasoning for managing distributed enterprises and their networks

*J. Frey*
*L. Lewis*
*Cabletron Systems*
*P.O. Box 5005, Rochester, New Hampshire, USA 03866*
*j.frey@ieee.org, lewis@ctron.com*

### Abstract

In this paper we describe our experiences and directions regarding the management of multi-domain Enterprise Networks (ENs). Our views are (i) the management of the EN is a distributed problem-solving activity and (ii) recent results in automated problem-solving in the distributed artificial intelligence and robotics communities can contribute to our understanding, development, and deployment of management software for the EN during the 90s and into the 21st century. We discuss a distributed, multi-level architecture for EN management and describe examples in fault management and business process management. Our examples demonstrate an integration of two off-the-shelf Network Management (NM) products: the Spectrum NM platform from Cabletron Systems and NerveCenter from the Seagate Corporation.

### Keywords

Distributed Artificial Intelligence, Enterprise Management, Network Management

## 1 INTRODUCTION

The idea of an intelligent, self-healing network has caught on. As in the robotics community, network researchers and managers foresee an autonomous, self-managed network that can negotiate problems with little human intervention. In the late 90's and into the 21st century, we should see a shift toward a view of the enterprise network (EN) as an intelligent, autonomous (or semi-autonomous) agent [1]. Where robots have sensors, sense-data, and effectors, the EN will have analyzers and monitors, network traffic and behavior data, and system commands to effect tasks such as failure detection/diagnosis/repair, upgrading software, dynamic reconfiguration, and controlling network devices. In the same way that researchers give robots instructions in the form of goals and the robot figures out how to achieve the goals, network managers will give the EN a set of goals and the network will figure out how to achieve those goals. This view of the EN opens the way for thinking about outstanding issues with intelligent architectures in general and their contributions to understanding and achieving good network management solutions.

One such issue is task analysis. What network tasks are better performed by reasoning with raw message traffic data? With network events? With network alarms? What tasks are better performed by reasoning with a symbolic model of a network? What tasks require reasoning that is distributed over multiple layers of network abstraction, as opposed to cooperative reasoning systems over a single layer of abstraction? What control mechanisms will be needed? These sorts of issues have received considerable attention in the AI and robotics communities, and similar issues have surfaced in recent research on the management of enterprise networks.

Our objectives in this paper are as follows. First, we provide a brief account of distributed, multi-level reasoning in robotics and AI research. Next, we consider the problems and goals of EN management and show how the achievements in the former disciplines map to an EN management architecture. Finally, we discuss working examples of the architecture in fault management and business process modeling. Our examples demonstrate an integration of the Spectrum network management platform from Cabletron Systems and NerveCenter from the Seagate Corporation.

## 2   DISTRIBUTED AI AND ROBOT REASONING

### 2.1 What is Distributed Artificial Intelligence?

A common approach to problem-solving is to decompose a task into subtasks and to perform the subtasks sequentially or assign them to individual agents in parallel. This approach assumes that a central controller exists which decomposes the task, assigns subtasks to appropriate agents, and synthesizes the results. In contrast, distributed artificial intelligence (DAI) is concerned with problem-solving for which a central controller is not present and for which subtasks are interdependent [2,3,4]. Specifically, DAI is concerned with problems that exhibit the following characteristics:

1. *Control and expertise required for solving a problem are distributed over a number of agents.*
2. *An agent is not always able to complete his subtask alone.*
3. *No single agent can solve the entire problem alone.*
4. *The solution to the problem requires cooperation and communication among agents in such a way that the expertise of each agent combines to solve the problem.*

It is clear that EN management is a DAI problem. A first-order categorization of network management tasks includes fault, configuration, accounting, security, and performance management [5]. However, it is not clear that this is the best or a complete categorization of network management subtasks, and less clear how and in what ways the specialists performing the subtasks depend upon the other specialists. A good way to pose the question is "What does X need from Y, and when and how often is it needed?", where X is a particular specialist and Y is the set of remaining specialists. A prior question, of course, is "What are the Xs?"

For example, problem-solving that involves fault, configuration, and performance management often requires specialists in each area who compare notes in order to solve the problem. In addition, another specialist in budget control and expenditures is consulted in order to find the most reasonable solution from a cost perspective.

As an aside, it is interesting to note that a problem that could turn out to be a DAI problem is the re-creation of intelligent behavior exhibited by a single human agent. For example, a controversial issue in the robotics community is that of determining what mechanisms are required to build an autonomous agent [6]. On the one hand, it has been argued that an agent requires a single controller who coordinates tasks such as perception, cognition, acting, planning, and learning. In this view, intelligent behavior is not a DAI problem since a controller delegates these tasks to sub-agents, where the sub-agents do their work independently of each other and the controller synthesizes their results. Alternatively, it has been argued that intelligent behavior of a single agent is itself an example of DAI. Perception, cognition, etc. are performed by specialists who negotiate among themselves, and *agency* is an emergent property or epiphenomenon of the multiple agents [7].

As with any sufficiently large multiple-agent enterprise, including EN management and possibly a single human agent, these questions are hard. A thorough study of network management within the DAI framework has not been forthcoming. Comparatively, there is more work and experimentation with AI paradigms applied to specific management tasks, e.g. paradigms such as neural networks, fuzzy logic, case-based reasoning, and expert systems. There is less work describing how these pieces fit together. The reason is that network management is plain difficult, and the understanding and characterization of DAI problems in general are equally challenging.

Here we do not try to characterize a complete network management architecture. In section 3 we discuss work on a piece of the total solution, fault management, and describe a working DAI architecture for achieving distributed fault management in the EN. Other efforts toward defining the subtasks and dynamics of integrated network management include [8,9].

## 2.2 What is Robot Reasoning?

The work on implementing intelligence in a robot since the early 80s is an interesting story. If our hypothesis that intelligent networks will come to be viewed as autonomous agents turns out to be true, then the uncovering of false starts and promising approaches in the robotics community could help us in understanding intelligent network management. In this section we describe two prominent approaches to the development of intelligent architectures for robots. For further detail, see [6].

### 2.2.1 The Single Loop Architecture
The single loop architecture (SLA) for robot intelligence is shown in Figure 1. Intelligent behavior starts with robot sensors and ends with instructions that are executed by robot effectors. Initially, sensory input is passed through several layers of abstraction, e.g. signals, signs, and symbols. Each layer filters noise and extraneous data out of the

data passed to it, and transforms the data into fewer, more informative chunks of data. When data becomes manageable, usually at the symbolic level, it is compared with pre-defined knowledge about what instructions should ensue. This operation is usually defined as "coordination" and may be implemented in a number of ways, including simple look up tables or expert systems. The output of the coordination module is a set of symbolic instructions. These instructions are decomposed down through levels of decomposition until they are suitable for processing by the robot's effectors, usually in the form of signals

The important ideas in the SLA are (i) the compression of sensor data through layers of abstraction, (ii) the coordination of highly abstracted sensor data and goals, and (iii) the decomposition of instructions through layers of abstraction into a form that is executable by effectors.

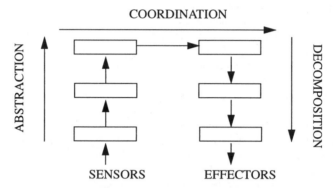

**Figure 1**  A single loop architecture for robot reasoning.

It is generally agreed that the SLA is unsuitable for dynamic, unstructured tasks. Sensor data must be abstracted into symbols before they can be coordinated with the goals of the system and initiate the decomposition of instructions for effectors. As a result, the processing time required to transform sensor data to control instructions through the loop is prohibitive. By the time the system figures out what to do in environment A, it is likely that environment A is obsolete and the instructions are no longer applicable. In other words, the rate at which the world changes over time is disproportionate to the amount of processing required for timely behavior.

## 2.2.2 The Multiple Loop Architecture
Robotics researchers observed that the coordination of abstracted sensory data and goals do not have to take place only at the symbolic level. Coordination could occur at any level of abstraction. Further, they observed that coordination at lower levels required increasingly less time, since upper levels would be by-passed. Therefore, problems which require a quick solution could be handled at lower levels of abstraction, and problems which require more time could be handled at upper levels of abstraction. This distinction is generally referred to as "reflexive" behavior for short-term solutions and "deliberative" behavior for long term solutions. Figure 2 shows a design that reflects these ideas, called the multiple loop architecture (MLA).

Each level of the MLA is a separate control loop that corresponds to a specific class of problems, where problems are partitioned and assigned to levels according to the amount of time and type of information required to solve them. For example, the short-term abstraction/coordination/decomposition loop at the lowest level provides quick reaction, bypassing upper level control mechanisms. In the network management domain, such tasks might include intelligent routing and temporary disconnection to a busy host. The medium-term loop provides reaction to more complex problems and operates on increasingly abstract input such as signs. Tasks of this sort might include alarm correlation in a busy network with multiple alarms, where some alarms are real and others are apparent, and the task is to distinguish the two and suppress all apparent alarms. The top level would provide reaction to problems that require more time. The classic example of a task of this kind is the reasoning involved in deciding to move a host from subnet A to subnet B because the majority of the host's clients reside on sub-net B, thereby causing increased traffic on the link between A and B.

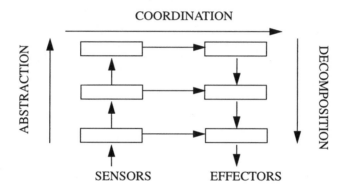

**Figure 2** The multiple-loop architecture for robot reasoning.

In comparison with the SLA, the MLA is intuitive and shows promise for a clearer understanding of intelligent network management. However, the primary unresolved issues of the MLA approach are (i) a methodology by which to classify and divide tasks according to information available and time required to solve them, (ii) the particular algorithms on each level that execute its class of tasks, and importantly (iii) how to synthesize activities on multiple levels when a task requires multiple levels of coordination. The latter issue is the classic DAI problem.

## 3  MANAGING THE ENTERPRISE NETWORK

With the ideas and insights of the preceding section under our belts, let us now turn to the practical matter of managing the EN. Three important truths regarding EN management motivate our distributed, multi-level architecture for EN management:

1. The EN is inherently a distributed, multi-domain entity. ENs typically are partitioned in ways that help administrators understand and manage the EN, e.g. with respect to geographical domains, functional domains, or managerial domains.
2. The tasks involved in managing distributed ENs are too complex for a single controlling agent, and thus the tasks themselves must be partitioned into distributed, cooperative agents.
3. The data types for analyzing and reasoning about EN behavior come in various forms, e.g. traffic data, symbolic models of the network, events, alarms, et al.

## 3.1 Spectrum's Distributed Client/Server Architecture

For the reasons above, the Spectrum NM platform was designed as a distributed client/ server architecture. The Spectrum servers, called SpectroSERVERs (SSs) monitor and control individual EN domains. The Spectrum clients, called SpectroGRAPHs (SGs) may attach to any SS in order to graphically present the state of that SS's domain, including topological information, event and alarm information, and configuration information.

   Importantly, each domain can be viewed from a single SG. See Figure 3. If SG-1 is attached to SS-1, but the user wishes to see the domain controlled by SS-2, the user can click on an icon in SG-1 that represents SS-2. Figure 3 shows the primary client/server attachment between SG-1 and SS-1, where virtual attachments between SG-1 and other SSs are indicated by dotted lines.

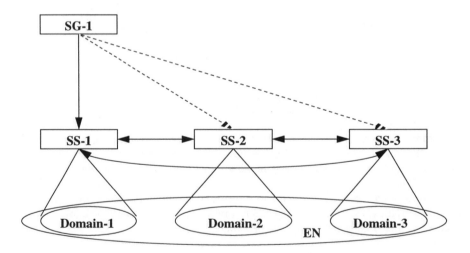

**Figure 3**   The SPECTRUM distributed client/server architecture.

### *3.1.1 Experiences with the Architecture in an Internal Testbed*
We have tested the architecture in an internal testbed to establish operational feasibility. A virtual enterprise network was configured using 250 SSs situated in our U.S. and

U.K. facilities. A three-layered hierarchical topology was used, with one master SS connecting to 14 SSs, each of which in turn connecting to 15-20 more SSs. Each end-node SS was given a database with several hundred manageable devices.

Cabletron recommends a maximum 7:1 ratio among SSs that are configured hierarchically. This ratio is derived from workstation operating system characteristics rather than communications traffic load among SGs and SSs. Note that load would be a strong, adverse factor if we allowed a one-to-one correspondence between SGs and SSs, or if we allowed heavy communication among SSs.

In our internal testbed, we purposely exceeded the 1:7 ratio in order to stress the architecture. Performance, accuracy, and reliability were monitored over a period of several days, during which time a variety of simulated failures were introduced and the resultant behavior analyzed. A few communication flaws at the physical layer were identified and corrected, but as a whole the test was successful and the test bed operated without incident. Subsequent passes through this test plan are being used to exercise new and enhanced distributed and fault tolerance functions.

These initial tests provided a good, empirical argument for the scalability of the distributed, client/server architecture. We consider each SS as an intelligent agent, capable of presenting management data on demand to any client SG. This keeps inter-SS communications to a minimum. Each SS "knows about" its peer SSs, but is prohibited from extensive communication with them. In Section 3.2, we show how SSs may communicate by intermediary agents who reside at higher levels of NM abstraction.

## 3.1.2 Customer Installations

The distributed version of Spectrum has been installed at dozens of customer sites worldwide, with setups ranging from a few (2 or 3) SSs to several hundred. For the most part, customer ENs are divided into geographical domains, using an SS at each facility or campus, with a central master SS at a headquarters location.

A particularly challenging project currently in progress is the management of a telecommunications network in Eastern Germany [10], deployed by Deutsche Telekom. This project poses unusual requirements because it is purely non-SNMP, using only a proprietary management protocol. Spectrum was chosen as the management platform because (1) it has a distributed, client/server design, (2) it has APIs for developing non-SNMP management applications, (3) it has APIs for configuring intelligent SS agents, and (4) it enables representation of both devices and services involved in the EN.

Consider (3). With multi-domain ENs with corresponding SS agents, polling-based management can be costly in terms of bandwidth load. By restricting SS polling (i.e. only using it for testing basic element presence/status) and instead having managed elements forward management data to the SSs via traps, in-band management traffic is reduced considerably. This was a requirement for the application. Note that a transition from polling to trap-based management and intelligent agents is considered by some to be the future of EN management [11].

Consider (4). Data collected via the management system is being utilized in two ways. First, network devices are represented topologically in order to monitor and control the operations of the telecommunications network as a whole. Second, a service-based representation is used to monitor/manage usage and repairs so that, for example, large business customers may be given relatively higher priority for repairs than residential customers. This secondary representation can be easily accomplished by giving relative weights to each managed element's alarms.

As of the end of 1996, the number of SSs deployed for this application was nearing 1000 (490 primary servers, each with a fault-tolerance backup server). The NM configuration is similar to our internal testbed save that (i) each end-node SS manages both devices and services, (ii) there are only two tiers (i.e. there is not yet a top-level master SS), and (iii) there are more end-node SSs per second tier SS (up to 40). Performance and capability results thus far have been excellent, and the project is on track towards a planned deployment of 4000 total SSs (first tier plus second tier). The success achieved thus far has already resulted in commitments for similar projects by telecommunications providers in several other countries.

## 3.2 Multi-Level Reasoning

Let us stop a moment to think about the management tasks that occur within network domains and management tasks that occur across domains. We will use fault management as our first example.

Fault management consists of event monitoring and filtering, event correlation, escalation of events to alarms, alarm correlation, and diagnosis/repair. The SSs that monitor individual network domains perform these tasks with Spectrum's inductive modeling technology (IMT) [12]. We may refer to this as *intra*-domain event/alarm correlation.

With multi-domain ENs, the requirement now is to perform the same function across domains. For example, a failed router in Domain-1 may affect the applications running in Domain-2. Conversely, the cause of an application failure in Domain-2 may be identified as the failed router in Domain-1. We refer to this as *inter*-domain event/alarm correlation. Since we have limited inter-communication among SSs, we need to find some other way to do inter-domain alarm correlation.

In light of our discussion of multi-level reasoning in Section 2.2, we can conceptualize the inter-domain event/alarm correlation task as shown in Figure 4. The figure, however, is somewhat misleading because it is in two dimensions. The bottom-most

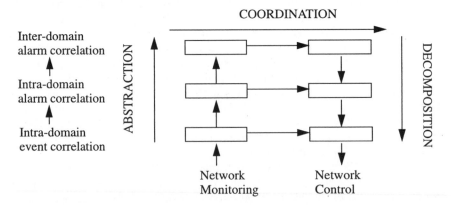

**Figure 4** The Multiple-Loop Architecture for multi-domain fault management.

two levels can be performed by SSs that monitor and control individual domains in the EN. The number of SSs is the (implicit) third dimension. The agent that resides on the top level collects alarms from multiple SSs and carries out inter-domain alarm-correlation, communicating with other SS agents as appropriate. Note that the SS agents communicate indirectly via the intermediary coordination agent on the top level.

What reasoning paradigm is appropriate for the coordination agent at the top-most level? Several reasoning paradigms are at our disposal, including simple look-up tables, expert systems, case-based reasoning systems, state-transition graphs, et al. Several commercial products that incorporate some one or other of these paradigms are available. At present, we are integrating NerveCenter (which uses the state-transition graph paradigm) with Spectrum, where NerveCenter is the top-most agent [13].

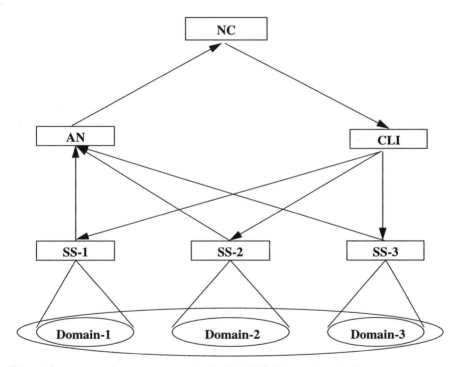

**Figure 5** An integrated architecture with SPECTRUM and NerveCenter.

A clearer picture of the integration architecture is shown in Figure 5 (we have left out the SG clients). The Spectrum Alarm Notifier (AN) is a client daemon that collects alarms from all domains in the EN. The AN can be configured to allow only select alarms to be passed to NerveCenter (NC). NerveCenter performs high-level reasoning over multi-domain alarms and if appropriate, communicates with other SS agents via the Spectrum Command Line Interface (CLI). Communications may include requests for further bits of information, or notification of inter-domain alarms.

At this juncture our integration with NerveCenter is in development/testing stage. Once complete, we wish to examine the ease with which one can encode knowledge in the top-level agent. In the AI community, this task is called the problem of knowledge acquisition. For example, we may ask: How easy is it to encode new knowledge in a state-transition graph? A rule-based expert system? A case-based reasoning system? Another important issue we are studying is the ability of the top-most agent to learn and adapt itself to new problems, given its past experience [14]. This latter task is quite ambitious, but it takes us closer to the realization of an intelligent network.

## 3.3 Business Process Modeling

In the final analysis, the network is simply an element in a larger picture which includes all aspects of an enterprise's business processes. These processes depend on the network and its various nodes in order to successfully complete their stated goals. When a network element is not functioning properly, fault management will help us to isolate it and correct the problem, but what else was affected? A router failure may cause a marketing forecast report to fail, or a file server crash might interrupt a nightly software distribution. Let's take a simple example and see how these higher-level functions can be managed via the architecture that we've been describing.

When a standard business function such as "Accounts Receivable" is described, we discover that it is, as any business process, composed of resources and workflows. Required resource components may include client workstations, compute servers, file servers, database applications, peripherals, and the voice and data networks connecting them all together. Workflows often consist of standard procedures, contingency plans, and data flows. Human operators are also involved, but since they lack a standard management interface, they will be ignored for purposes of this discussion. Computing systems and network elements have long been manageable, and applications are becoming increasingly so via proprietary interfaces and current standards efforts [15].

We find that Inductive Modeling techniques and multi-level reasoning can also be applied to business process management. Firstly, the systems, servers, applications, PBXs, peripherals, network devices, power supplies, (and operators, for that matter) can be modeled, along with the connection and dependency relationships between them. The objective is then to monitor the presence and health of these elements and roll the collective results up to a higher level representation of the "Accounts Receivable" business process. This monitoring data can also be fed to state machines (or other techniques) used to model the workflows. Since the elements are distributed across multiple domains in a large enterprise, we must consider that the information will be coming from several agents and must be correlated, perhaps by the means described in section 3.2. With this aggregate information, it becomes possible to focus upon mission critical common resources, understand the breadth of impact when a common element experiences troubles, and proactively manage problems on a business priority basis. This topic area is the subject of additional work by the authors [16].

# 4 SUMMARY AND CONCLUSION

We have examined the problem of EN management from two related perspectives. We looked at ongoing work on intelligent architectures in the distributed artificial intelligence community and in the robotics community. Our view is that there are elements in this work that will contribute to our understanding and implementation of EN management software as we approach and enter the 21st century. From a practical perspective, we described working architectures for solving the problem of intra- and inter-domain event/alarm correlation and business process modeling/management in multi-domain ENs which involved integrating the Spectrum NM platform and NerveCenter from Seagate Corporation.

## Acknowledgments

The authors wish to acknowledge the key engineer and architect responsible for distributed Spectrum: Eric Rustici. Additional thanks for conceptual support and contributions by David Taylor, Bill Tracy, Mike Soper, and A. J. Noushin. Special thanks to Frank Andrus and Michael Troitzsch for details on the Deutsche Telekom project.

# 5 REFERENCES

[1] L. Lewis. AI and Intelligent Networks in the 1990s and into the 21st Century. In *Worldwide Intelligent Systems*. Edited by J. Liebowitz and D. Prerau. IOS Press, Amsterdam. 1995.

[2] N. Sridharan. Workshop on Distributed AI. *The AI Magazine*. Summer 1987.

[3] M. Huhns (editor). *Distributed Artificial Intelligence*. Morgan Kaufman, London, 1987.

[4] K. Decker, E. Durfee, and V. Lesser. Evaluation Research in Cooperative Distributed Problem solving. Chapter 19 in *Distributed Artificial Intelligence* (Volume II) (edited by L. Gasser and M Huhns). Morgan Kaufman, London, 1989.

[5] *IEEE Communications Magazine. Special Issue: OSI Network Management Systems.* Edited by R. Pyle. May 1993 Vol. 31 No. 5.

[6] P. Maes (editor). *Designing Autonomous Agents: Theory and Practice from Biology to Engineering and Back*. MIT Press, Cambridge, Mass., 1991.

[7] M. Minsky. *The Society of Mind*. Simon and Schuster, Inc. 1985.

[8] K. Olesen. Network Management in Large Networks. In *Information Networks and Data Communications II*, D. Khakhar and V. Iverson (editors). Elsevier Science Publishers, North-Holland, 1988.

[9] J. Westcott. A Simple Model for Integrating Network Management. In *Information Networks and Data Communications II,* D. Khakhar and V. Iverson (editors). Elsevier Science Publishers, North-Holland, 1988.

[10] W. Weipert. The Evolution of the Access Network in Germany. *IEEE Communications Magazine*. February 1994.

[11] R. Oliveira, D. Sidou, J. Labetoulle. *Customizing network management based on application requirements.* Proceedings, First IEEE International Workshop on Enterprise Networking (ENW-96). Dallas, June 1996.

[12] W. Hamscher, L. Console, and J. de Kleer (editors). *Readings in Model-Based Diagnosis.* Morgan Kaufmann, San Mateo, 1992.

[13] L. Lewis and A. Noushin. Outline of the Spectrum/NerveCenter Integration. Technical Note lml-aj-95-10. Cabletron Systems. 1995.

[14] L. Lewis. *Managing Computer Networks: A Case-Based Reasoning Approach.* Artech House, Boston. 1995.

[15] J. Saperia, C. Krupczak, R. Sturm, J. Weinstock, *Definitions of Managed Objects for Applications,* IETF Application MIB Working Group draft, October, 1996.

[16] L. Lewis and J. Frey. *Incorporating Business Process Management into Network and Systems Management.* Third International Symposiaum on Autonomous Decentralized Systems, Berlin, Germany, April 1997.

# 2

# Integrated Customer-Focused Network Management: Architectural Perspectives

*L. Feldkhun, M. Marini, S. Borioni*
*Sodalia SpA*
*Via V. Zambra 1, 38100, Trento, Italy, Fax +39 461 316 401*

### Abstract

The industry-wide trend to decouple telecommunications services from the underlying network technology, together with increasing complexity of high-speed hybrid customer networks, has created the need for a new generation of integrated network management systems, dedicated to managing advanced heterogeneous customer networks as a service.

This paper presents business drivers for Integrated Customer-Focused Network Management (ICFNM) systems and introduces a number of key architectural perspectives for ICFNM implementation including: i) the position of ICFNM systems in the overall network management context, ii) the functional scope, and iii) the practical experience gained in designing and developing ICFNM using Object Oriented methodology and advanced commercial technologies (i.e., CORBA).

### Keywords

Integrated Network Management, CNM, Architecture, CORBA

## 1 INTRODUCTION: ROLE AND POSITION OF ICFNM

The ongoing rapid technological innovation and deregulation in the telecommunications industry creates business opportunities for telecommunications network and service providers. One area where these opportunities are greatest is in offering business customers an increasingly high level of control of network resources and services at decreasing costs.

This can be achieved by Virtual Private Networks (VPN): a service offering which allows business clients of Public Telecommunications Operators (PTO) to create corporate networks and define telecommunications services based on public network resources.

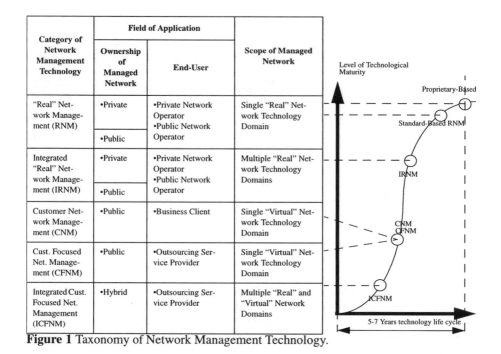

| Category of Network Management Technology | Field of Application | | Scope of Managed Network |
| --- | --- | --- | --- |
| | Ownership of Managed Network | End-User | |
| "Real" Network Management (RNM) | •Private | •Private Network Operator •Public Network Operator | Single "Real" Network Technology Domain |
| | •Public | | |
| Integrated "Real" Network Management (IRNM) | •Private | •Private Network Operator •Public Network Operator | Multiple "Real" Network Technology Domains |
| | •Public | | |
| Customer Network Management (CNM) | •Public | •Business Client | Single "Virtual" Network Technology Domain |
| Cust. Focused Net. Management (CFNM) | •Public | •Outsourcing Service Provider | Single "Virtual" Network Technology Domain |
| Integrated Cust. Focused Net. Management (ICFNM) | •Hybrid | •Outsourcing Service Provider | Multiple "Real" and "Virtual" Network Domains |

**Figure 1** Taxonomy of Network Management Technology.

Additionally, control can be offered through Outsourcing Services, which include a wide range of contractually defined services covering Operations, Administration, Maintenance and Provisioning of hybrid (in terms of ownership) corporate networks according to defined Service Level Agreements.

The growth of VPNs and Outsourcing business is strongly increasing because of:
• increasing role of telecommunication technology in business customers' competitive position
• rapid pace of technological innovation in telecommunications
• increasing technological sophistication and required skill level for telecommunications management
• increasing risk of technological obsolescence of privately owned telecommunications assets
• increasing level of customer focused control of publicly offered telecommunications resources and services (Bigaroni, 1992)
• decreasing costs of alternatives to private networks.

Provisioning for sophisticated services require network operators to introduce new generations of management systems and applications (Figure 1).

In this context we introduce the notion of an Integrated Customer Focused Network Management (ICFNM) system (Feldkhun, 1996). This system provides a particular view of network/service management, reflecting the perspective of a telecommunications services outsourcing contract established between a business client and an outsourcing service provider. The network and service management related information and controls, available to the end-user of ICFNM, are limited to the public network resources and services, which are contractually allocated to a particular customer of a PTO provider. This is in contrast to the amount of management information and the level of control available to the end-user of the "Real" Network Management (RNM) system of the public network. An ICFNM system generates and maintains a customer focused view of the status of resources/services in a multidomain (in terms of technology and ownership) network.

## 2   CHALLENGES FOR AN ICFNM SYSTEM

The objectives of ICFNM can be better understood by examining the context in which the system is positioned. Context analysis contributes to identify challenges for ICFNM and key architectural requirements for an effective system design.

### 2.1   Managed Networks

The need to deploy highly sophisticated communication services is reflected in the complexity of communication networks which are built on a number of interconnected transmission (e.g. PDH, SDH) and switching (e.g., X.25, Frame Relay, ATM) technologies. Each network technology is potentially offered by multiple suppliers.

Network and service operators are required to implement effective management to control and monitor network systems from a customer point of view in order to meet customers expectations and changing needs.

The main challenge is to provide operators with a complete and integrated view of the complex managed network environment. As an example of the network complexity, let's think of the number of equipment and network technologies which are crossed by an Internet Protocol (IP) packet generated at a desktop workstation. It crosses customer routers, it is encapsulated, for instance, into a Switched Multimegabit Data Service (SMDS) Protocol Data Unit, segmented into ATM cells, and it is then transmitted via a permanent Virtual Path Connection (VPC). The VPC may be physically laying on top of a PDH circuit multiplexed by access multiplexers, "SDH crossconnected" to finally reach a SMDS switch, which will terminate the ATM VPC and route the packet and so forth until the packet destination is reached.

From this complex and heterogeneous network stems the requirement for ICFNM to be capable of fulfilling operator needs for a simple-to-use tool; the tool allows for an integrated Operation Administration and Management (OA&M) of the end-to-end connectivity across multiple network technologies.

### 2.2   Management domains

Definitions of management domains have been extensively explored in the industry (Sloman, 1994).

The need to address multiple management domains by ICFNM is meant in terms of network ownership. Virtual Private Networks span customer premises and wide area networks (Figure 2), where customer premises equipment may be owned and managed by the customer or, alternatively, outsourced to PTO or other suppliers of outsourcing services.

Wide area networks are sectioned in different network domains under the control of different, sometime competing, network operators. Within one single PTO, different organizations may control different parts of the whole PTO network.

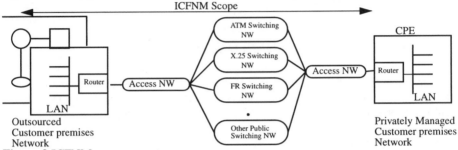

**Figure 2** ICFNM management scope.

While customer premises equipment and networks are fully dedicated to a given customer, network resources in the wide area network domain (e.g. access networks, multiplexers, switches, trunks, ATM Virtual Paths, SDH Virtual Containers) are generally shared among many of them. In the context of ICFNM, network resources are filtered and processed in order to find out those which are relevant to the customer being managed. From this perspective, for instance, a single ATM VP may belong to multiple virtual networks and therefore be monitored as a network resource virtually owned by multiple customers.

From these considerations originates the requirement for ICFNM to virtualize and integrate management domains consisting of resources under different ownership. The scope of ICFNM applications and the types and characteristics of available management operations are discussed in section 3.

## 2.3  Network Element and Network Management systems

Achieving an integrated view of the overall end-to-end network (from a customer perspective) requires ICFNM to interoperate with a set of Network Elements (NE) and Network Management Systems (i.e. "Real" Network Management System), which are designed and developed according to various criteria. This happens not only in case of different managed technologies, but also for managing equipment of the same technology but provided by different vendors.

Lack of standards for network management information models has caused a proliferation of proprietary solutions for the systems deployed in the market. Information models defined by ITU-T are available for a set of technologies, aimed at the public network side - a most prominent example is represented by SDH. Nevertheless, the late availability of the standards coupled with the time-to-market pressure has caused manufacturers to take proprietary approaches (e.g. adopting early drafts of the standards).

On the other side, Internet has defined, early on, a number of management "information models" (e.g. MIB 2 and extensions) complementary to the ITU-T ones, which have been successfully deployed in the field, but have limitations in covered network technologies and services.

As a result, a significant number of the network elements deployed in the field currently support proprietary management interfaces, and protocols. The need to integrate these networks has led, in some cases, to the development of SNMP and CMIP proxies providing views at either Network or NE level.

Peculiarities of management systems are not restricted to the aforementioned elements. Often NE and Network Management systems come with built-in limitations, mainly due to the fact that such systems have not been designed for further integration with other systems and/or applications.

Therefore almost every management system is restricted in some capabilities relevant to ICFNM, ranging from performance (e.g. response time to an external request), scalability (e.g. possibility to scale in order to serve a large number of application requests) and configuration possibilities (e.g. number of destination addresses for forwarding an asynchronous notification). For this reason the ICFNM architecture needs to address the differences in data models, protocols and functional capabilities offered by the interfaced managements systems.

## 2.4 Operation Support Systems

ICFNM capability to interoperate with a set of NE and Network Management systems is necessary, but not sufficient to achieve Integrated Customer Focused Network Management. Additional data, not available at the Real Network Management (RNM) level, is needed in order to model inter-network and service configurations.

From an architectural point of view ICFNM obtains additional information, complementing what is available at a network and network management level, from a set of Operation Support Systems (OSSs). Examples of OSSs relevant to the ICFNM application context are:
• Customer Management OSS (Contract OSS)
• Billing OSS
• Service Provisioning OSS
• Trouble Ticketing OSS (TT OSS).

For example, for an Outsourcing Business a Customer Management OSS will supply ICFNM with customer related data, including contractual data detailing Service Level Agreements, type, number, identifiers and characteristics of outsourced equipment. This information is processed by ICFNM in order to define the management scope for each customer and to accordingly configure management applications.

A Trouble Ticketing OSS is the destination to which ICFNM forwards notifications about fault occurrences which result from ICFNM diagnostics activity.

Challenges in interfacing external OSSs are similar to those found at the interface to RNM systems: the variety of data models, APIs, protocols, which require an architecture which would decouple ICFNM from the peculiarities and evolutionary changes of different OSSs.

## 2.5  Organization and processes

The pressure faced by telecommunication service providers to deal with global competition while rapidly developing new services, is reflected in the level of architectural flexibility required from ICFNM system. ICFNM aims to provide customer with a unique point of contact, the "front office", where help desk operators directly answer queries related to any service trouble.

In addition to the provisioning of a unique point of contact, ICFNM aims to off-load "back office" personnel by offering the front office personnel knowledge about service configuration, utilization parameters and faulty conditions while hiding complexity of networks and equipment. With powerful analysis tools help desk operators are able to direct technical personnel to the problem area in a majority of the cases.

ICFNM deployment configurations need to match PTO organizational processes. For that reason an architecture of ICFNM should foresee a hierarchy of operation centers distributed over the territory. Time-dependent management domains can be assigned to operation centers and individual operators, in terms of supported customers, managed network technologies and functional competencies.

## 3    ARCHITECTURE

This section describes the architecture of the ICFNM system in terms of functional capabilities, physical aspects and platform issues. The relevant choices with respect to the enabling technologies used are also discussed.

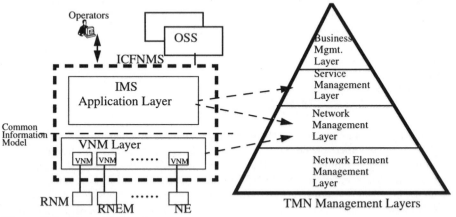

**Figure 3** Architecture layering.

## 3.1  Functional architecture

The functional architecture aims to describe and organize the functional capabilities of the ICFNM system by: defining the logical grouping of ICFNM functionality and mapping ICFNM functionality to a widely accepted telecommunication management framework (TMN).

The characterization of the context where ICFNM is positioned highlights the challenges to design a system capable of coping with the evolution of telecommunication services, technologies and organizations. This has lead to the main architectural choice made, i.e., the recognition of a three layered architectural approach (see figure 3):
- the RNM layer, with the purpose of managing the "real" NE and networks
- the VNM layer, with the purpose of applying customer focused filters to the overall set of management information available at the RNM layer for a particular network technology
- the IMS Applications layer, with the purpose of operating the set of virtual, integrated networks that provides telecommunication services to a specific customer.

The layered architectural approach allows for the handling of the system complexity by implementing the normalization of management data and virtualization of networks at the VNM layer while achieving integration of management data and functions at the IMS layer.

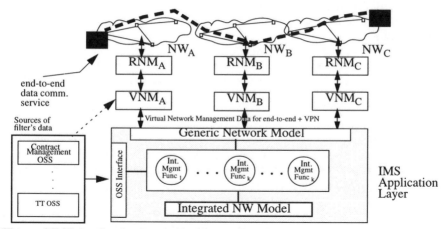

**Figure 4** IMS Application layer management functions.

## ICFNM: Virtual network management layer

This architecture layer fulfills two major objectives: normalization and virtualization with respect to network management information.

*Normalization:* in order to operate the integrated network, ICFNM needs to interface to different entities such as NEs, RNEM and RNM, thereby supporting a wide range of interfaces, generally characterized by different syntaxes, information models, services and behaviors.

Data available at the various RNM interfaces is not suitable for being directly used by the IMS Application layer because of the differences in the expressiveness of management data which are made available through their MIBs. Some MIBs are too fine grained, addressing details not relevant to ICFNM (e.g., network internal not relevant to service configurations). Other MIBs do not provide enough information for ICFNM (e.g., because no explicit association exists between network resources and customers using the resource).

In order to control the complexity of the managed systems, we decided to introduce the notion of a Virtual Network Management (VNM) layer. The VNM layer hides behavioral, protocol and modeling peculiarities of the different managed technologies by mean of:

• management protocol conversion to a common management protocol
• information model conversion to a Generic Network Model
• a set of management capabilities (e.g. threshold monitoring, polling, filtering) that apply to all network technologies and hide ICFNM applications from behavioral peculiarities.

*Virtualization*: the VNM layer offers to the ICFNM application layer (i.e., the IMS) a virtualization of the real network by generating virtual views of the management data available from the RNM.

The functions which generate virtual views use data typically provided by external OSSs (see also section 2.4). For example:

• Customer Ids to be associated with virtual network items provided by Contract Management OSS
• identifiers of network resources assigned to a customer (e.g., routers, interfaces, etc.) provided by Contract Management OSS and Service Provisioning OSS
• identifiers of network resources used by a given service (e.g., permanent virtual circuits) provided by Service Provisioning OSS.

When designing and tuning VNM, a key performance issue is deciding on the technical alternatives for filter application and construction of virtual views. This operation requires access to RNM MIBs, extraction of OSS data and computation of a large amount of information proportionally to the sizes of managed networks. Typical examples of architectural alternatives are:

• compute virtual network views on-demand as required by a management function. This alternative is feasible either when the response time is not a critical factor or when the OSSs and RNMs have close to real time response times. ICFNM applies this approach primarily for extracting up-to-date status information from the RNMs.
• store virtual network views persistently, by computing an initial value for it and keeping it up-to-date by handling asynchronous notification of RNM or OSS data changes. This alternative minimize the load on the RNMs and OSSs but requires them to support asynchronous change notifications. ICFNM adopts this approach for handling network configuration data.
• trigger periodical recomputation of the view and persistently store the result. ICFNM adopts this approach for collecting performance statistics on NE's and network resources.

Identification of the most suitable approach requires a trade-off analysis among the design simplicity allowed by the first option, the response time optimization offered by the second one and the increased computational load on the RNMs caused by the third one. Optimal balance between performance, load and complexity is a complex architectural issue. Indeed OSSs and RNMs services impose constraints thereby allowing only for sub-optimal solutions (for example, certain commercial RNMs do not support extensive polling of MIB attributes therefore requiring applications to perform limited polling and integrate this data with asynchronous notifications).

## ICFNM: IMS Application layer

The ICFNM Application Layer uses the services offered by VNM (see figure 4) to achieve the ICFNM main objective, that is to monitor, ensure, and report the Quality of Service offered to customers consistently with contractual data.

Functionalities in the area of Fault, Performance and Configuration Management are supported and summarized in Table 1. Network resources in the context of ICFNM Application layer are those filtered by VNM and relevant to the customer being observed.

Certain key management functions, for example, status monitoring of the end-to-end data communication service spanning across several network domains, cannot be achieved within a single network domain. At the single domain level, in fact, there is no notion of higher level services.

**Table 1** ICFNM Functional scope

| Mgmt Svc Area | Management Function |
|---|---|
| Fault Management | End-to-end monitoring of communication services across multiple technologies organized according to customer views.<br>Status monitoring of network resources.<br>Alarm collection, filtering, and correlation in order to identify faulty components during "alarms storms" caused by disruption of services spanning multiple networks.<br>Automated diagnostics in order to locate and isolate fault conditions.<br>Automatic generation and closure of Trouble Tickets.<br>Proactive fault management. |
| Performance Management | Monitoring, recording, and reporting on availability at service and network resource level.<br>Monitoring, recording, and reporting on end-to-end connection quality.<br>Identification of QoS level and variations. |
| Configuration Management | Generation and maintenance of the configuration of the integrated multi-domain customer network.<br>Population of the configuration data through network autodiscovery, access to external OSSs (e.g., Customer Contract OSS) and operator input.<br>Support for configuration queries and report generation. |

For example, the status monitoring of an end-to-end connection (see figure 4) will require the monitoring of the status of each NE and each service element belonging to the virtual view for that service, with respect to a particular customer.

Therefore, in order to manage customer specific telecommunication services provided on technologically diverse networks, the IMS Application layer integrates virtualized network data extracted by VNM from the RNMs.

The operation of mapping a service to its components available at VNMs layer can only be performed if topological information on the "integrated" network is available to the ICFNM system. This is accomplished by persistently maintaining an Integrated Network Model at the IMS Application Layer. The design of the Integrated Network Model is the key architectural issue of the IMS application layer and is a trade-off solution between:

• generality, to increase the system ability to easily integrate new networks
• expressiveness, to increase the system ability to support new management functions.

The chosen approach is based on an Object Oriented model of the integrated network. It achieves a balance between generality and expressiveness of the model by means of an interplay of classification and specialization techniques.

## 3.2   ICFNM physical architecture

The selection of a proper physical architecture needs to cope with different requirements, in terms of organizational processes, scalability and availability of the system.

The system needs to scale efficiently with the size of managed networks, be flexibly deployed into a distributed operational environment and be resilient to system faults. This permits optimized system deployment depending on networks size, characteristics of managed resources and traffic load patterns.

Different physical architectural options have been taken into account when designing ICFNM:

• Centralized ICFNMS: one Workstation(WS) hosts IMS and all VNMs
• Semi-distributed ICFNMS: a centralized Workstation hosts IMS, while VNMs are distributed over multiple WSs interconnected either locally with IMS or remotely with RNEM / RNM systems according to traffic patterns
• Fully-distributed ICFNMS: additionally to VNMs distribution, all different IMS applications are distributed over different interconnected workstations.

A thorough analysis concluded that a fully-distributed ICFNMS was the most suitable solution matching the driving requirements. In a fully-distributed ICFNM, VNMs, and IMS Applications may be freely distributed over multiple workstations interconnected via LANs and/or WANs thus permitting optimized and flexible system deployment.

## 3.3   ICFNM: Platform and enabling technologies

This section reviews the most critical components of the ICFNM development and run time platform in terms of relevant commercial products and enabling technologies chosen for the ICFNM system.

### Platform components

The design of a fully distributed system requires the selection of appropriate enabling technologies based on run-time and development considerations. For reasons described below, our choice was for the Common Object Request Broker Architecture (CORBA) (OMG, 1995). We selected IONA's Orbix implementation.

Specifically, CORBA Interface Description Language (IDL) was adopted to describe IMS interface to VNM layer as well as IMS internal object interfaces.

CORBA provides for a key part of our ICFNM communication infrastructure, complemented in the first release of the system by SNMP traps to support asynchronous communication. This was done because CORBA Event Services, while already defined by the OMG, had no product level implementation available at the design time for the first release of the ICFNM system.

Oracle Relational Database - interfaced by applications via an object encapsulation layer - HP OpenView Network Node Manager, X11R5, Motif were adopted as additional key elements for the definition of the ICFNM development and run time platform.

## *Rational for selecting the CORBA enabling technology*

The choice of CORBA has a significant architectural impact on the entire ICFNM system. The architectural perspectives presented below allow us to share practical experience of an early CORBA-technology adoption in the application domain of Integrated Service and Network Management.

a. *Information modelling perspective.*

The industry has acknowledged that Object Orientation provides the most suitable paradigm for managing network and system resources as well as for designing distributed applications. Notably OMG CORBA is gaining a growing consensus as a leading architecture for object distribution. CORBA technology can be effectively applied in the context of Network Management since it allows management systems to be aligned with the evolution of the computing world.

A joint OMG/ITU task force is defining a CORBA based architecture for TMN applications and addressing topics such as mapping GDMO and SNMP macros to IDL as well as mapping IDL into GDMO (Rutt, 1996).

Our experience proved CORBA IDL to be a powerful language for defining information models (comparable with GDMO and superior to SMI (Ashford, 1993)), and, in general, for application object interfaces.

The use of IDL allows for a clear separation between interface definition and implementation, thus completely decoupling the development of clients and servers. This allows the freedom, for instance, of using different languages like C++ and Smalltalk, and running on different operating systems. While this was not required in the first release of our system based on HP Workstations and HP-UX, we expect system evolution to require applications to be based on different platforms.

b. *Language implementation perspective.*

While separation between interface definition and implementation is also a key characteristic of GDMO / ASN.1, CORBA provides a much simpler solution to the same problem and it doesn't require designers to be familiar with complex abstract syntax notations or encoding/decoding schemes. IDL syntax is basically an extension to C++ and has been conceived with the goal of achieving a simple integration with implementation languages, allowing the mapping to C++ as well as to other languages.

CORBA also maintains the Object Orientation paradigm of TMN, supports Managed Object Classes and Managed Objects (the instances), and effectively supports inheritance to allow extensibility, evolution and specialization of object properties.

c.  *Communication perspective.*

From a communication perspective, CORBA allows a reduced communication overhead. Multiple naming contexts are endorsed where object locations are transparent to applications. This leaves the ORB charged with the localization of the proper object implementation, the implementation activation, the delivery of the method invocation and the reporting of results. Different activation policies allow a method base selection of persistency vs temporary activation.

Discovery of object services at run-time provided via a DII (Dynamic Invocation Interface) is an alternative to a static invocation of methods. The potential of DII is quite powerful in the ICFNM application context and allows the management of applications to be decoupled from the underlying technology evolution.

SNMP and TMN worlds foresee a unique management-agent paradigm, and recognize that different roles may be taken by the same application in different contexts.

Multiple communication paradigms are supported in addition to the client-server paradigm. Indeed publish-subscribe and peer-to-peer schemes are also available as third party commercial products.

d.  *Synergy between the development platform and the run-time environment through the product life-cycle.*

A main requirement for us was the ability to integrate effectively CORBA into our corporate software engineering computing environment. The OMT Object Oriented Analysis and Design methodology was applied to ICFNM system design, to describe application objects, to define the information models at the IMS-VNM border, the OSSs interfaces, the internal application interfaces and the database.

Adoption of Object Orientation was not limited only to analysis and specification stages, but was also applied to design, implementation, and support of ICFNM applications distribution.

A productive application of CORBA to the analysis and code construction phases was possible thanks to available commercial tools supporting the translation of OMT object models to IDL, and IDL compilers providing the mapping of object interfaces into C++ application skeletons.

e.  *Platform evolution strategy perspective*

CORBA technology is not dedicated to the telecommunication industry. It has been conceived and supported by the computing industry, resulting in lower technology costs at both development (libraries, development tools) and deployment levels (runtime licensing costs). The evolution of CORBA service definition and product availability (e.g., security) will provide a flexible platform on which applications targeting business needs can be built without the need to focus on support functions, possibly integrating software components from third party vendors.

# 4    CONSIDERATIONS ON ARCHITECTURAL QUALITIES

Deregulation, technology innovation and competition are motivating telecom operators to deploy timely and effectively new networking technologies and products.

Extensibility is a key architectural requirement for the ICFNM system, to allow for a simpler integration of new network technologies as well as for allowing the provisioning of new and modified management services. The Object Oriented approach, as well as the definition of a reusable and customizable virtualization layer (i.e., VNM) and CORBA are key architectural choices sustaining our reach for extensibility as seen in the previous sections.

An additional key architectural requirement for the ICFNM system is scalability. The system needs to cope with an increasing size of managed networks, number of customers, and services while offering adequate performance and response time to system operators, who are distributed across geographical areas. This objective is pursued by designing a fully distributed system, scaling efficiently with the size of managed networks which can be flexibly deployed into a distributed operational environment.

Support for distribution and scalability includes:
• clients of applications interfaced to the users execute on a local operator workstation
• flexible definition of a management domain at an operator level, (for instance, partitioning network domains according to managed Virtual Private Networks and functional criteria)
• definition of VNM classes, with VNM instances assigned to configurable domains and running on different workstations
• IMS application subsystems on different workstations. In case of computation intensive applications, such as of fault correlation and diagnostics processes, granularity of distribution may reach the individual process level, with one workstation allocated to one or several correlation processes.

The desired implementation of ICFNM has resulted in a system which may be freely distributed over multiple workstations interconnected via LANs and/or WANs, thus permitting an optimized deployment depending on networks size, characteristics of managed resources, and load patterns.

# 5    EARLY RESULTS

The first release of ICFNM system has been installed in the field for supporting Telecom Italia's growing needs for managing VPN and Outsourcing customers.

Benchmarks made on our deployed ICFNM system confirm the scalability, flexibility and performance of the system and indicate that CORBA is already a viable solution for distributed object-based telecommunication management systems. Detailed performance data is being collected from the field and processed in order to identify needs for VNM tuning as well as a practical feedback for the development of a new system release at the time of writing this paper.

A major result was the capability of designing and developing the system in an aggressive timeframe. CORBA support to object distribution has been the key factor in achieving this result. IDL proved to be immediately understandable to C++ software designers, allowing reduction of the ramp-up time for introduction of the new technology.

## 6    CONCLUDING COMMENTS

Telecommunications management business in the form of Outsourcing is a rapidly growing sector of the industry. Delivery of high quality outsourcing services to business customers at competitive costs requires a sophisticated organizational and technical infrastructure. This paper has introduced the concepts, drivers, and the overall role of Integrated Customer Focused Network Management System - an emerging management technology critical to the outsourcing business.

A technical approach for the development of ICFNM systems based on the object-oriented paradigm has been described. Specifically, this approach utilizes the CORBA development and run time enabling technology.

Our experience in advancing ICFNM technology suggests that the telecommunications industry is facing a number of challenges.

The integration of network management in general calls for an explicit recognition by standards bodies and industry associations (e.g., NMF) of the need to establish effective information models targeted for interconnected heterogeneous (from technology perspective) and hybrid (from ownership perspective) networks.

The further advancement of TMN concepts and standards, with respect to Service Management, is seen as an important enabler of reuse of information models in the development of ICFNM systems, as well as of interoperability among management systems and products within the overall TMN framework.

*Acknowledgments.*

We wish to thank V. Bigaroni and E. Pignatelli (Telecom Italia), G. Cortese, F. Macuglia, A. Marconi, and M. Peruso (Sodalia) for their valuable contributions.

## 7    REFERENCES

Ashford, C. (1993) Comparison of the OMG and ISO/CCITT Object Models. Joint NMF/OMG TF on Object Modeling.

Bigaroni, V. Calabrese, M. Long, D. Lumello, N. and Saracco, R. (1992) Toward open network supporting customer control services. Globecom '92. Orlando.

Feldkhun, L. Marini, M. Bigaroni, V. and Pignatelli, E. (1996) Integrated Customer Focused Network Management in a Heterogeneous High Speed Networking Environment (practical experience of Telecom Italia). Interworking'96, Nara, Japan.

Object Management Group. (July 1995) The Common Object Request Broker: Architecture and Specification, Revision 2.0.

Rutt, T. Mayne, R. and others. (1996) CORBA-Based Telecommunication Network Management System, Draft. OMG Telecom Special Interest Group.

Sloman, M. (1994) *Network and Distributed Systems Management.* Addison-Wesley.

# 3

# Customer facing components for network management systems

I. Busse, S. Covaci
GMD FOKUS, Hardenbergplatz 2, 10623 Berlin, Germany
Phone: +49 30 254 99 179, Fax: +49 30 254 99 202, Email: busse@fokus.gmd.de

## Abstract

The paper discusses the use of the WWW and Java in the provisioning of broadband connectivity to a big number of customers by a public network operator. Several options are analysed, ranging from a simple interface for service activation and usage to an integrated solution including outsourcing of customer network management by means of components dynamically loaded to the customer site. Besides the quick access to the service provided by the WWW, the paper describes the additional benefits drawn from the portability and extensibility of Java based solutions. Features like management functionality on demand, automatic release and update management of software, reduction in network load by moving functionality close to the data source are discussed in detail. The paper introduces an appealing alternative or complement to any network management solution. A prototype implementation currently under development is used for the validation of the proposed solutions in an intranet environment.

## Keywords

Broadband connectivity, network management, service management, outsourcing, mobile agents, java

## 1 MOTIVATION

Public network operators interconnect many customer premises networks via their broadband wide-area networks. While most of the customers already have network management solutions in operation to configure and monitor the state of their local networks they seldomly, if at all, get insight into public domains. Within public domains it is still very common that the setup and trouble shooting of connections is done by means of telephony and fax. This is a weak point for companies with distributed sites like for example distributed production plants that rely on wide-area connections.

A network operator is interested in providing his services with a minimum of human involvement in order to keep the cost down and to avoid human errors. Using network management solutions he can increase the availability of his network. The management solution can be offered to the customer or the provider can take over the task of network management also within the customer domain.

The customer wants to get an easy to use, reliable service at reasonable cost. This can only be achieved with an integrated network management solution including the national and international connections between the customer premises networks.

The following network management operations should be offered by a public network operator to the customers of his broadband connectivity service:

- subscription
- reservation, activation, modification, and deactivation of connections (if not done by inband signalling)
- testing of connections (prior to and during usage)
- receiving alarms
- trouble ticketing
- performance monitoring
- accounting.

There are several organizations working on interface specifications for this functionality:

Within the Internet community SNMP based specifications are developed [RFC1695]. The ATM Forum has defined management links between public and private networks in a protocol independent manner [M3] and adopted the Internet mib as a mapping to SNMP. They focus on the network management level, i.e., only connection set-up, performance monitoring, and fault alarms are covered. There are initial proposals for an extension to the service management level, e.g., the exchange of accounting information [ACCT96].

There are other initiatives also studying the customer-to-provider reference point, like the Network Management Forum and the Eurescom[1] projects. The Network Management Forum solution sets take into account the whole business process. The actual interface specification are provided in CMIP. The Eurescom work includes the service management level but is weak on the network management level that is currently restricted to virtual path configuration and alarm surveillance.

All the above mentioned work put the emphasize on the provision of a standardized management interface. Currently they use as management protocol SNMP and CMIP, and in the future Corba is considered an alternative.

For several reasons the provision of just a management protocol-based interface seems to be insufficient:

1. It is likely that the first contact to a connectivity provider will be made via a medium like the WWW. Here the potential customer can browse and compare offerings or let a broker facility select one according to its criteria. This is especially of interest if the connectivity provider is a player that is distinct from the network access provider, so that a customer is free to choose anyone according to his own preferences, e.g., the tariff.
2. Many customers, especially small companies, may not have management platforms installed and therefore no access to low level management protocol interfaces. In such cases it is useful to provide a ready to use graphical user interface for at least configuration and monitoring purposes. Also a help desk facility for the customer seems to be a necessary basic component.
3. Even if a management platform is available there is usually no support for the service management level. While there is a lot of off-the-shelf functionality to gather

---

1. Eurescom is an initiative of the European network operators.

network performance data and to collect and correlate alarms to isolate faults, the service management level functionality can hardly be used. So even for companies with installed network management solutions additional plug-ins or standalone solutions are necessary to enable access to the full functionality of the connectivity service.

4. If the connectivity service provider also offers to do the network management in the customer domain, additional functionality is required at the customer-to-service provider interface. The service provider must be able to install network management components at the customer site and enable an information flow for the network management level information in the reverse direction while a service management level information exchange still takes place as mentioned before. This scenario is often referred to as outsourcing of the network management activities.

5. Normally all software is updated sooner or later, thus requiring software distribution mechanisms for any kind of software that is downloaded and stored within the customer domain. This kind of functionality is currently not addressed at the considered management interfaces.

As we can see solutions (see Figure 1) range from a simple customer interface (a) and components for the integration into own legacy management platforms (b) to outsourcing of the network management (c).

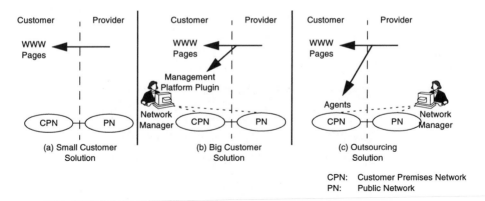

**Figure 1** Customer facing network management components.

In the subsequent sections we will show the usability of a WWW interface and of Java based mobile agents for the purpose of provisioning and management of an end-to-end connectivity service. The second section discusses a simple WWW based interface to access the broadband connectivity management. The third section describes the benefits we can draw from mobile agents in such an environment. Then we describe how agents can easily be implemented based on Java. The fifth section presents an example scenarios that we are currently prototyping. The paper is closed by describing the relation to other work in the area of client-server computing and mobile agents, and a final summary.

## 2 WWW BASED CUSTOMER-TO-SERVICE PROVIDER INTERFACE

As already indicated above the first contact with potential customers might take place via the WWW. A connectivity provider should be present with an appropriate service offering. To allow for comparison of competitive service providers, the service characteristics, e.g., pricing information, should be available and ready to use for brokerage services. This information is more or less static and can be retrieved using http from a freely accessible WWW server.

As the next step a subscription is required. It settles a contract between the customer and the provider and should enable the provider to bill its customer. Due to security issues this step still can't be done electronically. Nevertheless for the operation of the service some information, e.g., addressing information, has to be available in electronic form. This information can be entered by the customer and/or by a service administrator in the provider domain. The final goal should be to fully automate this procedure as soon as the security facilities and the legal base are available.

After a contract is "signed" the customer gains access to the relevant software components. During the contract negotiation or during the subsequent installation phase the necessary service components can be selected by the customer. The customer will be billed also according to the chosen components.

A simple application for the connectivity provisioning includes just a user interface for connection configuration. Additional packages may allow for alarm surveillance or performance monitoring. Since a customer might not have a legacy management platform we propose to offer an interface based on the WWW for this purpose. A prototype user interface is shown in the following Figure 2. All components share a common user interface as soon as they are installed by the customer.

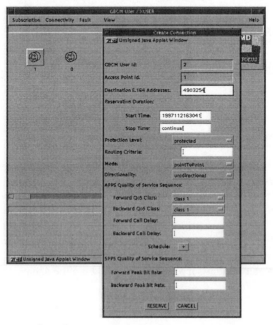

**Figure 2** Prototype user interface.

There are several options for communicating with the service provider domain. Due to the static nature of http in that it is hard to code state information this is implemented as a Java applet or application. The applet communicates with a dedicated server in the service provider domain (Figure 3).

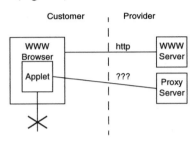

**Figure 3** Architecture and communication protocol options between customer and service provider.

At the customer-to-service provider reference point we can use the standardized interfaces based on SNMP/ CMIP, a private socket interface or Corba. To get a light solution it seems to be reasonable to settle on code that is included in the core Java distribution and is therefore already present at the customer site. SNMP and CMIP require special protocol implementations that have to be downloaded with the user interface since they are not included in the class libraries of the wide-spread browsers. For SNMP there are already two implementations, see [ADVENT] and [JASCA], that might be used. CMIP does not seem to be feasible at all due to its complexity.

A socket can be easily driven from within an applet and therefore provides a ready to use solution to any information exchange with the service provider domain. We decided to take a private socket based interface that carries the commands invoked at the user interface. This greatly reduces the required code at the customer site, leaving the mapping onto the necessary management interactions to the server inside the service provider domain. A number of platform providers like e.g. IBM and HP offer a string based API that can also very easily be used for this purpose [DERI]. With such an approach we might benefit from a standardized and thus portable string API to SNMP and CMIP.

For the future Corba [CORBA] and Java Remote Method Invocation [JRMI] are of interest since both will become part of the core Java and will thus be available in all Java enabled browsers. There is an increasing community promoting Corba in the standardization of service interfaces including the area of network management. The interfaces are developed separately or derived from existing mib specifications according to the SNMP SMI [RFC1157] or GDMO [ISO10165-4], making use of the mapping defined by the X/Open JIDM working group [XOJIDM]. Nevertheless, the Corba environment still lacks the management facilities provided by SNMP or CMIP agents. CMIS allows to perform operations on many objects or attributes specified e.g. by the means of scoping and filtering.

To keep applets from spying networks behind firewalls many browsers limit the destinations for socket set-up to the host from which an applet was loaded. Therefore the usability of the browser for the purpose of network management is poor since an applet cannot connect to an arbitrary network or network element manager. A workaround might be a proxy server for management requests within the provider domain.

In any of the afore mentioned solutions it is useful for such an implementation to be stored locally. This limits the number of downloads to the time of the subscription and the subsequent changes of the software configuration. Also service specific functions like logging of configuration requests and faults can be implemented more easily with the ability to save state information to the local disk. Unfortunately, the security restrictions of the available browsers do not allow any local disk access. It is desirable that a user of a browser can configure at least a single directory for being accessed by the downloaded applets.

An overview of the security restrictions is given in the following table. Note, that it makes a difference whether an applet is loaded over the network or from the local disk. Of course, the basic Java Virtual Machine (JavaVM) grants all permissions to an application.

**Table 1** Overview of security restrictions

|  | *Netscape Navigator* | | *AppletViewer* | | *JavaVM* |
|---|---|---|---|---|---|
|  | *net* | *local* | *net* | *local* | |
| file access | no | no | access control list | yes | yes |
| socket to any host | no | yes | no | yes | yes |

## 3 EXTENSIBLE AND MOBILE AGENTS - WHY?

An even more interesting approach is the possibility of using Java for the implementation of management agents. This allows to extend the functionality of an agent on demand. A further step would be to have mobile agents. Just a JavaVM needs to be in place. If a specific management task has to be fulfilled a roaming agent is sent around. It can provide interfaces supporting all protocols that are fully implemented in Java. Note, that a browser is not sufficient for this task, since agents should be able to do processing even if the user interface is not running.

There are several benefits we can draw from this approach:
1. In principle, there is no need for any running management agent if there is no manager that requires a specific information. As such the computing and communication overhead can be reduced by putting the agent first in place when a specific management action has to be performed and to destroy it when there is no longer a need for it.
2. Certain management tasks, like gathering usage data and computing charges, for example, can be located close to the real resources. Then we have only the reduced accounting data that need to be transported. This would reduce the communication overhead and thus the network load.
3. Sooner or later each piece of software is outdated, e.g., because a bug was discovered and fixed or because a new environment requires a different or extended functionality. An agent or the functionality within an agent can be replaced by a new

version of the software from remote. This is a basic functionality available in an agent execution environment.

4. The ability to move functionality to several places for execution allows for load balancing and results in a fault tolerant system. Agents can be directed to places with free computing power. In case an agent fails the functionality can be loaded in a different place on another host. Each host running the agent execution environment can act as such a back-up system.

5. A provider can react flexible on the changing environment in a customer domain within the already mentioned outsourcing scenario. A roaming agent can collect configuration data about the customer domain and return to the provider. Further agents are configured and sent to the customer domain according to the equipment installed and reported by the spy, and the management functionality required without an interaction between the customer and the provider. In case of special fault conditions additional agents for fault analysis can be sent to the customer domain performing analysis tasks.

## 4 EXTENSIBLE AND MOBILE AGENTS - HOW?

A suitable environment for extensible and mobile agents requires an execution environment that is available on a number of platforms. Code and state information needs to be portable from one execution environment to the next. Java seems to provide a good base for this purpose.

The JavaVM is available on all common computing platforms and provides a machine independent execution environment. It was designed for downloading and executing code from WWW servers therefore the basic functionality for moving code is available. A first step for moving the state of objects is provided by the object serialization facility [JOS] currently released by Sun. It allows to stream an object including all referenced objects with minimum effort. Based on object serialization the remote method invocation facility [JRMI] allows for passing arbitrary objects from JavaVM to JavaVM between two networked hosts.

Our agent execution environment has four basic components: agencies, agents, a name server and a class server. In the following we will roughly describe the purpose of each component.

### 4.1 Agency

An agency has three basic objects: a factory, a context and a security manager. The factory provides a remote interface. The method createAgent(Agent) implements a generic agent factory that is needed to create arbitrary agents from remote. Each factory can be uniquely identified with the hostname and the port number. It is registered at the central name server.

When an agent is created its object reference is stored within the context object in order to keep track of all agents at an agency. Then the agent is started within a separate thread executing the method run() of the agent object. Such an agent factory at each place provides the minimum infrastructure for mobile agents.

The security manager object will be based on the security feature of JDK 1.1 and is not yet fully implemented.

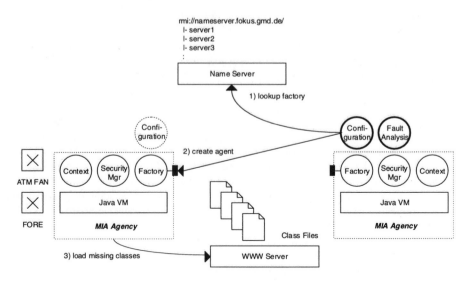

**Figure 4** Mobile intelligent agent (MIA) execution environment.

### 4.2 Agent

The class Agent implements the behavior that is common to all agents. It is abstract and cannot be instantiated. Within a subclass the specific behavior needs to be implemented first.

An agent has two basic attributes: agent identifier, that is globally unique, and class-base, to locate and load missing code over the network. Note, that there might be code missing even after the agent ran for many hours and changed its place several times. So from any place where code is detected to be missing the execution environment must be able to access the classbase of an agent. It can be either a WWW server (identified by "http://...") or another agent execution environment, i.e. a class server, (identified by "rmi://...") that also provides a mechanism for retrieving classes.

An agent needs to implement the runnable interface, i.e. a method run(). The method run() is executed whenever an agent is created or arrives at a new place. It implements the behavior of an agent.

The method move() allows an agent to change the place it is running in. It calls the method createAgent() at the new place and passes itself as the initial state. Then the local entry of the context is deleted. Since the JavaVM does not allow to continue directly after the call of move() on arrival at the new place, the method run() will be entered again from the very beginning. This is a drawback since it slightly complicates the programming of the behavior within the method run().

The usual method invocation can be used for the interaction between agents residing at the same place and thus within the same JavaVM. Remote method invocation is supported for interacting with agents residing at different places.

## 4.3 Name Server

A simple name server is provided by Sun within the remote method invocation facility. It is the java.rmi.Registry class. The implementation allows to resolve URLs to object references. Unfortunately, it provides only a flat name space.

## 4.4 Class Server

The class ClassServer provides only the method getClass() at a remote interface. It can be used to load missing code from the originating JavaVM of an agent. Alternatively a WWW server might be installed to act as the class base.

When passing an agent object via the remote method invocation interface to a new place, all local objects referenced directly or indirectly by the agent object are copied as well. Therefore an agent can be implemented as several Java objects. The object serialization algorithm will go over the entire graph of objects and write it to a stream, so that it can be completely reconstructed on the receiving side. It preserves identity, so that two object references which are pointing to the same object will also point to the same object in the new place. Furthermore cyclic structures are detected and reconstructed accordingly.

This is not done in case of a reference to a remote object. References to remote objects form the border of an agent. In fact they are lost when we don't care about them and an agent is moving from one host to the next as we will discuss under the list of open issues below.

The execution environment at certain place can have special features which are not coded in Java and are therefore not movable to other locations. Indeed this is an approach which is followed in some agent based network management solutions. The JavaVM is placed within a management platform which provides the communication stacks for speaking SNMP and CMIP to the outside world. For the Java program there is a special API for accessing these communication stacks. Here the agent facility is limited to download new code. This seems to be a good intermediate solution until implementations of communication protocols completely coded in Java become available. They are necessary for the agents to be usable in any place. At least for SNMP there already exist two implementations as already mentioned above.

The mobile agent technology is seen as a complement to client-server computing rather than as displacement for it. First of all the remote method invocation and thus client-server computing enables the mobile agent execution environment since it is used to pass state and code of agents between different hosts. Furthermore, agents can use the remote method invocation to communicate with remote agents providing a service.

## 4.5 Open Issues

Currently object addresses are used to reference remote objects. Thus each move of an agent that offers a service at a remote interface results in open references. There is no possibility to locate the agent afterwards without additional mechanisms. One solution is to keep a forwarding entry at each place which has been visited by an agent. Unfortunately, this would result in very big tables at each host and waste a lot of computing time for resolving a reference as soon as the number of agents and hosts increases. A better solution uses object names to store references. Object names can then be resolved at the central naming server.

It is not possible to transparently keep client-server relations via remote interfaces alive while an agent is moving from one place to another. Note, that it does not only affect the moving agents itself but also the clients that are connected to a service offered by the moving agent. Where a mobile agent is acting as a client it can simply close down the connection to its server and re-open it connecting to the same address from the new location. Where a mobile agent is acting as a server it must close the incoming connections and free the port. At the new location it has to choose a new port where it will offer its service. Then the service of the moved agent needs to be announced under the old name with the new address at the central naming server. This enables the clients which are bound to the service to re-open there connections after resolving the service name again via the central naming server.

It is possible to model relationships between agents. They are also modeled as remote objects. Thus the simple copy operation invoked during remote method invocation will stop when encountering a reference to a relation object. In such a case a new copy operation has to be provided that takes care of the relation graph and traverses it. Whether a referenced object will be copied or not is decided based on the attributes of the relation. This approach is taken in the COSCompoundLifeCycleService of Corba [COS]. It might be helpful to have a similar service in our environment.

A very important issue is security. Currently there are security restrictions neither at the remote interface for creating agents nor at the remote interface for retrieving classes. If the environment should be opened to a public network where we have to deal with malicious agents at least authorization and authentication mechanisms are required.

Mechanisms to make code and state persistent simplify the implementation of agents. Code and state persistence is of special interest if loading an agent over the network consumes a lot of resources and thus also takes a long time. State persistence is always necessary if the state can not be gathered from other resources. A good example are accounting records which cannot be generated again. Thus a persistence service is called for.

## 5 EXAMPLE SCENARIO

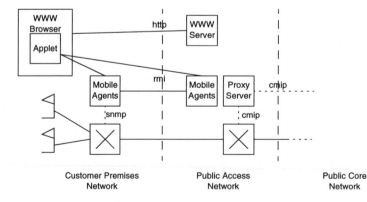

**Figure 5** Example scenario.

The connectivity service provider has set-up a WWW server as well as an agent execution environment in his domain. On the WWW server he provides freely accessible information about his service and a form for subscription. Furthermore a closed environment for configuration and supervision of the connectivity service.

In case of an outsourcing scenario the customer first subscribes to the service. Then the customer administrator has to download the basic agent execution environment. It has to be installed, at least once, in each of the customer premises networks. During configuration of the connectivity service the customer passes the addresses of these places to the service provider. This allows the service provider to send around an initial agent and discover the equipment that is installed in the customer domain. In our prototype we use ping and SNMP requests to detect hosts and ATM switches. Then initial agents for connection and IP configuration are sent to the customer domain.

As soon as the infrastructure is in place the customer gets access to WWW pages to configure and supervise its connections in its own and the public networks. He can set-up and clear down ATM virtual path and virtual channel connections. In addition monitoring of cell counters can be done in real-time. The graphical user interface can be accessed from each WWW browser, wherever the customer administrator is located.

As soon as a misbehavior is detected an alarm is raised and sent to the network administrator in the service provider domain. Furthermore a trouble ticket informs the customer about the problem and the estimated time of recovery. The service provider can now use additional agents to test the equipment in the customer or its own domain and track down the fault to a specific part of the network. As long as the problem can be solved without changes of hardware modules there is no need for the network manager to send out a technician.

Note, that the scenario still includes a proxy server to access equipment in the public domain via CMIP. This is required since the network and network element level interfaces as well as the cooperative interfaces in public domains are currently standardized based on CMIP. Nevertheless, towards the customer we are in favor of a WWW and Java based solution due to its minimum requirements concerning the hardware platform and its simplicity for a customer and the service provider.


## 6 RELATION TO OTHER WORK

There are many agent systems that are currently under development not only in research projects but also in industry. Telescript by General Magic is likely the most elaborated agent environment since the requirements of mobile agents were taken into account from the very beginning of its design. For example the execution at a new place starts directly behind the move statement what greatly simplifies programming. Others include Agent-Tcl by Dartmouth College or ARA by the University of Kaiserslautern.

Most of the well established agent platforms provide more elaborated features for agent programming than Java. Nevertheless, there is very little support for a global usage. We have chosen Java because of its market strength and the upcoming support on nearly all computing platforms. Furthermore, it is very easy to build a basic agent environment. There are several other agent platforms based on Java currently under way, e.g., Java-to-Go by the University of Berkeley and the Aglets library by IBM Tokyo.

Another technology of interest in our field is Corba. Corba provides the base for client-server computing. It allows to pass the state of objects. Unfortunately it is not possible to pass code since there is no virtual machine by default but the object behavior is compiled into native code. In order to achieve code portability one might restrict the implementation environment to a specific language that is compiled to a virtual machine, like Smalltalk or Java for example. Corba might be an alternative for doing the remote method invocation but since we have a homogenous Java environment we don't need the language independent interface description and coding. Nevertheless there are many services already specified and partially implemented which are of interest also to an agent based environment, e.g., the event service. Here we can definitely benefit from Corba compatible interfaces.

# 7 SUMMARY AND OUTLOOK

We have shown how the WWW and Java can be applied in the field of broadband connectivity service provisioning. This solution puts only minimum requirements on the computing platform at the customers site since Java is light weight and portable. Software components can easily be updated or extended by a service provider. As soon as Java will be part of operating systems, as it has already been announced by many computer vendors, like Microsoft, Sun, and IBM for example, it will be available from scratch on nearly all hosts. It is foreseen that such solutions enable the forthcoming network computing technology as drawn by Oracle.

Furthermore, the usage of WWW reduces the human interaction required on behalf of the service provider. Both the customer as well as the service provider can benefit from the simplicity of WWW based interfaces. Therefore the solution is likely to reduce cost at both sites. Via the WWW a customer can easily find a service provider that meets his requirements and a service provider can reach an increasing number of (potential) customers.

The basic idea of WWW and Java based network management can be used in any area of network and service management. For example it also provides a good base for intranet management solutions that can be sold to customers. In this case both the customer and the service provider can benefit from the fact that the implementation can run on any platform with a JavaVM. Furthermore software release and update management can be simplified by setting up a server that is contacted by the application on start-up or from within a running process without an interaction of a user.

Before such a solution can be applied at a public customer-to-service provider reference point the open security issue needs to be addressed. To protect both the customer as well as the service provider side from malicious attacks through the mobile agent facility. A simple WWW based interface can already be set-up since the security requirements can normally be satisfied by a firewall.

# 8 REFERENCES

[ACCT]      Greene, W., Heinanen, J., McCloghrie, K., Prasad, A.: "Managed Objects for Managing the Collection and Storage of ATM Accounting Information", Internet draft, IETF, April 1996, work in progress.

[ADVENT] Advent Network Management, Inc.: "Advent Network Java SNMP Package", Version 1.0 Beta, http://www.adventnet.com/snmp_api.html, 1996.

[CORBA] OMG: "The Common Object Request Broker: Architecture and Specification", Revision 2.0, July 1995.

[COS] OMG: "CORBAservices: Common Object Services Specification", Revision 1.0, March 1995.

[DERI] Deri, L.: "Network Management for the 90s", to be published.

[DMFA] Garcia-Lopez, E.: "Distributed Management Facilities Architecture", Version 2.0, TINA-C, March 1996.

[ISO10165-4] ISO/ITU-T: "Guidelines for the Definition of Managed Objects", ISO/IEC 10165-4, ITU-T Recommendation X.721, 1992.

[JASCA] Nikander, P., Wessman, P.: "Java SNMP Control Applet", http://termiitti.akumiitti.fi/nixu/, 1996.

[JOS] Sun Microsystems, Inc.: "Java Object Serialisation Specification", Revision 0.9, May 1996.

[JRMI] Sun Microsystems, Inc.: "Java Remote Method Invocation Specification", Revision 0.9, May 1996.

[M3] ATM Forum: "Customer Network Management (CNM) for ATM Public Network Service (M3 Specification)", Revision 1.04, October 1994.

[RFC1157] Case, J., Fedor, M., Schoffstall, M., Davin, J.: "A Simple Network Management Protocol (SNMP)", RFC1157, IETF, May 1990.

[RFC1695] Ahmed, M., Tesink, K.: "Definitions of Managed Objects for ATM Management Version 8.0 using SMIv2", RFC1695, IETF, August 1994.

[WEST] Westerkamp, E.: "Automatic Software Distribution of Java Applications".

[WREG] Wreggit, D. J.: "Software Agents Using Java", December 1995.

[XOJIDM] X/Open Company Ltd.: "Inter-Domain Management Specifications: Specification Translation", Second Sanity Check Draft, June 1996, work in progress.

4

# A VPN Management Architecture for Supporting CNM Services in ATM Networks

*J. T. Park[1], J. H. Lee[1], J. W. Hong[2], Y. M. Kim[3], and S. B. Kim[3]*

*[1]School of Electronic and Electrical Engineering*
*Kyungpook National University, Taegu, Korea*

*[2]Dept. of Computer Science and Engineering*
*Pohang University of Science and Technology, Pohang, Korea*

*[3]Telecommunicatoin Network Research Lab.*
*Korea Telecom, Taejeon, Korea*

## Abstract

As enterprises use ATM networks for their private networks and as these private networks use public ATM networks for wide area communication, the need for the customers to be able to manage both private and public networks is increasing. Currently, some standardization work is being done towards providing this capability to the customers. In this paper, we propose a virtual private network (VPN) management architecture for supporting integrated customer and public network management in ATM networks. The key component of this architecture includes an integrated customer network management (CNM) agent and public network manager. It also incorporates a CORBA-based shared management knowledge (SMK) system to provide a distributed processing environment for the exchange of management information between managers and agents.

## Keywords

Integrated customer and public network management, VPN Management, CNM agent, public network manager, TMN, ATM network, CORBA-based SMK system

# 1   INTRODUCTION

As private networks need to communicate each other (due to enterprises having multiple private networks throughout the country or throughout the world), these private networks often use public networks in order to exchange information between private networks, thereby, forming virtual private network. In such situations, network administrators of private networks generally would like to be able to monitor and manage their corporate network as well as public networks which their private networks use. Currently, this is a very difficult task mainly due to a couple of reasons: 1) administrative and 2) technical. The first reason is that public network administrators do not desire their networks to be openly managed by private network administrators. The second reason is that there is no clean mechanism available for the private network administrators to be able to manage any portion of the public networks. Recently, however, the trend is to allow the private network administrators to manage a portion of the public network essential to the operation of virtual private networks. Our work focuses on the technical aspect of allowing the public networks to be managed by private network administrators.

Some work has been carried out by standardization bodies such as ATM Forum, ITU-T and ETSI for providing the solution [4, 7]. The work mostly focuses on defining the interfaces for ATM network management and on the specification of management information base (MIB). ATM Forum has specified five management interfaces for the exchange of management information for ATM networks. Among them, M3 is a management interface between customer network management system (CNMS) and public network management system (PNMS), and M4 is the interface supporting network element level and network level management of public ATM networks. The M3 service is provided by the public network provider via customer network management (CNM) agents in the provider's network. CNM is a concept providing capabilities for network providers and their customers to exchange management information [1]. CNM services provide the customer using the provider's network with the capabilities to manage the portion of the public ATM network, which is critical to the operation of their virtual private networks.

Earlier work on CNM service provisioning stated the necessity of interworking between the CNM agents and public network managers [1, 5]. However, they did not provide solutions for the efficient interworking. Our work presented in this paper attempts to provide that solution. In particular, we propose an architecture for public network management system supporting CNM services. We have developed a public network manger which integrates CNM agents based on the M3 and M4 standard management interfaces. Our integrated network management system incorporates a CORBA-based Shared Management Knowledge (SMK) system [9, 12] which provides a distributed processing environment for the exchange of management information between managers and agents.

This paper is organized as follows. Section 2 discusses CNM and interworking requirements. Section 3 presents our proposed architecture for integrating CNM agent and public network manager. Section 4 presents a prototype implementation carried out as a proof of concept. Finally, a summary and possible future work are given in Section 5.

## 2    CNM AND INTERWORKING REQUIREMENTS

The ITU-T recommendation X.160 [4] defines an architecture for the Customer Network Management Services for public data networks. The CNM functional architecture (shown in Figure 1(a)) is composed of several function blocks that provide the functionality needed for CNM. The customer's management function provides the customer-related CNM functionality using the CNM functions provided by the service provider. The service boundary between the customer's management function and the CNM function is defined by the CNM reference point. Management information is exchanged between the customer's management function and the CNM function through the CNM reference point.

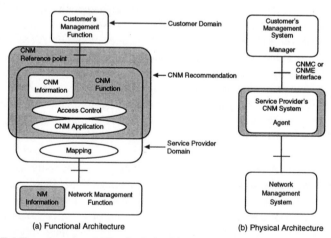

Figure 1  CNM Functional and Physical Architectures

As shown in Figure 1(a), the CNM function comprises several functional components: CNM information, access control, CNM application and mapping. The CNM functional architecture can be mapped onto the TMN functional architecture. The customer's management function and the CNM function correspond to the operations system function in TMN and the CNM reference point can be mapped onto the TMN X-reference point [1]. The CNM physical architecture(shown in Figure 1(b)) is composed of customer's network management system and a service provider's network management system. These two management systems are connected by the M3 interface.

CNM manager communicates with CNM agent within the public ATM network through the M3 interface. This communication takes place using SNMP over UDP. At the physical level, the communication may use an ATM UNI or dedicated circuit. The M3 specification is classified into two classes to allow public network providers to offer modular, incremental capabilities to meet different levels of customer's needs. The first class of M3 functions, Class I, requires that a public network provider to provide information on the configuration, fault and performance management of a specific customer's portion of a public ATM network. The second class of M3 functions, Class II, requires that a customer can request the addition, modification or deletion of virtual connections and subscription information in a public ATM network. In a nutshell, Class I provides monitoring capability only whereas Class II provides controlling capability as well.

Figure 2 shows an interworking structure between a CNMS and public NMS, in particular the interworking of a CNM agent and a public network manager. Recall that the public network management system is to provide either the Class I or Class II services to the customers. It is possible to provide the Class I service even if the CNM agent manages public network without interworking with the public network manager. However, it is absolutely necessary for the CNM agent to interwork with the public network manager if the Class II service is to be provided. This is mainly due to the control (i.e., add/modify/delete) capabilities associated with the Class II service provided to the customers who may potentially cause problems if they are not properly mediated by the public network manager.

Some work has been done for enabling the CNM agent to provide CNM services within the public networks. Most of these work thus far have proposed new information models or implementation methods to improve the performance of CNM manager and agent. However, the interworking mechanism between public network manager and CNM agent and the relationship between M3 MIB and M4 MIB in public

ATM network elements have not been studied. In the next section, we present our work on a framework which includes the interworking mechanism.

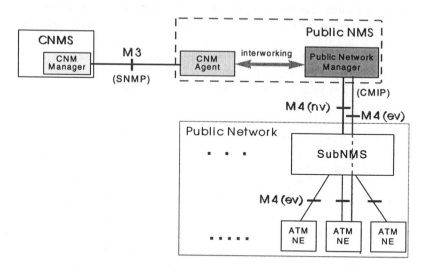

Figure 2  Interworking between CNM agent and public network manager

# 3   INTEGRATED PUBLIC NMS FOR SUPPORTING CNM SERVICES

Figure 3 illustrates the proposed architecture of a public NMS which integrates a public NMS and CNM agent. The integrated public NMS provides CNM services to CNM managers. The unique feature of the architecture is that it incorporates shared management knowledge (SMK) system [9, 12] to enable distributed management of CNM and public NMS.

In Korea, the government is trying to deregulate the communication service environment in a way that several regional public ATM network service providers are allowed to provide multimedia communication services. In this case, a customer may need to access several carrier management systems, i.e., public NM systems to manage its own VPN. Each public NM system may require access to several subordinate network management systems (i.e., subNMSs) to accomplish end-to-end management tasks. Consequently, VPN management requires a multi-level hierarchical manager-agent structure based on TMN where there are many-to-many relationships between managers and agents. This requires a distributed management architecture and an efficient interaction mechanism for managers and agents which are distributed across the network and interconnected in a hierarchical way. This distributed management functionality is accomplished by CORBA-based SMK system as shown in Figure 3.

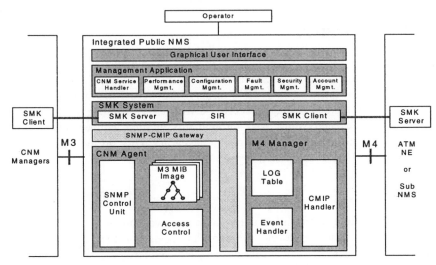

SIR : SMK Information Repository          SMK : Shared Management Knowledge

Figure 3  Integrated Public NMS Architecture

This architecture is composed of management applications, CNM agent, M4 manager, SNMP-CMIP gateway and SMK system. As shown in Figure 3, a CNM agent consists of M3 MIB image, Access Control, and SNMP Control Unit. This CNM agent realizes the CNM function modules described in Section 2. These major components of the integrated architecture are described in detail below.

## 3.1  CNM Agent

The CNM agent provides CNM services to CNM managers. As shown in Figure 2, the interaction between the CNM manager and CNM agent is defined by the M3 interface using SNMP as the management protocol. The SNMP control unit receives requests from a CNM manager and passes them to the M4 manager through the SNMP-CMIP gateway. It also handles replies and event reports from the M4 manager and forwards them to appropriate CNM managers. The access control is used to check the permission of management requests and controls the access of management information. There may be multiple CNM managers accessing a single CNM agent, information on these CNM managers is also stored here. Since the actual MIB is located in the managed objects of the public network, M3 MIB image is used to reflect the actual MIB and is maintained by the CNM agent.

## 3.2  SNMP-CMIP Gateway

All the management operation requests of CNM agent for public NM applications are performed through this SNMP-CMIP functional module. This module translates SNMP requests of CNM agent into understandable forms of CMIP and forwards it to the M4 manager. Also, when the M4 manager receives CMIP replies and event reports, those intended for CNM managers must pass through this module, converting them into those that can be handled by the CNM agent. Our earlier work on the CMIP/SNMP gateway function [13] has been applied here.

## 3.3  SMK System

The exchange of management information between two function blocks in TMN requires a common view about the protocol knowledge, management functions, managed object classes and their instances, and authorized capabilities. Such information is collectively referred to as the shared management knowledge (SMK) and is defined in ITU-T M.3010 [8]. In order to perform management functions, the knowledge on the specific role as a manager or an agent is necessary in addition to the specific options for each function. It is necessary to discover the managed object classes and their instances for each management communication interface pair.

In Figure 3, the SMK server in the integrated public NMS performs the functionality of providing the information on CNM agent which is related to M3 MIB image, access control and SNMP control unit to the SMK client. SMK Information Repository (SIR) exists in the integrated public NMS, and it is the place for the storage of SMK information. The SMK server can, with direct access to the M3 MIB image, obtain the management information without accessing the SIR. In case where the contents of M3 MIB change very often, and the size of M3 MIB is relatively small, it may be more efficient to directly access the M3 MIB to get the SMK. However, if the contents of M3 MIB are mostly static, and the size of MIB is large, it may be more efficient to store them in SIR in advance. This can be achieved by making the SMK server access the SMK information and store them in SIR when the agent process is activated. As the SMK client requests the SMK information, the SIR is accessed for the provision of required SMK information. In this case, M3 MIB is only accessed once each time the agent process is activated. The management of SMK information in a sense is similar to the view maintenance problem in distributed database management system. The suitable SMK management strategy may depend on the update frequency and size of M3 MIB, the complexity of M3 MIB access functionality, and of course the semantic correctness of SMK update operations [9]. We have designed and implemented this SMK management system which are described in more detail in Section 4.

## 3.4 M4 Manager

The M4 manager performs management interactions with public network management (M4) agents on behalf of the public network management applications as well as the CNM agent. Management requests from the public NMS come from the public NM applications directly whereas the CNM requests comes from the CNM agent via the SNMP-CMIP gateway. These requests must be distinguished by the M4 manager so that the replies and event reports from M4 agents can be delivered to appropriate managers. The LOG table is used to log all the interactions that take place within the integrated manager.

One of the important subcomponents within the M4 manager is the event handler. Event notification capability provides a customer with the ability to receive unsolicited events from public networks upon detection of abnormal conditions related to the customer's access. Event notifications could include changes in the status of ATM UNI configuration parameters, PVC configuration parameters, or access authorization failures. Event notifications are helpful to the customer for isolating failures in the customer's network. M4 agents report status changes and faults of specific customer-related NEs to M4 manager. When NMS receives an event report from an M4 agent, it determines whether the received event report relates to a specific customer's portion of public ATM network. If so, the event report is transmitted to the public network management application and the CNM agent. Thereafter, the CNM agent delivers it to the appropriate CNM manager by using an SNMP Trap. Figure 4 shows the procedure of event report handling.

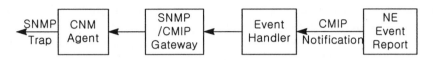

Figure 4   Procedure of event-report handling

Here, we describe the concept of M3 MIB Image which has been incorporated into the integrated public NMS in order to increase performance. Consider the case where a public network provider provides the M3 Class I service. When a CNM agent receives a SNMP GET request, a normal processing would require that this request be passed to the M4 manager via the SNMP-CMIP gateway. Then, the M4 manager would get appropriate information from an M4 agent. This process may take a long time for a simple GET request. Instead, in order to reduce the response time, we have installed the M3 MIB image within the CNM agent and it is updated periodically by the M4 manager. Basically, the M4 manager periodically polls the customer-related M4 MIB from the M4 agent and updates the M3 MIB image. When a SNMP request is received, the value of M3 MIB image can be returned immediately. When a specific event occurs which requires modification to the M3 MIB image, the M4 manager transfers the event to CNM agent through event handler directly. The procedure of the Class I service with the use of M3 MIB Image is illustrated in Figure 5.

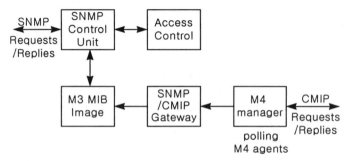

Figure 5   Procedure of Class I services

When a CNM agent receives a SNMP SET request for the update of management information, in the case that network provider supports the M3 class II service, it invokes the M4 manager for the provision of update services. The M4 manager, in response to the request from the CNM agent, stores service profiles in LOG table and requests M4 agents to update the related management information in ATM NE directly or via subNMS indirectly. The M4 manager is responsible for modification, addition and deletion of connections and subscription information in public ATM networks according to customer's requests. The response of M4 agent is transferred to a M4 manager and then to CNM agent based on the information stored in LOG table. The procedure of providing the Class II service is illustrated in Figure 6.

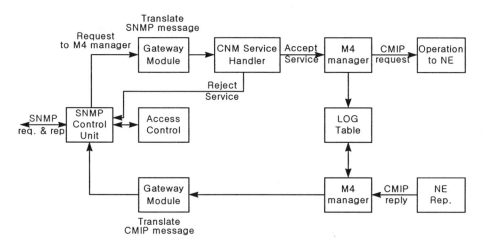

Figure 6  Procedure of Class II services

## 4    PROTOTYPE IMPLEMENTATION

Figure 7 illustrates the architecture of a prototype implementation of the integrated CNM agent and M4 manager, M4 agent and SMK system. We used the OSIMIS platform which was developed at UCL [10] for implementing manager-agent interactions. It provides an environment for the development for management applications which hides the details of the underlying management service through the object-oriented application program interfaces (APIs) and allows implementers to concentrate on the intelligence to be built into management applications rather than the mechanics of management service/protocol access. It also provides an implementation of the Internet SNMP over UDP using the ISODE ASN.1 tools and a generic CMIS to SNMP application gateway driven by a translator between SNMP and OSI GDMO MIBs.

In the SMK system, the CMIS scoping in the M4 interface and SNMP in the M3 interface is not a suitable technique for the identification of complete set of MO classes, and a more generic identification mechanism was needed [9]. Thus, we used the CORBA-based approach. The SMK client object of a management application is associated with the SMK server object at the managed system through the communication facility provided by the underlying ORB. The SMK information can be accessed by using the interface which is described in Interface Description Language (IDL). CORBA IDL and ORB implementation embedded in ORBeline [11] from PostModern Computing Technologies was used as a CORBA implementation. The communication mechanism of the CORBA ORB enables the location transparency between distributed objects to be supported, and this created the dynamic, efficient distributed processing environment.

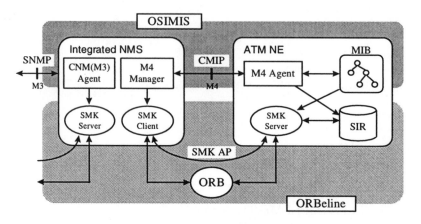

Figure 7  Prototype implementation using OSIMIS and ORBeline

The definition of SMK interface for a manager is shown in below. This SMK interface is one that a manager invokes to acquire SMK. A management agent implements SMK server and provides SMK information. A manager accesses information on SMK through this interface. SMK interface is also described in IDL and corresponds to a CORBA object. IDL interface has a following structure.

```
Interface <identifier> [<inheritance_spec>]
{
        <interface_definition;>*
};
```

<interface_definition:> includes the declaration of type, constant, exception, operation, and attribute. In the declaration of operations, the operation names and parameter types are specified below. The operation Get_ProtocolInfo identifies the type of management information protocol: CMIP or SNMP. The operation Get_AccessPolicy checks the access policy upon reception of request from the SMK client. The meaning of most

operational primitives are self-explanatory. The details of these specifications can be found in [9].

The SMK server process can be generated by forking the agent process. An agent generates a child process after initializing the MIB when running the program. The child process creates the SMK server object, and brings in the reference point to the MIB. The design algorithm for creation of SMK server object process at an agent is described below.

**Algorithm Server-Process (environment_variable)**

```
Begin
  initializeSyntaxes(syntaxes);
  agent.initialize(environment_variable, service);
  MO::initializeMIB(MIB_INIT_FILE);

  // create SMK server process
  if (fork() == 0) {
         // create SMK server object
         SMKObject *smkObj = new SMKObject(object_name);
         // get reference pointer to MIB
         MO **root = MO::getRoot();
         smkObj->GetMO(root->getWholeSubtree());
         CORBA::BOA::impl_is_ready();
  }
  coordinator.listen();
End
```

## 5  CONCLUSION

ATM network management system is absolutely necessary for providing the future communication services in ATM networks. The research in ATM NM is being carried out by many researchers as well as by standardization organizations. In this paper, we focused on the M3 and M4 interfaces for the purpose of CNM service provisioning and public network management.

As a solution to the problem of managing public ATM networks by the CNM systems, we proposed a VPN management architecture of an integrated public NMS supporting CNM services. The proposed architecture has functional modules for supporting both CNM functions and public network management applications. The prototype implementation of our design was made by using the OSIMIS platform and ORBeline.

For future work, we plan to investigate the mechanism of information exchange between public NMSs through the TMN X interface and interworking with CNM agents in the case where the customer's PVCs traverse multiple public networks. Another future work is to do a performance analysis of our prototype implementation. We plan to deploy our prototype in an operational ATM testbed of Korea Telecom.

## 6   REFERENCES

[1]   Jacqueline Aronsheim-Grotsch, "Customer Network Management CNM," Proceedings of NOMS'96, April 1996, pp. 339-348.

[2]   Open Management Group, "The Common Object Request Broker: Architecture and Specification," December 1993.

[3]   Henry J. Fowler, "TMN-Based Broadband ATM Network Management," IEEE Communication Magazine, 33(3):74-79, March 1995.

[4]   ITU-T Recommendation X.160, "Architecture for Customer Network Management Service for Public Data Network", 1994.

[5]   Michael Hinchliffe and Nigel Cook, "Customer Network Management and ATM Networks," Proceedings of NOMS'96, April 1996, pp. 155-164.

[6]   Miyoshi Hanaki, "LAN/WAN Management Integration using ATM CNM Interface," Proceedings of NOMS'96, April 1996, pp. 12-21.

[7]   ATM Forum M3 Specification Revision 1.03, "Customer Network Management for ATM Public Network Service," 1994.

[8]   ITU-T Draft Recommendation M.3010, "Principles of Telecommunication Management Network," 1995.

[9]   J. T. Park, S. H. Ha and J. W. Hong, "Design and Implementation of a CORBA-Based TMN SMK System," Proc. of NOMS'96, April 1996, pp. 64-74.

[10]  G. Pavlou, "The OSIMIS TMN Platform: Support for Multiple Technology Integrated Management Systems," Proc. of RACE IS&N'93, November 1993.

[11]  PostModern Computing Technologies, ORBeline 2.0 User's Guide, 1996.

[12]  J. T. Park and J. W. Hong, "Implementation and Performance of a TMN SMK System", Proc. of DSOM'96, L'Aquila, Italy, October 1996.

[13]  J. T. Park, Y. W. Choi and J. D. Kim, "The Integration of OSI Network Management and TCP/IP Internet Management using SNMP," Proc. of IEEE First International Workshop on Systems Management, April 1993, pp. 145-154.

# 7   BIOGRAPHIES

**Jong-Tae Park** received the BS degree from Kyungpook National University, Korea and the MS degree from Seoul National University, Korea. He received PhD from Computer Science and Engineering, University of Michigan, in 1987. From 1987 to 1988, he was at AT&T Bell Labs, working on network management and service provisioning. Since 1989, he has been working at the School of Electronic and Electrical Engineering at Kyungpook National University, Korea, where he is now an associate professor. His research interests include TMN, distributed DBMS, network and resource management in ATM networks, multimedia communication, and personal communication services.

**Jae-Hong Lee** received the BS and the MS degrees in Electronic Engineering from Kyungpook National University, Korea, in 1995 and 1997, respectively. He is currently working as a researcher at LG Electronics, An-Yang, Korea. His research interests include ATM networks and network management.

**James Won-Ki Hong** is an assistant professor in the Dept. of Computer Science and Engineering, POSTECH, Pohang, Korea. He has been with POSTECH since May 1995. Prior to joining POSTECH, he was a research professor in the Dept. of Computer Science, University of Western Ontario, London, Canada, where he worked on the CORDS project and MANDAS project. Dr. Hong received the BSc and MSc degrees from the University of Western Ontario in 1983 and 1985, respectively, and the PhD degree from the University of Waterloo, Waterloo, Canada in 1991. His research interests include network and systems management, distributed computing and multimedia systems. He is a member of IEEE and ACM.

**Young-Myoung Kim** is a Member of Technical Staff in Telecommunication Network Research Lab (TNRL) of Korea Telecom, Korea. He has been with Korea Telecom since 1989. He received the MS degree from KAIST in 1989. His research interests include Telecommunications Network Management. He is an associate member of IEEE Communications and Computer Society.

**Seong-Beom Kim** is a Principal Member of Technical Staff in Telecommunications Network Research Lab (TNRL) of Korea Telecom, Korea. He has been with Korea Telecom since 1984. Prior to joining Korea Telecom, he was a member of research staff in ETRI. He received the BSc and MSc degrees from Hanyang University, Seoul, Korea, in 1980 and 1986, respectively. His research interests include telecommunication network and system management, concurrency control in DBMS

## PART TWO

# Web and Java Approaches
## Chair: German Goldszmidt, IBM T.J. Watson Research Center

# 5

# Network Management using Internet Technologies

*Franck Barillaud*
*IBM CER*
*06610 La Gaude, France*
*email: fbarillaud@vnet.ibm.com*

*Luca Deri[1] and Metin Feridun*
*IBM Zurich Research Laboratory*
*8803 Rüschlikon, Switzerland*
*email: lde@zurich.ibm.com, fer@zurich.ibm.com*

## Abstract

As networks grow in size, speed and flexibility, the role of network management becomes increasingly important. Recent developments in Internet technologies might provide the capabilities that are well suited to the solutions of some very challenging, outstanding problems in network management. The increasing popularity of the World Wide Web, with its established user interface and the ability to run on almost any platform, offers a new way to provide wide access to complex software applications. The goal of this paper is to describe some key capabilities of Internet technologies and critically assess whether there is a match between these technologies and the needs in managing networks.

## Keywords

Network Management, World Wide Web, Java.

## 1 INTRODUCTION

Networks today are managed typically through the use of powerful and general purpose monolithic management platforms. These to some degree of success provide integration of management tools, but pose a number of disadvantages:

- Management platforms are expensive, both in terms of software as well as the cost of the hardware required.
- They are typically complex to install, run and maintain.
- Management platforms are based on the centralized paradigm of management: a small number of sites (typically a single one) collect data from the network and analyze them. This can create bottlenecks and thus delays in reacting to network problems; in addition, ability to scale becomes an issue.

---

[1] also at the University of Berne, Institut für Informatik, Neubruckstr. 10, CH-3012 Bern, Switzerland. Email: deri@iam.unibe.ch.

- In general, it is difficult to *remotely* access data and tools on the management platforms. The means available are typically primitive such as telnet, and do not match the capabilities offered on the consoles.

Recent emergence of Internet technologies such as the World Wide Web (Berners-Lee, 1992) and the Java language (Arnold, 1996) offers new means to overcome some if not all disadvantages of today's network management platforms.

The web browser as an end-user interface is available on almost every computer platform and is enjoying widespread use in low-cost access and integration of a broad range of services. The Java language on the other hand provides the means to create software applications that are portable across platforms and can be distributed, accessible through web browsers.

In this paper, we show two ways the web and Java technologies can be applied to the complex world of network management. We outline the effort required to integrate powerful tools into the Web, providing the ability to manage network resources efficiently using HTTP and Java. We also show how the integration of Java with mobile agent technology allow us to create a new generation of roaming applications to better solve problems that affect networks every day.

## 2   INTEGRATING THE NETWORK MANAGEMENT WORLD INTO THE WEB

The key driver for integration of web technologies into network management is the desire to have simple but powerful tools accessible from every platform. The web technologies (web browser, HTTP, HTML and web servers) provide in addition the following benefits:

- Web servers act as central repositories, reducing maintenance costs. For example, a management application can be changed at a server, and these changes propagate automatically through web browsers; there is no need to update the browsers.
- Facilities such as documentation and on-line help can be consolidated at servers.
- Web "pages" provide a natural environment to integrate multiple services.

A survey of the current trends in the use of the web technologies for network management can be found in (Jander, 1996).

### Webbin': HTTP-based Network Management

Webbin' is a research project which aims to simplify the way network management is performed. Webbin is based on the idea that the complexity of protocols such as CMIP or SNMP has to be hidden by the system and that the users have to rely on the services provided by the system and to reuse them every time a new application has to be developed instead of replicating them (craftsman paradigm, i.e. everything has to be custom built for a certain task). The core element of Webbin is a software application called *Liaison*[2] (Deri, 1996a) which allows CMIP/SNMP resources to be managed through HTTP (Figure 1). Liaison is based on a special type of software components called *droplets* that have the ability to be replaced and added at runtime allowing the dynamic modification and extension of the behavior of the application that contains them. Liaison comes with droplets that implement all CMIP and SNMP operations, a basic directory service and a metadata repository for SNMP. Additionally there are a couple of droplets that have the ability to query the metadata information contained inside the OSI stack. The idea is to implement a droplet for each management CMIP/SNMP operation and then cooperate with the existing droplets in order to reuse the services they provide especially with respect to the metadata access. This demonstrates how powerful software components are and how they allow the reuse of existing services and then the incremental building of applications instead of starting from scratch every time.

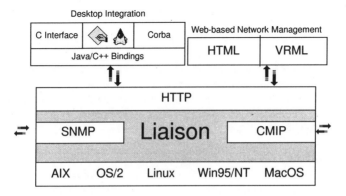

**Figure 1** Liaison Architecture

The basic Liaison configuration contains droplets for:
- browsing CMIP and SNMP resources using a Web-browser;
- displaying network topology in 3D using VRML (Virtual Reality Markup Language) (Deri, 1997b);
- performing batch operations which are used by *external bindings*, available in C, C++ and Java, which enables the creation of simple management applications by exploiting Liaison's services (Deri, 1996b);
- managing CMIP and SNMP instances from Corba (OMG, 1995) through *Corba Bindings* (Deri, 1997a), exploiting the external bindings.

The droplet paradigm allows one to combine services easily. For instance Liaison comes with a droplet that implements a directory service. This service is used by the discovery droplet that is responsible for locating the resources available on the network. Composition of services has several advantages. It keeps the application complexity low and allows service implementations to be replaced with new and more efficient ones without having to affect the users of those services. Liaison has the

---

2 Liaison can be freely downloaded from http://misa.zurich.ibm.com/Webbin/. It is currently available for AIX, MacOS, Linux and Win95/NT.

ability to transparently locate and exchange information with other Liaison's running on different hosts. This enables one to use the Liaison that is closest to the managed resource. In fact, thanks to the TCP/IP-based Liaison-to-Liaison communication services, a Liaison that has to deal with management resources running on a remote host, when possible delegates the request to another Liaison that is running locally with respect to the managed resources. This solution allows bandwidth to be saved because:

• local computation: all communications Liaison-managed resource are local,
• Liaison-to-Liaison communication uses a simple protocol that moves less data than an equivalent CMIP/SNMP request/response,
• Liaison-to-Liaison communication is always 1:1 (1 request/1 response), whereas CMIP is 1:n in the case of scoped operations.

CMIP/SNMP protocols have been mapped to URLs (Deri, 1996c). This mapping is very important because the entire architecture relies on it. It is based on strings, i.e. every value is mapped to a string, and it allows a URL to be mapped uniquely to a management operation and vice-versa. The URL is composed of 5 elements, http://<host>/<protocol>/<operation>/<context>?<parameters>, where: a) <host> identifies the host where the HTTP server runs, b) <protocol> specifies the protocol used c) <operation> specifies the protocol operation d) <context> specifies the context to use, if any e) <parameters> contain the operation parameters, if any. At the moment we are using HTML for basic network management. This is very useful for instance in situations where Liaison is installed on a managed device and users connect to the box via SLIP or PPP protocols and manage it using a conventional Web-browser. Whenever it is necessary to provide a more sophisticated application or when multiple operations have to be performed in batch mode, then users can write their own applets or applications using the external bindings. In case an interaction with Corba is needed, then the Corba bindings offer a way to manipulate CMIP/SNMP resources from a pure Corba world without having to deal with large amount of generated code.

## 3   NETWORK MANAGEMENT AGENTS

The management of high speed networks often require massive transfers of data. Such transfers can be inhibited by bandwidth constraints, storage limitations and data ownership considerations. Today's management systems are based on a platform-centered paradigm that separates applications from the required data and services. This centralized paradigm has been stretched to its limits by the emerging large scale, complex multi-domain networks. There is a growing need for new technologies to automate and distribute management functions to enable scaleable and robust operation.

Mobile agents are programs that can be dispatched and executed remotely under local or remote control. They can be used to move management functions to the data rather than move the data to these functions, thereby facilitating network management architectures for distributed automated management of networks of arbitrary scale and complexity.

In the traditional management environment, a program that monitors virtual circuits at an ATM switch must poll MIB tables and analyze them at the network management station. Both the data provided by the MIB as well as the control functions that it supports are rigidly built-in at MIB design time. The network management platform must poll and process large amount of MIB data (the virtual circuit MIB table can include thousands of entries) at a fast pace. On the other hand, a mobile agent can be delegated to the ATM switch to monitor and analyze the virtual circuit table . It can access the MIB and other relevant data directly by the instrumentation located in the hardware, more efficiently and reliably than a remote network management platform.

Health functions may also be used as examples to illustrate the need for mobile agents in the management of high speed networks. Health functions cannot be included as part of a static MIB design as they may vary from site to site and over time. Nor can they be usefully computed at centralized management platforms, as this can result in excessive polling rates and lead to errors due to perturbations introduced by polling. Mobile agents can support flexible and effective evaluation of health functions and threshold decisions at devices. For example, an agent in a LAN hub can check the status of a port and disconnect it when its load exceeds a threshold.

Use of mobile, intelligent agents can also enable managed devices to acquire autonomous management capabilities. For example, when communication with management entities is lost, a device may activate management programs, i.e. agents, that provide it with sufficient pre-defined management functions to allow autonomous operation of the unit.

## SMAP: Smart Management Applications

The goal of the SMAP research effort is the development of a distributed management paradigm to overcome the disadvantages of centralized network management. Smart Management Applications (SMAPs) are small, mobile network management tasks that can be executed at one or more locations in the network with the following key benefits:

- small footprint, as SMAPs are specialized to handle a small set of network management tasks and are activated only when necessary;
- localized problem resolution, as SMAPs can execute close to the managed equipment. thus providing a closer, detailed view of the network state as well as reducing network management traffic by consolidating collected data locally.

In the SMAP approach, the management applications (or at least parts of them) move closer to the probes (agents) to solve management tasks. The centralized manager or console is reduced to the role of presenting information and invoking management tasks as necessary.

### SMAP Architecture

The architecture of the SMAP platform is shown in Figure 2. The platform is based on the principle that management applications should only impose a minimal overhead on the managed system.

SMAPs are code objects which can be easily down loaded on demand to carry out management tasks. There are two categories of SMAPs:

* SMAPs which drive management tasks, for example, a management console application;
* SMAPs which when activated, carry out network management tasks such as monitoring packet delays at a network node.

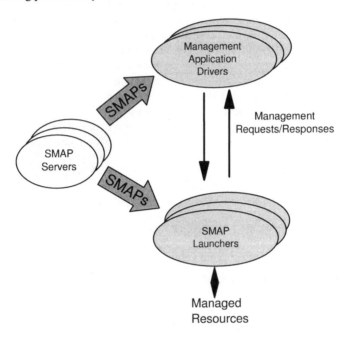

**Figure 2** SMAP Architecture

The SMAP platform consists of three components, distributed across the network:

* Management Application Drivers activate management tasks in the network and also provide, if necessary, the network operator interface.
* SMAP Launchers receive commands from management application drivers to carry out a management task, download appropriate SMAPs for the task, and start them. Activated SMAPs can communicate with other SMAPs or with the managed resources.
* SMAP Servers are SMAP repositories, used by both management application drivers and SMAP launchers.

The Java language has been chosen for the implementation of the SMAP platform. Java provides convenient facilities for the movement of code, a required feature of the SMAP architecture and is portable across many platforms.

SMAPs for driving and carrying out management tasks are coded as Java classes (as applications or applets) and are retrievable through HTTP servers across the network (Figure 3). The management application drivers consists of an appletViewer or a Java-enabled browser which loads the appropriate driver SMAPs as required. Once activated, these SMAPs connect to selected SMAP launchers.

In a launcher, the SMAP Launcher Java class instantiates an instance of the ApplicationHandler class, which in turn downloads appropriate SMAPs from an HTTP server and activates the management task. Loaded classes are removed when no longer needed.

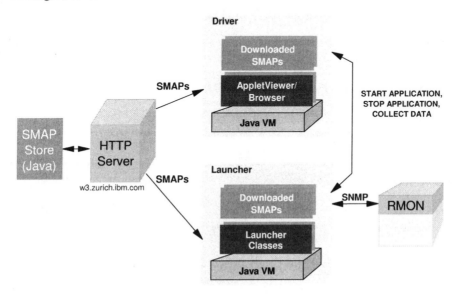

**Figure 3** SMAP RMON Example

## Example : LAN Traffic and RMON using the SMAP platform

Observation of packets on a LAN can prove helpful to network managers and designers. Traffic patterns between hosts or LAN segments provide useful information that can help produce more efficient LAN designs. Traffic data can allow network operators to pass some of the operating costs to network users based on usage patterns. Packet types observed on a network can also reveal potential problems that would otherwise be undetected by traditional network management software.

RMON (Remote Monitoring) MIB (Waldbusser, 1995) is an Internet standard for remote and distributed performance monitoring of ethernet and token-ring LANs. The statistics collected by an RMON probe is available through SNMP. Using customizable filters, an RMON probe can be used to count packets based on packet characteristics (protocol, size, ...). RMON also provides platform independent means to capture packets.

We applied SMAP concepts to LAN management using RMON probes. We selected an initial set of four management problems, based on monitoring IP traffic on LAN segments:

- segment usage: determine the number of packets local to, going out of, and coming into the LAN segment.
- segment traffic per protocol: collect and count IP packets based on the IP protocol type.
- HTTP server usage: collect and count HTTP packets on the same segment as the HTTP server to determine Web usage for selected LAN segments.
- segment traffic patterns: capture IP packets and determine the top 10 source and destination pairs based on the number of bytes/packets exchanged.

As an example, the segment usage application is implemented as follows: we use a Web browser to load the user interface code, a Java applet , from an HTTP server. From the user interface, we select the management task (segment usage), the destination SMAP Launcher, and initiate the application by pressing the start button. This causes two actions:

- first, a connection to the destination SMAP Launcher is established; and
- second, a START_APPLICATION command is sent.

The instance of the ApplicationHandler class on the launcher site loads Java class for this application. Using a network class loader, all required Java classes are automatically loaded (SegTrafficUsage, Filter, Channel, and all SNMP classes). The driver can request data by sending COLLECT_DATA commands to the SMAPs, and terminates the management task by issuing a STOP_APPLICATION command.

## Communicating SMAPs

The prototype implementation of the SMAP architecture uses a simple messaging scheme for inter-SMAP communications. This scheme is efficient, but it cannot support a richer form of communication between SMAPs. There are a number of alternatives. One is the use of Java facilities such as the Remote Method Invocation (RMI) API (Sun, 1996) or CORBA. In this approach, the communication between SMAPs is viewed as one between Java or CORBA objects, i.e., method calls. A second approach is to use intelligent agent technologies such as Aglets (Chang, 1996) to enable mobility as well as security.

## 4   CONCLUSION

Internet technologies such as the World Wide Web and the Java language will play an important role in the future of network and systems management. Already today, many in the industry are basing their next generation management applications on Web-based management (Jander, 1996).

In this paper, we have shown how the Internet technologies can be applied to network management. Webbin' allow networks to be managed through a web browser and provides a set of tools and programming interfaces to enable the development of network management applications. The SMAP approach supports distributed management, combining the web, Java and mobile agent technologies.

The pace at which Internet technologies change is remarkable; technologies change or new ones introduced almost every week. We expect that these developments will not only enhance how we do network management, but keep us researchers and developers busy for a long time.

**ACKNOWLEDGEMENTS**
The authors would like to acknowledge Lucas Heusler for his contributions to the SMAP work, Robert Akolk, Bela Ban, Dieter Gantenbein for contributions to Webbin', and the anonymous referees for their comments on the paper.

# 5   REFERENCES

Arnold, Ken and Gosling, James (1996). *The Java Programming Language.* Addison-Wesley, Massachusetts.

Berners-Lee, T., Cailliau, R., Groff, J. and Pollermann, B. (1992). World-Wide Web: The Information Universe. *Electronic Networking*, Vol. 1, No. 2, Spring 1992.

Chang, Dan and Lange, Danny (1996). Mobile Agents: A New Paradigm for Distributed Object Computing on the WWW. Proceedings of the *OOPSLA '96 Workshop on Toward the Integration of WWW and Distributed Object Technology*, San Jose, California. October 1996.

Deri, Luca (1996a). Surfin' Network Resources across the Web. Proceedings of the *IEEE Workshop on Systems Management*, Toronto, Canada. June 1996.

Deri, Luca (1996b). Network Management for the 90s. Proceedings of the *ECOOP '96 Workshop on System and Network Management*, Linz, Austria. July 1996.

Deri, Luca (1996c). HTTP-based SNMP and CMIP Management. *Internet Draft* (draft-deri-http-mgmt-00.txt), December 1996.

Deri, Luca and Ban, Bela (1997a). Static vs. Dynamic CMIP/SNMP Network Management Using CORBA. Accepted for publication at *IS&N'97*, Como, Italy. May 1997.

Deri, Luca and Manikis, Dimitris (1997b). VRML: Adding 3D to Network Management. Accepted for publication at *IS&N '97*, Como, Italy. May 1997.

Jander, Mary (1996). Web-based Management: Welcome to the Revolution, *Data Communications*, November 21, 1996 issue, 39-53.

Object Management Group (1995). *The Common Object Request Broker: Architecture and Specification*, Revision 2.0, July 1995.

Sun Microsystems (1996). *Java Remote Method Invocation Specification*, Prebeta Draft, Revision 1.1, November 1996.

Waldbusser, Steven (1995). *Remote Network Monitoring Management Information Base*. Internet RFC-1757.

# 6   BIOGRAPHIES

**Franck Barrillaud** is a member of the Network Management Product Design Team at IBM CER, La Gaude, France. He is working in the area of ATM network management development. Since 1996, he is leading the effort to introduce internet technologies into the network management products of the IBM Networking Division. His focus areas are in distributed and scalability aspects of network management and the introduction of artificial intelligence tols to solve network management complexity.

**Luca Deri** is currently a PhD student at the University of Berne, working part-time at the IBM Research Division, Zurich Research Laboratory. He received his degree in Computer Science with a thesis on Network Management from the University of Pisa. He worked at Finsiel S.p.A. as a consultant after the graduation, and was a research fellow at the University College of London. His professional interests include network management, software components and object-oriented technology. His Web page address is http://www.zurich.ibm.com/~lde/.

**Metin Feridun** is a research staff member at IBM Research Division, Zurich Research Laboratory since 1990. He previously worked for BBN Systems and Technologies (1987-1990), and GTE Laboratories (1983-1987). He received the S.B. degree (computer science) from Massachusetts Institute of Technology in 1978, the M.Eng. degree (computer engineering) from Rensselaer Polytechnic Institute in 1979, and the Ph.D. degree (electrical engineering) from Cornell University in 1983. His research interests are in distributed network and systems management.

# 6

# Using the World Wide Web and Java for Network Service Management

*M. C. Maston*
*Cisco Systems, Inc., WAN Business Unit*
*1400 Parkmoor Avenue, San Jose CA 95126 USA*
*408-525-2734 (Voice)*
*408-525-7414 (Facsimile)*
*mmaston@cisco.com*

### Abstract

This white paper discusses the potential applications of World Wide Web technologies to satisfy network service management requirements. Web-based tools can be applied in many areas of network management from simple element to highly sophisticated service management applications. The background issues leading to this approach, examples of solutions and the overall benefits of Web browsers and Sun's Java[*] are discussed.

### Keywords

World Wide Web, HTTP, CGI, Java, network management, wide area

## 1    INTRODUCTION

One of the most confounding issues faced by network equipment vendors, network service providers and end users today is achieving scaleable, reliable and distributed network management. These problems are most vexing in the wide area networking (WAN) space due to several factors including:

- immature or exceedingly complicated management standards
- extremely complex security considerations created by large global networks
- lack of coordination and compatibility among vendor network management equipment
- a myriad of complex management issues surrounding the expensive bandwidth consumed by management traffic

---

[*] Java is a registered trademark of Sun Microsystems, Inc.

- relatively poor understanding of the management issues across the LAN/WAN boundary
- monumental cost and support burden associated with administering management applications across many different operating systems and hardware platforms.

All of these issues contribute to relatively low quality network management functionality being the norm while the market continues to push the edge of the envelope for even more sophisticated management support. It is usually at this point in a technology cycle that a paradigm shift must occur in order to reset the standards and methodologies in place to a new starting point. From such a point, satisfactory solutions can begin to be delivered. Without this shift, products will continue to fall behind market expectations with little or no hope of catching up.

Presently, network management is no longer considered "icing on the cake" by serious network service providers. Rather, it is perceived as an integral component in a complete network design to assure initial and ongoing success. This change in attitude comes largely from a shift in what *users* have come to demand in the way of services from their service provider. Service providers then naturally drive vendors to add functionality in order to satisfy end-user requirements and allow their service to remain competitive in a cutthroat market. Continuous, reliable and high performance service is considered the minimum for what a service provider must offer and the requirements for more advanced management features escalate from there.

Today's service providers and their users expect features such as virtual private networks, customer network management (CNM), advanced reporting facilities and a plethora of other highly customized network management services. The bar has been raised for all service providers to bring bigger and better services to market faster and more often. More importantly, the expectation has been set that these services are delivered in a simple, "plug and play" way so that customers of the service provider can spend little time *learning* and more time *using* the service. Simply exposing an SNMP MIB for customer use no longer provides a complete solution and will only become less satisfactory in the future.

In order for these services to be delivered and continue to be augmented at the necessary pace, a paradigm shift as described before must take place. This change in the delivery of management services will most likely come from the technologies developed for the World Wide Web. Specifically, the HyperText Markup Language (HTML) and Sun Microsystems Java bring significant advantages to the task of deploying versatile, robust and powerful network management solutions. This white paper will discuss how these tools can be used to provide a fundamental change in the way network management applications are developed and delivered.

## 2    NETWORK MANAGEMENT ISSUES

The key issues that face most vendors and service providers in delivering network management applications are similar to those faced by most mass market software producers.  As might be expected, network management has a few twists that distinguish it from mainstream software, but in general the same rules apply.  The most prominent features and functionality required for management applications include:

- security of the network and of all user traffic on the network
- standards-based solutions; no access to information through proprietary interfaces
- support for all the equipment in today's heterogeneous networks
- a scaleable solution that can be deployed economically across the market continuum from enterprise to carrier networks
- affordable for large and small network service providers
- robust and reliable architecture
- high performance, responsive user interface
- "plug and play" installation, upgrades and maintenance
- ease of use for both the novice and "power user"
- availability on a wide range of operating systems and hardware platforms from desktop PC's to high-end workstations
- user interface that can be adapted to different operational models by the end user

Each of these requirements can be met exceptionally well by a combination of HTML and Sun's Java.  By exploiting the strengths of each of these tools, a very sophisticated network management product can be developed.  Before explaining how these technologies can be put to work for network management, however, it is necessary to separate network management into three classes:  element, network and service management.

## *Element Management*

In the simplest terms, element management is the process of monitoring, controlling and configuring any network entity.  An entity may be a workstation, router, backbone ATM (Asynchronous Transfer Mode) switch or other network-connected device. Potential management activities include reading statistical counters, issuing commands to control the flow of network traffic,  and configuring parameters such as the IP address of the entity or its network name.

## *Network Management*

Network management is the next step in the hierarchy of management from element management.  Where element leaves off in only controlling properties of individual network nodes, network management delivers an overall network view and is able to perform operations within or across the complete breadth of the network.  Administering PVC's, defining preferred routing and testing the integrity of trunks between nodes are just some of the duties of a typical network management application.

Typically, network management functionality is delivered by vendors on a network management station (NMS).  This allows users a single point of access to a set of information that is common to many or all of the network elements.  The management station also serves as a focal point for network-wide operations such as provisioning, diagnostics and statistics collection.

## Service Management

Service management embodies the processes required to monitor, command and control the actual services delivered by the managed elements. These services can range from LAN services such as TCP/IP to WAN services such as frame relay, ATM, video and voice.

These services can (and typically do) provide more than the ability to transfer network traffic from one point on the network to another. Service providers deliver guaranteed quality of service, advanced performance reporting, customer network management, and usage-based billing. These are value-added services above and beyond basic network transport access. Many service providers use these types of mechanisms as differentiators from their competition.

## 3     WORLD WIDE WEB TECHNOLOGIES

With each passing day there are new products being offered to enhance and expand the capabilities of the World Wide Web. Products which allow monetary transactions, database access and on-line information retrieval via the Web are common and continue to appear at a phenomenal rate. For the purposes of this discussion, the basic tools of the Web, HTML and Java, will be highlighted.

## HTML

HTML has been proven to be extremely easy to learn for many non-technical individuals. Perhaps even more impressive are the results these same people have produced given a little time and a simple text editor. Rich text content, along with audio, video and other multimedia data can be easily delivered with very little in-depth computer knowledge. All of this information is transmitted to client "browsers" such as Netscape Navigator™ from a Web server using the standards-based HyperText Transfer Protocol (HTTP). In essence, HTTP is simply a connection-oriented TCP/IP-based file transfer mechanism. Typically, a user asks for a particular page and the browser retrieves that page from the server. Once downloaded, the browser scans the contents of the page which may call out other files (such as bitmap images) located on the originating Web server or elsewhere. The browser then requests these files until the entire page content is loaded and rendered on the user screen.

What HTTP was not designed to do, however, is provide a continuous stream of updating information. A file is requested, downloaded and the connection is closed once the file has been retrieved. No further interaction between the browser and the server takes place. Obviously, this makes having any sort of dynamic information on a Web page difficult to accomplish. Fortunately, this is where Java steps in to save the day.

## Sun Microsystems Java

Sun Microsystems designed Java with the following criteria in mind:
- Simple, object-oriented and familiar

- Robust and secure
- Architecture neutral and portable
- High performance
- Interpreted, threaded and dynamic

Java is a programming language which is syntactically similar to C++, but has some of the less useful functionality removed. This tool picks up where HTML leaves off by allowing for small applications, called "applets", to be developed. These applets can draw arbitrary shapes and text, create data connections, accept user input and generally perform those functions necessary to produce dynamic, interactive content on Web pages.

Applets are called out in a standard HTML page just like a bitmap graphic or link to multimedia data. Any parameters which are important to the applet are described along with the applet citation. When the page is loaded, the applet is requested and downloaded from the server. Once retrieved, the applet is validated by the browser and begins to execute on the *client* machine. Normally, the applet will continue to run until it exits programmatically or a new page is loaded.

# 4    NETWORK MANAGEMENT WITH A WEB BROWSER

As previously mentioned, the technologies that have been driven by the explosive growth of the Internet and World Wide Web sites are well suited for network management applications. These tools provide a way for fairly inexperienced computer users to publish electronic information content with some truly impressive results. In this section, two practical applications of these tools are illustrated.

## *Element Management using HTML*
In order to perform element management, two key functions must be possible:

- the ability to read any and all configuration information about the element
- the ability to set the state of any and all configuration parameters for the element

The easiest way to gain this functionality is to leverage the existing network management protocols deployed in the elements. A sample system using SNMP as the managing protocol is depicted in Figure 1.

In this implementation, a Web server is actually integrated into the managed device. This Web server does not significantly differ functionally from the Web servers freely available via the Internet or those sold commercially. In addition to a set of predefined HTML *forms* that contain configuration templates for the managed device, a Common Gateway Interface (CGI) application is installed. The CGI program in this instance is referred to as the Command Processor since its sole function is to accept commands from the user and generate the appropriate configuration messages.

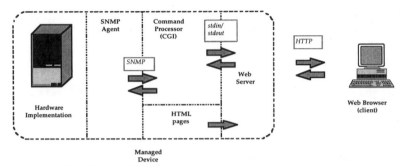

**Figure 1**       Web CGI-based Element Management

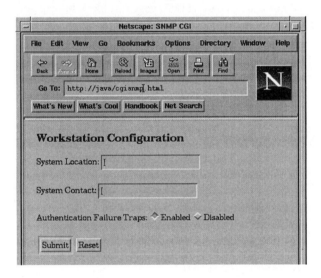

**Figure 2**    A simple HTML form

In this case, the Command Processor accepts name/value string pairs through the *stdin* logical device. It then parses these strings and generates the appropriate SNMP SET and GET commands. Any result codes and values returned from these commands can then be formatted into HTML and returned to the client browser (via the Web server) through the *stdout* device. The string pairs "posted" by the HTML form are really just the name and value of all the user input fields on the form. Figure 2 shows an extremely simple example of a form created to set some of the variables in a MIB for a UNIX workstation, the *sysLocation*, *sysContact* and *snmpEnableAuthenTraps*. Figure 3 shows the underlying HTML file the browser uses to build this page.

The form shown in Figure 2 allows a user to type in desired strings for the *sysLocation* and *sysContact* field using text input boxes. The *snmpEnableAuthenTraps* field is controlled by radio buttons (mutually exclusive). For the new values to be sent to the Web server and thereby set on the target device, the user must press the "Submit" button at the bottom of the screen. The "Reset" button initializes the form to its original state.

```
<HTML>
<HEAD>
<TITLE>SNMP CGI</TITLE>
</HEAD>
<BODY>
<H2>Workstation Configuration</H2><P>
<FORM ACTION="cgi-bin/cmdproc" ENCTYPE="x-www-form-encoded" METHOD="POST">
System Location:  <INPUT TYPE=text NAME=".iso.org.dod.internet.mgmt.mib-
2.system.sysLocation.0" ><P>
System Contact:  <INPUT TYPE=text NAME=".iso.org.dod.internet.mgmt.mib-2.system.sysContact.0"
><P>

Authentication Failure Traps:
<INPUT TYPE="radio" VALUE="1" NAME=".iso.org.dod.internet.mgmt.mib-
2.snmp.snmpEnableAuthenTraps.0" CHECKED=true> Enabled
<INPUT TYPE="radio" VALUE="2" NAME=".iso.org.dod.internet.mgmt.mib-
2.snmp.snmpEnableAuthenTraps.0"> Disabled <P>

<INPUT NAME="Button" TYPE="submit" VALUE="Submit">    <INPUT NAME="name"
TYPE="reset" VALUE="Reset">
</FORM>
</BODY>
</HTML>
```

**Figure 3**    HTML source file for form page

In Figure 3, the most important elements of the form are highlighted for clarity. The first line defines what action is to be taken upon the "Submit" button being activated:

<FORM ACTION="cgi-bin/cmdproc" ENCTYPE="x-www-form-encoded" METHOD="POST">

Here, the form will "POST" the entered data to the Web server using the "x-www-form-encoded" MIME encoding format. The exact details of this encoding are not important for this discussion. The "ACTION" field directs the server to launch a program called *cmdproc* and pass the posted data to it.

The next fields are the actual data entry fields:

System Location:  <INPUT TYPE=text NAME=".iso.org.dod.internet.mgmt.mib-
2.system.sysLocation.0" ><P>

System Contact:  <INPUT TYPE=text NAME=".iso.org.dod.internet.mgmt.mib-2.system.sysContact.0"
><P>

Authentication Failure Traps:
<INPUT TYPE="radio" VALUE="1" NAME=".iso.org.dod.internet.mgmt.mib-
2.snmp.snmpEnableAuthenTraps.0" CHECKED=true> Enabled
<INPUT TYPE="radio" VALUE="2" NAME=".iso.org.dod.internet.mgmt.mib-
2.snmp.snmpEnableAuthenTraps.0"> Disabled <P>

There are two "text" entry fields and a pair of "radio" buttons. As mentioned previously, each field has a name and a value. In order to allow the command processor to be generic in implementation, the "NAME" field of each input is actually set to be the fully qualified SNMP MIB name of the object to be set. By using this naming convention, the Command Processor application can be almost completely oblivious to the device being managed and can therefore be freely re-used for many products without modification.  The radio buttons, it may be noted, use the same name for each button. This is a common convention of HTML forms so that dependent controls can be grouped together.  More importantly, however, each button has a predefined value associated with it (1 and 2, respectively), which correspond with the acceptable values the MIB object can be assigned.  These values are not what are shown on the screen, however.  Instead, HTML allows more familiar labels to be used ("enabled" and "disabled"), but the numerical values are what actually get submitted to the Command Processor.

Finally, the line,

```
<INPUT NAME="Button" TYPE="submit" VALUE="Submit">    <INPUT NAME="name"
TYPE="reset" VALUE="Reset">
```

is used to create the "Submit" and "Reset" buttons.

Assuming that the user sets the form fields to the following:

System Location:  MyOffice
System Contact:  MCM
Authentication Failure Traps:  Disabled

the following string of characters concerning the form content will be submitted to the Web server and then passed on to the Command Processor:

```
.iso.org.dod.internet.mgmt.mib-2.system.sysLocation.0=MyOffice&.iso.org.dod.internet.mgmt.mib-
2.system.sysContact.0=MCM&.iso.org.dod.internet.mgmt.mib-
2.snmp.snmpEnableAuthenTraps.0=2&Button=Submit
```

Name/value pairs are concatenated and delimited using the "&" symbol so they can be easily parsed.  In addition, each name is separated from its corresponding value by an equals sign.  The reader may note that the name and value of the "Submit" button was passed as well.  This is normal and can (and should) be ignored by the Command Processor.

## *Element Management using HTML and Java*

For situations where only static information is to be displayed, the above example is adequate and versatile.  For more sophisticated element management, where information needs to be periodically or asynchronously refreshed, HTML and CGI alone are not sufficient.  The first reason for this, as noted previously, is that HTTP does not hold connections to the browser open and therefore has no way of sending updated data when it is available.  Secondly, HTML only really permits text and bitmap images to be specified and has no mechanisms for drawing arbitrary graphics such as lines, circles, etc. Together, these limitations make the creation of dynamically updating graphs, charts and text exceedingly difficult and cumbersome

Using a mixture of HTML and Java applets, however, quickly eliminates this dilemma. HTML code can be used to produce the static fields and fixed hyperlinks that would be useful for headers, links to on-line help and other relatively unchanging information. Java applets, alternately, can be used for a myriad of tasks including context-adaptive forms, real-time graphs and charts as wells as alarm lists. A simple example of a combination of HTML and Java applets can be seen in Figure 4.

Shown here is the shelf-level view of the StrataCom AXIS interface shelf, a high density device for aggregating user traffic into a broadband ATM network. HTML "frames", a way to divide the browser window into logical panes, have been used to provide the user with a toolbar of common functions. Each button on the toolbar can transport the user to different views of this element including alarms, configuration, diagnostics and statistics pages. The buttons themselves are simply bitmap images that are defined as hyperlinks to the appropriate supporting HTML pages.

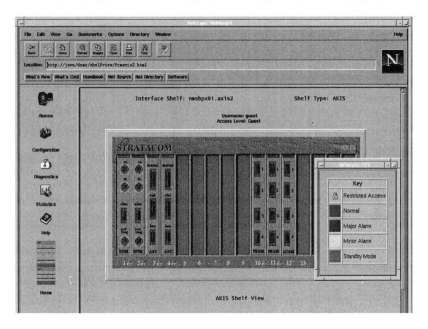

**Figure 4**    Java and HTML example

A Java applet has been used to build up the actual device view on the right side of the page. The applet scans the SNMP MIB of the AXIS device. It determines how many cards are installed, what slot they are installed in and what type of card they are. Empty slots are shown as blanks. After discovering the cards present, appropriate images of those cards are rendered in the correct positions along with the physical ports and their states. For instance, slots 10 and 11 have four port T1 FRSM (Frame Relay Service Module) cards installed. All four ports are shown in green to indicate they are active and operating normally. Slot 12 has an AUSM (pronounced *awesome*) interface, the ATM UNI Service Module, installed in it, again with all ports showing normal status. When the status of any port or card changes or when a card is inserted or removed, the applet will automatically update the state based on an SNMP trap, changing to red if in major alarm, for example.

Most importantly, each card and port (and even the shelf itself) shown is actually a hyperlink to pages about that object. As the applet builds the view of the card shelf, the page links are automatically created. When the user moves to a new page, the toolbar hyperlinks are updated to represent pages for alarms, configuration, statistics, diagnostics and help for the correct object. In this way, a cohesive, integrated user interface can be presented to user at every level of device management.

Figures 5a-c show the HTML pages responsible for producing this interface.

```
<HTML>
<head>
<TITLE>AXIS Shelf Home page</TITLE>
</head>
<FRAMESET COLS="15%,85%">
<FRAME SRC="buttons.html">
<FRAME SRC="axis.html">
</FRAMESET>
</HTML
```

**Figure 5a**      axisshelf.html - The main page

```
<HTML>
<CENTER>
<A HREF="alarms.html"><IMG BORDER=0 SRC="images/alarm.gif"></A><P>
<h2>Alarms</h2><P>
<A HREF="config.html"><IMG BORDER=0 SRC="images/netconfig.gif"></A><P>
<h2>Configuration</h2><P>
<A HREF="diags.html"><IMG BORDER=0 SRC="images/diags.gif"></A><P>
<h2>Diagnostics</h2><P>
<A HREF="stats.html"><IMG BORDER=0 SRC="images/report.gif"></A><P>
<h2>Statistics</h2><P>
<A HREF="help.html"><IMG BORDER=0 SRC="images/q_mark3.gif"></A><P>
<h2>Help</h2><P>
</CENTER>
</HTML>
```

**Figure 5b**      buttons.html - The toolbar page

```
<HTML>
<CENTER>
<H1>StrataCom AXIS Shelf</H1>
<P>
<APPLET CODE="AXISsnmp.class" CODEBASE=classes WIDTH=800 HEIGHT=300>
<PARAM NAME="LEVEL"
  VALUE="topshelf">
<PARAM NAME="MIBS"
  VALUE="axis.mib">
<PARAM NAME="STATE_NORMAL"
  VALUE="GREEN">
<PARAM NAME="STATE_MAJOR_ALARM"
  VALUE="RED">
<PARAM NAME="STATE_INACTIVE"
  VALUE="BROWN">
<PARAM NAME="UPDATE_BY_EXCEPTION"
  VALUE="true">
</APPLET>
</CENTER>
</HTML>
```

**Figure 5c**      axis.html - The shelf applet page

Figure 5a defines two frames, dividing the page vertically into sections that maintain proportions of 15 and 85 percent of the width of the page. The button toolbar is loaded into the smaller frame along the left edge of the window. The AXIS shelf applet page is loaded in the larger frame on the right.

Figure 5b shows the HTML code responsible for displaying the five button images and text labels in the toolbar frame page. Each is defined as a hyperlink to a page with more functionality related to each category.

The page shown in Figure 5c has some arbitrary header text and then the applet reference, signified by the "APPLET" keyword. Java applets have several standard parameters including the name of the class being used, where the class is located within the Web server directory and the dimensions of the applet on the page. These parameters are specified using the "CODE", "CODEBASE", "WIDTH" and "HEIGHT" keywords, respectively. There are other standard parameters, but they will not be addressed here. Applets can also have parameters defined for them by the applet designer which end users can then use to adjust the operation of the applet. The shelf Java class used here, *AXISsnmp.class*, has several useful parameters, some of which are called out in Figure 5c. The "LEVEL", "MIBS", "STATE_NORMAL", "STATE_MAJOR_ALARM", "STATE_INACTIVE" and "UPDATE_BY_EXCEPTION" parameters are used to tune the applet's functionality to the user's needs. These parameters are defined as follows:

- LEVEL - Starting view level of the applet when it is first run. In this case, the "topshelf" view is specified, indicating the uppermost shelf-level view should be used. Lower level views of specific cards/slots can also be specified.
- MIBS- The MIB file(s) to be loaded for use with this device. Any SNMP commands will be correlated with these MIBs to generate the appropriate messages. More than one MIB can be specified, multiple file names being delimited by the "|" character.
- STATE_NORMAL, STATE_MAJOR_ALARM and STATE_INACTIVE - These allow the user to set the color to be used when a device is in each of these states. Color names can be specified or RGB values. STATE_MINOR_ALARM also exists but is not shown. If values are not explicitly called out for each value, standard default color mappings are used.
- UPDATE_BY_EXCEPTION - Defines whether the display is updated by exceptions (traps) or if the device is periodically polled. Option POLL_RATE is used to defined the polling interval if this value is set to "false".


## 5   SERVICE MANAGEMENT WITH A WEB BROWSER

Using Web technology to perform service management functions is fairly straightforward. Most service management systems are actually based on the information collected in the network management layers: elements, ports, trunks, connections and all other manageable parameters. The service management system then manipulates this information and provides service *views* into it.

For example, a common requirement that can be fulfilled by Web technology is that of customer network management. Customers wish to be able to view and, in some cases, control the service they have purchased from a network service provider. To accomplish this, a service provider must first and foremost deliver an isolated view of the overall network, showing only those items in use by the particular customer. This information is usually available from vendor management stations in some form of database, but the actual delivery of this data to the customer can (and in most cases *has*) become the major stumbling block for most service providers.

Customer service data is typically offered to customers as access to an SNMP proxy MIB. Many customers of these services, however, have little or no interest in spending the time and money to integrate to this interface to get meaningful information. That is not to say no customers want a purely SNMP-based interface; many sophisticated corporate IT groups are willing and able to make the investment necessary to get a collective view of their local network and the purchased service. The vast majority of service purchasers, however, are only interested in running their business efficiently and have no time for generating customer management applications, but still require the information such systems deliver at their fingertips.

The dilemma that most service providers face, therefore, is that each customer wants something different in the way such a system would look and work. Many only want very simple performance charts they can review to assure themselves they have gotten what they have paid for. Others want alarms, topology views, near real-time performance statistics and control capabilities. And, of course, all of these features have to be very customizable, available on every hardware platform from low-end PC's to workstations and have to be "plug and play" for the least experienced users. Given the need to deliver the most cost-effective service, the intensely competitive market and staff shortages at most network operators, these requirements are nearly impossible to meet for the majority of service providers. Only technologies such as HTML and Java can allow many of these goals to be accomplished within the timeframe available.

## *Service Management using HTML and Java*
Many times, aspects of service management require the dynamic qualities provided by Java combined with HTML. As an example, Figure 6 shows a sample page from a fairly sophisticated customer network management system. This page has a connection topology map which shows the state of customer connections between endpoints around the country. The customer also has a form at the bottom of the page that allows the creation of additional connections within the network.

First, the topology map shows five connected cities: Portland, San Jose, Los Angeles, Boston and Atlanta. The fictitious customer has offices in each of these locations and different types of network connections including voice, frame relay and ATM running between them. The service provider actually has many other nodes, but the customer is only allowed to see those in use by their service. Further, there are several intermediate nodes between the endpoints shown (such as might exist between Los Angeles and Atlanta), but only the start and end of each *virtual customer* link is shown. A failure of one of these transitional elements will be shown as a link failure between the endpoints. The operational status of the links carrying these connections and the connections themselves are reflected on this map by the color of the nodes and interconnecting lines.

All of this information is gathered, updated and displayed in real-time by a Java applet communicating with a management server at the service provider location.

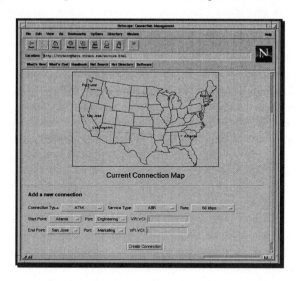

**Figure 6**     Service management with HTML and Java

Second, the user can add new connections to the service without having to directly involve the service provider by using a context-adaptive provisioning applet. In actuality, the service provider is always in control of the process since no commands issued by the user will be *directly* sent to the service network. Instead, these commands will be handled by an intermediate proxy located at the service provider facility which will determine if a customer has the appropriate privileges to execute any given commands. As can be seen from the figure, the customer can create ATM (CBR, VBR, ABR, etc.) as well as frame relay, voice and other types of connections. When the user chooses a particular connection type, the remainder of the input fields adapt to reflect information needed to complete the operation. In this case, the user need only select the endpoints of the connection, the desired port and an appropriate virtual path identifier (VPI)/virtual circuit identifier (VCI) pair. If this had been a frame relay rather than an ATM connection, the form would have changed to prompt the user or a DLCI (Data Link Connection Identifier) rather than the VPI/VCI needed for ATM.

The endpoints are limited by the applet to the cities predefined by the customer. Ports are given meaningful names such as "Engineering" and "Marketing" and are restricted to those ports on equipment already in use by the customer. By imposing these restrictions the service provider is assured that the customer cannot accidentally attempt to make changes to equipment not assigned to them. Additionally, customers will appreciate these limits since it not only helps reduce mistakes, but also gives them easily remembered names and only a few choices to make rather than having to recall potentially obscure node names and port numbers assigned by the service provider.

Finally, the customer can choose an appropriate data rate from a set of common values. Once all parameters are set to the user's liking, a request to set up the connection can be sent to the service provider by pressing the "Create Connection" push-button below the form. The actual implementation of what happens next is completely up the service provider. The information from the customer request may be sent to a service proxy, checked for validity, converted to the appropriate management protocol messages and sent to the network. Alternately, all of the information could be formatted and sent by electronic mail to a service provider human operator. The operator could then evaluate the request and manually create the network connection. How the actual connection is provisioned is completely within the discretion of the service provider and generally unimportant to the service user as long as the results are reasonably timely.

## 6    CONCLUSION

Clearly, the technologies driven by the World Wide Web are quite powerful and potentially useful for a variety of tasks, including network element and service management. This is not to say that it should be considered a complete replacement for the huge amount of management software and protocols that exist today. Nor by any means are HTTP and Java the right choice for every application. These tools are, however, uniquely qualified to meet a large number of the requirements of network and service management applications due to their ease of use, low cost, reliance on standards, flexibility, platform independence and security.

Web technology delivers on the promise that a developer can create one set of programs and content material which will run on virtually any machine and operating system. This functionality is what should be exploited by service providers most and wherever possible. Web servers and browsers should not be seen as a competing approach to the existing management protocols, but instead a way to deliver the information gained through such protocols to a wider audience. By building upon all the work that has already been done in the standard management interfaces, service providers can use the power of Web-based management to improve their internal processes and deliver  the highest quality services to their customers.

## 7    BIOGRAPHY

Michael Maston is a network management product line manager within the WAN Business Unit (formerly StrataCom) at Cisco Systems in San Jose, CA. His primary responsibilities currently include defining requirements for device management for Cisco's wide area ATM switches and Web-based network management solutions. Prior to working for Cisco, Mr. Maston spent 8 years as an electrical engineer developing hardware and software for Ford Aerospace, GTE Government Systems and Landis and Gyr Energy Management Systems. He holds a Bachelor of Science in Electrical Engineering from Santa Clara University in Santa Clara, CA.

7

# Distributed systems management on the web

*B. Reed, M. Peercy, E. Robinson*
*IBM Almaden Research Center*
*650 Harry Rd*
*San Jose, CA 95120*
*breed@almaden.ibm.com*

HTTP://www.almaden.IBM.com/cs/people/Breed/im97

**Abstract**

The need to manage a multi-platform enterprise and the advancement of web technologies have made it possible to build powerful cross-platform management applications. As more network and system management tools become web-enabled, the web browser becomes a convenient point of integration.

This paper describes the architecture and implementation of a web-based management console to an existing PC management application. Justification for using an embedded web server instead of the normal general purpose web server and CGI's is given. The limitations and solutions are discussed, as well as advantages over traditional GUI's.

**Keywords**

web-based management, embedded web server, CGI, WWW

## 1    INTRODUCTION

As the number of computer systems grows in an enterprise domain, administration of all the systems in the enterprise from a single console becomes more and more of a requirement. A number of systems management packages and tools exist to integrate, on a single screen, data from several sources throughout an enterprise network. These facilities have traditionally been designed for mainframe computers and UNIX workstations, both because these machines have controlled the critical resources of the enterprise and because the computational requirements of managing such resources demand such machines. In fact, systems management utilities often require a workstation as a dedicated platform for execution. However, as x86-based personal computers and PC servers obtain dominance in the market, the ability to monitor and administer these resources becomes extremely important.

A number of companies have developed products for the administration of PC resources. Among these products are IBM Tivoli's NetFinity, Compaq's Insight Manager, Intel's LanDesk, and Microsoft's SMS. Each of these products addresses parts of the hardware, software, and network domains of the PC computing universe. Nonetheless, prior to this research, none of them integrated all their features well into the enterprise systems management utilities being run on UNIX workstations to manage UNIX and mainframe resources.

Recently there has been a lot of interest in web-based management solutions (Wellens, 1996)(Bruins, 1996)(Mullaney, 1996). Switches and routers (McLean, 1996) are available that can be controlled and configured from a web browser. (Deri, 1996). In addition, Deri (Deri, 1996) has shown a way to access network management resources from the web. In this paper we show how to use the web for more complex systems management operations. By web-enabling systems management tools the web can provide a point of integration for enterprise management.

In this paper we discuss the problems and issues involved and describe our web-based solution. Section 2 describes the problem and our constraints. Section 3 provides background in the NetFinity product and general web HTTP/HTML serving. In Section 4 we discuss the solution, and how it was implemented. Section 5 explores issues surrounding the solution, particularly limitations and unexpected benefits. Section 6 continues with performance measurements of our embedded web server serving systems management data. We conclude with Section 7.

## 2   THE PROBLEM

In general, UNIX and mainframe systems and network management tools have a centralized management architecture with a heavy footprint requiring one or more powerful machines dedicated to the task of management. Many times the graphical user interfaces (GUI's) used to administer the network also have a large footprint and are intimately tied to the management machine. On the other hand, PC systems management tools such as NetFinity are designed to manage PCs in a peer-to-peer fashion without requiring dedicated management machines.

In order to integrate PC systems management into UNIX-domain systems management packages, we would like to integrate the PC's GUI's into the UNIX-domain's. However, these PC GUI's impose significant restrictions on scalability, screen real estate, and platform. Not only is direct porting of PC GUI's to the UNIX domain unpleasant because of the cross-platform implementation, but the PC GUI's are principly designed for single-system-at-a-time management. Thus scalability as well as screen real estate become extremely limiting issues. This integration required from us a different approach.

Our goal was to enhance NetFinity so that it would meet the following criteria:

1. the manager and the GUI would retain a small footprint,
2. the GUI would be cross platform, and
3. the GUI would easy to develop, change, and integrate with other tools.

Our solution to this problem was to embed a minimal web server into NetFinity so that any platform with a web browser could manage the NetFinity machine. By minimal we mean only the basic functionality needed to process HTTP requests without the overhead of general purpose file serving and CGI-bin style processing. Thus the GUI is automatically cross-platform, at least for all the platforms that have a capable browser ported to them. Furthermore, the embedded web server is very small, consuming few resources. Finally, since HTML is significantly easier to work with than traditional event-driven GUI code, development, maintenance, and integration is quite pleasant.

A minimal web server was implemented by encapsulating the hypertext transfer protocol (HTTP) in a C++ subclass of iostream. In this paper, we describe the processes which brought us to this design, describe the design, and offer some performance numbers of the embedded server delivering real-time systems management data.

## 3    BACKGROUND

### 3.1 Introduction to NetFinity

NetFinity is a completely distributed management tool for PC's. NetFinity agents are installed on every managed PC, and managing machines do not need to be dedicated to NetFinity management. The NetFinity machines communicate over a remote procedure call (RPC) layer that sits on top of NETBIOS, TCP/IP, IPX, or serial protocols. The RPC layer allows multiple hops through multiple protocols to be used to contact a remote host. Among the services supported are screen captures, process management, remote sessions, alert management, and system monitoring. The base executables that perform these services are small; in some cases they are loaded and unloaded on demand, and therefore do not require additional system resources. The user interfaces communicate with the bases over the RPC layer.

## 3.2 Introduction to the web

The World Wide Web consists of two main standards: hypertext transfer protocol (HTTP) (Berners-Lee, 1996) and hypertext markup language (HTML). HTTP is a stateless protocol used to transfer files to web browsers: a distinct HTTP request is made for each file requested. HTML is the usual file type that is transported by HTTP and is a markup language that describes the format of the document to the web browser. Along with simple document rendering, HTML also supports clickable links to other documents and conveniently arrayed forms in web pages. The standard order of events in forms processing is as follows:

1.  The web browser renders the page. Included in the page are form elements allowing user entry of commands or data.
2.  The user makes the entries he or she wishes and submits the form.
3.  The web browser generates a uniform resource locator (URL) consisting of the destination of the form submission and a number of variable/value pairs representing the entered data.
4.  The web server receives the request and services it according to the page and arguments in the request.

Furthermore, variable/value pairs may be hidden in the rendered document and thus used to maintain state between requests. Also, since all the documents we are generating change often, we send the documents to the web browser as "expired" to prevent them from being cached.

For security we used a publicly available Secure Socket Layer (SSL) (Freier, 1996) library to implement our embedded web server.  Most popular web browsers support SLL. SSL uses asymmetric keys to negotiate a symmetric key that is used to serve the HTTP request. For performance the symmetric key is cached for a short period of time. SSL provides encrypted communication as well as authenticating the server to the browser using an X.509.v3 certificate. Clients are authenticated using the NetFinity authentication RPC and basic web authentication.

## 4    PUTTING NETFINITY ON THE WEB

Because of the simplicity of HTTP, it is possible to create a minimal HTTP server with a very small footprint. We encapsulate the HTTP protocol in a class that inherits from iostream and includes methods to retrieve variables passed, the document requested, and authentication information. By inheriting from iostream, all of iostream's formatting and buffering are available to the server application code.

The HTTP class consists of the following:

1. methods for replying through the socket to the HTTP requestor,
2. methods for parsing the services and arguments present in the URL,
3. methods to ease formatting of URL data into the HTML forms, and
4. all the methods inherited from iostream.

The web systems management interface application listens on a socket for the HTTP request, in the form of a universal resource locator (URL), from the web browser. When it receives a request, the server spawns a thread with the URL and the socket instantiated in the HTTP class. The thread parses the request, using HTTP class methods, to determine the systems management service and the arguments desired. The thread then executes the RPC communication appropriate to perform the requested systems management service. The RPC's receive information from the backends, and the results are formatted dynamically and returned to the browser, again using the HTTP class. To avoid conflict with other web servers which may be running on the system, our server runs on a user-selected port whose default is different from the traditional port 80.

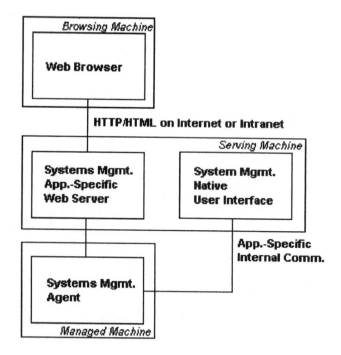

**Figure 1** Architecture of web interface daemon.

The result is a daemon that communicates with a web browser as if it were a web server and communicates with NetFinity backends as if it were a NetFinity GUI. Figure 1 shows the general architecture. Through this server architecture any web browser on any platform can manage a NetFinity machine. Implementing the daemon is very easy since it is totally procedural: many of the normal GUI event driven difficulties are avoided.

## 5   ANALYSIS

### 5.1 Limitations

Of course, there are tradeoffs to using the web as a management interface. Limitations include the use of forms as a means of interaction, the lack of client form validation, and basic formatting restrictions.

Being limited to forms as the only means of user interaction turned out not to be as limiting as one might imagine. In many cases, this restriction made the user interface less cluttered and easier to use, and, in general, it is a good fit for our systems management applications. While we did not have full control of screen layout, extensive use of tables (Raggett, 1996) improved the appearance greatly. In the end the limited control of page layout simplified the pages and allowed them to look good on a variety of platforms.

The one service that we could not do with forms was Remote Session. Remote Session offers the user a shell on the remote machine. Therefore, it must be fully interactive. We used Java to achieve this level of interactivity. Java is a very powerful yet simple language that enabled us to implement Remote Session in a few hundred lines of code. When the session is started, it makes an HTTP request to the daemon, and once the connection is made, the daemon and the Java applet use that connection as a full duplex channel.

The lack of form validation was not functionally limiting, but it does make forms somewhat more difficult for the user since he or she learns of errors in the form only after submitting it. Javascript from Netscape has the potential of fixing this. However, the lack of support, as well as security issues, prevented our use of it.

## 5.2 Unanticipated benefits

A number of unanticipated benefits arose from the using the web as the GUI. These include the ability to mail web pages as an alert-triggered action, the bookmark service offered by the web browser, links to other web resources, the find button of the browser, and rapid turn around of GUI modifications.

Various services in NetFinity and on NetFinity-managed machines generate alerts. These alerts can trigger pagers, e-mail, window pop-ups, etc. We constructed an alert action that sends a MIME-encoded e-mail of type "text/html" with links back to the machines involved in generating the alert. Thus, if the user's mail reader is properly configured, his or her web browser loads the page that was sent in the e-mail. E-mail is a convenient form of notification since people in general either watch, or have applications which watch, for new mail in their mail spools. Furthermore, since the e-mail is actually a web page, the links in the message allow the user to quickly access the machines causing and reporting the problem.

One of the difficulties in using any application with multiple GUI windows is finding a particular window on the screen or from one use to another. A native GUI's traditional means of changing context is by opening a new modal dialog box. In contrast, a web browser works in one window unless the user chooses to open another one. Also, when using a web browser, a bookmark can be set at any page and the user is able to return to that page at any time. In addition, instead of explaining complicated navigating instructions to another user who needs to find the window, a URL can be given or the page can be mailed. Also, pages can be printed straight from the browser. The find button is particularly useful on information packed pages. All these features come free to the web application developer but would be very costly to implement in native GUI's.

In summary, using a web browser as the front end and implementing a web server layer to translate service activity from the web domain to the application domain offers a plethora of benefits. The browser performs ancillary tasks, as a matter of course, that are only dreamed about in traditional GUI's. Traditional GUI's do not necessarily miss these resources, but they can be leveraged to provide a valuable service to the application user. In the case described in this paper, the user is the system administrator whose job becomes significantly easier with the portable and powerful web browser/server interface.

The use of HTML made GUI prototypes as easy as editing a web page. Changes to program code that did page layout were also simplified. Through the use of rapid prototyping and the ability to make quick changes to the code, the GUI evolved more rapidly. This often resulted in a better user interface than the native GUI.

## 6    PERFORMANCE

Today's web servers traditionally process forms through the use of CGI (Robinson, 1996) binaries or scripts. These are spawned by the HTTP server in response to an HTTP request to a specifically patterned URL. CGI bins are often written in C or C++. Also, Perl is a favorite language for these scripts.

Our embedded HTTP server does not spawn its form processing requests to a CGI program. We handle the form processing with the same executable as the server itself. In fact, the server connection, an http object as described previously, is passed to whatever handler is indicated by the requested URL. Each request spawns a new thread, and therefore operation is extremely lightweight compared to traditional CGI.

We compared the operation of the server implemented through the http class with a server implemented through spawning. License agreements for HTTP servers generally do not permit benchmarking the servers. This fact, plus the wish to make the CGI-spawning HTTP server as comparable as possible to our threaded server, led us to write our own spawning server. The spawning server simply accepts an HTTP connection, spawns a program, and directs the output of the program back through the connection. Since this is the absolute minimum that any HTTP server must do, it provides a good baseline.

The servers were run on a 100 MHz Pentium with 32 Meg of RAM running Windows NT 3.51. The HTTP requests were generated by a UNIX machine on the same LAN. Processor usage was gathered using the Performance Monitor built into Windows NT. SSL was not enabled. NetFinity RPC requests were made to the local machine. The page being served is the main page that lists the available services.

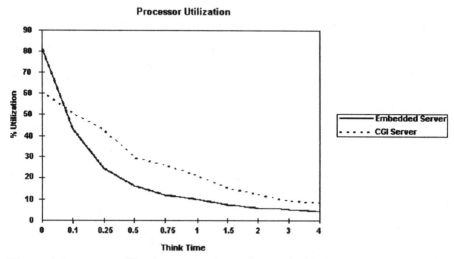

**Figure 2** Processor utilization comparison of an embedded server verses a CGI server.

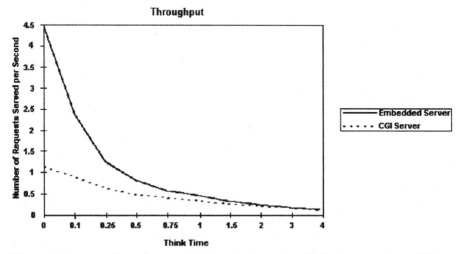

**Figure 3** Request throughput comparison of an embedded server verses a CGI server.

Figure 2 shows the processor utilization in relation to the think time for the two servers. Think time is the time from the end of the previous request to the start of the next request. It does not include the time to service the request. Figure 3 shows the throughput in relation to the think time. As think time increases, think time dominates the service time, and the throughput will tend toward the reciprocal of think time.

Note that even with a think time of 4 seconds the processor utilization of the CGI server is double that of the embedded server with approximately the same throughput. The contrast is even bigger with a think time of 250 milliseconds. The embedded server uses a little more than half of the processor utilization with twice the throughput. Only near zero think time does the utilization of the embedded server pass that of the CGI server; however, the throughput is nearly quadruple that of the CGI server. This difference is simply due to the overhead of spawning the GCI process.

In normal operations the time between HTTP requests to a server will be much greater than 4 seconds since generally the page will have information that will take time to be read by the user. However, as explained above, one machine can be used to administer other machines. This is especially useful to be able to manage non-IP machines (such as IPX and NETBIOS) from the Web. If a CGI server were used, a dedicated machine would be needed if the throughput were to get above 1 request per second. The embedded server uses less than 25% of the CPU at 1 request per second eliminating the need for a separate dedicated server for all but the busiest of managers.

It should be noted that the overhead of spawning is not the only performance advantage of using an embedded server: application initialization can also be very costly. In the case of the NetFinity RPC layer, communications must be set up and taken down each time a program is run. Using an embedded server, the initialization cost is paid only once; however, using CGI's requires the initialization to be run for each request.

## 7    CONCLUSIONS

Platform-independent user interfaces offer powerful flexibility in administration and integration of similar or disparate systems management tools. Web-based systems management provides a large segment of the overall picture which will bring single-image systems management to reality. Furthermore, because the HTML interface is as simple at the server end as it appears at the browser end, advancement in systems management features can proceed at a much greater pace than is possible with traditional GUI's. Aided by Java for functionality requiring a fully interactive interface, web-based systems management is definitely a step towards more powerful systems management.

## 6    REFERENCES

Berners-Lee, T., Fielding, R., and Nielsen, H., "Hypertext Transfer Protocol -- HTTP/1.0", RFC 1985, May 17, 1996

Bruins, B., "Some Experiences with Emerging Management Technologies", Simple Times, July, 1996

L. Deri, Surfin' Network Management Resources Across the Web, Proceedings sf $2^{nd}$ Intl IEEE Workshop on Systems and Network Management, Toronto, June, 1996

Freier, A., Karlton, P., and Kocher, P., "The SSL Protocol, Version 3.0" (Internet Draft), March 1996

McLean, M., "Browsers to Simplify Remote-Device Setup", LAN Times, May 13, 1996

Mullaney, P., "Overview of a Web-based Agent" Simple Times, July, 1996

Raggett, D., "HTML Tables", RFC 1942, May 1996

Robinson, D., "The WWW Common Gateway Interface Version 1.1" (Internet Draft), February 15, 1996

Wellens, C., and Auerbach, K., "Towards Useful Management", Simple Times, July, 1996

## 7    BIOGRAPHY

**Benjamin Reed** has been working since 1995 on system and network management solutions at IBM Almaden Research Center. He is currently working on his PhD in Computer Science at the University of California Santa Cruz.

**Michael Peercy** received his BS in Computer and Electrical Engineering from Purdue University and his MS and PhD from the University of Illinois. He has been a Visiting Scientist and Software Engineer at IBM Almaden Research Center since 1994, working in the area of systems management.

**Ed Robinson** has been working for Binary Consulting, Inc. on integrating web technology with modern compiler technology. He is a PhD candidate at the University of Houston. Prior to working at Binary Consulting, he was at IBM Almaden Research Center where he contributed to the work presented in this paper.

## PART THREE

# TMN - Today and Tomorrow
## Chair: Adarshpal S. Sethi,
## University of Delaware

# 8

# Integrated TMN service provisioning and management environment

*G. Chen and Q. Kong*
*CiTR Pty. Ltd.*
*P.O. Box 1643, Milton, Queensland, Australia 4064, Tel: +61 7 3259 2222, Fax: +61 7 3259 2259, {g.chen, q.kong}@citr.com.au*

## Abstract

The world wide competition and deregulation in the Telco industry has intensified the need to build a service management system, where the Telco service offering environment, service creation environment, service provisioning environment, service management environment, and service implementation (based on the network) environment are all integrated in a single architecture.

This paper addresses the issues of building such integrated system. We propose an integrated service provisioning and management environment with a Java-based user environment, a CORBA-based distributed service management environment and a TMN-based network management environment. The paper discusses the business requirements and technical requirements for such architecture. It also reports our first phase prototype implementation of the system. It focuses on the experiences of integrating CORBA, Java and OSI technologies to achieve integrated management.

## Keywords

Service management, service provisioning, CORBA, TMN, Java

## 1 INTRODUCTION

The world-wide deregulation of telecommunication industry has forced the telcos to re-organise the business process and adopt the new technology to manage its network and services. As opposed to the network management, which solves the cost reduction side of the equation, the service offering and service management activities deal with the revenue generation side of the equation. The technology framework in which the telco services are offered and managed becomes the major enabler for telcos to compete in the new business environment.

The telco industry has seen the following business trend and technology progression in recent years:

- Following the deregulation and increased competition, the customer care system integrated with the process re-engineering provides the maximum competitive edge to the telco industry.
- The distinction between traditional distributed computing environment and telecommunication environment is blurred following the use of computer technology. For example, some telecommunication service has little difference from a traditional financial service.
- An integrated environment for both service offering and service management is essential to telco's success. There is a chasm between the TMN service management and network management. The challenge facing the industry is to bridge this chasm by providing a reliable, scalable and extensible computing environment upon which telco services can be offered, managed and integrated with the network.
- In the choice of computing technology, the distributed object oriented technology shows its capability to bridge this chasm.

However, the telecommunication industry still needs evidence that the distributed object technology can be integrated seamlessly with the existing TMN environment, OSS environment and the user environment. Without this integration, the technology will have very limited role to play.

The Network Management Forum (NMF) has identified that Customer, Service Provider (SP) and Network are the major players and components in the integrated service management chain [7,13].

NMF is currently investigating the technology requirements, particularly from a customer's point of view, for a complete cycle of service offering and service management. The overriding requirement is the definition of an integrated environment where the telco services can be offered, controlled and managed. In the same environment, the network resources which provide the ultimate support for services can be integrated; the service providers can interact with other service providers and the customer care systems can be integrated to achieve required customer satisfaction and quality of services.

In addition to these, the advanced broadband transmission technology such as ATM has created more opportunities to extend the telecommunication services. This has put more requirements on the environment in which these new services are deployed and managed.

Three types of requirements are identified in the integrated service management chain. The customer care requirements include the CTI interfaces to the management of customer's ordering requests, quality of service requests and any other access requests. It also includes the customer network management requirements to provide interaction with different service providers. The service management interoperability requirements address issues such as: support of service level agreement; support of cross SP service provisioning and management; and support of customer network management interface.

The multi-technology integration requirements are addressing very important issues of using different technology to provide telco services. These requirements include:

- the support of the integration between technologies used by service management component and network management component

- the support of protocol independent object abstraction
- the support of service designing and creation environment.

This paper addresses the issues of the technology integration to provide this support.

## 2    SERVICE PROVISIONING

### 2.1    Service Management Life-cycle Scenario

The relationship between different components in the service management chain can be demonstrated by the following telecommunication service management life-cycle scenario.

Service ordering and provisioning is the basic activity in telecommunications. A simplified service ordering and provisioning process consists of:

- a customer issues a service order to a service provider
- the service provider collects and verifies customer information and starts the service ordering process
- service level agreement between the customer and the service provider is created or updated according to the negotiation between the customer and the service provider,
- services are installed and tested—this may involve actions at both service management layer and network management layer.

The following figure illustrates this process.

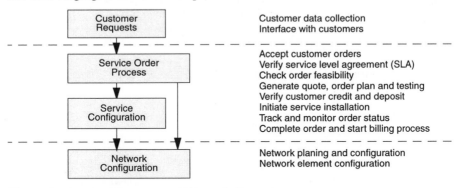

**Figure 1**    Service Management Life-cycle Scenario.

Depending on the type of services ordered, the processes, in particular the installation and test phase of the services, may not be automated.

### 2.2    Automation in Service Ordering and Provisioning

Automation in service provisioning process includes the following aspects: when a customer requires a new telecommunication service, a call centre, which is responsible for handling customer service requests, is contacted. Customer care system is used to provide a quick and reliable environment for the call centre. A telephony interface or on-line interface such as Web and Java can be used for such access.

To service the request, a process control task (such as a provisioning workflow) may be started, which controls the business rules and policies for the provisioning of such a service. The integration of customer care system and process control system provides the smooth management of the ordering process. This allows the tracking, monitoring, and reporting the customer requests as well as the processing of the requests. The workflow process integration customises the existing generic workflow management system to the specific telco business environment. In the cases where the consistency and correctness is crucial to the business process, the transactional capability can be supported by the workflow engine [2].

The processes involved in a workflow instance can be defined and deployed as distributed objects. They are supported by the distributed object environment. The environment contains a set of well specified generic object implementations which customise the distributed object environment to a specific business domain. These processes require a well-integrated management interface so that the service offering aspect and the management aspect are integrated in a single object environment which presents the TMN service management view.

Each service request can be mapped to a set of network requests and implemented on a potion of network and its elements. Network requests are sent to agents involved in the implementation of the service in TMN network management environment. This functionality is well supported by the network management environment.

The automated service ordering and provisioning scenario requires a great amount of integration between different technologies which offer different functionality. In the next sections, we focus on the integration aspects of technologies.

## 2.3  Technology Support

The following figure illustrates the technology support for service ordering, provisioning and process automation.

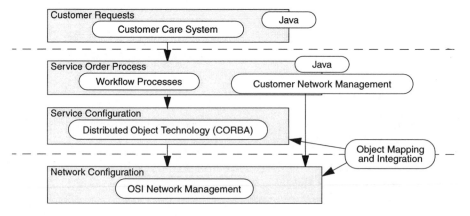

**Figure 2**    Automated Service Ordering and Provisioning Process.

As indicated in Section 2.2, technology required in supporting automated service ordering and provisioning process may include:

- Customer care system—the front-end contact point to customers
- Customer Network Management—the ability to manage inter-SP service configuration activities
- Java—a GUI architecture providing standard interface to operators and customers
- Workflow system—the back-end process control and monitoring system
- Distributed technology—the distributed platform, such as CORBA, to support distributed service management and to interface with the network environment
- Object mapping and integration—tools and environment providing static and automatic mapping from service management environment to network elements
- Network management—the basic technology, such as OSI NM, used to manage telecommunication management networks.

There exist many different technology choices to support the automated service ordering and provisioning process. In the next two sections, we present an architecture and discuss the experience of implementing such architecture.


# 3   ARCHITECTURE

In this section, we analyse the requirements for supporting the ordering and provisioning process in more detail and propose an architecture based on the integration of different technologies. The aim is to define an architecture that will meet the business requirements, survive the impact of technology evolution and migration and such architecture can be easily deployed to support telco's business processes.

## 3.1  Business Requirements

Although the trend towards adopting re-engineering using new distributed network and service management technology is very strong, across telecommunication industry, there are well-founded business reasons why people still have reservations of moving too quickly. The following concerns are among the first:

- availability and reliability—the ability to support telco business in a 7 by 24 way. This is overriding requirement that the industry will not use any technology that cannot provide reliable service
- integration with legacy systems—the strategy to protect the huge investment the industry has made. These systems include all existing network management and element management applications, communication stacks and protocols, and OSS applications which support telcos core business. A clear cut with the past, no matter how desirable it is, is not an option to the telco industry
- client GUI platform—the strategy to provide flexible and light-weight client GUI technology, which is well integrated with the distributed object paradigm and does not require a huge overhead of traditional client platform. With the popularity of

internet and Java technology, telcos need a cheap and flexible way to deploy services to customers quickly and allow easy customer access.

It is essential that the architecture for integrated service provisioning and management satisfy all these requirements.

## 3.2    Three Layered Architecture

The following figure illustrates the service provisioning and management architecture proposed here. It has three basic functional layers.

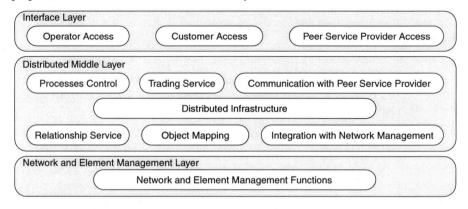

**Figure 3**    Management Architecture.

### *Interface Layer*

The interface layer supports different type of access to the system. It includes Java-based GUI for both operators and customers. The interface also includes APIs for peer service provider access to archive integrated peer-to-peer service provisioning.

This layer supports the separation of object presentation from their semantics. The semantics of the objects should be encapsulated behind an interface. Access can be permitted to the interface without compromising the semantics.

Java is a platform independent programming language. By using Java-based GUI, the presentation rules are encapsulated in Java objects, while the semantics are encapsulated in other system objects. In this architecture, they are encapsulated in CORBA objects supported by the distributed middle layer. This allows new applications to be developed in a manner that they present the information without interference with the core functionality. The development time for new applications can be reduced and the integrity of core functionality can be enhanced.

### *Distributed Middle Layer*

Distributed middle layer is the kernel part of the architecture. It consists of:

● service management components such as TMN management functions: fault, configuration, accounting, performance and security (FCAPS) [3]

- provisioning process control and automation components
- distribution of application components by supporting distributed infrastructure and services.

It also serves as the integration bus between different functional layers, different components and also between different service providers.

CORBA Object Request Broker (ORB) and CORBA services [11,12] are used as the distributed infrastructure. Among many services, the relationship service and the trading service are two important services in providing integrated service ordering and provisioning system. The relationship service offers functionality to model the component structure and topology in a provisioning process and the support for the process automation. The trading service helps the process control to be achieved in a distributed operating environment.

One major responsibility of this layer is to support interaction with the interface layer and the network and element management layer. The Java-CORBA gateway becomes the interaction point between the Java-based interface layer and CORBA-based middle layer. The interaction between the middle layer and the underlying network and element management layer requires the integration between CORBA and network management technology. The components such as object mapping, integration technology and relationship service all play important roles here. Section 4.1 discusses these integration issues in detail and Section 5.2 discusses our experiences of implementing the integrations.

### Network and Element Management Layer

The network and element management layer consists of a management platform and a set of network management functions. This environment provides the support for FCAPS functions at network and element level to deliver services. It interacts with the service management environment to ensure that the service requests are mapped to network operations and the quality of service requirements are satisfied.

OSI-based TMN network management and element management is the main component of this layer.

## 3.3 Technical Features

The architecture for integrated service management supports the following features:

- *Scalability*—the scale of the system can be extended when it is required by the business expansion. Distributed systems offer an opportunity to manage systems that are currently too large for a single host to manage by using multiple smaller hosts. One issue in the scalability is to allow hosts to be added into the system without affecting the functionality
- *Extensibility*—the functionality of the system can also be extended to meet customers' new requirements. The building block concept in distributed systems allows functional blocks to be added when required
- *Reliability*—continuous support for service is the major requirement of service providers. The architecture therefore must allow construction of systems that can

recover from hardware and software failure. Special effort is required to support different level of reliability and fault tolerance in a distributed system. Section 5.3 discusses this feature.

## 3.4  Deployment

The system can be deployed in a wide range of configurations. The primary partitioning of functionality between machines is the user interface and service machine. Typically a PC-based system would provide the user interfaces and a Unix-based system would provide the services and platform support. This configuration offers the advantage of using high performance and reliable systems for the servers that enforce the semantics of the system, while using lower cost systems for the user interfaces. The PC front end also provides an interface that is typically more advanced than Unix-based systems.

The Java user interface provides a mechanism for rapid development of user interfaces that are portable across multiple platforms. The CORBA infrastructure provides a basis for distribution and extensibility.

Java and CORBA are both available on a wide range of platforms (unix workstations, PCs, Macs etc.) and the platforms are able to interwork in a heterogeneous environment. The architecture may be implemented on different platforms depending of the performance required for a particular application. For example a low performance application may be implemented on a PC-based system, while a high performance high availability application may be implemented on a Unix-based system.

## 4    TECHNOLOGY INTEGRATION

The major challenge of providing an integrated telecommunication service management is the integration of existing technologies. The requirements for integration are caused by:

● different technologies are used to provide different functions
● different technologies are used by different service providers
● different technologies are used in today's systems which are supporting Telcos business processes. Such examples include the network management systems and OSS systems.

The architecture presented in the previous section requires the seamless integration of different technologies. These include:

● the integration between customer care system and the process control system
● the integration between GUI technology, such as Java to back-end distributed technology
● the integration of process control system to distributed environment
● the integration between CORBA-based service management components and OSI-based TMN network management components.

In this section, we focus on the integration between CORBA and Java, and between CORBA and TMN to provide an integrated management environment.

## 4.1   CORBA/TMN Integration

The integration between CORBA distributed object technology and the OSI network management technology provides the basis for the integration between telecommunication service management and network management layers [1]. Different object models are used in these two different technologies. Thus, the initial focus of the technology integration is the mappings between object models.

Three different integration approaches are used in our study. These are, *abstract mapping, close mapping* and *TMN (GDMO) factory.*

Using the abstract mapping approach, a CORBA object is used to represent a functionality which is usually modelled by a set of GDMO objects or an application entity. Such object presents a CORBA IDL interface and its implementation is based on CMIS operations and uses CMIP protocol. Since the CORBA object represents high level application semantics, the number of these objects, compared with that of the GDMO objects, is relatively small. Also, the GDMO containment structure and associated scope and filter operations can be embedded within the object implementation. This will reduce the complexity of modelling containment relationship within CORBA. These integration objects serve as the base level building blocks and more application semantics level CORBA objects can be defined.

The close mapping approach offers a more direct static mapping between two object models. A CORBA object is used to represent a GDMO object and a one-to-one translation between the attributes and operations can be achieved. Based on these CORBA objects (proxies) more application semantics level CORBA objects can be built. This approach needs to model OSI containment structure and scope and filter operations in CORBA which may present a performance problem.

TMN factory approach allows dynamic creation of CORBA proxies of GDMO objects at run time through a proxy service. This approach allows application to selectively create CORBA representation of GDMO objects and reduces the overhead of promoting large number of GDMO objects to CORBA space.

The experiences we gained from building a prototype of the system suggest a combination of different approaches depending on the way the services are modelled.

## 4.2   CORBA/Java Integration

Java and CORBA complement each other extremely well. Java can be used to develop small machine independent GUIs that may be downloaded across the web to the client's machine from a central server. The GUI can then use a CORBA implementation to make remote method calls on objects elsewhere on the network. This maps well to the real world organisation of distributed data.

This configuration allows the most suitable machines to be used for each component of an application and it also opens up the application to any machine where the Java client can run. No longer does the user have to be on a certain type of machine to be able to run the application.

However, the Java GUI components need to be well integrated with the CORBA environment in order to perform the GUI functionality. The following figure depicts the integration architecture:

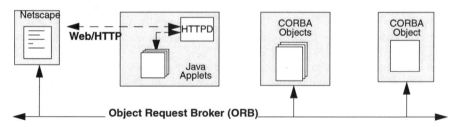

**Figure 4**    Java/CORBA Integration Architecture.

This architecture will enable true distributed computing with Java as the downloadable GUI front end. A Java object can also have a CORBA object interface so that remote objects can make computation requests on it.

## 5    IMPLEMENTATION

### 5.1  Service Management Environment

We have implemented a prototype of service management environment based on CORBA integrated with an OSI-based network management platform. In the first phase implementation we used Orbix 1.3 and OpenView DM 3.3. A Video-on-Demand (VOD) service is used as an example to demonstrate various concepts discussed in this paper. The following figure depicts the architecture of this environment:

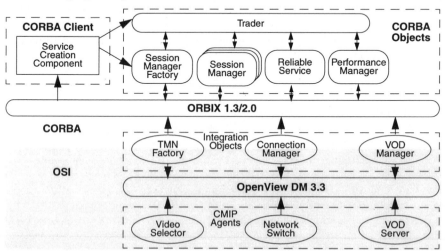

**Figure 5**    Service Management Environment (VOD Demonstrator).

The demonstrator has three sections—the TMN section, the CORBA section, and the integration section (CORBA and TMN). The TMN section consists of the CMIP agents for the Video Selector, the Network Switches and the VOD Servers. The CORBA section has a number of CORBA objects to implement the VOD Service.

The environment is supported by a CORBA-based platform integrated with a CMIP-based platform. The environment offers the existing CORBA distributed infrastructure and distributed services. A set of generic services to support service management and service offering environment is added to this platform.

In the above example, a service is triggered by a CMIP event modelling the video selector at the customer premises to select a video. This event is delivered to the CORBA service creation component which create session manager to provide the service. A trader is used for location transparency as it is vital to achieve scalability and reliability.

## 5.2  CORBA/TMN Integration

The CORBA objects TMN Factory, Connection Manager and VOD Manager are objects that integrate with CMIP environment. These objects have defined CORBA IDL interfaces and the implementation of these objects performs CMIP operations on the underline CMIP GDMO objects which model the elements and transmission network. The CMIP environment is implemented using HP OpenView DM.

The CORBA object Connection Manager is an abstract representation of the connection in the transmission network. This object does not map into an individual CMIP object but instead it models a connection which is constructed by CMIP operations. This object is a low level building block supporting application semantics. The implementation results indicate that this level of mapping offers the best overall system performance due to its higher level of abstraction [1] and is always preferred integration approach.

The VOD Manager is a CORBA object which models the VOD server in the CMIP space. This object represents the close mapping between CORBA and GDMO. With this object the CORBA environment can directly access CMIP objects to control and coordinate network elements.

The TMN Factory is a generic CORBA object which dynamically creates CORBA objects acting as proxies for CMIP objects. It provides a generic mechanism to allow access to TMN functionality that has not been encapsulated in an abstract object. It achieves this by providing a fixed interface that can be used to perform the standard CMIP operations on CMIP objects. An IDL object is created on demand and bound to a CMIP object. Remote clients can access the CMIP object via the IDL proxy. The TMN factory allows access to any CMIP object. Clients only need access to the CORBA infrastructure rather than the CMIP stack.

The TMN Factory provides an on-demand 1-to-1 mapping of CMIP objects into the CORBA domain. Due to the 1-to-1 mapping the service is not scalable. It is intended to provide access to rarely used CMIP objects.

## 5.3  Scalability and Reliability

Customers require systems that run continuously. The platform must therefore allow construction of systems that can recover from hardware failure. Some form of replication is required to allow another host to provide the services of a failed host.

Distributed systems offer an opportunity to manage systems that are currently too large for a single host to manage. Distribution allows multiple smaller hosts to be used in place of the single large host.

The demonstration offers scalability inherited from CORBA architecture. By using distributed trader, the service creation process can decide on which host the new service and management objects should run. The system is tested to handle 100,000 objects.

CORBA architecture does not have built-in reliability. In this demonstrator, we designed a generic reliability component which detects the fault in the management system and re-create the management objects by using trader and *virtual data store* (VDS) which is a component based on CORBA's persistent service. At any time during the running of the system, a host running the service management system can be un-plugged from the system. The reliability service detects the failure of communication with the objects on that host, recovers the state of the objects from the VDS and re-initiates the service from a different host.

## 5.4  Java Performance Manager

In a system where there could be a large number of users requiring the application services, there are significant problems in configuration management. For example, what happens when a new version is released. By adopting the Java and CORBA approach, the application client can be stored centrally on a few servers and then downloaded on request to the user.

To demonstrate the advantages of using Java as the user environment, a Java applet which monitors the performance of the VOD system was implemented. The applet is a performance monitor based on a simple bar chart display. Performance data from the CORBA and TMN sections of the application are gathered together into a single CORBA object. The Java applet then accesses the CORBA object in order to obtain the performance data and present it to the user. The bar chart consists of a count and a moving bar colour-coded to the current value.

## 5.5  Service Provisioning Environment

The work is continuing to construct a service provisioning environment. The initial effort is to have a GUI-based environment as the single access point for operators to deploy services required by customers to a set of network elements. The major components and activities in this environment include:

● A *Java service order applet* allows operator to enter customer information to a service order form. The ATM service provisioning is used as a demonstration scenario. Information required in an order form include customer name and address, bandwidth required, type of data to be transferred, service type, whether it is bandwidth on demand (pay as you use), etc. Service request is submitted by the operator on behalf of customers. The customer request is stored as the service agreement and is linked to the customer profile.

● A *Java resource monitor applet* allows network operators to view the status of network resource and to find the available resource to meet customer request. The Java/CORBA gateway supports the submission of such search request and returns results to the operator.

● A *Java configuration applet* provides the access point for the service deployment. It includes allocating the available network resource for the service; configuring network resources; associating the service with the set of network resources; testing the service; and finally, activating the billing system. A separate GUI may be used as the network manager's control console to display the connection between end points, or to allow network manager to re-configure the system.

Java/CORBA gateway allows service configuration request to be performed on service objects in the CORBA environment. The CORBA/TMN integration gateway allows the service requests to be mapped into a set of network configuration requests and implemented by network elements such as switches.

The following figure illustrates the interaction among system components in the service provisioning environment.

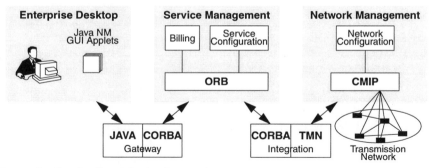

**Figure 6** Service Provisioning Interaction.

The components are being implemented as extended building blocks to the existing system to form an integrated demonstration.

## 6 SUMMARY

The paper has demonstrated an integrated service offering and management environment based on the integration of CORBA, Java and TMN network management technology. The prototype we built demonstrated that the architecture offers a very viable solution to telecommunication business processes.

The prototype offers a highly scalable, reliable and extensible environment to support telco business processes. It demonstrates the integration approaches between CORBA,

Java and TMN environments are sound and performing well. The resulting system provided very useful platform to evaluate issues relating to TMN service management.


# 7    ACKNOWLEDGEMENTS

The authors would like to thank the many people who have contributed to this paper—in particular, Michael Neville and Mike Sharrott.


# 8    REFERENCES

[1]    Q. Kong and G. Chen, Integrating CORBA and TMN environments, *Proceedings of the IEEE/IFIP Network Operations and Management Symposium*, Kyoto, Japan, April 1996.

[2]    Q. Kong and G. Chen, Transactional workflow in telecommunication service management, *Proceedings of the IEEE/IFIP Network Operations and Management Symposium*, Kyoto, Japan, April 1996.

[3]    ITU-T Recommendations M.3010 — Principles for a telecommunication management network.

[4]    W. Widi, *Standardisation of Telecommunication Management Networks*, Ericsson Review, No. 1, 1988

[5]    W. J. Barr, T. Boyd, and Y. Inoue, The TINA initiative, *IEEE Communication Magazine*, March 1993.

[6]    Q. Kong, G. Chen, and G. Holliman, Telecommunication service management, *Data Communications OpenView Advisor*, Volume 1, No. 8, August 1995.

[7]    Network Management Forum Smart Ordering Team, Ordering white paper, issue 1.1, January 1996.

[8]    E. K. Adams and K. J. Willetts, *The lean communications provider—Surviving the shakeout through service management excellence*, McGraw-Hill, 1996.

[9]    J. P. Chester and K. R. Dickson, Standards for integrated services and networks, *Integrated Network Management IV*, A. S. Sethi, Y. Raynaud and F. Faure-Vincent (eds.), Chapman & Hall, 1995.

[10]    M. Flauw and P. Jardin, Designing a distributed management framework—An implementer's perspective, *Integrated Network Management IV*, A. S. Sethi, Y. Raynaud and F. Faure-Vincent (eds.), Chapman & Hall, 1995.

[11]    Object Management Group, The Common Object Request Broker: Architecture and Specification, Revision 2.0, July 1995.

[12]    Object Management Group, CORBA Services: Common Object Services Specification, Revised edition, OMG document number 95-3-31, 1995.

[13]    Network Management Forum, OMNIPOINT V3 Specification, 1996.

# 9

# The use of allomorphism for the access control service in OSI management environment

*Ramos, A. M.*
*Universidade do Vale do Itajaí - Univali*
*University of Vale do Itajaí - Univali*
*Faculdade de Ciências Exatas - Curso de Computação*
*Rua Patricio Antonio Teixeira, s/n - Cep 88160-000 -*
*Biguaçu/SC/Brazil - Phone/Fax: 55 048 2314091-*
*E-mail: moraes@inf.ufsc.br*

*Specialski, E. S.*
*Universidade Federal de Santa Catarina*
*Federal University of Santa Catarina State - UFSC*
*Curso de Pós-Graduação em Ciência da Computação*
*Cx Postal 476 - Campus Universitário - Cep 88040-900 -*
*Florianópolis/SC/Brazil Phone 55 048 2319738 -*
*Fax 55 048 2319770 - E-mail: beth@inf.ufsc.br*

## Abstract

This paper proposes the use of Allomorphism for the access control service in OSI management environment, with no Access Control Function implementation (ISO10164-9, 1990), which is responsible for this service. The Allomorphism is a powerful SMI resource that makes the OSI Model very strong and flexible. This strategy reduces the overhead introduced by the access control function on OSI management systems without expose the management to arbitrary use of its resources. The Allomorphism allows to define different access views to the management resources according to security requirements. In this context, this work presents the security requirements in networks management systems, the OSI Access Control Function Model and its disadvantages, the idea of Allomorphism and, finally, how to implement it to access control.

## Keywords

Allomorphism, access control service, OSI management environment

# 1    INTRODUCTION

The scope of this paper is the OSI management environment, the access control service and its implementations alternatives using the access control function (ISO10164-9, 1990) and the allomorphism. The paper is designed to encompass the following objectives:

- To emphasize the vulnerabilities created for the network by the management systems and, consequently, the importance of access control service;
- To analyze the access control function (ISO10164-9, 1990) and their particularities;
- To analyze the allomorphism concept and its flexibility;
- To propose an alternative implementation for the access control service in OSI management environment which is based on the use of allomorphism;
- And, finally, to analyze the impact of the access control service implementation using the allomorphism.

Computer Network Management can provide benefits to the network system such as to control faults, to evaluate performance, to identify systems violation, to exercise control over configuration, to determine and allocate costs and charges for the use of the network resources, and others. However, some security conditions need to be implemented to control unauthorized people to access the network resources through the management system application itself.

Although management systems make the administration process of the network easier for the administrator, they also create a vulnerable point which can be used to manipulate extremely important information for the good functionality of the network. When management exists more points to attack the system and  more vulnerability are created, more possibility to compromise the net and its systems. These and many other weaknesses would not exist if a management system was not implemented (De Lucca, 1993).

So, the OSI management Model (ISO7498-4, 1989) defined the security functional area which aims to control security mechanisms, facilities and services in such a way that it will protect the network resources as well as (its own) the management application itself against system violations. In order to achieve all the requirements, three security management functions were defined to support the functional area; which are Security Alarm Reporting Function (ISO10164-7, 1991); Security Audit Trail Function (ISO10164-8, 1990); Objects and Attributes for Access Control (ISO10164-9, 1990).

The Security Alarm Reporting  Function notifies wrong operations in security mechanisms and services; and attempts to violate the network or management security. The Security Audit Trail Function records all potential events related to security in a log file for security audit. The Objects and Attributes for Access Control Function protects the management resources against nonauthorized access.

One or more access control mechanisms are implemented to ensure only access to specific resources by authorized users.

The alternative proposed in this article is resulted from the access control function (ISO10164-9, 1990) implementation in the UFSC's network management framework (Ramos, 1994).

## 2 OBJECTS AND ATTRIBUTES FOR ACCESS CONTROL FUNCTION

The Objects and Attributes for Access Control Function (ISO10164-9, 1990) has specific access control within the management system. Once an application management access is obtained, all associations among agent and management processes and all management operations on managed objects store in MIB are controlled by the Access Control Function and are managed by the Security Alarm Reporting Function and Security Audit Trail Function.

Many access levels may be needed to be implemented:

- Read/write access, read only or no access at all;
- Access only for some objects; and
- No permission to have a management association established.

Nonauthorized entities should not have access to management operations; and, all management warnings including, security alarm, should only be sent to authorized users.

In order to accomplish this, the security authority must define an access control policy for the management resources. These rules define: what kind of management information need controlled access; those who may manipulate them and the conditions to do so; and, what rules will control the access to management information.

Objects to be managed and their attributes model the management resources. Therefore when management objects and their attributes are access controlled so are the network resources automatically access controlled.

The management information which need access control is associated to another object, exclusive to access control, called target. The target has attributes to represent information and rules of access control to be applied to its associated object, which means the information that the object represents. For example, a target may represent a managed object, an attribute of managed object, an attribute value or even a management action.

The access control policy for resources management consists of specification of target objects that represent information and rules. There are two types of access control for OSI management:

- For association of management applications, to guarantee that non-authorized entities initiator will establish management associations;
- For management operations, to guarantee that operations will be made by authorized entities but with restrictions related to time, operation type, resources and information involved.

The Access Control Function has the basic model presented in Figure 1, with the following functions: Access Control Enforcement Function (AEF) and Access Control Decision Function (ADF).

The AEF function receives the access request made by an user, also called initiator, selects the access control parameters for that request and transfer these parameters to the ADF function.

The ADF function receives the access control parameters from the AEF function and decides if the access request is valid or not. For this validation, the ADF function compares the request parameters received from the AEF function with the rules of the access control policy defined for the network. The request is validated if it follows the rules. Otherwise, the request is rejected. The decision is passed back to the AEF function. If the request is rejected the AEF function will also receive the rejection instructions which can be just an access denied or even an instruction to abort the connection.

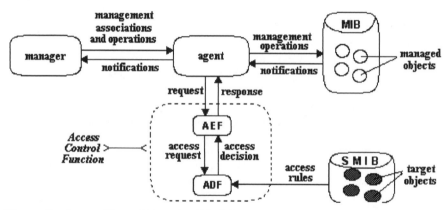

**Figure 1** - Access Control Function Model

The OSI Management Information Model is based on Object Oriented Paradigm where objects are logic abstraction of physical resources. Objects can be modeled using the paradigms: class, attribute, encapsulation, inheritance, multiple inheritance, and allomorphism as defined by SMI (Structure Management Information) (ISO10165-1,2,3,4). Objects are organized following a hierarchy in a Management Information Base (MIB).

The information of access control is also stored with object oriented paradigm using objects of access control called target objects. Each management resource, or management object, which needs security is associated to a target object. All target objects are in a Security Management Information Base (SMIB) defined at conceptual level. The SMIB can be integrated partially or totally to MIB.

## 3. ALLOMORPHISM

Allomorphism allows an instance of a class to behave like an instance of one or more other classes. This feature provides more power and flexibility to the OSI Management Model (Pavlou, 1993).

In general, the OSI Information Model allows members of arbitrary classes, derived or not, to be related allomorphically if they have a compatible behave. The object that emulates completely behave and characteristics of a member of another management object class is said to be allomorphic. Allomorphism is used by management systems to allow a better interoperability of manager and managed objects.

For example, when the management system is updated to a new version, the objects needed to be extended to include the new features. By definition, allomorphic classes must provide management of new systems and accommodate their new additional features in the current system.

Allomorphic classes, when derived, are called allomorphic subclasses and present similar behave to their superclasses. Different versions of a resource is an example o allomorphic class. Allomorphic subclasses are a type of specialization of superclasses where the inheritance characteristics from the superclass must be restriction in the allomorphic subclass (ISO10165-1, 1991).

Allomorphic inheritance shows restrictions to attributes, actions, and notifications. Regarding attributes, any acceptable value in the subclass must also to be acceptable in the superclass. The following restrictions apply for the inherited actions of the subclass:

- New optional and response parameters may be added to the action argument if the superclass is able to be extended;
- New mandatory parameters may be defined for the action argument if default values are defined to be assumed when nothing is specified;
- New necessary parameters may not be added - because they must be present at the execution time of the action;
- New actions can be defined; however, they can never be invoked by a system that doesn't accept the new class.

Restrictions of notifications for inherited events refer to inclusion of new response parameters and extension of value limits for existent arguments that are neither one allowed.

Figure 2 shows the inheritance tree that starts with the root class which is the superclass of all other classes. Net and system are subclasses of the root class. Net has only one subclass which is sub-net. On the other hand, system has a subclass equipment with 3 other subclasses: switch, router and hub.

Figure 3 shows the structure when the router subclass is defined as allomorphic to its superclass equipment. One class can be allomorphic to one or more classes. In the allomorphic class is important to have an attribute (allomorphsClassesDefault) that identify the default classes which has an allomorphic behavior. At the moment of the managed object creation, the allomorphs attribute of TOP Class receive the value of this attribute (allomorphsClassesDefault).

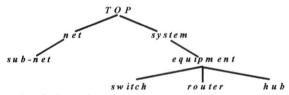

**Figure 2 -** Example of Hierarchy

```
router MANAGED OBJECT CLASS
        DERIVED FROM              equipment;
        CHARACTERIZED BY
        routerPackage                     PACKAGE
                ATTRIBUTES
                        routerID GET-REPLACE,
                        model             GET-REPLACE,
                        address           GET-REPLACE,
                        nameServer        GET-REPLACE,
                        route             GET ADD REMOVE,
                        routePrivate      GET ADD REMOVE,
                        allomorphsClasessDefault   REPLACE-WITH-DEFAULT
                        DEFAULT_VALUE
                        Attribute-Module.defaultallomorphsClasses GET-REPLACE;
                ACTIONS
                        addRoute,
                        deleteRoute,
                        changeRoute,
                        queryRoute,
                        addRoutePrivate,
                        deleteRoutePrivate,
                        changeRoutePrivate,
                        queryRoutePrivate;
                NOTIFICATIONS
                        routeAdded,,
                        routeDeleted,
                        routeChanged,
                        routePrivateAdded,
                        routePrivateDeleted,
                        routePrivateChanged;
        REGISTERED AS  at&tdcs52rot1;
        default values definition
        defaultAllomophsClasses::={at&tdcs52}
```

**Figure 3 -** Definition of the Router Class

The router subclass is at the same time derived and allomorphic to its superclass equipment. One instance of the class router can behave like an instance of the equipment class as well as an instance of the router class itself.

If the router subclass identifies equipment and system superclasses as members of its allomorphsClassesDefault, it is possible to operate an instance of the router class as an instance of the system and equipment classes. However, the allomorphic resource is not transient. If the router class is allomorphic to the equipment class, and the if equipment class identifies the system class in its allomorphic set attribute, it doesn't mean that one instance of the router class can behave like an instance of the system class.

Allomorphism also allows instances of arbitrary classes to present same the behave. One example is shown in Figure 4 where the switch class is defined as allomorphic to the router class. This hypothetical case would be used to have a switch imitating a router in tests of a network system.

It is important to remind, however, that Allomorphism and Polymorphism have different concepts. Polymorphism is an object oriented paradigm and allows a function to behave in different ways. Allomorphism, on the other hand, is a resource of OSI Management Information Model which allows an instance of a class to imitate the behave of an instance of one or more other classes.

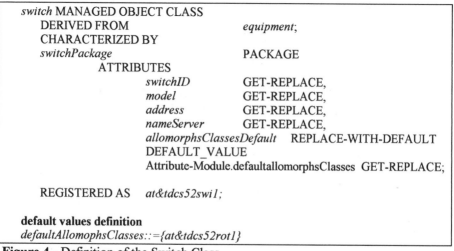

**Figure 4** - Definition of the Switch Class

## 4. ALLORMORPHISM AND THE ACCESS CONTROL FUNCTION

The Access Control Function (ISO10164-9, 1990) guarantees associations and management operations at run time. When an OSI management system is activated, the access control function will be called by the application process in order to have an access decision every time a) an association among application processes (agents or mangers) is established; or, b) a management operation (create, delete, action, set and get) is made on an object stored in the MIB.

The access decision process calls the AEF function which calls the ADF function. The later makes inference on the SMIB security base. It accesses the target object associated, compares the access request with the protection rules defined to the target object, and returns the decision access to the AEF function which will repass the decision to the agent application processed as shown previously in Figure 1.

It is noticeable that the Access Control Function Model requires a well organized structure, however, flexible and of easy maintenance of the SMIB base in order to make a fast decision access.

The management system already makes an overhead operation to the network when it adds the security service. When more traffic is created, regular and management applications processes will become slower. The access decision for a local process doesn't generates more traffic in the network, but the OSI Model is based in CMIP (Common Management Information Protocol) protocol, which means that it will be waiting for an answer. The OSI Model is connection oriented; so, there will always be a remote application process, the manager, waiting for a management operation or a management association.

This paper proposes the use of allomorphism for the access control service instead of OSI Model Access Control Function (ISO10164-9, 1990). Allomorphism is a property  of the OSI Management Model which allows the network manager to define relationships among classes and release them to users according to the management system needs.

According to (Klerer, 1993), allomorphism allows view definitions (or masks) of managed object classes. The network manager may hide some attributes and events to a group of users. An allomorphic class can be defined from a managed object class with some characteristics left apart. The user will access only the redefined characteristics in the allomorphic class.

Figures 5, 6 and 7 show a new representation of the equipment class presented in Figure 2. The equipment super-class will have two new classes: OperatorRouter and UserRouter, to implement the router equipment in the access control service.

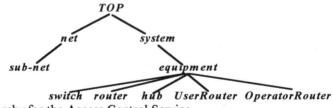

**Figure 5** - Hierarchy for the Access Control Service

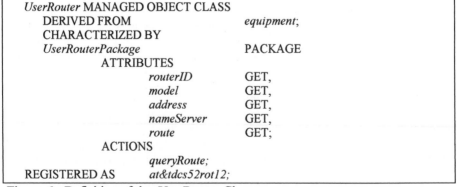

| | |
|---|---|
| *UserRouter* MANAGED OBJECT CLASS | |
| DERIVED FROM | *equipment*; |
| CHARACTERIZED BY | |
| *UserRouterPackage* | PACKAGE |
| ATTRIBUTES | |
| *routerID* | GET, |
| *model* | GET, |
| *address* | GET, |
| *nameServer* | GET, |
| *route* | GET; |
| ACTIONS | |
| *queryRoute;* | |
| REGISTERED AS        *at&tdcs52rot12;* | |

**Figure 6** - Definition of the  UserRouter Class

```
OperatorRouter MANAGED OBJECT CLASS
        DERIVED FROM           equipment;
        CHARACTERIZED BY
OperatorRouterPackage      PACKAGE
            ATTRIBUTES
                    routerID GET-REPLACE,
                    model            GET-REPLACE,
                    address          GET-REPLACE,
                    nameServer       GET-REPLACE,
                    route            GET;
            ACTIONS
                    addRoute,
                    deleteRoute,
                    changeRoute,
                    queryRoute;
            NOTIFICATIONS
                    routeAdded,
                    routeDeleted,
                    routeChanged;
    REGISTERED AS         at&tdcs52rot11;
```

**Figure 7** - Definition of the OperatorRouter Class

One instance of the router class can behave like an instance of one of the two classes: OperatorRouter or UserRouter. But the router class definition needs to contain the OperatorRouter and the UserRouter values in its allomorphsClassesDefault attribute to hold the allomorphic characteristic as seen in Figure 8.

Router, OperatorRouter and UserRouter classes model the same physical resource (Figure 9). Each class represents one different type of modeling for different users.

One given user of the router resource will access only the resources of the modeled object referred by the class defined for him/her. The network operators will only manipulate the instances of the modeled classes for the Operator group. Regular users of the management system will manipulate instances with behave like the ones in the User group class. The network manager will manipulate, on the other hand, all instances and they may have allomorphic behave of all kinds of groups.

The router class, as seen previously in Figure 8 , belongs to the Manager group in Figure 9. The instances of this class with all their attributes can be totally seen by the manager group, and the instances can also behave like their allomorphic classes.

The Operator group has the OperatorRouter class associated to its group and can see the router through the modeled class. Close to all resources associated to the router can be available to this group, except the information that manipulates the private routes which are exclusive to the network manager group.

The UserRouter class is the simplest. This class only allows the User group to execute the action QueryRoute. No other action will be available for this group. In addition, this group has neither authorization to receive notification or to alter attribute values for this router.

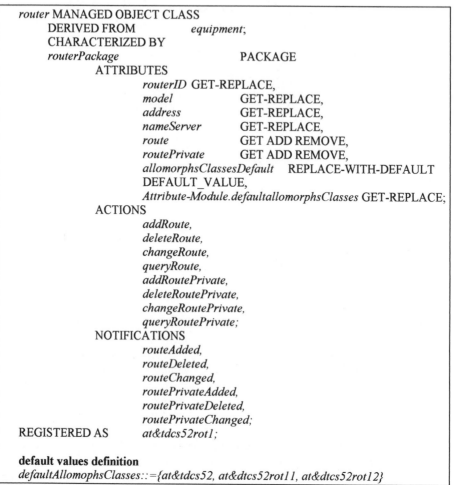

```
router MANAGED OBJECT CLASS
      DERIVED FROM            equipment;
      CHARACTERIZED BY
      routerPackage                      PACKAGE
                  ATTRIBUTES
                        routerID GET-REPLACE,
                        model              GET-REPLACE,
                        address            GET-REPLACE,
                        nameServer         GET-REPLACE,
                        route              GET ADD REMOVE,
                        routePrivate       GET ADD REMOVE,
                        allomorphsClassesDefault    REPLACE-WITH-DEFAULT
                        DEFAULT_VALUE,
                        Attribute-Module.defaultallomorphsClasses GET-REPLACE;
                  ACTIONS
                        addRoute,
                        deleteRoute,
                        changeRoute,
                        queryRoute,
                        addRoutePrivate,
                        deleteRoutePrivate,
                        changeRoutePrivate,
                        queryRoutePrivate;
                  NOTIFICATIONS
                        routeAdded,
                        routeDeleted,
                        routeChanged,
                        routePrivateAdded,
                        routePrivateDeleted,
                        routePrivateChanged;
      REGISTERED AS          at&tdcs52rot1;

      default values definition
      defaultAllomophsClasses::={at&tdcs52, at&dtcs52rot11, at&dtcs52rot12}
```

**Figure 8** - Definition of the Router Allomorphic Class

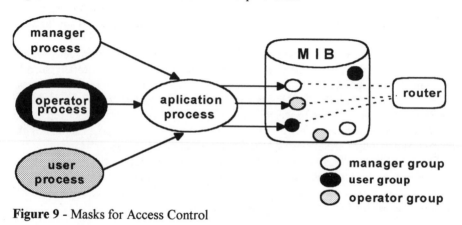

**Figure 9** - Masks for Access Control

A management application process must provide its group and its access level to the agent application process before executing an operation on a *router* class instance. The management application process uses a filter to choose the appropriate access mask that will be applied. One can conclude then that an instance of the router class will exhibit allomorphic behavior of an instance of the class compatible with the management application process and all its access rights.

The access control service can be implemented by defining several modeled classes of the same resource ( access masks: router, OperatorRouter, UserRouter, etc.), one for each group of users.

Definition of user groups and classes of the managed objects for these groups are rules in the access control policy which is defined by the network manager and is implemented in the management system. Information about the different access control policy models for management systems can be found in (Ramos, 1994).

Access control policies can be applied to one user (Individual-based), user groups (Group-based), or through security levels associated to management information which need protection (MultilevelSecurity).

When implementing the access control service by allomorphism, the one responsible for the system security must consider the following regarding the access control policy:

- Identify elements in the management system that need access protection; such as: application processes at association level and managed objects at management operation level;
- Identify the best access control policy model to help the system needs; that means: decide if the policy must be applied to each user individually, to a group of users, or to create security levels for accesses;
- Create resource views (using masks) for different group of users protecting important information.

The access control service implemented with allomorphism concepts in an OSI management system eliminates the Access Control Function (ISO10164-9, 1990) need because there is no access control at operational level in the management system. This process is transparent to the user since it is implemented in the system. A user or a group of users will see only a set of management information that is defined by the access mask defined for him.

When a user attempts to make an access not defined in his/her domain, the object will reject the access through its behavior because it does not accept such resource and its characteristics are not extended. This is guaranteed with the filter mechanism.

Although the access control service is implemented at the system management implementation, it is not static for the whole system life due to its allomorphism that provides flexibility in the system. It is possible to define a new access control policy while creating new user groups and new masks for resources. To do so, new allomorphic classes are defined according to system protection needs.

## 5. CONCLUSION

Since computer network became an important and strategic process in private and public institutions, it is of extremely importance to present a system with security and efficiency. To maintain a functional network system at acceptable level, it is necessary to implement a network management system applications.

Management system provides many benefits to a network administration, however, like any other computing system, it is subject to security violation and requires a security service.

Application management and security services are extra processes in the network and cause some overhead in its performance. They need to be implemented together to guarantee the network security.

Not only authentication mechanisms, cryptography and passwords control are important in a security service, but specially an access control system. The access control function is fundamental as a control in the management system with protection of management resources.

The Access Control Function Model (ISO10164-9, 1990) guarantees associations and management operations at run time. As presented in this paper, this function presents a complex with many messages exchanges between the AEF and ADF functions, which cause an overhead in the management application, and makes many times its use not acceptable. Due to this overhead, many administrators deactivate the access control function to release the system for better performance, however, they expose the system to violations.

This paper presents another option which is very practical and uses the allomorphism to implement the access control service in a flexible and simpler way and with less impact in the OSI management.

Another advantage of this implementation is a reduction in the overhead with the access decision process. When we define the access control policy and the different access masks, we transfer the access decision to the management system, while in the OSI function they are made at execution time.

The responsibility is then is passed to the one who implements the system who is also responsible for creating the allomorphic classes. And so, he/she will guarantee that a user and group of users will access only those resources defined in their domain.

Allomorphism also allows different access control policies in management systems (Individual-based, Group-based, Multilevel-based, etc.). Using definition of new allomorphic classes, it is possible to create new user groups, new access masks, define new access control policy according to protection  of access in the OSI management systems.

# 6. REFERENCES

CCITT X.720. (1992) CCITT - Recommendation X.720 - Information Thecnology - Open Systems Interconnection - Structure of Management Information: Management Information Model.

CCITT X.722. (1992) CCITT - Recommendation X.722 - Information Thecnology - Open Systems Interconnection - Structure of Management Information: Guidelines for the Definition of Managed Objects.

ISO7498-4. (1989) ISO/IEC 7498-4 - Information Processing Systems - Open Systems Interconnection - Basic Reference Model - Part 4: Management Framework.

ISO10164-7. (1991) ISO/IEC DIS 10164-7 - Information Technology - Open Systems Interconnection - Systems of Management - Part 7: Security Alarm Reporting Function.

ISO10164-8. (1990) ISO/IEC DIS 10164-8 - Information Technology - Open Systems Interconnection - Systems of Management - Part 8: Security Audit Trail Function.

ISO10164-9. (1991) ISO/IEC CD 10164-9 - Information Technology - Open Systems Interconnection - Systems of Management - Part 9: Objects and Attributes for Access Control.

ISO10165-1. (1991) ISO/IEC DIS 10165-1 - Information Thecnology - Open Systems Interconnection - Structure of Management Information: Management Information Model.

ISO10165-2. (1989) ISO/IEC DIS 10165-2 - Information Thecnology - Open Systems Interconnection - Structure of Management Information: Definition of Management Information Model.

ISO10165-3. (1989) ISO/IEC DIS 10165-3 - Information Thecnology - Open Systems Interconnection - Structure of Management Information: Definitions of Management Attributes.

ISO10165-4. (1991) ISO/IEC DIS 10165-4 - Information Thecnology - Open Systems Interconnection - Structure of Management Information: Guidelines for the Definition of Managed Object.

KLERER, M. (1993) Systen Management Information Modeling, IEEE Communications Magazine.

DE LUCCA, J. E. (1994) Arquitetura para Segurança em Gerência de Redes, Trabalho Individual - CPGCC UFSC, Santa Catarina.

PAVLOU, G. et alli. (1993) A Generic Management Information Base Browser, The ISO Development Environment, Versão 8.0, University College London.

RAMOS, A. M. (1994) Interface de Controle de Acesso para o Modelo de Gerenciamento OSI, Dissertação de Mestrado - CPGCC UFSC, Santa Catarina.

## 7    BIOGRAPHY

RAMOS, ALEXANDRE MORAES, M.Sc., is an Assistant Professor Graduate of Computer Science Department at University of Vale do Itajaí in Santa Catarina, Brazil. At the same time, he works on network management at Water Resources and Meteorological Integrated Centre of Santa Catarina, Brazil. His research interests include network management, telecommunications management, and information technology.

SPECIALSKI, ELIZABETH SUELI has been working on computer network research and development since 1980, mainly in network management. She has been teaching on Computer Science Pos-Graduation Program at Federal University of Santa Catarina, Brazil since 1989. She heads a workgroup which designs and develops a network management platform and provides network consulting services. She also heads a group which works for Human Resources formation in Internet administration.

# 10

# Experiences on building a distributed computing platform prototype for telecom network and service management

*Rahkila, S., Stenberg, S.*
*Nokia Research Center, Distributed Computing Platforms,*
*P.O.Box 422, 00045 Nokia Group, Finland, phone: +358-9-43761*
*sakari.rahkila@research.nokia.com*
*susanne.stenberg@research.nokia.com*

## Abstract

Open object-oriented distributed computing will be essential for the telecommunications business in the future. The most promising approach for the object-oriented distributed computing is the Common Object Request Broker Architecture (CORBA) technology standardized by Object Management Group (OMG). CORBA forms the foundation for building applications constructed from distributed objects and for interoperability between applications in homogeneous and heterogeneous computing environments.

The World Wide Web has grown to become one of the most popular services on the Internet. An integration of the Web and CORBA enable the benefits of distributed object computing while using existing Web resources.

The integration of the two recognized network management protocol standards CMIP and SNMP, and CORBA technology, allows management applications to take advantage of distributed object computing as well as the standardized network management protocols.

Nokia Research Center has launched a project to build a platform prototype on an object-oriented distribution infrastructure. The purpose of the prototype is to provide a framework which supports the creation, management and invocation of distributed telecom services.

## Keywords
Network Management, Service Management, World Wide Web, CORBA

## 1.   INTRODUCTION

Telecommunications networks are continuously growing in scale and complexity, and the number of equipment and services they provide is increasing. Integrated, distributed management of heterogeneous networks and services is increasingly important. There is thus a need for a consistent approach to integrate management solutions. Additionally it would be desirable to use off-the-shelf (buy vs. build) components and to leverage existing investments by integration of 'legacy' management applications. Distribution in management applications is needed for the scalability and for the cost/performance benefits.

The CORBA [1] technology has gained credit solving distribution, interface and integration problems. However, CORBA does not provide a network management architecture; it provides a distributed object computing architecture.

In the telecommunications industry the Telecommunications Management Network (TMN) [2] paradigm is widely established. In the Internet community, on the other hand, SNMP [3] based network management has gained widespread acceptance due to its simplicity of implementation. Thus, TMN and Internet management will co-exist far in the future.

Currently the Web based technology is changing and evolving very rapidly, and has attracted a lot of attention also for network management solutions. However, the Web development has been very closely based on the use of the Hypertext Markup Language (HTML) and the Common Gateway Interface (CGI), which are very limited in functionality. Therefore, new approaches are needed to support utilization of resources at the client's end and distribution of services over the network. This also includes the use of 'mobile code', such as JavaSoft's Java for easy integration of existing Web browser and the object-oriented CORBA technology.

Below we present the TMN, SNMP, CORBA and Web concepts, and describe the effort in the Nokia Distributed Computing Platform (DCP) prototype project to integrate these technologies.

## 2.   INTRODUCTION OF TECHNOLOGIES

### 2.1. Network Management

The goal of TMN is to enhance interoperability of management software and to provide an architecture for management systems. A TMN is a logically distinct network from the telecommunications network that it manages. It interfaces with the telecommunications network at several different points to send/receive information to/from it and to control its operations.

The TMN information architecture is based on an object-oriented approach and the agent/manager concepts that underlie the Open Systems Interconnection (OSI) systems management [4]. For a specific management association, the management processes involved will take on one of two possible roles: a manager and an agent. A managed object is the OSI abstract view of a logical or physical system resource to be managed. All management exchanges between a manager and an agent are expressed in terms of a consistent set of management operations and notifications. A Managed Object is the OSI abstract view of a logical or physical system resource to be managed. These operations are realized through the use of the CMIS [5] and the CMIP [6].

SNMP is a protocol suite developed for the management of the Internet. SNMP was designed to be an application level protocol that is a part of TCP/IP protocol suite.

The SNMP framework is based on the principle of minimally simple agents and complex managers. Internet uses the term network element to describe any object that is managed. The network element consists of the managed entity and the managed entity's agent [7]. The SNMP structure of information uses simple two-dimensional tables as it's basic containment structure for managed objects.

## 2.2. CORBA

The Object Management Architecture includes the Object Request Broker (ORB), Application Interfaces, Common Facilities, Object Services, and Domain Interfaces.

Application Interfaces are specific to end-user applications, they are not defined in the OMA architecture.

Domain Interfaces are vertical application domain-specific interfaces, e.g. domain interfaces for telecommunication applications.

Common Facilities [8] are horizontal end-user-oriented facilities applicable to most application domains (e.g. Distributed Document Component Facility).

Object Services [9] is a collection of services with object interfaces, which provides basic functions for all objects to be shared by all applications. The Object Services are intended to be modular so that clients in the client/server environment are free to use as many or as few as necessary to accomplish a task.

The Object Request Broker provides the basic mechanism for transparently making requests to - and receiving responses from - objects located locally or remotely. The ORB supplies delivery services, activation and deactivation of remote objects, method invocation, parameter encoding, synchronization, and exception handling. For interoperability between ORBs, CORBA 2.0 specifies a mandatory message format, called General Inter-ORB Protocol (GIOP), which can be hosted on any network transport in theory, however, TCP/IP is mandatory. Hosted on TCP/IP the protocol is called Internet Inter-ORB Protocol (IIOP).

The CORBA Interface Definition Language is used to define the interface to a CORBA object. IDL is programming language independent. It is mapped to the implementation language, i.e. the programming language in which the object implementation code is written. IDL compiler output binds clients and object implementations to the ORB.

## 2.3. Web and Java

The Web is largely founded on the use of two Internet standards, HTTP and HTML. The HTTP is an application-level protocol, which defines both message formats and a mapping to TCP/IP. HTTP is a stateless protocol, in which headers are transferred as (ASCII) text. The HTML defines the data formats for creating documents (pages) that are available via servers. HTML is not really a programming language. It specifies the use of tags to code information, which tells the Web browser how to display the documents' text elements. The documents are identified by an Uniform Resource Locator (URL), which can point to a static document, i.e. a file, or to a dynamic document. A dynamic document is one which is generated by a program which is identified by the URL. This mechanism is provided by CGI. The CGI is a standard for external gateway programs to interface with information servers such as HTTP servers.

The Web architecture has scalability problems, e.g. all interactions to a remote system go through one HTTP server. Another problem is the client interactions with the HTTP server, in which the HTTP protocol opens and closes a new connection for each request.

The integration of the Web and CORBA alleviates some of the scalability problems in the Web architecture. The use of CORBA removes the Web server as a single focal point for all client requests, and compared to HTTP, the IIOP is a far more promising protocol for reaching resources over TCP/IP. The Web and CORBA applications can be integrated using a gateway approach. The definition of a CORBA IDL mapping for HTTP enables interoperability between ORB/HTTP domains. Another simple approach is to use the CGI gateway. A CGI program can be associated with a CORBA client.

Java is an object-oriented programming language created by JavaSoft. Java is interpreted and features automatic garbage collection. The Java compiler generates bytecode instructions which are executed using interpreters provided for a variety of platforms. Thus, Java code is considered architecture-neutral. If additional performance is required, the Java bytecode can be translated into machine code for a particular CPU.

Applets are Java programs that follow a set a conventions that allow them to run within a Java enabled browser, i.e. a browser including the Java interpreter. A Java applet is made up of one or more Java objects. Java applets are embeddable into HTML documents, which means they can be down-loaded on-demand from a server and executed in the Web browser. In a sense Java represents mobile code.

For Web development, Java applets provide far better facilities than HTML forms. The applets are more dynamic and are reusable.

The natural way to integrate Java and CORBA is to provide a Java mapping  from CORBA IDL. This allows applets to appear as CORBA objects. A Java and CORBA integration supporting the Web is to provide an IIOP package written in Java. This effectively extends the Web browser to support IIOP.

## 2.4.  Integration of Technologies

The Joint Inter-Domain Management (JIDM) working group sponsored by X/Open and Network Management Forum (NMF) is seeking to enable inter-operability between CMIP, SNMP, and CORBA. JIDM concentrates on CMIP/CORBA and SNMP/CORBA inter-working.

The NMF has produced specifications that enable interworking between SNMP and CMIP environments. The SNMP/CMIP interworking package consists of five specifications. Three of the specifications address the relationships among object definitions in the two models, e.g., the translation of the Internet MIBs to GDMO and vice versa.

As stated in the mission statement of the OMG Telecommunications Domain Task Force, it provides a forum for the exploration, specification, and application of object technology within the telecommunications industry. The Task Force has outlined an architecture for a CORBA-based telecommunication network management system, using ORBs to facilitate the monitoring and control of telecommunications network elements, networks and services.

## 3.   NOKIA DCP PROTOTYPE

### 3.1. Overview

The purpose of the Nokia Distributed Computing Platform prototype is to provide an extensive framework which supports the creation, management and invocation of distributed telecom services. The framework is built on an object-oriented 'software bus' that facilitates both service development and service integration. The technology providing the software bus is CORBA. Objects form the components of all services, in this framework which integrates the technologies described above (Figure 1 ).

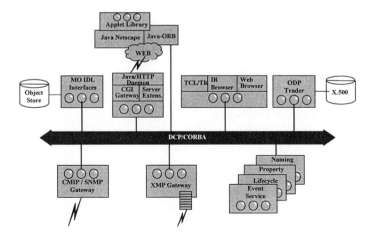

**Figure 1**  Nokia Distributed Computing Platform Prototype.

The following chapters describes the various parts of the DCP prototype.

### 3.2. GDMO/ASN.1 to IDL Compiler

The DCP prototype includes implementation of GDMO/ASN.1 to IDL compilers in accordance with the JIDM Specification Translation and related class libraries. The compilers utilize a persistent metadata repository for storing the knowledge about the GDMO and IDL definitions. The metadata repository stores the information about the GDMO objects using the Management Knowledge Management [10] metadata object definitions. The metadata repository uses an object database for persistence.

Tools for browsing the persistent metadata repository are also produced. These tools include facilities to browse the GDMO metadata using a Web browser such as Netscape or Mosaic and UNIX shell invocation using Tcl/Tk windows.

## 3.3.  String PDUs

It is possible to represent all the structures (syntax) used to build CMIS(P) messages using a string representation. This allows the building of management entities, which send and receive string messages. The approach used in the DCP prototype follows the one outlined in [11].

The string API is based on ASN.1 and hence allows any syntax defined in ASN.1 to be passed. Each message is built as a string which includes three basic pieces of information. These are the ASN.1 module name, the ASN.1 type name, and a string containing the value.

The DCP prototype contains a string PDU package which is used for building, parsing and passing CMIP string PDUs using CORBA as transport mechanism. The architecture of the package can be divided into three separate areas: ASN.1 data type classes, CMIP data type classes and builder classes. The ASN.1 classes are used to represent the basic ASN.1 data types such as Integer, Real etc. The CMIP classes represent the CMIP datatypes. These classes use the ASN.1 classes to compose a more complex data types. The builder classes are used to create CMIP String PDUs.

## 3.4.  CMIP/SNMP Gateway

The CMIP/SNMP gateway enables management of SNMP devices by using CMIS services. The gateway supports sending and receiving of CMIP and SNMP protocol events and timer events, and dispatches methods to managed objects also residing in the gateway.

The managed objects of the OSI/Internet management gateway communicate with Internet agents by using the SNMP protocol to carry out the operations invoked through their CORBA IDL interfaces. The managed objects derived from base classes in the DCP prototype represent the NMF translation of Internet MIBs to ISO/CCITT GDMO MIBs.

The naming service of the DCP prototype is used by the gateway to register managed objects and to enable scoping. The X.500 names-library allows conversion of X.500 names to the CORBA names. The gateway uses the CORBA Life Cycle Service of the DCP prototype to create a managed object corresponding to the managed-object class (object identifier) received as a parameter of CMIS M-CREATE request. CMIP protocol messages are received and sent as string PDUs over CORBA, as described in the previous chapter.

The stateless approach used in the gateway does not maintain the Internet MIB's data. Instead, for each received OSI management service request, the gateway generates one or more Internet management service requests (i.e. SNMP requests) to the Internet agent in order to achieve the function of the OSI management service request. A problem with this approach is that certain OSI management notifications which result from a change of state in the MIB cannot be emulated, since many such changes don't cause the Internet agent to send a trap, and a comparison of states in the gateway is impossible because the Internet MIB data is not replicated.

The stateful approach would require the gateway to replicate the Internet MIB locally, and to send periodic requests to SNMP agents to keep the replicated MIB current. The gateway would then try to fulfill each incoming CMIS request by using locally-replicated MIB data, instead of sending appropriate requests to the SNMP agent. The stateful approach could usually provide better response time, but has the drawback that the data retrieved might not be current. The poll frequency used to update the locally-replicated MIB has a significant effect on the accuracy of the response.

The stateless approach was selected because it is far less complex and it is also the approach used in the ISO/CCITT and Internet Management Coexistence (IIMC) documents of NMF.

## 3.5. Nokia XMP++

The Nokia XMP++ [15] classes enable building of C++ based applications on top of the X/Open XMP API [16]. The XMP API provides an industry-standard API to CMIP and SNMP management protocols. A new version of XMP++, having a more generic nature, has been integrated into the DCP prototype. This XMP++ gateway allows building of OSI/CORBA gateway applications.

The XMP API library implementation, which has been used, is the one provided by the HP OpenView Distributed Management (OVDM) platform.

There are three interface layers in XMP++, represented by three groups of classes: netHandle, xmpPlatform, and xmpWkSpace and xmpSession.

The xmpWkSpace and xmpSession classes interface directly with the XMP API; respectively they correspond to the XMP API Workspace and Session concepts. The xmpPlatform class replaces the XMP API Workspace and Session concepts with a single management connection concept. The netHandle and its derived classes are the interface classes for the network management applications to access the network. The netHandle class uses the xmpPlatform class. It also uses the event handle class (Reactor class) from the ADAPTIVE Communication Environment (ACE) library V3.3 [17] for receiving messages from the network and for catching signals from the system.

The XMP++ gateway provides a conversion between the CMIP protocol messages which are received and sent as string PDUs over CORBA, and the information of the XMP++ interface.

The class model of the gateway is based on a MIT actor-dispatcher design [18]. The dispatcher provides for demultiplexing of timer events, signal events and I/O events received through communication channels. It defines an interface for registering  actors  for certain events, and dispatches the incoming messages to the pre-registered actors.

The actors are derived from CORBA objects. The actor (base) class declares virtual methods for handling different events that it can register for. The derived actor classes implement the event handling in a specific manner. The proxy actors represent managed objects and take care of the mapping of operations.

The DCP_Gdmo_Disp accepts incoming messages, and forwards them to the appropriate XMP_Proxy_Act  (Figure 2 ).

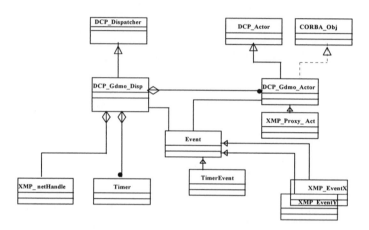

**Figure 2**  Actor class framework.

The XMP_Proxy_Act performs the conversion and calls the DCP_Gdmo_Disp which sends the message to the network using the XMP_netHandle class. The DCP_Gdmo_Disp keeps track of pending operations and supports e.g. retries.

The Name Service finds the desired CORBA object reference to the XMP_Proxy_Act  based on OSI distinguished names.

## 3.6. DCP Prototype Services

The Naming Service is the principal mechanism for objects in the DCP prototype to locate other objects. The COSS Naming Service defines operations how a client can get an object reference by a name. The Naming Service is designated to be built on top of existing name and directory services such as ISO X.500 [12]. Through the use of a names-library, name manipulation is simplified and names can be made representation-independent thus allowing their representation to evolve without requiring client changes. The DCP prototype contains an implementation of a specially designed X.500 names-library allows conversion of X.500 names to the CORBA names.

The concept of a Trading Service [13] provides a mechanism for dynamically finding services. The object which realizes the Trading Service is called trader. A trader is an object that enables clients to find information about suitable services and servers. Service providers can advertise or export services to a trader. Such advertisements are known as Service Offers. They describe a service using a set of service properties (name, value, mode triples). The service provider or object acting on behalf of the service provider is called an exporter. Potential service users, importers, can import from the trader, which means obtaining information on available services and their accessibility. The trader tries to match the importer's request against its Service Offers. After a successful match, the importer can interact with the service provider.

Traders can share information about Service Offers among each other. This is realized by a concept known as federation of traders. That is, traders can act recursively as clients to other traders.

The Property Service [14] provides a set of interfaces for dynamically manipulating the properties of objects. A CORBA object supports an interface which consists of operations and attributes. The interface is defined in IDL. Properties are intended to be the dynamic equivalent of CORBA attributes.

The Property Service defines operations to create and manipulate sets of name-value pairs or name-value-mode tuples. The names are simple IDL strings and the values are IDL anys. The use of type *any* is significant in that it allows a property service implementation to deal with any value that can be represented in the IDL type system. As with CORBA attributes, clients can get and set property values. However, with properties, clients can also dynamically create and delete properties associated with an object. In addition, clients can create and manipulate properties and their characteristics, such as the property mode. For example, a management application can create an expire date property for a particular service and set the mode of property to be fixed read-only. This means that other clients can read the property (and use the service if it is in use), but can not write or delete the expire date.

The Life Cycle Services defines services and conventions for creating, deleting, copying and moving distributed objects.

To create a new object, a client must find a factory object, which is an object that knows how to instantiate an object of the desired kind. The Life Cycle service defines a GenericFactory interface, which supports a general `create_object` operation. The LifeCycleObject interface has operations to move, copy and remove objects. The FactoryFinder defines an interface for finding factories.

The DCP prototype includes an implementation of a generic factory with GenericFactory interface as defined in COSS [9]. The job of the generic factory is to match the creation criteria specified by clients with offers of implementation specific factories which actually assembles the resources for objects. Objects of the DCP prototype support the LifeCycleObject interface.

## 3.7. Web User Interface

Web based user interfaces offer some benefits for network management applications. Web browsers provide for cost-benefits, the user interfaces are easy to use, and enables mobile management in that remote access only requires access to the Internet and a Web browser [19].

Java applet based UIs allow new versions of the user interfaces to be deployed faster, as they are downloaded from the server. The DCP prototype provides a Java applet library for accessing the distributed services provided by the platform. This includes a set of applets, which provide a user interface to the network management operations. The set contains applets for manipulating tree structures, e.g. MIBs, naming trees etc. The Java applets communicate with the DCP prototype both trough the HTTP-IIOP gateway described below and directly by IIOP using a commercial Java-ORB implementation.

## 3.8.  HTTP - IIOP Gateway

The CORBA IDL mapping of the HTTP produced by ANSA treats HTTP as a set of methods with arguments and results. This mapping has been used to provide gateways that can convert between HTTP and CORBA IIOP representations. Interoperability between the DCP prototype and the Web is achieved by using the HTTP-IIOP (H2I) and IIOP-HTTP (I2H) gateways from ANSA. This allows Web clients to use objects (clients) developed on the DCP prototype.

The IIOP-HTTP gateways between the DCP prototype and the Web require a trader to locate new services and to map between URLs and CORBA object references for the services accessible through the platform.

For future development of the DCP prototype the HTTP-IIOP and IIOP-HTTP gateways will not be used, since using Java-ORBs provides a much more natural integration of the Web and CORBA based services.

## 4.    CONCLUSIONS AND FUTURE ENHANCEMENTS

We believe that CORBA technology has the potential to become the basis for a paradigm shift from centralised management to distributed management of networks and services.  We also believe that future management solutions benefit from the integration of CORBA and Web technologies.

In the first phase of the DCP prototype development the main emphasis was on 'proof-of-concept', i.e. to prototype the integration of Web, Java applets, CORBA objects with a legacy service. In the next phase of the project such issues as security, fault-tolerance, multi-threading, and support of real-time processing will be considered.

Several papers [20], [21], [22], [23], [24], [25], [26], [27] have treated CORBA/OSI management integration with encouraging conclusions, but also bringing forth the drawbacks involved.

Notwithstanding, it seems that the next generation commercial network management platforms will support native CORBA communication [28]. In our case it means that the DCP prototype objects can communicate directly with services included into e.g. such systems as HP's future Synergy technology [29] based platform.

During the project it has become evident, that even though current commercial ORBs are not optimized [30], the CORBA technology is mature enough to be used as a base for the telecom software platform development.

The different architecture of open distributed telecom systems, as opposed to monolithic systems, still poses the main difficulty. Even though we have found the learning curve of CORBA usage moderate, the combination of Web/Java/CORBA requires patient learning. The emerging Java-ORB technology has the potential to make life easier. We have also found the use of object databases very useful in conjunction with the adoption of object-oriented technology such as Java and CORBA.

## 5.   REFERENCES

[1]     Object Management Group, The Common Object Request Broker: Architecture and Specification, Revision 2.0,  July 1995.

[2]     ITU-T M.3010, Principles for a Telecommunications Management Network, Draft June 1995.

[3]     A Simple Network Management Protocol, RFC 1157, May 1990.

[4]     ISO 10040, Information technology - Open Systems Interconnection - Systems management overview, November 1992.

[5]     ISO 9595, Information technology - Open Systems Interconnection - Common management information service definition.

[6]     ISO 9596, Information technology - Open Systems Interconnection - Common Management Information Protocol.

[7]     Black, U., Network Management Standards, The OSI, SNMP and CMOL Protocols, McGraw-Hill, 1992.

[8]     Object Management Group, Common Facilities Architecture, Revision 4.0, November 1995.

[9]     Object Management Group, CORBAServices: Common Object Services Specification, March 1996.

[10]    ISO/IEC 10164-16, Information Technology - Open Systems Interconnections - Systems Management - Management Knowledge Management Function.

[11]    IBM Corp., CMIP Run!, Vol. 2 No. 4, 94.

[12]    ISO 9594-1, The Directory - Overview of Concepts, Models, and Services.

[13]    Object Management Group, Trading Object Service, RFP 5  Submission, May 1996.

[14]    Object Management Group, Object Property Service, OMG TC Document, June 1995.

[15]    S. Rahkila, S. Stenberg, Nokia XMP++: An Object-Oriented Approach to TMN Application Development, Proceedings TINA '95, Volume 1, Melbourne February 1995.

[16]    X/Open, Management Protocols API (XMP), CAE Specification, Mar 1994.

[17]    The      ADAPTIVE      Communication      Environment      (ACE), *http://www.cs.wustl.edu/~schmidt/ACE.html*

[18]    ABCL : an object-oriented concurrent system, ed. by Akinori Yonezawa , MIT Press, 1990

[19]    J. Reilly, The World-wide Web and Programming Future Broadband Network and Service Management Applications, Presented at the NWPER'96, Aalborg May 1996. http://misa.zurich.ibm.com/Consortium/doc/papers

[20] Luca Deri, Network Management for the 90s, Submitted to ECOOP'96, Linz July 1996. http://misa.zurich.ibm.com/Consortium/doc/papers

[21] Della Torre, C., A Generic Distributed Service Management Test Bed Integrating CORBA and the XMP API, Proceedings ECOOP '94, July 1994.

[22] O'Sullivan, D., CORBA and Telecoms Management - can they live in perfect harmony?, Proceedings DOOC '95, October 1995.

[23] Davis, J., Du, W., Kirshenbaum, E., Moore, K., Robinson, M., Shan, M-C., Shen, F., CORBA Management of Telecommunications Network, Proceedings DOOC '95, October 1995.

[24] Park, J-T., Ha, S-H., Hong, J.W., Song, J-G., Design and Implementation of a CORBA-Based TMN SMK System, Proceedings NOMS '96, April 1996.

[25] Dittrich, A., Höft, M., Integration of a TMN-based Management Platform into a CORBA-based Environment, Proceedings NOMS '96, April 1996.

[26] Kong, Q., Chen, G., Integrating CORBA and TMN Environments, Proceedings NOMS '96, April 1996.

[27] Usländer, T., Brunne, H., Management View upon CORBA Clients and Servers, Proceedings ICDP '96, February 1996.

[28] Hewlett-Packard,                     Press                     Release, http://hpcc998.external.hp.com:80/nsmd/ov/whatisov/telecom.txt

[29] Herman, J., The Sorry State of Enterprise Management, Distributed Computing Monitor, March 1996.

[30] Gokhale, A., Schmidt, D., Measuring the Performance of Communication Middleware on High-Speed Networks, Submitted to SIGCOMM Conference, ACM, August 1996.

## 6.    BIOGRAPHY

Susanne Stenberg, Project Manager, Nokia Research Center (Finland), Distributed Computing Platforms technology area.
Sakari Rahkila, R&D Manager, Nokia Research Center (Finland), Distributed Computing Platforms technology area.

# 11

# Active Objects in TMN

*A. Vassila, G. Pavlou, G. Knight*
*Department of Computer Science, University College London,*
*Gower Street, London WC1E 6BT, UK*
*Tel: +44-171 419 3679, +44-171 380 7366, +44-171 380 7215*
*Fax: +44-171 387 1397*
*e-mail: n.vassila, g.pavlou, g.knight @cs.ucl.ac.uk*

### Abstract

Telecommunications Management Network (TMN) systems use the object-oriented information modelling techniques and communication facilities provided by OSI Systems Management (SM). TMN interfaces are specified in terms of rather passive Managed Objects (MOs) with prespecified behaviour. In this paper, we propose the concept of Active Managed Objects (AMOs) as the means to specify and express arbitrary management functions (including those specified in [1]) through a suitable scripting language. AMOs may be delegated to a TMN application in agent role and function close to other managed objects they access. Such a facility increases the intelligence and autonomy of TMN applications and enables the expression of management functions with arbitrary intelligence. Also, since it uses the normal TMN mechanisms for information modelling and access (GDMO, CMIS/P), it could be potentially standardised. In this paper, we describe the AMO concept, examine the information model and scripting language aspects and present our implementation experience.

### Keywords

TMN, OSI Management, Active Managed Objects, Management by Delegation

## 1 INTRODUCTION

This paper describes work at UCL* which is aimed at increasing the intelligence and autonomy of components of TMN[2] systems. TMN systems normally make use of the services and protocols defined within OSI management[3]. The OSI management information model[4] is based on an object-oriented database paradigm with rather passive 'Managed Objects' (MO) encapsulating management information on a managed system. MOs are accessed by applications in a managing role with the access being mediated by an 'agent'

---

* This work is partially funded by the ACTS project MISA (AC080).

function at the managed system, which provides database-like functions enabling target MOs to be selected in a flexible and data-dependent way.

Our work is motivated by the following considerations:

- There are some exceptions to the general passive nature of MOs, for example the *discriminator*[5] objects which take an active role in the handling of asynchronous notifications, and the *monitor metric*[6] and *summarisation*[7] objects which calculate various statistics based on values obtained from other local MOs. We have found these more 'active' MOs to be powerful tools in management - in particular, they enable much management activity to remain local to the managed system which would otherwise require interaction with a remote managing application. However, their 'managing intelligence' is of static nature, parameterised only through MO attributes and actions. We have sought, therefore, to generalise the active MO concept by providing a framework for a MO whose precise behaviour is specified in a program or script downloaded at runtime.

- Moving management intelligence close to the resources being managed should eventually improve the overall performance of the management system as it reduces network traffic, it enables autonomy of managed systems and at the same time improves fault tolerance of the whole system as constant communication between managing and managed entities is no longer substantial.

- The OSI management model has little to say about the managing function beyond specifying the operations it is allowed to invoke. However, the TMN architecture involves a hierarchy of applications in a manager role in which high level ones control the behaviour of their subordinates. We have developed a system in which the behaviour of a system in a manager role is specified in a runtime script. It is convenient to represent a subordinate managing system and its script as a MO which can be manipulated by a high level managing system.

An answer to the above considerations is to have a script executed by an interpreter that performs management operations. This approach has also been followed by other research groups: In Columbia University the Management by Delegation paradigm[8] has been defined, and a system implementing it has been developed. In addition, in INRS Telecommunications Canada[9], another delegation framework is being developed for application management, as well as for systems administration. [10] also describes a prototype implementation of an 'intelligent agent' system for application management using CORBA. Other work includes the scotty implementation package[11]. But, what we have identified, is that the execution of the script should be controlled by a higher level (probably remote) management entity. This entity should be able to control every aspect of the script. Given the TMN context for the work  this control must be through the normal TMN mechanisms. This means that the invocation of a script must be represented as a MO; the higher level managing system then interfaces with the script by manipulating the MO's attributes, actions and notifications. We call such MOs 'Active Managed Objects'[12] (AMO). A similar approach has been followed by ISO and has been documented in [13].It has to be noted that our primary interest is  in prototyping the idea of AMOs and assess their usability and effectiveness, rather than presenting a fully implemented system.

In Section 2 of this paper we clarify the AMO concept and in Section 3 we present a detailed system design for an AMO. This design is, in fact, independent of the scripting language used

- however, the capabilities of this language are clearly important. In Section 4 we discuss language issues in general, while in Section 5 we give a detailed description of the OSI definition of the AMO. A pilot implementation of an AMO has already begun using the OSIMIS TMN platform; in Section 6 we describe this work.

## 2 THE AMO CONCEPT

Figure 1 shows the main AMO system components. The 'Managed System' can be any OSI managed system, complete with the usual collection of MOs representing the resources being managed. In addition to these MOs, there is an AMO representing a script. From the point of view of OSI management there is nothing special about an AMO; for example, it presents a conventional interface to the management agent, which can invoke the normal range of operations (get, set etc.) upon it. The 'Remote Managing System' in Figure 1 can manipulate the AMO by sending CMIP request PDUs to the managed system.

In Figure 1 the Remote Managing System is shown as a stand alone component. However, it may itself be implemented as a script and AMO in another managed system which is itself

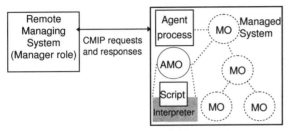

controlled by a higher level managing system. The interpreter in Figure 1 interprets a language which includes functions which perform management operations. We have investigated with two kinds of functions: those which provide access to local MOs and those which provide access to MOs on remote managed systems.

**Figure 1** The Active Managed Object Model.

### 2.1 Local MO Access

This can be used to extend the capabilities of managed systems so that they can accept and interpret delegated programs, expressed in a suitable scripting language. The latter enables

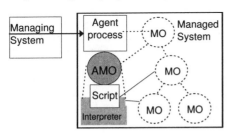

access to other managed objects in the local system through the full range of operations available at the object boundary i.e. get, set, action etc. The delegated logic assumes that all the other objects are local and accesses them through their distinguished names in that system e.g. *{subsystemId=nw, entityId=x25}*. The results, collected by the AMO, are then accessible to a managing application via the agent process in the normal way (Figure 2).

**Figure 2** Access to local MOs

Within a TMN system this capability can be used, for example, to perform some standard or proprietary TMN management function and make the results available to an application in a manager role. It can also be used to provide autonomous control within a managed system.

## 2.2 Access to Remote Managed Systems

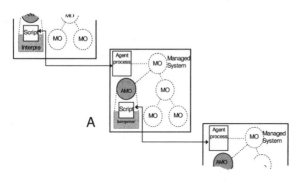

An example of this is shown in Figure 3. Here the managed system marked 'A' in the centre of the figure contains an AMO which controls an interpreter offering the full CMIS client service[14]. This service is being used to manage a subordinate managed system. At the same time, 'A' is itself being managed by a higher level managing system.

**Figure 3** Hierarchical Management via AMOs.

This gives fully hierarchical management controlled by scripts which can be downloaded at runtime. Strictly, it is not correct to describe the systems in Figure 3 as 'managed systems' since they embody the managing and the managed roles. For remote access all MOs must be addressed through global rather than local names. The global name of a MO consists of the managed system identifier (that can be resolved to its address using the Directory), followed by the distinguished name of the MO to be accessed. The scripting language and underlying infrastructure may provide location transparency so that local and remote object accesses are indistinguishable. Local accesses will retain the CMIS properties but without using the CMIP stack. In TMN terms, local accesses will occur through a q reference point while remote accesses through a Q interface.

It should be noted that the AMOs in Figure 2 and Figure 3 are identical from the informational point of view. Their management definition remains the same, while the difference lies solely in the capabilities of the scripting language. In the first case, operations within the local managed system are supported - one can expect these operations to be quick and cheap as there probably is a single address space. In the second case, operations may be performed on **any** managed system. Access is via an OSI management agent offering the normal facilities available through CMIS such as multiple object selection, filtering, access control, etc.

Although the systems in Figure 3 are shown performing remote operations there is no reason why they should not access local objects by directing CMIS requests at their local agent processes. This way, there will be no need for the script programmer to distinguish between local and remote objects, but the advantages of using local operations will be lost.

# 3   THE RUNTIME ENVIRONMENT

After having introduced AMOs, we now investigate their functionality in some detail. Our aim is to construct a language independent system so that the AMO can work with more than one language and interpreter.  Therefore, we must identify the language independent aspects of the system and define an information model for an AMO which encapsulates these. The identified purposes of AMOs can be summarised as:

- To provide management access to scripts; i.e. to  store information on scripts and identify the interpreters that will execute them.
- To initiate and terminate execution of scripts according to managing systems requests and local rules.
- To communicate with a running script for the purpose of passing parameters and notifications to scripts and retrieving results.
- To provide/present/structure results for managing applications.

## 3.1 AMO Functionality

The script and its invocation are the real resource that is managed by the managing application via the AMO. Therefore, the AMO must represent all the aspects of the script that could be of interest to the managing application and that are likely to be changed during its lifetime.

The first requirement is to deliver the script to the managed system and store it there. This is done by storing the script as an attribute in the AMO, and delivering it using regular CMIS operations. It is important to note that we assume that our system will be persistent, that is, all the AMO components will survive any system failure.

Once a script is started, it represents a separate thread of  control within the managed system. This thread will exist until either a managing application requests termination, the script voluntarily terminates or the managed system forces termination (perhaps because the script is consuming too many resources). Typically, a script will use an event-driven programming style. The events of interest being, for example, notifications from other MOs within the managed system, receipt of communication from other managed systems, timer events and receipt of signals from the manager application or the local agent process. The execution of the thread will normally be suspended whilst it waits for an event.

It is important that the managing application is not *obliged* to access the AMO and script during its execution for reasons other than an emergency, or to retrieve results and statistics. This means that the AMO must carry all the knowledge that concerns the execution of the script as directed by the managing application. However, although, as stated above, the managing application is not obliged to access the AMO during script execution it may do so in order to vary aspects of script execution - for example, to adjust polling frequency.

At the very least, a script must be able to communicate with its local environment and with its managing application. One role for a script could be to provide sophisticated, autonomous processing of local events. Such events will engender notifications from MOs which the AMO must pick up and pass to its script through some sort of asynchronous signalling mechanism. We propose that the AMO class inherits from the discriminator class which will enable it to receive notifications and filter them before generating signals. Normally the filter will be provided by the managing application which created the AMO and can be changed at any time. A signalling mechanism can also be employed to allow asynchronous communication from the managing application. This is invoked using the CMIS M-ACTION service.

It is evident that a managed system with AMOs and scripts embedded within it has many of the characteristics of a multiprocessing, multi-user operating system. It is unwise in such a system to assume that all scripts are benign; it is important that the managed system can defend itself against rogue scripts. This defence, one can envisage, could be exercised through pre-emptive scheduling mechanisms parameterised by some sort of priority value. Furthermore, an interpreter for a safe language would provide a major defence against malign scripts, by forbidding all access other to the script's own state variables and the MIB.

It is important to note that any management operations invoked by the script must behave exactly as if they were invoked directly by a managing process, because the AMO acts on behalf of a specific managing application. The normal access control mechanisms are applied to the attributes of the AMO and to operations on managed objects invoked by the script. The issues involving the mechanism by which this is done are under investigation. It is evident, however, that by identifying the managing application that created the AMO in every management operation, access control can be performed in a uniform fashion.

From the requirements above, we can summarise a number of features which must be present in the AMO information model:

- **Actions which can be used by the managing system to start, stop execution and to signal the script thread.** The result of the first action will be the creation of a new thread of execution for the script. The second action will result in a previously created  script thread being terminated. The side-effects of this must be further investigated, as there is the problem of destroying a script while executing, which can cause corruption of management information. The third action will cause the AMO to deliver a signal to the script. The exact mechanism through which such signals are handled within a script are language dependent. Typically, the effect will be to cause the script to exit from a waiting state and execute the appropriate code.
- **A notification of script termination and exceptions.** The AMO should be able to notify the managing system about the script execution (mode of completion, execution error etc.). asynchronously.   The normal mechanisms are employed with the AMO issuing a notification which triggers an event report through a discriminator.
- **A notification to be invoked from within the script.** It is important that the script should be able to trigger notifications voluntarily. Thus the script programmer can ensure that a notification (and hence an event report) is issued when a situation that requires further involvement by a managing application is identified. Again, the precise way in which this facility is invoked is language dependent.
- **A filter.** This is used to filter notifications coming from within the managed system and determine whether these should result in signals being passed to the script thread.
- **An interval timer mechanism.** This can be used in order to simulate polling procedures. The AMO will signal the script at specified intervals - minimising the need for managing application intervention. Note that polling can also be implemented by calling a timeout function from within the script. However, a mechanism controlled through the AMO, has the benefit of allowing the managing application to adjust operations at runtime.

- **An attribute to identify the script to be run and the controlling managing application**. The first will contain a string representation of the script. It will be set by the managing entity when the AMO is created. The second attribute is essential for the operation of the AMO system because it has to appear in all communications with local or remote systems, as the script inherits all the permissions that the issuing managing application owns.

## 4  FACILITIES OF THE SCRIPTING LANGUAGE

As stated above, the AMO has been designed so as to impose minimal constraints on the language employed for the script. We envisage scripts implementing anything from one line statistical calculations to complex control policies - so a range of languages may be deployed to suit these different applications. In this section we examine two aspects of the facilities these languages must provide. First we look at the facilities which are the counterpoint to the generic facilities provided by the AMO - those that implement API 1 in Figure 4. Next we outline the requirements for a general purpose language suitable for the TMN environment. This must provide the full functionality offered by the OSI model - API 2 in Figure 4.

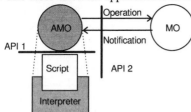

**Figure 4** Script APIs.

## 4.1 The interface to the AMO

It is useful to think of the script executing in conjunction with an AMO as analogous to a user space process executing in a general multitasking operating system. Interactions take place in two directions. The script must be able to read and write certain AMO attributes so that, ultimately, it may communicate with the remote managing application (similar to Unix ioctl() calls). It must also be able to cause the AMO to emit notifications and react to signals from the AMO (similar to the Unix signal()/Kill() mechanism). The script can be started during AMO creation or through a subsequent activation action while a termination signal will be passed to it at deletion in order to terminate gracefully.

In the following example script, the scheduling primitives can be seen, as well as an example of the CMIS extensions to the scripting language (described in Section 4.2).

```
RegisterTimers (900000, scollect)                          Initialisation code
ifnumber=Get systemId=grappa@interfacesId=NULL ifNumber;
UpTime=Get systemId=grappa@internetSystemId=NULL sysUpTime;
proc scollect () {                                         Callback function
  date = getSystemDate();
  upTime=Get systemId=grappa@internetSystemId=NULL sysUpTime;
  if (upTime<Uptime) issuenotification(systemId=grappa "has been reset!!");
  for (j==1, j<=ifnumber, j++) {
    dn="systemId=grappa@interfacesId=NULL@ifEntryId=SnmpInteger="j;
    inucastpkts = Get (dn "ifInUcastPkts");
    outucastpkts = Get(dn "ifOutUcastPkts");}
  SetOutputInfo (date, UpTime, inucastpkts, outucastpkts);
}
```

In the above example, the procedure `scollect()` is executed every fifteen minutes. Several values are retrieved from the local MIB-II, written in the `OutputInfo` AMO attribute and a notification is emitted if needed. This notification can be sent as an event report or logged, depending on the actions of the managing application (creation of a discriminator).

## 4.2 A general purpose language for TMN management

In this section we examine the required facilities of the scripting language in order to enable the expression and realisation of management policies. The base language facilities with respect to the available data and control structures need to be comparable to those of compiled programming languages. Object-oriented aspects, i.e. classes, inheritance and polymorphism, are necessary for structuring more complex scripts. Scripting languages that can be thought as potential candidates are Tcl [15]/ Scheme[16] and their object-oriented extensions and of course Java[17]. These languages need to be extended with management facilities that will enable interaction with the TMN environment and the MIB. It is these extensions we consider in this section.

A script may access managed objects that are 'local' to the AMO i.e. part of the same Management Information Tree (MIT) across a TMN interface, or 'remote' as part of another MIT and TMN interface. Remote object accesses may incur an increased latency as an external representation of the 'method call' will travel across the network, involving protocol overhead etc. As such, the script should be aware of the system in which it executes through the name of the top MIT object so that it can distinguish between local and remote access. On the other hand, the syntactic aspects of object access should be in principle the same for both local and remote objects. Global distinguished names may be used to provide location transparency while local distinguished names will default to the local environment.

It appears at first that this unification is difficult: local access can be modelled as managed object method invocation, in the same fashion as in any object-oriented environment. On the other hand, remote access should reflect the facilities of CMIS, where access is mediated by an agent. The difference is crucial and, in a first consideration, it appears to have an effect on the language extensions. As it has been shown in [18], higher level abstractions on top of CMIS are possible, providing the illusion of direct object access. In fact, it is possible to unify the two types of extensions, assuming that CMIS facilities such as scoping and filtering are also available for local access. In fact, such facilities increase the available flexibility and simplify the script logic, as well as optimising the management traffic for remote access. Managed objects may be addressed through their distinguished names (see example script in section 4.1), or through (implementation dependent) object references i.e. handles or pointers. Unifying local and remote access means of course that association establishment has to be completely transparent to the script i.e. hidden.

The style of interaction should be both synchronous and asynchronous. A synchronous style of interaction has method call semantics and results in natural, linear program flow but blocks the thread of execution until the call returns. An asynchronous style of interaction has message passing semantics and requires the management of state since the result will be returned through a 'callback'. Asynchronous facilities are of paramount importance in single threaded environments as they prevent blocking the whole application for remote accesses with increased latency.

# 5   AN INFORMATION MODEL FOR AN AMO

In the light of the analysis in Sections 3 and 4 we can now give a more formal specification of an AMO which inherits from the eventForwardingDiscriminator managed object class.

**ATTRIBUTES**

| | |
|---|---|
| Id/Name | Identifies the AMO and name it in the OSI MIT. |
| Authorisation (a DN) | Identifies the managing system that created the AMO. It will contain the distinguished name of the managing which will be used as input to access control functions. |
| Script (string) | Holds the delegated script. |
| Timer (sequence of integers) | Used to change the intervals between script executions without having to alter the script itself. |
| Filter (CMISFilter) | A filter for notifications from local MOs. Notifications which pass the filter result in signals to the script. |
| Parameters (name-value pairs) | To be provided to the script when it starts. The syntax is a set of pairs that indicate the name of a parameter and its value. |
| InputInfo (name-value pairs) | Settable by the managing application, readable by the script every time it is changed. |
| OutputInfo (name-value pairs) | Gettable by the managing application, writeable by the script to indicate some results from its execution. |
| Priority (integer) | Indicates the priority that should be given to the script execution. |

Apart from the standard notifications or actions that any managed object can include, the AMO will also support the following. The need for these was discussed in Section 3.1.

**NOTIFICATIONS**

| | |
|---|---|
| TerminationInfo | Triggered when the script terminates. |
| InformManager | Triggered as a result of the issuenotification() function. |

**ACTIONS**

| | |
|---|---|
| ActivateSThread | The AMO activates the script thread upon receipt of this action. |
| DestroySThread | The script thread is deleted. |

# 6   IMPLEMENTATION EXPERIENCE

In this section we present considerations regarding the implementation of AMOs. The purpose of this presentation is twofold: to explain how the AMO concept has been implemented in our environment i.e. the OSIMIS TMN platform; and to identify the necessary requirements in order to implement AMOs on other TMN platforms.

The first important consideration is related to the choice of the underlying scripting language. In most TMN platforms, APIs like those depicted in Figure 4 are implemented in C/C++, so the first important requirement for the scripting language is to be extensible and able to interface easily to C/C++.

One of the reasons for choosing Tcl as opposed to e.g. Scheme as the scripting language to verify the AMO concept, apart from the existence of the Safe-Tcl interpreter, was the ease with which it interfaces to C/C++. The interface between Tcl (or any similar scripting language) and the 'encapsulating' C/C++ environment is in two directions:

- from Tcl to C/C++, for accessing local or remote managed objects and for manipulating the AMOs own attributes and emitting notifications
- from C/C++ to Tcl, for starting the AMO script, receiving notifications from local or remote managed objects and receiving input from actions invoked on the AMO (including setting its attributes, deletion etc.).

The next important consideration has to do with combined event handling in the scripting language and the encapsulating C/C++ environment i.e. the TMN application in agent role. Both these systems need to deal with events with respect to communication endpoints and timer alarms. In our case the starting point is the OSIMIS system which implements a managed system as a single Unix process without thread support. As an AMO script will execute in the same operating system process with (part or all of) the encapsulating management agent, it is necessary to be able to combine their events so that there is a central listening and dispatching loop, serving one or more AMO scripts and the surrounding agent. There are two different possibilities for accomplishing this:

1. through a scripting language interpreter that can be interrogated about the endpoints and alarms it deals with so that they can be combined with those of the underlying TMN C/C++ system; or
2. through a scripting language that can be extended with the endpoints and alarms of the underlying TMN C/C++ system.

In both cases, the assumption is that the target central mechanism can be extended with 'foreign' descriptors and alarms. In our implementation we have followed the second approach as it suited both the nature of Tcl and the OSIMIS process coordination mechanism. An extension of the OSIMIS event handling mechanism passes control to Tcl and integrates with it the OSIMIS C++ endpoints and timer alarms.

Finally, the most important implementation consideration is related to the TMN platform APIs for accessing other managed objects, local and remote, and interacting with the AMOs own attributes, reacting to actions and emitting notifications. A key feature of scripting languages such as Tcl (or Scheme) is that the main data types are the string and list while other complex types can be emulated through these. An important requirement for any underlying TMN platform is that its APIs should support the manipulation of attributes, actions, notifications and their values through string representations. The same applies also to distinguished names, filter expressions and other CMIS-level parameters. The OSIMIS APIs support string expressions in addition to the native C/C++ types, so it has been straightforward to map these onto Tcl language extensions. When accessing local managed objects, scoping and filtering may be supported in addition to accessing objects on a one-by-one basis; this provides additional flexibility and is particularly important for presenting the same access paradigm for both local and remote objects, as it was explained in Section 4.2. In addition, access control functionality is necessary as the script / encapsulating AMO assumes the identity of the managing application that 'owns' it. In OSIMIS, the local managed object access API allows the evaluation of scoping and filtering parameters. In addition, an object instance modelling the Access Decision Function (ADF) is globally available and this allows for the evaluation of access control rights in order to be able to grant or deny access. When accessing remote managed objects, these facilities are available through the OSIMIS RMIB manager support infrastructure. Local and remote managed object accesses can be distinguished through the global name prefix: the scripting language knows about the environment in which it executes through the name of the top MIT object e.g. {c=GB, o=UCL, ou=CS, cn=ATM-CM-OS, networkId=ATM}.

In summary, the key requirements in order to be able to implement AMOs on any TMN platform with C/C++ APIs are the following: a flexible scripting language that integrates easily with C/C++ and that is safe for the local system; the possibility for a combined process coordination mechanism that integrates the scripting language and C/C++ TMN platform events; and flexible managed object access APIs for local and remote objects that support scoping, filtering and access control locally, as well as accept string parameters.

## 7   SUMMARY AND FURTHER WORK

In this paper we have identified the need for programmable management facilities within the TMN environment and have investigated how the execution of such programs may be controlled using existing TMN mechanisms. We have identified two APIs which need to be available to program scripts; one provides access to the local environment in which the script is embedded, the other provides access to MOs in local and remote systems. These APIs have been analysed in some detail and a scheme for their implementation within the UCL OSIMIS system has been presented.

The control of the script execution is effected by representing the script and its execution as a MO - a concept we have named an 'Active Managed Object'. The requirements for the attributes, notifications and actions of an AMO have been studied and our conclusions have been presented.

Currently a pilot implementation which implements a subset of the facilities of the two APIs above exists, using the Tcl scripting language. This implementation will be used to assess the effectiveness of the AMO concept and will gradually be extended to include more of the required facilities, as well as to assess the efficiency of the chosen language.

## 8   REFERENCES

[1]    ITU/CCITT Recommendation M.3400 - TMN Management Functions, October 1992.
[2]    ITU/CCITT Recommendation M.3010 - Principles For A Telecommunications Management Network,        October 1992.
[3]    ITU-T X.701, Information Technology - Open Systems Interconnection - Systems Management  Overview, June 1991.
[4]    ITU-T X.720, Information Technology - Open Systems Interconnection - Structure of Management   Information - Part 1: Management Information Model, January 1992.

[5]     ITU-T X.734, Information Technology - Open Systems Interconnection - Systems Management - Part 5: Event Report Management Function, 1992.
[6]     ITU-T X.738, Information Technology - Open Systems Interconnection - Systems Management - Part 11:      Metric Objects and Attributes, 1994.
[7]     ITU-T X.739, Information Technology - Open Systems Interconnection - Systems Management - Part 13:      Summarisation Function, 1994.
[8]     G. Goldszmidt and Y.Yemini, Evaluating Management Decisions via Delegation, in *IFIP International    Symposium of Network Management*, April 1993.
[9]     Jean-Charles Grégoire, Management with Delegation, in *IFIP AIPs Techniques for LAN and MAN       Management,* Paris France, 1993.
[10]    P. Steenekamp and J. Roos, A Framework for Policy-based Agents: Implementation of an Application Management Scenario, in *IFIP/IEEE DSOM Workshop*, Italy, 1996.
[11]    J. Shönwälder and H. Langerdörfer, Tcl Extensions for Network Management Applications, in *3rd Tcl/Tk Workshop*, Toronto, 1995.
[12]    A.Vassila and G.Knight, Introducing Active Managed Objects for Effective and Autonomous Distributed       Management, in *Bringing Services to People, IS&N Conference*, Heraklion Greece, 1995.
[13]    ISO/IEC DIS 10164-21 - Information Technology - Open Systems Interconnection - Command Sequencer for Systems Management, 1996.
[14]    ITU-T X.710 - Information Technology - Open Systems Interconnection - Common Management  Information Service/Protocol, Version 2.
[15]    J.K.Ousterhout, Tcl and the Tk Toolkit, Addison-Wesley, 1994.
[16]    William Clinger and Jonathan Rees (eds.), Revised[4] Report on the Algorithmic Language Scheme, November 1991.
[17]    Sun Microsystems, The Java[TM] Language: A White Paper, 1995.
[18]    G.Pavlou, G.Knight, K.McCarthy  and S.Bhatti, The OSIMIS Platform: Making OSI Management Simple,  in *Integrated Network Management IV pp.480-493*, Chapman and Hall, 1995.

## 9     BIOGRAPHIES

**Anastasia Vassila** graduated from the Computer Science Department of the University of Crete in 1993. Currently she is a PhD student in UCL under the supervision of Graham Knight. She does research on integrated network and systems management, emphasising on autonomous, event-driven management in the OSI/TMN framework.
**George Pavlou** received his Diploma in Electrical and Mechanical Engineering from the National Technical University of Athens in 1982 and his MSc in Computer Science from University College London in 1986. He has since worked in the Computer Science department at UCL mainly as a researcher but also as a lecturer. He is now a Senior Research Scientist and has been leading research efforts in the area of management of broadband networks and services.
**Graham Knight** graduated in Mathematics from the University of Southampton in 1969 and received his MSc in Computer Science from University College London in 1986. He has since worked in the Computer Science department at UCL mainly as a researcher and lecturer. He is now a Senior Lecturer and has led a number of research efforts in the department. These have been mainly concerned with two areas: network management and ISDN. Currently he is leading the UCL effort in three EU-funded projects in the areas of network and systems management and broadband networks.

## PART FOUR

# Standards, Protocols and Interoperability
## Chair: Douglas N. Zuckerman, Bellcore

# 12

# Joint Inter Domain Management:

# CORBA, CMIP and SNMP

*By the JIDM taskforce and as editors:*
*Dr. N. Soukouti*
*SMILE, Sarl.*
*E-mail: soukouti@smile.fr*
*Postal: 29, rue Viala*
*75015 Paris*
*France*
*Tel: :+33 1 40 59 09 00*
*Fax: +33 1 40 59 09 01*

*Dr. U. Hollberg*
*IBM European Networking Centre*
*E-Mail: hollberg@heidelbg.ibm.com*
*Postal: Vangerowstr. 18*
*D-69115 Heidelberg*
*Germany*
*Tel: ++49 (6221) 59-4223*
*Fax: ++49 (6221) 59-3300*

**Abstract**
In this paper, we present an overview on the work of a joint activity of the X/Open and the Network Management Forum known as *"Joint Inter Domain Management (JIDM)"*. The activity of this group has been split into two phases, called *"Specification Translation"* and *"Interaction Translation"*. In this paper, we present the basic concepts of the *Specification Translation*, which has been defined to allow for the migration of information models between different domains of management technology. We will also present in short the problems to be solved during the ongoing *Interaction Translation* phase, which aims at enabling the inter-working between management software across the different domains: CMIP, CORBA and

SNMP. The Specification Translation document is currently in final review as X/Open preliminary specification. The Interaction Translation document is expected to be released in the second Quarter of 1997.

The intention of the paper is to make the JIDM activity known and to raise interest in its results; the interested reader is referred to the X/Open preliminary specification.

### Keywords

CORBA, OMG, COSS, OSI, CMIP, GDMO. SNMP. ASN.1, SMI, MIB

## 1 INTRODUCTION

The *Joint Inter-Domain Management (JIDM)* group is jointly sponsored by X/Open and the Network Management Forum (NMF). The group was initiated in response to the perceived need to provide tools that can enable interworking between management systems based on different technologies. The JIDM group has focused on the three key technologies: CMIP, SNMP, and CORBA, and is seeking to enable inter-operability between them both within a single organisation and between organisations. SNMP has a large established base in the general purpose computing market, CMIP is mandated in the telecommunications arena by the TMN standard and CORBA is recognised as the emerging standard covering distributed object oriented programming. Each technology has its strength; thus full inter-operability will enable designers to select the most appropriate technology to apply to any given problem. The *ISO-Internet Management Coexistence (IIMC)* group of the Network Management Forum has addressed SNMP/CMIP interoperability. Thus, JIDM has chosen to concentrate on CMIP/CORBA and SNMP/CORBA inter-working.

To enable inter-working it is necessary to be enable the mapping between the relevant information models and to provide a mechanism to handle protocol conversion on the domain boundaries. This allows objects in one domain to be represented in the other domain and the interactions can be governed by the domain of choice rather than by the domain in which the target object is implemented. For example, an object in the CORBA domain should be able to interact with a GDMO object as if it were in the CORBA domain, ideally without having to know that the target object is in a different domain. Naturally the reverse is also desirable, that an OSI Manager should be able to manage CORBA objects as if they were defined in GDMO (this requires the reverse mapping).

A major advantage of the JIDM approach is that the strength of CORBA (object oriented system with well-defined APIs which are aimed at standardising and simplifying the task of creating distributed applications) can be combined with the strength of CMIP (powerful protocol with strong wire compatibility allowing integration of multi-vendor hardware) to give the best of both worlds. The implementor would have an effective environment in which to implement manager or agent functionality and yet be able to easily integrate components from multiple vendors. Figure 1 on the next page illustrates the main interoperability scenarios identified for OSI, SNMP and CORBA domains.

Section two gives an overview of JIDM's activities, the *Specifications Translation* and the *Interaction Translation*. Section three deals with *Specifications Translation,* it explains in four subsections the principles of the translation processes from ASN.1 and GDMO to CORBA IDL and vice versa, and from SNMP MIBs to CORBA IDL. The paper closes with hints on related work, conclusions and references.

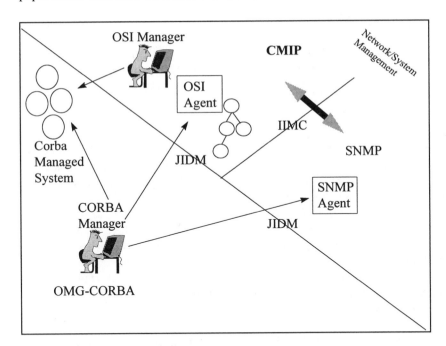

Figure 1: Inter-Working between CORBA, CMIP and SNMP

## 2    JOINT INTER DOMAIN MANAGEMENT

### 2.1  Specification Translation

The *Specification Translation* covers the process by which specifications are translated from one management domain to another.   Algorithms for the static translation between GDMO/ASN.1 specifications and CORBA IDL interfaces, and for the static translation from SNMP MIBs to CORBA IDL have been defined. These algorithms may need to be extended later on to generate additional  material for use in the Interaction Translation.   When translating from GDMO/ASN.1 to IDL, trade-offs are encountered between enabling access to the  full power of CMIP and generating simple IDL representations which simplify the application programmer's task.   Wherever possible, these have been resolved according to the principle of keeping it simple in the most number of cases at the expense of making lesser used constructs more complex.

### 2.2  Interaction Translation

The *Interaction Translation* covers the process by which interactions in one management domain are transformed into corresponding interactions in the other domain.  A *gateway*, the entity responsible for transforming interactions, may receive a CMIP PDU and must map this into one or more requests or replies on CORBA IDL interfaces.  For example, if a scoped and filtered CMIP GET request is received, the gateway would have to identify the set of objects matching the filter within the scope and invoke the appropriate operation on each of those objects.  The results would be collated and formatted into one or more CMIP PDUs in reply.  Interaction Translation must cover initialisation identifying how the gateway is initialised and populated, and how it identifies the existing object population and what other service instances it may need to use.  The gateway will probably make use of existing standard services in the CORBA domain, e.g. OMG Name Service to resolve Distinguished Names, OMG Lifecycle  Service to create new object instances and the use of OMG Event Channels for event distribution.
Interaction Translation requires that the interactions be captured by a gateway which converts them in accordance with the mapping rules.   Thus, in the OSI/CORBA scenarios, the gateway must:

- receive any incoming CMIP SET/GET/ACTION request and translate it into one or more invocations of methods supported by the addressed object(s),
- receive any event generated by an application object and translate it into a CMIP EVENT-REPORT request to be forwarded to those remote systems which had

registered their interest in receiving that kind of events,

- receive incoming method invocations and forward them as CMIP SET/GET/ACTION requests to the appropriate OSI agent,
- receive CMIP EVENT-REPORT requests and forward them as CORBA events to interested parties in the CORBA domain,
- receive any incoming CMIP CREATE/DELETE request and translate it into an invocation to a method being supported by an object (e.g. a factory object),
- receive method invocations for creating or deleting objects in a remote system and forward a CMIP CREATE/DELETE request to the addressed OSI agent.

This protocol conversion is complicated by such things as the need to map identifiers due to the differences between GDMO and IDL scoping and case-sensitivity, to map between GDMO Distinguished Names and CORBA Object References and to handle CMIP scoping and filtering requests which may require one CMIP request to be mapped to multiple sequences of IDL operations.


## 3   A SCENARIO

A common scenario is that a set of objects are defined in GDMO. The GDMO document is statically translated into IDL interfaces by applying the Specification Translation algorithm as explained in this text. A manager implemented as a set of CORBA objects, would manage objects supported by an OSI agent as if they were CORBA objects (i.e. via the generated IDL interfaces). These interfaces would be supported by a gateway supporting the Interaction Translation algorithm. This would dynamically translate IDL requests into CMIP PDUs based on the original GDMO specification. The Interaction Translation needs to bi-directionally translate between CMIP PDUs originating from the agent and the appropriate IDL requests and replies. If the Manager is an OSI manager, it generates CMIP PDUs exactly as if the objects were supported by an OSI agent. The CMIP protocol is transformed by the gateway into IDL interactions on the CORBA implemented object; the results of the IDL interactions are transformed back into CMIP PDUs.

For the rest of this paper, we will focus on the specification translation and implementation issues.

## 4    JIDM SPECIFICATION TRANSLATION

### 4.1  Assumptions and Principles

The algorithms have been designed using a number of guiding principles:
*   **Completeness**.  The aim was to provide a complete mapping as so far as it was possible.  Rules have been provided for all cases regardless of their frequency.

*   **Simplicity**:  ASN.1 allows many constructs that are difficult to map into IDL.  Many of these are not frequently used.  In the light of the completeness principle, it was decided to select the simplest mapping for 80% of the cases allowing the remaining, more obscure cases, to be more complicated if necessary.

*   **Reuse of OMG services**:   The CORBA domain does not have a network management architecture; it provides a distributed processing environment.  It is populated by an increasing number of Common Object Services such as naming and events which are useful building blocks.  The JIDM group is trying to exploit these services where possible e.g. event services are used for OSI notifications.

*   **Freedom of Implementation:**   The JIDM group refrains from defining or constraining implementation unless it is absolutely necessary.

In order to translate between GDMO and CORBA IDL, (hereinafter referred to simply as IDL), or between SNMPv2 and IDL, it is necessary to be able to map the basic type definitions (i.e. mapping between ASN.1 types and IDL type definitions).  This section describes the translation  by addressing the mapping of the primitive types and of the constructed types.
There are two versions of ASN.1 defined, X.208 and X.680.  Since GDMO explicitly builds on X.208 and all new GDMO will provide X.208  versions (at least in the short term), JIDM focused on that version.  However, translation is also provided for all the basic types (e.g. BMPString and UniversalString) from X680 as a step towards migration to X.680.
ASN.1 has a much more complex type system than IDL.  As a result, the translation necessarily risks to lose some information; e.g. tag values, value range constraints or compound type constants.  Capturing this information for subsequent use in the run-time system is a key issue.  Several schemes have been proposed including the use of string constants and  #pragma directives, but there is no published solution yet.  In some cases, the complexity of some data types makes it desirable to define operations for manipulating their values.  In CORBA, the standard technique is to define pseudo

IDL (PIDL), which allows a fairly tight definition of the operations but with the implication that these operations have library implementations and can only be invoked locally. For example, this is used to provide access methods to support manipulation of the BitString data type.

## 4.2   ASN.1 to IDL translation

This section provides the translation of ASN.1 types in an ASN.1 module to IDL types in an IDL module.

Translate each ASN.1 module in the GDMO document (input file) to an IDL module.

Ignore ASN.1 macros and the export clause.

At the beginning of the module, translate each imported type in the ASN.1 module into a typedef of the corresponding imported IDL type and an #include statement for the module to be imported from.

Map each value assignment as a const declaration or method of the *ConstValues* interface of the generated module according to the rules.

Generate named IDL types for anonymous types and use the named type in the rest of the IDL. Repeat this recursively until all types are either named constructed types or one of the base types. This includes applying the rules for "COMPONENTS OF Type" and OPTIONAL and DEFAULT for sequences and sets.

Rearrange the order of generated types such that there are no forward references.

While generating the final output, resolve the conflict in ASN.1 identifiers due to case-insensitive nature of IDL identifiers and different rules for name scopes.

Note that the mapping is not direct in the sense that one ASN.1 type may be translated into several IDL types and constants. Some ASN.1 constants are not translated into IDL constants but into IDL operation.

## 4.2   GDMO to IDL Translation

GDMO to IDL translation is important from several points of view, e.g.

- It allows to implement GDMO MIBS as CORBA objects and thus implementing distributed OSI agents.
- It allows to give a CORBA manager an IDL view of the GDMO managed objects.

In this section, the Specification Translation process is described in terms of inputs and outputs and a rough outline of the process is given. The process can be implemented by a compiler which operates on a set of input files and produces some output files. Since IDL definitions are processed in terms of files which determine their granularity and reusability, it is necessary to specify which definitions are generated in which files.

In addition, GDMO adds some complexities since GDMO specifications use full text names e.g. "CCITT Rec. X.721 (1992) | ISO/IEC 10165-2 : 1992".  Since such names are used to import parts of other specifications, there must be a way for the translation process to access the files containing these specifications.  In addition, it is desirable to be able to associate the resulting  IDL files with the original GDMO to facilitate browsing and reuse.  Both will be achieved by providing a *nickname database* which maps from the unique registered name of the GDMO document (or relevant Object Identifier) to a short nickname suitable for use as a filename base. This nickname will be used to find imported files and to control the names of the generated IDL files.

### *4.2.1    Outline of Translation Algorithm*

- Translate ASN.1 types and constants according to the ASN.1 to IDL translation algorithm.
- Translate each GDMO managed object classes template into an IDL interface as follows:
    1. Retain the inheritance relationship of the class for the interface,
    2. Unwind the packages of the managed object class:
    3. Translate each attribute to operations on the interface in accordance to its associated property lists.
    4. Translate each action to an operation of the interface; in addition, generate interfaces for the support of multiple replies to be used with OMG event transmission mechanisms.
    5. Translate each notification to an operation to be used by OMG event services; in addition, generate interfaces for the handling of notifications in both, the *Push* and *Pull* model.
- Translate Parameter templates into IDL type definitions.
- Translate Behaviour templates into comments.
- Ignore Name Bindings templates.


Figure 2  summarises the translation process of GDMO documents.

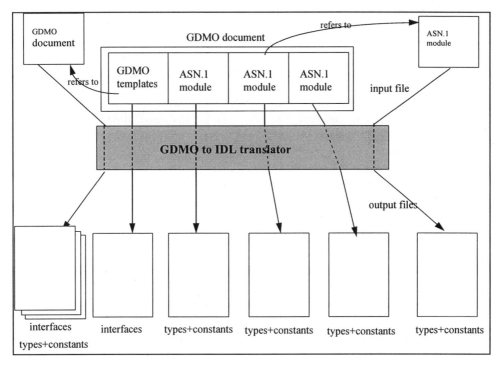

Figure 2: GDMO to IDL translation process

GDMO managed object classes are all derived directly or indirectly from the class *top*, as defined in X.722. In turn, *top* will be derived from a CORBA interface called *ManagedObject* which will hold generic information. For instance, operations on groups of attributes will be supported by the inherited interface *ManagedObject*; it will also offer operations for the support of multiple attribute access.

## 4.3  IDL to GDMO Translation

The CORBA-IDL to GDMO translation is important from several points of view, e.g.:

- It allows to manage CORBA applications from an OSI manager once the IDL interfaces are translated into GDMO/ASN.1.

- It allows to develop applications based on a CORBA IDL information model and to translate the model later on into GDMO/ASN.1, to be used to build OSI agents.

The basic guidelines of the translation algorithm are the following:

- Translate each IDL module into one GDMO document.
- Translate each IDL interface into one GDMO managed object class with one inlined package template.
- Translate exceptions into error parameters.
- Translate IDL attributes into GDMO attributes.
- Translate IDL operations into GDMO actions.
- Translate IDL types and constants into ASN.1 types constants.

## 4.4  SNMP-MIB to IDL Translation

The SNMP-MIB to IDL translation is important from several point of view, e.g.:

- It allows to implement an SNMP MIB as CORBA objects, thus opening up access to such an SNMP agents MIB to any CORBA client.  Besides, such a kind of implementation will be very flexible and allow for easy extensibility.

- It allows to manage SNMP MIBs from a CORBA manager as soon as the SNMP MIB definition has been translated into CORBA-IDL.

The translation algorithm can be summarised as follows:

- Translate each SNMP information module into an IDL module:
- Process any ASN.1 type and value as described earlier in the ASN.1 to IDL section.
- Translate each Table entry into an interface where the entry components are attributes in this interface.
- Translate single-instance variable of a group into one interface where each variable is mapped into one attribute in this interface.
- Translate notifications (NOTIFICATION TYPE macro) into operations on a separate interface which will be used  with OMG event channels.

This algorithm is compatible with the IIMC recommendations of the NMF for the SNMP to GDMO translation.

## 5    RELATED WORK

A taskforce called *telecom* sponsored by the Object Management Group is defining currently a TMN based on CORBA and going to use JIDM work.   The ISO-ODMA(Open Distributed Management Architecture) group would use JIDM work to provide support for OSI management on top of ISO-ODP (Open Distributed Processing).  A working group of the NMF and X/Open, called "*High Level CMIS API* " is very actively working towards a C++ embedding of GDMO and ASN.1 defined information models, as well as an embedding of ACSE and CMISE services.  Their work will become especially important during the interaction translation phase, it has been of some impact on the specification translation phase. It should be noted, that the translation from GDMO/ASN.1 to C++ is less constraint than that of GDMO/ASN.1 to CORBA IDL.

## 6    CONCLUSION

The JIDM *specification translation algorithms* make it possible to bridge across the gaps between the different management domains, CORBA, CMIP and SNMP. Available information model in GDMO and ASN.1 can be translated into CORBA IDL specifications, and vice versa. Also SNMP MIB definitions can be translated into CORBA IDL.   This allows to *reuse specifications* in a different domain, to *build gateways* between different domains, e.g. between CMIP and CORBA.   The latter potential will be explored during the second, the *interaction translation* phase of JIDM.

- JIDM allows CORBA programmers to write OSI managers and OSI agents without needing to know GDMO, ASN.1, CMIP and its programming interfaces.
- JIDM allows GDMO and CMIS programmers to access CORBA implemented resources, services or applications without knowing CORBA.
- Implementing manager applications using CORBA provides a uniform view of the resources independent of whether they are basically managed via GDMO, SNMP, or even proprietary resources when the latter are encapsulated into CORBA objects. This is an important step toward integration of management systems.
- Once the JIDM interaction translation is finished, complete management platforms

can be based entirely on CORBA and access the non-CORBA world through gateways.

# 7   ACKNOWLEDGEMENT

The work described in this text has been done jointly by the JIDM team during the last three years. The authors wishes to thank the other JIDM task force members, mainly Subrata Mazumdar and Tom Rutt (both Lucent Technologies), Tim Roberts (BNR), Juan Hierro (Telefonica), and  Martin Kirk (X/Open) for their ideas, work and co-operation.

# 8   REFERENCES

X/Open Preliminary Specification: "Inter Domain Management. Specification Translation". Final Draft,  November 1996.

X/Open Preliminary Specification: "TMN/C++ Application Programming Interface", May 1996, with three parts: ASN/C++, May 1996, CMIS/C++, Aug. 1996 (planned) and GDMO/C++, Dec. 1996 (planned).

# 9   BIOGRAPHY

Nader Soukouti is senior engineer at Smile. He received his Ph.D. from the university Paris VI in 1995 and the computer engineer degree from HISSAT, Damascus, in 1988. Nader has been member of JIDM task force since its creation in 1993. His main domain interest is Network Management and CORBA.

Ulf Hollberg is senior consultant at the IBM European Networking Center in Heidelberg with architectural responsibility for IBM's  TMN product components. Ulf received his Ph.D. from the University  of Bonn. Ulf Hollberg is member of the JIDM team since 1993.

# 13

# A service engineering approach to inter-domain TMN system development

*T. Tiropanis, D. Lewis, R. Shi*
*Computer Science Department, University College London*
*Gower Street, London WC1E 6BT, United Kingdom*
*Tel: +44 171 419 3687, +44 171 391 1327, +44 171 419 3249*
*Fax: +44 171 387 1397*
*E-mail: {ttiropan, dlewis, rongshi}@cs.ucl.ac.uk*

*A. Richter*
*GMD-Fokus*
*Hardenbergplatz 2, D-10623 Berlin, Germany*
*Tel: +49 30 25499 287, Fax: +49 30 25499 202*
*E-mail: richter@fokus.gmd.de*

### Abstract

The deployment of service management systems in a multi-service environment opens a whole area of issues. Requirements for openness and reusability can be satisfied by following a service engineering approach where the service management functionality is decomposed to reusable service components. Such an approach that was based on TMN technology is presented in this paper and conclusions are drawn about the potential of other available technologies with regard to service management.

### Keywords

TMN, TINA, Service Management, Service Components, Distributed Processing

## 1 INTRODUCTION

In recent years much work in the area of distributed and telecommunication-based systems has been underpinned by an assumption that the resulting standards, interfaces and products will have to operate in a global open market for telecommunication services and software. Such a market will be characterised by a more crowded and diverse arena of players than has been apparent in the traditional telecommunications industry structured around large state-owned enterprises. The increased competitiveness of this global market will produce ever increasing pressures to reduce costs and to increase service quality to the customer. Open interfaces have an important part to play in this both in reducing the burden on the service provider in dealing with multiple proprietary interfaces and in increasing choice to the customer by encouraging competition on the quality and cost of service rather than ties to specific interface technologies.

An area seen as being critical to cost reduction and customer satisfaction is that of management. This covers the management of the providers' resources to ensure their efficient use, the management of the delivery of the service to the customer and the integration of these activities to maximise the effectiveness of both.

The increasing management requirements on the network level, due to the use of the more complex ATM technology, and the demand for a number of elaborate service management areas such as subscription and accounting management, set an extra requirement for software reusability in the construction of network and service management systems. Principles of Object Orientation can find use in the decomposition of network and service management software to reusable components. Issues regarding the integration of these components are also considerable.

This paper presents work in the practical application of the TMN recommendations to the area of inter-domain service management by grouping part of the management functionality to reusable components. This work, undertaken in the RACE II PREPARE project [PREPARE], and later continued in the ACTS project Prospect, was conducted in an environment specifically designed to impose many of the requirements of an open service market on the development of a system of inter-working service and network management components; in this case managing multimedia tele-services over a broadband network. The requirement for openness justified the use of a service engineering approach. The next section outlines some of the standards that influenced this work. Section 3 describes the specific context in which this prototyping work was conducted. Section 4 details the design approach taken, the components used and the experiences gained before some conclusions are drawn.

## 2   CURRENT ARCHITECTURES

The most mature open architecture applicable to service management, and the basis chosen for the PREPARE project, is the ITU-T M.3000 series of recommendations on TMN. This standard has been widely accepted by the telecommunications industry in Europe and now increasingly in North America and Japan. The core architecture recommendation [M.3010] structures management functionality into a set of layers starting at the network element layer, through the network layer to the service layer and finally the business layer. TMN compliant standardisation efforts have led to a wide range of interfaces being developed at the network element level and more recently at the network level. Open interfaces at the service level are only now starting to emerge, primarily intended for operation at the inter-domain interface identified in TMN, i.e. the X interface. The NM Forum is particularly active in this area through its SMART group. This group is addressing specific areas of management functionality identified in a common Business Process Model [NMF-BPM], such as trouble ticketing, service ordering and billing, where industry agreements on inter-work are required.

The process of defining open interfaces in TMN concentrates on the development of information models defined in terms of Managed Objects (MO) that can be accessed using management information retrieval protocols, e.g., SNMP and CMIP. The TMN Interface Specification Methodology [M.3020] provides guidelines for defining interfaces in terms of management interfaces that are decomposed into Management Service Components (MSCs) which are in turn decomposed into Management Functional Components (MFCs). However, in practice MSCs and MFCs are not often utilised as units of reusability for interface specifications, this being hampered further by a lack of a strict notation for these components. MOs are frequently reused between specifications, however without a clear mechanism for reusing MOs in groups forming MFCs the real value of this reuse tends to be more in the interface syntax of an MO than in the functional semantics it may offer. A further, large factor restricting widespread reuse of components in TMN is the current lack of a common API for management platforms to encourage the development of platform independent components implementations. This area is currently being addressed by the X-Open and the NM Forum [SPIRIT].

A different approach to component reusability has been taken by the Open Management Group (OMG) in developing its Common Object Request Broker Architecture (CORBA) [CORBA]. Drawing its membership from the wider data processing industry, as opposed to the telecommunication industry sector that influenced the development of TMN and the NMF, the OMG aims to apply to the area of distributed processing the object oriented techniques common in software engineering today, e.g. use of C++ classes as units of functional reuse. This is performed by providing mechanisms to support distribution transparencies, that have been identified as important in separating the design of a distributed system from the mechanisms providing, and the problems presented by distribution.

Significantly, the ODP [X901] recommendations go beyond outlining a technique defining interface specifications for distributed objects, and propose a general approach to defining distributed systems. This is based on the five ODP viewpoints: the enterprise viewpoint aiming largely at requirements capture; the technology viewpoint for defining the technological constraints placed on a system; the information viewpoint addressing the information to be processed in the system; the computational viewpoint addressing how the system is decomposed into computational objects and the engineering viewpoint describing how information and computational models are implemented as engineering objects, e.g. C++ classes.

Though the adoption of ODP techniques may encourage the development of a clear, object-oriented information model, the availability of object oriented platforms providing the appropriate distribution transparencies, e.g. a CORBA implementation, leads to the use of the computational objects of the computational viewpoint as the primary units of system decomposition and eventual reuse in the engineering model.

The reusability of such computational objects, and the corresponding engineering objects, is aided by the ODP engineering concept of clustering which allows for the grouping of engineering objects into larger units of functionality. This provides system designers with a wider and more continuous range of granularity in defining reusable components than it seems to be practised in TMN.

Though the ODP standards and the OMG architecture are applicable to any distributed processing problem, the Telecommunications Information Networking Architecture Consortium (TINA-C) have attempted to apply these principles to the needs of the telecommunications industry [TINA-018]. This has involved the adoption of ODP as the basis for their architectural approach resulting in the definition of an abstract Distributed Processing Environment (DPE) [TINA-005]. To provide continuity with existing telecommunications industry standards TMN principles have been adopted for the management aspects of this architecture while the influence of the IN standards [Q1204] is seen in the approach to functional decomposition and reuse. By bridging the gap between the emerging distributed processing industry and the well established telecommunications industry in this way, TINA provides a path for TMN-based management architectures to be integrated with the distributed systems techniques in a manner better suited to the requirements of the open service market. The following sections give an example of how this path could be followed based on some of the practical experiences gained from the PREPARE project.

## 3  OPEN SERVICE MARKET CONTEXT IN PREPARE

As stated in the previous section, the PREPARE project offered an opportunity to investigate the development of management services in an open service market context and in an environment that imposed requirements down to the implementation level. This consisted of an enterprise model imposing requirements to manage broadband networks and multimedia tele-services [Bjerring-94].

The enterprise model chosen consisted of a value chain of service providers. At the lowest level multiple public network operators (PuNOs) provide an ATM Virtual Path service over their respective network domains, and conspire to provide this service, spanning several domains, via a single management interface. This management interface is used by a Virtual Private Network (VPN) provider, in concert with the management of private network resources of ATM LANs owned by private network operators (PrNOs) to provide end to end management services connecting hosts on remote ATM LANs. This VPN service is in turn used by providers of multimedia tele-services, namely a multimedia conference (MMC) service and a multimedia mail (MMM) service in the provision of these tele-services to a customer organisation between internationally distributed ATM LAN sites.

The decomposition of management functionality between these various organisational stakeholders was performed initially at a very low level of granularity and in line with the TMN functional architecture. This involved identifying one TMN Operations System Function (OSF) for each stakeholder at the service level with additional OSFs, if required, at the network level and the network element level via Q-adapter functions. The aim was to implement this TMN architecture as a set of interoperating Operations Systems (OSs) mapped simply to the OSFs. Figure 1 shows schematically how these OSs were distributed between organisational stakeholders and mapped over a broadband demonstration network, spanning London, Berlin and Copenhagen, and consisting of ATM LANs and ATM cross-connects (XCs). The X and Q3 interfaces developed between these OS are also identified.

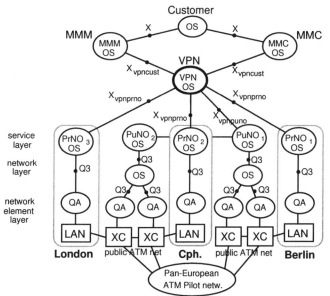

**Figure 1**    Overall TMN Physical Architecture for PREPARE.

## 4    APPLYING SERVICE ENGINEERING PRINCIPLES TO TMN OSF DESIGN

The diversity of the offered services in PREPARE added a requirement for openness and reusability to the design and specification of the reusable service management components. The need that the development of these components had to be shared between different project partner organisations emphasised this requirement.

### 4.1    Decomposing services into service components

Several OSs  were derived from the enterprise specification in PREPARE. For the service management support for the services, corresponding service operation systems (S_OS) were defined. The communication between S_OSs in different domains was accomplished over the TMN X interface. A common information model in GDMO and ASN.1 specified the information to be exchanged over the X interface. A set of event-trace diagrams defined the X interface interactions between the S_OSs for a set of different service management scenarios.

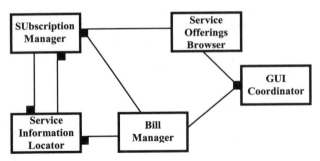

**Figure 2**      Relationships diagram for the generic Service Components.

The functionality of the S_OSs was decomposed to Service Components (SC). These SCs were categorised to *generic SCs* and *service specific SCs*. All the generic SCs are the reusable service management modules. The service specific SCs dealt with requirements for service specific functionality. The integration of management and service specific components is an idea borrowed from TINA.

The following generic SCs were identified for supporting the management of the services in PREPARE:

- The *Service Offerings Browser* (SOB) *SC* offers functionality for browsing through details for services on offer. The details are stored as entries on the X.500 directory.
- The *Subscription Manager* (SUM) *SC* implements the functionality needed for managing existing subscriptions to services. It employs a unique identifier for each service subscription and which is also used for de-multiplexing subscription specific service requests exchanged between SCs.
- The *Service Information Locator* (SIL) *SC* provides functions for locating service management agents on other customer or provider domains. For a given subscription ID, this SC can return the address of the customer or provider service management agent by using the combination of X.500 and X.700 described in section 4.3.2.
- The *Bill Manager* (BM) *SC* offers the utilities required for collecting, browsing, and paying of bills between customers and providers for a specific subscription.
- The *Graphical User Interface coordinator* (GUI_C) *SC* implements the functions for integrating the GUIs of the several service components installed and running on the same workstation function within a common front-end.

We mapped each SC to one Computational Object (CO) [TINA-HV][X901]. The relationships between the COs are shown in Figure 2 using TINA notation.

The GUI_C SC was used by both the SOB SC and the BM SC in order to register their graphical user interfaces within the same front. The SUM SC was used by the SOB SC in order for the latter to perform on-line subscription to a service on behalf of a user, and it was also used by the SIL SC in order to authorise the requester of the location of a service management agent. The SIL SC was used by the SUM SC in order to find the contact agent of a service provider when a subscription was requested by a user, and it was also used by the BM SC when sending or paying bills.

## 4.2 Building-blocks and contracts

SCs can be grouped together according to the dependencies between them. Grouping of SCs allows for their better management, i.e., they can be treated as an entity that can be moved or copied to other locations and plug into other applications by offering a combined set of interfaces to them. The notion of a building block (BB) in TINA is that of a group of computational objects that has these properties. In this document the term *contract* is used for the interfaces offered by building blocks.

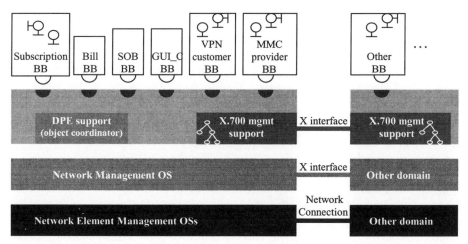

**Figure 3** Structure of TMN S_OSs with building blocks and contracts.

The relationships between the SUM and the SIL in Figure 2 indicate the strong dependencies between them. For this reason they were put together in the Subscription BB. The other SCs were organised into separate BBs. The Bill BB includes the BM SC, the service offering browser (SOB) BB includes the SOB SC and the GUI coordinator (GUI_C) BB includes the GUI_C SC.

Most of the above BBs include an extra *coordinator SC* that is responsible for the initialisation, termination, activation and deactivation of the contracts of the BB. This component is also responsible for other generic building block management functions.

The service specific components were organised into separate BBs. There are two types of BBs for each service: the customer BB and the provider BB, e.g., for the VPN service, there are the VPN service customer BB and the VPN service provider BB. The same applies for the MMC and MMM services.

A picture of all the BBs organised on the site of the MMC provider is given in Figure 3. This combination of BBs comprises the VPN customer OS and the MMC provider OS and shows the case of being the MMC provider and a VPN customer at the same time. The communication between the BBs was accomplished over a distributed processing support infrastructure (DPE support) that was developed in PREPARE, which resembles the TINA DPE. The inter-domain communication which was performed over the TMN X interface was supported by an X.700 management support module which was integrated with the DPE. Both the DPE support and the X.700 management support entities are explained in detail in section 4.3.

## 4.3    Key service components

Since no CORBA or TINA compatible platforms were used in PREPARE, two custom service components were developed. The object coordinator was to support TINA DPE-like functionality. The X.700 management object was to support an X.700 management gateway to the DPE. It was used as a special DPE service by other service components that utilised the TMN X interface. The TMN X interface was used only for inter-domain transactions in PREPARE as shown in Figure 3.

*The object coordinator*
Although in Figure 3 the object coordinator appears to be one layer underneath the other service components, it can also be considered as a special service component. It provides location transparency to all the contracts on a global level, by means of a unique global name it assigns to each of them.

The object coordinator offers an interface, with a well known reference, for the other contracts to use. It consists of the following methods:

• *Register*. This method is invoked by a contract of a BB when it registers with the object coordinator.
• *Deregister*. This method is invoked by a contract when it wishes to de-register.
• *Reregister*. This method is invoked by a contract when it wishes to amend its registration details, such as location. This can provide object mobility support that can be used for personal or terminal mobility.
• *Call*. This method is invoked by a contract when it wishes to call a method provided by another contract.

For scalability reasons, many object coordinators can be instantiated across the network. They can set up connections between them and via special interfaces they can:

• *Locate* contracts registered with any other object coordinator.
• *Ensure the uniqueness* of a global name when a contract registers.
• *Invoke* methods on contracts registered with any of the object coordinators.

On a high level, an object coordinator resembles a DPE or an Object Request Broker (ORB), while the communication between different object coordinators resembles the concept of the Kernel Transport Network (KTN) in TINA. In PREPARE, the object coordinator was used only for intra-domain access. All the transactions between different domains were made over the TMN X interface. Therefore, the inter-coordinator communication was restricted to the boundaries of each administrative domain.

The object coordinator proved to be a core component for applying and trying out the TINA principles on our design. Although we cannot claim that it substituted the functionality a DPE can provide, it fulfilled our requirements for a DPE support component.

### The X.700 management object

The X.700 management object is a component that was developed as a DPE service. It offers an X.700 management API to other components which can handle X.700 operations, actions and event reports. In PREPARE, it was used by contracts that utilised the TMN X interface for inter-domain management interactions.

The implementation of this component that was used in PREPARE, offers access to the X.500 directory as well. It is based on the integration of the X.500 directory and X.700 management which provides location transparency for X.700 agents that have an entry in the X.500 directory [PREPARE]. Each entry includes the address of the corresponding X.700 agent. Due to similar naming conventions for X.500 data objects (directory objects) and X.700 data objects (managed objects), the concatenation of these two provides a unique name for each X.700 data object regardless of which agent is hosting it.

## 4.4    Technology used

The object coordinator and all the service components were implemented in Tcl which satisfied requirements for flexibility and simplicity when developing service management logic. Tcl is an interpreted scripting language. As a high level language it can be used by service managers who are not necessarily experienced programmers. Since it is embeddable into C or C++ code, it is also possible to reuse already existing software and integrate it with Tcl code.

The Tcl Object Coordinator (TOC) is the implementation of the object coordinator component. It is based on Tcl (version 7.4) [TCL-TK], the object-oriented Tcl extension ObjectTcl (version b1.1) [OTcl], and the distributed programming Tcl extension Tcl-DP (version 3.3b1) [Tcl-DP].

There are two implementations of the X.700 management object. One is based on tcl-rmib which is part of the OSIMIS 4.0 management platform [OSIMIS][TCL-RMIB]. The other one is based on tcl-idmis which is part of the ANDROMEDA [Dittrich-95d] platform, which, in turn, is based on OSIMIS and ISODE.

Both, tcl-rmib and tcl-idmis, offer a Tcl-based X.700 management interface, where management operations are invoked by Tcl commands. The parameters passed to and the results returned from the X.700 management object are strings. To support the asynchronous reception of event reports they are mapped to Tcl call-backs. Tcl-idmis integrates X.500 and X.700 under the same API and therefore offers location transparency for X.700 management agents.

## 5   CONCLUSIONS

Though the PREPARE open service market context is a fairly limited example of potential enterprise situations that could evolve, it gives us some insight into the problems that a management system designer may face while working in this area. Such a context places new requirements on management systems as they encompass the integration of service and network management. One key requirement is the need for modularity of management systems at an engineering level. This is seen as essential in order to obtain the level of implementation reuse and the resulting fast time to market that competitive pressures and changing customer requirements will demand in the open service market. TMN recommendations offer the possibility for defining interfaces in a modular way, e.g. by using MFCs. However the mapping of modularity in interface specification to modularity in implementable components is outside the scope of TMN. This coupled with the slow emergence of an effective platform independent TMN API has meant that there is currently no commonly practised approach of modular TMN system design.

ODP, with its five view-points, supports implicitly the mapping modularity in the specification of functional interfaces to modularity in an engineering model. ODP has been taken by the TINA Consortium and applied to the telecommunications arena, though to date practical application of this architecture has been limited to a few prototypes. The work presented in this paper is an example of such a prototype, where some of the aspects from the TINA overall architecture have been applied to a home-grown DPE offering location transparency between engineering objects. By integrating this DPE with an existing TMN platform, the engineering aspects of applying ODP and TINA principles to TMN have been investigated to some extent. This is an important area, since even if TINA DPE or CORBA implementations are taken up increasingly for management applications there is a large investment in TMN-based management interface specification and an growing base of installed TMN agents and applications that will have to be accommodated by the new platforms.

The ACTS project, Prospect, has been building on the PREPARE and other related work in setting up further management networks for integrated multi-domain services in which an common engineering approach can be exercised. In PREPARE, the inter-domain interactions were supported over the TMN X interface and only the intra-domain interactions were done over a distributed processing platform. The use of CORBA platforms for uniform communication among components across domains in Prospect [Lewis-97] was considered beneficial to the approach followed in PREPARE. While TMN is still used on the network management layer in Prospect, due to the availability of such systems and to the previous investment in this area, a CORBA-TMN gateway is being developed to bridge the gap between CORBA-based service management components and the TMN-based network management components.

## 6   ACKNOWLEDGEMENTS

All partners in the PREPARE project and in particular Sven Krause, Ingo Busse and others at GMD-Fokus, Lennart Bjerring of L.M. Ericsson, as well as George Pavlou and David Griffin of UCL. This work was conducted under the partial funding of the EU through the RACE II PREPARE project (contract R2004) and of the ACTS Prospect project (contract AC052). The views expressed in this document do not necessarily reflect those of the PREPARE consortium or of the Prospect consortium.

# 7  REFERENCES

[Bjerring-94] Bjerring, L.H., Schneider, J.M., End-to-end Service Management with Multiple Providers, Proceedings of the 2nd International Conference on Intelligence in Broadband Services and Networks, Aachen, Germany, September 1994, pp 193-206, Springer-Verlag, 1994.

[CORBA] *The Common Object Request Broker: Architecture and Specification*, OMG Document Number 92.12.1, Revision 1.1, Object Management Group, 1992.

[Dittrich-95d] A. Dittrich, *The ANDROMEDA Platform: an object-oriented development and run-time environment for management services*, Proceedings of the 6th IFIP/IEEE International Workshop on Distributed Systems: Operations & Management (DSOM '95), 1995.

[Lewis-97] D. Lewis, T. Tiropanis, C. Redmond, V. Wade, *Inter-domain Integration of Service and Service Management*, Proceedings of IS&N '97 Conference in Como, Italy, May 1997.

[M.3010] *Principles for a Telecommunications Management Network*, ITU-T Recommendation M,3010, 1992.

[M.3020] *TMN Interface Specification Methodology*, ITU-T Draft Revised Recommendation M.3020, 1994.

[NMF-BPM] *A Service Management Business Process Model*, Network Management Forum, Morristown, NJ, 1995.

[OTcl] *Object Tcl (version b1.1)*, http://www.x.co.uk/devt/ObjectTcl/, IXI Limited, 1995.

[OSIMIS] G. Pavlou, G. Knight, K. McCarthy and S. Bhatti, *The OSIMIS Platform: Making OSI Management Simple*, in Integrated Network Management IV, ed. A.S. Sethi, Y. Raynaud and F. Faure-Vincent, pp. 480-493, Chapman & Hall, 1995.

[PREPARE] J. Hall (Ed.), *Management of Telecommunication Systems and Services*, Springer-Verlag Berlin in Heidelberg, 1996.

[Q1204] *Intelligent Network Distribution Functional Plane Architecture*, ITU-T Recommendation Q.1204, 1993.

[SPIRIT] *X/Open Consortium Specification. SPIRIT Platform Blueprint*, SPIRIT Issue 2.0, Network Management Forum, Morristown, NJ, 1994.

[TCL-TK] John K. Ousterhout, *Tcl and the Tk Toolkit*, Addison-Wesley Publishing Company, 1994.

[TCL-RMIB] T. Tin, G. Pavlou, *Tcl-MCMIS: Interpreted Management Access Facilities*, Proceedings of the Sixth IFIP/IEEE International Workshop on Distributed Systems: Operations & Management (DSOM '95), Ottawa, Canada, October 1995.

[Tcl-DP] B. C. Smith, L. A. Rowe, *Tcl Distributed Programming (Tcl-DP) (version 3.3beta1)*, ftp://mm-ftp.cs.berkeley.edu/pub/multimedia/Tcl-DP, June 1995.

[TINA-005] Graubmann, P. and Mercouroff, N., *Engineering Modelling Concepts (DPE Architecture)*, TINA Baseline document TB_NS.005_2.0_94, December 1994.

[TINA-018] Chapman, M. and Montesi, S., *Overall Concepts and Principles of TINA*, TINA Baseline document TB_MDC.018_1.0_94, February 1994.

[TINA-HV] Natarajan, N., Dupuy, F., Singer, N., Christensen, H., *Computational Modelling Concepts*, TINA Baseline document TB_A2.HC.012_1.2_94, February 1995.

[X901] *Information technology - Open Distributed Processing - Reference Model - Part 1: Overview*, ITU_T Draft Recommendation X.901/ISO/IEC Draft International Standard 10746-1, 1995.

# 8   BIOGRAPHIES

**Thanassis Tiropanis** graduated from the department of Computer Engineering and Informatics at University of Patras, Greece, in 1993. Since 1994, he has been working as a Research Fellow at UCL. Thanassis was involved in the European projects PREPARE and Prospect investigating the deployment of service management systems for the open services market. He has lead a team that investigated the introduction of mobility to the Prospect multi-service environment. Architectural aspects of mobile services is an area of particular interest on which he is doing a Ph.D.

**David Lewis** graduated in Electronic Engineering at the University of Southampton in 1987 and received an M.Sc. in Computer Science from UCL in 1990, since when he has worked as a research fellow. He has worked primarily on the EU funded projects PREPARE and Prospect, in which he has been responsible for planning high speed international test-bed networks and for leading teams developing integrated, multi-domain service management systems. He is also working on a Ph.D., researching a service management development architecture for the open services market.

**Rong Shi** graduated from the department of Computer Science at Tsinghua University, Beijing, China, in 1987 and received an M.Sc. at the same department, in 1989. Since 1994, she has been working as a Research Fellow at UCL. Rong Shi was involved in the European ESPRIT MIDAS project and RACE II PREPARE project. Now, she is working on the ACTS MISA project investigating the inter-domain routing algorithm for the SDH and ATM integrated networks. She has lead one activity in the MISA project on the system high level specification.

**Alexander Richter** received his diploma in Computer Science from the Technical University Berlin, Germany. During his studies he joined the Research Center for Open Communication Systems (FOKUS) of GMD in Berlin where he is still working in the Department for Management in Open System (MINOS).

Mr. Richter worked for the RACE II project PREPARE. After this project was finished he has been working for the ACTS Project PROSPECT where his scope is the management of services in the open service market. In this area the mobility aspects of service usage and provision are of interest.

# 14

# TMN/C++: An object-oriented API for GDMO, CMIS, and ASN.1

*T.R. Chatt*
*Vertel*
*21300 Victory Blvd., Woodland Hills, CA 91367, USA*
*Phone: +1(818)227-1400     Fax: +1(818)598-0047*
*e-mail: tom-chatt@vertel.com*

*M. Curry, TCSI Corporation, mikec@tcs.com*

*J. Seppä, Nokia Telecommunications*
*juha.seppa@ntc.nokia.com*

*U. Hollberg, IBM European Networking Center*
*hollberg@heidelbg.ibm.com*

## Abstract

The TMN/C++ API offers a standardized object-oriented application programming interface (API) for telecommunications management applications. The architecture comprises three modular, layered APIs. The ASN.1/C++ API provides a role-independent interface to the data itself and its encoding. The CMIS/C++ API provides a service-oriented interface for the basic management information model (MIM) services—creating and deleting objects, getting and setting their attributes—in an agent or manager role. The GDMO/C++ API provides a MIM-oriented framework for accessing and implementing managed objects in a containment tree and accessing CMIS services in an agent or manager role. The API set supports two distinct application types: "specific" applications which efficiently implement a static MIB specification, and "generic" applications which can dynamically interpret new MIBs. By the end of 1997, these standards documents are expected to be published as X/Open preliminary specifications and adopted by the NMF.

## Keywords

TMN, GDMO, CMIS, ASN.1, C++, API, telecommunications, OSI, network management, agent, manager, MIM, NMF

## 1     PROJECT BACKGROUND

By 1994, the notation, service model, and protocol to be used in telecom management software had been fairly well standardized by organizations such as the International Telecommunications Union (ITU, formerly the CCITT), and the Joint Technical Committee (JTC1) of the International Organization for Standardization (ISO) and the International Electrotechnical Commission (IEC). The Guidelines for Definition of Managed Objects (GDMO) established a notation and an object-oriented framework for describing management information.[1] The Common Management Information Service (CMIS) specified the way in which such information would be exchanged across a network.[2] And Abstract Syntax Notation One (ASN.1) standardized a machine-independent notation for representing data and a practical means of exchanging it.[3] This level of "meta-standardization" provided a framework for standardizing information exchange for particular types of equipment and services. Once the framework was established, industry organizations such as the ITU-T and the Network Management Forum (NMF) could focus on standardizing the information required to manage equipment (such as SONET or SDH network elements) and enable services (such as electronic bonding). The combination of standard information and a standard means of communicating it enabled network interoperability between various manufacturers' equipment and the management software of various service providers and carriers.

With the adoption of the TMN model and the obvious advantages it offered came a need for a significant amount of complex software to implement it. Though the applications to be built on the TMN model may be quite heterogeneous, the commonality inherent in the standard information model and transport allowed the development of reuseable software for any TMN application. The software market responded by providing a variety of TMN software tools, generally comprising compilers which translated the standard GDMO and ASN.1 abstract notation into some specific programming language form, and "platforms" or "toolkits" which provided a CMIP engine in an "agent" or "manager" role. While these tools facilitated the development of TMN software applications, their application programming interfaces were non-standard and often complex. The learning curve invested in developing a TMN application using a particular vendor's TMN tools was often steep, and was not applicable to another vendor's TMN tools. Thus, application builders incurred repeated learning curve costs to use various tools as best fit particular applications, or else they were locked in to one vendor for a "one tool fits all" solution.

---

[1]  GDMO is specified in ITU-T Rec. X.722 I ISO/IEC 10165-4.
[2]  CMIS is specified in ITU-T Rec X.721 I ISO/IEC 10165-2.
[3]  ASN.1 was standardized in 1990 as ITU-T Rec X.208 I ISO/IEC 8824.
Significant extensions to the notation caused it to be re-standardized in 1994 as ITU-T Rec X.680, X.681, X.682, and X.683 I ISO/IEC 8824-1 through 8824-4.

In response to this need to standardize an API for TMN applications, X/Open promulgated XMP and XOM, which provided a C language API for managing objects, and a method for mapping GDMO/ASN.1 onto the XMP/XOM domain.[4] The wide acceptance of this standard, and its incorporation into many of the commercial TMN platforms demonstrated the need for such a standard. However, the utility of this standard in the TMN domain was hampered by its high degree of abstraction. It is generally perceived as difficult to use, and most TMN platforms which boast XMP/XOM compliance also offer more easy-to-use API layers which hide it. Moreover, XMP/XOM was a C API, while C++ was gaining increasing acceptance as a more natural language for implementing object-oriented methodologies such as GDMO.[5]

This was the industry context in which the TMN/C++ API was conceived. At the Network Management Forum general meeting in autumn of 1994, a group of member companies initiated a collaborative effort to design a portable, easy-to-use, object-oriented application programming interface (API) for telecommunications management applications using the TMN standard model. The project, which got fully underway in 1995, was co-sponsored by the Network Management Forum (NMF) and X/Open, and enjoyed participation and support from a variety of organizations:

- telecom service providers and equipment manufacturers, such as Nokia, Motorola, Ericsson/HPT and Hitachi Telecom
- hardware platform vendors, including IBM, Sun, and HP
- TMN software platform providers, such as Vertel and TCSI
- software tool vendors, such as OSS, Opening Technologies, and DSET
- other industry and software standards organizations, such as the Open Software Foundation (OSF), and the International Organization for Standardization (ISO)
- other (non-telecom) ASN.1-based service providers, such as 3M Health Care.

The three-year project will produce a series of related standards documents to be adopted by the NMF and published by X/Open. The standards, which are coherent but loosely coupled, are being published separately as they become available and approved. The first two standards, TMN/C++ and ASN.1/C++, are technically approved and ready for publication at the time of this writing. It is anticipated that the entire series will be completed by the end of 1997.

---

[4] See X/Open Documents C315 OSI – Abstract Data Manipulation API (XOM) and C306 Systems Management – Management Protocol API (XMP)
[5] By early 1995, telecom companies including Nokia, NTT, and Ericsson/HPT, who had developed in-house TMN application platforms had done so using C++.

## 2    DESIGN GOALS AND OVERALL ARCHITECTURE

The working group agreed upon a set of goals for the API, with "simple, intuitive, and easy to use" being at the top of the list. The most natural expression of the object-oriented TMN model in a programming language was desired. Source code level portability was a second key goal, such that application developers would only have to invest in one learning curve for a common API and would not be locked in to particular vendors and tools. It was agreed that the complete functionality of GDMO, CMIS, and ASN.1 should be provided for, rather than trying to restrict the API to some subset of the model. The diversity of applications to be built upon the API—from network element agents which may be deployed on embedded processors to service management layer applications likely to be on large corporate information system platforms—demanded that the API be flexible and optimizable enough to meet these various needs. Finally, the commercial side of those present required that the API not be so grandiose as to be unimplementable in a reasonable market timeframe.

The C++ programming language was chosen both for its natural strength in expressing an object-oriented model, and for its increasing deployment in the telecom industry and in the software industry in general. The use of C++ class inheritance hierarchies to model ASN.1 and GDMO objects was adopted as the most intuitive way to translate these abstract notations into a programming language. Rather than limit the TMN/C++ API to the commercially available features of C++ as of 1995 (which would have looked a bit stilted in a TMN/C++ standard published in 1997), the decision was taken to target the emerging ANSI standard for C++ as a programming language baseline.[6] All features were made use of, including virtual functions, templates, and exceptions. In particular, extensive use was made of the ANSI standard libraries, including strings and containers/iterators (the "standard template library" or STL).

The API was developed to accommodate two rather different software application models, which were termed the "specific" and the "generic" models. In the "specific" application model (see Figure 1), the application is designed around a particular set of management information which is established before development occurs. This information, fixed in a particular set of GDMO/ASN.1 definitions, is passed through a translation process (ie., a compiler) and specific C++ classes are produced which model the information. In this model, a TMN/C++ API implementation would provide a base library and header files with C++ classes representing the basic types and objects defined in GDMO and ASN.1, and the CMIS services. The API implementation would also provide a compiler which would take GDMO and ASN.1 inputs and generate specific classes for that particular information model. The application developer would then develop the application using these base classes, plus the compiler-generated classes (which are

---

[6] See ANSI X3J16/95-9987 (April 1995), Draft International Standard for Information Processing: Programming Language: C++.

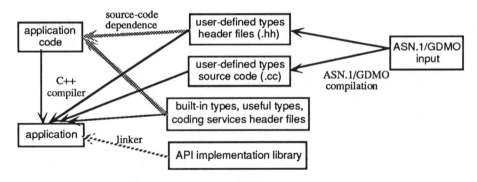

**Figure 1**   The "Specific" application model.

fixed at application development time). The specific model offers the benefits of a type-safe and easier-to-use interface by producing C++ classes tailored for a specific application. This model is also likely to produce more efficient code in the final application.

In contrast, the "generic" model (see Figure 2) allows knowledge of the management information definition to be postponed until runtime. In this model, a TMN/C++ API implementation provides only a base library and header files defining C++ classes representing the basic types and objects of GDMO and ASN.1, along with the ability to construct user-defined GDMO objects and ASN.1 types at runtime (eg., by loading and interpreting new or revised GDMO and ASN.1 specifications). The application developer would then develop an application using generic classes, and would construct management information definitions "on the fly". The generic model involves an exposed interface to the "meta-data", that is, the data which describes the structure of the management information. This model opens up the powerful possibility of an application which discovers new information definitions as it runs, and adapts to handle these new definitions. A generic application runs in an interpretive, rather than a compiled, mode.

Since the generic and specific application models differ, different APIs are needed to address them. Both the generic and specific APIs can stand alone, and TMN/C++ API implementations may decide to offer only one or the other. However, the

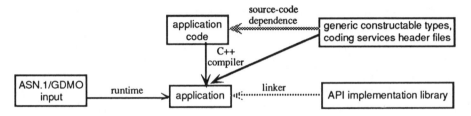

**Figure 2**   The "Generic" application model.

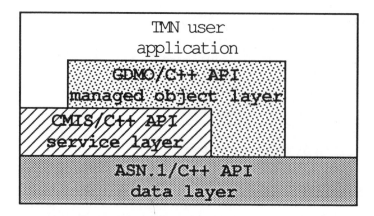

**Figure 3**    TMN/C++ layered architecture.

overall API was designed such that both generic and specific APIs together form one coherent API, and are not completely distinct. This allows one application to easily support a combination of compiled and dynamically interpreted information, and also simplifies both the understanding and the implementation of the complete TMN/C++ API.

The overall project of a TMN/C++ API seemed most naturally decomposed into three relatively independent, layered APIs. (See Figure 3.) The foundational layer is the interface to the management information (data) itself. Thus, the ASN.1/C++ API was conceived to provide a mapping from data syntax specifications (expressed in ASN.1) onto C++ classes and objects for actual programmatic manipulation of the data. This layer deals strictly with data access and encoding, and is independent of the use of this data in the context of an X.700 management information model in an agent or manager role. Both "specific" and "generic" interfaces are offered in this layer, to allow for the efficient implementation of a static specification, and the flexible implementation of a dynamically interpretive application. Recognizing that ASN.1 has much wider applicability outside of TMN, the ASN.1/C++ API was designed to be a "stand alone" API, independent of the other two TMN/C++ API components.

The service layer, CMIS/C++ API, provides a programmatic interface to the services required to operate an agent or manager in a management information model (eg., creating and deleting objects, getting and setting their attributes). This layer provides the framework for an agent or manager "engine", and thus provides symmetric interfaces for the agent and manager roles (though the two roles may be coherently combined in one application). However, because this layer is independent of any particular management information base (MIB), it is by its nature only "generic", and has no "specific" components. The CMIS/C++ API is actually broader than just CMIS. It also includes the services defined in ACSE for initiating, responding to, and controlling associations between agents and managers.

The managed object layer, GDMO/C++ API, provides a programmatic interface for navigating both inheritance trees (class definition meta-data) and containment trees (instances of management data). It also integrates the functionality of the service layer, representing it as methods associated with managed objects. This layer, which is the most closely tied to the X.700 model, has different (and not entirely symmetric) interfaces for agent and manager roles. (As with the service layer, agent and manager APIs may be coherently combined in one application.) And since this layer maps GDMO specifications onto C++ classes and objects, it has both "specific" and "generic" variants.

## 3    ASN.1/C++ API

The ASN.1/C++ API defines a mapping from ASN.1 abstract entities, such as types and values, onto C++ entities such as classes and objects. For each ASN.1 type, there is a corresponding C++ class.For each ASN.1 value, there is a corresponding C++ object. ASN.1 modules, which encapsulate ASN.1 definitions, are translated as C++ namespaces, which scope the corresponding classes and objects. The goal was to define a simple, easy-to-understand correspondence between the ASN.1 model and the C++ produced.

An abstract base class, `AbstractData`, heads the C++ class hierarchy which is used to model all ASN.1 types. (See Figure 4.) The `AbstractData` class provides all the functionality which is inherent in any item of ASN.1 data: it can be encoded and decoded, transformed to or from ASN.1 value notation, tested for validity against its subtype constraints, queried to discover its type information, and compared against other values. This class is abstract in the C++ sense (ie., contains pure virtual functions) and cannot be instantiated directly. Rather, it serves to provide a useful abstraction through which instances of its subclasses may be manipulated.

For each ASN.1 built-in type, a C++ subclass of `AbstractData` is defined to provide a type-specific "natural" representation. For instance, the C++ class `BOOLEAN`, which represents the ASN.1 type BOOLEAN, derives from `AbstractData`, but otherwise has the look and feel of the C++ `bool` built-in type. It is freely constructable from, assignable from, and castable to the C++ `bool` type. Similarly, the C++ class `INTEGER`, which represents the ASN.1 type INTEGER, has the look and feel of a C++ `int` type.[7]

For user-defined ASN.1 types, the derivation of the corresponding C++ classes follows the type definition. For instance, given an ASN.1 type 'Rate ::= INTEGER', the corresponding C++ class `Rate` derives from the C++ class

---

[7] This is actually a simplification. The actual specification provides for "huge" integers (ie., those which exceed the capacity of a given machine's native 'int'), as well as multiple INTEGER classes to make most efficient use of short and long ints.

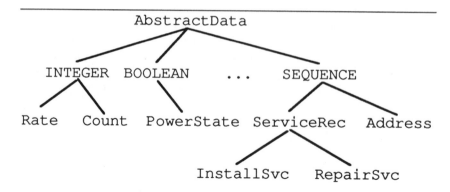

**Figure 4**   ASN.1 Base Class Hierarchy.

INTEGER. And given an ASN.1 type 'InstallSvc ::= ServiceRec', the corresponding C++ class InstallSvc would derive from the C++ class ServiceRec.

The application interacts with the ASN.1/C++ API across four basic kinds of interface: abstract data, information objects, encoded buffers, and coding services. (See Figure 5.) The abstract data interface is through the AbstractData class hierarchy described above. In the specific ASN.1/C++ API, the interface includes type-safe generated C++ classes to represent specific ASN.1 user-defined types. In the generic ASN.1/C++ API, the application manipulates data using the base classes and generic access methods (eg., to access components of structured types by field name or position). Similarly, the application may manipulate STL containers which represent ASN.1 information object sets, either through type-safe generated classes or generically.

The coding services are provided in the form of template functions for "encode" and "decode", which take an AbstractData subclass object as an argument. Following the style of an STL algorithm, the encode and decode functions may operate on any buffer which provides an octet (ie., unsigned char) iterator. This allows the implementation of buffers to be vastly open-ended. A buffer may

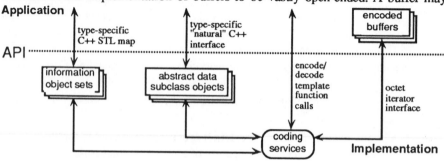

**Figure 5**   ASN.1 Application Interfaces.

be as simple as an `unsigned char*`, or it may be any C++ class which can "look" like one. An implementation may even provide "smart" buffer classes which can provide an efficient bypass around the one-octet-at-a-time public interface.

Where C++ code is provided or generated to represent specifications expressed in ASN.1, the names of the C++ classes and objects are—to the greatest extent possible—made the same as the names of the ASN.1 entities they represent. Case is preserved and hyphens are mapped to underscores. Where limits on the length of names are needed, a standard algorithm for name shortening is provided. Rules are also provided for generating names where no names exist in the ASN.1 (anonymous fields and types), as well as for resolving keyword conflict and disambiguation issues.

The mapping of ASN.1 entities onto C++ causes some issues to arise, such as compatibility for assignment and comparison, which are not explicitly addressed in ASN.1 itself. The ASN.1/C++ class hierarchy implements a general rule of compatibility within any class inheritance subtree with a concrete common base class. For example, all INTEGER subtypes are compatible, and all subtypes of a particular SEQUENCE or SET type are compatible, but different SEQUENCEs and SETs are not compatible. This provides an easily understandable rule which is naturally coherent with the ASN.1/C++ class hierarchy.

For several reasons, including the convenient storage of ASN.1/C++ objects in STL containers, it was deemed useful to define comparison rules within and among all types. These comparison rules are semantically sensible where feasible (eg., string types are lexicographically ordered), and arbitrary where not.

The realization of the abstract notation into C++ also introduced the need for directives in an ASN.1 specification, where supplemental information is needed or desireable to guide the translation process. Directives may control C++ name generation, memory storage issues, and implementation variations. Thus, a standard directive format, which is a "distinguished" ASN.1 comment, was specified. The directive notation is position-independent and relocatable. By design, directives may enhance the translation process, but are never mandatory, so that "pure" ASN.1 is always translatable.

## 4    CMIS/C++ API

The CMIS/C++ API provides a programming interface to the OSI application-service-elements (ASEs) that are used by all management applications, namely the CMISE and the ACSE. The software components of the API are:
- a set of C++ classes that model the application-service-elements, and their service primitives and parameters;
- objects for referring to outstanding operations (invocation handles), and mechanisms for applications to receive indication and confirmation service primitives (callback and queue classes);
- a "convenience" API for automatic association management.

The basic idea behind the CMIS/C++ API design was to give the application developer a *service-oriented* (as opposed to protocol-oriented) view of the communications functions used by a management application. This was because a user's basis for understanding the communications functions provided by the ACSE and the CMISE are the service definition documents, not the protocol specifications. The service definitions document the *meanings* of the service interactions, their parameters, and the relationships between the parameters. These semantics are primarily what an application developer cares about.

Moreover, the service definitions use a very simple and consistent style for documenting service primitives and their parameters. It would have been possible to model the relatively more complex APDUs of the protocol definitions as C++ objects (and indeed, as ASN.1/C++ objects), but mapping between service primitives and APDUs, while necessary, is the proper function of a protocol machine, not of application logic. The higher level of conceptual abstraction found in the service definitions made the service view the more appropriate basis for an API design. The mechanics of constructing or interpreting APDUs is left to the API implementation.

The basic design approach for the CMIS/C++ API was to use object oriented techniques to build explicit and detailed models of the CMISE and the ACSE, and their service primitives and service parameters, as C++ objects. The main reason to model these service elements explicitly was so that the service definition documents, along with typical textbook explanations of the OSI application layer structure, could in large measure stand as primary sources of reference documentation on the structure and semantics of the API. It was felt that the closer the parallels between these explanations of the service elements and their C++ embodiment in the API, the smaller the conceptual leap that a prospective user would need to take to understand how to use the API.

Another compelling reason for fine-grained modeling of all of an ASE's features was discovered during the design discussions for the ACSE interface. The initial thinking was to provide only a simplified interface for association establishment. As discussions progressed, however, it became apparent that to provide any less than the full set of capabilities offered by the ACSE definition would lead to a design that failed to support certain applications' requirements. Presenting the ACSE's capabilities only in some pre-packaged "convenience interface" form amounted to imposing a policy on the way in which those capabilities could be used. The convenience of such an interface would be apparent only in applications whose requirements accorded with that policy: a convenience interface is only convenient if it does what you need.

Conceptual simplifications which departed from a close modeling of the OSI service elements were avoided. For instance, since CMIS is inherently asyncronous, the CMIS/C++ offers only an asyncronous interface. (A syncronous interface is offered at the higher-level GDMO/C++ interface.) However, C++ modeling techniques enabled the construction of user-friendly semantic-oriented classes. The service parameters are a good example.

The CMIS and ACSE service definition documents are purposefully silent on the exact form of service-defined parameter values, concentrating instead on their semantics. This allowed some degree of abstraction in the design of the C++ service parameter classes, in that there was no *a priori* requirement, from the service perspective, that the parameter types be literal ASN.1 data objects as represented in protocol. In the CMIS/C++ API, the interfaces presented by the parameter classes are defined primarily in terms of their semantics at the service interface, rather than their structures in protocol.

A simple example of how such an abstraction is modeled in the CMIS/C++ API is the Scope class. Part of the definition of this class is:

```
class Scope {
public:
    enum SimpleStyle { baseObjectOnly, wholeSubtree };
    enum ParameterizedStyle { nthLevel, baseToNthLevel };

    Scope(SimpleStyle style = baseObjectOnly);
    Scope(ParameterizedStyle style, unsigned int n);
    // ...
};
```

This abstraction aligns more intuitively with the semantics stated in the CMIS service definition document than does the ASN.1 type used to transmit the value (a CHOICE with three INTEGER alternatives).

Another powerful abstraction provided by the API is the use of C++ operator overloading to simplify the construction of "filter" parameters using a symbolic paradigm. For example, an expression like:

```
(administrativeState == AdministrativeState::locked &&
supersetOf(packages, availabilityStatusPackage_singleton))
```

can be used to produce a filter parameter value with a fraction of the effort that would be required to construct the same value using only the ASN.1 types provided by the protocol specification.

## 5    GDMO/C++ API

The GDMO/C++ API provides programmers with a "managed object"-oriented API to CMIS services and to access and to implement managed objects. In the OSI management model, an agent role application contains managed objects— abstractions of managed resources—which it reveals to be managed by remote systems. A manager role application in turn issues management operations on managed objects contained within one or more agents. The GDMO/C++ API supports both manager and agent roles. Using the GDMO/C++ API, the user works with C++ classes and objects which closely model the managed objects described in the GDMO for the MIB being implemented. This provides a natural realization of the OSI management model, where managed resources are modeled with managed objects.

The main goals of the GDMO/C++ API are:

- Provide an easy-to-use programming interface.
- Provide high level abstraction close to OSI management model.
- Hide as much of the complexities of communication as possible.
- Provide for strong compile time type checking when ever possible.
- Allow incremental development of management applications.
- Preserve all necessary information from GDMO specification.

The GDMO/C++ API provides a framework for management applications. This framework provides a context in which managed objects can be easily managed, by maintaining a context for each managed object instance, and automatically routing management operations to the appropriate managed object instance context. The framework validates all incoming and outgoing service requests, and provides default processing to handle all normal and error cases. In this manner, much of the complexity of communication is hidden.

Whereas the ASN.1/C++ API provides data types used in TMN, and the CMIS/C++ API provides services for communicating data, the GDMO/C++ API provides a complete application framework in which the application developer only need fill in the particular behavior of the managed objects being implemented. The API framework establishes a default implementation (usually returning a CMIS error) for all operations, which facilitates incremental application development. Programmers only need to define those parts important for the application, while everything else has been covered by API implementation. The framework thus enables faster and easier management application development.

Managed object classes defined in GDMO are modeled by "managed object handle" (MOH) C++ classes for manager role applications, and by "managed object" (MO) C++ classes for agent applications. The MOH C++ classes provide a communications interface for a manager role application, through which it can issue management operations on managed objects and receive responses to the operations. The MO C++ classes provide the basic structures for agent applications to implement managed objects.

The frameworks for manager role applications vs. agent role applications differ in several notable ways. The most important is the organization of the managed objects. On the agent side, the management information tree (MIT) model (also known as the "containment tree") is closely followed, with each managed object represented by one MO instance in the tree. The agent role framework provides functionality to implement and manage this tree, including default automatic processing of scope and filter.

On the manager side, instead of imposing a tree structure, the notion of MOH collections is introduced. There are enumerated collections, in which the user may insert or delete any arbitrary MOH. There are also rule-based collections, which are constructed based on the result of scope-and-filter operations. The manager role user may arrange any number of MOHs and MOH collections as suits the application. On the manager side, multiple MOHs may provide an interface to the same managed object.

Like the ASN.1/C++ API, two different software models, generic and specific, need to be addressed by this API. The specific API is based on a fixed set of managed objects known at compile time, as opposed to the generic API which is able to handle new GDMO and ASN.1 definitions dynamically. In the specific model, each GDMO managed object class is translated or "mapped" onto a specific C++ class which represents it. The specific C++ classes provide type-safe interfaces based on the GDMO and ASN.1 definitions, and thus can provide compile-time type checking. In the generic model, a base C++ managed object class is used to represent an instance of any GDMO managed object class. Each instance, however, is dynamically linked to type information for its particular managed object class. In this way the validity of any operation issued on a generic managed object instance can be checked at runtime.

Both manager and agent role frameworks offer an application shell with complete default behaviour which the user can specialize and override as necessary. Two basic mechanisms for this specialization are offered. A callback mechanism, essential for the generic model, allows the developer to register callbacks for particular events at various contexts (eg., an object instance, an object class, the framework). A C++ virtual function mechanism allows the developer to derive specialized classes from the mapped (or base) managed object classes, and to provide specialized implementations of the virtual methods. The latter method is more suited to the specific model, and works in concert with an "object factory", which is used by the framework to automatically instantiate objects when needed. The factory also uses a virtual function mechanism which allows the framework to instantiate user-defined classes.

The framework offers further interface simplifications. OSI management operations are inherently asynchronous. An operation consists of four phases—request, indication, response and confirmation—which CMIS models as service messages. The GDMO/C++ API provides management operations in two modes: asynchronous and synchronous. In the synchronous mode of operation the four phased operation is abstracted to a blocking function call. This model makes developing simple management applications easy, and is also useful for multi-threaded applications. Not all applications can accommodate to synchronous mode (eg., single-threaded user interface applications, which need to perform other activities while a communications operation is being processed), and thus asynchronous mode is provided as well.

The framework also simplifies application development by providing a cache for selected managed object attribute values. On the manager side, the cache feature can be combined with an automatic tracking feature, which allows the manager application to treat its MOHs as C++ objects which are remotely implemented, offering a highly simplified CORBA-like interface.

## 6    SUMMARY

By exploiting the object-oriented modeling power of the C++ language, the TMN/C++ project has created an ensemble of highly modular APIs which greatly simplify the development of TMN management applications. By structuring the project as coherent but loosely-coupled APIs, a wide variety of different application domains can be suited. Lower-level message-oriented applications can have the greatest degree of control over their network interface by using ASN.1/C++ by itself. Applications which want explicit control at the CMIS service and association control level will find that CMIS/C++ offers an easy-to-use embodiment of the service model. And ultimately, applications which want to operate at a high level of abstraction will find that GDMO/C++ offers a good combination of simplicity, flexibility, and power. Simplicity is achieved at each level by close C++ modeling of the semantics of ASN.1, CMIS, and GDMO and MIM, combined with leveraging of established C++ technology such as the Standard Template Library. Common design idioms are deployed across all the APIs to achieve a consistent and coherent ensemble, furthering ease of use.

Several NMF member companies have already begun to prototype and even productize components of this emerging standard. Early reports indicate that it is commercially feasible to develop, and that users of the APIs find their application development significantly expedited.

## 7    REFERENCES

ANSI Working Paper X3J16/95-9987 (April 1995), "Draft Proposed International Standard for Information Processing: Programming Language: C++".

ITU-T Recommendation X.680 "Information Technology - Abstract Syntax Notation One (ASN.1): Specification of basic notation." (1994), also ISO/IEC 8824-1 (1995).

ITU-T Recommendation X.721 "Information Technology - Open Systems Interconnection - Common management information service definition" (1991), also ISO/IEC 9595.

ITU-T Recommendation X.722 "Information technology - Open systems interconnection - Structure of management information - Guidelines for the Definition of Managed Objects" (1992), also ISO/IEC 10165-4.

Network Management Forum.

TMN/C++ Application Programming Interface. NMF 039.

ASN.1/C++ Application Programming Interface. NMF 040.

CMIS/C++ Application Programming Interface. NMF 041.

GDMO/C++ Application Programming Interface. NMF 042.

## 8    BIOGRAPHY

**Tom Chatt** is a Principal Software Engineer at Vertel in Woodland Hills, CA, with architectural responsibility for Vertel's TMN Manager and Agent Development Environment products. **Michael A. Curry** is a Principal Engineer for the TCSI Corporation in Berkeley, CA.    **Ulf Hollberg** is a Senior Consultant at the IBM European Networking Center in Heidelberg, with architecural responsibility for IBM's TMN product components. **Juha Seppä** is a Project Manager at Nokia in Tampere. He has several years background in practical TMN implementations for GSM networks.

All of the authors have been active participants in the NMF High-Level API Working Group which is developing these standards.

# 15

# SNMP and TL-1: Simply integrating management of legacy systems?

*A. Clemm*
*Ericsson Fiber Access*
*1525 O'Brien Drive, Menlo Park, CA 94025-1451*
*Tel. (415) 463-6722, fax (415) 463-6755*
*E-mail: aclemm@eur.ericsson.se*

### Abstract

One of the challenges associated with open telecommunications network management is the integration of legacy systems. SNMP is as open management paradigm of growing importance in telecommunications. This raises the question whether it could also form an appropriate basis for an integrated management of TL-1 based devices. This paper takes a closer look at this possibility. TL-1 and SNMP are compared. Technical issues involved in managing TL-1 based systems using SNMP are investigated. Because of the differences in power and intended operating paradigms, it is argued that the suitability of an according management integration depends very much on the application.

### Keywords

SNMP, TL-1, TMN, legacy systems, management protocols, management gateways, integrated network management, telecommunications network management

## 1 MOTIVATION

Open management architectures and standard management protocols have reached maturity and gained widespread acceptance. This is no longer true only for data but increasingly also for telecommunications. The main driving force for this is the desire to achieve an integrated network management, both in terms of components to be managed and of functionality. With the Telecommunications Management Network (TMN) framework in place and commercial tools available, an increasing number of Q3 interface implementations can be expected in the near future. SNMP (Simple Network Management Protocol) management, although not originally designed for use in the telecommunications arena, plays an important role as well.

However, for any of the new standard compliant management systems, it must be taken into consideration that they do not exist in a vacuum but that they belong to a history of network management which cannot be ignored. Two main challenges are associated with this:

- The new, integrated management systems must offer at least the same level of functionality that users were provided by legacy management systems. Reducing the functionality that they previously had is simply not acceptable from the users' point of view. Such legacy management systems tend to be highly specialized and very well geared for a particular management function or for the management of a particular product or technology. Extracting this functionality in an integrated, generic way that allows reuse of the existing code base rather than having to recreate it from scratch is in general far from trivial.
- The need for integration does not pertain only to new systems. Rather, the value of an integrated management depends just as much on the ability to include legacy systems that coexist with systems based on open standards. Otherwise, in the users' perception a new, "integrated" management system means just another system in addition to those that they already have to operate and maintain, adding to the complexity of their network operations instead of doing just the opposite.

In this paper, we deal with the second challenge. In particular, we explore what an integration of (legacy) TL-1 based systems into (open) SNMP management environments entails.

Despite (or rather, because of) its primitive capabilities, SNMP (Simple Network Management Protocol) has started playing an important role also in the telecommunications area. This is underlined by work of industry consortia, such as the ATM Forum and its SNMP MIB definitions for management of ATM devices, and by industry projects [2]. TL-1 (Transaction Language 1), on the other hand, is a Man Machine Language (MML) introduced by Bellcore. These systems have very high commercial relevance in the telecommunications area as they are widely deployed throughout telecommunications networks especially in North America.

SNMP is praised for its simplicity, ease of agent implementation, and flexibility. Most importantly, rich and relatively inexpensive off-the-shelf management tools and implementation environments are available. As a consequence, there is a growing trend to consider SNMP as a pragmatic solution to achieve management integration also of systems until now managed using TL-1. In this paper, we take a closer look at this conception and point out what technical challenges are involved. It is shown that such an integration, while possible, does not come without tradeoffs. Depending on the application, it may not be as simple as it may perhaps initially appear. Consequently, "open" does not in every case mean "preferable". We focus on the original SNMP as opposed to SNMPv2 [7-9], as it is currently the much wider accepted of the two. Most of our results apply however also to SNMPv2, as can be inferred by the reader.

In section 2, TL-1 and SNMP based management are compared and contrasted on the basis of their communication, information, organisation, and function models. This forms the basis for understanding what commonalities can be exploited and what differences have to be overcome to integrate TL-1 and SNMP based management. Section 3 discusses different possibilities for integration of TL-1 and SNMP based management. Section 4 explores technical issues involved in the realization of one of the alternatives, namely the introduction of a management gateway between SNMP managers and TL-1 agents which holds the promise of using off-the-shelf SNMP management platforms for managing TL-1 systems. Drawbacks of trading off the relative openness of SNMP against the legacy management framework of TL-1 are pointed out, as TL-1 is the more powerful of the two. From this, it is inferred which kinds of applications are good candidates for SNMP based management integration, and for which kinds drawbacks may offset the potential advantages. Conclusions are offered in section 5.

## 2 TL-1 AND SNMP: A COMPARISON

TL-1 and SNMP differ entirely in their communication, information, function, and to some extent even in their organisation models. They are compared in the following section. Because more readers will be familiar with SNMP than with TL-1, TL-1 is treated in a little more detail than SNMP.

## 2.1  TL-1

TL-1 is a MML which was standardized by Bellcore [11-13]. Communication between managers and network elements (i.e. agents) takes place by exchange of operations application messages, essentially simple text strings that follow a particular (TL-1) format. The management services are defined by contracts which specify the messages to be exchanged. A contract may specify an (input) command and an (output) response constituting a management operation, or an autonomous message.

A large number of different contracts exists, in particular for management operations. They can roughly be classified with respect to the kind of function they provide, and with respect to what aspect of the network element respectively what Managed Object (MO) they refer to. As they are rather closely related to the particular function and often the type of MO they refer to, a clear separation of information, communication, and function models is difficult. Whenever there is a requirement for a new operation, it is simple and legal to add a new (proprietary) transaction. This has resulted in many proprietary TL-1 dialects. Essentially, they all have in common services to enter, edit, delete, and retrieve various kinds of management information as well as to invoke special functions such as the execution of tests.

Contracts refer to object entities which are addressed using hierarchical "access identifiers", or AIDs. An AID reflects at the same time the type of object, e.g. a linecard, and the instance itself. The hierarchies are based on containment, closely reflecting physical reality, and are addressed that way. The following is an example for an AID in an Ericsson LOC (Loop Optical Carrier) system realizing FITL (Fiber In The Loop) technology: DFM-2-ONU-15-LC-5. It denotes the linecard in slot 5 of Optical Network Unit 15 on the Passive Optical Network (PON) off Distribution Fiber Module 2 of a particular Host Digital Terminal. An AID can be used to refer not only to one particular but also to groups of entities, subject to the individual contract definitions, which makes scoping of management operations across several MOs possible (e.g. DFM-2-ONU-15-LC would refer to all linecards in that particular ONU). Also associated with object entities is usually a service state, indicating their ability to perform their functions. A state machine ensures that no operations are enacted when it is illegal to do so.

The following examples of an input command and an output response illustrate what typical TL-1 messages look like.

```
RTRV-EQPT:SYS001:HDT:20;
```

This is a typical command to retrieve parameters from a particular device. RTRV-EQPT ("retrieve equipment") constitutes the command code, with RTRV being the command verb which specifies the action to be performed, and EQPT being a command modifier expressing the constraint that the action is performed against a piece of equipment (versus for instance a line). SYS001:HDT:20 constitute the staging parameters. SYS001 is the target identifier. It identifies the network element to receive and process the command and can be thought of as the agent identifier. HDT is the AID. It identifies the entity within the network element against which the retrieve is to be performed, in this case the "host digital terminal", a piece of equipment used in FITL systems. 20 is the correlation tag ("ctag"); it is sent back in the response so that the manager can correlate responses it receives against the respective commands. TL-1 commands may in addition (not shown here) contain a general block following the correlation tag with data parameters. The data parameters allowed differ for different command codes and AIDs.

The following is a possible response for the above TL-1 command:

```
SYS001 96-4-15 11:30:45
M   20 COMPLD
    "HDT::GENERIC=rel2.0,TMG=SLV1&SLV5&INT,DIAGINTVL=4,
DIAGALARM=2,STATINTVL=48,SAMCHAN=LOCAL,LBO=266:IS-NR,SX"
```

The first line contains the identifier of the system issuing the response, followed by a date and time stamp. The M in the following line indicates the mcode which specifies the type of output message. In this case, it is the response to an input command. The value of the ctag of the input command follows, 20 in this case. COMPLD indicates the command has successfully completed. The following lines contain the response data block, i.e. parameter values of the HDT that were requested. IS-NR refers to the HDT's service state: in service, normal. An error response would have contained e.g. a DENY instead of a COMPLD, followed by an error code and error text.

Autonomous messages look similar to response messages, indicating the type of autonomous message and level of severity. Instead of a ctag, they contain an automatic message tag (atag) which is automatically incremented for each new autonomous message, except for messages indicating the clearance of an alarm condition which contain the atag of the message in which the alarm was first reported.

## 2.2 SNMP

SNMP is a management protocol originally designed for the management of TCP/IP capable data communications network devices. The guiding principle for its design was simplicity, to ensure agent implementations could easily be made available. The management information base consists of objects (not in the object oriented sense) representing variables. Both object types and object instances ("SNMP variables") are viewed as parts of the Internet registration tree, with instances forming the leaves. Columnar and non-columnar objects are distinguished, depending on whether or not they are conceptually considered part of a table entry. The management protocol itself offers a set of operations to retrieve and set values of particular object instances, to traverse tables, and to send simple traps, i.e unsolicited events from the manager to the agent. The SNMP management framework is in general considered polling based, trap directed. The reader unfamiliar with SNMP concepts is asked to refer to the literature [5,6].

Similar to TL-1, new SNMP object types are easy to define and add, leading to a wide variety of proprietary MIBs (Management Information Bases), i.e. collections of object type definitions. However, two aspects make the dealing with proprietary aspects in SNMP easier than in TL-1: the clear separation of the management information from the other aspects of the management framework which are commonly shared, and the availability of standardized MIBs for many purposes.

## 2.3 Differences between TL-1 and SNMP

The following tables summarize the most important differences between TL-1 and SNMP.

**Table 1** Communication model

| SNMP | TL-1 |
|---|---|
| Small set of general, basic communication services (get, set, get-next, trap), parametrized as to to which SNMP object (i.e. variable) to apply | Large set of specialized communication services, parametrized to a lesser extent (different services for semantically analogous operations on different types of objects) |
| Unreliable as based on UDP | Unreliable, however, message sequencing aids in detecting lost messages |
| No scoping | Scoping to some extent |
| Target for read operations not necessarily required ("Get-next" for MIB-discovery) | Target required |
| ASN.1 encoding | ASCII strings |
| Minimal security (community name as unencrypted password) | Very basic security (log on and off) |
| Connectionless, datagram oriented | Session oriented |
| Polling based, trap directed | Polling and event based |

**Table 2** Information model

| SNMP | TL-1 |
|---|---|
| Simple information model using tables and variables plus grouping mechanism | No explicit information model, implicit information model underlying addressing scheme (AIDs), operations and parameters but no clean separation of information, communication, and function models |
| Clear separation of information from communication model | |
| No state model | Inherent state model |
| Objects abstracted away from physical world | Objects reflect physical world directly |

**Table 3** Function model

| SNMP | TL-1 |
|---|---|
| Design for monitoring of devices in TCP/IP based data networks | Design for operations, administration, maintenance, and provisioning in telecommunication networks |
| No dedicated function model | Extensive function model, e.g. testing, protection switching, alarm status retrieval, etc. |
| No dedicated state and lifecycle management | Extensive state and lifecycle management (e.g. inventory of equipment before it is installed) |

**Table 4** Organisation model

| SNMP | TL-1 |
|------|------|
| Manager-agent paradigm | Manager-agent paradigm; variation possible with a mediation device involved in certain management functions |

To summarize, TL-1 is actually the richer and more powerful of the two, while SNMP is better structured and typically the easier to understand. The differences in the goals of their design clearly show: SNMP for the monitoring of devices in TCP/IP based networks, TL-1 for the OAM&P (Operations, Administration, Maintenance, and Provisioning ) of very specialized telecommunications systems. Provisioning for instance is a concept virtually unknown in the management of data networks. In the following sections, the issues involved in applying SNMP on this unusual terrain are inspected more closely.

## 3 ALTERNATIVES FOR INTEGRATION OF SNMP AND TL-1 BASED MANAGEMENT

There are different possiblities for the integration of SNMP and TL-1 based management:

- Multi-protocol (or rather, multi management paradigm) management agents, i.e. addition of SNMP capabilities to TL-1. This alternative is not flexible enough as basis for management integration, as changing existing agent implementations in the field may be impossible.
- Multi-protocol (or rather, multi management paradigm) managers. The actual management protocol that is used is hidden inside a layer within the management system. This layer offers an internal unified model which is as homogeneous as possible to management applications which need not be aware of differences in management protocols. This alternative mandates extensions to existing management platforms, which arguably some are equipped to accomodate (e.g. PMI (Portable Management Interface) adapters in SunSoft's Solstice [10]).
- Management gateway between SNMP manager and TL-1 agent (figure 1). Different possibilities exist for where to place the system boundaries. The gateway could be realized in the network element itself (as indicated in the figure) or outside, e.g. in a mediation device or even in the management system itself, making this alternative similar to the previous one.

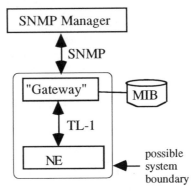

**Figure 1** Management gateway as integration means.

Because the last alternative illustrates best the issues involved in the integration of SNMP and TL-1 management, it is the one which shall be investigated further. It also holds the promise to be able to use standard SNMP management platforms without further modifications as basis for integration and therefore deserves a closer look. Gateway approaches have been successful and are well accepted to achieve interoperability in the management arena, e.g. for interoperability between CMIP and SNMP [1, 3, 4]. On a broader note, the considerations in building a gateway contrast also the SNMP and TL-1 based management paradigms rather nicely. We focus on conceptual issues. The actual design of a gateway is beyond the scope of this paper.

From section 2 follows that a gateway means that TL-1 agents will be managed using a less powerful paradigm than the one used before. Since this may involve not being able to use the full power these agents offer, at first sight this may appear like something that should be avoided. However, there are several advantages to be gained which makes this alternative well worth exploring. Most importantly, management which is integrated and homogeneous is preferable in many scenarios over one which is not, even if a subset of the devices to be managed could be managed more efficiently differently. This is especially true for applications not taking advantage of the full capabilities TL-1 might offer. SNMP tools available today hold the promise for rapid development cycles if the devices to be managed can be managed using SNMP.

## 4 BUILDING AN SNMP-TL-1-MANAGEMENT GATEWAY

A management gateway introduces a management hierarchy. It fulfills management operation requests from the upper-level management system in terms of management operations enacted upon lower-level management systems. As SNMP is less powerful than TL-1, in general the mapping between them is fairly straightforward, as the gateway does not have to fill a gap in functional power between the two. This means that besides keeping track of the different addressing schemes, the gateway will not have to keep much state which keeps it relatively simple.

As TL-1 does not really separate communication, information, and function models, it has to be investigated what transactions are provided by TL-1 and determined how the same information and operations can be conveyed using SNMP. We do this by drawing on examples from the set of TL-1 transactions used for the management of Ericsson's LOC systems. In subsection 4.1, we describe how communication and function model can be mapped. In 4.2, we look more closely at the mapping of the information model. This provides the basis for understanding the following discussion in 4.3 as to where applications are likely to benefit from the integration of TL-1 based systems in SNMP environments and where its practical limitations are.

Not discussed are aspects dealing with the use of a gateway to also implement a management information hierarchy in which the most recent information is collected at the leaves and propagated to some degree into the higher levels. Implementing caching strategies to keep static information in the gateway is desirable from a performance standpoint but adds complexity because state and replicated management information have to be maintained. In the context of this paper, we consider this a design issue which is beyond its scope.

## 4.1 Mapping TL-1 contracts and SNMP operations

As explained in section 2, TL-1 messages start with a command code, followed by lists of various parameters. The command code consists of an (action) verb and one or two modifiers that state the kind and instance of "MO" the operation is to be applied to. Accordingly, TL-1 transactions can be grouped along the type of actions (that are being applied to different entities). Different modifiers will be reflected in what parts of the MIB the respective SNMP operation(s) refer to.

## Memory administration contracts

Memory administration contracts allow the configuration, i.e. the changing, deleting, adding and querying of network element parameters.

- *Enter*: Corresponding to a "create", these contracts are not directly mappable. However, they can be simulated if the "MO type" being entered is modeled as an SNMP table, using set operation(s) to implicitly create a new table entry (compare with [8]).
- *Retrieve*: Correspond and can be mapped to SNMP get operations. Scoping is not available in SNMP. To imitate scoped retrieves à la TL-1, multiple gets or sequential get-next operations have to be issued. The use of SNMP here is less efficient.
- *Edit*: In general, the changing of parameters can be mapped to set operation(s).
- *Delete*: Analogous to the case of *Enter,* these contracts are not directly mappable to SNMP. However, they can be simulated if the type of the MO to be deleted is modeled as an SNMP table, using a set operation to delete the table entry (compare with [8]).

## Network maintenance contracts

Network maintenance contracts provide the means for surveillance, alarming, recovery, and state control. They are fairly hard to imitate in SNMP as it is they which define to a large extent the function model for which SNMP has no counterpart. Where TL-1 contracts define "actions", these have to be simulated by means of SNMP variables whose setting to certain values implies the side effect that a certain operation is executed. The SNMP get response returns the success indication. If the action has failed, an error is reported. In cases where more detailed information has to be provided, a dedicated SNMP variable can be introduced whose value is to be retrieved separately. Some examples for network maintenance contracts follow.

- *Message reporting control*: The contracts allow for the enabling and inhibiting of message reporting. This constitutes "actions" for which there are no corresponding SNMP counterpart operations. They can be simulated by introduction of an SNMP variable whose value implies whether message reporting is enabled or inhibited. To differentiate between different logged on users, message reporting control requires the use of an actual table instead of a simple variable.
- *Alarm and event status retrieval*: TL-1-based systems typically maintain status information of any outstanding alarms and events, which can be retrieved through these contracts. In SNMP, this has to be simulated by introducing a table to maintain alarm information. The table can be retrieved using get and get-next operations. TL-1 parameters identifying the entity an alarm or event pertains to have to be translated into SNMP MIB counterparts. The table needs to include the alarmed entity as part of the index, otherwise random access to alarm information pertaining to a particular piece of equipment is not possible. Complete table traversal may still occasionally be necessary as random access across several dimensions is not possible in SNMP tables.
- *Performance monitoring*: TL-1 allows retrieval of certain performance related parameters. This translates in a straightforward manner, mapping to SNMP get operations on MIB variables introduced for that purpose.
- *Testing*: These contracts allow the remote execution of a wide variety of tests on equipment, e.g. on leakage, capacitance, or the presence of dial tone. Notice that the execution of some tests may span longer time intervals during which service supported by the system may be affected, which requires also the means for aborting tests in progress. There is no counterpart for any of this in SNMP, as this resembles actions. The corresponding actions can be simulated by introducing dedicated SNMP variables whose setting to a certain value implies that a test is to be executed. Action parameters

will have to be placed into additional SNMP variables. The response to the test result can be placed into yet additional MIB variables which will have to be retrieved separately. A special value can be used to indicate that a test has not been finished. The variable(s) will have to be reset after a test result has been retrieved, or have to be reset as a side effect of the setting of the variable resembling the action. Again, SNMP concepts are being stretched a bit.

Many other network maintenance contracts exist that each require their own mapping, for instance for protection switching control, loopback control, alarm error tracing, or the backing up of network element state information.

### Security management
Unlike TL-1, SNMP does not require logging on and off managed devices. A management gateway can just log on as a registered user before the first SNMP command is issued.

### Autonomous messages
Facilities for transmitting autonomous messages are much richer in TL-1 than in SNMP. In general, for autonomous messages to be reported, an SNMP enterprise specific trap will have to be used, with some information to be retrieved manually from the SNMP agent. For this purpose, an "alarm MIB" with variables to keep information on all standing alarms has to be introduced. Events which are reported only once and are not subject to clearing have to be investigated separately.

## 4.2  Management information

TL-1 messages refer to an implied model of the network, as stated in the modifiers of the command code and the AID used to address the particular "MO instance" that the TL-1 message refers to. The AID is not just an address but along with the command code indicates the type of MO that is being referred to. The underlying TL-1 model has to be translated into an explicit SNMP model. Figure 2 depicts an excerpt of the model which is implied by the AIDs used in the TL-1 transactions for the management of Ericsson's LOC system. Indentation depicts hierarchical dependency.

> *HDT Host Digital Terminal*
> > *TGM Timing Generator Module*
> > *DFM Distribution Fiber Module*
> > > *ONU Optical Network Unit*
> > > > *PAD Packet Assembler Disassembler*
> > > > *PM Power Module*
> > > > *LC Line Card*
> > > > *BP Binding Post*
> > > > *CPT Contact Point*
> > *OFT Office Feeder Termination*
> > > *DG Digroup*
> > > *CH Channel*

**Figure 2** Excerpts of LOC object model as implied by TL-1 transactions.

A specification of a methodology to be used to derive the SNMP model from the TL-1 transactions is beyond the scope of this paper. Here, only a few aspects are mentioned: different types of objects have to be translated into different SNMP groups. The individual parameters of transactions indicate which variables have to be provided as part of the objects. Objects that may occur more than once in a LOC system have to be translated into tables - in this example anything at a lower hierarchy level. TL-1 does not really make use of complex data types so the mapping of data types is straighforward. One restriction exists in the case of enumerated types of which there are several, the most prominent example being the system state. Here, values can be transferred as graphic strings which will have to be interpreted in the management system.

## 4.3 Benefits and limits of TL-1 management integration based on SNMP

Management applications communicating with TL-1 devices through the gateway may constrain themselves to the same applications which are typical for the management of TCP/IP based networks: occasional polling of devices to determine operational state, processing of traps for that same purpose, plotting of performance-related attributes over time, remote inspection of system configurations, etc. As was shown, the mappings for these communications exchanges are very straightforward. It is reasonable to expect the gateway approach to work very well here and achieve management integration with minimum effort, both development and computationwise.

However, in general the desire for management integration in a telecommunications environment does not stop here. Rather, in addition to these applications, all the other kinds of applications that previously TL-1 was applied to are expected to carry over into the SNMP environment. After all, the same level of functionality as before still has to be supported. In many cases, integration of SNMP and TL-1 based management does not only have to be concerned with the mere mapping of SNMP onto TL-1 at the protocol level. It also needs to be concerned at an application level with the paradigms according to which TL-1 is used. Where SNMP lacks the facilities to support these paradigms, the now SNMP-based applications need to make up for these deficiencies by adding something on top. It is here where the technical challenges come in and where the limits of this approach lie. They extend beyond the mapping of TL-1 network maintenance contracts and autonomous messages. The challenges include the following (see also figure 3):

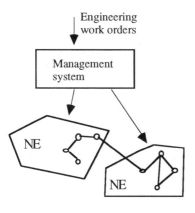

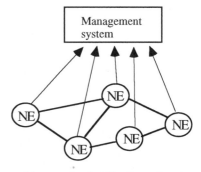

(a) few complex NEs, flow of static information from top to bottom

(b) many simple NEs, flow of static information from bottom to top

**Figure 3** Typical management scenarios: (a) TL-1, (b) SNMP.

- Control of management information. In SNMP, it is typically the agent who owns and controls all management information. The manager discovers new agents and new pieces of equipment as it polls the network. In TL-1 based management, it is common for the manager to be in control of the static management information and retrieve only dynamic state information and alarms from the network element. An operator inventories equipment of the network, the information of which is obtained for instance from engineering work orders. The associated management information is communicated to the agent. Only from this point in time is it possible to actually monitor the equipment. Differences between the management view and the real world can be resolved through discrepancy reports or sophisticated reconciliation procedures. This reflects that typically telecom operators have much more complete and central control over the network than their datacommunications counterparts. The case where network elements simply pop up or disappear, common in data communications networks where users simply plug, unplug, and move their machines, simply does not occur in a telecom environment.

- Complexity and distribution of managed systems. TL-1 based systems are typically more complex and may be more geographically distributed than their usual SNMP counterparts. This results in a higher number of tables as opposed to non-columnar objects, making management information more cumbersome to access and manipulate.

  Also, graphical topology display is no longer automatically obtained by using a management platform, as much of the topology of interest is defined not through the different SNMP agents that are out there but rather within the MIB tables of a single agent.

## 5 CONCLUSIONS

Will SNMP based systems form the next generation of legacy systems in the telecommunications industry, soon obsoleting efforts to integrate TL-1 legacy systems into SNMP management? Quite possible, especially considering the limitations inherent in SNMP and the advent of TMN. Why, then, should integration of SNMP and TL-1 based management be considered? The answer to this is manyfold. First, comparing the two provides insight about the strengths and weaknesses of each, so there is an educational benefit. Second, as of today, SNMP is undoubtedly the most prevalent approach to open management, with other more powerful open management technologies still to be proven. Third, because of its importance, mappings from other technologies to and from SNMP exist (e.g. OSI management / CMIP and CORBA). Accordingly, management of TL-1 devices through SNMP automatically opens them also up to management by other paradigms through use of multi-stage gateways - concerns about performance aside.

The mapping of SNMP to TL-1 is straightforward in areas where the traditional strengths of SNMP lie. It is here where an integration of TL-1 based systems into SNMP environments is sound and well-suited. Beyond that, integration of TL-1 based systems into SNMP environments is not trivial. The attempt to carry over TL-1 concepts into SNMP reveals in many cases SNMP's limitations in this terrain. SNMP concepts have to be stretched to be able to achieve if the full management functionality provided by the richer TL-1. Certain features that TL-1 offers with respect to better management efficiency cannot be used, like scoping. Network maintenance contracts can often only be simulated through MIB extensions which make for very complex MIB semantics. Autonomous messages and alarming may require a two-phase concept. However, a mapping is certainly possible.

In general, the best use of SNMP is when it is constrained to those applications for which it was designed, namely polling based monitoring of network equipment. This is by comparison relatively simple to support. It is also where SNMP based integration of TL-1 based systems provides the most value. Use of SNMP for applications to manage TL-1 capable devices beyond that is certainly possible although it may at times be awkward in the described telecom environment and offset the desired benefits. Here, the SNMP management applications are no longer off-the-shelf and take more effort to develop than under the legacy paradigm. Therefore, an integration of TL-1 based systems into SNMP based management has to be carefully considered on an application by application basis. Nevertheless, management gateways, along with adaptation layers, are important and in general very suitable techniques to cope with the integrated management challenge posed by legacy systems.

## ACKNOWLEDGEMENTS

For valuable discussions and comments on earlier versions of this paper which undoubtedly helped improve it, I want to thank my colleagues at Ericsson Fiber Access, in particular David Hood, Steve Cutcomb, Alice Bedoyan, and Ron Reedy.

## REFERENCES

[1] S. Abeck, A. Clemm, U. Hollberg. Simply Open Network Management - An Approach for the Integration of SNMP into OSI Management. *3rd IFIP/IEEE Int. Symposium on Integrated Network Management,* San Francisco, CA, April 1993.

[2] M. Ho, A. Farrell, J. De Wiele. SNMP Management of Telecommunications Carrier Networks. *IFIP/IEEE Network Operations and Management Symposium,* Kyoto, Japan, April 1996.

[3] K. McCarthy, G. Pavlou, S. Bhatti, J. Neuman de Souza. Exploiting the power of OSI management for the control of SNMP-capable resources using generic application level gateways. *4th IFIP/IEEE Int. Symposium on Integrated Network Management,* Santa Barbara, CA, May 1995.

[4] Network Management Forum CS341. *CMIP/SNMP Interworking.* Includes Forum 026: Translation of Internet MIBs to ISO/CCITT GDMO MIBs, Issue 1.0, October 1993, and Forum 028: ISO/CCITT to Internet Management Proxy, Issue 1.0, October 1993.

[5] RFC 1155. *Structure and Identification of Management Information for TCP/IP-Based Internets.* Internet Engineering Task Force, May 1990.

[6] RFC 1157. *A Simple Network Management Protocol (SNMP).* Internet Engineering Task Force, May 1990.

[7] RFC 1902. *Structure of Management Information for Version 2 of the Simple Network Management Protocol (SNMPv2).* Internet Engineering Task Force, January 1996.

[8] RFC 1903. *Textual Conventions for Version 2 of the Simple Network Management Protocol (SNMPv2).* Internet Engineering Task Force, January 1996.

[9] RFC 1905. *Protocol Operations for Version 2 of the Simple Network Management Protocol (SNMPv2).* Internet Engineering Task Force, January 1996.

[10] *Enterprise Manager 1.1.1 Video Workbook.* Order No. 802-3197-02, SunSoft, 1995.

[11] TR-TSY-000831. *Operations technology generic requirements (OTGR): Operations application messages - Language for operations application messages.* Issue 2, Bellcore, February 1988.

[12] TR-TSY-000833. *Operations technology generic requirements (OTGR): Operations application messages - Network maintenance: Network element and transport surveillance messages.* Bellcore, Issue 5 Revision 1, April 1993.

[13] TR-1093. *Generic State Requirements for Network Elements.* Issue1, Bellcore, September 1993.

## BIOGRAPHY

Alexander Clemm has been actively involved in network management research and development since 1990. Currently he is responsible for the systems engineering of Network Management Systems at Ericsson Fiber Access in Menlo Park, California. He has published about a dozen papers and holds a Ph.D. in computer science from the University of Munich and an M.S. from Stanford University.

## PART FIVE

# Management Paradigms
## Chair: Bruno Vianna, Telebras

# 16

# Determining the Availability of Distributed Applications

**Gabi Dreo Rodosek**
**Thomas Kaiser**
Leibniz Supercomputing Center
Barer Str. 21, 80333 Munich, Germany
Email: {dreo, kaiser}@lrz.de

## Abstract

Distributed applications can be seen as complex pieces of software which are distributed across various heterogeneous end systems in a network. Mostly, they rely on the provision of other applications as well. Adequate methods for testing the availability of distributed applications do not exist yet. Availability must be determined based on the availability of the involved end systems and network nodes. In view of this, we propose (i) a service graph for the description of functional dependencies, (ii) extend it to a parameterized availability graph to describe the involved components, instantiate the graph, and (iii) give calculation rules for determining the availability of applications. Although most dependencies are described during the design phase, some of them can be recognized only during operation. To deal with this, user trouble reports about unavailable services are used to refine the availability graph and improve the availability calculations.

**Keywords:** Distributed Applications, Service graph, Availability graph

## 1 INTRODUCTION

During the past decade worldwide network and system capabilities have rapidly advanced to meet the challenge of the information age imposing high requirements on network and systems management ([HeAb 94], [Slom 94]). The pace has been further fueled by customer demand for a variety of innovative services that require the support of high-quality reliable networks. Availability of provided services, as one of the most important Quality of Service (QoS) parameters, is another customer demand. Until recently, availability is mostly referred to the availability of particular links, network nodes or end systems. However, when talking about the availability of applications, it is necessary to deal with complex dependency relations between several components which are necessary for the provision of an application. Obviously, this increases the complexity substantially. Since users perceive degradations in terms of used services, it becomes crucial to solve this topic. To clarify the terminology, a service provider uses distributed applications (e.g., MHS) to provide services (e.g., email) to users (customers) with certain QoS parameters, as agreed in service-level agreements (SLAs).*

In spite of the relevance of this topic, an adequate solution – being applicable also in a production environment – has not yet been found. The difficulty is to recognize the dependencies, and thus, to identify the involved components for the provision of an application. In general, it is necessary to describe (i) functional dependencies between services, and (ii) environmental dependencies (i.e., how services are realized in a concrete environment). With a service description, it is possible to identify the constituent functional building blocks, and thus recognize functional dependencies. Existing approaches (e.g., [ISO 10746], [ITU-T Q.1201], [TINA-C 95], [ITU-T M.3010], [Gosc 91], [KPM 96], [Dreo 95]) refer to a service description with respect to certain requirements (e.g., for trading, for correlation of trouble tickets). Dependencies between resources are for example modeled using relationships ([Kaet 96]) to determine the root cause of a fault. However, availability aspects have not yet been tackled.

The paper proceeds as follows: Section 2 illustrates the complexity of the problem area, especially functional and environmental dependencies, and clarifies the term availability from the user and the service provider view. Section 3 describes the proposed methodology, including three steps: firstly, we propose a service graph for the description of functional dependencies and extend it to a parameterized availability graph to describe environmental dependencies as well as give calculation rules for determining the availability. Additionally, the instantiation of the availability graph is discussed. Secondly, we answer the question how to test the availability of end systems and network nodes hereby referring to the testing capabilities of management tools. Thirdly, in order to cope with the dependencies which are not known during the design phase, we discuss some enhancements of the proposed approach to calculate the availability as precise as possible. Implementational aspects, namely the realization of the proposed methodology in a production environment, are described in Section 4. Finally, Section 5 gives some concluding remarks and outlines further work.

## 2   PROBLEM DESCRIPTION

The complexity of determining the availability of distributed applications results from various points such as functional dependencies between applications, their distributed realization on various end systems (e.g., servers) and the used communication infrastructure. To give an idea about the complexity, we describe a relatively simple example of a WWW proxy depicted in Figure 1.

The example starts with a user requesting a WWW page. The WWW client requests first the IP address of the proxy server for which a request is sent to the DNS (domain name service) and served by a corresponding DNS server of the provider (steps 1-4). Afterwards, the WWW client sends its request for a WWW page to the proxy server (step 5). The proxy server itself requires the IP address of the remote WWW server and therefore makes a request to the DNS server (steps 6-9). The WWW proxy either retrieves

---

*The terms service and distributed application are used interchangeably.

the page from the remote WWW server (steps 10-11) or provides the WWW page from the cache (steps 10'-11'). If all steps are successful, the proxy server sends the reply to the WWW client (step 12). This example simplifies the real situation. For example, we assume that there is only one configured DNS server in the resolver part. Besides, we have not discussed the functionality of the WWW client (e.g., the possibility of internal caching). Additionally, we assume (IP) connectivity between all involved devices, that the WWW proxy server has enough resources (e.g., memory, CPU), and all required processes are running. Finally, we have demonstrated only some functional dependencies between applications. For example, that a WWW application requires only the provision of a DNS. We have not assumed that the WWW application can be distributed over a distributed file system. As illustrated, the concrete environmental dependencies are very complex and cannot all be foreseen.

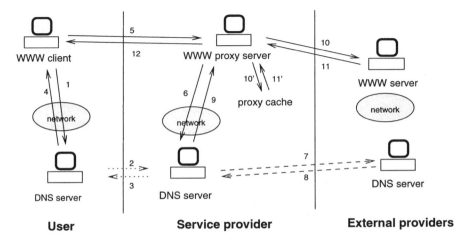

**Figure 1** Example: WWW proxy

Since the provision of services in today's distributed environment is a complex task, many applications rely on a hierarchy of underlying services. A distributed application (e.g., remote printing) depends on client, server and gateway processes, which themselves depend on system software and hardware. Besides, networking devices and communication links must be properly configured and in operating state. For example, the usability of a WWW application depends on whether the DNS application is available, whether the DNS server can resolve the stated name, whether the WWW server workstation is in operation and the network connectivity provided. In general, the availability of a distributed application depends on the availability of (i) other applications, such as DNS, the availability of (ii) end systems (e.g., workstations) where the software is running, and availability of the (iii) connecting communication infrastructure, the network.

As depicted in Figure 1, it is necessary to distinguish between different views of availability. A service is available for a user if he can use it at a certain time with the agreed QoS parameters at a predefined access point. It is the obligation of the service provider to assure the correct provision of services at this access point. We define the interface between a user and a service provider as a service access point (SAP) where the usage and provision of services take place. If the provision of a service relies on the provision of services from an external service provider, we define this interface between providers as SAPs as well. The QoS parameters at the SAP are handled within SLAs, too.

To illustrate the various availability views, let us consider the Leibniz Supercomputing Center (LRZ) as a service provider of network connectivity and computing power for the Munich Universities. The Department of Computer Science of the University uses the provided IP connectivity from the LRZ to realize a distributed file system. In our example, the connecting router interfaces are the SAPs. If one of these interfaces is down, the availability of services such as NFS mounts within the department is 0% for the time of the outage of the interface. However, from the service provider's viewpoint, only one of 250 interfaces is down. On the other side, the LRZ itself uses services from other service providers. In such a case, the LRZ acts as a user.

Besides, the term availability of distributed applications itself is imprecise. Is an application available, for example, if the end systems and the network nodes involved in its provision are available or if only n-1 of n building blocks of an application are available or if there is no user trouble report?

## 3   PROPOSED APPROACH

Beside the specification of an appropriate testing interval, the question is how to test whether a complex distributed application, like DNS, is available at the SAP. A straightforward method could be to test the usage of a service also by a service provider with the maximal possible request rate for a service. This is of course unrealistic. Another unrealistic approach is to document user trouble reports about unavailabilities of services, and perform availability calculations solely on these reports.

Availability is closely related with faults. Because a distributed application is realized with components (network nodes, end systems), a straightforward definition is that an application is available if there are no faults in a distributed system. Faults in our context are breakdowns of devices or wrong configurations of software which could have an influence on the availability. From the user's view, performance degradations may also be seen as faults, which we assume are handled with SLAs.

The proposed methodology consists of the following three steps.

### 3.1   Identifying the involved components for the provision of distributed applications

In order to recognize the involved components for the provision of an application, we need to describe (i) generic functional dependencies between applications, and (ii) specific environmental dependencies (i.e., the realization of a distributed application in a concrete environment).

We propose to describe the functional dependencies between services in terms of a layered *service graph* where the nodes represent the services and the edges represent the relationship *is_functional_dependent*, as depicted in Figure 2. For example, the provision of the WWW service depends on the provision of the IP connectivity service. In terms of a service hierarchy, we may distinguish between application services (e.g., email, backup) as well as basic and communication services, which are used for the provision of application services. In order to determine the availability of distributed applications, it is

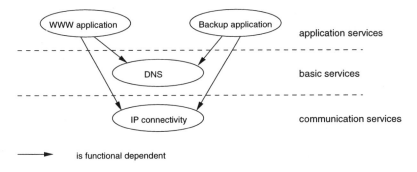

**Figure 2** Layered service graphs

necessary to (i) recognize the involved components, and (ii) to determine the availability of these components. For this purpose, we use the service graph as a basis to generate a *parameterized availability graph* (Figure 3). Some ideas for the design of the availability graph have been adopted from [JeVa 87].

The availability graph serves two purposes: firstly, it describes dependencies in the system, and secondly, it gives calculation rules for determining the availability of applications.

The nodes of the availability graph represent the parameterized service descriptions, such as IP connectivity(client, server), whereas the edges are used for calculation purposes. As depicted in Figure 3, we distinguish between AND and OR edges.

The skeleton of the availability graph (refer to Figure 3) represents the functional dependencies between services (e.g., the WWW service requests the provision of the DNS). For example, the WWW service is available if the DNS service AND the IP connectivity service are available. Referring to Figure 3, the service WWW(client, server, file) is available if WWW_client(host) AND DNS(client, server, request) AND WWW_server(host) are available. In other words, the application is not available if there is a fault(WWW_client(host)) OR a fault(DNS(client, server, request)) OR a fault(WWW_server(host)).

From this skeleton, we further refine the nodes. The refinement of the DNS node is shown in Figure 3. The DNS service is available if either one of the two name servers is available (an example of the OR operation).

Parameters are used to represent functional and environmental (i.e., implementational) aspects. Functional aspects, like WWW(client, server, file), refer to the functionality of a service (e.g., the functionality of the WWW service which is to transfer files from server to client). Environmental aspects are used to describe, for example, that a WWW server may be realized over a distributed file system (AFS) or local disk. In such a case, we extend the node WWW server with the graph on the lowest layer. For example, the WWW server runs on a host with minimum requirements on cpu power and disk space, which are additional parameters of the node host(process, cpu, disk). This means that the WWW server on the host(process, cpu, disk) is available if the *process* is running, the currently available cpu is over the limit *cpu*, and there is at least *disk* space available.

To summarize, during the design phase the availability graph is generated from the service graph and extended/refined to deal with known environmental dependencies. In case new dependencies or "hidden" side-effects are encountered during operation, the availability graph needs to be extended. An example of such a "hidden" side-effect is the unavailability of a service due to performance problems of another workstation caused by NFS timeouts. We have the possibility to extend the graph by including either new AND and/or OR edges and/or new parameters.

The next step is the instantiation of the parameterized availability graph, which means to replace parameters with concrete values. Afterwards, the instantiated availability graph is the basis for the calculation of the availability of a service, because the leaves of the graph specify the involved components for the provision of a service and the dependencies in terms of AND and OR. By testing the correct operation of the involved components, it is possible to calculate the availability of applications.

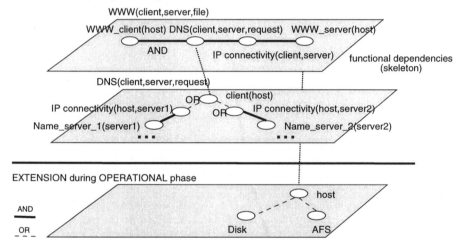

**Figure 3** Parameterized availability graph

A topic deserving more attention is the instantiation of the parameterized availability graph. For example, host(www process, cpu, disk) is instantiated to *sunmanager(httpd, 10MIPS, 10MB)*. The problem herewith is that a complete automatic instantiation is not possible due to the complexity of the environment. A semi-automatic instantiation can be achieved if the (i) distributed system is described with relevant attributes (i.e., workstation with a specific amount of memory, CPUs), and (ii) if there exist certain rules (e.g., that a DNS server workstation needs to have certain attributes.) These rules may represent a part of the organizational policies. Combined with inventory tools to recognize the devices and their attributes in a system, such instantiation can be performed automatically to a large extent. Despite this, certain steps during the instantiation need to be performed manually.

An open question is the granularity of the availability graph. The refinement of the graph should stop if the leaves of the instantiated availability graph can be either tested with existing testing methods or they point to other services, as shown in Figure 3 and Figure 4.

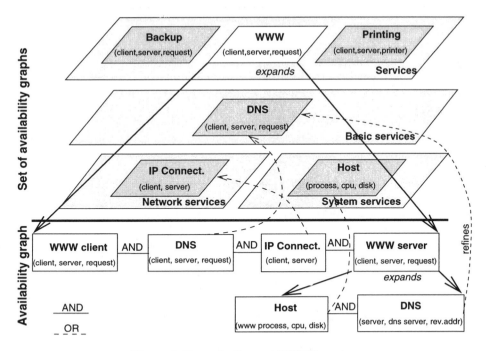

**Figure 4** Example: Intranet WWW service

## 3.2 Testing and calculating the availability of involved components

After having identified the involved components, it is necessary to analyze how to test the availability of these components. Rather then describing various testing methods, we analyze this problem by calculating the availability of the IP connectivity service in our environment. In this case, the components we have to deal with are IP routers with IP interfaces. Because there exists a relatively simple built-in IP test (e.g., "icmp echo request"), this is considered to be a simple service.

To determine the availability of components we need to repeat tests for each device every $\Delta t$, which needs to be short enough to recognize relevant faults, i.e. there is no degradation of the service perceived. Thus, the availability A of an IP interface I within a time period $[t_1, t_2[$ is defined as

$$A(I)_{t_1,t_2} := 1 - \frac{\sum_i \{\Delta t \mid t_1 \leq \tau_i < t_2 \ \wedge \ \text{fault}(I, \tau_i)\}}{t_2 - t_1}$$

where $\tau_i = t_1 + i * \Delta t$ for every $i \geq 0$ and fault(I, t) means fault of I at time t.

For the provision of the IP connectivity service, a backbone of IP routers is required. The SAPs are the interfaces of the routers. The task is to determine the availability of the IP connectivity service between these SAPs. However, to test the connectivity of $n$ interfaces in a shared medium backbone, we need to perform $n \times (n-1)$ tests. Due to the enormous effort it is necessary to reduce the number of tests. Therefore, the testing is performed centrally from one interface (i.e., the interface $I_P$ where the provider has its workstations) to any other interface. In this case, we do not measure the availability $A(I_1, I_2)$ of IP connectivity between the two interfaces $I_1$ and $I_2$, but the availability $A(I_P, I_1)$ and $A(I_P, I_2)$ combined to $A(I_P, I_1, I_2)$. We can assume that $A(I_P, I_1, I_2) \leq A(I_1, I_2)^{\dagger}$. We also assume that all interfaces have the same priority for the provider.

For the IP connectivity service, the following availability calculations are of interest:

- Availability $A(I)$ for one user access point (i.e., interface I) which can be calculated as stated previously.
- Assume that a user (e.g., the Department of Computer Science) runs, for example, a distributed file system over the Munich Network. Thus several SAPs (i.e., $I_1, I_2, \ldots$) are of interest. The availability $A(I_1, I_2, \ldots)$ of all relevant IP interfaces for a user is reduced by the sum of time intervals in which at least one interface is down.
- Availability $A(I_{all})$ for all IP interfaces of the service provider is the average of the availability of all interfaces. This result is relevant for the service provider to gain information about the reliability of the used devices.

Determining the availability of the IP connectivity service is relatively simple because the service itself is simple. The availability of the IP service depends only on the availability of the IP network devices. Besides, there exists a simple testing mechanism.

If we want to determine the availability of end systems like hosts for our applications, there exists no easy test mechanism. We do not test only the connectivity, but need to test at least whether all required system processes are running and whether there are enough free resources. Existing management tools enable us to perform some of these tests. However, without describing some dependencies, we cannot conclude that an application running on some end systems is working correctly. In order to calculate the availability of distributed applications calculation rules and adequate testing methods are necessary.

To improve and to refine the availability graph as well as to recognize what are the most relevant parameters in a production environment, we introduce another step proposing a learning approach.

## 3.3   Improving the methodology by refining the availability graph

It is impossible to specify all dependencies in a large production environment during the design phase. Some of them cannot be foreseen and are only recognized during operation. Therefore, it is necessary to extend and refine the availability graph.

For this, we propose a "learning" approach by analyzing user trouble reports about unavailable services. When a user reports an unavailable service, and a fault (the cause of the reported unavailable service) has not been diagnosed by a provider, a discrepancy has occurred. In general, there may be three reasons for this:

- either the testing methods are imprecise for this specific case, or
- the existing availability graph for this specific problem is not complete, or
- the problem description of the user was imprecise or has pointed to another fault.

---

$^{\dagger}$In practice a partial outage only between $I_1$ and $I_2$ is rare in a shared medium backbone.

Since user trouble reports are in general imprecise, we developed the so-called "Intelligent Assistant", a specialized tool which supports the user during the report and the localization of a fault in a predefined way. When the cause of a fault is identified, we can either (i) improve our testing method, or (ii) extend the availability graph, or (iii) change the decision trees in the "Intelligent Assistant". Let us illustrate this on an example. A user reports a trouble ticket with "WWW service unavailable", but a fault could not be recognized by the provider by testing the components as identified by the instantiated availability graph. After diagnosing the fault, we may recognize that the reason for the degradation of the service was, for example, an overload due to NFS problems. Thus, we need to extend our availability graph to describe a dependency with NFS and to install or develop an appropriate testing method.

## 3.4   Assessment of the solution

To summarize, we have proposed a methodology which consists of the following steps:

- build generic service graphs for each service, extend the service graphs to parameterized availability graphs to describe the involved components, instantiate the availability graphs in a concrete environment,
- perform testing of involved components and calculations based on the given calculation rules,
- refine the availability graphs with additional parameters and nodes to improve the availability calculations.

A relevant point is that a core set of service graphs and parameterized availability graphs can be specified. Such a core set can be applicable also for other service providers. Only the instantiation of the availability graphs is environment dependent.

The proposed solution – giving us the possibility to determine the actual and statistical availability of applications – is evaluated with respect to the following criteria: (i) scalability, (ii) manual effort, and (iii) dynamic changes.

There are two types of scalability we have to consider. Firstly, with respect to the number of services and nodes in an availability graph, and secondly, to the number of instantiations. Adding a new service requires a specification of a new service graph which can be considered to scale well. The same appears also for nodes in the availability graphs. The increasing number of instantiations of the derived availability graph is approached by adding parameters.

With respect to the manual effort, the main part consists of the definition of the service and the availability graphs. As already said, it is possible to specify a core set of generic service graphs and derived availability graphs. The effort to incorporate specifics is considered to be acceptable with appropriate tool support.

Dynamic changes refer to the instantiation of the availability graph as a consequence of changes in the system or network configuration. An approach to deal with new configurations is a version controlled database. Because network and system configurations should be stored in such a version controlled database, the additional storage of the availability graphs can be considered to be acceptable.

# 4   IMPLEMENTATIONAL ASPECTS

Due to being in a production environment, we started from a bottom-up approach to configure and implement a tool set for determining the availability of our IP router backbone (3.2). To supervise our network nodes, we take two different approaches: firstly, we use a developed tool to test IP connectivity which allows us to ping 250 nodes in parallel in a few seconds. Secondly, special configurations of HP OpenView Node Manager are used to poll all important network nodes every 5 minutes. We use both approaches to tune our tests. In recent years we adopted and refined tests of the former OpC, now IT/Operations, to supervise our server systems, and automatically generate trouble tickets. We used these tools as a basis for our own developments. As depicted in Figure 5, we

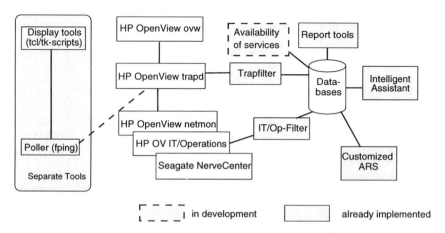

**Figure 5**   Tool support in our production environment

developed several tools to implement the proposed methodology. For the evaluation of the events and to store the detected faults in a database, we have implemented an application which hooks in HP OpenView event processing and stores them after some preprocessing in an ARS schema. With this tool, we have implemented the availability calculations of Step 2. Report tools have been developed to display the calculated data. Besides, tools for the specification, modification and usage of the availability graphs are under development (Step 1). These tools aquire availability data out of several databases.

To test end systems more precisely, and to obtain more precise information about the status of the network, we specified rules to correlate network and end systems alarms. Due to the fact that the correlated alarms generate trouble tickets in ARS, and user trouble reports are documented in the ARS as well, we apply the proposed approach in [Dreo 95] for the correlation of device-oriented and service-oriented symptoms. This is the realization of Step 3. Fault diagnosis is performed according [DrVa 95].

As seen from Figure 5, almost all necessary building blocks (i.e., tools) to develop the tool to monitor the availability of services have already been implemented and are in production use. So far, we are testing the semi-automatic realization of the instantiation of the availability graph. We extend the attributes of the documented components in our inventory database as well.

Being in an open university environment means that our customers do not have service subscription profiles. Therefore, we use the "Intelligent Assistant" to enable a user to specify his used services, and start the availability calculation for these services.

# 5  CONCLUSION AND FURTHER WORK

The rapid growth of complex, heterogeneous and distributed systems as well as distributed applications impose new complexity to IT management. In particular, management of distributed applications is a challenging topic. Distributed applications can be seen as complex pieces of software which are distributed across various, heterogeneous end systems, the network, and rely on the provision of other applications.

This paper proposes a methodology for determining the availability of distributed applications. We approach this by illustrating the complexity of the dependencies between involved components on an example and discuss the various views of availability. If we want to determine the availability of a complex distributed application (e.g., DNS), we have to deal with (i) dependencies between applications, (ii) non-existent testing mechanisms (e.g., providing information whether the DNS service is operating correctly), and (iii) not knowing which devices are involved in the provision of which applications. Solutions for this are presented in the first step. In the second step of the methodology, we clarify the term availability, and present the availability calculations for the simple IP connectivity service. This is possible because (i) simple testing methods do exist (e.g., icmp echo request), (ii) there are no dependencies with other applications, and (iii) it is clear what devices to test. Besides, in practice, we have to deal with dependencies which are not known at design phase. Therefore, we need further to improve and extend the availability graph during the operational phase by analyzing user trouble reports (step 3). In view of this, we propose a service graph to describe the functional dependencies which are known at design phase, and extend it to a parameterized availability graph to identify the involved devices. The availability of an application is calculated based on the availability of devices and the rules as given by the availability graph.

With the availability graph, we have a structure to describe complex dependencies in a distributed heterogeneous environment. Because we are a large service provider, experiencing the mentioned problems, we are focusing on completing the availability graph as precisely as possible according to our expertise. Such an availability graph can be applicable also for other service provider environments, as well as other applications, such as event correlation. Another aspect worth noting is that the availability graph – and the described dependencies – gives the basis to analyze bottlenecks in the system either on the service or device level. Our further work will therefore concentrate on the optimization of the availability graph, and the development of methods to recognize bottlenecks (i.e., weak points) in the system.

## Acknowledgements

The authors wish to thank the members of the Munich Network Management (MNM) Team for helpful discussions and valuable comments on previous versions of the paper. The MNM Team is a group of researchers of the Munich Universities and the Leibniz

Supercomputing Center of the Bavarian Academy of Sciences. It is directed by Prof. Dr. Heinz-Gerd Hegering.

## 6  REFERENCES

[Dreo 95] G. Dreo, *A Framework for Supporting Fault Diagnosis in Integrated Network and Systems Management: Methodologies for the Correlation of Trouble Tickets and Access to Problem-Solving Expertise*, PhD thesis, University of Munich, 1995, Verlag Shaker.

[DrVa 95] G. Dreo and R. Valta, "Using Master Tickets as a Storage of Problem-Solving Expertise", In [INM-IV 95], pages 328–340.

[Gosc 91] A. Goscinski, *Distributed Operating Systems – The Logical Design*, Addison-Wesley, 1991.

[HeAb 94] H.-G. Hegering and S. Abeck, *Integrated Network and System Management*, Addison-Wesley, September 1994, Reprint 1995.

[ICDP 96] A. Schill, C. Mittasch, O. Spaniol and C. Popien, editors, *Proceedings of the IFIP/IEEE International Conference on Distributed Platforms: Client/Server and Beyond: DCE, CORBA, ODP and Advanced Distributed Applications*, IFIP, Chapman & Hall, 1996.

[INM-IV 95] Y. Raynaud, A. Sethi and F. Faure-Vincent, editors, *Proceedings of the 4th IFIP/IEEE International Symposium on Integrated Network Management, Santa Barbara*, IFIP, Chapman & Hall, May 1995.

[ISO 10746] "Open Distributed Processing – Reference Model", IS 10746, ISO/IEC, 1995.

[ITU-T M.3010] "Principles for a Telecommunications Management Network", Recommendation M.3010, ITU-T, October 1992.

[ITU-T Q.1201] ITU-T, *Principles of Intelligent Network Architecture*, Q.1201, 1993.

[JeVa 87] E. Jessen and R. Valk, *Computer systems: Basics of Modelling (in german)*, Springer Verlag, 1987.

[Kaet 96] S. Kätker, "A Modeling Framework for Integrated Distributed Systems Fault Management", In [ICDP 96], pages 186–198.

[KPM 96] A. Kuepper, C. Popien and B. Meyer, "Service Management using up-to-date quality properties", In [ICDP 96], pages 448–459.

[Slom 94] M. Sloman, *Distributed Systems Management*, Addison-Wesley, June 1994.

[TINA-C 95] TINA-C, "TINA-C Deliverable: Service Architecture Version 2.0", Technical Report, TINA Consortium, 1995.

## 7  BIOGRAPHY

**Gabi Dreo Rodosek** received B.S. and M.S. degrees in computer science from the University of Maribor, Slovenia, and the degree of a Dr.rer.nat. in 1995 from the University of Munich, Germany. **Thomas Kaiser** received his diploma (Diplom-Informatiker, M.Sc.) in computer science in 1989 from the Technical University of Munich, Germany.

Both are research staff members at the Leibniz Supercomputing Center and members of the Munich Network Management Team, being engaged in several network and systems management projects.

# 17

# Secure Service Management in Virtual Service Networks

*Hai Qu*
*Computer and Information Science Department*
*University of Delaware, Newark, DE 19716, USA*
*Tel: +1 619 6511543  Fax: +1 619 6582243*
*Email: qu@cis.udel.edu*

*Tuncay Saydam*
*Computer and Information Science Department*
*University of Delaware, Newark, DE 19716, USA*
*Tel: +1 302 8312716  Fax: +1 302 8318458*
*Email: saydam@cis.udel.edu*

### Abstract

This paper extends our discussion and treatment of security of service management applications. After a brief discussion of the key players within the secure service management environment, it presents in detail the security service protocol and application programming interface (API) to facilitate request and response between a service management application and its security server. Guidelines on defining service management protocol and extending current CMIP's security management functions are also given. Conclusions encapsulate the results so far achieved as well as the future work and directions of our study.

### Keywords

Service Management, Network Management, Security Services, Security Service Protocols, Security Server, ASN.1.S, SP, ACSE_S, CMIP, Security Management, Security MIB.

## 1  INTRODUCTION

Service Management is an emerging topic on how to make network users to have friendly, flexible and secure control over QOS of their communications. In this area, security of Service Management is still not fully developed, and more study needs to be done on that

[PRI93] [PRI94]. Our problem to solve is to protect service management communications from various security threats by providing the following security services: Authentication, Data Confidentiality, Data Integrity, Access Control, Security Audit and Security Alarm.

In [Qu96], we designed the extensions of ASN.1, Presentation Layer and ACSE, which were named ASN.1.S, SP and ACSE_S respectively, in order to provide a generic approach to serve the common needs of protected communications among application processes. In this paper, we will try to solve our problem using this generic approach as the basis, by defining Security Service Protocol and API (Application Programming Interface), giving guidelines for designing Service Management Protocols, and specifying the extensions to be done for CMIP's Security Management functions.

## 2   EXTENSIONS TO ASN.1, PRESENTATION LAYER AND ACSE

Here we will only briefly introduce how the extensions to ASN.1, Presentation Layer and ACSE were done. The details are in [Qu96].

### 2.1   ASN.1.S: ASN.1 Extended for Security

We extend the current ASN.1 by introducing new data types as follows. The extended ASN.1 is called **ASN.1.S**. The extended encoding rules are called **BER.1.S**, which is BER.1 plus the new encoding rules for these new data types.

### *ENCRYPTED Type*

This new data type is defined as having tag UNIVERSAL 11, which is not used in current ASN.1 standard. The format for this data type is as follows.

```
ENCRYPTED
    { algorithm          OBJECT IDENTIFIER,
      keyType            KeyType,
      toBeEncrypted      ANY }

KeyType ::= ENUMERATION
              { noKey(0),            -- some checksum needs no key
                secretKey(1),        -- the symmetric secret key
                senderPrivateKey(2), -- the asymmetric private key
                receiverPublicKey(3),-- the asymmetric public key
                sessionKey(4),       -- the symmetric temporary key
                CAPrivateKey(5)      -- the assymetric private key }
```

*algorithm* indicates the encryption algorithms to be used for this data type. The encryption key to be used is indicated by *keyType*, which can be the symmetric secret key, sender's asymmetric private key, receiver's public key or a session key dynamically generated during connection establishment. In the definition of this data type, the actual value for *algorithm* and *keyType* may or may not be fixed. They can be assigned different values each time the application tries to send encrypted data.

The data to be encrypted is *toBeEncrypted* which can be any ASN.1.S type. The Presentation Layer first encodes it according to BER.1.S into a BIT STRING. Then it encrypts this BIT STRING into another BIT STRING. The final data to be transferred in PPDU includes object identifier value for *algorithm*, enumeration value for *keyType* and the latter BIT STRING. The encoding for ENCRYPTED data type is just like the encoding for a SEQUENCE data type, as follows.

```
[UNIVERSAL 11] IMPLICIT SEQUENCE
                    { algorithm        OBJECT IDENTIFIER,
                      keyType          KeyType,
                      encrypted        BIT STRING }
```

## *CHECKSUMMED Type*

This new data type is defined as having tag UNIVERSAL 12, which is not used in current ASN.1 standard. The format for this data type is as follows.

```
CHECKSUMMED
     { algorithm              OBJECT IDENTIFIER,
       keyType                KeyType,
       toBeChecksummed        ANY }
```

Most features are the same as in ENCRYPTED type. The encoding of this data type is like the encoding of the following.

```
[UNIVERSAL 12] IMPLICIT SEQUENCE
                    { algorithm        OBJECT IDENTIFIER,
                      keyType          KeyType,
                      toBeChecksummed  ANY,
                      checksum         BIT STRING }
```

## 2.2   SP: Presentation Layer Extended for Security

There is no new Presentation Layer service primitives to be introduced. But we need to add more parameters into P-CONNECT primitives as shown in Table 1. (In the tables of this paper, M means Mandatory, C means Conditional, O means Optional and = means the same value.)

**Table 1** New Parameters of SP-CONNECT Primitives

|                          | request | indication | response | confirmation |
|--------------------------|---------|------------|----------|--------------|
| secretKey                | O       |            | O        |              |
| initiatorPrivateKey      | O       |            |          |              |
| responderPublicKey       | O       |            |          |              |
| responderPrivateKey      |         |            | O        |              |
| initiatorPublicKey       |         |            | O        |              |
| suggestedSecrecyInfoList | O       | O          |          |              |
| suggestedIntegrityInfoList | O     | O          |          |              |
| resultSecrecyInfoList    |         |            | O        | O(=)         |
| resultIntegrityInfoList  |         |            | O        | O(=)         |

## 2.3   ACSE_S: ACSE Extended for Security

ACSE_S supports authentication itself and also supports negotiation of other security
parameters by pass-through to SP. In SA-ASSOCIATE.request, we need to add some
new parameters to the existing A-ASSOCIATE.request. Also, a new group of service
primitives, SA-AUTHENTICATE, are introduced for both parties to authenticate the
other party any time during a connection by requesting certificate, password or challenge
number.

**Table 2** New Parameters of SA-ASSOCIATE Primitives

|                          | request | indication | response | confirmation |
|--------------------------|---------|------------|----------|--------------|
| initiatorDN              | O       | O(=)       |          |              |
| recipientDN              | O       | O(=)       |          |              |
| initiatorAuthenticator   | O       | O(=)       |          |              |
| responderAuthenticator   |         |            | O        | O(=)         |
| secretKey                | O       |            | O        |              |
| initiatorPrivateKey      | O       |            |          |              |
| responderPublicKey       | O       |            |          |              |
| responderPrivateKey      |         |            | O        |              |
| initiatorPublicKey       |         |            | O        |              |
| suggestedSecrecyInfoList | O       | O          |          |              |
| suggestedIntegrityInfoList | O     | O          |          |              |
| resultSecrecyInfoList    |         |            | O        | O(=)         |
| resultIntegrityInfoList  |         |            | O        | O(=)         |

## 3   THE PLAYERS AND THE PROTOCOLS IN THE SOLUTION

In Figure 1, we show all the players and protocols used in the solution.

## 3.1   Service Management Application and Local S_MIB

Service Management Application is a special network communication application that
makes service management functions (value-added services) usable to the users of com-

munication services, e.g. bandwidth management, connection management and QOS management in VPN (Virtual Private Network) and UPT (Universal Personal Telecommunication) [PRI94]. The application processes are running at SMS (Service Management System), NMS (Network Management System) or CPN-NMS (Customer Premises Network NMS). They communicate with each other using P1, the Service Management Protocol, which will be defined later.

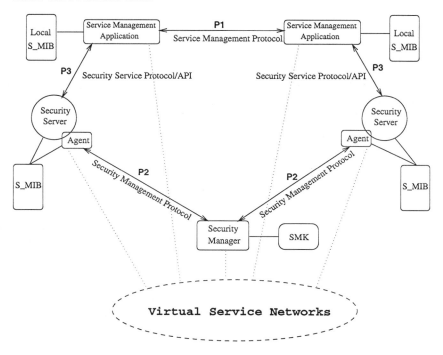

Figure 1  Security of Service Management

Our principal concern here is how to make these application processes to communicate in a secure and controlled way. The Local S_MIB (Local Security MIB) at the Service Management Application will keep some local security information that the application can easily access without contacting the Security Server. The Security Server in the same domain with the application will manage the application's Local S_MIB, using P2, the Security Management Protocol, in a hierarchical management framework, or using some other proprietary management protocol where the Security Server acts as the proxy agent on behalf of the applications which are indirectly managed by the Security Manager.

## 3.2   Security Server and S_MIB

Security Server logically stores all the security information for its domain in the Security MIB, i.e. S_MIB, and provides security services to Service Management Application upon request (e.g. providing authentication service) [Muf93]. The application process acquires security service using P3 API which will utilize P3 protocol, the Security Service Protocol, if the Security Server is remote to the application process. If the security server is in the same system as the application process, no open communication protocol activities are involved. A given Security Server can serve more than one application process which are normally in the same domain (e.g. a local area network). S_MIB will contain managed objects such as security policies and domains, security services, security mechanisms, security algorithms, security audit trails, security alarms, etc.

## 3.3   Security Manager and SMK

Security Manager is the manager role functionality of the security management (i.e. management of security). The corresponding agent role functionality is at the Security Server. The manager and agent communicate using P2, the Security Management Protocol, which is one of the five Specific Management Functional Areas in CMIP [Sta93]. The manager needs a local database SMK (Shared Management Knowledge) that keeps such information as the structure of all S_MIBs it can manage, cached information about the roots of subtrees in S_MIBs, information about domain objects, security policies for management, and so on, which will facilitate the manager to perform management operations over security MIB objects on security servers and the applications.

A manager can manage more than one Security Server. The manager can remotely manage the S_MIB objects that correspond to the security service configuration parameters in the Security Server. The management functions include security object creation/deletion/modification, policy/domain management, authentication and access control management, security audit and alarm management, etc.

## 3.4   P1: Service Management Protocol

Service Management Protocol is used to carry management information between peer service management entities. There are quite a few studies which explore these service management functionalities [Say95]. Since P1 needs to be defined from the scratch, it would be better to define the PDUs using the new ASN.1.S types, ENCRYPTED and CHECKSUMMED, to make selective field data secrecy and integrity possible [Qu96].

## 3.5   P2: Security Management Protocol

The following security management functions are already defined among the thirteen Systems Management Functions of CMIP which are within the Security Management Functional Area [Sta93]. They are Security Alarm Reporting, Security Audit Trail and Access Control Management. In addition to the above functions, we still need the following functions to make Security Management in CMIP to be complete: Authentication Management, Confidentiality Management and Integrity Management. The formal definitions of the managed objects for these functions will be studied in the future.

## 3.6    P3: Security Service Protocol/API

One of the main objectives of this study has been the definition of Security Service Protocol/API (P3) to facilitate the security service request and response between service management applications and Security Servers. No matter whether the security server is local or remote to the service management application process, a set of API functions of a programming language, e.g. C or C++, is to be defined for the application to access security service. The actual entities in the application that will call these API functions are SP, ACSE_S and ACSE_S user. If the security server is remote to the application, the API functions will be responsible for acquiring corresponding services from security server if it cannot provide the security services and/or parameters in its Local S_MIB.

Some Internet documents have already defined a Generic Security Service API (GSSAPI) [RFC1508] [RFC1509]. Its goal is to provide a portable programming interface which is independent of underlying security mechanisms. The user can only have pointers or handles for security data. The user simply calls a sequence of API functions and sends opaque *tokens* to peer user through in-band or out-of-band channel. There are some drawbacks in GSSAPI.

- User data can be signed, can be sealed (integrity plus confidentiality), but cannot be encrypted without integrity checksum.
- The security tokens are opaque to users, i.e. they are only byte strings to users. But the way that user sends data over an application protocol makes it necessary to know the structure of tokens in order to embed the tokens into protocol data. This is especially true for applications defined using ASN.1.
- Selective field confidentiality and integrity can not be achieved. There is no way to indicate parts of data to be protected.

Our purpose of P3 API is different from GSSAPI. P3 API presents to user a concrete method to access security services flexibly. The user still has choices to use different mechanisms but has more direct control over the process of security service. These tangible security services are building blocks for more abstract services. These functions will overcome the drawbacks of GSSAPI.

### API Functions in C

Here we specify the C function prototypes for the API functions and also describe the functionalities the application may need to access by explaining what the API functions will accomplish. All the following functions return an integer value where 0 means success and other values mean failure.

(1) *int GetSessionKey ( int \*length, char \*\*sessionKey );*

Given the requested length of session key to be generated, this function will generate a session key or get a session key from security server and set the actual length of the key when returning. This function will be called by SP to generate a session key during connection establishment time.

(2) *int CryptoRequest ( enum Algo algo, char \*parm, char \*key, int size, char \*data, int \*resultSize, char \*\*resultData );*

Given the cryptographic algorithm, the parameters, the cryptographic key and the data, this function produce new data to be pointed to by *resultData* by performing cryptographic algorithm on the original data. This function will be called by SP.

(3) *int GetRandomNumber ( int \*length, char \*\*randomN );*

Given the requested length of random number to be generated, this function will generate a random number or get a random number from security server and set the actual length of the result when returning. This function will be called by ACSE_S in order to generate a random challenge number for authentication purpose. SP will also call this function if it is performing some cryptographic algorithm and is in need of a good source of random numbers that is not available in local system.

(4) *int AuthCheck ( char \*myDN, char \*peerDN, AuthData \*authData );*

*myDN* and *userDN* are the DNs (Distinguished Names) in the X.500 Directory Service format. An application process can claim a specific user identity among the possible more than one identities it may act for. Given the DN that API caller is claiming, peer user's DN and the authentication data that peer user provides, this function checks the authenticity of this peer user. This function is called by ACSE_S user to verify if his peer's password or certificate is correct or not.

(5) *int AccessControl ( char \*myDN, enum Op op, char \*subjectDN, char \*targetDN );*

Given the operation requested, the subject, i.e. the requester of an operation, the target on which the requested operation will be performed, this function checks them against the access control rules that may be stored locally or in security server, and rejects or accepts the requested operation. This function will be called by ACSE_S user.

(6) *int SecurityAlarm ( int type, int cause, char \*detector, char \*provider, char \*additionalInfo );*

When an entity, either SP, ACSE_S or ACSE_S user, detects a security attack, it can generate a security alarm that the event discriminator will send to security manager or security server who will forward this alarm to security manager. This function will be called by SP, ACSE_S or ACSE_S user.

(7) *int SecurityAuditTrail ( int type, int cause, char \*additionalInfo );*

This function generates a security audit trail which may be logged by the security server. This function will be called by SP, ACSE_S or ACSE_S user.

(8) *int GetMyAuthData ( char \*myDN, char \*peerDN, Authenticator \*auth );*

Before it establishes application association with another user with identity *peerDN*, the

API caller claiming *myDN* as its identity calls this function to get information that is necessary for it to make authentication data which will be sent to peer user during association establishment. This function will be called by ACSE_S user.

(9) *int GetCertificate ( char \*userDN, Certificate \*cert );*

The main purpose of this function is for the sender of data to get the public key of peer indicated by *userDN*, for encrypting data to be sent, or for decrypting signed data. This function will be called by ACSE_S which will pass the public key to SP.

(10) *int GetPrivateKey ( char \*myDN, int \*length, char \*\*privateKey );*

The API caller claiming *myDN* calls this function to get its asymmetric private key. The keys can be acquired by means that is out of the scope of our study, or from Security Server through Security Management exchange. This function will be called by ACSE_S.

(11) *int GetSecretKey ( char \*myDN, int \*length, char \*\*secretKey );*

The API caller claiming *myDN* calls this function to get its symmetric secret key. The keys can be acquired by means that is out of the scope of our study, or from Security Server through Security Management exchange. This function will be called by ACSE_S.

## P3 Service Specification

P3 uses ACSE_S directly for association control and authentication services. All other services as defined below will be provided through SP-DATA service primitives. We list the parameters for one group of service primitives as an example, which are basically corresponding to the arguments in API functions described earlier.

| SS-CRYPTO | request | indication | response | confirmation |
|-----------|---------|------------|----------|--------------|
| algo | M | M(=) | | |
| parm | O | O(=) | | |
| key | M | M(=) | | |
| data | M | M(=) | | |
| result | | | M | M(=) |
| resultData | | | C | C(=) |

## P3 Protocol Specification

There is no P3 PDU for association establishment, which will be done by using SA-ASSOCIATION, SA-RELEASE, SA-ABORT and SA-P-ABORT primitives directly. SA-AUTHENTICATE primitives could also be used for extra authentication during the association.

There is a simple correspondence between the service primitives and the PDUs for P3. The request and indication primitives for a service are mapped to one PDU and the response and confirmation are mapped to another PDU.

Below, we only list the names of the PDUs and how the service primitives are mapped into PDUs.

**Table 3** P3 PDUs

|  | req/ind | resp/conf |
|---|---|---|
| SS-SESSION-KEY | SESS-RQ | SESS-RE |
| SS-CRYPTO | CRYP-RQ | CRYP-RE |
| SS-RANDOM-NUMBER | RAND-RQ | RAND-RE |
| SS-AUTH-CHECK | AUTH-RQ | AUTH-RE |
| SS-ACCESS-CONTROL | AC-RQ | AC-RE |
| SS-ALARM | ALM-RQ | ALM-RE |
| SS-AUDIT | AUDIT-RQ | AUDIT-RE |
| SS-AUTH-DATA | AUDA-RQ | AUDA-RE |
| SS-CERTIFICATE | CERT-RQ | CERT-RE |

The ASN.1.S definitions of these PDUs are to be studied in future. Basically the ASN.1.S definitions are corresponding to the parameters in the service primitives. Here we give the PDU definitions for CRYP-RQ and CRYP-RE as examples. We suppose the data integrity function in SP is either used for each whole PDU, or never used in a connection. Hence, in the ASN.1.S definitions, CHECKSUMMED data type is not needed. The security server and the application will, at connection set-up time, negotiate on whether data integrity will be used or not. Encryption is based on selective field approach, therefore ENCRYPTED data type appears in the definitions.

```
CRYP-RQ ::= [APPLICATION 2] SEQUENCE
                { algo        [0] OBJECT IDENTIFIER,
                  parm        [1] ANY OPTIONAL,
                  secretData  [2] ENCRYPTED
                                  { DES,
                                    secretKey,
                                    SEQUENCE
                                       { key   OCTET STRING,
                                         data  OCTET STRING } } }

CRYP-RE ::= [APPLICATION 16] SEQUENCE
                { result      [0] Result,
                  resultData  [1] OCTET STRING }

Result ::= ENUMERATION
               { success(0),
                 algoNotAvail(1),
                 invalidParameter(2),
                 invalidAuditType(3),
```

```
invalidUser(4),
noAccessRight(5),
wrongPassword(6),
...  }
```

Most likely the keys used for both encryption and checksumming will be the symmetric key, because the symmetric keys are easier to manage for systems that are in the same domain than the public/private key pairs. Also, if the Security Server and the Service Management Application are located physically very closely, these two security services will not be used for better performance so that the application could get security services from its Security Server efficiently.

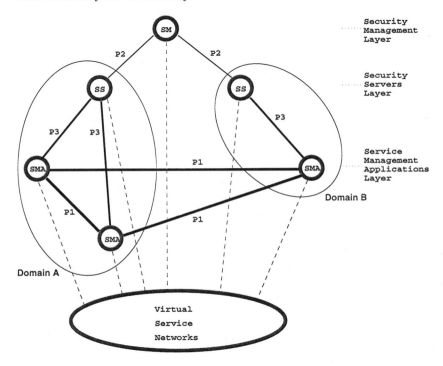

Figure 2   Hierarchical Organization of Security Entities and Domains

## 4   HIERARCHICAL ORGANIZATION OF THE SOLUTION

As a hint to specifying the solution more formally in the future, here we present the hierarchical organization of entities and domains [Slo94] in our solution as in Figure 2.

SMA, the Service Management Application process, can communicate with other SMAs in the same security domain or in different domains. The SMAs constitute the Service Management Applications Layer in this logical organization. SS, the Security Server process, will help provide security services to SMAs in the same domain. These SS's constitute the Security Servers Layer. SM, the Security Manager process, which is at the Service Management Layer, will manage SS's directly and SMAs indirectly. The SS's themselves are also organized in another hierarchy, e.g. the hierarchy of certification authorities, which reflects the hierarchical web of trust among the Security Servers. So, domains A and B in Figure 2 may form a bigger domain.

## 5   CONCLUSIONS

In this paper, we discussed the functions of the players in the solution to the problem of Secure Service Management. Service Management Application processes communicate with each other using Service Management Protocol, which is designed to utilize the security services provided by the underlying SP and ACSE_S. Its security information is managed by its Security Server in the same domain. The Security Server provides security services to SP, ACSE_S and the application process through a newly designed P3 API and P3 protocol which is protected by the underlying SP and ACSE_S. The security information in Security Server is managed by the Security Manager.

Because we utilized the generic security service approach involving ASN.1.S, SP and ACSE_S, there was no duplicate effort done for providing security services to the applications. This unique, comprehensive and flexible underlying approach helps solve our problem very efficiently. Also, P3 API provides a portable programming interface that can eliminate the need for the application to know whether the Security Server is on local system or on a remote system.

Another result we achieved is to have solved the problem of security of OSI network management, i.e. security of CMIP. We don't need to redefine CMIP PDUs and services, but need to implement CMIP entity on top of SP and ACSE_S. Access control function has been partially defined in CMIP; that is, the managed objects for access control have been defined, but how to actually achieve access control has not been defined. We can use Distinguished Names as the identifiers of subjects and targets. Other security management functions to be extended in CMIP were also suggested in this paper.

Our future work will include the perfection of the logical structure of the problem, the complete formal specifications of the three protocols and security-related managed objects definitions in CMIP, and object-oriented modeling [Col94] [Say95] of the solution.

## 6   REFERENCES

[Col94] Derek Coleman, et al. (1994) *Object-Oriented Development: The fusion method*, Prentice Hall

[Muf93] Sead Muftic, et al. (1993) *Security Architecture for Open Distributed Systems*, John Wiley & Sons Ltd.

[PRI93] *Service Management Reference Configuration*, RACE Project 2041 PRISM, 1993

[PRI94] *VPN and UPT Service Management: Second Case Study Report*, RACE Project 2041 PRISM, 1994

[RFC1508] J. Linn (1993) *Generic Security Service Application Program Interface*, Zolot Associates

[RFC1509] J. Wray (1993) *Generic Security Service API : C-bindings*, DEC

[Qu96] Hai Qu and Tuncay Saydam (1996) *Security of Service Management: A Generic Approach to Secure Communications within OSI*, draft paper to be submitted

[Say95] Tuncay Saydam, et al. (1995) *Object-oriented design of a VPN bandwidth management system*, pp344-355, Integrated Network Management, IV, edited by Adarshpal Sethi, et al., Chapman & Hall

[Sch96] Bruce Schneier (1996) *Applied Cryptography: protocols, algorithms, and source code in C*, John Wiley & Sons, Inc.

[Slo94] Morris Sloman, editor (1994) *Network and Distributed Systems Management*, Addison-Wesley Publishing Company

[Sta93] William Stallings (1993) *SNMP, SNMPv2 and CMIP: The Practical Guide to Network-Management Standards*, Addison-Wesley Publishing Company

# 7  BIOGRAPHY

Hai Qu is a PhD student in Computer & Information Science Department, University of Delaware. He got B.Sc. and M.Sc. degrees in Computer Science from Fudan University, Shanghai, China. Since his graduate study, he worked on research and development of computer network protocols and applications. He was a Lecturer in Fudan University doing teaching, researching and network application development. He worked in National Computing Center, UK, on OSI protocol testing. He started his PhD study in 1994 working on network management and service management related issues.

Tuncay Saydam is a professor of computer science and computer networks at the University of Delaware since 1979. He has received Dipl.Ing., M.S. and Ph.D. degrees from Istanbul Technical University and the University of Texas. He has been an invited research professor at the Swiss Federale Institute of Technology (ETHZ) and Ecole Polytechnique Federale de Lausanne, where he has been active in research. He has more than 20 years of academic and intensive consulting experience. Author of more than 50 research papers, Prof. Saydam's current research interests include service management, QoS management, virtual networking and telecommunications software engineering.

# 18

# Intelligent Agents for Network Fault Diagnosis and Testing

*Garry Grimes*[+]
*Euristix Ltd.*
*Century Court, 100 Upper St. Georges St., Dun Laoghaire, Co. Dublin*
*Tel: +353 1 288 4788   Fax: +353 1 288 5392     email: gary@euristix.ie*


*Brian P. Adley*
*Department of Electronic and Computer Engineering &*
*    Telecommunications Research Centre*
*University of Limerick, Limerick, Ireland*
*Tel: +353 61 202609     Fax: +353 61 202572   email: Brian.Adley@ul.ie*

### Abstract

In modern complex networks remote control testing and diagnosis is necessary for efficiency. In this paper we propose the use of Intelligent Agents to carry out these tasks. These autonomous intelligent or programmable agents allow for full flexibility in unplanned conditions. They may be instructed to perform tasks at the resource thus reducing network loads in times of trouble and reducing the burden on the managing system.

The framework of existing network management standards is sufficient to implement such intelligent agents using only MIB descriptions based on existing standards. Tests may consist of one or several managed Intelligent testing agents and should be managed according to the situation. In this paper we look at distribution of testing scripts and control of the agents as well as the storing and returning of test results, in particular from distributed tests. We also present several practical implemented uses of the technology.

### Keywords

Intelligent Agents, Test Objects, SNMP, Test Manager

## 1    INTRODUCTION

In order to accommodate unpredicted changes in the way networks are used and in their composition including both hardware and software components, flexibility is needed in the management system. In addition complex monitoring and control operations may be needed that involve multiple co-operating management objects on different platforms.  Furthermore, in large networks a network management application may be topologically distant from  the multiple objects that are performing a specific management operation. These requirements support the need for intelligent dynamic and autonomous agents to perform management operations. Intelligent Agent technology is being   investigated in many distributed application situations and in this paper we investigate its application to Network Management. In particular, we have implemented and analysed  the use of intelligent agents for fault analysis within current network management standards and architectures. This paper presents the results of this work and concludes that current management standards can support the use of Intelligent Agents.

## 2    CURRENT NETWORK MANAGEMENT APPROACHES

Standardised network management systems are implemented based on an object based manager-agent model,

---

[+] This work was completed while at the University of Limerick

(Ashford, 1989) & (Gering, 1993). The manager requests or updates information on the state of a resource from an agent which represents that resource in the form of managed objects held in a managed information base, MIB.

The system is thus managed through a distributed database. The data is held in MIBs which are maintained by agents. The agents running on the nodes are in effect database interfaces. They update their own MIBs depending on what is happening with the real resource that the MIB represents and they allow the managing system to query this information. The agents cannot make decisions or initiate any management activities, this is left completely to the managing system. Where information is to be gathered from the MIBs of several agents and compiled for analysis it is the managing system which will perform all the operations involved; querying the agents, receiving the results and compiling them.

## 3    MANAGING TESTING AND DIAGNOSIS FUNCTIONS

### 3.1   Problems with the established architecture

In most cases only a single agent per resource will be required to execute the test function selected by a manager, for example (Noto, 1996). In the case of end-to-end path testing however, tests may be required that necessitate the co-operative participation of several distributed management agents. This can lead to the following problems (Vassila, 1995) & (Koojiman, 1995):

- Many management operations are necessary in a complex test situation. With the manager performing so many management operations on several geographically distant agents, network bandwidth usage is concentrated at the management system. Also testing functions on several agents may produce large quantities of resulting data to return to a manager. This is not good, especially where such tests are being carried out in serious failure situations.
- Where all control is centralised to one managing station this station (and the link to it) becomes of critical importance where the carrying out of a test must be orchestrated by the manager which initiated it. This would be the case where agents simply store and convey information. Should the manager or its link experience a failure, all operations by the agents, including any tests-in-progress must terminate or suspend until normal operation resumes.
- Manager processing power is limited. Greater loads on managing systems, such as from processing test result data from several agents means less available for other management tasks.

The processing power and operational resources available to agents of modern managed devices are reaching the stage where they are capable of far more complex activities than ever before. The SNMP RMON MIB (Waldbusser, 1991) allows for this to a certain degree. This standardised MIB allows for fault management to the extent that a manager may request data on network interface traffic from managed devices and summarisations from the agent which has been monitoring the network traffic. This information is very useful for fault management, diagnosis, and testing, but only standard information is available. Should more complex or specific testing be required this approach lacks the flexibility to adapt to requirements.

### 3.2   Improvements to the established model: The Intelligent Agent

Much work has been done, including (Vassila, 1995), (Magedanz,1995), and (Meyer, 1995), into solutions to the problems of centralised network management and improvements in flexibility to extend the functionality of current managed systems. Such approach is shown in figure1. The central idea here is the delegation of tasks which would normally be explicitly controlled by a management station, to the managed agents. This is achieved by incorporating a facility into the agents which allows them to download scripts of instructions and to interpret and execute them. With this architecture agents become capable of performing tasks autonomously with or without the co-operation of other managed systems. Should managers require several agents on a remote system or network to work together to perform a task, the managing system can delegate responsibility for control of the task to one of the managed systems. The manager thus instructs the agents involved in the tasks they are to perform. It should also be possible to indicate a certain course of action to be taken in the case of exceptional circumstances arising, such as failure of a system involved in a test.

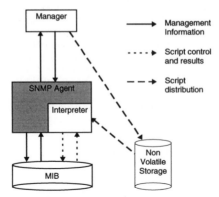

**Figure 1** Architecture of an Intelligent Agent.

With such a scenario the performance of the managed network would be increased as the gathering and processing of data is moved nearer to the source of that data. Full flexibility is achieved where agents can be instructed 'on-the-fly' to carry out specialised tasks. Also many of the weaknesses in the established model of use are resolved:

- Where sets of management instructions are performed autonomously by an agent, management functionality is delegated away from the manager. The only management information flowing between the managing system and the agents is the initial set of management instructions and the final results of the management operations which have been carried out by an agent or agents. Network traffic concentration at the managing station is therefore greatly reduced.
- Once the management instructions have been communicated to the agent or agents involved, then faults in the management station or in the link between it and the agents will not be catastrophic as the agent processes can continue to function independently. When the operation results have been compiled they will be available to the manager as soon as the situation has been resolved and proper functioning has resumed.
- The processing load on the manager is greatly reduced where the bulk of the data gathering and manipulation is done by the agents rather than the manager. A managing system may therefore initiate many such operations at no extra overhead and wait to be informed by an agent when it has completed all the necessary operations and the final results are available. It may also perform intermediate checking of results to ensure that all is proceeding as it should.

## 4.0   TESTING BY INTELLIGENT AGENT

### 4.1   Manager-Agent relations in testing

Using an Intelligent Agent system for testing and diagnosis will involve different manager-agent relations depending on the type and extent of the tasks to be carried out. Different manager-agent relations for a distributed test have been discussed by Ohta et al. in (Ohta, 1993). In the paper the authors state that the optimum form of management will depend completely on the application scenario. Greater system processing loads favouring delegation of tasks away from the manager.

With a test involving distributed co-operating agents we found that delegation of manager status to one participating (or non-participating) agent works best. Using a test-manager agent to co-ordinate a test and retrieve and process results cuts down on processing load of and network traffic to the main manager system. Test control in the event of necessary intervention by the main manager is simplified. Should the main manager have to suspend or terminate a test then control of all testing agents is effected by communication with the delegated test-manager agent. The diagrams in Figure 2 and Figure 3 below show the manager-agent relationship as direct, Figure 3 and with a test-manager in Figure 2. A test using the architecture of Figure 3 with *n* testing agents will require *n* sets of management communications for each command to begin, suspend, resume or terminate the test. The manager must also have a understanding of the functions of each testing agent and know in which order commands to agents should be sent. For example, in the tests described in Section 8.1 the agent producing test traffic should be suspended before the traffic analysis agent.

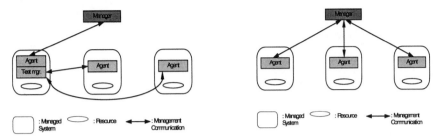

**Figure 2** Manager-Agent relations with a delegated test manager agent

**Figure 3** Manager-Agent relations with no test manager agent

In the case when a test can be done using one managed agent, such as with the Broadcast Storm source locator function (see Section 8.3) it was found that a direct manager to agent relationship works best. When using an Intelligent Agent for testing there is nothing to be gained by delegating a further test-manager agent expressly for controlling one testing agent or retrieving and processing results. Such functions can be incorporated into the testing agent script itself.

## 4.2   Structure of a test-manager

Each intelligent agent has the necessary functionality to take on the role of test manager. In our particular implementation delegation was arbitrary, however in practice there will be external conditions which affect the decision as to which particular agent should be given this responsibility. There are many issues to consider; including base processing power, and average load on each agent's processor, as well as the average traffic through its network interfaces. Such discussion is outside the scope of this paper but should be considered for future work.

The test manager agent has various tasks to perform regarding test set up and control, as well as its own testing operations. These various steps which must be taken by the test manager could be coded as part of the agent itself. This however makes testing less flexible with the setting up, running and completion of tests being of a set procedure, and reporting of results having the same format. In order to introduce enough generality into this method of control it would be necessary to provide large quantities of test related information by setting MIB objects. Many tests will require specific information for set up and dealing with exceptional circumstances. Statically defining and creating specific MIB objects to store all this information is not desirable. Most of the more specific variables will remain unused in many circumstances and it is not realistic to assume that all test conditions and criteria can be anticipated and catered for in this way. Having said this, it is useful to provide some general 'lowest-common-denominator' test information as MIB objects. General object descriptions for test functions have been detailed in (ISO, 1993) & (NMF, 1992).

In contrast to a fully MIB specified approach to the architecture of a test manager, a more flexible and straightforward method is to develop a script for any agent which will deal with the set up and control of tests. Supplying this script to an agent which will execute it effectively makes that agent into a test-manager. In order to do this a special control script is sent to the delegated test-manager. This instructs the test manager in how the test is to be set up among the agents involved, how unusual circumstances are to be dealt with and when the test should be initiated and completed. It should also instruct how to obtain and process results from each agent which has been involved in the test and can notify the main manager when final results are available.

## 5.0   IMPLEMENTATION

### 5.1   Why use a Standardised Network Management Protocol ?

There are many benefits to be gained from standards-based integrated network management; more rapid deployment of new services and integration of these services into existing systems and a common management platform across heterogenous environments (Shrewsbury, 1993).

As Intelligent Agent Testing and Diagnosis is a service which may span diverse systems we were interested to see if an Intelligent Agent System could be incorporated into an existing network management framework and to this end decided to base an implementation on an existing management protocol.

After considering SNMP, the Simple Network Management Protocol (Rose, 1990), (Case, 1991) and CMIP, Common Management Information Protocol (ISO, 1991) network management platforms we decided to use SNMP. The CMIP standard would have allowed for a very capable and well defined system, SNMP however supported our requirements with less overhead, allowing more time to investigate the issues involved.

## 5.2  The Scripting language

Because of its network management extensions and its widespread use and support the Intelligent Agent System was implemented using Scotty. This is the Tool Command Language developed by John Ousterhout (Ousterhout, 1990) with extensions developed by Juergen Schoenwaelder (Schoenwaelder, 1995).

The security and integrity of a network node becomes an issue when it provides the capability to execute scripts remotely. In the Intelligent Agent System this is somewhat resolved by not allowing 'dangerous' commands to directly access the underlying operating system of a device. Such commands are either not allowed or put into 'wrappers' which allow only certain 'minimum-damage' functionality. Also scripts are allowed to write to and read from files in only one local directory, this works in a way similar to the HTTP access of World Wide Web home sites.

## 6    DESIGN OF THE MIB

The requirement was to develop a MIB which would allow the downloading of scripts to the agent, the control of this script and the returning of results when the script was complete.

### 6.1  Script downloading

There were several options available as regards script downloading. Among these HTTP, FTP, TFTP, SMTP and SNMP itself.

The function of distributing scripts to intelligent agents requires a mechanism for transporting text files between the managing and managed systems. Several efficient TCP/IP protocols have been developed which can serve this purpose; HTTP (Berners-Lee, 1996), FTP (Postel, 1985), TFTP (Scollins, 1992), SMTP (Postel, 1982) and SNMP itself.

We wished to implement a script transfer mechanism using only SNMP to investigate whether the protocol would be sufficient for such a purpose, thereby removing the requirement for supporting additional ones. For the purposes of comparison we also decided to implement script downloading using a protocol developed specifically for ASCII text transfer. We selected HTTP for this purpose as it is supported by Scotty and is based on a simple model; set up connection, transfer data, close connection, which exactly fits the requirements of script transfer.

### 6.2 Script transport and the MIB

Using SNMP for script downloading requires breaking up scripts into lines which will then be performed by executing a specific line or by concatenating several lines together if they belong to the same script. This is necessary because of the message size limitation in the SNMP PDU. Scripts are broken into consecutive lines and are transported to the agent by the manager in separate Set commands. These scripts are stored in the MIB Table. Separate scripts are stored as conceptual tables within the main table. When a script is completely downloaded it is available for execution. To facilitate re-use the agent may be instructed to transfer a script to some non-volatile storage. To this end, the MIB contains an action object which on changing value (via a SET) instigates the I/O operation. To allow later retrieval of stored scripts the agent creates and stores a permanent script handle.

Transport of scripts with HTTP is more straight-forward and requires only two MIB variables as it is not stored in the MIB at all for transfer (though this could use a process similar to the SNMP download but from non-volatile storage). One variable will be SET to indicate the uniform resource locator (URL) of the location of the script to be downloaded, the other will hold any error messages that may occur during transfer. The managing system which supplies the script must have a HTTP server process running which is capable of responding to requests by the agent. When the script URL variable is set the agent issues a HTTP get command and stores the script on non volatile storage. In our implementation which is file system based, this involved extracting the local filename from the URL as a handle for storage and retrieval. Depending on the existence of non-volatile storage and the type available, this operation should be performed appropriately by the agent in a manner transparent to the managing system.

In this investigation into methods of script transport we found that both SNMP and HTTP have the functionality to deal with the downloading of scripts to agents. It is possible to implement an intelligent agent system without recourse to a protocol other than SNMP. However, attempting to use a management protocol like SNMP to perform a task for which FTP, TFTP, and HTTP were developed is subject to several inefficiencies. The drawback lies with the increased overhead on the agent process. It must create and fill MIB table rows with each script line. These lines must then be extracted from the MIB and concatenated together to form the original script. The use of a more suitable protocol for transport such as HTTP, allows direct receiving and storing of a downloaded script without the use of the MIB as an intermediary. The extra process of  table row creation, setting, extraction and concatenation is not required. If it is possible to include support in an intelligent agent for script retrieval via HTTP, or another such lightweight file transport protocol, we believe this to be far more efficient than the use of SNMP for the same purpose.

## 6.3  Control of the Script

As indicated by the specifications in (NMF, 1992) for a test management service, tests may be both implicitly or explicitly controlled. Implicit control may be script based, where an agent is instructed to monitor  conditions in its controlling script. It may issue commands to other agents which are co-operating with it in the running of the test using explicit control commands. Also provided is the facility to specify when a script should begin to iterate and when to end, intervals after which it should iterate and number of iterations.  Explicit control is provided by a SET on the action variables which control script execution.

### 6.3.1  Scripts are pre-prepared for execution

To use a downloaded script it is necessary to provide information for its use. A table is maintained with all this information, and the same script may be used in different ways depending on this information. The MIB has a scriptTable for this purpose. Each row of this table holds as well as the script name; its status, owner, number of iterations of execution and interval between each, and also input (arguments) to the script. As scripts need to be reusable under different conditions they are written to accept arguments. These are most likely to specify the names or addresses of other network devices when scripts are for path testing purposes and will be different from host to host.

### 6.3.2  Execution monitoring and control

Described in (ISO, 1993) is a model for managing the invocation of tests on remote resources. These tests may be controlled and subject to monitoring, suspension and resumption during application or uncontrolled where results are returned on test completion. This remote control test specification is intended for an OSI system and the corresponding CMISE network management system with its more complex GDMO MIB description format.

The Network Management Forum produced OMNIPoint1 which provides a similar specification for a Testing Management Function. This specification though with ISO standards in mind is not explicitly intended to be implemented in GDMO.

These models have not addressed distributed co-operating agents within tests but do suggest a framework for any generic test that should be required to be performed. Because of this genericity they were particularly applicable to the use of testing by intelligent agents for which testing is not restricted to any one function or set of functions.  Though we developed the intelligent agent to use SNMP as its management protocol with its MIB written using ASN1 the ITU-T and NMF test management models where not GDMO specific are still applicable to an SNMP MIB. In particular X.745 and OMNIPoint2 specify the testing procedure as requiring several states: *Not Initialised, Idle, Initialising, Suspended, Testing, Terminating* and *Disabled*. Testing actions are also specified in X.745 as *Initiate a test, Suspend, Resume* or *Terminate a test.*

When a script in the scriptTable is to be executed its ID is SET to the current script id variable. Control of the current script is by the RMON method of setting 'action' variables.

To control execution the currentScriptAction variable may be set to one of five values:

- Static - The initial value, also if there is no script or the script has not yet been been run.
- Running - The current script is to execute.
- Suspended - The current script is to be Suspended before the next iteration.
- Resumed - The suspended current script is to resume operation where it was suspended.

- Terminated - The current script is to cease execution after the current iteration.

The script condition variable shows the current state of the script. This will indicate whether the script is: *static, executing, about to suspend, suspended, resumed from suspension, about to terminate* or *terminated*.

## 6.4  Return of Results

Intelligent agents which perform management tasks that are delegated to them by a managing system often do not require that results be returned. Research work thus far on the use of intelligent agents, including their use in performing network management tasks has not addressed this. Standardised specifications which address testing issues assume well defined procedures and well defined results.

When considering testing using downloadable scripts it is essential that results be available to management, even where a testing system is quite autonomous it is always necessary that a management station be able to monitor results.

Where tests are performed using downloaded scripts where there can be no precise standardisation, there can be no pre-arranged structure for storing or reporting results. Every script is different and performs a different function and will produce different results.

Two solutions exist within our framework for result reporting of scripted tests:
1.  Output results to a file and export via HTTP to the manager.
2.  Place results in a MIB and allow them to be retrieved via SNMP.

It is a simple matter to write out results to a file as they are obtained and indicate the result filename or URL for retrieval by the manager. The results file may then be displayed or parsed by a management application.

To make results available via SNMP it is necessary to store them in a MIB. However MIBs are developed for a particular purpose. Though it is possible to develop a MIB 'on the fly' specially prepared for storing script results and download it along with the script itself, this is not a realistic solution. SNMP agents are not intended to load MIBs at any time other than on initialisation. If a new MIB is loaded the agent must be reset and all previously obtained information will be lost. The Intelligent Agent System therefore uses a configurable result table in which script results may be stored, regardless of format. The table rows are created to store what ever type or format of result is required and can retrieved via SNMP GETs or SNMPv2 GET-BULK.

## 6.5  Presenting Test Results to a User.

A problem with retrieving data from MIBs generally is that unless the retriever knows the MIB structure and the purpose of each variable and their relationships, it is difficult to transform the data into easily understandable information. This problem is compounded by the returning of results in an unknown format by an intelligent agent.

To allow for more 'user-friendly' reporting of results the Intelligent Agent System allows for the downloading of a Result-Display-Script by the (test) script developer to the individual agents or to a controlling agent which will gather final results.

This script is available to a managing system for uploading when the script has results available to specify exactly how result data should be presented to a human user. These Display Scripts are exported to the agent in the same manner as the Scripts that the agent runs. Uploading of them has been provided for by HTTP and by loading into a Display Script Table maintained by the Agent and from which the Display Script may be retrieved using SNMP GETs.

The IETF has taken into consideration the uploading of user interfaces written in Java (Gosling,1995) for management by Web browsers, and indeed our display scripts may be written in any interpreted language which provides an SNMP interface. However, the Display Scripts developed during this project use the Tcl extensions Tk with the extra Scotty functionality.

## 7.0  THE INTELLIGENT AGENT PROGRAM

The agent program uses the network management extended Tcl interpreted language Scotty. Scotty is based on event driven programming. Typically a Scotty application will load an initialisation script which installs some event handlers. In the case of an SNMP agent, these event handlers are configured to respond to SNMP PDUs received and interface with the MIB database to update or return MIB variables. Once initialisation is complete the application enters an event loop. The event driven approach works well in most cases however one problem is the lack of threads. New processes must be explicitly started to deal with SNMP requests otherwise the listening process will block until the request has been completed. In the Intelligent Agent System Event handlers deal with making available table contents, retrieving and setting MIB variables and performing explicit script control actions.

The Scotty extensions allow for very comprehensive network tesing and include interfaces to TCP/IP services like UDP Datagrams, TCP message passing, ICMP services as well as HTTP and SNMP. All these are available to downloaded scripts.

## 7.0 MANAGEMENT STATION APPLICATION

The management can be performed by using the Scotty command line interface for control and this was adequate during development. However the Tk toolkit that forms a part of the Tcl suite of extensions makes all such commands available for control by GUI. A manager application was developed which really provided a graphical front end to the Scotty SNMP GETs and SETs but specifically developed for the control and monitoring of the Intelligent Agent by its MIB.

## 8.0   TESTS IN PRACTISE - DESIGN OF TEST SCRIPTS

Four applications of the Intelligent Agent System to Network Fault Diagnosis and Testing were developed to better investigate the use of fault and performance testing with an intelligent agent system. These functions were: A distributed one way UDP through-put test, a distributed loss rate test, a broadcast storm locator, and resource utilisation over time.

### 8.1   Distributed one way through-put test (UDP)

The purpose of this test is to determine the maximum UDP datagram throughput of a link. It is a distributed function using two testing agents; the datagram traffic generator at one end of the link and the datagram receiver at the other end. In addition there is a test manager agent which runs the distributed function at regular intervals until the required results have been obtained. These results are then retrieved by the test manager agent and summarised (average throughput over time), the main manager is notified and can retrieve the final results.

It works as follows: A script is downloaded to an agent at the test traffic source to send as many datagrams as possible within a certain time period (*n* seconds). The agent at the test traffic destination point is instructed to receive as many datagrams as possible from the time of receipt of the first datagram until the test time ( *n* seconds) has elapsed. A test manager agent is delegated which sets off these tests at regular intervals, a main manager request to suspend, resume or terminate the test is directed only to the controlling test manager agent which also gathers the end results from the datagram source and destination agents. It then calculates the throughput from *datagrams received / time*.

This method works well for in-service testing as test traffic has to travel only one way to be examined, thus reducing test traffic to half of loopback. Importantly, this also allows true one-way throughput examination rather than duplex. Explicit manager control of the test and retrieval of results is by communication with the delegated test-manager. Test scripts for datagram traffic generation and receiving are shown in figures 4 and 5.

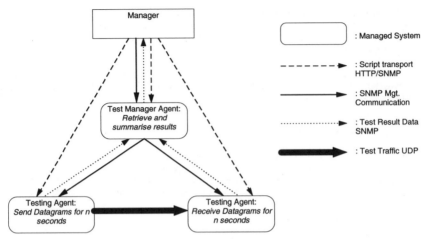

**Figure 3** Architecture of the distributed UDP throughput test

## 8.2  Distributed loss rate test

This was mainly developed with ATM testing in mind and was inspired by the Ohta et al paper though it was (by necessity) implemented over TCP/IP using UDP datagrams. The test involved setting up a test traffic insertion point, a drop and analysis point and several intermediate monitoring points. All of these functions are distributed amongst agents along the path to be tested and are controlled by a Test-Manager-Agent which starts the test at various intervals and which collects results from all of the (sub) test-agents. This Test-Manager-Agent is so due to its own controlling script giving it that function. Explicit control of all agents (i.e. the whole distributed function) is effected by control of the Test-Manager-Agent. This test is distributed so that one way testing is possible, this is necessary in assymetrical point-to-point ATM links where regular loop back tests are not possible.

## 8.3  Broadcast Storm source location

Monitors network traffic, at packet level and logs all broadcast traffic on a sliding window basis over the last $n$ packets. Should broadcast packets reach unacceptably high (configurable) levels it archives the last $m$ packets (since just before the storm began) and makes them available for examination by a managing station. It also examines the contents of the archived logs to determine the source of the initial broadcast packet of the storm. This however only works in the situation where the broadcast storm is of RESENDs (which have identical data contents).

As a broadcast storm will often cut off communications the agent is completely autonomous and will continue monitoring, logging and processing relevant data until communication with the manager has been restored and the results can be retrieved.

## 8.4  Simple resource utilisation over time

An agent on each host monitors CPU utilisation though this may be easily extended to include monitoring of any quantifiable device value. There is no interfering packet transfer while the monitoring goes on, information may be obtained from the management station whenever required. This was really developed to test the uploadable display script functionality. All results from each CPU testing agent can be presented together in graph format from the results compiled by the central controlling agent which holds the display script.

## 9.0  RESULTS AND CONCLUSIONS

In this paper we investigated the application of Intelligent Agent technology to remote network testing and diagnosis functions. We have found that the use of intelligent agents provides complete flexibility to develop and implement precise testing functions to deal with unforeseen situations and requirements. In distributed testing where several agents are involved in a test it was found to be very useful to delegate a test-manager agent to control testing and retrieve results. This reduces test management traffic (control and result data), main manager processing load, and also allows for an autonomous testing system.

We also found that it is not useful to develop a specific agent to deal with control of intelligent agents for testing. We find that fully configurable tests require fully configurable management, and that a scripted test manager is the best approach.

Configurable tests require configurable result storage which cannot be planned for in advance. We used a result table with agent configurable rows to solve this. The main manager application must be able to interpret these results if they are to be of use. We investigated a method of including a test specific 'display script' with the script of the delegated test manager agent. This can be uploaded by the manager and presents results in a human-readable format with text and graphics.

In this investigation we were interested to see if a complete Intelligent Agent system to meet our requirements could developed within the framework of an existing network management architecture such as SNMP. Such a system is not intended to replace the static agent architecture but to work with it. Either as an extensible agent or simply listening for SNMP communication on a different UDP port as was implemented here.

We found that it is possible to implement such a system within the SNMP framework with no changes to the protocol. A MIB was developed according to the SNMP specifications (Stallings, 1993) with objects to hold all the required test information (including test script) and objects to control test functioning. Though script distribution was found to be possible using only SNMP it is more reliable and efficient to use HTTP. This was provided for in our implementation.

Due to our requirement to base the Intelligent Agent system on an existing network management standard we chose SNMP over CMIP mainly because of the widespread use of SNMP and the complicated and heavy-weight nature of CMIP. Scotty proved an excellent development platform for the system and provided excellent support for SNMP, HTTP, UDP and TCP. However an SNMP intelligent agent based platform is rather strained and has its drawbacks. The nature of actions (e.g. script control) not being an integral part of SNMP, they must be implemented by changing MIB object values. Action parameters in SNMP must be implemented by MIB objects which are explicitly SET before actions can be performed. Retrieval of large amounts of data is less efficient than CMIP and 'round-about' methods must be used for data reduction. The presentation of result data in table object format also seems rather tortured when compared with the GDMO Test results record object class described in X.745 (ISO, 1993).

In the course of this work we used Tcl to create our test scripts. This proved adequate for the purposes of testing but it would be useful to investigate other interpreted languages such as telescript, Perl and Java. Our test scripts were relatively small and involved minimum complexity, they were quite applicable to the simple unstructured form of Tcl. Large scripts are very unwieldy when written in Tcl and become very difficult to develop, maintain, and debug. The safety features of Tcl (latest version) with respect to agent scripts are very useful, however other languages, such as Java appears to address this area better.

## 10 ACKNOWLEDGEMENTS

The authors would like to acknowledge the financial support of Ericsson System Expertise Ltd. and Teltec Ireland without whom this work would not have been possible.

## 11 REFERENCES

Ashford, C. (1993) The OSI Managed Object Model. *ECCOOP '93.*
Berners-Lee, T. Fielding, R. Frystyk, H. (1996) Hyper Text Transfer Protocol, Internet Working Draft, February 1996.
Case, J. Fedor, M. Schoffstall, M. Davin, J. (1991) RFC1213: A Simple Network Management Protocol.
Gering, M. (1993) CMIP vs SNMP. Integrated Network Management III, 347-359.
Gosling, J. McGilton, H. (1995) The Java Language Environment: A White Paper, Sun Microsystems.
ISO/IEC DIS10164-12 (1993) Test Management Function.
ISO/IEC9596 Information Technology (1991) Recommendation X.711 - Open Systems Interconnections - Common Management Information Protocol Specification.

Koojiman, R. (1995) R. Divide and Conquer in network management using event-driven network area agents, Technical Univerity of Delft, The Netherlands.

Magedanz, T. (1995) On the Impacts of Intelligent Agent Concepts on Future Telecommunications Environments, *IS&N '95 Conference Proceedings*, 396 - 414.

Meyer, K. Erlinger, M. Betser, J. Sunshine, C. Goldszmidt, G. Yemini, Y. (1995) Decentralizing Control and Intelligence in Network Management, *4th International Symposium on Integrated Network Management*, *Santa Barbara, CA*.

NMF (Network Management Forum): Forum 012 (1992) Application Services: Testing Management Function, Issue 2.0.

Noto, M. and Tesink, K. (1996) Definitions of Tests for ATM management, *IETF Internet Draft*.

Ohta, S. and Fujii, N. (1993) Applying OSI systems management standards to virtual path testing in ATM networks. Integrated Network Management III, 629-640.

Ousterhout, J.K. (1990) Tcl: An Embeddable Command Language, *USENIX '90 Conference Proceedings*.

Postel, J. Reynolds, J. (1985) RFC959: File Transfer Protocol.

Postel, J.B. (1982) RFC821: Simple Mail Transfer Protocol.

Rose, M. McCloghrie, K. (1990) RFC1155: Structure and Identification of Management Information for TCP/IP-based Internets.

Scollins, K. (1992) RFC1350: Trivial File Transfer Protocol.

Schoenwaelder, J. and Langendorfer, H. (1995) Tcl Extensions for Network Management Applications, Technical University of Braunschweig, Germany.

Shrewsbury, J.K. (1993) TMN in a nutshell.

Stallings, W. (1993) SNMP, SNMPv2 and CMIP: The practical guide to Network Management Standards, Addison-Wesley.

Vassila, A. and Knight, G.J. (1995) Introducing Active Managed Objects for Effective and Autonomous Distributed Management, in *IS&N '95 Conference Proceedings*, 415 - 430.

Waldbusser, S. (1991) RFC1271: Remote Network Monitoring Management Information Base.

## 11   BIOGRAPHY

**Garry Grimes** obtained the BSc in Computer Systems from the University of Limerick in 1994. He was awarded the MEng by research and thesis in telecommunications in 1996 from the University of Limerick. This was awardedon the basis of research into the application of Intelligent Agent mechanisms to standardised network management architectures. This work was mainly directed at building on current fault and performance management techniques in TCP/IP managed networks. Garry is now based at Euristix Ltd., Dublin, Ireland, where he is involved in the design and development of enterprise network and element management solutions.

**Brian Adley** has been member of the Department of Electronic and Computer Engineering faculty at the University of Limerick from 1992. Between 1975 and 1992 he has held a number of communications related positions in engineering and management including Hewlett Packard Inc., Apollo Computer Corp, Protein Databases Inc., ICL and Marconi Space and Defence. He holds a B. Sc. (Hons) in Computer Science from De Montfort University (1975) and a M. Sc. Computer Science from Boston University (1991). His current interests are distributed management and image processing.

```
##
## Send a result back to IAgent
##
proc setresults { str1 res1 } {
  global rf
  udp send $rf { [ list str1 res1 ] }
}

proc udp_send { file host delay } {
  global msg stat size
  if {[catch {udp send $file $msg} err]} {
          catch {
                udp close $file
          }
          return
  }
  incr stat(send,$host) $size
  if {$delay > 0} {
     after $delay "udp_send $file $host $delay"
  } else {
     after idle "udp_send $file $host $delay"
  }
}

proc udp_summary { file host secs } {
  global stat sdt
  udp close $file
  set send $stat(send,$host)
  set o_speed [expr $send * 8.0 / 1024 / 1024 / $secs]
  puts [format "%6.3f MBit/s send" $o_speed $host ]

set info "udpsend output results line"
## Create the output summary line
set sn "SN:udpsend.tcl"
set fdt "FDT:"
append fdt [exec date]
set ds "DS:"
append ds $send
set dur "DUR:"
append dur $secs
set spd "SPD:"
append spd $o_speed

set summ [list $sn $sdt $fdt $ds $dur $spd]
puts $summ
## Update the resultTable with output
setresults $info $summ
}

##
## Connect a udp file handle to the test port on host and
## send datagrams
## to it. Calculate the send rate in kB per second.
##
proc udp_test { host secs len delay } {
  global msg stat size testport
  set size $len
  set testport 5556
  set msg ""
```

```
  for {set len 0} {$len < $size} {incr len} {append msg "+"}
  set stat(send,$host) 0
  if {[catch {udp connect $host $testport} f]} {
           puts stderr "$host: $f"
           return
  }
  after [expr $secs * 1000] "udp_summary $f $host
$secs"
  udp_send $f $host $delay
}

##
## Parse the command line arguments and call udp_test
## for every host given on the command line.
##
global sdt
set secs 10
set delay 10
set bytes 1024
set newargv ""
set parsing_options 1
while {([llength $argv] > 0) && $parsing_options} {
    set arg [lindex $argv 0]
    set argv [lrange $argv 1 end]
    if {[string index $arg 0] == "-"} {
        switch -- $arg {
            "-d" { set delay [lindex $argv 0]
                   set argv [lrange $argv 1 end]
            }
            "-t" { set secs [lindex $argv 0]
                   set argv [lrange $argv 1 end]
            }
            "-s" { set bytes [lindex $argv 0]
                               set argv [lrange $argv 1 end]
            }
            "--" { set parsing_options 0 }
        }
    } else {
        set parsing_options 0
        lappend newargv $arg
    }
}
set argv [concat $newargv $argv]
set sdt "SDT:"
append sdt [exec date]
if {$argv == ""} {
    puts stderr {usage: udpspeed [-d delay] [-t seconds] [-s
size] hosts}
} else {
    set time 1
    foreach host $argv {
        after [expr {$time * 1000}] "udp_test $host
$secs $bytes $delay"
        incr time $secs
        incr time 1
    }
}
```

**Figure 4** Script to send datagram traffic.

```
##
## Send a result back to IAgent
##
proc setresults { str1 res1 } {
   global rf
   udp send $rf { [ list str1 res1 ] }
}

proc udp_receive { file host } {
   global stat size
   udp receive $file
   incr stat(received) $size
}

proc udp_summary { file host secs } {
   global stat sdt
   fileevent $file readable ""
   udp close $file
   set received $stat(received)
   set i_speed [expr $received * 8.0 / 1024 / 1024 / $secs]
   puts [format "%6.3f MBit/s received " $i_speed $host ]

## Prepare results summary info
   set info "connection analysis output iteration summary"
   set sn "udprecv.tcl"
   set fdt "FDT:"
   append fdt [exec date]
   set ds "DS:"
   append ds 0
   set dr "DR:"
   append dr $received
   set dur "DUR:"
   append dur $secs
   set spd "SPD:"
   append spd $i_speed
   set summ [list $sn $sdt $fdt $ds $dr $dur $spd]
   puts $summ
## Update the resultTable with output
   setresults $info $summ
   udp close $f
   exit
}

##
## On receipt of the first datagram, start the timer.
##
proc firstreceipt { f host } {
   global secs
   fileevent $f readable "udp_receive $f $host"
   after [expr $secs * 1000] "udp_summary $f $host
$secs"
}

##
## Connect a udp file handle to the test port on host and
```

```
## receive datagrams
## from it.
##
proc udp_test { secs len } {
   global msg stat size testport
   set size $len
   set host [exec hostname]
   set testport 5556
   set stat(received) 0
   if {[catch {udp open $testport} f]} {
         puts stderr "$host: $f"
         return
   }
   fileevent $f readable "firstreceipt $f $host"
}

##
## Parse the command line arguments and call udp_test
## for every host given on the command line.
##
global sdt rf
set secs 10
set bytes 1024
set newargv ""
set parsing_options 1
while {([llength $argv] > 0) && $parsing_options} {
   set arg [lindex $argv 0]
   set argv [lrange $argv 1 end]
   if {[string index $arg 0] == "-"} {
         switch -- $arg  {
            "-t" { set secs [lindex $argv 0]
                  set argv [lrange $argv 1 end]
            }
            "-s" { set bytes [lindex $argv 0]
                  set argv [lrange $argv 1 end]
            }
            "--" { set parsing_options 0 }
         }
   } else {
         set parsing_options 0
         lappend newargv $arg
   }
}
set argv [concat $newargv $argv]

set sdt "SDT:"
append sdt [exec date]

## Create a UDP file handle for the result
set rf [udp open 5000]

## Begin the test
udp_test $secs $bytes
```

**Figure** 5  Script to receive datagram traffic.

# TRACK II

## PART ONE

# Broadband Network Management

Chair: Jan Roos,
University of Pretoria and Sun Microsystems

# 19

# Layered Bandwidth Management in ATM/SDH Networks

*Tai H. Noh*
*Lucent Technologies, Bell Labs Innovations*
*Room 1B-413, Crawfords Corner Rd., Holmdel, NJ 07733*
*Phone: (908)949-5184, Fax: (908)949-5908,*
*e-mail: tnoh@lucent.com*

## Abstract

ATM technology offers greater flexibility than STM technology, but in some applications this flexibility has to be compromised to realize a network that performs at least as well as the current STM implementations to achieve guaranteed Quality of Service and favorable economics. This paper presents a layered bandwidth management scheme that exploits the strengths of both STM and ATM technology to realize a more manageable and cost-effective network than is possible by applying only one of these technologies in the integrated ATM and SDH transport network.

## Keywords

Bandwidth Management, ATM, STM, SDH, Transport Network Evolution

## 1. INTRODUCTION

The recent trends in ATM/SDH broadband technology have led to high-speed fiber transport links, high-capacity network systems, and multimedia services. These trends have increased the importance of the efficient utilization of bandwidth with guaranteed Quality of Service (QoS) but also the vulnerability to network failures. This paper discusses a bandwidth management strategy for the capacity efficiency in the integrated ATM and SDH transport network. Various end-to-end network protection options are described in the reference [Noh, 1996].

Bandwidth management implies rearranging the current network bandwidth configuration to enhance network adaptability to both expected and unexpected traffic variations. Configuration Management functions in the TMN execute bandwidth management control actions, for example, rearranging VP bandwidth when a reconfiguration control parameter, such as VP use, or network traffic exceeds a predefined threshold.    A bandwidth reconfiguration interval is based on the traffic behavior in ATM networks. The behavior of an ATM traffic source can be decomposed into four levels:  *network level, call level, burst level,*  and *cell level* [Hui, 1988]. An important attribute of each level is its rate of bandwidth reallocation, which differs substantially from one level to another [Figure 1]. Bandwidth management belongs to the *network level.*   Most of bandwidth reconfiguration schemes proposed to date deal with SDH/PDH networks[Gopal ,1990] [Hasegawa, 1987] and ATM networks [Gerla, 1989] [Monteiro, 1993] [Lee, 1993] [Sato, 1990] separately.  There have been very few studies that address overall strategy of bandwidth management to utilize the strengths of both ATM and STM technology.

We divide the *network level* into two layers- the STM path layer and the ATM VP layer - - to introduce the layered bandwidth management concept into the evolving transport network. Sufficient transport bandwidth should be allocated for each level to bound the amount of blocking occurring at the next level.

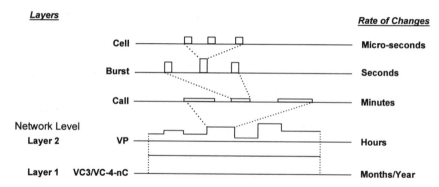

• The facility network should be designed  so that the probability of failure  to set up an VC-4-nC is small.
• An VC-4-nC should be allocated so that the probability of failure to  set up VPs is small.
• A VP bandwidth should be allocated so that the probability of failure  to use the VP for setting up a call is small.
• A call should be admitted only if the probability of failure due to transporting bursts is small.
• A burst should be transported  only if the probability of cell loss is small (The CBR does not have a burst layer).

**[Figure 1]  Layered Bandwidth Allocation and Rate of Changes**

The layered bandwidth management can be supported with either STM cross connection and ATM cross connection as two separate network elements (NE), or via a hybrid STM/ATM cross connection that provides a smooth path in the evolution toward ATM in the transport network.   Understanding the technical strength   and economics of alternatives will help both network service and system  providers  develop the most cost-effective evolution planning strategy.

This paper is organized as follows. Section 2 provides transport options for ATM traffic. Section 3 discuses transport network evolution scenarios toward ATM. Section 4 introduces layered bandwidth management. Section 5 provides applications of the layered bandwidth management. Finally, section 6 presents some concluding remarks.

## 2. TRANSPORT OPTIONS FOR ATM TRAFFIC

There are three options to transport ATM traffic over SDH networks. They are embedded ATM transport, hybrid transport, and pure ATM transport within a single SDH tributary.

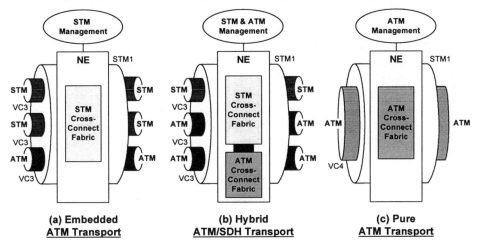

[Figure 2] Transport Options for ATM Traffic

### Embedded ATM transport

This option corresponds to (a) in Figure 2. ATM cells are mapped to the concatenated mode of SDH payloads by using conventional SDH Network Elements (NE). The SDH NEs provide transport of ATM cells from the customer ATM CPE to the ATM switch and back to the customer ATM CPE. The ATM cells are completely transparent to the SDH NEs. The STM fabric cross-connects VC3 paths under STM management only.

### Hybrid Transport

This option corresponds to (b) in Figure 2. The STM and ATM traffic are combined onto the same SDH pipe over STM tributaries. The NEs have visibility of the ATM cells. The ATM cross-connect fabric in the hybrid NE is used to aggregate ATM traffic to achieve better fills of the concatenated mode of STM tributaries carrying ATM cells. The VC3 paths containing ATM traffic are dropped at the ATM fabric while STM fabric provides VC3 cross-connects. Therefore, this NE requires both STM and ATM management.

## *Pure ATM Transport*

This option corresponds to (c) in Figure 2 and carries all ATM and STM traffic as ATM cells in a single SDH tributary. STM traffic is circuit emulated (if necessary) into ATM cells to  maximize the bandwidth utilization of a given SDH pipe by using ATM technologies.  The single tributary VC4 in the STM-1 pipe is utilized for ATM traffic under ATM management only.

## 3. TRANSPORT NETWORK EVOLUTION TOWARD ATM

A layered transport network architecture is composed of three layers: the circuit, path, and transmission media layers [ITU-T G.805 ,1992]. The circuit is an end-to-end connection established/released either dynamically or by short-term provisioning.  The transmission media network that interconnects nodes is the physical connection based on long-term provisioning.  The path layer bridges these two layers and provides logical connections between terminated node pairs.  This path layer plays an important role in constructing flexible ATM/SDH networks.

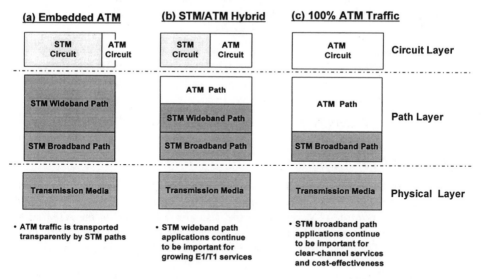

[Figure 3]  Transport Network Evolution toward ATM

The path layer is further divided into three different paths: STM wideband, STM broadband, and ATM VP in the evolving transport network [Figure 3]. The STM wideband and broadband path layers have been constructed in the existing SDH networks by using SDH cross connect systems or Add/Drop Multiplexers (ADM). The Virtual Path (VP) extends the concept of path layers to ATM networks. To enhance network reconfiguration capability, selecting the right network element to support the three different paths is one of the key concerns in the evolving network from the perspective of guaranteeing QoS and favorable economics. Thus, it is very important to examine both circuit and path layers in the network evolution scenarios to minimize the cost and QoS degradation associated with the circuit emulation between STM and ATM networks.

In the embedded ATM network [(a) in Figure 3], SDH paths provide transparent transport of ATM traffic. Switching of ATM traffic is performed by ATM edge switches located in the access or CPE network and ATM hub switching systems in the junction or interoffice network. An ATM Service Access Multiplexer (SAM) can be used at the edge of the public ATM network to provide ATM interfaces and adaptations for customer services to help reduce the transport inefficiencies associated with the conventional SDH hierarchical tributary structures.

Service providers may want to add ATM functionality to the SDH NEs for the improvement of network bandwidth and flexible introduction of new services with the inherent benefits of ATM technologies. ATM layer grooming is performed by the ATM path to reduce the number of expensive ATM switching systems [(b) in Figure 3]. STM wideband paths applications continue to be important for the growing E1/T1 services. To utilize the inherent strength of the ATM VP technology, E1/T1 services can be circuit emulated. The E1/T1 traffic would be converted to ATM at the edge, connected as ATM through the network, and converted back to circuits on the other side. There is a potential to reduce the overall network equipment cost because ATM connections may be less expensive than STM wideband connections of the same rate, since there does not need to be tributary processing in different paths.

There are important problems however. First, circuit emulation causes a delay. Second, the circuit emulation cost can not be negligible for large STM traffic. There will be a 13% bandwidth penalty for circuit emulation (ATM cell overhead). These issues will be more significant in the area where most traffic is added/dropped from/to the local office [Wu, 1994].

As ATM traffic demands increase, the transport option in the transport access network will be changed to pure ATM transport with a single SDH tributary such as STM-1 and STM-4c [(c) in Figure 3]. The ATM traffic in the single SDH tributary will be terminated at the junction network for grooming at the ATM layer. STM broadband path applications play an important role as a container for the ATM traffic.

# 4. LAYERED BANDWIDTH MANAGEMENT

## 4.1 Description

Capacity efficiency can be achieved by two-layered bandwidth management in ATM/SDH networks: the STM path layer and the ATM path layer [Figure 4]. An important attribute of each layer is the operating time scale as described in Figure 1.

Configuration Management in the TMN executes the ATM path layer management [Layer 2] for the short term traffic variations and STM path layer management [layer 1] for the long term service growth.

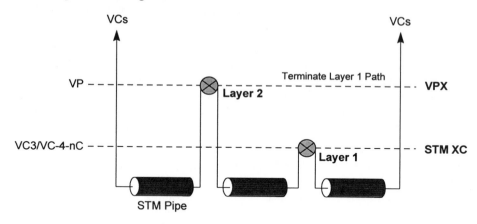

**[Figure 4] Layered Bandwidth Management**

At the ATM layer, the route and bandwidth of VPs are defined independently. The route is provisioned in the database of the VP terminators and cross-connects. The bandwidth reconfiguration of the VP can be achieved simply by changing database values in the VP terminators and cross-connects. This inherent benefit provides more dynamic bandwidth reconfiguration. However, without additional control, cell loss and fluctuation of transmission delay are inevitable due to statistical store and forward transmission. Therefore, the service class that guarantees no cell loss and no intra-network delay jitter is supported by some additional control with extra cost.

In SDH networks, a digital path is established by assigning a time slot for the TDM frame at each cross-connect on the path. Thus, path route establishment and bandwidth assignment are interdependent. Fixed bandwidth digital paths can be established hierarchically. STM cross connect systems are very cost-effective with non-blocking operation [Eng, 1990]. The use of the STM path management between network nodes eliminates the problem of processing cells at gigabit-per-second rates in the high speed backbone network.

The concatenated mode of SDH tributaries such as VC-4-nC provides a container for ATM traffic. For some applications in ATM networks where the bandwidth reallocation unit for traffic demand is VC3/VC-4-nC, switching entire VC3/VC-4-nC ATM containers is more desirable than switching individual ATM cells. The high-speed transport network with multiple ring interconnections is an example in the ATM/STM hybrid network. The interconnected ring networks can be dynamically reconfigured by adding or dropping the entire VC3/VC-4-nC tributaries. SDH will be terminated only when grooming at the ATM layer is needed. Characteristics of each layer management is summarized below.

- *Layer 2: ATM layer paths -VP*
    - *—ATM path traversing ATM VP Cross-Connects (VPXs)*
    - *—Bandwidth management for bundle of ATM VCs*
    - *—Reduce the number of intermediate ATM switch hops along the path*
    - *—Flexible bandwidth allocation*
    - *—Efficient bandwidth utilization for service growth and network restoration*
- *Layer 1: STM layer paths - VC3/VC-4-nC*
    - *—STM path traversing ADM/STM Cross-Connects*
    - *—Bandwidth management for bundle of VPs*
    - *—Reduce the number of intermediate ATM hops along the path*
    - *—Reduce buffering delay and nodal processing overhead*
    - *—Reduce the number of SDH path termination*

## 4.2 Examples and Discussions

To illustrate the operations of each layer management, consider the simple network model that has three ATM Switch (ATM-SW) nodes. A cross-connect system interconnects them for the bandwidth management. The physical links between the cross-connect system and ATM-SWs are STM-1s. The model network accommodates two types of services (class 1 and class 2). No spare capacity is reserved for the network protection. The class 1 is a video service that is coded at 10 Mb/s and the transport is rate-controlled at 10 Mb/s. The class 2 traffic is a 10 Mb/s native LAN traffic. Each service class is assigned to different VPs for the segregation of different QoS requirements. The network management system is located at an administrative center that communicates with the ATM-SWs to collect the data of existing demands, from which the required bandwidth for the forecast demands is calculated [Hui, 1988] [Logothetis, 1995].

Figure 5a shows an initial configuration of this model that provides the capability of the layer 1 management. Customer AB transports three class 1 and three class 2 services between the switch pair A&B. Customer AC and BC transport two class 1 and two class 2 services between the switch pair A&C, B&C respectively. The customer AB would require two VC3s. Surplus bandwidth on each VC3 is about 8 Mb/s and 28 Mb/s respectively since a VC3 provides the customer an available bit rate of 48 Mb/s. The customers BC and AC would require a VC3 with 8 Mb/s surplus bandwidth.

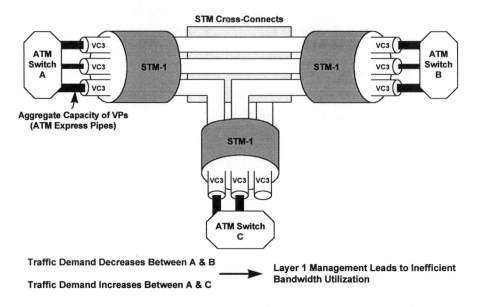

[Figure 5a]: Layer 1 management

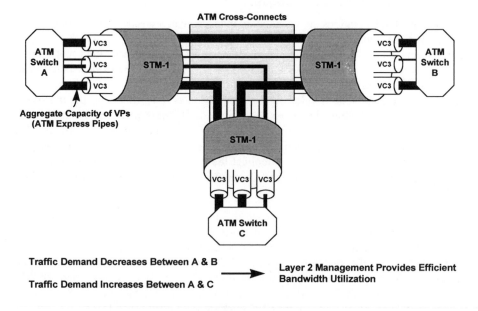

[Figure 5b]: Layer 2 management

Now, consider the situation where the forecast traffic demands change. The customer AB stops subscribing to a class 1 service and the customer AC wishes to transport two more class 2 services. However, the customer AC can not use the surplus bandwidth on a VC3 between the ATM-SW A and the STM cross connect system that handles VC3 cross-connects. This leads to inefficient bandwidth utilization. This inflexibility issue associated with the boundary of the VC3 tributaries can be resolved by using ATM technology. In Figure 5b, the STM cross-connect system is replaced by an ATM VP cross-connect system for the layer 2 management. Customer AB and AC could now share the same VC3 to transport their ATM traffic.

As ATM traffic demands increase, the transport option in the transport access network will be changed to pure ATM transport with a single SDH path as described in the section 2 and section 3. Figure 5c shows that bandwidth utilization can be maximized by using a single tributary VC4. There is no barrier to use surplus capacity.

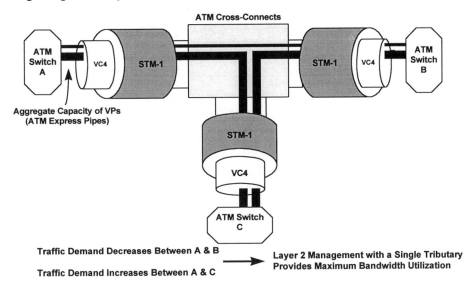

[Figure 5c]: Layer 2 management with a single SDH tributary

*The discussion of this example leads to the following conclusions:*

- *Layer 2 Bandwidth Management is more flexible and bandwidth efficiency can be maximized with the single STM tributary for ATM traffic.*

- *Layer 1 Bandwidth Management leads to inefficient bandwidth utilization due to the hierarchical structure of the STM tributaries. However, this is no longer an issue if the individual STM express pipe already has significant traffic aggregation.*

# 5. NETWORK APPLICATIONS

The ATM transport network is divided into the transport access network, the transport junction network, and the transport high-speed backbone network. The Telecommunication Management Network (TMN) integrates all operation, administration and maintenance functions for these network components.

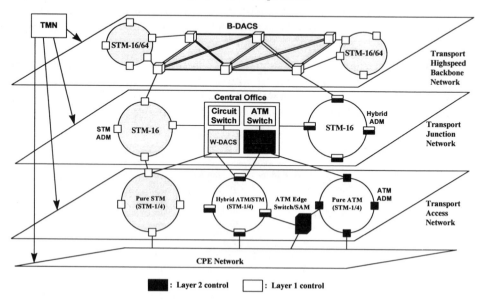

**[Figure 6]:  Layered Bandwidth Management at the Transport Network**

The transport access network collects traffic from a set of end-user CPE network elements and multiplexes it to one or more designated nodes in the network.  SDH/ATM rings connect businesses to the network and can be used for residential access as well. For ATM services the transport access network   provides bandwidth management dynamically by using SAM and /or ATM edge switches  to best utilize the bandwidth [Layer 2 management].

The junction transport network provides grooming, consolidation, and segregation of different facilities and services in the transmission network between switching systems and between carrier terminals. Grooming is a function that allows use of both incoming and outgoing facilities by the cross-connections of tributaries. Various types of DACSs have been deployed for these network requirements. The importance of this grooming function continues to increase as networks evolve to ATM.  Figure 6 shows ATM VPX for the layer 2 management and wideband DACS (W-DACS) for the STM wideband service grooming. The role of W-DACS continues to be important for the growing E1/T1 services as described earlier in the section 3. The ATM/STM hybrid ADMs can be effectively utilized to accommodate the transport access network traffic in the high-speed optical transmission lines.

The transport backbone network uses very high capacity fibers that are administered at the broadband granularity [Layer 1 management]. STM based high speed STM-16/64 rings and broadband SDH cross connects are used in this network to provide STM express pipes that are filled by the Layer 2 management at the access and junction networks.

*Applicable Areas of each layer of management in the ATM/SDH transport networks are summarized as follows:*

*Layer 2 Bandwidth Management is appropriate for dynamic bandwidth reallocation*

- *Edge Network*

    - *ATM Service Access Multiplexer (SAM) and/or ATM Edge Switch*

    - *ATM/STM hybrid rings for low ATM traffic demand and pure ATM rings for high ATM traffic demand*

- *Junction Network*

    - *ATM mesh network (ATM VPX) and ATM/STM hybrid rings*

*Layer 1 Bandwidth Management is appropriate for quasi-static bandwidth reallocation*

- *SDH rings and ATM/STM rings for ring interconnections between transport backbone rings(STM-16/STM-64) and transport access rings (STM-1/STM-4)*

- *Broadband STM mesh network (B-DACS)*


## 6. CONCLUDING REMARKS

We have discussed a bandwidth management strategy in ATM/SDH networks using the strengths of both STM and ATM technology. We have proposed the applicable areas of each layer of bandwidth management from the perspective of long-term transport network evolution toward ATM.

*The recommendations are summarized as follows:*

- *Current STM cross-connect applications continue to be important.*

- *Layer 1 control is appropriate for quasi-static bandwidth reallocation and applicable to high speed backbone network.*

- *Layer 2 control is appropriate for dynamic bandwidth reallocation and applicable to the grooming at edge and junction network.*

- *Layer 2 management may replace wideband STM path cross-connects if stringent cost targets can be met and operations issues addressed for applications not sensitive to delay.*

- *Terminate STM path only when the ATM processing is needed. If the entire network is controlled by the layer 2 management, the SDH path will be terminated at every node even if ATM processing is not needed.*

## ACKNOWLEDGEMENT

The author would like to thank Stuart Waldman and Dennis Doherty for their reviews and comments. A number of others provided useful comments on the earlier drafts, including Bill Goers, Jack Davis, Fred Feldman, Mark Wilson, Patrice Lamy and Kevin Sparks.

## REFERENCES

**K.Y. Eng, R.D. Gitlin and M.J. Karol (1990)** A Framework For A National Broadband (ATM/B-ISDN) Network," IEEE Globecom'90, pp. 308.7.1-308.7.6

**M. Gerla, S. Monteiro, and R. Pazos (1989)** Topology Design and Bandwidth Allocation in ATM Networks," IEEE JSAC, SAC-7(8), pp. 1253-1262

**G. Gopal, C. Kim, and A. Weinrib (1990)** Dynamic Network Configuration Management," ICC'90, Vol. 2, pp. 302.2.1-7

**S. Hasegawa, A. Kanemasa, H. Sakaguchi, and R. Maruta (1987)** Dynamic reconfiguration of Digital Cross-Connect Systems with network control and management," Globecom'87, pp. 1096-1100

**J. Y. Hui (1988)** Resource Allocation for Broadband Networks, IEEE J. On Select. Area in Communication, Vol.6, No.9, pp. 1598-1608.

**ITU-T G.805 (1992)** Architectures of Transport Networks Based on the Synchronous Digital Hierarchy (SDH)

**M. Lee and J. Yee (1993)** A Design Algorithm for Reconfigurable ATM Networks," IEEE Infocom, pp. 2a.1.1-2a.1.8

**M. Logothetis, Michael and S. Shioda (1995)** Medium-Term Centralized Virtual-Path Bandwidth Control Based on Traffic Measurements," IEEE Transactions on Communications, Vol. 43, No. 10, October 1995

**S. Monteiro and M. Gerla (1993)** Topological Reconfiguration in ATM Networks," IEEE Inforcom, pp. 267-271

**T. Noh (1996)** End-to-end Self-healing ATM/SDH Networks," IEEE Globecom'96, pp. 1877-1881

**K. Sato and I. Tokizawa (1992)** Flexible Asynchronous Transfer Mode Networks Utilizing Virtual Paths," ICC'90, pp. 318.4.1-318.4.8

**T. Wu, J. Bartone, and V. Kaminisky (1994)** A Feasibility Study of ATM Virtual Path Cross-Connect Systems in LATA Transport Networks," IEEE Globecom'94, pp. 1421-1427

## BIOGRAPHY

**Tai H. Noh** is a Member of Technical Staff at Bell Laboratories/Lucent Technologies, where he is currently involved in the SDH/ATM integration, end-to-end network and network systems evolution planning, and end-to-end self-healing SDH/ATM network architecture. He joined Bell Laboratories in 1987. He holds a B.S. from Seoul National University, an M.S. from Rensselaer Polytechnic Institute, and a Ph.D. in Computer Information Engineering from Stevens Institute of Technology.

# 20

# Unified Fault, Resource Management and Control in ATM-based IBCN

*S. Sartzetakis*
*Institute of Computer Science, Foundation of Research and Technology - Hellas, Heraklion, GR71110, Crete, Greece tel:+30 81 391727, fax:+30 81 391601, e-mail: stelios@ics.forth.gr*

*P. Georgatsos*
*Algosystems S.A., 4 Sardeon St., N. Smyrni,17121 Athens, Greece tel: +301 9310281, fax:+301 9352873, email: pgeorgat@algo.com.gr*

*G. Konstantoulakis*
*NTUA, Computer Science Department, Telecommunications Lab, Heroon Polytechniou 9, 157 73 Athens, Greece tel: +301 772 1494 fax: +301 772 2530, email gkonst@telecom.ntua.gr*

*G. Pavlou, D. P. Griffin*
*Department of Computer Science, University College London, Gower Street, London WC1E 6BT, UK. tel: +44 171 419 3687, fax: +44 171 387 1397, email: {G.Pavlou, D.Griffin}@cs.ucl.ac.uk*

## Abstract

In this paper we present the initial specification of a system that covers both the control and management planes of network operation, with emphasis on fault, performance, and configuration management. The system incorporates rapid and reliable ATM network layer self-healing mechanisms, intelligent load balancing, dynamic routing and resource management functions all interworking together with the overall goal to ensure cost-effective network performance and availability under normal and fault situations. Restoration mechanisms in the ATM network layer are integrated with the control and management plane functionality, aiming at providing an integral and network-wide treatment to the problem of fault recovery. The system is designed and is being implemented adopting the emerging TINA framework and using a CORBA-based distributed processing environment. The validity and effectiveness of the system are assessed and demonstrated through international ATM field trials.

## Keywords

Integrated ATM control and management. Fault management and network survivability. Dynamic routing, self-healing, OAM. TINA-C, CORBA DPE.

# 1   INTRODUCTION

In order to minimise the cost of building and maintaining wide area ATM networks, efficient and cost-effective resource management techniques are needed to avoid the over-allocation of resources. This requires building a network that has the intelligence to monitor, control, and enforce its own QoS agreements, guaranteeing the high availability expected without the unnecessary cost of over-resourcing

Research in resource and routing management has led to a number of optimising algorithms, tackling the problem of network availability and performance in isolation, not considering the interactions between the control and management functions. It does not take explicitly into account the diversity of performance and bandwidth requirements of the many service classes supported by the networks. Furthermore, the work to date does not take an integrated approach to resource control, routing and alarm management from the perspective of network reliability and availability.

Recognising the need for combining the functional capabilities of the control and management planes for the benefit of network operation, this paper introduces the REFORM (REsource Failure and restORation Management in ATM-based IBCN) (REFORM, 1997) system with the purpose to ensure network performance and availability within acceptable levels under normal and fault conditions. The specifications of this system both from the functional and architectural aspects is the main theme of this paper.

Our system integrates control and management plane functionality for providing a complete, network-wide solution to the problem of network availability under normal and fault conditions. Specifically, control plane functions such as route selection, Operation Administration and Maintenance (OAM), and self-healing mechanisms are integrated with higher level network-wide routing and resource management functions.

We treat the problem of network availability in ATM-based IBC networks in an integral manner. The term "integral" refers to the ability to:

- cover the complete failure management cycle, that is prior, post failure, and failure normalisation phases;
- involve both control and management plane functionality.

While research has already been conducted in the area, it is the first time that a complete system is introduced that provides:

- **an integrated approach to the problem of network performance and availability**, considering the whole failure management cycle and integrating the OAM and control mechanisms with the network management functions in the performance and fault management areas;
- **intelligent load balancing, OAM restoration, dynamic routing and spare resource management** mechanisms all **interworking together**, explicitly taking into account the multi-service nature of the network environment.

The REFORM system has been designed according to the TINA architectural framework, using a CORBA-based distributed processing environment with all the advantages this brings and which are explained later in this paper. While the distinction between the control and management planes is maintained at a conceptual level, this distinction is relaxed during the system design and implementation phases. Part of control and management plane functionality will be modelled and implemented based on a DPE platform, using a uniform set of DPE services. The REFORM system is one of the first systems to provide such integrated solutions.

In the following section we provide briefly some background information we consider helpful in understanding the scope of our system. Section 3 presents the REFORM system specifications, and section 4 presents the REFORM architecture. Following that, we describe a realistic environment scenario that is used to assess the functionality and performance of the

REFORM system as whole. Finally Section 6 summarises the paper and describes aspects of future work.

# 2    BACKGROUND INFORMATION, STATE-OF-THE-ART

Considering networks as multiple-resource, multiple access systems, an obvious issue is how access to the resources is granted. **Resource management** refers to the necessary means and functions for resolving such contention problems with the overall objective to obtain solutions satisfying certain cost-effective criteria but constrained by general performance requirements. ATM networks by their very asynchronous multiplexing nature and the flexibility of the VPC layer that they offer, provide several degrees of freedom in traffic multiplexing. Although this flexibility eases the task of traffic admission and multiplexing it may have a disastrous effect on network performance if it is not managed properly. The problems and indeed the solutions of resource management are related to those of routing management - routes in ATM networks are defined in terms of VPCs and VPCs are defined in order to support routing.

**Routing** functionality is spread over the control and management planes of the network and the network elements. Route selection is part of the call control functions running at the network switches, whereas route definition functionality belongs to the management plane or it could itself provided by the control plane, on the basis of suitable distributed routing protocols (e.g. PNNI). Different sets of route may be defined for different connection types according to their performance requirements.

Successful **fault management** requires that information is gathered from network elements in a standardised way. Practical experience on existing networks shows that some fault scenarios involve alarm burst phenomena (e.g. uncontrolled pouring-in of numerous redundant alarm notifications) which tend to overload the management system.

Network robustness can be achieved with the deployment of autonomously operating **restoration mechanisms**. These restoration mechanisms reside at different network levels and different sub-networks and inter-work with each other through escalation schemes. The autonomously operating restoration mechanisms, provided by the network need to interact with the network management plane, in order to achieve the goal of network robustness. The network management system can simply monitor the restoration process or it can also actively participate in this process in order to complement and optimise the effect and operation of the self-healing actions taken at the control plane. Fast control plane restoration (self-healing) mechanisms as well as medium or longer term restoration mechanisms, in the management plane must be accommodated.

There are two possible self-healing techniques that can be used for restoration management. The first is the **pre-assignment of VPCs**, that are characterised as back-up VPCs; the second is a *flooding* based dynamic route search scheme. The pre-assignment of the back up VPCs yields several benefits compared to the flooding based dynamic route search scheme used in most existing self healing schemes. Its primary advantages are restoration rapidity and realisation of path restoration between path termination nodes. In this scheme there is the need for the management of back up VPCs as well as of original VPCs. In particular this restoration mechanism requires a message transmission system which may be based on OAM flows or other standardised protocols (e.g. PNNI), a spare resource management algorithm which is based on iterative algorithms in order to achieve optimum utilisation of network resources and also a well defined failure management cycle which can be realised as components of a management system.

The ATM-based networks operation and maintenance is organised in a layered fashion. Five hierarchical **OAM** layers and the relative OAM information flows have been defined (F5-1: Virtual Channel, Virtual Path, Transmission Path, Digital Section, Regenerator Section). The two first OAM layers (F4-F5) are in the ATM layer and the rest three (F1-F3) in the physical layer. The importance of the *integration* between *management system and OAM* has been identified by ITU (1992). Recommendation I.610 proposes an interaction scheme between the TMN and OAM. This scheme involves both X & Q interfaces for the entire path starting from the customer premises, passing the access local exchange up to the core network.

# 3    REFORM SYSTEM SCOPE, APPROACH, AND FUNCTIONAL SPECIFICATIONS

Adopting a network operator's viewpoint and by taking into account the broadband and multi-service nature of IBC network environment, the scope of the REFORM system is to provide the necessary means and functions for ensuring network performance and availability within acceptable levels under normal and fault situations. *Ensuring network performance* means the network will be able to maintain the performance of the existing connections within acceptable levels. *Ensuring network availability* means the network will be able to set-up connections guaranteeing their performance requirements. To achieve this, the REFORM system offers:

- fast fault detection, alarm indication and report functions;
- rapid and reliable network self-healing mechanisms for resource and service restoration, spread across both the control and management planes of the network operation;
- efficient dynamic and/or static resource and routing management schemes and algorithms with inherent load balancing functionality, coping with fault conditions; and
- efficient resource migration algorithms.

All these will interwork together with the overall goal to ensure the:

- cost-effective network performance and availability in normal conditions; and
- cost-effective, reliable and robust network recovery in the performance of existing connections as well as of the availability for new connections, from fault situations.

The term **self-healing**, as used above, refers to the ability of the network to reconfigure itself around failures quickly and gracefully with the goal of restoring service availability within acceptable levels for existing as well as future connections. Self-healing implies resource restoration by means of a distributed mechanism, as opposed to centralised schemes. Self-healing mechanisms mainly reside in the control plane (one node), while higher level decisions can be escalated to the management plane. The REFORM system provides self-healing mechanisms at both the control and management planes. The latter is provided as part of suitable distributed routing and resource management schemes, dynamically adapting to fluctuating traffic load and in cases of resource failure. Centralised protection is also provided through management activities for defining suitable sets of routes and for allocating spare resources, in a cost-effective manner. These activities are adaptive in a quasi-static form regarding network resource failures, redesigning new sets of routes or reassigning spare resources. When failures occur and the lower-level self-healing capabilities of the network cannot succeed in restoring network availability under the current traffic conditions these activities may also be triggered. The REFORM system integrates all these functions through a carefully designed hierarchical architecture. Aspects of this architecture are discussed in Section 4.

To facilitate hierarchical designs and functional decomposition, we adopt the distinction between the management and control planes in the operation of networks. Such a distinction is necessary since it maps to the different domains and actors' interfaces currently found in the telecommunications arena; those of network equipment manufacturers, network operators and management system providers. In general, the control plane is required for the operation of the network, including the base functionality of signalling, Call Control (CC), and Operation

Administration and Maintenance (OAM). The management plane is required for the optimal operation of the control plane, by tuning and appropriately managing various operational parameters in the control plane.

With respect to fault management, the approach adopted is that the control plane encompasses the functionality of fault detection and fast-responding self-healing mechanisms through the OAM capabilities that it offers. The management plane encompasses resource and routing management functionality for tuning and optimising the operation and effect of the control plane actions as well as for guaranteeing cost-effective allocation and usage of the spare resources in the network. Re-routing is also provided for restoring network availability to future connections. It is performed through a mixture of distributed and centralised logic, with the purpose to avoid damaged network areas while at the same time relieving the load from the back-up network resources. If such activities fail to restore network performance and availability given the current traffic conditions, other management activities will be activated (e.g. for defining new sets of routes, service migration). While the control plane functionality provides a fast reaction to a fault situation, the management plane provides the required short, medium or longer term reactions, depending on the severity of the failure and the actual traffic conditions, complementing the effect of the self-healing actions taken at the control plane. We broaden therefore the scope of restoration management beyond the level of self-healing mechanisms for service restoration at a local level. By incorporating effective resource and routing management functionality -suitably coping with fault conditions- our system restores service availability for existing and new connections, at a network-wide level. We build upon the results of the RACE II projects ICM (1997), IMMUNE (Nederlof, 1995), and TRIBUNE, and on other proven results in the literature (Fujii, 1994), (Griffin, 1995).

## 3.1    System Functional Specifications

We are concerned with failures caused in or escalated to the ATM network layer. For example, when a physical link failure has occurred and cannot be recovered by the restoration mechanisms in the physical layer(s), the fault is escalated to the ATM layer. This type of failure will cause a large number of alarms, as many as the VPLs defined on this link. REFORM therefore will provide for alarm correlation functionality in order to determine the original link failure. Faults in the ATM layer are also concerned with equipment related failures e.g. VP cross-connect, VPC TP (Termination Point) failures. The failures will be identified through the standardised OAM alarms. Restoration mechanisms and escalation procedures in the network levels below the ATM layer are not addressed.

The REFORM system incorporates rapid and reliable ATM layer self-healing mechanisms, intelligent load balancing, dynamic routing and resource management functions all interworking together with the overall goal to ensure cost-effective network performance and availability under normal and fault situations. Cost-effective solutions are pursued, on one hand by means of considering suitable objective functions in the optimisation problems corresponding to the route definition, spare resource allocation and management algorithms, and on the other hand by considering different *survivability service classes*.

In the rest of the section we outline the scope of the functionality of our system per each of the network phases with respect to failures. We concentrate on describing how control and management functionality is integrated to fulfil the objectives of our system.

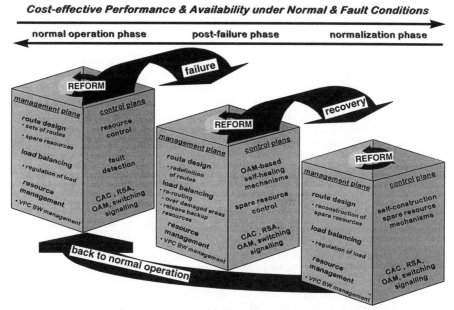

**Figure 1** The REFORM system during the network operation phases.

During **normal operation** of the network, the system incorporates intelligent routing, load balancing and resource management functions, which tune the operation of the control plane through appropriate management actions. All this functionality is provided through a suitable hierarchy corresponding to different levels of abstraction and time-scales. The overall aim during this phase is to guarantee effective utilisation of network resources and cost-effective network availability within acceptable levels, while ensuring that the performance requirements of the various service classes are met. The *route design functions* are responsible for the design of appropriate admissible sets of routes per connection type, based on previously defined sets of VPCs, taking into account the performance requirements of the different connection types that the network supports. The route design functions are also responsible for the allocation of appropriate spare resources in the network according to certain cost-effective objectives and an overall protection strategy to provide as many options as possible for bypassing potential faults. Additionally, they should provision for adaptability to cater for unpredicted traffic fluctuations but according to the constraints of cost-effectiveness. The *load balancing functions* (which may distributed or centralised) operate on the defined sets of routes trying to influence the routing decisions taken in the network switches, by conveying to them network-wide information. Through their activities load distribution may be regulated over the whole network. Note that balanced networks have the potential benefit of minimising disruptions of existing connections in case of network eventualities (node, link failures). The control plane encompasses the base functionality of CC, including routing and CAC, signalling and OAM. The routing functionality in the control plane is responsible for taking the actual routing decisions by means of suitable route selection algorithms. The control plane routing algorithms may also perform load balancing at a local level (i.e. at the vicinity of a node). The resource management functions are responsible for the management of the bandwidth of the VPCs according to actual network conditions. The system employs such functions both in the control and management plane, differing in terms of time-scale and abstractions. They try to ensure effective usage of VPCs and to avoid congestion at the cell and the connection level.

During the **post-failure phase** the system employs fast-responding self-healing functions in the control plane through the OAM facilities that it offers. The system follows a self-healing scheme based on pre-assigned back-up resources, rather than based on flooding messages. The standardised OAM flows are used to provide a fast and reliable means for inter-node message exchange. In addition, the adaptive, distributed and/or centralised routing, load balancing and resource management functionality of the system, with their inherent self-healing capabilities, will be activated for tuning the operation and complementing the effect of the control plane actions, to a network level. The overall goal during this phase, is to ensure a rapid recovery of network performance and availability, taken into account the actual traffic conditions. The entire system functionality is properly integrated through the levels of the REFORM hierarchy. The system combines the merits of distributed and centralised control schemes during the restoration process. Once a failure has been detected, the control plane will react through its OAM-based self-healing procedures and the associated automatic re-routing mechanisms. Spare resource control is also required for resolving competition as spare resources are normally shared. The load balancing and resource management activities in the management plane will also be in effect, through their own self-healing capabilities for providing a first level network-wide treatment to the restoration procedure. The load balancing functions will influence routing decisions so that future connections avoid routes traversing damaged paths, while at the same time taking care not to overload the back-up routes. The resource management activities will try to ensure that sufficient bandwidth on the back-up resources can be allocated. All these activities will take a network-wide view and the actual network traffic conditions. The system monitors network performance and availability with a wide perspective during the restoration process, in order to assess the effectiveness of the actions taken so far. If network performance is found unacceptable, the centralised routing functions will be activated for assigning new sets of routes and VPCs as necessary. The system allows for different *survivability classes* for prioritising access to network resources. If required, the restoration process may also be extended to higher levels for performing appropriate service migration functions, an extension of the system currently under study.

The **normalisation phase** is concerned with the reconstruction of routes and spare resources to cater for the predicted traffic and to build in flexibility to allow a rapid response to subsequent failures. The overall goal is to enter the normal operation mode by causing the minimum of disruptions to existing connections. The route design functions will determine new sets of routes and spare resources according to a general strategy satisfying certain cost-effective criteria and taking into account network usage predictions. Different policies for spare resource reconstruction, like for example distributed self-reconstruction mechanisms are under investigation. The load balancing and resource management functions operate as in the normal operation case taking into account the newly resulted configuration. The control plane OAM and routing functionality will also be activated according to the reconstruction directives given by the management plane.

The detailed description of a functional model decomposing the above functionality into individual components and specifying the information exchanges between them is beyond the scope of this paper.

# 4   SYSTEM ARCHITECTURE

The key aspect of the REFORM architectural approach is the adoption of the emerging TINA framework with the purpose of evaluating its suitability and assessing its potential benefits when applied to network management, while comparing it as well to established technologies

(OSI-based TMN, Internet SNMP). We use a commercial CORBA-based platform for providing the Distributed Processing Environment (DPE) for designing and implementing both the control and management plane functionality. Existing TINA-C architectures in the area of Resource Management are taken into account. Also existing TMN architectures for configuration and performance management, such as the ICM Virtual Path Connection and Routing Management (Griffin, 1995), (ICM, 1997), constitute the starting point for the management plane architecture; these are enhanced and modified in the light of the TINA-C developments and ported on the CORBA-based DPE. They are complemented by new functionality, not yet addressed by TINA and not available in existing TMN architectures.

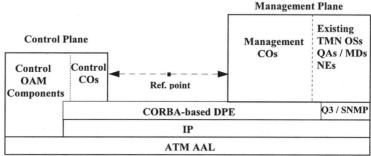

**Figure 2**    Overall REFORM system architecture.

The overall system architecture is shown in Figure 2. There exist two major domains, related to management and control plane respectively. The major underlying assumption is the use of TINA-C engineering principles in the form of a CORBA-based DPE.

The detailed architecture of the management plane takes into account state of the art work in this area i.e. in TINA-C and the RACE-II ICM project. The management plane functionality is specified as Computational Objects (COs) for Fault, Performance, Connection and Resource Configuration Management. Existing TMN applications from previous work - Operations Systems (OSs), Q-Adaptors (QAs) - are integrated in the REFORM system. Also, Network Elements (NEs) are accessed through TMN Q3 or SNMP interfaces. The integration of those elements and applications in the TINA-based REFORM environment raises issues of migration from TMN to TINA, which are discussed below. The architecture of the management domain has both the hierarchical aspects, as dictated by the TMN layered architecture which is adopted by TINA, and also the flat peer-to-peer aspects which are facilitated by the use of a DPE-based computational decomposition. Federation of activity between management domains is also considered.

The control plane functionality will be supported by relevant control logic and OAM transport protocols operating directly over the ATM adaptation layer. The state of the art in terms of projects such as IMMUNE and TRIBUNE together with the relevant ongoing ITU-T work is the starting point. The integration between the control and management plane domains will be accomplished through the interaction of computational objects over the DPE. As such, the control plane domain includes computational objects for the purpose of interfacing to the management plane. Such interactions are expected to be bi-directional.

All functions and operational components of our system in the leftmost box in Figure 2 are regarded as local inside a node, and will be implemented according to the particular capabilities of the network elements. We take into account the low level interfaces that provide connection with the hardware subsystem. Thus, direct message exchange will be feasible.

Overall, there are three types of transport in the REFORM system:

- the ATM AAL, which supports the control plane interactions; the direct use of the ATM AAL ensures high-speed interactions but the relevant presentation and distribution facilities are minimal;
- the TMN Q3 using RFC1006/TCP/IP over the ATM AAL; this supports management plane interactions for the existing TMN applications; and
- the CORBA-based DPE using TCP/IP over the ATM ALL; this supports the majority of the management plane interactions and the interactions between the control and management domains.

There is also a possibility to use the CORBA-based DPE directly over the ATM AAL. This depends on developments on CORBA mappings in the lifetime of REFORM.

In order to be able to test our system under stress and worst case conditions all OAM protocols are integrated within the testing systems we use for our experimentation. The test tool is attached to the switches and monitors the data streams, OAM flows and collects statistical information. On the other hand we are able to generate fault in the OAM flows and collect statistical information. We are also able to generate faults in the OAM streams in order to cause predefined error profiles.

## Migration from TMN to TINA
REFORM tackles aspects such as how to integrate or migrate towards the TINA environment, using in particular the DPE engineering model and the problem decomposition from the enterprise, information and computational viewpoints.

Given the fact that network elements will support TMN Q3 or SNMP management interfaces while existing TMN management applications (OSs) with Q interfaces may be reused, issues in migrating these components to the TINA environment will have to be tackled. In addition, existing TMN architectures for ATM resource management (e.g. ICM) will be reworked according to the new methodologies and the architectural input from TINA-C in these areas.

In the key area of integrating elements or applications with Q3 interfaces, the X/Open Joint Inter-Domain Management (XoJIDM, 1996) task force has been investigating the similarities and differences between GDMO/ASN.1 and CORBA IDL object models and proposes generic translations in both directions. It should be mentioned that elements with SNMP interfaces may also be seen as Q3-capable elements through generic translation: the ICM project has contributed to the relevant NMF specifications and has developed a generic Internet Q-Adaptor (IQA) which is available to REFORM.

Given the fact that fault and performance management need fine-grain event reporting/logging and sophisticated query and retrieval facilities, REFORM will assess the suitability of OMG Common Object Services (COS) to support such functionality. Specialised facilities in addition to the OMG common services will also be investigated, such as management brokers that enable multiple object access and sophisticated information retrieval, sophisticated event reporting and logging facilities, metric monitoring and summarisation facilities etc. In general, the full expressive power of the OSI Management / TMN framework needs to be available over the CORBA-based DPE.

## 5    AN EXPERIMENTATION SCENARIO

Our work is centred around a number of trials and experiments to validate the system in a realistic environment. The field trials are based on European National Host (NH) platforms (ACTS, 1997). The NHs are consolidated advanced platforms comprising communications infrastructure, services and generic applications. The NHs are made available to projects interested in performing operational trials or experiments of leading   edge applications, services and management that involve real networks, services and users.

As an example of a typical, experimental scenario, the following describes a fault in the delivery of video services and the associated activity of the REFORM system to resolve the problem. Consider a video distribution service consisting of a video server which distributes a video stream to the users of the service on demand. (Figure 3). Faults can affect this service in a number of ways. For example: a user may be unable to access the video distribution server to request a new video stream; an existing video stream may be interrupted; or the video server itself may collapse. In the following a fault caused by a link failure is discussed.

On failure there are a number of ways of restoring the service. These can be at many levels of the REFORM architecture: from the reception of OAM alarms and the invocation of fast, localised self-healing mechanisms through to network wide, off-line redesign of the network routes available to the video services. The goal of our system is to enable these mechanisms to work together in an integrated way. By the co-operation of algorithms and functions in each of the architectural levels the network can be made operational as quickly as possible through the immediate response of localised algorithms embedded in the network elements while the higher level management algorithms restore efficiency in the network as a whole over a longer time period.

Figure 3 shows a simplified view of a sample network configuration formed by the EXPERT testbed (involving the Swiss NH and the Dutch NH) and the Norwegian NH testbed. A video distribution server is attached to the LATEX node of the Swiss NH, there is a single user group attached to the NT2 node of the same network and a second user group located in the Dutch NH.

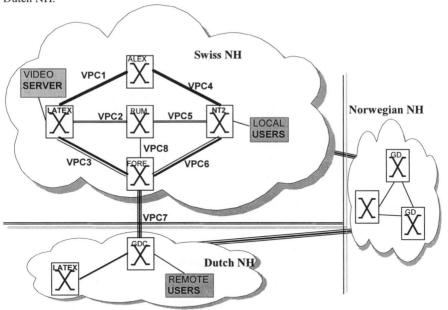

**Figure 3**  The NH set-up for the REFORM experiments.

Video streams are transmitted via VCCs which are set up on demand between the users and the server and routed over previously installed VPCs. In this example, two possible routes have been assigned to video traffic between the video server and the local users on the Swiss NH: over VPC1 and VPC4; or over VPC2 and VPC5. Similarly, two routes have been assigned to the remote users at the Dutch NH: over VPC3 and VPC 7; and over VPC2, VPC8 and VPC7.

The fact that two routes have been assigned for each source-destination pair is implemented in the network by alternative route selection entries in the VC switches of the network nodes. When a connection request is made, Route Selection Algorithms (RSAs) determine which route should be used (Georgatsos, 1996). The route selection entries are defined initially by the route design algorithms in the upper levels of the REFORM architecture (note that this is an example of the co-operation between the management and control levels of the REFORM architecture).

During normal operation of the network, the RSAs select appropriate routes for new connections according to their algorithms (Georgatsos, 1996). The load balancing functions manage the route selection criteria used by the RSAs to balance the load over the network as evenly as possible by altering the selection priorities associated with each route selection entry.

In this scenario we now introduce a fault in the link between the LATEX and ALEX nodes in the Swiss NH, so putting VPC1 out of service and disrupting the video streams which were allocated to VCCs routed over this VPC.

On failure of this link, the LATEX node is notified by a control plane signal (OAM alarm) that the link is out of service. On receipt of this signal, OAM-based restoration protocols will attempt to restore the failed link. Because this may take a large amount of time in the case of physical failures requiring human intervention, the local control plane self-healing algorithms will prohibit the RSA from selecting this VPC for future connections. As a secondary response the control plane algorithms will try to obtain as much bandwidth as possible for the alternative route (VPC2 and VPC5) from the common pool bandwidth which is continually updated by the higher level bandwidth management functions which redistribute capacity to existing routes and VPCs according to measured and predicted traffic levels.

As new connections are established they are routed over the only remaining route and it is likely that it will become overloaded leading to connection rejections. Before this situation arises, the Load Balancing component, in the management plane, will register a high traffic load compared to the network wide average and attempt to resolve this problem. However, there are no alternative routes available from the LATEX node, and Load Balancing is unable to resolve the potential problem.

At this point, Load Balancing informs its superior component in the REFORM management hierarchy, Route Design, of an unbalanced network which it is unable to resolve (Griffin, 1995). Route Design then attempts to redesign the VPC network and the routing plan to cope with the new physical topology. In this case potential new routes are considered (VPC2, VPC8 and VPC6; or VPC3 and VPC6) according to existing and predicted traffic and the quality constraints of the video service (for example there may be a maximum number of links and nodes video traffic may be routed over due to maximum allowable delay figures) and a new route is established if possible. The management decisions are implemented in the network by updating route selection tables in the RSAs at the corresponding nodes. As time progresses Load Balancing will update the route selection priorities for the new and the already existing VPCs, to balance the load in the overall network and in the control plane the RSAs will make all new routing decisions according to the new routing table entries.

The above is an outline of the functionality of the REFORM system highlighting the co-operation required in an integrated system covering the control and management planes. The overall problem of fault, resource management and control is handled by a number of layers. Each layer has a well defined limit of responsibility with operational parameters defined and managed by the layer above.

# 6 CONCLUSIONS, FUTURE WORK

We introduced an integrated framework for ensuring network performance and availability in multi-class ATM networks under normal and fault conditions. We integrated control and management plane functionality, covering the complete failure management cycle (prior, post, normalisation). We combine in one system:

- rapid and reliable network self-healing mechanisms, spread across the control and management layers of the network operation;
- intelligent routing coping with fault conditions, taking into account the multi-service environment;
- intelligent load balancing functionality with inherent self-healing capabilities;
- cost-effective spare resource allocation and appropriate management schemes

We provide a number of generic and specific software components to offer solutions to the problems of fault, and routing, management for ATM networks. These components can be exploited by the existing network operators, or provided to new network operators as part of a packaged solution.

Our goal is also to make substantial contributions to TINA-C especially in the fields of fault and performance management as well as to the definition of the migration path from currently used management frameworks, like TMN, to emerging ones, like TINA. Since we adopted the emerging TINA framework based on a commercial CORBA-type platform, we will evaluate, validate and demonstrate through experimentation in real network environment the suitability of such approaches to network management.

We plan to carry out field trials promoting in parallel the use of advanced networks and services as they are provided by NH environment incorporating real user communities in Europe. This way we demonstrate how the functional capabilities of the developed system can improve performance and reliability of end-to-end connections.

Much of the above work is still on-going, and experimentation is planed for this and the next years. Up to date information is widely available on line (REFORM, 1997).

## Acknowledgements

REFORM is a research project (July 96 - June 99) of the ACTS programme partially funded by the Commission of the European Union. The authors wish to thank all their project colleagues who contributed in many ways to the formation of the ideas described here. Without them this work and our co-operation for these three years would never have started.

# 7 REFERENCES

REFORM (1997) "REFORM System Requirements and Analysis", Deliverable D1, A208/Bell/WP1/DS/P/002/b1. On-line information on the REFORM system is available under: http://www.algo.com.gr/acts/reform

ACTS (1997) On-line infomation: http://www.uk.infowin.org/ACTS/

Fujii H. and Yoshikai N. (1994) "Restoration Message Transfer Mechanism and Restoration Characteristics of Double-Search Self-Healing ATM Network", IEEE Journal on Selected Areas in Communications, Vol.12, No.1.

Georgatsos P., Griffin D. (1996) "A Management System for Load Balancing through Adaptive Routing in Multi-Service ATM Networks", INFOCOM 1996.

Griffin D., Georgatsos P. (1995) "A TMN system for VPC and routing management in ATM networks", Integrated Network Management IV, Proc. of 4th. ISINM 1995, ed. A.S.Sethi, et al., Chapman & Hall, UK, 1995.

ICM (1997) "Integrated Communications Management of Broadband Networks", ed. Griffin D., Crete University Press, Heraklion, Greece, ISBN 960 524 006 8.

ITU-T (1992) Recommendations: M.3010, Principles for a Telecommunications Management Network, Geneva, October 1992. / M.3200, TMN Management Services: Overview, Geneve, October 1992. / M.3100 Generic IM / I.610 / ITU-T SG4 "TMN Architecture and Principles, Management Services, Generic Object Model" / ITU-T SG11 "TMN Protocols and Interface Models"

Nederlof L., et al (1995), "*End-to-End Survivable Broadband Networks*" IEEE Communications Magazine, September 1995.

TINA-C (1994) Documents: Overall Principles and Concepts of TINA (1995), Fault Management and Resource Configuration Management, Connection Management Architecture, Management Architecture, Network Resource Information Model Specification.

XoJIDM (1996) (X/Open Joint Inter-Domain Management task force) NMF-X/Open; Proposal "Inter-Domain Management: Specification Translation" (from ASN.1 to IDL).

# 8    BIOGRAPHIES

**Stelios Sartzetakis** received his BSc degree in Mathematics from Aristotelian University of Thessaloniki in 1983, and his Masters in Systems and Computer Engineering from Carleton University of Ottawa, Canada in 1986. He joined ICS-FORTH in 1988 where he has been responsible for FORTH's telecommunications infrastructure at large, principal in the creation of FORTHnet, and participated in a number of EU funded projects. Today he is senior telecommunications engineer in the networks group (www.ics.forth.gr/~stelios), responsible for research projects in broadband telecommunications networks and services management.

**Panos Georgatsos** received the B.S. degree in Mathematics from the National University of Athens, Greece, in 1985, and the Ph.D. degree in Computer Science, with specialisation in network routing and performance analysis, from Bradford University, UK, in 1989. Dr. Georgatsos is working for Algosystems SA, Athens, Greece, as a network performance consultant. His research interests are in the areas of network and service management, analytical modelling, simulation and performance evaluation. He has been participating in a number of telecommunications projects within the framework of the EU funded RACE and ACTS programmes.

**George Konstantoulakis** was born in Chania, Greece at 1968. He received the Degree in Electrical Engineering from NTUA in 1992. In 1991 he joined the Telecommunications Laboratory of NTUA where he finished his thesis and now finishing his Ph.D. His research interests are in the area of broadband communication networks, high speed - real time architectures and algorithms. He has over ten publications in the above areas. Member of the IEEE and the Technical Chamber of Greece by which he took the reward of the new engineer with the best graduate thesis (1993) at national level. He is reviewer of IFIP TC6.

**George Pavlou** received his Diploma in Electrical and Mechanical Engineering from the National Technical University of Athens in 1982 and his MSc in Computer Science from University College London in 1986. He has since worked in the Computer Science department at UCL mainly as a researcher but also as a lecturer. He is now a Senior Research Scientist and has been leading research efforts in the area of management of broadband networks and services.

**David Griffin** received the B.Sc. degree in Electronic, Computer and Systems Engineering from Loughborough University, UK in 1988. He joined GEC Plessey Telecommunications Ltd., UK as a Systems Design Engineer working on TMN architectures and ATM traffic experiments. In 1993, Mr. Griffin joined ICS-FORTH as a Research Associate on the EU RACE II ICM project. He joined UCL in 1996 and is currently employed as a Research Fellow working on a number of EU ACTS projects in the area of resource management for TINA systems covering performance, fault, configuration and accounting management.

# 21

# Performance Management of Public ATM networks - A Scaleable and Flexible Approach

*R. Davison, M. Azmoodeh*
*BT Laboratories,*
*Martlesham Heath, Suffolk, UK,*
*rob.davison@bt-sys.bt.co.uk, manooch.azmoodeh@bt-sys.bt.co.uk*

### Abstract

ATM technology promises a flexible, multi-service network that will support the broadband future. It will need to be robust and efficient in delivering quality of service and this will not be achieved by network protocols alone. Performance management systems will be needed.

The flexibility required from future networks will also be required from their management systems. Approaches derived from the developing technologies of distributed object-oriented computing (CORBA) and distributed artificial intelligence (agents) could allow management systems to be implemented as sets of small-grain, co-operating distributed objects. Such systems will be more flexible, more robust and easier to modify .

This paper presents a framework being developed for understanding performance management functions and an approach to their implementation based on agent technology and the use of CORBA platforms.

### Keywords

performance management, ATM, agents, virtual path management

## 1. INTRODUCTION

ATM technology offers the opportunity for constructing networks which can carry many types of service including services beyond those considered when the networks were designed. This will be achieved through the flexibility ATM offers to support diverse bandwidths and quality of services. There are still technological as well as economic uncertainties on how ATM will be developed in the communication networks of the future, however it is already clear that the new flexibilities offered by

ATM will imply a need for more complex management and control than for conventional single service networks. This is crudely illustrated in figure 1. Each graph represents a space bounded by the number of service types available in a network, the network's efficiency and the quality it offers. Figure 1(a) illustrates how for a conventional network the network design process results in network which can carry few service types with a bounded range of efficiency and quality. The role of network management can be seen as restricting that range further to give the desired quality and efficiency. Figure 1 (b) shows how the possibilities covered by a designed ATM network are much larger and hence network management has to do more to restrict the network to the required behaviours.

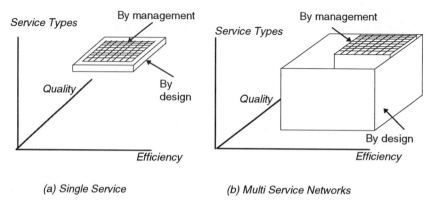

(a) Single Service                (b) Multi Service Networks

**Figure 1** The Behaviour Range of Conventional Networks  vs. The Behaviour Range of ATM Networks.

The functional area most greatly affected by the move to multi-service ATM networks is performance management. This is responsible for ensuring that the multiple service types receive their contracted QoS. Sections 2 and 3 of this paper discuss the necessity of performance management functions and the reasons for their complexity. Section 4 describes architectural ideas used to classify and organise the necessary functions to manage the performance of a large ATM network.

The flexibility offered by ATM networks will only be of benefit if it can be matched by flexibility in its management systems. Existing management systems are expensive to maintain and difficult to change, so a major challenge is to devise a management platform which can adapt to the evolving nature of future networks. The fields of distributed object-oriented computing and distributed artificial intelligence could offer benefits here and are discussed in section 5. As part of our ongoing research we have built a management system that manages ATM virtual paths and was designed and implemented using the concepts expounded in this paper. This is discussed in section 6.

The work described in this paper is based upon a scenario deployment of ATM to support a large, public ATM network carrying a diversity of services including both data and delay-sensitive services. The network will have call control using signalling and its own protocols for traffic and congestion control. Our concern here is the management of the ATM overlay network and not with interactions with lower (e.g. SDH) or higher (e.g. IP) layer management and control functions.

## 2.    THE NEED FOR ATM PERFORMANCE MANAGEMENT SYSTEMS

It is often claimed that future networks will be 'self managed'. This term is not properly defined. However, it is usually derived from the 'Internet' style of networks where 'seemingly' there is little management control. Although, strictly speaking this assertion is not true, the intended meaning is that the network protocols (such as IP or ATM protocols) will detect and resolve all performance management issues. We believe that this view ignores two key aspects :

- that control actions are needed at many different time-scales including some that may span hours, days or even weeks and are appropriately addressed by management systems;
- that network providers need the ability to modify a network's behaviour, for instance to increase utilisation, for external business reasons which again requires management functions.

It is certainly true that developing network controls are expanding the range of circumstances that can be coped with by control alone, but there will always be a need for management systems. This does not imply a need for large, centralised systems. The functionality required could be implemented in a highly distributed manner as will be described in section 5.

A second statement that has been made is that due to development of new technologies, bandwidth as well as switch processor speed and buffer capacities will be so plentiful that performance management is unnecessary. It is fairly obvious that for lightly loaded networks, congestion is less likely to occur, so a suggestion would be that as loading increases the resources (bandwidth, buffer capacities, processor's speed) available should be increased to avoid congestion problems. As expanded in (Jain, 1992) and (Jain, 1990), this is likely to move the point of congestion (for instance from an overloaded circuit to an overloaded buffer). It may be argued that a balanced re-configuration - where all nodes and links are upgraded -would avoid this problem. However, the burstiness of traffic sources will mean that simultaneous bursts can occur, resulting in worst case scenarios that are dramatically different from normal conditions. A network designed to support these extreme cases is unlikely to be economically viable, so some form of management is necessary to cope with these cases.

An interesting area of study is the trade-off between the cost of implementing management systems and the costs of network hardware. Currently, network utilisation is a key factor for providers. However it is possible that as network hardware costs fall, it will be cheaper to accept lower network utilisation in return for less active performance management systems.

## 3.    SOURCES OF COMPLEXITY IN ATM PERFORMANCE MANAGEMENT

Having justified the need for ATM performance management systems we move on to address why ATM performance management poses new challenges. Several fundamental aspects of ATM have a significant impact on the functionality required from network management systems. The most important are:

### *Shared physical resources and virtual reservation*
In the PSTN, when a call is set up, physical resources are reserved to support that call during its lifetime. In contrast, in ATM networks virtual resources (VPCs) are reserved and the underlying physical resources are shared. An important consequence is that rather than the result of congestion being call barring as it would be in today's PSTN, the result of congestion will be downgrading of the quality of existing connections. This is a very different behaviour that will impact upon network management.

### *Diverse connection types in a single network*
A number of different connection types are being defined for ATM networks each able to support different kinds of services. Examples are Constant Bit Rate connections that require the same bandwidth throughout their life and Available Bit Rate connections that use flow control techniques to match the bandwidth they can use to the bandwidth available in the network. The different connection types all share the same cell structure and can be carried by the same network. However, their behaviours are fundamentally different in terms of how they use resources, the traffic contracts they will have and how they act in the event of congestion. So a single ATM network will contain connections that behave in very different ways, will impact upon each other and yet will need to be managed as part of a single system.

### *Managing switch buffers*
The asynchronous nature of ATM is enabled by placing buffers in switches that cope with short term variations in load. The size of the buffers and the strategies used to partition and empty them will impact upon quality of service and upon the efficiency that can be produced from the network. Although it is unclear how dynamically these will need to be changed, it seems likely that some management intervention will be needed.

## *Complex traffic patterns make prediction difficult*

An important benefit of ATM is that it provides the ability for users to send highly variable bit-rates into the network providing they stay within agreed parameters such as, for example, mean and peak transmission rates. This does mean that functionality that depends on being to able to predict network usage, such as accepting new connections or preventing congestion by re-routing, are complicated since such predictions will be statistical with potentially large inaccuracies.

Standards bodies such as ATM-F, ITU and ETSI have, so far, dealt primarily with near term issues in the deployment of ATM, such as connection management and fault management. The challenging area of performance management has yet to be adequately addressed by these groups. (ATM-Forum)(ITU-T, I.371)

## 4. A FRAMEWORK FOR DESIGNING ATM PERFORMANCE MANAGEMENT SYSTEMS

This section describes architectural ideas we have been using to understand, specify and start to implement ATM performance management systems.

### 4.1. Two aspects of performance management functions

The actions taken by performance management systems can be divided into two categories: those actions taken in response to impending network congestion or QoS problems and those taken to modify the behaviour of the network system as a result of some business decision. Figure 2 represents this division by representing the two categories as 'direction setting' and 'direction enforcing'.

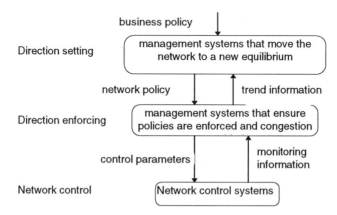

**Figure 2** The two aspects of performance management functionality.

Direction setting functionality would be responsible for converting a business policy - such as providing a new connection type or increasing network utilisation - into its impacts upon different parts in the network. These impacts are then passed on as network policy statements to the direction enforcing functions which use them to configure and manage network controls and resources.   Our work to date has concentrated on direction enforcing management functions which are the topic of the rest of this paper.

## 4.2.   Different layers of traffic variation

It has been widely recognised that cell, burst, and call level traffic variations need to be studied and controlled separately (Key)(Hui).  (Arvidsson) takes this approach further and defines further traffic variations more of interest to management functions. We believe the use of time-scales to aid understanding is useful and have derived our own time-scales as defined in table 1.

**Table 1** Time-scale of events occurring in an ATM network

| Layer | Variation | Example |
|-------|-----------|---------|
| Permanent Environmental /Network | Changes due to permanent change in user population/sub-population behaviour and changes due to permanent change in the network | reduced tariffs, new service types |
| Temporary Environmental /Network | Changes due to temporary change in user population/sub-population behaviour and changes due to temporary change in the network | phone-in, periodic change such as weekday v weekend |
| Connection | Change due to use of a connection | shutting it down, changing its QoS |
| Activity | Changes due to an application level change in the use of a connection | change of scene in a video or downloading next page in a document reading application |
| Burst | changes due to variations in bursts at either the  user terminal or inside network due to a particular combination of bursts | packetisation effects, higher level flow control |
| Cell | changes due to cell level variations | segmentation into cells, shaping and policing functions |

The time-scales were used to capture the different causes of traffic variation in an ATM network.  These causes could then be grouped together to form the problems that a particular network management function should solve. This use of time-scales does not necessarily imply that those functions aimed at solving problems at a particular time-scale have to operate within that time-scale. For instance, a source policing algorithm would have to work at speeds to match cell rates however a management function that set parameters for such an algorithm would work on a much longer time-scale considering trends in traffic measurements. This example illustrates why the time-scales are used to organise network problems rather than being applied to structuring network management functions and is a key difference from (Arvidsson)

## 4.3.   Analysis vs. tuning

A further structuring concept that we have found useful is to break some identified functions into an analytical part and a tuning part. The analytical part takes a more global view and uses analytical techniques to produce a new solution, whereas the tuning part takes a more local, dynamic view and tries to tune parts of a solution. The algorithms used in the analytical part can take in data from a large part or the whole of the network and can have time and space to process this data to produce near-optimal solutions. However we believe that there is a need for a more dynamic form of management that is able to respond more quickly if less optimally. The tuning part only monitors and changes a small part of the network so is able to respond and to maintain a satisfactory state. Eventually, in all cases, variations will be sufficiently great and will require analysis.

The example in section 6 illustrates this approach.

## 5.   TECHNOLOGY TO RE-ARCHITECT NETWORK MANAGEMENT

The flexibility offered by ATM networks will only be of benefit if it can be matched by flexibility and adaptability in its management systems. Existing management systems are expensive to maintain and difficult to change, so a major challenge is to devise a management platform which can adapt to the evolving nature of future networks. As part of our work we have been considering two technologies to aid in this goal - object-oriented distributed computing platforms and distributed artificial intelligence.

### 5.1.   CORBA-based network management

The TMN architecture (ITU-T, M3010) has provided an implementation architecture offering benefits due to structuring and distributing functions. However, since its conception, distributed computing technology has developed significantly enabling :
- a finer grained distribution with benefits of re-use, maintainability and scalability;
- a higher level API, rather than a protocol interface, for application programmers to use.

The developing CORBA standard is an attempt to provide a standardised approach to distributed computing and is seen by many as a key technology in the future of communication management. (OMG)(TINA-C).   Designers of network management systems can progress from an implementation where the unit of distribution is a large system offering many managed objects at its interfaces to one where the unit of distribution is  a small object which offers a few interfaces to manage perhaps as little as a single resource. Figure 3 illustrates this possibility and is derived from (ITU-T, X.703) and (OMG). It shows how a single, large Operations Systems Function in a TMN could be replaced by many, smaller systems in a CORBA-based solution.

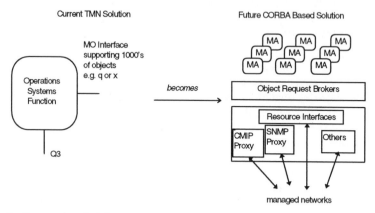

**Figure 3** Conventional Management Systems vs. a CORBA based Management System

This picture is a possible future rather than an implementable present. Concerns still exist over CORBA technology in areas including performance, scalability and security, but we believe this technology represents a key part of the future.

## 5.2.  Distributed artificial intelligence

The field of Distributed Artificial Intelligence (DAI) and particularly agent technology has suffered from producing more hype than content, however it does contain genuinely useful ideas which could benefit the network management domain. The previous section describes how management systems could be built of small management applications, each encapsulated as an object on a distributed CORBA platform.   DAI ideas and techniques could extend this idea to move from a hierarchically structured architecture such as TMN to one based upon peer-to-peer interaction.   This would give a more loosely coupled system that could be more flexible and adaptable.

This technology is at an early stage but early applications to complex domains such as air-traffic control (Ljungberg) and telecommunications (Azarmi) suggest fruitful possibilities

# 6.  APPLICATION TO VIRTUAL PATH BANDWIDTH MANAGEMENT

As part of our work we are implementing and experimenting with demonstrators which are being designed and constructed using the ideas described above. At the time of writing, the best example is a system produced to perform virtual path bandwidth management. Its overall aim is to ensure that the set of virtual paths set up in a network are able to deliver both the appropriate quality of service to individual connections whilst maintaining the connection acceptance rates desired by the provider. This is a non-trivial task because it depends upon the call patterns of users and their traffic patterns during calls, both of which are complex and difficult to estimate. The system acts in the Temporary Environmental/Network time-scale defined in section 4.2 dealing with variations over a number of users and hence traffic variations over hours or days. Examples may be different call patterns at different times of the day or week.

A conventional approach to the problem might be to design and implement a single algorithm that takes in the estimate for traffic volumes and produces a complete VP network. Such a system would go against our aims of highly distributed and flexible systems. Instead we use the ideas of analysis vs. tuning described in section 4.3 to produce a system that consists of a design algorithm that is a long-term function with a global view and a set of tuning functions that work quickly and flexibly with only local views. Although both are looking at problems caused in the 'temporary environmental/network' time-scale, the tuning functions operate on the network on a frequent basis whilst the design algorithm operates on a weekly or monthly basis.

The VP bandwidth management system comprising the VP network design function and tuning agents is illustrated in figure 4.

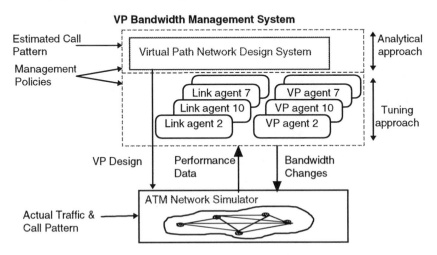

**Figure 4** The organisation of the VP bandwidth management system

The ideas described in section 5 were used to implement the tuning functions as  a set of agents each responsible for a different resource in the network and only having dynamic knowledge or control of that resource. By communicating with each other they are able to  modify VP bandwidths in the network avoiding potential congestion. It is important to notice that, in an agent-like manner, there is no type of agent that does VP bandwidth management , rather the behaviour of VP bandwidth management emerges from their distributed interaction.

The agents were constructed to control a public ATM network simulator produced by the RACE ICM project (RACE ICM). This simulator provides near-real-time simulation and provides a management interface allowing network data to be extracted during simulation and limited management controls to be applied. Experiments so far suggest that the agents described do achieve the goal of increasing call acceptance although they have yet to be properly bench-marked.  The graph in figure 5 illustrates this and is based on a simulated network of 4 ATM nodes, interconnected by 5 155 Mbps transmission links supporting 11 VPCs. Three connection types were supported: 1Mbps CBR, VBR with mean of 3 Mbps and peak of 5Mbs and VBR with mean of 4 Mbps and peak of 6 Mbps. Traffic volume in the network consists of order of 50 users for each source-destination pair, generating calls based on call pattern with mean holding time of  110sec and mean silence time of  130 sec. For each connection, traffic is generated as a set of bursts with rates based on truncated negative exponential distribution.

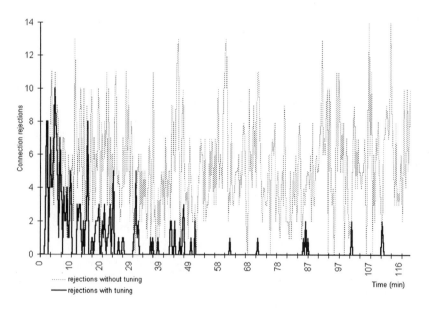

**Figure 5** Connection Rejection with and without tuning agents

Each agent contains a very simple rule-based system that provides its behaviour. An expected benefit of this approach was that the required behaviour would be provided by many, small systems which should be easier to maintain and modify. However in practice we found that although the 'micro' aspect of the agents was simpler, the 'macro' aspects were much more difficult owing to unfavourable interactions such as deadlock, infinite loops, and inappropriate actions. This highlighted the need for development tools that cope with distribution. A second factor that needs further research concerns the granularity of agents. The system that we built does have very small agents which may result in higher levels of messaging overhead between the agents and a greater configuration problem. There is a design trade-off to be understood which balances the benefits of smaller agents against the resulting overheads.

## 7.   CONCLUSIONS

ATM performance management is a vital area that needs to be addressed if ATM is to provide a flexible, multi-service network platform. Network controls alone will not be sufficient.

A time-scale based framework has been presented that categorises network problems and events according to the traffic variation they cause. This framework is being used to capture the functions that are needed to manage ATM networks and has been found to be useful.

Communication management systems are already becoming the major barrier to change, preventing rapid deployment of new services and functions. For this situation to be changed, new approaches are needed that will result in smaller and more flexible systems that are cheaper to build, change and discard. Initial work on using the technologies of DAI and CORBA offers promise in achieving this goal.

The ideas described in this paper are being developed and enhanced as part of an ongoing programme of work aimed at producing proven specifications and designs for a new breed of ATM performance management systems.

*Acknowledgements*
The authors would like to thank Sagar Gordhan of BT for his support in producing software to monitor the ATM simulator and for his experimental analysis of the VP bandwidth management demonstrator.

## 8.     REFERENCES

Arvidsson A. (1995) "On the usage of virtual paths, virtual channels, and buffers in ATM traffic management", 12$^{th}$ Nordic Teletraffic Seminar NTS12, Finland 22-24 Aug 1995.

ATM-Forum Traffic Management Specification V4.0, ATM-F/95-0013R11

Azarmi N., Nwana H, Smith R (1996)  Special Edition on Intelligent Software Systems, BT Technology Journal, vol. 14 no. 4 October 1996

Hui J Y (1988) "Resource allocation for broadband networks", IEEE journal of selected areas in communication, 6, 1988.

ITU Draft Recommendation X.703, Open Distributed Management Architecture

ITU-T Recommendation I.371, "Traffic Control and Congestion Control in B-ISDN", Perth, November 1995

ITU-T Recommendation M3010 Principle for a Telecommunications Management Network

Jain R. (1990) "Congestion control in computer networks : issues and trends" IEEE Network Magazine, May 1990, pp24-30

Jain R. (1992) "Myths about congestion management in high speed networks", Internetworking: Research and Experience, Vol 3, pp 101-113.

Key P. (1995) "Connection Admission Control in ATM networks" BT Technology Journal, 13, 3, July 1995.

Ljungberg and Lucas (1992) "The Oasis air-traffic management system", Proceedings of the Second Pacific Rim Conference on AI, 1992

OMG Telecom Special Interest Group (1996) "CORBA-based Telecommunication Network Management" White Paper, Draft 3.0, Feb 1996

RACE Project R2059 ICM, http://gryphon.elec.qmw.ac.uk/~icm

TINA-C (1994), Overall Concepts and Principles of TINA, TINA Del. TB_MDC.018_1.0_94

Rob Davison graduated from Bristol University, UK,  in Computer Systems Engineering and joined BT Labs in 1988. He has worked in the field of communications management since 1990 with a particular emphasis on the application of AI techniques.  He currently leads a team of researchers addressing network management with focus on novel architectures for network management, ATM performance management and the application of AI .

Manooch Azmoodeh graduated from the University of Manchester, UK, with a Ph.D in computer science. He spent five years at the University of Essex researching intelligent databases and query languages and then joined BT Labs in 1988.  He has researched various aspects of communication management systems, with emphasis on the application of AI and database techniques. His current interests are new network management architectures based on emerging software technologies and their application to broadband networks.

# 22

# DIVA: A DIstributed & dynamic VP management Algorithm

*S. Srinivasan and M. Veeraraghavan*
*Bell Laboratories, Lucent Technologies*
*101 Crawfords Corner Road, Holmdel, NJ 07733*
*E-mail:* {cheenu,mv}@bell-labs.com

### Abstract

The concept of preestablishing Virtual Path Connections (VPCs) in ATM networks offers a number of advantages, such as simplified on-demand connection setup and fault management. However, preallocation of bandwidth resources to VPCs minimizes resource sharing and leads to poor resource utilization. This is especially true if the VPC bandwidth allocations are computed allowing for some uncertainty in traffic characterization. In this paper, we propose DIVA, an algorithm for distributed and dynamic VPC bandwidth management. DIVA alleviates the problem of poor VPC resource utilization by dynamically adjusting VPC bandwidth allocations, thus minimizing the effect of traffic uncertainty. This algorithm is proposed for hierarchical networks based on the ATM Forum's PNNI routing standard. Using hooks provided by PNNI routing, network nodes monitor VPC bandwidth usage, make dynamic VPC bandwidth/buffer modifications, and set up and remove VPCs dynamically. This algorithm also handles the additions and failures of network elements such as links and switches.

### Keywords

ATM networks, virtual path management, dynamic/distributed algorithm, hierarchical networks, PNNI

## 1 INTRODUCTION

Provisioned Virtual Path Connections (VPCs) in ATM networks offer a number of advantages: (a) reduced on-demand connection setup time, (b) fast rerouting of bundles of Virtual Channel Connections (VCCs), useful in networks with high reliability requirements, (c) potential reduction in "switching" costs by using crossconnects without expensive call processing software in parts of the network instead of more expensive switches, and (d) in particular application areas, such as wireless ATM networks, where provisioning VPCs between adjacent pairs of base-stations allows mobile handoffs to be simplified (Srinivasan and Veeraraghavan, 1996).

On the other hand, a drawback of using provisioned VPCs with preallocated capacity is that network utilization (transmission efficiency) is reduced since link and node resources are partitioned (rather than shared) (Veeraraghavan et al., 1996). This effect is especially true if the resource allocations to VPCs are made allowing for uncertainty in traffic characterization. The effect of varying traffic patterns can be sharply reduced by using a scheme which dynamically adjusts VPC resource allocations based on usage. The presence of such a scheme will reduce the transmission inefficiency factor of using provisioned VPCs with preallocated capacity. Besides being dynamic, the VPC management scheme needs to be distributed for the reason that a centralized solution does not scale well with the size of the network, and is also very poor from a fault tolerance perspective. In this paper, we propose a *dynamic and distributed algorithm for monitoring and managing VPCs.*

Dynamic management of VPC routes and resource allocations can be done by continuously monitoring the network and reacting to repeated congestion patterns and topological changes caused by failures and addition of network elements such as links and nodes. The standardization of the Simple Network Management Protocol (SNMP) (Stallings, 1993) and the ATM Management Information Base (MIB) (Ahmed and Tesink, 1994) provides one method to achieve this. However, constantly reading and writing MIB variables for this purpose is typically too burdensome to the switch agents and is also too slow. An alternative to gathering network state information by reading MIB variables is provided by the ATM Forum's PNNI routing protocol (ATM Forum, 1996), which enables information about the network to be gathered and disseminated in a scalable manner. Hence, DIVA is proposed for hierarchical networks based on the PNNI routing standard. Using hooks provided by PNNI routing, network nodes monitor VPC bandwidth usage, make dynamic VPC bandwidth/buffer modifications, and set up and remove VPCs dynamically. Note that, currently, there is no solution for dynamic VPC management in the PNNI standard. One approach to realize changes in VPC routes/allocations is by using the switch signaling software. However, the use of sequential node-by-node setup and configuration procedure can result in excessive VPC adjustment delays. This approach also requires additional software to monitor VPCs and initiate adjustments, on already overloaded switches. The Parallel Connection Control (PCC) scheme (Veeraraghavan et al., 1996), originally proposed for on-demand connection set up, provides solutions for both these drawbacks. First, PCC recognizes the inherent parallelism in the connection establishment procedure and executes actions at a number of switches in *parallel* rather than in sequence. This concept can be used for fast VPC setup/modification. A second aspect of PCC is that it separates some of the connection management functions into *connection servers*, distinct from switches. Hence, the additional software for monitoring VPCs and initiating adjustments can be moved to the connection servers.

DIVA utilizes these advantages of PCC and the hooks provided by PNNI routing to provide a scalable solution for the VPC management problem. In other words, DIVA is targeted at networks that use PNNI and PCC for switched connection management. Specifically, it addresses the following issues.

- Where should VPCs be laid and what are the parameters needed?
- How should VPCs be realized? To answer this, we need to answer the following two questions: (a) How should VPCs be routed? (b) How much bandwidth and buffer resources should be allocated to each VPC?
- When should adjustments be made to VPC routes/allocations?
- How are changes to VPC configurations accomplished in reaction to link and node (a) failures, and (b) additions and restorals?

Past approaches for the creation and maintenance of VPCs are the following.

- Centralized optimization based schemes (Siebenhaar, 1994, Cheng and Lin, 1994, Faragó et al., 1995, Logothetis and Shioda, 1995): These are based on the solution of a non-linear optimization with non-linear constraints. Such optimizations are computationally intractable and approximate models are used which retract from the optimality while still not simplifying the schemes sufficiently to allow them to be run often enough, further compromising on the optimality of the assigned bandwidths. The centralized nature of these approaches has drawbacks from the points of view of scalability and fault tolerance.
- Distributed heuristic (Shioda and Uose, 1991): The drawbacks of this scheme are the following. (a) Only one VPC is allowed between each pair of switches which is unreasonable in practice since the required bandwidth to accommodate all the traffic between a pair of switches may not be available on a single route. It is also not acceptable from a fault tolerance perspective. (b) The route of each VPC is precomputed and fixed. In practice, network state is constantly changing and so does the "best" route between a pair of nodes. Hence, a VPC management scheme should allow for VPCs to be rerouted periodically.

None of the above schemes addresses the issue of dynamically handling network state changes such as link and node *failures* and *additions*. Lastly, and perhaps most importantly, any solution to the VPC management problem needs to address the practical issue of interworking with existing standards and/or systems. There is no attempt in this direction in (Siebenhaar, 1994, Cheng and Lin, 1994, Faragó et al., 1995, Logothetis and Shioda, 1995, Shioda and Uose, 1991), which detracts from their practicality.

In our work, we aim for a solution to the VPC bandwidth/buffer/route management problem that has the following properties.

- The solution should be scalable with respect to the size of the network. This precludes any centralized approaches.
- The solution should be robust and be capable of handling network state changes such as network element additions and failures.
- It should not assume fixed precomputed routing of the VPCs. VPCs should be routed according to the "best" paths, which could change with time as the network state changes.
- It should be able to take advantage of existing standards and interwork with them.

The rest of the paper is organized as follows. Section 2 defines some terms used in the rest of the paper. Section 3 describes the details of DIVA, our algorithm for VPC management, and Section 4 summarizes the contributions of the paper.

## 2 TERMINOLOGY AND DEFINITIONS

We model the call arrival processes as *Markov Modulated* processes (Shroff and Schwartz, 1996, Choudhury et al., 1996). Each source is characterized by the tuple $[M, R]$, where $M$ is the $K$-state transition matrix, and $R$ is a $K \times 1$ vector of source intensities.

When cells from multiple sources arrive at a switch, the cells are stored in a *buffer* till the destination port/link is free. Cells may be lost due to *buffer overflow* of finite

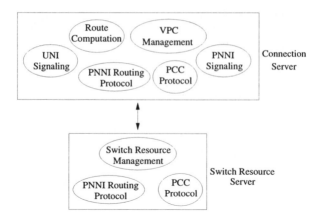

**Figure 1** Distribution of functions between connection server and switch resource server.

sized buffers and the stochastic nature of the sources. An important QoS criterion for a connection is the *Cell Loss Ratio* (CLR), which is the ratio of the total number of cells that are dropped due to buffer overflow to the total number of cells that arrive at the switch. Cells also experience a *Cell Delay Variation* (CDV), which is the time spent in a buffer, before being switched out. We make a distinction between the end-to-end CLR and CDV constraints for a connection and a VPC. We refer to the former simply as the end-to-end CLR/CDV constraints and to the latter as the end-to-end VPC CLR/CDV constraints.

The *equivalent bandwidth* of a set $S_n$, of $n$ sources, is the constant rate $c_{eq}(b, \theta, S_n)$ at which a buffer of size $b$ cells should be emptied so that the CLR experienced by the source is less than $\theta$. Expressions for $c_{eq}(.)$ exist in the literature, for instance in (Guérin et al., 1991).

Calls arriving at a VPC are modeled as Poisson processes with a mean rate $\lambda$. The call departure process is also Poisson with mean $\mu$. The ratio $\rho = \frac{\lambda}{\mu}$ is called the *offered traffic*. A call gets blocked on the VPC when the resources available on the link are inadequate to support the call. The probability that this happens, is termed as the *Call Blocking Probability* (CBP).

VPCs are of two types, *homogeneous* if they only support calls of a single traffic class, or *heterogeneous* if they support multiple classes of traffic. For a homogeneous VPC capable of supporting $\nu$ calls, the CBP is a function of $\nu$ and $\rho$. For a heterogeneous VPC with $k$ classes capable of supporting $\nu_j$ calls of class $j$ with offered traffic $\rho_j$, $j \in \{1, \cdots, k\}$, the CBP is a function of $\nu_1, \cdots, \nu_k$ and $\rho_1, \cdots, \rho_k$ (Kaufman, 1981, Labourdette and Hart, 1992, Choudhury et al., 1995).

## 3   DYNAMIC AND DISTRIBUTED VPC MANAGEMENT

In this section, we describe the network architecture assumed, the procedure for exchange and dissemination of routing information inside the network, and our algorithm, DIVA, for managing VPCs in a hierarchically organized network.

## 3.1   Network architecture

DIVA is designed for networks that use PNNI and Parallel Connection Control (PCC) (Veeraraghavan et al., 1996) for switched connection management. The network nodes are gathered into hierarchical *Peer Groups* (PGs) for scalability reasons. Nodes are of two types, switches and *connection servers* (CSs). CSs collect and disseminate routing information, compute routes, and coordinate parallel connection setup inside a PG. Switches run *switch resource servers* (SRSs) which perform switch resource management functions, consisting of (a) Connection Admission Control (CAC) for buffer and bandwidth resource allocations, (b) selecting incoming and outgoing VPIs/VCIs for VCCs, or incoming and outgoing VPIs for VPCs, (c) configuring the switch fabric by setting Port/VPI/VCI translation tables for VCCs or Port/VPI translation tables for VPCs, and (d) programming parameters for various runtime (user-plane) algorithms, such as cell scheduling, per VC queuing or per-VP queuing, priority control, rate control, etc. Software implementing UNI (User Network Interface) signaling (The ATM Forum, 1994), PNNI signaling (for PG to PG connection setup), PNNI routing (for communicating current topology state information), and PCC (for intra-PG connection setup) are distributed between the SRSs and CSs, as shown in Figure 1.

The function of VPC monitoring and adjustment initiation, required for VPC management, is also assigned to connection servers, which initiate VPC modifications/setups in response to topological changes reported via the PNNI routing protocol messages. For this added functionality, a *primary connection server* is assigned to each switch. Any of the other connection servers in the same PG act as *alternate connection servers* in case of failures. (These details are described in Section 3.3).

Each PG in the network has a CS designated as the Peer Group Leader (PGL). A PGL of a PG at level $L$ represents this PG at level $L+1$ as a *logical group node* (LGN). Two LGNs may be connected to each other by a *logical link* which is an aggregate representation of one or more links between the corresponding PGs at the lower level. PGLs obtain routing information about the switches in their peer groups and propagate a condensed version of this information to its peers in the higher-level peer group.

Figure 2 shows the PNNI/PCC network architecture assumed for our VPC management algorithm.

## 3.2   Inter-node routing information exchange

As described in Section 3.1, DIVA utilizes hooks in the PNNI routing protocol to gather periodic updates about network topology and state. This section describes how the PNNI routing protocol is executed in the network architecture shown in Figure 2 with switches and connection servers.

PNNI routing messages are of two types, *Hello* packets and PNNI Topology State Packets (PTSPs) (ATM Forum, 1996). Nodes in the network run a modified version of the *Hello protocol* (ATM Forum, 1996). *Hello* packets are exchanged across each link that comes up and the nodes attached to the link exchange information to establish their identity as well as their PG membership. Each switch also informs each of its neighbors belonging to a different PG the identity of its primary connection server. This piece of information propagates to the connection servers in the adjacent PG and is used for setting up connections spanning multiple PGs (as will be explained later). Neighboring

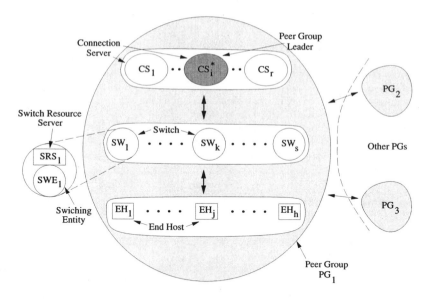

**Figure 2** Architecture of a peer group.

nodes also exchange periodic *Hello* packets which serve as "heartbeats" to monitor the health of the neighboring node.

Switches in the network create PNNI Topology State Elements (PTSEs) about their *local* state, namely about their own identity and capabilities. Switches *do not* maintain topology state information and hence do not exchange such information with each other. Instead these PTSEs are encapsulated into PTSPs and sent to the primary connection server of the switch. These PTSEs also contain topology state parameters describing the state of links associated with the node in question as well as all the VPCs originating at the nodes. This information is used by the connection server to perform VPC bandwidth management functions.

The connection servers in a PG exchange topology state and reachability information with each other in order to synchronize their topology databases. Each connection server reliably floods the connection servers in its PG with detailed information about the state of the nodes for which it is the primary connection server as well as links associated with these nodes. This information is used by the connection servers to perform detailed route computation inside the PG. Also, the PGL condenses this information and floods it in its PG. The PGL also receives such condensed information about other PGs and floods this information among all the nodes in its child PG if this is not a lowest level PG and among the different connection servers in a lowest level PG.

PTSEs in the topology databases of the connection servers are subject to aging rules and deletion just as in PNNI based networks.

## 3.3   DIVA: Algorithm for VPC management

In this section, we present the details of our VPC management algorithm, DIVA.

## *Where should VPCs be laid and what are the parameters needed?*

The following parameters have to be determined by or specified to the connection server which sets up the VPC. (a) VPC type, namely heterogeneous or homogeneous, (b) the source characteristics, as the set of $[M, R]$ tuples for each source, (c) the end-to-end CLR and CDV constraints $\tau$ and $\theta$, for each class of traffic supported by the VPC, and (e) the number of calls of each class of traffic, either directly specified or indirectly specified by giving the CBP constraints.

The pairs of switches $i$ and $j$ to be connected by VPCs can be determined by one of the following methods.

1. **Approach based on traffic monitoring:** As explained in Section 3.1, one of the functions performed by the connection servers is to receive UNI signaling messages requesting SVC setup and determine routes within its PG to set up SVCs. Thus each connection server $CS$ knows the number of on-demand connections set up between each switch $SW$ for which it is the primary connection server and other switches in the PG. If there is no VPC between switches $SW$ and $SW'$ and the number of connection setup requests over a certain window of time $\tau_{monitor}$ exceeds a threshold $r_{max}$, $CS$ decides that $SW$ and $SW'$ need to be connected by one or more VPC.

   Besides determining the node pairs that should be connected by a VPC, some of the parameters of the VPC can also be determined from the SVC setup and release data, namely, source characteristics, end-to-end VPC CLR and CDV constraints, the CBP constraint on the VPC and the offered traffic. The details of how these parameters are determined are deferred to a later paper. The type of VPC to set up as well as the CBP constraint, either system wide or for each node pair, have to specified to the connection servers.

2. **Rule based VPC approach:** For certain applications, we may decide to have VPCs between certain pairs of nodes from the very beginning, based on a certain set of prespecified rules. For example, in IP-over-ATM networks we can avoid connection setup to transfer connectionless IP packets by connecting each pair of IP routers by a VPC.

3. **VPCs requests from an administrator/manager:** Each such request has to specify all the parameters enumerated above, for every pair of switches, $i$ and $m$, that have to be connected by one or more VPCs.

The above three approaches are not mutually exclusive. While a given approach may work better for certain types of VPCs, it may not be applicable for others. The complete answer to the question of where to lay the VPCs is a combination of all three approaches.

## *How are VPCs realized?*

The two questions related to the realization of VPCs are: (a) How should VPCs be routed?, and (b) How much bandwidth and buffer resources should be allocated to each VPC? Let us assume that a connection server needs to realize a sufficient number of VPCs between two switches $i$ and $m$ with the given set of parameters defined above. Connection servers and switch resource servers are involved in the VPC-realization phase. The following steps are involved in setting up a VPC.

**Step 1:** First, the connection server needs to *determine the number of calls* that must

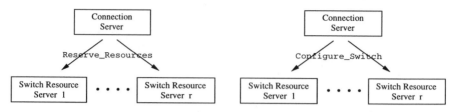

**Figure 3** VPC setup within a PG.

be accommodated on these VPCs. This data may be directly provided in the request or it may need to be inferred by the connection server from the specified call blocking probabilities and offered traffic. The number of calls can be computed using the Erlang-B formula (Jagerman, 1974) for homogeneous VPCs and using a multiservice CBP method (Kaufman, 1981, Labourdette and Hart, 1992, Choudhury et al., 1995) for heterogeneous VPCs.

**Step 2:** Next, the connection server performs *route computation* using routing information (reachability and the state of different elements) previously collected via PNNI routing protocol messages. It uses an estimate of the required bandwidth derived from the source descriptions to determine the "best" (most likely to succeed) "shortest path" routes for the VPCs. It then attempts to set up the VPC along one of these routes.

**Step 3:** For each selected route, the connection server needs to *apportion the end-to-end VPC CLR and CDV*, for traffic class $j$, $\theta_j^{tot}$ and $\tau_j^{tot}$ respectively, among the switches on the route. This is a complex problem (Nagarajan, 1993), especially if switches from multiple vendors are to be accommodated. One simple solution, based on estimating the number of switches $n_s$ the VPC passes through, is to apportion the end-to-end class $j$ VPC CDV constraint $\tau_j^{tot}$ equally among all the switches giving a per-switch constraint of $\frac{\tau_j^{tot}}{n_s}$. Similarly, if the end-to-end class $j$ VPC CLR constraint $\theta_j^{tot}$ is equally distributed among the $n_s$ switches, the CLR constraint on each switch is $1 - (1 - \theta_j^{tot})^{\frac{1}{n_s}}$.

**Step 4:** VPC setup is done using parallel setup of switches within a PG and sequential setup from one PG to the next. Connection setup within a PG consists of two phases: a *resource reservation phase* and a *fabric configuration phase*, as shown in Figure 3. In the *resource reservation phase*, the connection server sends a Reserve_Resources message to the SRSs (Switch Resource Servers) on the switches of the selected route, *in parallel*, requesting them to set aside the necessary resources. Each SRS that receives such requests pushes them into a queue and processes them sequentially. This allows multiple connection servers to simultaneously contact an SRS without causing any deadlocks (Veeraraghavan et al., 1996). Each Reserve_Resources message specifies the parameters $[M_j, R_j]$, $\theta_j$, $\tau_j$, and $\nu_j$ for each service class $j$. Each SRS then performs CAC (Connection Admission Control) functions to determine if it can satisfy the request. To do this, it must determine the required bandwidth and buffer resources from the parameters specified in the Reserve_Resource message. Buffer and bandwidth allocations are performed simultaneously by the SRS on each switch within the constraints of (a) the available switch buffer and bandwidth resources, and (b) the cell loss and delay variation constraints on the connection. If the SRS determines that it has sufficient resources to meet these requirements, it sets aside these resources and marks them as unavailable. It also selects VPIs to be used for this VPC. It then sends a `Resources_Available` message back to the requesting connection server. If the requested resources are not available, it responds with a `Resources_Unavailable` message.

**Step 5:** If the first phase is successful, responses carrying actual delay guarantees are received by the connection server which uses these to compute the residual constraints to be satisfied by switches in PGs further down the route. Then the connection server (A.2.5 in the example) enters the second phase by sending a `Configure_Switch` message in parallel to all the switches on the route, as shown in Figure 3. An SRS receiving this message sets port/VPI translation table entries. In other words, it configures the switch fabric to realize the VPC. It also sets buffer allocations for the VPCs (assuming per-VP queuing in the nodes) and provides the allocated bandwidth to the switch scheduler for enforcement.

For multi-PG VPCs, a connection server in each PG performs this two phase setup approach for the switches in its PG. PG to PG setup proceeds sequentially using PNNI signaling messages for communication between connection servers in different PGs.

Once the above 5 steps are completed successfully in the last PG along the route of the VPC, appropriate data tables are created at the VPC terminating switches, by the connection servers in the PGs of these switches, to enable them to treat each other as "logically connected" neighbors. □

## *When and how should adjustments be made to VPC routes and resource allocations?*

This determination is done using one of two approaches: *by monitoring call blocking probabilities* or *by monitoring VPC cell usage*. Using the periodic updates received from each switch $SW$ for which it is the primary connection server, each connection server $CS$ monitors the state of each VPC associated with the switches. If, in the previous timeframe, either (a) the observed CBP $\beta_{meas}$ on a VPC is not within a specified range $[(1 - \rho_l) \times \beta, (1 + \rho_u) \times \beta]$ where the target CBP for the VPC is $\beta$, or the average occupied bandwidth $c_{occ}$ is not in the range $[min \times c_{actual}, max \times c_{actual}]$, the primary connection server of one of the terminating switches of the VPC takes corrective action.

When $\beta_{meas} < (1 - \rho_l) \times \beta$ or $c_{occ} < min \times c_{actual}$, it attempts to reduce the allocated bandwidth. The connection server uses the two-phase approach described in Section 3.3 to realize this adjustment of lowering the assigned bandwidth (expressed as a percentage in the `Reserve_Resources` message). The exception to this bandwidth reduction step is when the occupied VPC bandwidth is smaller than a minimum $c_{min}$. In such a case the connection server marks the VPC as "frozen". No more connections are admitted on this VPC and it is allowed to empty out. When the last call completes, the VPC is torn down and the resources recovered.

On the other hand, when the bandwidth is inadequate to support the required CBP and $\beta_{meas} > (1 + \rho_u) \times \beta$ or $c_{occ} > max \times c_{actual}$, the connection server attempts to increase the bandwidth assigned to the VPC. This is also done using the two-phase approach described in Section 3.3. In the first phase, the connection server contacts all the switches in parallel requesting an increase in bandwidth (again, expressed as a percentage). In the second phase, it sends a commit request. If the attempt to increase bandwidth fails, the connection server attempts to set up a new VPC between this pair of switches on a new route.

## How do we accommodate link and node additions and failures?

Lastly, we briefly describe the actions that happen in the network when its state changes in any of the following ways: addition of a switch, connection server or link, and failure of a switch, connection server or link.

**Addition of a link:** The switches on either side of the newly added link inform their respective primary connection servers about the new link. These connection servers in turn flood this information among other connection servers in the PG. This information also passes up the hierarchy if the link is between PGs. Thus the new link is added to the topology databases of all the necessary connection servers in the routing domain. These connection servers then start using the link for routing future connections.

**Addition of a switch:** As soon as a switch $S$ fires up (it may be a new switch or a failed switch being restored after repair), it starts the *Hello* protocol over each of its outgoing links. Simultaneously it requests a list of connection servers in its PG from its neighbors. Once it receives this information $S$ sends a `Register` message to a chosen connection server $CS_c$ informing it of its status as a newly added node and requesting that it be assigned a primary connection server. $CS_c$ chooses the least loaded connection server*, in its estimation, $CS_p$ in the PG. It then forwards the `Register` message to $CS_p$. $CS_p$ then establishes a direct logical signaling link with $S$ and informs $S$ about it. Once this link is established, $S$ starts sending periodic PTSPs on this link to $CS_p$. The other connection servers in the PG know about $S$ and topology state information associated with $S$ from the PTSPs flooded periodically by $CS_p$ in the PG.

**Addition of connection server:** When a connection server, $CS_n$ comes up, it contacts its neighbors (which may be switches or other connection servers) to determine the address of other connection servers in the PG. It then proceeds to contact them and establish logical links to some or all of them and performs a topology database copy. It now starts receiving PTSPs from other connection servers and proceeds to do the same. Other connection servers in the PGs also use a "handshake" protocol to off-load some of the switches they are currently responsible for to the newly added connection server which is lightly loaded.

**Failure of a link:** Failures, both *intermittent* and *permanent*, may be detected by Operation and Maintenance (OAM) functions executed at different levels of the network hierarchy, namely, the physical layer OAM, ATM VP and VC layer OAM (ITU-T, 1993). *Hello* packets running at the PNNI routing protocol layer also aid in detecting failures. For example, when a link goes down, the nodes on either side of the link stop receiving *Hello* "heartbeat" packets from the node across the link and pass this information up to their primary connection servers. Correlating the alarms and failure indication messages received from different levels of the hierarchy and determining their cause is a complex problem. The connection servers could execute one of the algorithms from the literature (see (Katzela and Schwartz, 1995) for one such algorithm) to determine the failed network element. For example, using the information about missing *Hello* packets, the connection servers could determine which link has failed. This information is then flooded inside the network and eventually reaches the the topology databases of necessary connection servers which then do not use the failed element in computing future routes. Also, when a connection server $CS$ determines that a link $l$ has failed, it reroutes any

---

*Real-time loading information is assumed to be exchanged between the CSs in a PG in the PTSPs.

VPCs that it is monitoring, which use this link. The reuse of signaling software for VPC setup/modification and PCC for fast connection setup are the two aspects of DIVA that enable fast restoration of VPCs.

**Failure of a switch:** The failure of a switch is equivalent to the failure of all the links around it. The node is marked unreachable by the connection servers and is not used for routing connections.

**Failure of a connection server:** When a connection server $CS_1$ fails, each switch $SW$ for which it was the primary connection server goes through the procedure followed by a switch that is newly added to the network (as explained above). Switch $SW$ chooses another connection server $CS_2$ as its primary connection server, and starts sending their PTSPs to $CS_2$.

## 4 SUMMARY

Virtual Path Connections (VPCs) provide numerous advantages in ATM networks. They however suffer from the disadvantage of poor resource allocation due to preallocation of bandwidth and lack of sharing. In this paper, we addressed the issue of making VPCs resource efficient using dynamic VPC bandwidth management. We proposed **DIVA**, a distributed and dynamic algorithm for VPC monitoring and maintenance for the ATM Forum's Private Network-Network Interface (PNNI) standards based hierarchical networks. We described the approach used by DIVA to answer the following questions about VPC management.

- Where should VPCs be laid and how are the parameters to set up VPCs determined?
- How should VPCs be realized, *i.e.*, (a) how should VPCs be routed, and (b) how much bandwidth and buffer resources should be allocated to each VPC?
- When and how should adjustments be made to VPC routes/allocations?
- How are changes to VPC configurations accomplished in reaction to link and node (a) failures, and (b) additions and restorals?

## 5 REFERENCES

Ahmed, M. and Tesink, K. (1994). *RFC-1695: Definition of managed objects for ATM management using SMIv2*.

ATM Forum (1996). Private Network-Network Specification Interface v1.0.

Cheng, K.-T. and Lin, F. Y.-S. (1994). On the joint virtual path assignment and virtual circuit routing problem in ATM networks. In *Proc. IEEE Globecom*, pages 777–782.

Choudhury, G. L., Leung, K. K., and Whitt, W. (1995). An inversion algorithm for computing blocking probabilities in loss networks with state-dependent rates. In *Proc. IEEE Infocom*, pages 513–521.

Choudhury, G. L., Lucantoni, D. M., and Whitt, W. (1996). Squeezing the most out of ATM. *IEEE Trans. Comm.*, 44(2):203–216.

Faragó, A., Blaabjerg, S., Ast, L., Gordos, G., and Henk, T. (1995). A new degree of freedom in ATM network dimensioning: Optimizing the logical configuration. *IEEE J. Selected Areas Comm.*, 13(7):1199–1206.

Guérin, R., Ahmadi, H., and Naghshineh, M. (1991). Equivalent capacity and its application to bandwidth allocation in high-speed networks. *IEEE J. Selected Areas Comm.*,
9(7):968–981.

ITU-T (1993). *B-ISDN Operation and Maintenance Principles and Functions.* Rev. 1,
Geneva.

Jagerman, D. L. (1974). Some properties of the Erlang loss function. *Bell System Technical J.*, 53(3):525–551.

Katzela, I. and Schwartz, M. (1995). Fault identification schemes in communication networks. *IEEE Trans. Networking.*

Kaufman, J. S. (1981). Blocking in a shared resource environment. *IEEE Trans. Comm.*,
29(10):1474–1481.

Labourdette, J.-F. and Hart, G. W. (1992). Blocking probabilities in multi-traffic loss systems: Insensitivities, Asymptotic behavior, and Approximations. *IEEE Trans. Comm.*,
40(8):1355–1366.

Logothetis, M. and Shioda, S. (1995). Medium-term centralized virtual-path bandwidth
control based on traffic measurements. *IEEE Trans. Comm.*, 43(10):2630–2640.

Nagarajan, R. (1993). *Quality-of-Service Issues in High-Speed Networks.* PhD thesis, Department of ECE, University of Massachusetts, Amherst.

Shioda, S. and Uose, H. (1991). Virtual path bandwidth control method for ATM networks: successive modification method. *IEICE Trans.*, E74(12):4061–4068.

Shroff, N. and Schwartz, M. (1996). Improved loss calculations at an ATM multiplexer.
In *Proc. IEEE Infocom*, pages 561–568.

Siebenhaar, R. (1994). Optimized ATM virtual path bandwidth management under fairness constraints. In *Proc. IEEE Globecom*, pages 321–329.

Srinivasan, S. and Veeraraghavan, M. (1996). Virtual paths in hierarchical wireless ATM
networks. In *Proc. $3^{rd}$ Int. Workshop Mobile Multimedia Comm.*

Stallings, W. (1993). *SNMP, SNMPv2, and CMIP: The practical guide to network-management standards.* Addison-Wesley Publishing Company.

The ATM Forum (1994). *ATM User-Network Interface Specification v3.1.*

Veeraraghavan, M., Kshirsagar, M., and Choudhury, G. L. (1996). Concurrent ATM connection setup reducing need for VP provisioning. In *Proc. IEEE Infocom*, pages 303–
311.

## 6   BIOGRAPHY

SANTHANAM SRINIVASAN received his B.Tech. degree in Electronics & Communication Engineering from the Indian Institute of Technology (Madras) in 1991 and M.A.
and Ph.D. degrees in Electrical Engineering from Princeton University in 1993 and 1995,
respectively. He is currently at Bell Laboratories in the Networking Research Laboratory,
where he is engaged in research in network management for IP and ATM networks.

MALATHI VEERARAGHAVAN is currently a Distinguished Member of Technical Staff
at Bell Laboratories in the Networking Research Laboratory. Her research interests include mobility management, signaling and control of networks, and network management.
Dr. Veeraraghavan received her B.Tech. degree in Electrical Engineering from the Indian
Institute of Technology (Madras) in 1984, and M.S. and Ph.D. degrees in Electrical Engineering from Duke University in 1985 and 1988, respectively. She served as an Associate
Editor of the IEEE Transactions on Reliability from 1992-1994.

## PART TWO

# Service Management
## Chair: Roberto Saracco, CSELT

# 23

# Customer Management and Control of Broadband VPN Services

M.C. Chan, A. A. Lazar and R. Stadler
Center for Telecommunications Research
Columbia University, New York, NY 10027
{mcchan, aurel, stadler}@ctr.columbia.edu

## Abstract

We present an architecture for customer management and control of a broadband VPN service. The architecture is aimed at giving the VPN customer a high level of control over the traffic on the VPN, such that end-to-end requirements for the customer's enterprise network can be met. We describe how different control and management objectives can be achieved with this architecture. Its design includes a generic resource controller, which can be specialized in order to realize a large class of control schemes, following a customer's specific requirements. We have implemented a prototype of this architecture on a high-performance emulation platform. The prototype allows us to validate the management and control functionality of the customer control system and to demonstrate the performance characteristics of different realizations of the architecture.

## Keywords

Virtual Private Networks, Management of Broadband Services, Customer Control, Management Architectures, Prototyping

## 1. INTRODUCTION

Broadband technology has the potential to change corporate networking in major ways. Broadband networks are aimed at providing quality-of-service (QOS), thus making it possible to support real-time services like voice and video communication, in addition to best-effort data delivery. Due to their ability to integrate different services on the cell-level, they provide a promising platform for distributed multimedia applications that are emerging today. Furthermore, the advent of broadband technology will enable the integration of today's separate corporate networks (voice network, data network), which often rely on different public services (e.g., leased lines for voice traffic and LAN interconnection, frame-relay service for low-volume data exchange) into a single enterprise network, using a single Virtual Private Network (VPN).

Corporations want to control and manage their enterprise networks according to their own control objectives and management strategies. This implies that a corporate customer, using a VPN service, needs the capability to control and manage its traffic on the VPN--possibly in cooperation with the provider. For the designer of an enterprise network, the question arises, which part of the control functionality is executed in the customer's domain and which part in the provider's domain. More precisely: which functions are performed by the customer alone, which by the provider alone, and which in the form of customer-provider cooperative control.

There are strong reasons for *customer control*, i.e., for running traffic management functions in the customer domain. First, different customers pursue different control and management objectives while running their enterprise networks. For example, customer requirements concerning the traffic carried on in a VPN are very diverse with respect to supporting multimedia traffic with different performance characteristics and performance requirements. Some customers may want to operate a multiclass network with several traffic classes for both real-time and non real-time traffic; others may want to support just one class of traffic with peak rate allocation. Some might want to implement a call priority scheme which enables calls of higher priority to pre-empt those of lower priority when the network is congested; others may want to apply other control schemes in case of congestion. Providers face difficulties in their efforts to accommodate such diverse requirements. Customers who know their requirements better than the providers may be in a better position to execute control according to their objectives. Also, operations under customer control can be executed faster than those performed in cooperation with the provider, since no negotiation is required. For example, setting up connections over a VPN can be done by the customer in a distributed way, based only on local information. This allows customers to engineer or configure their traffic control systems in such a way that short connection set-up times can be achieved, which is required by some applications.

Second, customers want provider-independent control in order to meet special requirements for the enterprise network [ZER92]. For example, usage collection that permits billing at a level of detail beyond the provider's capability, such as billing at an application level, may be needed. Furthermore, the partitioning of the VPN by the customer may be required to implement sophisticated access control mechanisms, which prevent unauthorized access to certain partitions of the network. Also, automatic fall-back mechanisms may be desirable for critical applications that need high network reliability.

Third, moving the responsibility for VPN traffic management from the provider to the customer accelerates the introduction of Broadband VPN services. Specifically, public VPN services based on Constant Bit Rate (CBR) Virtual Paths (VPs) can be provided efficiently today [ATS93, FOT95]. However, such a service requires resource control by the customers, since they will be billed based on allocated bandwidth--even if they do not use it.

In this paper, we present an architecture for customer-based management and control of a broadband VPN service. We outline how different control and management objectives can be achieved with this architecture. An element of this architecture is the design of a generic resource controller, which can be specialized in order to realize a large class of control schemes, following a customer's specific requirements. Further, we present a prototype implementation of this system in a high-performance emulation environment. The prototype allows us to demonstrate performance characteristics of the customer control system and to validate the management and control functionality.

The paper is organized as follows. Section 2 describes different broadband VPN services from a customer's perspective, specifically a VPG-based VPN service, which gives the customer a high level of control. Section 3 discusses customer control and management objectives for a VPN. Section 4 presents our architecture for a customer operated control and management system for a VPG-based VPN service. Finally, Section 5 describes our experience with a prototype implementation of the architecture and the emulation platform we use for prototyping.

## 2. BROADBAND VPN SERVICES

A broadband virtual private network (VPN) is a service that provides broadband transmission capability between islands of customer premises networks (CPNs) (Figure 1). It is a central building block for constructing a global enterprise network (EN) which interconnects geographically separate CPNs. A VPN service involves several administrative domains: the customer domain, the domain of the VPN service provider--also called "value added service provider" (VASP)--, and one or more carrier domains [SCH93]. As a result, it is necessary to address the aspects of multi-domain management in the context of VPN service management and provisioning ([LEW95], [TSC95]). The scope of this paper is limited to the customer domain and the interaction between the customer domain and the VPN provider domain.

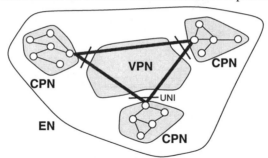

**Figure 1** Customer's view of a virtual private network.

Traditionally, leased line circuits based on STM (SDH/SONET) technology have been used for providing VPN services [YAM91]. The speed of the circuit can be changed by customer-provider cooperative control. However, dynamic bandwidth adjustment for leased line circuits is inefficient and costly compared to ATM-based services, which place no restriction on the line speeds the customer can choose from [HAD94].

Service providers are beginning to offer broadband VPN services using ATM transport networks. Two common approaches are Virtual Circuit (VC)-based VPN services ([SAY95]) and VP-based VPN services [ATS93]. These services provide ATM logical links between separate CPNs. In the case of a VC-based VPN service, the customer requests a new VC from the provider for every call to be set up over the VPN. Bandwidth control and management between customer and provider is performed per VC. In the case of a VP-based VPN service, customers can perform their own call and resource control for a given VP, without negotiating with the

VPN provider. Bandwidth control and management between customer and provider is performed per VP. VC-based and VP-based VPN services replace today's leased line services. They offer customers more flexibility in dynamically requesting adjustments in the VPN capacity. Since networks typically exhibit a dynamic traffic pattern, such a technique of rapid provisioning will result in lower cost for the customer, because pricing is expected be based on the VPN capacity per time interval allocated to the enterprise network. A VPN is accessed via common user-network physical interfaces (UNIs).

A Virtual Path Group (VPG)-based VPN service has been proposed to enhance customer control over the VPN [CHA96a]. The Virtual Path Group (VPG) concept has been introduced to simplify virtual path dynamic routing for rapid restoration in a carrier network [HAD89]. In a VPG-based VPN service, a VPG is defined as a logical link within the public network provider's ATM network. Figure 2 shows a VPG-based Virtual Private Network connecting 3 CPNs. A VPG is permanently set up between two VP cross connect nodes or between a VP cross connect node and a CPN switch that acts as a customer access point for the VPN service. A VPG accommodates a bundle of VPs that interconnect end-to-end customer access points. The VPN provider allocates bandwidth to a VPG, which defines the maximum total capacity for all VPs within the VPG. A VPG-based VPN consists of a set of interconnected VPGs.

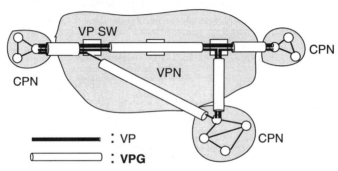

**Figure 2**   A VPG-based virtual private network.

VPs and VPGs are set up by the network management system of the VPN provider during the VPN configuration phase. Only the network management systems must know about the routes of the VPGs, their assigned bandwidth, and the VPs associated with them. The use of VPGs has no impact on cell switching, as cells are transmitted by VP cross connect nodes based on their VP identifier. In order to guarantee cell-level QOS in the carrier's network, policing functions (Usage Parameter Control) are required at the entrance of each VPG.

The VPG concept enhances the customer's capability for VP capacity control. It allows transparent signalling and dynamic VP bandwidth management within the customer domain. A customer can change the VP capacities, within the limits of the VPG capacities, without interacting with the provider. As a result, the VPG bandwidth can be shared by VPs with different source-destination pairs. Furthermore, customers can independently achieve the optimum balance between the resources needed for VP control and the resources needed to handle the traffic load.

## 3. CUSTOMER CONTROL AND MANAGEMENT OBJECTIVES

From the perspective of traffic control, the customer wants to achieve two sets of objectives. The first set relates to end-to-end QOS requirements for the traffic on the enterprise network, which translates into QOS objectives for the traffic that traverses the VPN. QOS objectives on the cell level are usually expressed in terms of bounds on end-to-end delays and error rates; on the call level QOS objectives include call blocking constraints and bounds on call set-up times. The second set relates to efficient use of VPN resources, primarily trunk bandwidth.

Efficient use of the VPN bandwidth can be achieved by exploiting statistical multiplexing at the cell- and/or the call-level. On the cell-level, multiplexing gains among calls with the same source-destination pair (with respect to. the VPN) can be achieved using the schemes described in [HYM91, ELW93]. Cell multiplexing among calls with different source-destination pairs can be performed based on the contract region concept [HYM94]. On the call-level, schemes for VP control (e.g. [OHT92]) can be used to exploit multiplexing among calls with the same source-destination pairs. Finally, the techniques described in [FOT95] and [CHA96a] can be used to multiplex calls with different source-destination pairs. Depending on the type of VPN service the provider offers, the customer can choose to implement one or more of the above multiplexing schemes in the customer control system.

In terms of managing the enterprise network, customers want the capabilities to control the bandwidth cost of the VPN service, define QOS objectives and set preferences and priorities for resource allocation to deal with congestion situations. These management objectives apply to the customer domain only and are different from customer to customer. They define the policies according to which the customer control system operates. Management capabilities can be realized by tuning controllers in the customer control system (Section 4.3). For illustration purposes, we describe below some of the management capabilities we have implemented in our prototype system.

Cost management allows the customer to define the maximum average cost of the VPN communication resources over a specific period of time. This capability is realized by setting constraints on the negotiation of VPN bandwidth between the customer and the provider. VP management allows the customer to manipulate VP bandwidth directly. Operationally, the control of the VP bandwidth can be executed either automatically by the customer control system or under direct control of the operator of the enterprise network. The operator can allocate a fixed amount of bandwidth to a VP, which must be respected by the control system. QOS and priority management operations define how calls are handled in the enterprise network. In our implementation, every call is characterized by a performance class and a priority class. The performance class of a call determines its QOS requirements. QOS management deals with managing the level of service provided to different performance classes. In particular, the customer can modify the blocking objectives of calls belonging to a performance class. The level of priority determines the relative importance of a call. In our scheme, a high priority call can pre-empt a call of lower priority in case of congestion. The customer can enable and disable priority control and can set blocking objectives for priority classes. The concepts of QOS and priority class are independent in the sense that a call that demands stringent QOS requirements can have low priority and vice versa. Finally, the above described management capabilities are orthogonal in the sense that they can be applied independently of one another.

## 4.  A CUSTOMER CONTROL AND MANAGEMENT ARCHITECTURE

Figure 3 shows the systems involved in the provisioning and operation of a VPG-based VPN. In the provisioning phase, information concerning the VPG topology, the VP topology and the mapping between them is exchanged and stored in the management systems of the customer and the provider. Knowledge about the VPGs is also required in the provider's control system, which performs Usage Parameter Control (UPC) per VPG. The use of VPGs has no influence on cell switching and transmission, since cells are switched according to the VP identifiers in their headers. Figure 3 also shows the organization of the control system according to time-scales. The customer control system contains three classes of controllers: VP admission controller, VPG controller, and VPN controller. These controllers operate on different time-scales and run asynchronously.

We illustrate the interaction among these controllers with an example. Assume that one of the VPs experiences a sudden increase in traffic load. The VP admission controller associated with this VP admits calls as long as there is sufficient capacity. If there is not sufficient capacity available, calls are blocked. On a slower time scale, the VPG controller detects the congestion in this particular VP and attempts to allocate additional bandwidth to it. If the increase in traffic load is transient and, therefore, the demand for bandwidth drops after some time, the interaction stops here. Otherwise, if the congestion persists, the VPN controller, which runs on a slower time-scale, will request additional VPN capacity from the provider.

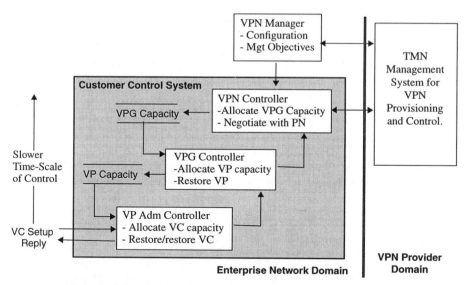

**Figure 3**  A functional model of the customer control system.

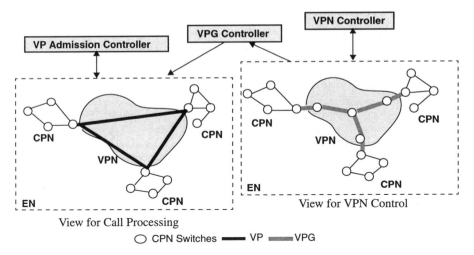

**Figure 4** Network views the controllers operate on operate on.

For the purpose of dynamic bandwidth control, a VPG-based VPN can be compared to an ATM network in which the link size can be varied. Therefore, controllers in the customer domain operate on two views of the network (Figure 4). The view on the left side of Figure 4 shows a network of end-to-end VPs which connect a set of CPNs. The view on the right shows a VPG network, which connects the same set of CPNs. The relationship between VPs and VPGs defines the mapping between both views.

The VP admission controller, which participates in call setup and release in the enterprise network, operates on the left view. The controller decides whether a call can be admitted into the VPN, based on the VP capacity, its current utilization and the admission control policy. The VP admission controller always ensures that enough capacity is available, such that cell-level QOS can be guaranteed for all calls that are accepted. The controller runs on the time scale of the call arrival and departure rates (seconds or below). There can be one VP admission controller per VP, or one for a set of VPs. The VPG controller operates on both views. Depending on the state of the VPs (in particular, traffic statistics and VP size) and the control objectives, it dynamically changes the amount of VPG bandwidth allocated to associated VPs. This controller enables customers to exploit variations in utilization among VPs that traverse the same VPG, allowing bandwidth between VPs of different source-destination pairs to be shared without interacting with the provider. In order to guarantee QOS, the sum of the VP capacities must be less than or equal to the capacity of the VPG link. The controller runs on a time-scale of seconds to minutes. The VPN controller operates on the right view. It is the only controller which interacts with the provider, and it runs on the slowest time scale of all the controllers (minutes or above). The VPN controller dynamically negotiates the bandwidth of the VPG links with the provider, based on traffic statistics and control objectives (e.g., minimizing the VPN cost), while observing the customer's QOS requirements.

## 4.1  Controller Design

Figure 5 shows the functional design of a VP admission controller and a VPG controller according to our implementation. In this design, the VP admission controller includes two objects: a VC capacity allocator and a coordinator. The allocator receives requests from a VC connection manager in the customer domain. The coordinator changes the capacity of the VP upon request from the VPG controller. It changes the capacity of the VP only when the bandwidth requirements of the active calls in the VP do not exceed the new capacity. The VPG controller includes four objects. The trigger object periodically initiates the VP capacity allocator to run the VP allocation algorithm. The coordinator sends the new VP capacities to the coordinators of the associated VP admission controllers, using a synchronization protocol. Finally, an estimator object collects statistics from the VP admission controllers. This data is used by the capacity allocator.

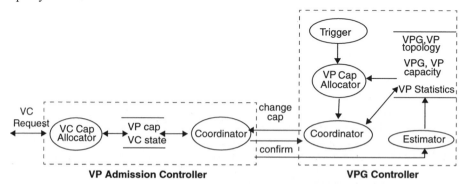

Figure 5  Functional model of a VP admission controller interacting with a VPG controller

Obviously, there exist many ways of realizing the above design, with respect to control algorithms, mechanisms for trigger realization, synchronization protocols, and centralized or distributed implementation of the controllers. For example, the control system may include one VP admission controller per VP or one centralized controller for the whole VPN. The same applies for VPG control. Also, VP admission controllers can send bandwidth requests to VPG controllers, triggered by a pressure function, or a VPG controller can periodically recompute the VP capacities and distribute them to VP admission controllers. Similarly, the synchronization protocols between the VP admission controller and the VPG controller can be realized in different ways. One possibility is that the VP admission controller, upon receiving a request to change the VP size, checks whether the current utilization is above or below the new size. If the utilization is below, the VP size is changed and a confirmation is sent to the coordinator of the VPG controller. If it is not below, the VP size remains the same and a failure reply is sent instead. In another possible implementation, when the attempt for changing the VP size is not successful, the VP admission controller waits and blocks further calls from being admitted. Then, the utilization of the VP can only be decreased, as calls can leave but no new calls are admitted. When the utilization drops below the new size, the VP size is updated and the reply sent to the VPG controller. A customer's choice for a specific design of the control system is based upon its control objectives and requirements for the control system, which relate to system size, expected traffic and signalling load, efficiency of resource control and robustness of the control system. In order to enable the realization of a large class of control objectives and control schemes, we have designed a generic controller as one of the building blocks of a customer control system. This generic controller enables many interaction patterns among controllers and is constructed in a modular way.

Figure 6 shows a functional model of the generic controller, which includes two sets of subcontrollers in a symmetrical design. One set of subcontrollers regulates the access to the resource, and the other set controls the size of the resource. The two sets of subcontrollers cooperate by accessing a shared data object, the resource graph. Each set of subcontrollers is made up of three functional components: trigger, allocator, and coordinator. The trigger decides when a computation should be done. The allocator performs the computation, which can be initiated by an external controller or by the trigger. The allocator that controls the access to the resource computes the amount of the resource that should be given to a particular request. The allocator that controls the size of the resource determines the resource capacity. A change of the resource capacity is coordinated by the coordinator object, which facilitates the interaction with other controllers. In particular, it implements the synchronization protocol needed to ensure that state changes among distributed controllers do not violate a set of resource constraints. The resource graph is modeled as two sets of weighted graphs, one representing the resource allocation and statistics, the other the resource capacity. Interfaces are provided to access and modify the relationship among these graphs.

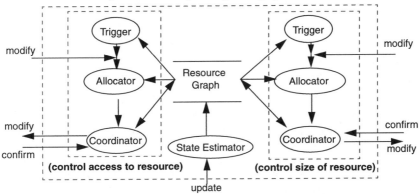

**Figure 6** Functional model of a generic controller.

In our implementation, a generic controller is realized as a container class in C++, which includes as base classes the subcontrollers, trigger, allocator, coordinator, etc. Interfaces offered by these subcontrollers are implemented as virtual functions that are overloaded for a specific realization of the controllers.

The design of the generic controller shown in Figure 6 has brought us the following benefits. First, it was possible for us to design and implement all three classes of controllers --VP admission controller, VPG controllers, and VPN controller-- as a refinement of the generic controller class. For example, the VP admission controller in Figure 5 has two "non-trivial" controller objects --the VC resource allocator and the coordinator-- and five "trivial" controller objects. (Trivial controller objects can be thought of as objects which perform no action except that of forwarding data to another object. They are not shown in Figure 5). The VPG controller contains four non-trivial controller objects and three trivial objects. Second, based on the generic controller design, we were able to realize different control schemes that attempt to achieve different control objectives for the customer control system. Realizing different control schemes is often possible by exchanging a set of subcontrollers in the system. For example, we implemented two classes of VC capacity allocators, realizing different VP admission schemes. One scheme aims at achieving call blocking objectives related to performance classes. The other scheme realizes call pre-emption in case of congestion, taking into account the priority of a call.

## 4.2  Enabling Management Objectives

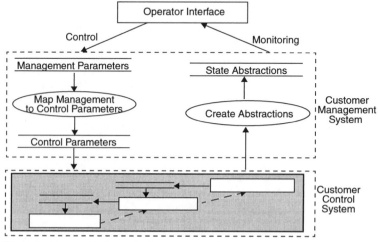

**Figure 7**  Framework for customer management.

The customer operates a management system to control and monitor the traffic on the enterprise network. A part of this system manages the traffic over the VPN, which is the focus of this paper. Examples of management capabilities that are related to the VPN service include controlling the bandwidth cost of the VPN service, VP bandwidth management, and QOS management (see Section 3). In the following, we describe how the management objectives outlined in Section 3 can be realized. Figure 7 shows our framework for implementing management capabilities. In this framework, management parameters, which directly relate to management objectives, are mapped onto control parameters, which influence the behavior of the controllers, and are subsequently distributed to the controllers in the customer control system [PAC95]. In our implementation, management parameters are made available to the operator of the enterprise network through the management console (Figure 8). A management parameter can be mapped onto control parameters for one or more classes of controllers. For example, cost management operations affect only the VPN controller. Allocating a specific capacity to a VP through a VP management operation affects both the VPG and the VPN controllers. QOS management operations, such as setting call blocking objectives, generally affect all classes of controllers. In response to a change in blocking objectives, the VP admission controller adjusts its admission policy, the VPG controller changes the VP allocation strategy, and the VPN controller negotiates the VPG sizes according to the new bandwidth requirements.

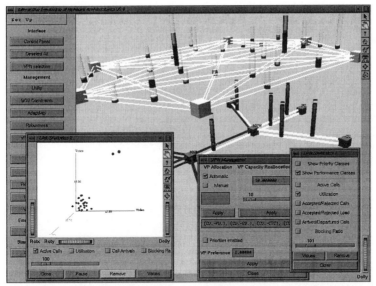

**Figure 8** Management console for a VPG-based VPN service. The upper layer represents the VP network, the lower layer the VPG network. The vertical bars on the VP network indicate the utilization, the vertical bars on the VPG network the allocation of VPG bandwidth to VPs.

Figure 8 shows the screen of the management console that we have implemented for customer management of a VPG-based VPN service. Both layers of the VPN are visible. The upper layer represents the VP network, the lower layer the VPG network. The vertical bars on the VP network show the current utilization of the VPs. The three segments of a particular bar correspond to the three traffic classes supported in our particular system. The outline of the cylinders indicate the currently allocated VP capacities. The vertical bars on the VPG network give the allocation of the VPG bandwidth to the VPs. A "cloud view" on the lower left corner shows the number of active calls in the VPs. Each axis corresponds to a traffic class. In this specific snapshot, one can see that two of the VPs experience a much higher load than the others. The interface in Figure 8 allows an operator to perform management operations and observe the reaction of these operations on the global state of the system.

## 5. PROTOTYPING AND EVALUATING THE ARCHITECTURE

Two main tasks are involved in the development of a network architecture: the development of a software system and the design and analysis of algorithms. The first task focuses on software engineering aspects to satisfy the system requirements. The second concentrates on developing control functions that meet performance objectives. A thorough evaluation of the performance characteristics of a network control system has to take into account both of these aspects. Our approach to evaluating a target architecture is to build a software prototype, designed according to this architecture, which runs the intended algorithms [CHA96b].

The emulation platform consists of four building blocks: parallel simulation kernel, emulation support, real-time visualization and interactive control, and emulated system (Figure 9).

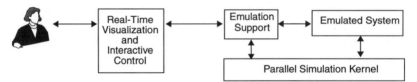

**Figure 9** Building blocks of the interactive emulation platform.

 The module for real-time visualization and interactive control contains an interface which provides 3-D visual abstractions of the system state. The emulation support module coordinates the exchange of control and monitoring messages between the graphical interface and the emulated system. It reads the states of the emulated system, and performs filtering and abstraction operations before making the information available for visualization. Control information from the user is mapped onto a set of control parameters that are interpreted by the emulated system.

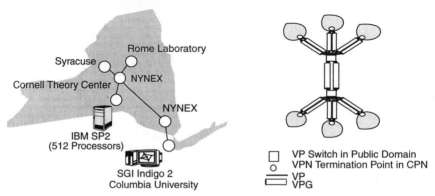

**Figure 10 (a)** Hardware configuration of the interactive emulation platform.

**Figure 10 (b)** Network topology used in the evaluation.

In our implementation, both the emulated system and the simulation kernel (coded in C++ and MPI) run on a SP2 supercomputer located at the Cornell Theory Center (CTC) in Ithaca, New York. The real-time visualization and interactive control module resides on an SGI Indigo2 workstation at Columbia University (Figure 10(a)). It is written using Open Inventor, a 3D graphics tool kit based on Open GL. The emulation support module is distributed on the two machines. These machines communicate through NYNET, an ATM network that connects several research laboratories in New York State.

## 5.1   Evaluating the Customer Control and Management System

In our evaluation, we implemented several versions of the VPN control system on the emulation platform. We have also built implementations of the management capabilities discussed in Section 3, namely, VP management, QOS management and priority management. Management operations are performed through the graphical interface (Figure 8). In the following, we summarize one of the results of the evaluation, in which we studied the effectiveness of resource control under constant traffic load. A more complete description of the experimental results can be found in [CHA96a]. The performance of the customer control system was evaluated in a scenario based on the topology of the NYNET testbed. In this scenario, a VPN service interconnects 6 CPNs. The VPN contains 14 unidirectional VPGs which support 30 unidirectional VPs, connecting the 6 CPNs in a full mesh topology. The two VPGs in the middle carry 9 VPs; the remaining VPGs carry 5 VPs each (Figure 10(b)).

There is one VP admission controller per VP, executing a complete sharing policy. A centralized VPG controller periodically recomputes the capacities that are allocated to the VPs, by estimating the expected utilization of the VPs for the next control cycle. The VPG controller distributes the bandwidth to the VP admission controllers using a two phase protocol. This protocol ensures that the sum of the capacities of the VPs within a VPG does not exceed the capacity of the VPG. No VPN control is performed, i.e., the VPG link capacities remain constant during the course of the experiments.

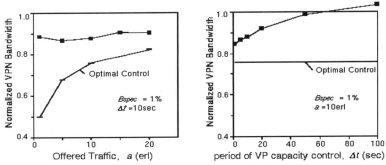

**Figure 11** Performance Evaluation of the customer control system.

We model the network traffic load, the processing time of the controllers and the time delay to send a message from one controller to another. The network traffic is composed of two classes with different bandwidth requirements. A class 1 call needs one unit of bandwidth, while a class 2 call requires 10 units of bandwidth. The holding time of the calls of both classes is exponentially distributed with a mean of 100 seconds, and call arrivals are modeled as Poisson processes. We vary three parameters in the experiments: the number of VPs in the VPG link ($n$), the offered load ($a$), and the control period for changing the VP capacities ($\Delta t$). All VPs in the VPG link experience the same offered load.

We define the normalized VPG capacity as the ratio of the VPG capacity needed to attain a specific call blocking probability (*Bspec*) with VP capacity control over that without control (fixed VP capacities). The plot on the left side of Figure 11 shows normalized VPG capacities for 5, 10 and 20 VPs. It indicates that the control effect is especially large when the offered traffic per VP is small and the number of VPs multiplexed in the VPG is large. The figure also shows the necessary VPN bandwidth for different traffic loads; the figure on the right side of Figure 11 gives the necessary VPN bandwidth for different control periods. The VPN capacity is computed as the sum of the VPG capacities. The figures also contain the lower limits for VPN bandwidth, which are calculated assuming complete VPG bandwidth sharing by all calls in the VPN. They approximate the performance of an optimal control scheme. The distance between the curves for the optimum control and our scheme in the left figure suggest that there is room for improving the algorithms and protocols we are running, specifically in the case where the offered traffic is low.

## 6. REFERENCES

[ATS93]   T. Aoyama, I. Tokizawa, and K. Sato, "ATM VP-Based Broadband Networks for Multimedia Services," IEEE Communications Magazine, April 1993, pp. 30-39.

[CHA96a]M.C. Chan, H. Hadama and R. Stadler, "An Architecture for Broadband Virtual Networks under Customer Control," IEEE NOMS, April 1996.

[CHA96b]M.C. Chan, G. Pacifici and R. Stadler, "Prototyping Network Architectures on a Supercomputer," HPDC-5, August 1996.

[FOT95]   S. Fotedar, M. Gerla, P. Crocetti, and L. Fratta, "ATM Virtual Private Networks," Communications of the ACM, vol. 38, no. 2, Feb. 1995.

[ELW93]   A.I. Elwalid, D. Mitra, "Effective Bandwidth of General Markovian Traffic Sources and Admission Control of High Speed Networks," IEEE/ACM Transactions on Networking, Vol. 1, No. 3, pp. 329-343.

[HAD89]   H Hadama, and S. Ohta, "Routing control of virtual paths in large-scale ATM-based transport networks," Trans. of IEICE, vol. J72-B-1, no.11, pp 970-978, 1989 (in Japanese).

[HAD94]   H. Hadama, T. Izaki, and I. Tokizawa, "Cost Comparison of STM and ATM Transport Networks," NETWORKS'94.

[HYM91]   J. M. Hyman, A. A. Lazar, G. Pacifici, "Real-Time Scheduling with Quality of Service Constraints," IEEE Journal on Selected Areas in Communications, September 1991.

[HYM94]   Hyman, J.M., Lazar, A.A. and Pacifici, G., ``VC, VP and VN Resource Assignment Strategies for Broadband Networks", Proceedings of the 4th International Workshop on Network and Operating System Support for Digital Audio and Video, Vol. 846, Springer-Verlag, 1994.

[LEW95]   D. Lewis, S O'Connell, W. Donnelly, L. Bjerring, "Experiences in Multi-domain Management System Development," in IFIP/IEEE ISINM, Santa Barbara, 1995, pp. 494-505.

[OHT92]   S. Ohta, K. Sato, "Dynamic Bandwidth Control of the Virtual Path in an Asynchronous Transfer Mode Network, "IEEE Trans. Comm. Technol., 40, 7, pp. 1239-1247.

[PAC95]   G. Pacifici and R. Stadler, "Integrating Resource Control and Performance Management in Multimedia Networks," in Proceeding of the IEEE ICC, 1995.

[SAY95]   T. Saydam and J.P. Gaspoz, "Object-Oriented Design of a VPN Bandwidth Management System," in IFIP/IEEE ISINM, Santa Barbara, 1995.

[SCH93]   J.M. Schneider, T. Preuss, and P.S. Nielsen, "Management of Virtual Private Networks for Integrated Broadband Communication," in Proc. of ACM SIGCOMM '93, pp. 224-237.

[TSC95]   M. Tschichholz, J. Hall, S. Abeck, R. Wies, "Information Aspects and Future Directions in an Integrated Telecommunications and Enterprise Management Environment," Journal of Network and Systems Management, Vol. 3, No. 1, 1995, pp.111-138.

[YAM91]   T. Yamamura, T. Yasushi, N. Fujii, "A Study on an End Customer Controlled Circuit Reconfiguration System for Leased Line Network," ISINM, 1991, pp. 383-394.

[ZER92]   T.G.Zerbiec, "Considering the Past and Anticipating the Future for Private Data Networks", IEEE Communication, March 1992, pp.36-46.

## Acknowledgments

This research was supported by the Department of the Air Force, Rome Laboratory, under contract F30602-94-C-0150. It was conducted using the resources of the Cornell Theory Center.

# 24

# ATM Network Resources Management using Layer and Virtual Network Concepts

*G. Woodruff, N. Perinpanathan, F. Chang, P. Appanna, A. Leon-Garcia*
*Department of Electrical and Computer Engineering, University of Toronto*
*Toronto, Ontario, M5S 1A4. Canada*
*Phone: (416) 978-4764     Fax: (416) 978-4425*
*gillian@comm.toronto.edu / appanna@comm.toronto.edu / alg@comm.toronto.edu.*

### ABSTRACT

This paper describes the organization of a resource management system for an ATM network within an integrated operations environment. The system encompasses both traffic and network management aspects within a single framework. A management hierarchy is described based on the concept of layer networks, and a new Virtual Network layer is defined for fairly allocating network resources amongst competing network and service applications. Each Virtual Circuit (VC), Virtual Path (VP), Virtual Network (VN) and Transmission Path (TP) layer network is assigned resources which can be accessed by and assigned to other client layers where applicable for the routing and admission of transport connections. Examples of general layer network management entities and an object-oriented information model capturing required resource and performance information are presented to illustrate the power of re-using concepts at each management layer.

**Keywords**: Virtual Network, Layer Network, Resource Management, Information Model.

## 1. INTRODUCTION

One of the key transport technologies within the future broadband network will be ATM (Asynchronous Transfer Mode). ATM standards include aspects related to connection control and network management. There is also now an enormous body of theoretical literature addressing related ATM traffic and resource management issues and requirements, some of which are far from resolved; and although their algorithmic aspects would not be subject to standardization, the transport network operator's connection control and management system must have access to a standard set of relevant network state information.

The objective of this paper is to bring together what we perceive as some essential elements of an ATM *Network Resources Management* (NRM) system within a management framework encompassed by standards bodies and consortia. Our initial focus here is on a system which is to operate within a single administrative domain of a public network. An NRM system, in our definition, is responsible for all aspects related to the fair allocation of transport network resources to competing users; as such, it includes: i) traffic management (ATM cell-level flow controls and connection set-up/release); and ii) network management (performance monitoring and configuration management). Controls of type i) would traditionally be proprietary and fall under the responsibility of the network access and switching equipment, whereas ii) is traditionally the responsibility of the network management system, with standardized interfaces now being defined within the ATM Forum and the ITU-T's Telecommunications Management Network (TMN) [M.3010].

The Telecommunications Information Networking Architecture Consortium (TINA-C) provides the framework for developing these traditionally separate NRM controls on a common object-oriented software platform in a distributed processing environment, allowing the sharing of common network resource information and management functionality. In this paper we give an overview of a possible

NRM architecture for ATM, based on concepts from TMN, TINA-C, the ATM Forum, and current ATM traffic management research. We start in Section 2 with a summary of ATM's *layer networks* and in Section 3 with an overview of what resources are to be managed and how these resources can be organized using the layer network and *Virtual Network* concepts. The virtual network is a logical collection of switching and transmission resources, and simplifies management of these resources. This is followed by an overview of the *NRM architecture* in Section 4, the *managing entities* required to support basic NRM functionality in Section 5, and some essential elements of the *NRM Information Model* to represent resource states in Section 6.

The objective is to define general concepts of ATM resources and NRM management functions which can be re-used at different levels of a management hierarchy to provide a rich and versatile "toolkit" applicable to a wide range of NRM needs. This effort is part of a larger project whose aim is to develop an integrated resource management architecture for large-scale, integrated services ATM networks. There are five projects across several Canadian universities addressing different facets of the larger project.

## 2. ATM CONNECTION AND LAYER NETWORKS

ATM has the ability to support two types of connections (referred to as "trails" by ITU-T and TINA-C): Virtual Circuit connections (VCCs) and Virtual Path connections (VPCs). Each VCC/VPC consists of a concatenated set of VC/VP links (VCLs/VPLs) connected by nodes supporting VC/VP switching. Each VC/VP link makes use of an identifier (VCI/VPI) which is used for switching and is translated into the identifier for the next link. VCLs can (but need not) be carried within VPCs; this simplifies VC connection set-up procedures, as less VC switches need be involved, augments the connection-carrying capacity of the network (limited by VCI address space), and provides a means to partition groups of like users.

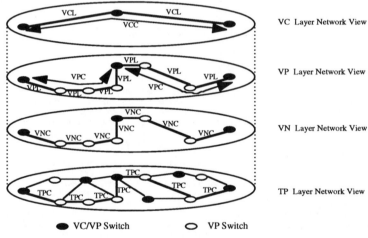

Figure 1: The NRM layer networks.

The layer network concept [G.805][TINA95] can be applied to ATM to view the connection topology and resources at the VC, VP and Transmission Path (TP) layers, as shown in Figure 1. The VP layer is considered a *server* to the VC layer; the TP layer can be considered a server to both the VP layer and the VC layer (when VCCs are not routed through VPCs). In the NRM context, a server layer can allocate some of its resources to the *client* layer (e.g. some transmission capacity assigned to a server VPC can be allocated to a client VCL). In addition, a server layer can define the restricted topology over which routing procedures may be carried out for a client layer connection, e.g. a VCC could be routed over the server VP topology, making use of information regarding VPC resource availability.

The layer network concept is an important one for the NRM system, as each layer provides a simplified logical network view of a specific portion of network resources, within which management functions can be carried out. We have extended the layers to include a *Virtual Network* (VN) layer, as illustrated in Figure 1 and motivated in the following section.

## 3. THE NEED FOR RESOURCE MANAGEMENT: ATM SERVICE GUARANTEES

### 3.1 Levels of resource allocation and Virtual Networks

A resource management system is necessary due to the need to allocate limited resources amongst competing users and to monitor resource usage. In the ATM transport network, competition for resources will occur at several levels.

**At the cell level,** VC or VP connections being multiplexed or switched at Network Elements (NEs) compete for buffer space and transmission time and consequently may experience cell loss and variable cell transit delays; fairness can be judged by providing cell-level Quality-of-Service (QoS) guarantees for each connection, and is realized through the Usage Parameter Control (UPC) or Network Parameter Control (NPC) of agreed connection traffic flow characteristics (Traffic Descriptors or TDs), buffer management, and scheduling policies at these NEs.

A large body of the ATM literature has focused on the development of models relating these policies and the cell-level QoS, which depend on the traffic flow characteristics, buffer and bandwidth resources, and switching architecture. We believe that a standard "black box" representation of specific switch and multiplexer performance capabilities is necessary if connection management functions are to be unbundled from the switching elements. This representation would be used to determine whether a new connection with specific TDs and QoS needs can be supported. From the resource management perspective, it is highly desirable to transpose this procedure into one where the NE capabilities and the connection requirements can be represented as a set of additive "virtual" resources to be allocated or released.

The *equivalent bandwidth* concept is one promising approach [ElWal93][Kesid93] in which the common resource "currency" is transmission bandwidth. The development of other capacity representations which account for multipoint switching or specific scheduling implementations with limited buffering remains an open issue. For our purposes here, we assume that an NE is capable of supporting a connection's TD and QoS requirements when the required equivalent bandwidth can be allocated to the associated terminating VCL or VPL. This provides a convenient link-oriented representation of NE resources required to provide cell-level QoS, which can be assigned to server layer networks and allocated amongst their clients.

**At the connection (VC or VP) level,** requests for connectivity may be denied due to a lack of NE resources to support cell-level QoS requirements (e.g. equivalent bandwidth), or due to a lack of VCI/VPI *addresses* or other resources such as *processing* required to perform per-connection UPC, flow control or scheduling, and buffer management; fairness at this level can be judged by providing connection-level Grade-of-Service (GoS) guarantees such as probability of blocking, and is realized through the "virtual" dedication of pools of these resources throughout the network to designated classes of users making connection requests.

We refer to this subset of network resources for a class of users as a Virtual Network or VN; within this network, requests for VC or VP connectivity or an increase in their resource allocations may be treated on an equal, first-come-first-served basis or on some other scheduling schemes. This concept of VN is more general than that of a collection of VPCs, as it provides a means for fair allocation of total TP layer resources amongst competing classes of transport applications. The resources allocated logically to the VN are managed as a set. All the VPCs, VCCs inside a VN may be managed as one entity instead of individually. Instead of a VN, if we use a collection of VPCs, then each VPC has to be managed or monitored singly as opposed to a group. Additionally, allocating resources to the VN will simplify resources management as now, only a subset of the network resources is to be managed as opposed to the entire set of network resources. For example, if VNs are defined for a particular class of traffic, the allocation/de-allocation of resources are eased because it is much simpler to allocate resources to similar traffic (homogeneous) than heterogeneous traffic. The logical segregation of resources will increase blocking probability but this is counterbalanced by multiplexing gain that are produced by the sharing of resources by the homogeneous traffic of the VN. For the large-scale network under consideration, the multiplexing gain should be significant.

The VN concept has only lately been appearing in the ATM literature under various nomenclature [Dziong96][Farago95], and we believe it will be a powerful and versatile resource management tool for large-scale networks. For example, VNs may be dedicated for a particular QoS/service/traffic class (e.g. video traffic), for an overlay network dedicated to a particular application (e.g. IP overlay network), or for a particular customer group with specific network management requirements (e.g. as a fault-protected Virtual Private Network); VNs could be nested within other VNs; and significant gains in bandwidth efficiency are possible by utilizing VP connections for fast routing only, with VC connection resources drawing upon the total VNC (see definition below) resource pools rather than individual VPC [Dziong96].

For this reason, we have added an additional VN layer network between the VP and TP layers in Figure 1. This Figure illustrates the physical topology of a VN, which corresponds to that of the TP layer. Since the VCCs and VPCs resources are drawn from the VN, it is logical to put the VN layer between the TP and the VP layer, instead of between other layers. We adopt the same layer network terminology, such that VPCs at a client VP layer are routed over Virtual Network connections (VNCs) at the VN layer. VNCs are shown routed in a "single hop" over the TP topology (and therefore consist of a single VN link - VNL); each VNC is assigned a portion of the TP connection's (TPC's) equivalent bandwidth. We assume that each VNC is also allocated a pool of VCI/VPI addresses to be associated with VCL/VPL terminations. Other NE resources, such as processing, are also allocated to switches at each VN layer.

Note that the routing over a VN (its logical connectivity) can be further restricted by defining a VN routing topology. This would imply a representation where VNCs are routed over multiple hops of the TP topology. However, since switching only occurs at VC and VP levels, no actual connectivity at the VNC level would be implied. The physical VN representation is necessary to capture information regarding the location of VC/VP switching nodes and resources associated with switching and terminating VCL/VPLs.

**At the VN layer,** there will also be competing requests for the limited resources of the TP (physical) network; it is therefore possible for VN requests to be denied. The way that fairness is judged at this level, and the associated sharing policies, are open research topics. However, all layer networks can be seen to share the pattern of connectivity requests (for a specified QoS and TDs) which are satisfied by allocating server layer resources. These resources need to be provisioned in such a manner as to meet the layer network's service guarantees. This repeating pattern motivates our definition of general layer network resource management functional entities discussed in Section 5.

## 3.2  QoS management per Virtual Network

The management of multiple QoS classes can be handled by establishing a VN for each class, and for illustrative purposes in this paper we will assume this approach. Resources allocated to a VN are then at the disposal of all connections which will be satisfied by the VN's QoS constraints. This can be achieved by establishing an end-to-end QoS reference connection and allocating QoS limits to each NE switch and link; a link's QoS allocation then equals its server layer's end-to-end connection QoS.

For example, a VPC would be routed over the VN layer network which supports the appropriate QoS class. The end-to-end QoS of the VPC would be the "sum" of the component QoS values associated with the NEs performing VP switching and the VNC "hops", and can be bounded by limiting the number of hops allowed over the VN layer. Similarly, a VCC routed over the VP layer would have an end-to-end QoS equal to the "sum" of the component QoS values associated with the NEs performing VC switching and the VPC "hops". The VCC end-to-end QoS can be similarly bounded by restricting the number of hops over the VP layer.

## 4. THE NRM ARCHITECTURE

The NRM architecture consists of a structured set of interacting managing systems to realize real-time traffic control and network management. The architecture is aligned with TINA's functional layering, and the NRM system resides within TINA's resources layer and TMN's Network Management Layer (NML) and Element Management Layer (EML). The architecture captures the strong interactions among the managing systems to perform admission control, routing, resource allocation, resource reconfiguration, connection set-up/release, and performance monitoring functions necessary to effectively manage the ATM network's resources.

The managing systems that form the NRM architecture are shown in Figure 2. These managing systems are defined based on the computational modeling concepts specified by the TINA-C [TINA95a]. They include the Connection Session Manager, a set of Layer Network Managers (LNMs), and a set of Element Managers, which are organized and structured based on the ATM transport network functional architecture defined by ITU-T [I.326] and the management functional layering principle defined by TMN [M.3010]. Interactions among these managing entities consist of invocations, responses, and notifications.

Connection requests initiated at TINA's service layer, as depicted in Figure 2, are directed to the Connection Session Manager. The Connection Session Manager identifies the Virtual Network (VN) within which the connection is to be established, and directs the request to the appropriate LNM. The LMNs are defined for VC, VP, VN and TP layers, and perform dynamic resource management within their respective layer. For example, a VCC request with a particular QoS class is directed to the VC manager of the appropriate VN reserved for that QoS class.

The Element Manager translates the physical NE state information into vendor-independent logical resource information for the LNMs. Each Element Manager is delegated an individual NE or a set of NEs. The Element Manager is notified by the NEs of faults, which are also polled to retrieve performance related parameters. Other functions of the Element Manager include: monitoring and reporting facility usage to the LNMs; interacting with local NEs to configure network connections upon the LNM's command; initiating continuity checks and traffic loop-back tests; and maintaining a local database to keep updated information about the local transmission path, link, and node states under its purview.

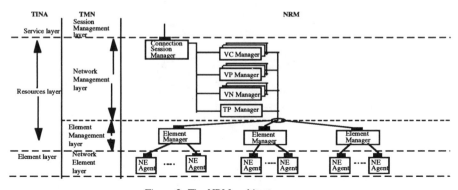

Figure 2: The NRM architecture.

## 5. THE LAYER NETWORK MANAGER (LNM)

Each Layer Network Manager performs similar traffic management and network management functions (such as connection management, resource reconfiguration management, and performance management) based on the resource states of its layer network. Therefore, to achieve high software reuse[1], it is desirable to define a generic LNM that can be specialized to yield a Virtual Connection (VC) Manager, Virtual Path (VP) Manager, Virtual Network (VN) Manager, and Transmission Path (TP) Manager. Note that a VC, a VP, and a VN Manager may exist for each Virtual Network defined over the Transmission Path layer network. However, only a single TP Manager is required irrespective of the number of VNs that are defined.

Such a generic LNM can be specified as a building block that consists of a set of managing entities, which in turn can be implemented using Computational Objects (COs) that interact with each other in order to make management decisions pertaining to the layer network. This approach is shown in Figure 3, which illustrates a generic LNM building block along with its managing entities and possible interactions (indicated by directed arcs), both within and external to the LNM. As shown in Figure 1, the VC layer network is considered the "highest" layer network while the TP layer network is considered the "lowest" layer network; therefore, lower-numbered LNMs can act as a server to any higher-numbered LNM and as a client to any lower-numbered LNM. For example, LNM(N) can act as a client to LNM(N-1) and as a server to LNM(N+1).

The managing entities defined within the generic LNM include the Admission Controller, the Routing Manager, the Resource Manager, the Connection Manager, the Configuration Manager, and the Network Information Manager. The Admission Controller, and the Configuration Manager act as clients to appropriate managing entities within lower-numbered LNMs. The Routing Manager and the Resource Manager act as servers to appropriate managing entities within higher-numbered LNMs. The Network Information Manager oversees the storage and monitoring of the layer's information base, which presents the necessary network state information for use by these managing entities.

Functions performed within each managing entity are transparent to the other managing entities. However, a managing entity may invoke another managing entity through appropriate interfaces to obtain its management services. Section 5.1 illustrates the sequence of interactions that take place among the above managing entities to set up a new VCC over the VP layer network. Section 5.2 then generalizes the functions performed by each of these managing entities so that they can be applied to a variety of network resource management tasks.

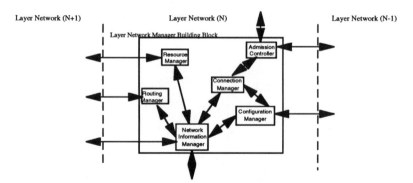

Figure 3: The Layer Network Management building block.

## 5.1 Example: VCC set-up over VP layer

Figure 4 shows the interaction among the managing entities within the VC, VP and VN Managers to set up a new VC connection. The Connection Session Manager sends a VCC request to the VC Admission Controller (1). The VC Admission Controller then invokes the server (in this case, the VP layer) to find an appropriate route over the server layer topology (2) and to reserve the necessary transmission

---

[1] Software re-use is achieved by instantiation of objects from the same object classes and by specialization that allows the addition of attributes and operations to an existing object to yield completely different objects. An object encompasses both function codes and data structures.

resources (3). Switching resources at each VC switching point are reserved through invocation of the VN Resource Manager (4). The VC Admission Controller then invokes the VC Connection Manager (5) to perform the VC connection binding operations, and the VC Network Information Manager then instantiates the VCC and Termination Point objects within the VC layer's information base (6). The VC Network Information Manager then invokes the Element Managers for each VC switching node (7) to set the UPC parameters and the translation table entries with specified VCIs. Once the Network Information Manager receives a confirmation from the Element Manager, an acknowledgment is conveyed to the Admission Controller, which indicates to the Connection Session Manager that the new VCC has been set up.

Note that the set of interactions among the LNMs to set up a VPC over the VN layer network, or to set up a VCC over the VN layer network are similar to those shown in Figure 4. In the former case, the VP Manager acts as a client to the VN Manager, while in the latter case the VC Manager acts as a client to the VN Manager. This generality also extends to the case of setting up a VNC over the TP layer network.

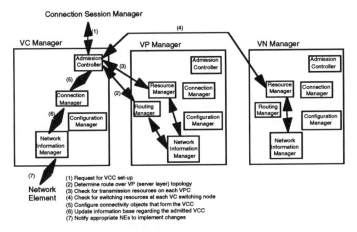

Figure 4: Interactions among the VC, VP and VN LNM managing entities

## 5.2 Description of the LNM's generic managing entities

This section gives a summary of the generic functions performed by each managing entity, and how these functions specialize at specific layers.

Admission Controller

The Admission Controller is invoked by the Connection Session Manager following a connection/topology request from the service layer, and coordinates the set of steps to decide whether a new connection/topology can be admitted into the network. A connection/topology specializes, for VC and VP layers, as a point-to-point or multipoint connection; for VN layers, as a network topology; and for the TP layer, as a "physical" network (which may itself be a layer network e.g. within SONET).

The Admission Controller interacts with the server Routing Manager and Resource Manager to determine whether a route and the necessary resources are available within the server layer network for connection/topology admission. Based on the responses received from these managing entities, the Admission Controller may decide to accept the connection/topology, reject the connection/topology, or optionally allow for re-negotiation in case of rejection. The number of rejected requests is stored by the Network Information Manager to track service levels being provided.

## Configuration Manager

The Configuration Manager is invoked internally, and coordinates a similar set of steps as the Admission Controller to dynamically modify the layer network's topology and resource states, either on a scheduled or on-demand basis, in order to track projected or sudden changes in traffic demand or topology. These functions specialize, for VC and VP layers, as changes in connection resource allocations, re-routing in the event of failure, or adding/dropping leaf terminations; for VN layers, as new resource allocations or topologies; and for the TP layer, as changes in switching nodes or transmission capacities within the ATM network domain. Changes in resource allocations may arise due to performance degradation at the specific layer (e.g. QoS degradation at the VC and VP layers, GoS degradation at the VP and VN layers, resource shortage at the TP layer) or due to dynamic adaptation to measured and projected traffic demand.

## Routing Manager

The Routing Manager determines an appropriate route based on its layer network's topology and resource states (e.g. estimated or measured utilization of resources). The Routing Manager may be invoked by a client layer's Admission Controller during a new connection set-up, or by a client layer's Configuration Manager to perform re-routing in the event of failure or adaptive reconfiguration . The Routing Manager interacts with the Network Information Manager to obtain topology and resource-related information for route selection, which may in turn store a set of alternative routes for future use. Since routing is not performed at the VC layer network, a Routing Manager does not exist within the VC Manager.

## Resource Manager

The Resource Manager regulates the allocation of its connection hop resources (such as effective bandwidth and addresses) and Network Element resources (such as processing and virtual resources) to satisfy the client layer's traffic demand and performance constraints. Network Element resources reside at VN and TP layers, while resources for a connection "hop" can reside at all layer views. The Resource Manager may be invoked to allocate or re-negotiate resources for a new connection, or to re-assign resources allocated to already existing client layer connections. As with the Routing Manager, the Resource Manager does not exist within the VC Manager.

## Connection Manager

The Connection Manager binds connection and Termination Point objects during connection set-up, and unbinds them during connection release. Functions performed by the Connection Manager include logical configuration of the switch, i.e. setting up translation table entries from incoming to outgoing ports. The Connection Manager may be invoked by the Admission Controller to bind a new connection, or by the Configuration Manager to bind a modified or restored connection. The Connection Manager in turn invokes the Network Information Manager to instantiate and update the set of objects that form the connection.

## Network Information Manager

The Network Information Manager monitors the Management Information Base which captures the necessary topological and resource state information for layer management. Managing entities within a LNM invoke the Network Information Manager to update the network state information each time a management decision is made. The Network Information Manager in turn notifies the appropriate Element Manager(s) regarding changes in connections or resource states in order to physically effect the logical changes within the appropriate NEs. The Network Information Manager also receives performance-related parameters from the Element Manager. In addition, the Network Information Manager triggers the Configuration Manager to take appropriate actions whenever it detects a degradation in its layer network's performance.

The next section will discuss some aspects of the information model that are used by these managing entities for resource management.

# 6. NRM INFORMATION MODEL (NRM-IM)

The NRM Information Model (NRM-IM) captures the necessary ATM network resource and performance information to enable the resource management functions. The NRM information model (NRM-IM) is an ATM-specific and protocol-independent information model that abstracts an ATM network into a logical view. This logical view contains all network architecture components, connection objects and related resource and performance information identified in previous sections of this paper. The NRM-IM establishes a standard view of the network among all managing entities so as to allow sharing of common network information.

The NRM-IM consists of two parts: a network view (nv) and an element view (ev). The element view focuses on modeling individual network elements, while the network view models the aggregation of the network elements and resources and provides a network-wide view. The element view Management Information Base (MIB) is used for communication between the NEs and the Element Managers at the EML. It is also used by the Element Managers at the EML to interact with the Layer Network Managers at the NML. The network view MIB is used for interactions among Layer Network Managers at the NML.

The NRM-IM (ev) is adopted from ATM Forum's M4 MIB element view [ATMF-M4ev] and Bellcore's TA-1114 [TA-1114]. The NRM-IM (nv) is based on ITU-T's functional architecture of the transport network, both generic [G.805] and ATM-based [I.326], ATM Forum's M4 MIB network view [ATMF-M4nv], and TINA's Network Resource Information Model (NRIM) [TINA95b].

Figure 5 shows a subset of the NRM-IM (nv) object classes which contain the network resource and performance information for resource management. The object classes are specified using the widely used object model of Rumbaugh [Rumb91]. The object classes shown in this figure are generic classes. All object classes in each layer network **inherit**[2] from these generic classes. For instance, the `Trail` class can be **specialized**[3] to become a VCC, VPC, VNC and `TPC` in the VC, VP, VN and TP layer network respectively and the `LinkConnection` class can be specialized to become a VCL, VPL, VNL, and `TPL`. The inheritance and specialization principles of object-oriented technology enable the creation of objects that capture the attributes and operations at different abstraction level. They also allow code re-use. The Connection Termination Point (CTP) and the Trail Termination Point (TTP) are similar to the Network Connection Termination Point (NWCTP) and the Network Trail Termination Point (NWTTP) in TINA's NRIM, respectively.

The ATM network resource and performance information required for resource management is captured by the attributes of these object classes. The information or the attribute values would be gathered by assignment, estimation or measurement depending on the nature of the information. Network resources that we show are address space (VCI, VPI range), equivalent bandwidth and reserved switch processing. Parameters characterizing connections include traffic descriptors (TDs), QoS class and bandwidth utilization. Performance information is captured by the QoS performance parameters and GoS. These attributes are shown in Figure 5 and are listed below. Managing entities within each Layer Network Manager would manipulate these attributes while carrying out their corresponding resource management functions.

`Address space` -- Attribute of `TTP`.
The `Address space` can be a VCI range or VPI range or both. It indicates whether a VCI or VPI is available for a new connection.

`Equivalent Bandwidth` -- Attribute of `Trail` and `LinkConnection`.
This is an abstract representation of a connection's resource requirements at termination points based on the traffic descriptors and the QoS performance parameters of that connection.

`Switch Processing` -- Attribute of `Subnetwork`.
This is a quantitative representation of processing capacity available for a switch to handle new connections.

---

[2] Inheritance is the sharing of attributes and operations among classes based on a hierarchical relationship.

[3] Specialization: Each subclass inherits all the properties of its superclass and adds its own unique properties.

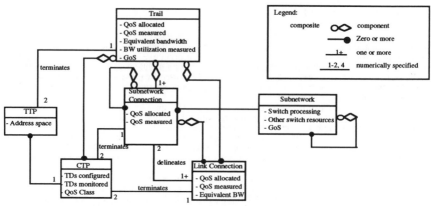

Figure 5: The NRM-IM (nv) object classes

Traffic Descriptors Configured -- Attribute of CTP.
This characterizes expected connection traffic flow. These TDs include Peak Cell Rate (PCR), Sustainable Cell Rate (SCR), Burst Tolerance (BT) and others, as defined by the ATM Forum and ITU-T.

Traffic Descriptors Monitored -- Attribute of CTP.
This indicates the UPC/NPC violation counts and number of successfully passed cells at each CTP by aggregating the parameters obtained from the UPC/NPC Monitoring Data object class in the element view.

Bandwidth Utilization Measured -- Attribute of Trail
This reflects measurements of specific parameters describing actual traffic flow e.g. TDs, average and variance of cell rates, etc.

GoS -- Attribute of Trail.
This indicates the connection-level Grade of Service such as probability of blocking.

QoS Class -- Attribute of CTP.
The QoS Class can be specified or unspecified. Several QoS classes are specified by the ATM Forum to provide different types of services. A set of default values will be assigned to the QoS performance parameters if a specified QoS Class is assigned.

QoS -- Attribute of Trail, SubnetworkConnection and LinkConnection.
This is a set of QoS performance parameters that conveys the cell-level QoS performance information for the given connection segment. These parameters are defined by both the ATM Forum [ATMF-v4] and ITU-T [I.371]. There are two types of QoS performance parameters. The QoS allocated is the assigned QoS contribution allowed for that connection or connection segment, based on the assumed reference connection. The QoS measured is the actual cell-level performance obtained through measurement. For instance, the QoS performance values are typically measured by sending OAM cells intrusively or non-intrusively through the segment of interest and applying loopback at the far end. The values of these attributes would then be calculated by studying the received OAM cells.

Typically, all objects are contained in either ATM NEs or layer networks. Although this provides a clean partitioning of the objects, object relationships that relate objects in different layer networks and views (ev/nv) are still required to maintain the integrity of the information model:

- Objects with inter-layer relationship: LinkConnection and Trail classes are involved in the inter-layer relationship. Based on the functional architecture of the transport network, a link connection at the client layer will be served by a trail at the server layer. The relationship is illustrated in Figure 6.

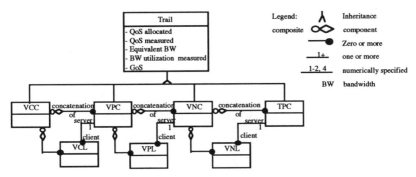

Figure 6: Client-server relationship of connection objects between adjacent layer networks.

- Objects with inter-view relationship: CTP, TTP, Subnetwork and SubnetworkConnection are the object classes that exist in both network view and element view of the information model. These objects carry the network state measurement information from individual network elements and forward it to the network view for aggregation, filtering and interpretation. In other words, these classes are the only element view aspects required to construct the network view. The CTP and TTP classes appear the same in both views. The Subnetwork and the SubnetworkConnection network view object instances at their smallest granularity are equivalent to the ATMCrossConnectControl and the ATMCrossConnect at the element view respectively, as suggested by TINA's NRIM.

# 7. SUMMARY

This paper has given an overview of how Virtual Network, layer network, computational modeling and information modeling concepts can be applied for resource management within an ATM network. We have stressed the power of the *layer network* concept, which leads to a hierarchical structure of similar, interacting *Layer Network Managers* responsible for the management of their layer network state information and the allocation of their resources to other client layers where applicable. We have also introduced a *Virtual Network* layer, which we believe will be an important concept for the fair allocation of network resources amongst competing network services and applications. The implementation of traditionally separate traffic and network management functions within a repeating layered management structure provides a rich and flexible foundation for future resource management systems.

Such systems will likely take advantage of the advances made in the emerging distributed computing environment, e.g. CORBA. This environment will offer a platform for the flexible distribution and expansion of the management capabilities presented here, and brings with it new challenges regarding implementation for good "signaling and delay" performance, and design for scalable and stable operation.

## 8. REFERENCES

[ATMF-v4]   ATM Forum Traffic Management Working Group, Draft ATM Forum Traffic Management Specification Version 4.0, December 18, 1994.

[ATMF-M4ev] "M4 Interface Requirements and Logical MIB: ATM Network Element View", Version 1, ATM Forum, October 1994.

[ATMF-M4nv] "M4 Interface Requirements and Logical MIB: ATM Network View", Version 1 (Draft), ATM Forum, 1995.

[Dziong96]  Z. Dziong, Y. Xiong and L. Mason, "Virtual Network Concept and its Applications for Resource Management in ATM Based Networks", Broadband Communications '96, Montreal, Canada, April 23-25, 1996.

[ElWal93]   A. I. ElWalid and D. Mitra, "Effective bandwidth of general Markovian traffic sources and admission control of high speed networks", *IEEE/ACM Trans. on Networking*, vol. 1, no. 3, June 1993, pp. 329-343.

[Farago95]  A. Farago, S. Blaabjerg, L. Ast, G. Gordos and T. Henk, "A New Degree of Freedom in ATM Network Dimensioning: Optimizing the Logical Configuration", *IEEE JSAC*, vol. 13, no. 7, pp. 1199-1206, September 1995.

[Fowler95]  H. Fowler (Bellcore), "TMN-Based Broadband ATM Network Management, *IEEE Communications Magazine*, March 1995.

[G.805]     ITU-T Recommendation G.805 (Draft), "Generic Functional Architecture of Transport Network".

[I.326]     ITU-T Recommendation I.326 (Draft), "Functional Architecture of Transport Networks Based on ATM".

[I.371]     ITU-T Recommendation I.371, "Traffic Control and Congestion Control in B-ISDN", March 1993.

[Kesid93]   G. Kesidis, J. Walrand, C. S. Chang, "Effective bandwidths for multiclass Markov fluids and other ATM sources", *IEEE/ACM Trans. on Networking*, vol. 1, no. 4, August 1993, pp. 424-428.

[M.3010]    ITU - CCITT Recommendation M.3010, "Principles for a Telecommunications Management Network", October 1992.

[Rumb91]    J. Rumbaugh, *Object-Oriented Modeling and Design*, Prentice Hall, 1991.

[TA-1114]   TA-NWT-001114, "Generic Requirements for Operations Interfaces Using OSI Tools: ATM/Broadband Network Management", Bellcore Technical Advisory, Issue 2, Livington, N.J., October, 1993.

[TINA95]    "Overall Concepts and Principles of TINA", TINA Document TB_MDC.018_1.0_94, February 1995.

[TINA95a]   "Computational Modelling Concepts", TINA Document TB_NAT.002_3.1_94, December, 1994.

[TINA95b]   "Network Resource Information Model Specification", TINA Document TB_LR.010_2.0_94, December, 1994.

Biography

G. Woodruff and P. Appanna are Ph.D. students at the University of Toronto.

N. Peripanathan and F. Chang completed their M.A.Sc. degree in 1996 and are currently with Nortel.

A. Leon-Garcia is a professor in the department of electrical and computer engineering at the University of Toronto.

# 25

# Management of New Federated Services

*Bhoj, P., Caswell, D., Chutani, S., Gopal, G., Kosarchyn M.*
*Hewlett-Packard Laboratories, Palo Alto, CA 94304, USA*
*Email: {preeti,caswell,chutani,gita,maka}@ hpl.hp.com*
*Telephone:+1 415 857 3797; Fax: +1 415 857 5100*

## Abstract

The explosive growth of the Internet, widespread use of the World Wide Web, and a trend towards deployment of broadband residential networks are stimulating the development of new services such as interactive shopping, home banking, and electronic commerce. These services are *federated* since they depend on an infrastructure that spans multiple independent control domains. Managing federated services and providing effective support to the customer of these services is difficult, because only a small part of the environment can be observed and controlled by any given authority. We characterize different dimensions of this problem, using our experience with the deployment of a system that gives the home consumer broadband access to community content as well as to the Internet. This type of system is referred to as Broadband Interactive Data Services or BIDS. We then focus on diagnosis and describe a customer support tool that was developed to partially automate diagnosis in BIDS. We use the experience with this tool to derive a blueprint for a general architecture for managing federated services. The architecture is based on service contracts between control domains.

## Keywords

Internet, Federated Systems, Federated Service Management, Broadband Interactive Data Services, Fault Diagnosis, Customer Support, Service Contracts.

## 1. INTRODUCTION

The explosive growth of the Internet and the widespread acceptance of the World Wide Web have caused a corresponding growth in new offerings of networked services such as Web-based interactive shopping [Online, Pathfinder] and home banking [Wells]. Several companies are also experimenting with Web-based inter-business solutions [Business, 96] since the standards-based multi-service Internet infrastructure offers a cost-effective alternative to proprietary Electronic Data Interchange (EDI) systems [Li] for inter-business transactions. In addition, cable and telephone network operators, as well as Internet service providers, are making a concerted effort to sign up residential customers with a variety of IP-based data services. Access technologies such as cable modems and

Asymmetric Digital Subscriber Line (ADSL) [Minoli, 95] offer the potential of providing millions of residential customers with*broadband* access.

While a great deal of excitement exists regarding the potential of Internet-based services, the operational characteristics and *management* of these services have received very little attention. These Internet services all cross administrative boundaries. Therefore, components in each of these domains must cooperate and function correctly for the services to work. Management of these services, which we refer to as *federated service management*, is a challenge since only a small part of the environment can be observed and controlled by any given authority.

Customer support is a critical but often neglected aspect of management that is particularly difficult in a federated environment. Our experience in working with a distributed system to deliver high-speed data services to the home has been that good customer support determines the customer experience with complex Internet services, particularly for relatively unsophisticated users. Borenstein et al., [Borenstein, 96] observed in their experience of running  First Virtual Holdings, an Internet-based electronic commerce service company, that the biggest unexpected problems centered on customer service,  and that "an Internet-savvy customer service department is an absolute prerequisite for anyone providing commercial services to the net".  While good customer support is essential,  economics for the mass market dictate that customer support departments cannot rely solely on human expertise to handle the growing complexity as well as increasing numbers of users. Our research is aimed at identifying technological support to help simplify the complex problem of customer support for federated services.

The goal of this paper is to articulate the challenges facing service management in federated systems, detail the issues surrounding customer support, and describe our work to date in this area.  In Section 2, we discuss examples of federated systems in everyday use, such as the telephone network and the Automated Teller Machine (ATM)/Point of Sale (POS) network, and contrast why the challenges are more severe in the kinds of systems being envisioned for the future.  In Section 3, we describe a type of federated system being deployed today to provide broadband interactive data services (BIDS) to residential customers.  In Section 4, we outline our experience in providing customer support for a trial deployment of  BIDS. We have implemented a system to aid in testing and diagnosis for customer support; this system is also briefly described in Section 4. We used this experience to  derive a blueprint of an architecture and requirements for a federated management system, which is discussed in section 5. We end in Section 6 with a summary and future work.

## 2. FEDERATED SYSTEMS

We first define the terminology used in this article. A *federated system*  is defined to be a system composed of different administrative entities cooperating to provide a service. A *service* is an application with a well defined interface and functionality. *Federated service management* is the management of services that span multiple heterogeneous control domains, and which rely on correct functioning of components across those domains. A *control domain* is defined to be an administrative domain that is managed by a single administrative entity, typically a business. There are several successful examples of operational federated systems offering networked services, including the telephone network, the ATM/POS network, EDI systems for inter-business movement of supplies and products, and the Federal EFT network [Juncker, 91]. In this section we focus on the public, open systems such as the ATM/POS network and the telephone network, rather than the private, closed EDI systems, and extract the characteristics that render these systems manageable and reliable. We then contrast these characteristics with those of the future Internet-based services.

The federated systems that are currently operational typically have a small number of different types of business entities that participate in the end-to-end service. For example, the different control domains for an ATM/POS network are the retailer's store, the retailer's headquarters, the retailer's bank, the switching organization such as Interlink which serves as an intermediary to route transactions, and the customer's card-issuing bank [Perry, 88]. Standards mandate all aspects of the system from message formats to physical thickness, size and embossing of the cards. The American Bankers Association (ABA) publishes interpretations of the standards for POS debit systems, along with the responsibilities in a POS system of card-issuing banks, manufacturers of terminals, retailers, retailers' banks, and switching organizations. Similarly for the telephone network, the entities that collaborate to set up a call such as the Local Exchange Carriers (LECs) and the Inter-Exchange Carrier are well defined. All interactions between various control domains and pieces of equipment have been specified in documents such as the LATA Switching Systems Generic Requirements (LSSGR) [LATA, 96]. In addition to detailed requirements, there are stringent certification procedures for equipment suppliers who build to these interfaces.

Both of these federated systems are highly reliable and provide a high degree of customer satisfaction with respect to support. They share a set of characteristics that make this possible, or at least easier to provide, than the Internet-based services that are now becoming available.

Each of these systems was created for the purpose of providing a *single service*[1].

- All the system components and their interfaces were designed and implemented with a priori knowledge of the single service they enable.

- A common understanding of semantics exists in all interactions between domains.

- Standards prevail at all levels in the system, including the application level.

*Regulatory policy* guides the development and operation of each of these types of systems.

---

[1] At least as originally designed. The phone network has grown over the years to offer many more services than a basic phone call, but the new services are derived from or provide enhancements to the original service.

- The legal responsibilities are clearly defined for all participants. Bodies such as the ABA, specify these responsibilities for the ATM/POS network. A host of policy-making bodies (federal and state regulators, legislatures, and courts) make decisions that collectively form a set of publicly-known policies for the phone companies.
- A very high level of reliability, and by implication testing, is imposed on these systems.
- There are strict certification procedures for the suppliers of system components. This limits the total number of vendors in the market place.

These systems were developed during a time of *minimal  competition* for the provision of services.

- They had the luxury of a fairly long maturation process that lead to a large and stable system core.
- The cost of poor reliability was deemed to be sufficiently high to make high reliability a design goal of the system.

In contrast, the kinds of federated systems that are being deployed today to support services such as Web-based interactive shopping, inter-business electronic commerce, and digital on-line photo-finishing, exhibit characteristics that differ radically from those listed above.

These systems are *multi-service systems*, where the exact mix of services being provided changes with time.

- Typically Internet-based services reuse an existing general purpose infrastructure for new applications.
- Although standards exist at the lower levels, such as  the Internet Engineering Task Force (IETF) and Worldwide Web Consortium (W3C) standards for transport level protocols and Web interaction protocols, there are no application level standards, nor commonly accepted application semantics.
- The configurations of participants change and grow very rapidly as new business opportunities are explored for offering new services. For example, participants offering services such as Web hosting, or search engines are relatively recent phenomena.

The use of Internet is *completely unregulated.*

- There are no legal precedents for service contracts among participants. It is  not even clear who can serve as an appropriate regulatory body to oversee compliance with service contracts.
- No testing and reliability requirements are being enforced  so far on service/component providers.
- The myriad choices for equipment and vendors, with general purpose, off-the-shelf equipment being configured into new and complex systems, impede certification procedures.

The pace of evolution of Internet-based services is frenetic and marked by *intense competition.*

* The opportunistic nature of the application domain results in transient services and service providers.
* The rapid pace often results in very short product life-cycles and low levels of testing due to time-to-market pressures.
* Reliability is not yet considered a primary design goal.

We believe that the remarkable dynamism of the Internet era precludes the use (at least in the short term) of  careful specification and engineering, and standards based construction and operation of service delivery systems that characterized federated systems in the past. Fundamentally, Internet services lack a coherent, top-to-bottom, end-to-end system architecture to serve as a blueprint for evolution. On the other hand, these new services, which started out as novelties, are rapidly becoming mission critical. Therefore, the businesses trying to sell these services to consumers must provide high levels of predictability and ease-of-use, to prevent customer frustration and rejection. We must concentrate research and development efforts in improving the operational characteristics of this new class of federated systems, and in providing solutions for federated service management.

## 3. BROADBAND INTERACTIVE DATA SERVICES (BIDS)

The imminent deployment of BIDS and BIDS-like systems has motivated our research into federated systems. The BIDS systems are emerging due to two phenomena: an ongoing effort to connect homes via broadband networks, and services motivated by the availability of high bandwidth, such as video-on-demand, videophones, home shopping, and lately high speed Internet access and World Wide Web access. These services are federated since the underlying system is composed of multiple independently managed systems that cooperate amongst themselves to furnish a service.  Since these services are targeted towards non-technical customers, effective and efficient operational support is critical to their success.

### 3.1 BIDS Architecture

The general architecture of a BIDS system is shown in Figure 1. Independent domains are drawn as boxes with dotted lines. A subscriber interacts with the system via an access network. This can be an HFC cable TV network, an ADSL or ISDN network provided by the local telephone company, or a Satellite or Wireless network. The access requires hardware such as PCs and modems and appropriate protocols to talk to the headend (the terminating point for the access network). The user is connected via a local access network to a server complex  (shown as the Application Service Provider Domain in Figure 1). The server complex has the necessary infrastructure for managing access to the access network, which includes managing subscription, billing, and security for the subscribers. It addition, the server complex can also provide content to the subscribers (such as a community Web server or bulletin board), and access to the Internet, other networks (online service providers (OLSP)), and services provided by those networks.

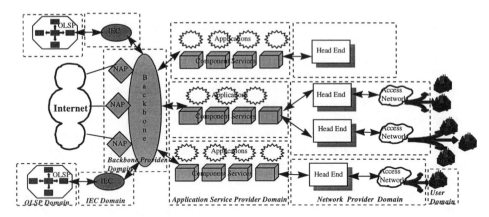

**Figure 1**   The Architecture of BIDS.

A customer accesses the system via applications such as a Web browser or an email application. Successful functioning of these applications is dependent on multiple system components. For example, to look at a Web page on a Web server on the Internet, a user goes through the following steps. The user's browser application (which must be installed prior to use) connects to the login server via a modem, an access network, and the server complex high speed network, to obtain the right credentials for access. Subsequently, the browser accesses the Internet by going through the Internet gateway and the firewall server. Problems with any of these components can cause the browser to fail to access the requested page.

From the architecture of BIDS it can be seen that BIDS is a federated system that spans the Internet and the online service providers. It is multi-service, unregulated, and the services are subject to intense competition since anyone can furnish a service and try to compete with the existing providers of a similar service.

One such system, called HP-BIDS,  has been developed by Hewlett-Packard and deployed in a trial. It is a representative example of federated systems that are likely to emerge in the future to provide high bandwidth data services to homes. A recent trial deployment of the HP-BIDS system has given us the opportunity to study the problem of customer support for a federated service, namely broadband access to the Internet. The trial has been going on for approximately 18 months at this point, and the experience gained from it has yielded several useful insights into the issues of diagnosing customer reported problems and dealing with multiple control domains in the process.

## 4. MANAGEMENT OF BIDS

Managing services in an Internet-based federated system such as BIDS is a challenge. In what follows we limit ourselves to the problems faced in diagnosing faults and doing customer support in BIDS and BIDS-like federated systems. This choice is deliberate since the success of the BIDS infrastructure will depend on providing satisfactory customer service to the end-users who are not necessarily technologically savvy. It is also an issue that has been largely ignored by the Internet community with a few exceptions [Borenstein, 96].

### 4.1 Difficulty of Diagnosis

Diagnosing services provided by a BIDS-like system is difficult because the underlying system has a complex configuration with respect to the mix of components, there is no prior operational experience, and little understanding of failure modes. Due to complex interaction between different parts of the system, the failure modes that are observed are very diverse, unexpected, not intuitive, and show no single dominating factor responsible for malfunctions.[2]

Since the mix of components is varying and not under central control, interactions between them will continue to be unpredictable. The practical implication of such interactions is that the symptoms of a problem can appear far away from the problem itself, both in space and time, and may have no apparent correlation with the problem. On the other hand, the complaints are stimulated by the failure of services being used by a customer. Diagnosing these services requires information about all the components that a service relies upon, which may not be available if the components are in a domain that does not export this information outside the domain. Yemini et al., [Yemini, 96] have discussed similar phenomena in the context of large scale heterogeneous networks, focusing on the problem of alarm correlation.

We give two examples that illustrate the spectrum of problems that were encountered. One of the problems that customer support had to diagnose was a call from a customer whose FTP software (a utility for transferring files across machines) stopped working after he had installed a new version of the FTP on his PC. The problem was traced to a temporary network failure, but not before customer support had checked the PC configuration, the cable modem, the new version of the FTP that was installed, the state of the subscriber account, and the availability of the servers to which the user wished to connect. Sometimes this had to be done on machines in different administrative domains, which required phone calls to the managers of those domains to get the right information.

A second problem that occurred quite often was a deterioration in performance of applications that used TCP on the HFC access network between a particular PC and the

---

[2] The faults observed in the initial deployment of HP-BIDS over a period of 3 months ranged from network problems, application problems such as email, user errors, problems with online service providers, PC configuration problems, and several others outside those categories.

servers in the server complex. It was traced to noise ingress from another user's home that resulted in packet loss which was perceived as a result of congestion by TCP. This caused TCP to slow down resulting in decreased throughput. The lower throughput propagated to the application level and the applications started timing out. This was seen as a disruption of service by the user.

## 4.2  Diagnosis in the HP-BIDS Trial

In the deployment of HP-BIDS trial, there is one point of contact for customer support for all the services that are accessed by a customer via the access network. This is desirable from the customers point of view, since a customer does not want to be confronted with the scenario where each service provider passes the responsibility of problems to other service providers in the system, whom the customer must call in turn. From the point of view of support it is also more effective for diagnosis  to have an organization that has an overall view of the system as opposed to a view of just one control domain. This provides an end-to-end view of  the federated service. A customer calls this support organization when an application that accesses the network in any way fails to perform according to customer's expectations. The customer support contact tries to identify and diagnose the problem by asking the customer questions, testing different pieces of the whole infrastructure, including the PC configuration and customer access rights. Testing can done directly by customer support if they have the necessary access, or on behalf of customer support by the administrator of the domain within whose purview lies the component.

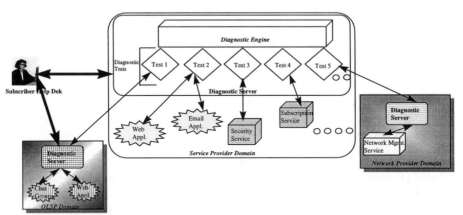

**Figure 2** Details of the Diagnostic System.

Since the diagnosis procedures employed to deal with such problems are manual, error prone (such as asking the user to read the parameters from the PC configuration file), and time consuming (and hence difficult to scale to a large customer base), we developed a diagnostic system that helped automate some of the diagnosis and trouble shooting procedures in HP-BIDS. This is a good short term solution for small to medium size systems and consists in automating the invocation of diagnostic tests within and outside the local domains. The experience gathered with the HP-BIDS diagnostic system is used to derive requirements for a more general service-contract based architecture that facilitates diagnosis.

The goal of our diagnostic system is to isolate the source of a problem rapidly and efficiently without requiring administrative rights to all the components. The diagnostic system does so by remote testing of different components of HP-BIDS using agreed upon interfaces exported by each component.[3] The meanings of the test results are agreed upon beforehand. The choice of the tests and the order in which they are to be executed is determined by the study of the history of problem call records.

The customer support engineers use their desktop computers to interact with a remote diagnostic server (Figure 2). An engineer orders a set of diagnostic tests to be run and the diagnosis server invokes management interfaces in each of the components relevant to the requested test. The results of each test are returned back to the support engineer.

The diagnostic server has two parts, a diagnostic engine that receives requests from the support engineer and dispatches the necessary tests, and the system specific tests themselves. The diagnostic engine understands the relationships between the tests and their input and output data, and can order the tests to be run such that the output of one test can be used as input for a later test.

It should be noted however, that a customer support engineer must understand the system well enough to know which tests are relevant to a customer's problem. In other words, a human troubleshooter remains a critical component of the diagnosis process. Furthermore, the diagnosis server does not address the problem of automated diagnosis across domain boundaries. Nonetheless, even such a simple tool is an improvement over manual procedures and will cut down enormously the average time required for customer support.

# 5. A GENERAL ARCHITECTURE FOR DIAGNOSIS AND CUSTOMER SUPPORT

## 5.1 Facilitating Diagnosis

If BIDS-like systems become commonplace, more and more federated services will be deployed over domains (such as the Internet) over which the service providers have no control. Organizations that provide support for these federated services will find themselves charged with diagnosing problems with these domains as well, in addition to diagnosing the services they provide directly. This dilemma arises from the desirability of providing one point of contact for customer support, irrespective of the underlying component services, and the inability to consolidate control of those components into one domain.

---

[3] These interfaces can also export the ability to monitor or control the domain.

We are developing an architecture that will reduce the complexity of the diagnosis process. In what follows, we give a very high level overview of this architecture and our rationale for the design choices. Even though this work is preliminary and the details are still being worked out, it offers useful guidelines for design of federated services to facilitate diagnosis and customer support, where none exist so far. We continue to validate and refine this architecture against our experience of HP-BIDS and similar systems.

Any such architecture must take into account the unique nature of the new federated systems introduced in Section 2. These factors motivate the following design decisions.

1. Focus on diagnosing individual services as opposed to diagnosing the entire system. This implies separating the diagnosis requirements of a service from those of the entire system.
2. Separate out the requirements of diagnosis and customer support, which are similar for various services, from the details of the functionality being offered by the services, which can vary a lot.
3. Specify the expectations of a service in the form of contracts. This delineates the exact responsibility of the service towards its users and is specified per service. These contracts allow us to deal across administrative boundaries.
4. Provide an infrastructure for verifying compliance with these contracts and a trusted third party that can arbitrate in case of conflicts.
5. Furthermore, make the diagnosis technique recursive so that the components that are used to construct a service can use the same techniques internally to localize diagnosis.

## 5.2  Representing Services in BIDS

The principal ideas of our architecture are brought out by the following example. A service in BIDS is represented by an abstract model, Figure 3, which is elaborated below. At one level there is a service such as the Web Service (referred to as a top-level service) that is provided to a user. A top-level service is the entry point for the user into the system and user interacts with the system by the means of this service. This service itself is put together by active collaboration and participation of other services (referred to as component services), which offer more primitive services. In case of the Web service, these are the local access network, the high speed network in the server complex, the session manager, the Web server, and perhaps even the PC operating system. The user is not exposed to these component services and does not care what these services are and how they affect the top-level service. Furthermore, each of the component services can itself be recursively composed of other more basic component services and a federator for those services. Thus for example, the Web server can be composed of a jukebox, a server machine, and a high speed bus that connects the two together.

Providing effective customer support requires coordination across all component services. This task is accomplished by the federator (a generalization of the HP-BIDS diagnostic server). The role of a federator can be assumed either by one of the service providers, by the manager of any of those systems, or a third party. The federator is only required to know the dependencies of a top-level service on its component services (the next level down) and to have access to the contract verification procedures, and possibly contracts themselves. These dependencies are expressed by an AND/OR dependency graph [Luger, 93]. In Figure 3, solid edges represent the dependency graph and the broken arrows represent access by a federator via the contract verification interfaces.

The rest of this discussion shows informally how the model impacts the process of provisioning a service, deploying it, and providing customersupport for it.

### 5.3 Service Provisioning

A service provider or the system integrator must locate (or construct) the suitable components, understand their functionality, compose them and configure them to construct a new top-level service, and deploy the service making it available to users. This task is arbitrarily complex if the components are numerous, varied, and not originally conceived for interoperability with each other. The integrator must also provide the missing functionality where needed. Thus in Figure 3, a Web service is provisioned by integrating multiple networks, servers, session managers, firewalls, and gateways.[4]

Each component explicitly specifies in the form of contracts what is being offered by that component (type and quality of service), what requirements must be met to access that service, the mode of billing, and some tools to verify and possibly enforce compliance with the contracts. These contracts are not a complete description of the service but an abbreviated model that captures the characteristics that are useful from the point of view of diagnosis. The contracts are represented as auxiliary interfaces offered by each component service to the federator. In time, the core of these contracts can become standardized for certain types of services, with a customizable component for special requirements.

Thus for example, if a server such as a Web server is managed independently, there is a service contract between it and other components that use it, that specifies the availability and the reliability of the server, and the average response times. It also furnishes information (such as abbreviated system logs) to verify compliance with its

---

[4] Some aspects of provisioning can be automated. The problems of locating the right services and understanding what they offer has been looked at in several trading and brokerage models in the context of object management [ODP, 10746-3] and service architectures for multimedia [Nahrstedt, 95]. Cohrs and Miller [Cohrs, 89] have looked at the problem of specifying and verifying the configuration of components in a federated system. These provisioning issues are beyond the scope of this paper but are mentioned for sake of completeness.

service contract.   The details of our contract based architecture will be described elsewhere and are outside the scope of this paper. A similar notion of contracts has been developed under ANSA [Hoffner, 93], though not from the point of view of diagnosis and customer support.

The service provider must also construct a structural model of the top-level service that gives the dependencies of the top-level service on its component services. This dependency graph (shown in Figure 3 by solid arrows) is an AND/OR graph. From this graph it is clear that the Web service depends on the access network, the Web server, etc., for it to function correctly. Same style of reasoning can be applied recursively to each of these components.

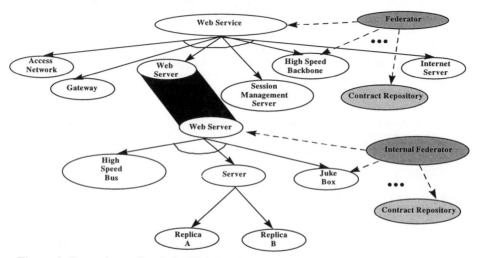

**Figure 3**  Dependency Graph for Web Service.

## 5.4  Customer Support and the Federator

The underlying assumption of the model is that problems arise if a component service does not or cannot fulfill its contract or due to circumstances that are outside the purview of the component services. Thus when a federator receives a customer report of a problem, it tries to identify the service with which the customer is experiencing problems.  Then it tries to determine if the customer is meeting the requirements of access (has the right security and access profile, has paid the bills etc.). If the customer meets the required constraints, the federator systematically verifies that each component service is complying with their contracts. This requires executing the contract compliance verification procedure provided by the service that hides the internal details of the service. If a component is found to be in violation of the contract, the federator reports the details of the failed compliance test and the particulars of the problem to the entity responsible for the faulty component. If the federator has the necessary rights and understands how that problem can be fixed, it can fix it itself.

Otherwise, it delegates repair and any further diagnosis and tracks its progress through a trouble-ticket like system. The contract verification process is driven by the knowledge of how the service is composed and statistics on distribution of faults, the historical experience with the component services, or the reliability data on component services.

The faulty service component (more precisely its internal federator) can repeat the same process to localize the fault to its component services. For example, a top-level federator may notice that the network is dropping too many packets. If that network is in a different administrative domain, it notifies the entity responsible for managing the network of the problem. This entity subsequently tries to localize the source of the problem (whether it is the hardware, protocols, or something else). The top-level federator monitors the progress of the problem/repair and informs the user when the problem is fixed.

## 6. SUMMARY

The emerging Internet-based federated systems span multiple control domains, lack any existing service level standards or system architecture, and offer a dynamic mix of services. Diagnosing faults and providing support to the customers of federated services is critical to their success but extremely difficult to accomplish, since only a small part of the environment can be observed and controlled by any given authority. We have characterized this problem, using our experience with the deployment of HP-BIDS. The HP-BIDS experience established the importance of a single point of contact for providing customer support to users and for doing diagnosis. It also revealed the desirability of automated diagnostic procedures that can be invoked across control boundaries.

We then described a customer support tool that was developed to automate some aspects of diagnosis in BIDS, and used the experience with this tool to derive requirements for a general service contract based architecture that can help ease the problem of diagnosis and management of federated services. We introduced the notion of a federator in this architecture and showed how the federator can diagnose the federated services by checking each component service for compliance with their contracts. This technique can be used recursively within each control domain to achieve the level of detail required in diagnosis.

Work on the service-contract based architecture is ongoing. Open questions include different levels of transparency of service contracts, complexity of describing dependencies and fault models for large systems, and guaranteeing consistency of contracts. Other aspects of the HP-BIDS diagnostic server are also being generalized, including testing support, alarm correlation for service management, and automatic generation of fault models for federated services based on knowledge capture.

The original contributions of this paper consists in formalizing the distinction between the new federated systems and the pre-existing systems, documenting the experience of managing BIDS-like systems from the point of view of diagnosis and customer support, and deriving a set of architectural requirements for federated service management based on this practical experience.

## 7. ACKNOWLEDGEMENTS

We thank Gary Herman, Rich Friedrich, and Ed Perry for comments and feedback, and Ellis Chi for helping with diagrams and editing.

## 8. REFERENCES

Borenstein, Nathaniel S., et al. (1996) Perils and Pitfalls of Practical Cyber Commerce. *Communications of the ACM.*

Cohrs, D.L., Miller, B.P. (1989) Specification and Verification of Network Managers for Large Internets. *ACM SIGCOMM89.*

Gifford, D., Spector, A. (1985) The Cirrus Banking Network. *Communications of the ACM, Volume 28, Number 8.*

Hoffner, Y., Linden, R., Beaseley, M. (1993) The Compatibility of Objects in Distributed Systems. *ANSA APM. 1066.00.02.*

Invoice? What's an Invoice?.*Business Week, June 10, 1996.*

Juckner, G.J., Summers, B.J. (1991) A Primer for Settlement of Payments in the United States. *Federal Reserve Bulletin.*

LATA Switching Systems Generic Requirements (1996),http://www.bellcore.com.

Li, M.S. EDI - An Overview. *IEEE Colloquium on Standards and Practices in Electronic Data Interchange, Digest No, 106.*

Luger, G., Stubbelfield, W. (1993) Artificial Intelligence: Structures and Strategies for Complex Problem Solving.*The Benjamin/Cummings Publishing Company, Inc.*

Minoli, Daniel (1995) VideoDialtone Technology.*McGraw-Hill, Inc. 1995.*

Nahrstedt, Klara, Smith, Jonathan M. (1995) The QoS Broker. *IEEE Multimedia.*

Online Shopping Network, http://www.ll.net/osc/Virtshop.htm

Open Distributed Processing Reference Model.*Technical Report ISO/IEC 10746-3.*

Pathfinder, http://www.pathfinder.com

Perry, T. (1988) Electronic banking goes to market. *IEEE Spectrum.*

Wells Fargo, http://www.2digm.com/resume/wells2.

Yemini, S.A., Kliger, S., Mozes, E., Yemini, Y., Ohsie, D. (1996) High Speed and Robust Event Correlation. *IEEE Communications Magazine.*

## PART THREE

# Management and Control in ATM Networks

Chair: Rolf Stadler, Columbia University

# 26

# A general framework for routing management in multi-service ATM networks

*P. Georgatsos*
*Algosystems S.A., 4 Sardeon St., N. Smyrni, 17121 Athens, Greece.*
*tel: +30 1 93 10 281, fax: +30 1 93 52 873, email: pgeorgat@algo.com.gr*

*D. P. Griffin*
*Department of Computer Science, University College London,*
*Gower Street, London WC1E 6BT, UK. tel.: +44 171 419 3687,*
*fax: +44 171 387 1397, email: D.Griffin@cs.ucl.ac.uk*

### Abstract

This paper presents a framework for routing management in ATM networks supporting guaranteed quality connections. It discusses the rationale behind the decomposition of a routing management service into a hierarchical system comprising both the management and control planes of the network. The concepts and ideas behind a set of algorithms for implementing the desired functionality are developed and discussed.

### Keywords

ATM, TMN, load balancing, performance management, route design, routing, routing management, VPC, multi-class environment.

## 1 INTRODUCTION

The overall objective of a routing policy is to increase the network throughput in terms of call admissions, while guaranteeing the performance of the network within specified levels. The design of an efficient routing policy is of enormous complexity, since it depends on a number of variable and sometimes uncertain parameters. This complexity is increased by the diversity of bandwidth and performance requirements of different connection types in a multi-class network environment. Furthermore, the routing policy should be adaptive to cater for changes in the network: topological changes due to faults or equipment being taken in and out of service; and changing traffic conditions.

Routing in Asynchronous Transfer Mode (ATM) networks is based on Virtual Path Connections (VPCs), a route is defined as a concatenation of VPCs. It has been widely accepted that VPCs offer valuable features that enable the construction of economical and efficient ATM networks, the most important being management flexibility. Because

VPCs are defined by configurable parameters, these parameters and subsequently the routes based on them can be configured and re-configured on-line by a management system according to network conditions.

To date, the majority of research work in the area of routing has been concerned with routing algorithms in isolation, rather than considering both the requirements and capabilities of the networks and the interplay with other resource management functions. One of the first attempts to design an integrated approach to the problem of routing management in multi-service ATM networks was presented by Griffin (1995).

This paper describes a general framework for tackling routing management in ATM networks. The framework encompasses both the control and management planes in a hierarchical manner assuming that route selection at call set-up time is handled by control plane functions, but according to parameters set by routing management. The routing management system is part of the management plane, and it is itself a hierarchical system operating in parallel to the network and its embedded control functionality.

The paper is organised as follows: Section 2 introduces the management service and describes our general functional model for routing management. Section 3 describes and analyses the issues behind the algorithms necessary for realising the components of the functional model. Finally, Section 4 draws conclusions on the work presented in this paper and identifies the scope for future work in this area.

## 2   DESCRIPTION OF THE MANAGEMENT SYSTEM AND GENERAL FUNCTIONAL MODEL

It is assumed that the network being managed is a public ATM network offering switched, on-demand services which are composed of a number of unidirectional connections. Each connection falls into a particular connection type or class, denoted throughout this paper by the term Class of Service (CoS). The CoS definition characterises the connection type in terms of bandwidth and performance requirements. We assume that it is in the responsibility of the network to guarantee the performance of the supported CoSs. Such performance guarantees are provided in terms of upper bounds on parameters characterising CoS performance in the network. A more detailed treatment of the assumptions made on the network can be found in (ICM, 1997) (Griffin, 1995).

The objective of the Routing management service is *to manage the network routing functions so as to maximise network availability whilst guaranteeing the performance requirements of current and future connections as specified in their CoS definitions.*

In the following paragraphs, we present an implementation independent functional architecture for the Routing Management service, by decomposing it into a number of distinct functional components. This architecture was first introduced by Griffin (1995) and it is further elaborated here. The basic dimensions of reasoning for the decomposition, are as follows:

- The definition of routing information is made on the basis of traffic predictions of anticipated network usage.
- Predictions may vary over time, hence routing information needs to be revised.
- Traffic estimates may prove to be inaccurate, furthermore actual traffic load may vary within a statistical range around the predicted values.

From the above it is apparent that there are basically two levels at which adaptivity to traffic variations should be provided; one at a level of traffic prediction changes and one

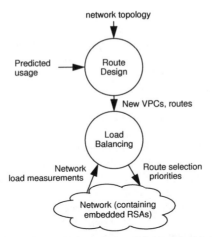

**Figure 1 Routing management - a hierarchical approach**

at the level of actual traffic fluctuations around the predictions. Therefore, we propose that routing management is achieved through a two level hierarchy (see figure 1).

The higher level of the routing management hierarchy, the Route Design component, operates at epochs where traffic predictions change, producing new sets of routes per CoS. Suitable sets of routes are constructed using, as input, predictions of network traffic at a connection level, per CoS and per s-d pair. Whenever traffic predictions change, the sets of routes may need to be reconstructed. The level of reconstruction depends on the significance of the changes (Griffin, 1995).

The lower level, the Load Balancing component, operates within the time-frame of traffic predictions and within the set of VPCs and routes defined by the higher level. Its objective is to manage the Route Selection Algorithms (RSA) embedded within the network switches according to actual traffic conditions. This level is introduced to compensate for inaccuracies in network usage predictions and short-term fluctuations of the load around the predictions. The Load Balancing component warns the higher level component of undesirable trends in network availability based on actual usage measurements.

This hierarchical approach to the problem exhibits fair management behaviour whereby initial management decisions taken with a future perspective are continuously refined in the light of current developments. Apart from its fairness, this behaviour provides a desirable level of adaptivity to network conditions.

The proposed routing management system implies a semi-dynamic type of routing policy, combining the merits of centralised and decentralised (based on local information) routing policies. Semi-dynamic routing policies were introduced in traditional data networks and improvements in network performance has been confirmed under such schemes (Rudin, 1976) (Yum, 1981).

An essential point underlying the proposed system is that it explicitly views route definition and route selection management functionality as management plane functionality. As a result, the routing functions in the control plane embedded in the network switches need only incorporate the actual route selection functionality; the management of which is left to the overlying management components.

One of the critical issues associated with routing has always been the assessment of the impact of routing decisions, at a particular instant, on future network performance. The proposed routing management model tackles this issue through its hierarchy. The higher level reserves the resources for routing based on estimates per source-destination (s-d) pair and CoS which refer to a long-term time period. According to the information supplied by the higher level, the lower level may be able to make predictions regarding shorter-term network usage and influence accordingly the routing decisions. With the proposed hierarchical structure the impact of routing decisions is time bounded and routing is always made on the basis of anticipated traffic.

Finally, it is worth-seeing how management of guaranteed performance connections is achieved in the proposed model. It is in the responsibility of the higher level routing component - Route Design - to apply suitable route design algorithms that preserve the performance targets of the supported traffic by defining appropriate sets of routes. The lower level component manages the performance of the different classes in an indirect manner by sharing the network resources between the competing classes, through appropriately influencing the routing decisions. Various prioritisation policies can be realised through this component in cases where network availability is limited, favouring certain CoSs at the expense of others.

## 3   FUNCTIONAL ASPECTS

### 3.1   Route Selection Algorithm (RSA)

Route selection is achieved in the control plane by means of a RSA which, without loss of generality, is assumed to operate on the basis of *route selection parameters* associated with the available routes. Following the ideas on the taxonomy of routing algorithms (Rudin, 1976), (Schwartz, 1980), several types of RSAs can be distinguished according to the selection method they employ, the information they utilise and the degree of adaptivity they offer. According to the selection method employed, RSAs may be:

- *Deterministic:* route selection is made according to a predefined order. In this case, a priority is assigned to each alternate route and higher priority routes are selected first.
- *Random:* route selection is made based on probabilistic criteria. Routes are assigned probabilities or frequencies and the selections are made to guarantee the frequencies.
- *Locally Adaptive:* route selection is made based on a policy taking into account the current load on the VPCs, as seen locally (e.g. select the least loaded VPC).

Apart from the routing information available at the network switches, RSAs may base their decisions upon purely local information (e.g. congestion at the VPCs originating from that switch), network-wide information (e.g. congestion on the routes to specific destinations), or upon no information at all. According to their degree of adaptivity, RSAs may be: static (not adaptive at all) or dynamic. A significant parameter associated with adaptivity is how adaptivity is provided; it can be provided: locally at connection acceptance/release times; through inter-node exchange, periodically or at exception; or from the management system periodically or at exception.

From the many algorithms that may be derived according to the above taxonomy, locally adaptive RSAs incorporating adaptivity provided by a network-wide routing management component (semi-dynamic routing schemes) seem to result in better network performance compared to other adaptive schemes, as indicated in the literature for data networks (Rudin, 1976) (Yum, 1981). However, the optimality of particular RSAs should not be judged in isolation but in conjunction with the algorithms incorporated in the higher routing management components.

## 3.2 The Load Balancing Component

The scope of the Load Balancing component is to:
- manage the RSAs in the network nodes according to network-wide traffic conditions;
- monitor the network with the purpose of providing warnings on deterioration in route availability and load deviations which would result in an unbalanced network.

Through its actions Load Balancing makes routing decisions network-state adaptive. Network-state adaptive routing has been recognised as a useful merit of routing algorithms as proved by the huge quantity of literature on the subject (Girard, 1983), (Eshragh, 1987), (Rudin, 1976), (Schwartz, 1980). Moreover, through RSA management, distribution of network load may be regulated; therefore enabling network load balancing. Balanced networks have been widely accepted as a valid objective of network design and routing policies (Gelenbe, 1994), (Kershenbaum, 1991). Apart from its active role in routing management, the Load Balancing component also contributes to preventive management. By taking a future perspective, it notifies the higher-level routing management component (Route Design) of undesirable trends in network availability. Thus, appropriate actions to increase network availability may be taken before the network availability deteriorates below acceptable levels.

The essence of any route selection management algorithm is to assign a figure of merit to each route, and to influence the RSAs so that traffic is routed over those routes with higher figures of merit. This view is in accordance with the traditional view where routing schemes are variants of shortest path algorithms (Schwartz, 1980). In connection-oriented networks, such as ATM, the figure of merit should refer to the *potentiality of the route* to accommodate new connections.

Route potentiality is calculated for all possible routes from a given switch to a particular destination node and for each CoS. It is calculated in terms of a metric associated with each VPC which reflects the potentiality of the VPC to accept connections of a particular CoS given its current load; this metric is referred to as *VPC acceptance potential*. Taking into account that on a particular VPC originating at a switch more than one route to a particular destination for a given CoS may originate, the potentialities of all these routes can be accumulated, giving rise to a figure of merit for selecting this VPC, referred to as *VPC selection potential*. Route selection management algorithms then grade VPCs at each switch according to their selection potential and configure the RSAs accordingly. This is achieved by setting the route selection parameters so that VPCs with a higher figure of merit have advantage over those with lower figures of merit.

To achieve network load surveillance, network availability for new connections can be estimated by extending the notion of potentiality to the node level. *Node potentiality* to a given destination for a particular CoS reflects the potential of the network to establish new connections from this node to the given destination node for a particular CoS, taking into account actual network traffic conditions.

Specific route selection management algorithms can then be distinguished according to the way they calculate route potentiality, the location in which they execute their algorithms for determining VPC selection potentials, and the way they perceive changes in VPC selection potentials and then trigger appropriate management actions.

Load balancing algorithms along the previously introduced concepts have been presented by Georgatsos (1996) and ICM (1997), and improvement in network performance has been verified assuming a deterministic type of RSA.

## 3.3　The Route Design Component

The scope of the Route Design component is to design and redesign - whenever necessary - a network of VPCs and a set of admissible routes per (s-d) and CoS satisfying predicted traffic requirements, given the constraints imposed by the network model and the performance targets of the CoSs. The following terminology and notions are introduced:

*ClassRoute network:* For a given CoS, the ClassRoute network is a sub-network of the VPC network consisting only of the VPCs that belong to routes of that CoS.

*SDClassRoute network:* For a given CoS and a given (s-d) pair, the SDClassRoute network is the sub-network of the ClassRoute network consisting only of the VPCs that belong to the routes interconnecting the given (s-d) pair.

*SDClassPath network:* For a given CoS and a given (s-d) pair, the SDClassPath network is the sub-network of the physical network consisting of the nodes and links that appear in the paths of the given (s-d) pair and CoS.

A *design problem*, defined by its literal meaning, is characterised by its objectives, its constraints and by its uncontrolled variables for which solutions are sought. The objectives of a design problem are usually defined in qualitative terms (e.g. optimum solutions) rather than in quantitative terms. Associated with a design problem, a *design alternative* is defined as an arrangement of the problem's uncontrolled variables, satisfying the problem's constraints.

A design problem is said to be a *well defined design problem* if there exist two functions of the problem's uncontrolled variables one corresponding to the cost (*cost function*) and the other corresponding to the benefits (*benefit function*) of alternative designs. Therefore, each design alternative of a well defined problem can be explicitly evaluated in terms of the cost that it implies and the benefits that yields. Then, the following are defined, where $c(.)$, $b(.)$ are the problem's cost and benefit functions respectively and D is the set of all design alternatives:

A *least-cost design* of a well defined problem is that design alternative which has the minimum cost among all other possible designs, irrespective of benefits. Design $d' = (\underline{x}')$ is a least-cost design if and only if $c(\underline{x}') \leq c(\underline{x})$, $\forall d = (\underline{x}) \in D$.

A *maximum-benefit design* of a well defined problem is that design alternative which achieves the maximum possible benefits among all other possible designs, irrespective of cost. Design $d' = (\underline{x}')$ is a maximum-benefit design if and only if $b(\underline{x}') \geq b(\underline{x})$, $\forall d = (\underline{x}) \in D$.

A *cost-effective design* of a well defined problem is that design alternative which achieves the maximum possible benefits at the least cost among all other possible designs. Design $d' = (\underline{x}')$ is a cost-effective design if and only if the following hold:

$$\frac{c(\underline{x}') - c(\underline{x})}{c(\underline{x}')} \leq \frac{b(\underline{x}') - b(\underline{x})}{b(\underline{x}')}, \; \forall d = (\underline{x}) \in D \text{ such that } c(\underline{x}) \leq c(\underline{x}') \text{ and } b(\underline{x}) \leq b(\underline{x}')$$

$$\frac{c(\underline{x}) - c(\underline{x}')}{c(\underline{x}')} \geq \frac{b(\underline{x}) - b(\underline{x}')}{b(\underline{x}')}, \; \forall d = (\underline{x}) \in D \text{ such that } c(\underline{x}) \geq c(\underline{x}') \text{ and } b(\underline{x}) \geq b(\underline{x}')$$

there are no $d = (\underline{x}) \in D$ such that $c(\underline{x}) \leq c(\underline{x}')$ and $b(\underline{x}) \geq b(\underline{x}')$

It can be proved that a cost-effective design corresponds to the solution of an optimisation problem with the objective function being the ratio of the problem's cost by the benefit functions, i.e. $c(.)/b(.)$, subject to the problem's constraints.

The existence of *optimum designs* (least-cost or maximum-benefit or cost-effective designs) to a well defined design problem depends on the existence of solutions to the corresponding optimisation problems (i.e. on the topological properties of the problem's state space as constrained by the set of constraints and on the mathematical properties of the problem's cost and benefit functions).

The essence of the proposed approach is to tackle the Route Design problem within the previously introduced conceptual framework of problem alternative designs. This entails decomposition of the overall Route Design problem into a number of distinct problems and tackling each of the problems by means of optimum designs i.e. defining appropriate cost and benefit functions and seek for optimum designs. Specifically, the approach is to decompose the overall problem of Route Design as follows:

- Map traffic predictions into max (maximum) flow requirements for all (s-d) pairs and each CoS. The max flow requirements are in terms of Virtual Channel Connections (VCCs) that the network must be able to establish at any instant, given the max connection rejection tolerance of each CoS.
- Determine SDClassPath networks, i.e. paths over which connections will be routed so that CoS performance is guaranteed. That is, given the max flow requirements determine suitable paths per (s-d) and CoS so that to satisfy the delay, jitter and cell loss constraints.
- Determine a suitable VPC network, based on the identified SDClassPath networks. That is, given the set of paths per (s-d) and CoS, determine a suitable VPC network.
- Map the SDClassPath networks to the VPC network to derive the SDClassRoute networks. That is, map the set of paths to the derived VPC network to obtain the set of routes per (s-d) and CoS.

## Determining max flows

It is assumed that traffic predictions characterise the number of connection requests for a specific interval, in which statistical convergence is achieved. By modelling the network with an appropriate queuing system, it is possible to obtain the (minimum) number of VCCs which the network should provide at any instant so that the blocking probability is less than or equal to the maximum tolerable rejection ratio per CoS. The derived number also denotes the maximum flow per (s-d) and CoS within the physical network.

## Determining SDClassPath networks

The Route Design component guarantees upper bounds on delay, jitter and cell loss ratio per CoS by defining appropriate set of paths per (s-d) and CoS. Each path per (s-d) is assigned a *performance quality* corresponding to an upper bound on the delay, jitter and cell loss that a connection traversing this path would experience.

Various policies for allocating (s-d) pairs of a specific CoS to paths offering specific performance qualities can be identified. The following *performance sharing policies* are proposed:

*Complete sharing:* a path of a specific performance category is shared exclusively by the CoSs of the same performance category.

*Compatible sharing:* a path of a specific performance category is shared only by CoSs of the same or superior performance category.

*Non-compatible sharing:* paths of a specific category can be shared by any CoS.

In order to meet the cell loss requirements, the cell loss performance targets - determining the level of multiplexing at the cell level - of the Connection Admission Control (CAC) algorithms deployed in the network switches need to be determined and then configured. It is assumed that the CAC algorithms corresponding to VPCs defined on the same link will have the same cell loss performance target. The allocation of cell loss performance targets for each link can be achieved according to the following three *link performance assignment policies.*

*Complete sharing:* Links are assigned performance targets in such a way so that paths of any cell loss category can be established on them.

*Complete partitioning:* Links are assigned performance targets in such a way so that paths of specific cell loss categories can be established on them.

*Complete sharing with initial reservations:* Links are assigned performance targets in such a way so that paths of specific cell loss categories can be established on some links while paths of any cell loss category can be established on the remainder of the links.

For a given performance sharing and link performance assignment policy, there may be many paths for a given (s-d) and CoS that satisfy the performance constraints of each CoS. The number of paths is however finite, constrained by the number of paths that can be found for each (s-d) pair in the physical network. So the following questions arise: are there any paths per (s-d) and CoS that satisfy the performance and max flow constraints? If yes, which paths should be selected per (s-d) and CoS?

The above questions state the design problem of the *SDClassPath configuration.* The uncontrolled variable of this design problem is the topology of the SDClassPath networks. The constraints of the problem are: the max flow constraints ensuring that the max flow requirements per CoS and (s-d) should be accommodated within the defined SDClassPath networks and path sharing constraints corresponding to the adopted performance sharing policy. The problem is a well defined design problem as explicit cost and benefit functions can be defined. For example for a given design, i.e. SDClassPath networks' configuration, the cost and benefit functions could be the respective cost and benefit functions of the related optimum flow assignment problem.

Tightly coupled with the SDClassPath definition problem is the *optimum flow assignment* problem concerned with the optimum distribution of the max flow commodities (a commodity is the tuple: (s-d) and CoS) along the defined paths, given the topologies of the SDClassPath networks. The uncontrolled variable is the flow per commodity along each path or equivalently their distribution frequency at the source nodes. The constraints of the problem are the flow balance constraints in and out each node, the capacity constraints for each link and the max flow constraints. The cost function may be related to the cost of used network resources (the utilisation of the links) and the benefit function may be related to network load balancing (the discrepancy of a link's utilisation from the average link utilisation). By considering appropriate cost or benefit functions, optimum flow configurations can be established as solutions to appropriately defined optimisation problems.

## Determining the VPC network

The definition of the VPC network encompasses the following tasks:
- the definition of the topology of the VPCs,
- the definition of the (required) bandwidth of each VPC.

which are viewed as design problems for which optimal designs can be sought.

As far as the *VPC topology design* problem is concerned, the uncontrolled variable is the set of *VPC cut nodes*. A VPC cut node is defined to be a node where a VPC is terminated. VPC cut nodes are connected with (at least) as many VPCs as the number of physically different paths in the SDClassPath networks that connect them. The paths connecting two different VPC cut nodes are termed as *VPC platforms* upon which VPCs may be defined depending on the adopted *bandwidth segregation policy* (see below). Therefore, the definition of VPC cut nodes in effect determines the minimum number of VPCs which may exist in the network. Note that each VPC platform is characterised by its required bandwidth which can be determined by the expected number of commodities flowing between its edges. The mapping between the anticipated flow to the required bandwidth should be made taking into account the multiplexing scheme as applied at the cell level by the network CAC algorithms as well as taking into account multiplexing at the connection level. The anticipated flow along the VPC platforms was determined during the definition of the SDClassPath networks. At one extreme a VPC topology design treats all VC/VP switches as VPC cut nodes, whilst at the other extreme only the access nodes are VPC cut nodes. Optimum VPC topology designs could be sought for, considering explicit cost and benefit functions in terms of the VPC cut nodes. For example, for a given design - selection of VPC cut nodes - a cost function could reflect the cost of managing the VPCs (the number of VPC platforms) and a benefit function could reflect the benefits of reduced set-up times (inverse of the average number of hops on VPC cut nodes from source to destination over all SDClassPath networks). Appropriate optimisation problems can be formulated to determine the optimum VPC topology designs.

Tightly coupled with the VPC topology design problem is the *VPC bandwidth assignment problem*. The VPC assignment problem can be viewed as a design problem with uncontrolled variable the bandwidth assigned to the VPCs. The constraints of the problem are the link capacity constraints. The following policies can be identified for VPC instantiation, corresponding to different *bandwidth segregation policies*. Considering a simple model consisting of a single resource (bandwidth available on a VPC platform) for which a number of different classes (corresponding to the (s-d)-CoS commodities) compete, the following resource sharing policies can be identified:

*Complete sharing:* the resource is shared among all competing entities on the basis of their arrival.

*Complete partitioning:* the resource is appropriately segmented and access by the competing classes is restricted to a segment. Access between the classes competing for the same segment is done on the basis of their arrival.

*Compete sharing with initial reservations:* similar to the complete partitioning policy, however a portion of the resource is also defined for being shared by all classes on the basis of their arrival. The optimality of complete sharing with an initial reservation scheme has been experimentally verified (Sykas, 1991).

Optimum designs for VPC bandwidth assignment can then be sought for, provided that explicit cost and benefit functions can be instituted. For example, for a given design i.e. a given set of VPCs and their allocated bandwidth, a cost function may reflect the increased cost in using network resources (increase in link utilisation) caused by bandwidth segmentation on the links compared to the case where no VPCs have been defined. The utilisation of links is measured assuming the optimum flow distribution in the network of VPCs. An example of a benefit function could be related to the gains in call blocking probability attained by the defined VPC set of the particular design as opposed to the case where no VPC has been defined.

## The dynamic part

The dynamic part of the Route Design could be achieved in a quasi-static way by rerunning the static part algorithms. However, we envisage a more efficient way of dealing with asynchronous triggers for the purpose of ensuring management operations are as economical as possible in terms of management overhead and reaction response. In particular, we envisage a hierarchical approach (Griffin, 1995) towards handling asynchronous triggers, as indicated below:

- Map the traffic predictions to max flow requirements and determine the optimum designs of the involved design problems corresponding to the new flow requirements.
- Determine whether the new flow requirements can be accommodated in the existing VPC and SDClassRoute networks, keeping the same VPC capacity.
- If not possible (either not feasible or flow distribution optimality constraints are violated), the bandwidth of the existing VPCs is appropriately modified (within link capacity constraints), keeping the same SDClassRoute networks to accommodate the new flow requirements.
- If not possible (either not feasible or flow distribution optimality constraints are violated), SDClassRoute networks are modified either by adding/removing existing VPCs and by appropriately modifying their bandwidth.
- If not possible (either not feasible or flow distribution optimality constraints are violated), the VPC network is appropriately modified (by creating/deleting VPCs) and by appropriately setting VPC bandwidth and defining SDClassRoute networks so that to accommodate the new flow requirements.
- If not possible (either not feasible or flow distribution optimality constraints are violated), triggers to higher level management functions (network planning, service migration) are emitted.

## 4  CONCLUSIONS AND FUTURE WORK

In this paper we presented a general framework for routing management in multi-service ATM networks. The framework encompasses both the control and management planes of the network, specifically covering the RSAs embedded within the network nodes and the management functions required to manage the RSAs. This hierarchical separation between the control and management functions allows network elements to be as simple as possible, reducing their complexity and hence their cost. At the same time the capabilities of the routing mechanisms are not compromised because the overlying management functions are able to influence the operation of the local routing algorithms by means of route selection parameters. By building the routing management intelligence into a parallel environment (the management plane) to that of the network itself (control and user planes) the complexity of the management algorithms is not constrained by the capabilities of the network elements.

In addition to the control plane/management plane hierarchy, the paper discussed the rationale behind the decomposition of the routing management service into a hierarchical system of management components fulfilling the roles of Route Design and Load Balancing. One of the main advantages of this separation of functionality is that management decisions taken on a longer term basis (by Route Design) can be continually refined (by Load Balancing) in the light of recent developments without the disadvantage of continually invoking computationally-intensive management processes.

Following the implementation independent analysis of the functions required for routing management, the resulting components were mapped to the Telecommunications Management Network (TMN) architecture (ITU1, 1992) by Griffin (1995) and ICM (1997) as a framework for implementation. The resulting system was implemented in the RACE II ICM project and the architecture and concepts described in this paper were validated through experimentation. The Load Balancing component, in particular, was subject to extensive experimental work (ICM, 1997) and at the time of writing, Route Design algorithms are being implemented following the concepts and ideas of section 3.

An important conclusion is that in our opinion the TMN framework is mature enough for realizing complex systems such as routing management. It was necessary to extend the ITU M.3020 methodology (ITU2, 1992) to give guidance on the decomposition of TMN systems (ICM, 1997), but having done this, the development of a large scale TMN system was facilitated through a well defined approach to system design and development.

An important aspect of our future work is concerned with the development of a more resilient routing management scheme by integrating the management service presented in this paper with fault management facilities. Our future work in this direction will be according to the TINA architecture. One of the advantages of the TINA approach is that there is no longer a distinct separation between control and management plane functions - a common framework is used for all telecommunications software in a distributed processing environment. This approach will have many benefits when interworking is required between route design, load balancing and control functions, and will facilitate interworking with fault management which also spans both the management (alarm handling and correlation) and control planes (self-healing OAM techniques, etc.).

In this direction there are a number of open issues related to the compatibility between TMN design methodologies and TINA modelling techniques. Work currently being undertaken in the context of the ACTS project AC208 REFORM (REFORM, 1997) concerns the relationship of the routing management system presented here with the resource configuration management, connection management and fault management subsystems of the TINA management architecture (TINA, 1994).

# 5 ACKNOWLEDGEMENTS

This paper describes work undertaken in the context of the RACE II R2059 ICM project and work currently in progress in the ACTS AC208 REFORM project. The RACE and ACTS programmes are partially funded by the Commission of the European Union.

# 6 REFERENCES

Eshragh, N., Mars, P. (1987) "Study of dynamic routing strategies in circuit-switches networks," 3rd UK comp. and telecomm. perf. engin. workshop, Edinburgh.

Gelenbe, E., Mang, X. (1994) "Adaptive Routing for Equitable Load Balancing," ITC 14, Elsevier Science B.V.

Georgatsos, P., Griffin, D. (1996) "A Management System for Load Balancing through Adaptive Routing in Multi-Service ATM Networks", proc. of IEEE INFOCOM'96.

Girard, A., Hurtubise, S. (1983) "Dynamic Routing and Call Repacking in Circuit-Switched Networks," IEEE Trans. on Comm., Vol. 31.

Griffin, D., Georgatsos, P. (1995) "A TMN system for VPC and routing management in ATM networks", Proc. of 4th ISINM, Chapman & Hall, UK.

ICM (1997) "Integrated Communications Management of Broadband Networks," ed., Griffin, D., Crete University Press, Heraklion, Greece, ISBN 960 524 006 8.

ITU-T Rec. M.3010 - Principles for a telecommunications management network, 1992.

ITU-T Rec. M.3020 - TMN interface specification methodology, 1992.

Kershenbaum, A., Kermani, P., Grover, G. (1991) "MENTOR: An Algorithm for Mesh Network Topological Optimization and Routing," IEEE Trans. on Comm., Vol. 19.

Rudin, H. (1976) "On routing and delta routing: A taxonomy and performance comparison of techniques for packet-switched networks", IEEE Trans. on Comm., Vol. 24.

REFORM (1997) "REFORM System Requirements and Analysis," Deliverable D1, A208/Bell/WP1/DS/P/002/b1.

Schwartz, M., Stern, T.E. (1980) "Routing techniques used in computer communication networks," IEEE Trans. on Comm., Vol. 28.

Sykas, E., Vlakos, K., Protonotarios, E., (1991) "Simulative Analysis of Optimal Resource Allocation and Routing in IBCNs", IEEE JSAC, Vol. 9.

TINA-C (1994), "Management Architecture," Version 2.0, Document label TB_GN.010_2.0_94.

Yum, T.P. (1981) "The Design and Analysis of a Semi-Dynamic Deterministic Routing Rule", IEEE Trans. on Comm., Vol. 29.

**Panos Georgatsos** received the B.S. degree in Mathematics from the National University of Athens, Greece, in 1985, and the Ph.D. degree in Computer Science, with specialisation in network routing and performance analysis, from Bradford University, UK, in 1989. Dr. Georgatsos is working for Algosystems SA, Athens, Greece, as a network performance consultant. His research interests are in the areas of network and service management, analytical modelling, simulation and performance evaluation. He has been participating in a number of telecommunications projects within the framework of the EU funded RACE and ACTS programmes.

**David Griffin** received the B.Sc. degree in Electronic, Computer and Systems Engineering from Loughborough University, UK in 1988. He joined GEC Plessey Telecommunications Ltd., UK as a Systems Design Engineer, where he was the chairperson of the project technical committee of the EU RACE I NEMESYS project while working on TMN architectures and ATM traffic experiments. In 1993, Mr. Griffin joined ICS-FORTH in Crete, Greece as a Research Associate on the EU RACE II ICM project. He was the leader of the ICM group on performance management case studies and TMN systems design. Mr. Griffin joined University College London in 1996 and is currently employed as a Research Fellow working on a number of EU ACTS projects in the area of resource management for TINA systems covering performance, fault, configuration and accounting management.

# 27

# Switchlets and Dynamic Virtual ATM Networks

*J.E. van der Merwe and I.M. Leslie*
*University of Cambridge, Computer Laboratory*
*New Museums Site, Cambridge CB2 3QG.*
*Telephone: +44 1223 334650. Fax: +44 1223 334678.*
*email:* {jev1001,iml}@cl.cam.ac.uk

## Abstract

This paper presents a novel approach to the control and management of ATM networks, by allowing different control architectures to be operational within the same network, and on the same switch. The resources available on an ATM switch are divided into **switchlets**, each of which encapsulates a subset of the physical ATM switch resources. A set of switchlets on different ATM switches combine to form a virtual ATM network. Each virtual network created in this way can potentially use different control and management mechanisms which are collectively called a control architecture. For example one virtual network might run ATM Forum signalling, another might implement an IP switching control architecture, while yet another can be reserved for an in house secure control architecture. In this manner a new control architecture can be introduced into a network without disrupting existing services and applications, thereby facilitating change management in an elegant way. Switchlets and a control architecture from a well known set can be created on demand to allow the dynamic construction of virtual ATM networks of predefined type. Alternatively the control architecture can be supplied (by a 'user') after the virtual network has been created, thus allowing the dynamic creation of virtual ATM networks of arbitrary functionality. Finally, the paper presents a proof of concept implementation as well as ongoing work in this area.

## Keywords

ATM, Open Switch Control, Virtual Networks, Network Control and Management, Change Management

## 1 INTRODUCTION

The inadequacy of existing ATM control and management strategies to meet the demands of for example multimedia applications, has been widely acknowledged. In addition to the more conventional approaches by standards bodies and communities (ATM Forum, 1993), (Cole, 1995), several other solutions to the ATM control problem have been proposed over an extended period of time. (Both (Crosby, 1995) and (Stiller, 1995), present good comparisons of different approaches to signalling and control, as well as some historical perspectives on how the standards process evolved.) More recent approaches such as (Iwata, 1995), (Lazar, 1996b), (Herbert, 1994) and (Ipsilon Networks, 1996) have proved that although there is general acceptance of the superior data transfer capability of ATM, ATM control can still not be considered a solved problem. In this paper, the mechanisms that constitute a particular control and management approach, are collectively called a **control architecture** and an instantiation of a control architecture is called a **control domain**. For example a UNI/NNI signalling implementation is an example of a control architecture, as is the IP switching architecture from Ipsilon Networks(Ipsilon Networks, 1996).

Despite showing some obvious advantages, the uptake by standards bodies of ideas from these new control architectures have been slow if at all. One of the main reasons for this situation is that a new control architecture is normally proposed as a **replacement** of an existing one. Naturally this leads to a reluctance to move to the new untested control architecture, even if the existing one is known to be non ideal. One of the major contributions of this paper is a solution to this problem by allowing different control architectures to be operational simultaneously in the same network and on the same switch.

An underlying and related problem is that of the control interface provided by the ATM switch. This is the interface used by the control architecture to manipulate the switch hardware in order to perform its control and management functions. An example of an operation performed through this interface is the manipulation of bits in a routing table in order to set up a connection through the switch. The proliferation of control architectures mentioned above means that a definite requirement for this interface is that it be **open**, so that different control architectures can be developed to make use of it.

One approach to the open switch control problem is to define a simple low level interface which can be used within any control architecture to exercise control of the physical switch. This approach, followed within the DCAN project (Herbert, 1994), has lead to the design of an open switch control interface, which is described in Section 2. The DCAN approach still has the limitation mentioned above that only one control architecture can be operational at any particular moment in time.

Section 3 addresses this problem by showing how a subset of the ATM switch resources can be presented to a particular control architecture as a **switchlet**. The term switchlet is used in favour of say, 'virtual' switch, to emphasize the fact that real resources are allocated to the switchlet. The switchlet presents the same open switch control interface to its control architecture, which means that the control architecture is oblivious of the fact that it is not in control of the whole physical switch.

Switchlets are combined into virtual ATM networks, each of which can potentially use a different control architecture, i.e. be of a different **type**. This means that switchlets permit the introduction of new control architectures into an existing network in a very elegant and controlled manner.

In the first instance the action of creating switchlets and combining them into virtual networks, is performed by human operators. The process can however be automated so that virtual networks can be **dynamically** created. This process is considered in Section 4. The control architecture instantiated on the newly created virtual network can be one of a predefined set of well known control architectures. Alternatively, an 'empty' virtual network can be created in which the control architecture is provided by the 'user' or the entity that requested the creation of the virtual network. This allows the implementation of an open multi service network (OMSN) (Van der Merwe, 1995), in which 'any user' can construct a network and become a 'network service provider'.

In Section 5, a proof of concept implementation is presented and discussed, together with some indication of ongoing work. Section 6 briefly considers related work and the paper ends with a conclusion.

## 2   DCAN APPROACH TO OPEN SWITCH CONTROL

Figure 1 illustrates the control of an ATM switch, by control software forming part of a particular control architecture, through an *Ariel* * **open switch control interface**. The premise of the DCAN approach is that switch control be opened up by providing a simple, generic, low level switch control interface on the switch. Switch control software running on a general purpose workstation, invokes operations on the *Ariel* interface in order to control and manage the switch.

The switch control software and the *Ariel* interface have a client-server relationship. The relationship is highly asymmetric because the server is always very simple and lightweight, while the client (the switch control software) is potentially very complex. For this reason, the control software is assumed to run on a general purpose workstation, while the *Ariel* server is simple enough that it can be implemented on very simple switches. The relationship is also one-to-one in that a single switch controller talks to a single *Ariel* interface. (Depending on implementation, the relationship could also be one-to-many, but not many-to-one.)

The switch control software is responsible for performing all the functions required by a specific control architecture, such as setting up virtual channel identifier/virtual path identifier (VCI/VPI) mappings, call acceptance control (CAC), resource allocation etc. A control architecture (or strictly speaking control domain) is not limited to the implementation of a single signalling protocol, for example an implementation capable of both UNI 3.0 (ATM Forum, 1993), and Spans (FORE Systems, 1995) signalling would still be a single control architecture.

Even though communication between the control software and the *Ariel* interface is based on client-server principles, this does not imply or require the facilities of a general purpose distributed processing environment (DPE). Rather communication between the control client and *Ariel* server is considered a 'local affair', for example on a default VCI/VPI pair. In fact the *Ariel* interface specifies the functionality required by an open switch control interface, and implementations based on different mechanisms can be (and have been) done.

### 2.1   The *Ariel* Interface

The purpose of the *Ariel* interface is to provide an open, generic switch control interface with a useful set of functions. In particular the *Ariel* interface should be useful even if detailed knowledge of the controlled switch is not available. In the ideal case *Ariel* should provide sensible control of a switch of unknown type.

The *Ariel* control interface consists of the following interfaces:

**Configuration** - the configuration interface is primarily used to **find out** what the configuration of a switch is as opposed to configuring the switch.

**Port** - a port interface is provided for each port on the switch and deals with a port as a complete entity, e.g. port up/down.

**Connections** - the connections interface is responsible for basic VCI/VPI mapping, and deals with quality of service (QoS) issues through a context index obtained (by the controller) through the context interface.

**Context** - the context interface bundles QoS abstractions into a single interface and is explained below in more detail.

**Statistics** - the statistics interface allows the controller to obtain switch statistics and accounting information.

---

* Credit is due to Sean Rooney of the Computer Lab for suggesting Shakespeare's *The Tempest* as the basis of our name space.

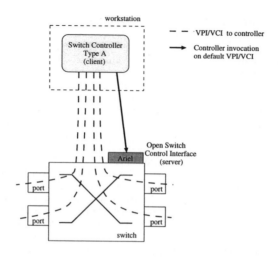

**Figure 1** Switch control through the *Ariel* interface

**Alarms** - the alarms interface allows the controller to be informed when certain events take place on the switch.

QoS aware connections are set up through the *Ariel* interfaces, by first creating a context, and then associating that context with the VPI/VCI mapping during the actual connection setup. This means that all QoS issues are effectively taken out of the Connections interface, which can be kept simple.

Another reason for separating the Connections and Context interfaces, comes from the realization that the Context interface provides a way to 'allocate resources' on the switch, and the Connections interface provides a way to 'use the resources', and that the two need not be tightly coupled. For example, a control architecture can allocate a certain amount of resources on the switch by means of the Context interface, and then 'over commit' these resources by allowing more connections to be created than the resources would suggest, because it has some external knowledge about the behaviour of the connections in question. Note that most current commercial ATM switches will not allow this separation between the allocation and usage of resources.

Dealing with QoS issues in a generic fashion at a low level is very difficult, and may not be possible. The reason for this is that the QoS **capabilities** of a switch are determined by the queueing and scheduling policies employed. These, in turn, are what differentiate one switch from another. It is therefore unlikely that all switch vendors would be willing to make such details about their switches available to be included in an **open** interface, such as *Ariel*.

One way of avoiding this problem is to hide the switch queueing and scheduling policies behind a generic interface. The ATM service categories identified by the ATM Forum (Sathaye, 1995), provide the means for such an abstraction. This approach is reasonable because it can be expected that switch manufacturers will build switches with queueing and scheduling mechanisms which will support a subset of these services. This approach is followed in *Ariel*. It must be noted that should lower level details about switch internals be made available by certain switch vendors, this knowledge can still be used in favour of the abstraction based on ATM Forum service categories.

Five service categories are (currently) defined by the ATM Forum, namely:

**UBR** - unspecified bit rate
**CBR** - constant bit rate
**rtVBR** - real time variable bit rate
**nrtVBR** - non real time variable bit rate
**ABR** - available bit rate

These service categories are parametrised by (in total) four QoS parameters and six traffic parameters, a subset of which must to be specified for each service category. Hiding the switch details behind this abstraction obviously leads to a loss of information. However, enough information is still available to perform functions such as call acceptance control (CAC) outside the switch in the controlling workstation.

All that is required to perform CAC outside the switch is the **resource mapping function** that is used by the switch to map QoS and traffic parameters to switch resources, as well as knowledge about the resources available on the switch. Such a resource mapping function constitutes significantly less sensitive information than the mechanisms used to implement it. It can therefore be expected that switch manufacturers will be more willing to provide such information. If switch specific CAC functions are not available, generic CAC functions could be used by control software. As long as the generic CAC functions err on the conservative side, this will result in useful (albeit not ideal) switch control.

## 3 SWITCHLETS

The approach described in Section 2 allows a switch to be controlled in an open fashion and allows different control architectures to be designed to utilise the *Ariel* interface. This allows much more openness and flexibility than is currently the case, but it still means that at any given time, a single control architecture is operational on a switch and within a network.

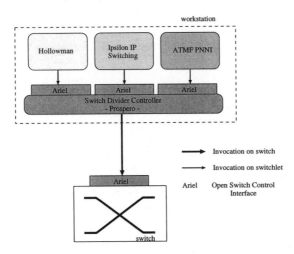

**Figure 2** Creating switchlets

Figure 2 shows how the *Ariel* interface on a physical switch can be used by a **Switch Divider Controller** to create several **switchlets**. (In keeping with our name space the Switch Divider Controller is called *Prospero*.) The *Prospero* Switch Divider Controller allocates a subset of the physical switch resources into a switchlet, and makes this available to switch control software through an *Ariel* interface. Switch control software, of a particular type, will control the switchlet by invocations on the switchlet *Ariel* interface, in exactly the same way as it would on a physical switch. As an example, Figure 2 shows three possible control architectures namely Hollowman (Rooney, 1997), IP Switching (Ipsilon Networks, 1996) and the ATM Forum's PNNI (ATM Forum, 1996).

Switchlets can be combined into virtual networks of a certain type, or control architecture. This is depicted in Figure 3, which shows a network of five switches, on which three virtual networks of different type are deployed.

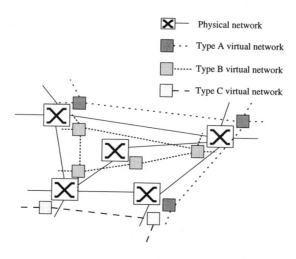

**Figure 3** Virtual ATM networks with different control architectures

The switch is completely oblivious of the fact that several control architectures are operational on it, and maintains the one-to-one relationship between an *Ariel* server on the switch and a single Switch Divider Controller. The *Prospero* Switch Divider Controller provides *Ariel* interfaces to the switchlet controllers. Except for the fact that less resources are available to it, the switch(let) controllers are therefore also unaware of the presence of the Switch Divider Controller. The Switch Divider Controller polices invocations on the switchlet *Ariel* interfaces to ensure that switchlet controllers do not utilise resources not allocated to it, or in any other way interfere with the functioning of other switchlets.

The switch divider controller is analogous to the small kernel in the Nemesis operating system (Leslie, 1996), which is responsible (amongst other things) for allocating system resources to scheduling domains, and for policing the actual usage of allocated resources. Indeed the proposed model has some similarities with the Nemesis operating system in that real resources are made available to an 'application', the control architecture in the network model, which is then allowed to use these resources according to some internal policy.

## 3.1 The *Prospero* Switch Divider Controller

Partitioning of switches into switchlets require the **specification** of a switchlet in terms of a subset of the physical resources available on the switch. Resources on a switch that need consideration include: ports, VPI/VCI space, bandwidth, buffer space, and queueing and scheduling policies.

Ports and VPI/VCI space constitute the connections resources of a switch and can be partitioned at various levels of granularity:

**port level** - whereby certain ports within a switch are allocated to a switchlet

**VPI level** - whereby certain VPI ranges on certain ports are allocated to a switchlet

**VCI level** - whereby certain VCI ranges on certain VPIs on certain ports are allocated to a switchlet

Partitioning at the VCI level, being the most general of the three possibilities, is the approach taken in specifying switchlets.

Bandwidth, buffer space, and queueing and scheduling policies combine to represent the **switching capacity** of the switch. As explained in Section 2.1, the approach taken with the *Ariel* interface is to hide QoS details behind the five ATM service categories. The same approach is followed in specifying switchlets, whereby a certain percentage of the resources for a particular service category will be 'marked' as belonging to a certain switchlet. The control architecture operating on the switchlet can then employ the same resource mapping function mentioned in 2.1, on its subset of resources to perform for example CAC.

A switchlet specification could therefore consist of the number of ports required, and then for each port the following information:

- The range of VPIs required
- The range of VCIs per VPI required
- The service categories required
- The capacity per service category required (Until a better understanding of the problem of dividing switch resources have been developed, service capacity will be specified as a percentage of what is available on the physical switch.)

The *Prospero* Switch Divider Controller has to **know** the capacity of the physical switch, and only allow switchlets to be created until this capacity is exhausted. Allocating resources on the switch to a switchlet does not involve any invocations (or allocations) on the physical switch. Rather, *Prospero* notes the allocation in its internal representation of the switch capacity, and uses that to police future invocations on a switchlet *Ariel* interface. Once connections have been established in a switchlet (and switch), *Prospero* has to rely on in-band policing mechanisms in the physical switch, to ensure that connections from one switchlet do not adversely affect that of other switchlets. This requires nothing new on the physical switch, since this functionality is needed in switches any way.

In the first instance *Prospero* provides a configuration interface, which can be used by human operators to create switchlets and virtual networks. Of more interest though, is the dynamic creation of virtual networks (on demand) by other software systems. Such a **Virtual Network Service** is considered in the next section.

Remote access to the *Prospero* interfaces are required for both static (i.e. done by a network administrator on a long time scale), or dynamic (i.e. done by software on an on demand basis) virtual network creation. Therefore both of these actions presuppose the existence of a **bootstrapping (virtual) network**, implementing a **bootstrapping control domain**. The bootstrap control domain is therefore any control architecture, contained within its own virtual network, which provides the required addressing, routing and other facilities to enable communication between the *Prospero* instances and other software entities.

In the prototype implementation, an existing IP-over-ATM (virtual) network is used as the bootstrap control domain. This solution has allowed progress to be made, but is considered too heavyweight because of its reliance on conventional ATM control. On the other hand the use of IP in the bootstrap network has attractive properties, and an alternative implementation based on the much simpler IP switching mechanisms (Ipsilon Networks, 1996) is therefore currently under investigation. Note that the bootstrapping problem is present in all ATM (and other) networks, and is not unique to the environment described in this paper. The bootstrapping facility has however been generalised into something that can provide more sophisticated services.

## 4   VIRTUAL NETWORK SERVICES

As mentioned in the previous section, sets of switchlets can be combined into virtual networks. The control domains for these different virtual networks could be instantiations of the same control architecture, or a different control architecture could be operational in each virtual network, or any combination of these.

The ability to have different control architectures in the same physical network allows for a very elegant way of introducing new control software into an existing operational network. The new control architecture can namely be made operational in its own virtual network while existing users and applications operate undisturbed in the original (virtual) network. After the new control architecture has been tested, users and applications can be migrated to it and the partitioning of switchlet resources can be modified to reflect the fact that the new virtual network is to become the default (or only) virtual network.

A more interesting possibility is to provide an on demand virtual network service, in which switchlets are dynamically created and merged into virtual networks. If this facility is combined with the services provided by a distributed processing environment (DPE), virtual networks become a service which can be offered, traded and manipulated like any other service. Such a DPE can be one of the facilities provided within the bootstrap virtual network, and is assumed in the following discussion.

The use of a DPE (in the bootstrap virtual network) does not mean that all control architectures have to be implemented by means of a DPE, or even be aware of the existence of a DPE. Indeed, a major strength of the approach presented here is that a conventional control architecture (e.g. an ATM Forum UNI/NNI compliant control architecture) can be instantiated in and confined to its own virtual network. On the other hand, new control architectures are being developed that can make use of the DPE facilities or can even be implemented in a DPE environment. The latter approach might lead to some simplifications in the control architecture. For example, because of the existence of the bootstrap virtual network it might not be necessary for a particular control architecture to implement its own bootstrapping procedures but instead rely on the bootstrap virtual network to provide such services.

## 4.1   Creating a Virtual Network

This section describes the system services required for the dynamic on demand creation of virtual networks. Figure 4 shows the interaction between the services that are involved in this process.

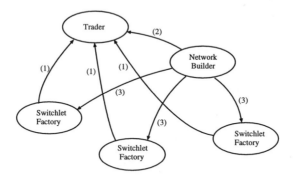

**Figure 4** Dynamic Virtual Network Services

The **Switchlet Factory** service encapsulates the *Prospero* Switchlet Divider Controller presented in Section 3. The Switchlet Factory informs a **Trader** service about its existence as well as its switch capacity. (Interaction (1) in the Figure.) Trading is the standard way of matching service providers and consumers in a DPE (APM, 1993). The Switchlet Factory also exports interfaces which allow the creation and destruction of switchlets. When a switchlet is created the Switchlet Factory also exports an *Ariel* switch control interface.

In order to create a virtual network, a **Network Builder** service is provided with the 'specification' of the desired network. This network specification is in terms that hitherto made little sense after network equipment have been commissioned. An example specification could be,

- UNI 3.1 signalling
- CBR capacity of 15 Mbps (Note that in the case of CBR a percentage of what is available is trivially converted to bandwidth and vice versa.)
- Between A and B
- With redundancy
- For three hours

Or could be as simple as 'Cheap network between A and B'.

The network specification could be the output of another service, or be provided by a human being.

The Network Builder has knowledge about existing virtual networks, and coordinates the creation of virtual networks so that, for example, the VCI space allocated to two switchlets in the same virtual network, and in adjacent switches will overlap. The Network Builder contacts the Trader and asks for all switches with the required capabilities and capacity according to the supplied network specification. The result of the query, invocation (2) in Figure 4, is a list of 'interface references' to Switchlet Factories matching the criteria. Using the supplied network specification and topology information obtained from some topology service, the Network Builder determines which switches should form part of the virtual network. The Network Builder then invokes required operations on appropriate Switchlet Factories to create the switchlets.

Two possibilities exist in terms of the type of the control architecture which will be instantiated on a newly created virtual network:

A predefined control architecture from a well known set can be started up when the virtual network is created. An example would be the creation of an ATM Forum UNI/NNI compliant virtual network. In DPE terminology this would be called a traded typed virtual network. In this case the appropriate software entities will be started up by the network builder as soon as the virtual network has been created.

Alternatively, a blank virtual network or virtual network without a control architecture can be created. In this case the control architecture is supplied or filled in by the entity that requested its creation. This would be called an 'anytype' virtual network in DPE terms. In this case the network builder will not start up the control architecture, but rather will return an interface reference to each switchlet to the entity that requested creation of the virtual network. Since this involves the services of the DPE in the bootstrap control domain, these control architectures will normally be required to make use of the DPE. In this manner a new control architecture can be **composed** out of **base components**. This means that by adding or modifying base components virtual networks with control domains of arbitrary complexity and functionality can be constructed. For example, the composed control architecture can use all base components but replace the routing component with a special purpose one for its particular application.

The newly created virtual network then proceeds to perform its own initialisation, bootstrapping and operation, all of which is confined to its 'own' switchlets and control domain. Note that in the case of a conventional control architecture, with its own bootstrap procedure, the operations of the control architecture can be truly confined to its own virtual network. This is not true if the control architecture use facilities provided by the bootstrap virtual network, or rely on the DPE in the bootstrap virtual network for its communication. This is an important issue in terms of the resources, both network and processing, which have to be allocated to the bootstrap virtual network.

Following the creation of a switchlet, the Switchlet Factory has to update its available capacity in the Trader, or potentially remove its trader entry if it has no capacity left. The decomposition of a virtual network happens when the requested time period expires and the Switchlet Factory claims back switchlet resources, and updates its Trader entry. Alternatively, if an undefined time period is required, the Network Builder will be responsible for periodic 'keep alive' invocations on the Switchlet Factory to keep the virtual network intact.

Something that has not been considered in the above discussion is how potential users of the new virtual network get to know about its existence. It is assumed that creation of the virtual network is the result of a request of potential users, or that potential users will be able to obtain this information through some external mechanism.

## 5  PROOF OF CONCEPT IMPLEMENTATION

In order to proof the feasibility of the Switchlet and Dynamic Virtual Network concepts, an implementation has been done on one of the ATM networks at the Cambridge University Computer Laboratory. The only switch resources that were taken into account for this implementation were switch ports and VPI/VCI space. (i.e. none of the capacity and QoS resources mentioned in Section 3.1 were considered.)

The part of the service ATM network used consists of three Fore Systems ATM switches (one ASX-100 and two ASX-200s), which interconnect a number of file servers, workstations, one router and several ATM video adapters (Fore Systems AVA-200s). These switches are an essential part of the Computer Laboratory's infrastructure, and as such minimal disruption and downtime were of key importance. The approach taken was therefore to utilise the existing IP-over-ATM network provided by means of Spans and UNI Signalling in the Fore Systems Switches as the bootstrap control domain. The VPI/VCI space available to this bootstrap control domain was however limited, which means the remainder of VPI/VCI space was made available to a *Prospero* Switch Divider Controller implementation. This arrangement is illustrated in fig 5.

The communication mechanism between *Prospero* and the switch depends on the switch model. Specifically, in the case of the ASX-100, an *Ariel* server implementation runs on the switch. For comparison purposes, several *Ariel* impersonations using different mechanisms were implemented. This included both message passing as well as remote procedure call (RPC) based implementations. For the two ASX-200 switches, invocations are made by means of SNMP (over IP provided by the bootstrap control domain) with a SNMP daemon running on the switch. Communication with the ASX-100 can also use the SNMP mechanism, however the *Ariel* implementation is more efficient and architecturally cleaner, and is therefore to be preferred.

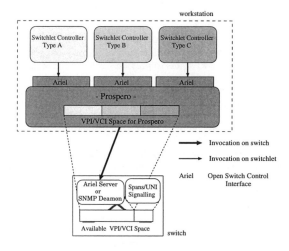

**Figure 5** Proof of concept *Prospero* implementation

The Dynamic Virtual Networking Services were implemented using an implementation of the Distributed Interactive Multimedia Architecture (DIMMA) to provide its DPE (Li, 1995). DIMMA is a framework Object Request Broker (ORB), which provides a common base for the construction of domain specific ORBs.

As mentioned in Section 4 the network builder keeps track of all current virtual networks in the system. The network builder is also the only entity which is allowed to create and destroy switchlets. This allows the network builder to perform garbage collection on virtual networks for which the control architecture malfunctions. Because users are allowed to supply 'their own' control architectures, this is an important function to ensure the integrity of the network as a whole. The current implementation also provides a trivial access control mechanism, whereby a list is kept of users who are currently allowed to request creation of virtual networks. In similar fashion, the current implementation obtains topology information from a manually constructed topology database.

The implementation described in the above paragraph clearly has some limitations. In particular the use of SNMP in some cases to talk to the switch is suspect from a performance point of view. More seriously however from an architectural point of view, is the fact that the Spans/UNI signalling is still running on the switch and is not using the *Ariel* interface. The security provided by the trivial access control mechanism described above, while adequate for an experimental environment, will also need to be significantly extended in a real world implementation. However, the fact that the implementation has been done on an existing service network with minimal disruption is testimony to the flexibility of the switchlet concept.

The virtual network environment described above has been used (and is still being used) in the Computer Laboratory to implement several novel control architectures (Rooney, 1997), (Van der Merwe, 1997). Work is currently being undertaken to implement more conventional control architectures in this environment.

## 6 RELATED WORK

A speculative application programmers interface (API) for the control of ATM switches was presented in (Lazar, 1996a). This API seems to provide similar functionality to the *Ariel* interface presented in Section 2.1 but contains no interface for obtaining switch statistics or for receiving alarms. The generic switch management protocol (GSMP) (Newman, 1996) is another low level switch control mechanism. GSMP closely match the functionality provided by *Ariel*, but has a much simpler notion of QoS. In fact, since no QoS issues were taken into account for the current implementation, GSMP is one of the *Ariel* impersonations which the current *Prospero* implementation can use.

Virtual ATM networks based on virtual paths (VPs) have been proposed as a mechanism to segregate different traffic types into more homogeneous (and thus more easily manageable) groups (Fotedar, 1995), (Farago, 1995). In (Gupta, 1995) the partitioning of resources to form virtual networks in a packet network is presented. In this case the purpose of the resulting virtual networks is to reduce connection setup times for 'real-time' connections.

In the Xbind project (Làzar, 1996b) networking devices such as switches are abstracted into virtual entities which export DPE interfaces. Since different algorithms are allowed to operate on these high level abstractions, some notion of virtual networking is possible within the Xbind environment. However, since it is impossible to reduce a control architecture, such as for example a PNNI implementation (ATM Forum, 1996), to a mere set of algorithms, this provides a fairly limited virtual networking environment.

## 7 CONCLUSION

The paper presented the concept of ATM **switchlets** whereby a subset of physical ATM switches are presented to a control architecture to manipulate as it sees fit. The switchlets are presented to the control architecture as an *Ariel* open switch control interface. The *Prospero* **switch divider**

**controller**, controls the physical ATM switch by means of an *Ariel* interface on the physical switch, and polices invocations made by the different control architectures operational on the switch.

Since the partitioning of switch resources is done at a very low level, very few restrictions are being imposed on control architectures implemented in the switchlet environment. This allows both conventional control architectures based on message passing protocols, as well as control architectures implemented using DPE methodologies to be accommodated.

It was shown how the switchlet concept can be used to introduce new control architectures into an existing network in a flexible and non disruptive manner. Indeed this has been proven by means of a proof of concept implementation.

Finally, the paper showed how the switchlet concept can be used to create **virtual ATM networks** of arbitrary topology on demand and to run arbitrary control architectures in these virtual networks. This means that virtual networks becomes a service which can be offered, and traded like any other service in a DPE environment. The control architecture instantiated in these virtual networks can be known a priori or can be supplied by the entity requesting network creation. This allows users to supply and manipulate their own control architecture in the created virtual network.

## 8 REFERENCES

APM Limited (1993) The ANSA Model for Trading and Federation. Tech. Rep. AR.005.00, APM Limited, Castle Park, Cambridge, UK.

ATM Forum (1993) *ATM User-Network Interface Specification - Version 3.0.* Prentice Hall.

ATM Forum (1996) Private Network-Network Interface Specification Version 1.0 (PNNI 1.0). ATM Forum document: af-pnni-0055.000 .

Cole, R.G., Shur, D.H. and Villamizar, C. (1995) Ip over atm: A framework document. Available from: http://ietf.cnri.reston.va.us/.

Crosby, S.A. (1995) Performance Management in ATM Networks. Tech. Rep. 393, University of Cambridge, Computer Laboratory, UK.

Farago, A. et al (1995) A New Degree of Freedom in ATM Network Dimensioning: Optimizing the Logical Configuration. *IEEE Journal on Selected Areas in Communication*, vol. 13, pp. 1199–1206.

FORE Systems (1995) SPANS UNI: Simple Protocol for ATM Signalling. Release 3.0. FORE Systems Inc., 174 Thorn Hill Rd, Warrendale PA, USA.

Fotedar, S. et al (1995) ATM Virtual Private Networks. *Communications of the ACM*, vol. 38, pp. 101–109.

Gupta, A. and Ferrari, D. (1995) Resource Partitioning for Real-Time Communication. *IEEE/ACM Transactions on Networking*, vol. 3, pp. 501–508.

Herbert, A. et al (1994) Scalable Distributed Control of ATM Networks. Project proposal, University of Cambridge, Computer Laboratory, UK. Project overview available from: http://www.ansa.co.uk/DCAN/index.html.

Ipsilon Networks (1996) IP Switching: The Intelligence of Routing, the Performance of Switching. Available from: http://www.ipsilon.com/productinfo/techwp1.html.

Iwata, A. et al (1995) ATM Connection and Traffic Management Schemes for Multimedia Interworking. *Communications of the ACM*, vol. 38, pp. 72–89.

Lazar, A.A. and Marconcini, F. (1996a) Towards an Open API for ATM Switch Control. Available from: http://www.ctr.columbia.edu/comet/xbind/xbind.html.

Lazar, A.A. et al (1996b) Realizing a Foundation for Programmability of ATM Networks with the Binding Architecture. *IEEE Journal on Selected Areas in Communication*, vol. 14, pp. 1214–1227.

Leslie, I.M. et al (1996) The Design and Implementation of an Operating System to Support Distributed Multimedia Applications. *IEEE Journal on Selected Areas in Communication*, vol. 14, pp. 1280–1297.

Li, G. (1995) DIMMA Nucleus Design. Tech. Rep. APM.1551.00.05, APM Limited, Castle Park, Cambridge, UK.

Newman, P. et al  (1996) Ipsilon's General Switch Management Protocol Specification Version 1.1. *Internet RFC1987.*

Rooney, S. (1997) An Innovative Control Architecture for ATM Networks. IM'97, San Diego.

Sathaye, S.S. (1995) ATM Forum Traffic Management Specification Version 4.0. ATM Forum Technical Committee - Contribution 95-0013.

Stiller, B. (1995) A survey of UNI Signalling Systems and Protocols for ATM Networks. *ACM Computer Communications Review*, vol. 25, pp. 21–33.

Van der Merwe, J.E. and Chuang S.C. (1995) Support for Open Multi Service Networks. Regional International Teletraffic Seminar, Pretoria, South Africa. Available from: http://www.cl.cam.ac.uk/users/jev1001/.

Van der Merwe, J.E. and Leslie, I.M. (1997) Service Specific Control Architectures for ATM. In preparation.

## 9  BIOGRAPHY

Kobus van der Merwe received the B.Eng. and M.Eng. degrees from the University of Pretoria in 1989 and 1991 respectively, and is currently working towards the Ph.D. degree at the University of Cambridge Computer Laboratory, Cambridge, U.K. His current research interests are in network control and management.

Ian Leslie received the B.A.Sc in 1977 and M.A.Sc in 1979 both from the University of Toronto and a Ph.D. from the University of Cambridge Computer Laboratory in 1983. Since then he has been a University Lecturer at the Cambridge Computer Laboratory. His current research interests are in ATM network performance, network control, and operating systems which support guarantees to applications.

# 28

# The Hollowman
# an innovative ATM control architecture

*S. Rooney*
*University of Cambridge, Computer Laboratory*
*New Museums Site, Cambridge CB2 3QG.*
*Telephone: 44 1223 334650. Fax: 44 1223 334678.*
*email:* Sean.Rooney@cl.cam.ac.uk

### Abstract

The current implementation of out-of-band control in ATM networks inhibits their successful exploitation. The confusion in signalling protocols between application services and their resource requirements results in the loss of one of the key advantages of ATM which is the ability of applications to decide the requirement of their connections. The control being immutably built into the switches results in switch vendors, rather than service suppliers, defining the management policy which best suits those services.

The World Wide Web has recently become one of the most important services on the internet but was unimagined five years ago. Clearly history should teach us of the impossibility of predicting the services that will be in common use in the near future. It is not at all obvious how one can define a priori the required control policy for these as yet unknown services.

This paper presents an innovative control architecture called *Hollowman*, which devolves control from the ATM switches into an application level distributed processing environment.

### Keywords

ATM, delegated control, distributed processing

## 1 INTRODUCTION

As (Crosby, 1995) points out, current ITU-T signalling protocols, such as Q.2931 (ITU-T, 1994), are flawed due to the lack of a clear distinction between application level services and their communication requirements. A signalling protocol should specify how to establish connections with arbitrary requirements and let applications decide which connections they need in order to implement services, rather than trying to specify a complete set of connection types within the signalling protocol itself.

It is clearly impossible to try and define all future services; this means that signalling protocols will constantly need to be modified as new services are required and network users will be frustrated in their desire to implement these innovative services. These two points combined will seriously inhibit the introduction of ATM. The limitations of the standards are evidenced by work such as *OPENET*, (Cidon, 1995), which is seeking to extend the Private Network to Network Interface (PNNI) (PNNI, 1994), in order to make it usable for intra-networking.

Orthogonal to this is the fact that currently the control policy is typically implemented in software running on the physical switch. Just as service suppliers should be able to decide on the resource requirements of the connections used in their services, so they should be able to define the policy to control those connections. At the moment this is not possible.

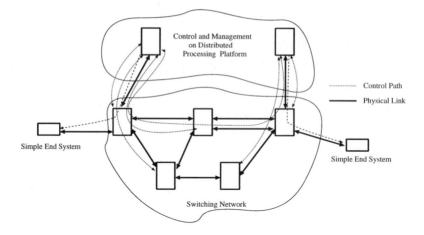

**Figure 1** ATM Network with devolved management and control.

What we propose is to devolve control out of the physical switches into a distributed application level control architecture. Figure 1 shows the relationship between the control architecture and the controlled network. This control architecture opens up the management of the ATM switches and devices and enables service providers greater flexibility in the management of their services. At the Cambridge Computer Laboratory we have built an ATM control architecture called the *Hollowman,* based on these concepts.

Such an approach is similar in principle to that of the TINA consortium (Barr, 1993) in as much as they both use the techniques of distributed computing. However as (Crosby, 1995) points out TINA remains compatible with existing control and management standards when it is their rigidity which is one of the key problems to overcome.

The *X-Bind* architecture described in (Lazar, 1996) is closer to that described here. Although there are key technical differences between the Hollowman and *X-Bind* which will be evidenced in the rest of this paper, the major difference is philosophical. We believe that there will be no one single, ubiquitous ATM control architecture. Our work is geared towards the development of a necessary infrastructure - called the *Tempest* - which allows the simultaneous execution of many different control architectures over the same physical network. Although the Hollowman is a fully functioning control architecture in its own right, we view it more as a set of components, techniques and algorithms which can be used and reused in other control architectures within this framework rather than as a future standard. The mechanics of how control architectures can be made to coexist is detailed in (van der Merwe, 1996), the rest of this paper concentrates on the Hollowman.

## 1.1   Terminology

A *Domain* is a logical node of the controlled network, which is defined by the set of resources that it contains. An *Application* is a schedulable entity within some *Domain* to which resources e.g. CPU time, may be assigned.

A *Service Type* is a well defined task that a given *Application* can carry out on behalf of another. An instance of a *Service Type* is called a *Service*. A *Service Offer* is the means by which the existence of a *Service* is advertised within the control architecture. A *Service Offer* defines: the type of the *Service*; the location of the *Service*; the protocols which may be used to access that *Service*.

The process of *Trading* is the act of matching the requirements of a service user for a *Service* with the set of available *Service Offers*. The process of *Reification* is the resolution of a *Service Offer* into a access point for that *Service*. *Reification* involves the reservation of sufficient resources within the control architecture to permit an *Application* to use the service.

A *Connection* is a set of resources allocated to two or more applications across the network in order to exchange information. A *Connection Type* is defined by the nature, amount, time period and location of the resources that need be allocated to a *Connection*.

## 1.2 Overview

First, the concept of *trading* is discussed and some extensions within the the control architecture are presented.

Second, the *soft switch* which is both the interface to the physical switch and the encapsulation of a switch control policy is explained.

Third, the *host manager* which manages the resources within a given domain is introduced and the means by which the host manager allocates one resource - virtual circuit identifiers - to applications is detailed.

Fourth, the means by which connections are created across the network by the *connection manager* is explained. The concepts of *caching* frequently used connections and *lazy evaluation* of connection end points in order to achieve better set-up times are introduced.

Fifth, the novel concept of an application specifiable *call closure* is presented. The interest of applications being able to take advantage of their high level knowledge in order to be able to optimise their resource usage is motivated.

Finally, we describe the set of Application Programmer Interfaces (API) that the Hollowman offers.

## 2 TRADER

A *Trader* is an application that maintains a set of available service offers for a given domain and whose location is well known within that domain. Domain applications register themselves with their trader at start-up and in doing so they obtain a name which is unique within the scope of their domain.

Trading is hierarchical within the control architecture, i.e. an attempt is made to find a match for a service request first of all within the same domain as the requestor and if this fails, within the scope of a higher level. Thus the use of service providers which are, in some sense *close*, is favoured; this is particularly advantageous if a domain corresponds to a single work-station, in which case optimised forms of communication between the service provider and user are employed.

Within the current implementation of the control architecture two levels of trading exist: a trader for each domain and a trader which federates all these traders. For convenience we term a trader that maintains offers for a given domain as a *domain trader* and the single encompassing trader as the *federating trader*. The domain traders maintain service offers for general applications

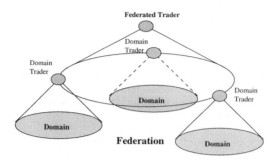

**Figure 2**  Example of three federated domain traders.

and the federated trader maintains domain trader service offers. Figure 2 shows a federation of three domain traders. At regular intervals the federated trader ensures that the domain traders are still active and if not removes them from the federation. All offers registered at a failed domain trader implicitly are no longer available within the federation.

Since domain traders are recorded as service offers, they may themselves be the subject of service requests. In particular it is advantageous for a given domain trader - trader *T1* to establish a connection with another, trader *T2* if the service offers of *T2* are often required in the domain *T1*.

Trading is a well understood concept and has existed in Distributed Processing Environments such as ANSA (APM, 1995) for quite some time; the federation of traders and the garbage collecting of service offers for failed applications is a natural extension of the basic trading concept. The originality of trading in the control architecture comes from the use of trader knowledge for domain resource management (this is detailed in Section 4) and the employment of trader service offers in ATM signalling (this is detailed in Section 5).

## 3  SOFT SWITCH

The *Soft Switch* has the following roles within the control architecture: it defines a set of logical control interfaces to the switch; it implements a control policy for the switch and it encapsulates the precise method of interacting with the physical switch.

The relationship between the physical switches and the soft switches is one-to-one. A soft switch contains a representation of the physical switch's resources as well as a set of switch control services. The complete control policy for a given switch is partitioned across these switch services. The different control policies that each switch service implements are independent. This separation permits different aspects of control policy to be kept distinct and allows different control functionality to be manipulated using a specific and dedicated interface. In what follows a *switch* denotes the combination of the physical switch and its associated soft switch.

A soft switch holds state about the physical switch in order to perform its control functions. For example the connection management service needs to know about the current resource usage on the physical switch in order to determine if a demand for a connection should be satisfied or not. In consequence, the soft and physical switch have to resynchronise in the event that either of them is stopped and restarted.

*X-Bind* uses CORBA/OMG (OMG, 1991) as the underlying platform for communication between *all* entities including exchanges between the management layer and managed network elements themselves. In our opinion running CORBA/OMG on switches is unnecessary and restrictive. The Hollowman communicates with the physical switch using the *Ariel* (van der Merwe, 1996) switch management interface. An *Ariel* server runs on the physical switch and an *Ariel* client runs in the same address space as the soft switch. The interface is defined by a set of services. A minimal set of services is defined for a switch but different switches may extend/enhance this set. The precise services that the physical switch supports is determined by the soft switch at start-up time. Figure 3 shows a schema of a switch.

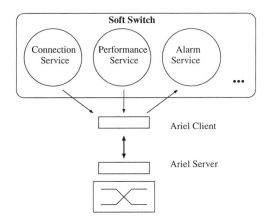

**Figure 3** Example showing a soft switch containing three switch services.

It is important to stress that *Ariel* does not define a single wire representation for communication between the *Ariel* client and server. Many different mappings between the *Ariel* interface description and an underlying communication mechanism are possible e.g. RPC/UDP, CORBA/IIOP, SNMP/UDP. In the last case an SNMP daemon running on the switch takes the place of the *Ariel* server. A more detailed account of *Ariel* is given in (van der Merwe, 1996).

## 4  HOST MANAGER

The *Host Manager* is an extension of ideas developed within the Nemesis real-time operating system (Leslie, 1996). Nemesis allows applications to manage the resources allocated to them at a very fine level of granularity and thus enables them to make precise guarantees about their behaviour.

The host manager is an entity which allocates resources to applications within a given domain. It is common that this scope corresponds to a single work-station but is not a constraint of the model. However all applications are resident in one and only one domain. We reserve our discussion of host managers to the features which are associated with the control of the ATM network, but it should be noted that the host manager is a much more general concept than presented here.

A connection is the means by which one application sends information to a set of others across

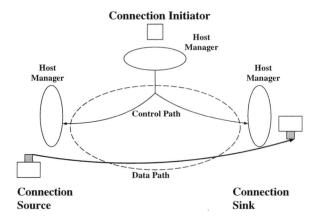

**Figure 4** Example showing third party connection set-up.

the network. All connections have one sending application, called the source, and $N$ receiving ones, called sinks.

The host manager has two interfaces: one which applications use to initiate connections and one which the connection manager uses to inform a host manager that another entity wishes to establish a connection with an application in the host manager's domain.

An application makes one of four requests in order to establish a connection:

**(1)** connect a sink service offer to a source service offer;
**(2)** connect itself as sink to a source service offer;
**(3)** connect itself as source to a sink service offer;
**(4)** connect a source service access point to a sink service access point.

In all four cases the host manager will check if the sink and source are actually both local and hence do not require a network connection in order to communicate. All but *(3)* may require joining a branch to a existing connection, in which case the host managers and connection manager recognise this and join at the appropriate place within the multi-cast tree. For the applications this is transparent.

It may be possible for applications to learn of the existence of an available service other than by using the control architecture trading mechanism which is why the *(4)* type request is supplied. For *(1)* and *(4)* the initiator of the request can be the sink or the source of that connection or neither. Figure 4 shows an example of third party connection set-up.

One managed resource is the virtual circuit identifiers (vci) that a given domain has at its disposal. The host manager at start-up obtains the set of Service Access Points (SAP) that have been allocated to that host within that control architecture. The SAP is the handle through which both applications and the host manager manipulate vci's. Host managers do not assign vci's directly to applications, rather they allocate them SAPs which the connection manager maps to a vci during connection creation.

An application wishing to establish a connection asks the host manager for a free SAP. If one exists the host manager sets the state of the SAP to *Reserved* and accounts it to the application.

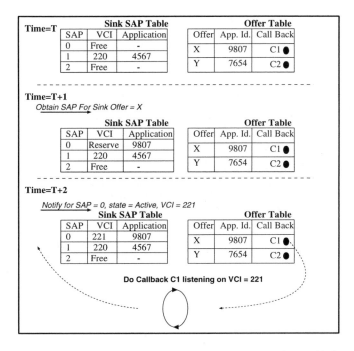

**Figure 5** Example showing the host manager state changes during the establishment of a connection.

The SAP can be viewed as a token for a vci which will redeemed during connection creation; the connection manager having a more complete knowledge of the current network connections can decide what is an appropriate vci to map that SAP to.

Each time that a connection is established to an application within the scope of a host manager, the state of the SAP is set to *Active* and the SAP/vci table mapping is updated. The freeing of a *Reserved* SAP simply sets its state back to *Free*, the freeing of an *Active* SAP causes the host manager to ask the connection manager to release the connection and to notify the other host managers involved about the change in state. If an application fails before releasing a SAP then at some subsequent moment the domain trader will realize that the application has disappeared and tell the host manager to release any resources associated with that application.

An application registering a service offer associates a call-back with that offer in the host manager. The call-back is invoked when an attempt is made to reify that offer. After a successful invocation of the call-back the service provider application should be able to guarantee the service. Before an attempt is made to reify an offer no resources at all may have been allocated to it. Figure 5 shows a simplified version of the sequence of events which occur when a host manager receives a request to reify an offer as a sink in a connection.

The connection manager requests the host manager to find a free SAP and to reserve it for the offer $X$. The host manager verifies $X$ is an offer present in its offer table, finds the application that supports $X$ and allocates it SAP = 0. At a later stage the control manager notifies the host manager that a connection has been established to SAP = 0 and that the vci associated with it is 221. The host manager finds out that $X$ is the offer waiting on SAP $= 0$, and executes the call-back for $X$.

## 5   CONNECTION MANAGER

Within the control architecture the *Connection Manager* has responsibility for establishing and tearing down connections between applications in distinct domains. In order to achieve this the connection manager has knowledge of the topology of the ATM network. It acquires this knowledge during the bootstrapping of the network or the network elements. From the point of view of the connection manager the network is made up of: host locations, switch locations, device locations, e.g. a camera.

The initiators of requests to the connection manager for connection creation are always domain host managers. The connection manager identifies the sink and source domains of the connection and determines a sequence of switches which constitutes a route between them. It then uses the host manager of the sink and source domain and the soft switches in order to reserve the required resources and establish the connection.

In the general case many routes may exist between two domains and the connection manager uses a routing algorithm to distinguish a 'best' route. Currently the default algorithm is a variant of the weighted spanning tree algorithm. The weights associated with each of the switches in a route are defined as a function of the current resource usage on a switch and the necessary resources required for a connection. The exact formula used to turn these two pieces of information into a weight is switch dependent and an intrinsic part of the control policy of the connection soft switch. However, within the framework of the Hollowman any of the diverse techniques described in (Lee, 1995) for resource constrained routing could be used.

The connection manager frees the resource associated with a connection when a host manager asks it to do so. The host manager may have been explicitly asked by an application or it may have decided that the application involved in the connection had failed.

The connection manager may decide not to remove a given connection between two domains if there are frequent requests for connections of that type between those domains, i.e. the connection may be *cached*. The connection can then be re-used the next time an appropriate connection request is received, thus reducing the latency in the set-up time. It should be noted that only the Hollowman makes a distinction between connections that are in use and connections that are cached; as far as the physical switch is concerned they are indistinguishable. Which connections are likely to be reused is highly application specific. This makes the use of connection caching problematic within a generic control architecture, but highly promising for application specific control architectures.

The demand for connection creation involves modifying state in at least four places: the source host manager, the sink host manager, the connection manager and the switches. It is possible that the attempt to create a connection fails after state has been already updated in one or more of the above. In this case all the updates must be undone and the original state before the creation restored. Thus a connection creation is in fact a distributed transaction which can be rolled back if the creation fails at any stage. This problem is made more complicated by the fact that the state in the connection and host managers should be locked for as short a time as possible to optimise concurrency and so we cannot, for example, simply lock the whole of the connection manager during an operation.

We have experimented with making the communication with the switches for creation and deletions of connections asynchronous in order to minimise the amount of time an operation occupies in the control architecture. Since the control architecture has a complete view of the state of resources in the switch, once the control architecture has decided that, say, a create operation is valid, then the only way the switch can refuse the connection is due to switch failure. The network connection is only marked complete when each switch has returned successfully. An application after asking for the creation of a connection will be forced to wait until the connection has been marked complete, however the application will not cause another application to block because the network connection belongs to it alone. In addition the create operation is executed in parallel on all the switches that it involves as well as with the modifications in state within the host and control manager, further optimising the connection creation time. The price of this is the introduction of asynchronous communication and hence some additional complexity. The above type of connection creation/deletion we denote as *lazy* in analogy to lazy evaluation within programming languages (Bird, 1988). We have experimented with this technique within the Hollowman and noted the expected factor of decrease in connection set-up time, as the $N$ switches do their processing in parallel. The drawback of this technique is that it supposes that once a connection has been authorised by the control architecture that the physical switch cannot refuse it, if the switch does refuse it then recovering from the failure is further complicated.

The control architecture has complete knowledge of the topology of the network and maintains information about the current resource usage in its nodes. When the network is interconnected with a larger network it is neither desirable or feasible to have such information for the unified network. The interconnection of a network managed using the control architecture described here and other networks outside its control is a subject of on-going research.

## 6   CALL MANAGER

There are advantages to being able to group logically associated connections together into a higher conceptual entity. For example, it is likely in the bi-directional communication case that if one connection fails then the other should automatically be removed as well. We term this logical grouping of connections a *Call*.

Applications are free to associate any group of connections into a call. This avoids attempting to define all call types that all applications will ever want. We have adopted a similar approach to that defined in (Minzer, 1991), having a language in which an application may define a complex mesh of connections for, say, a video-conferencing application. What makes the concept of a call powerful is by allowing an application to create the control behaviour that is to be used within it. The associations of a group of connections with the control behaviour to manage those connections we term a *Call Closure*. Using call closures, applications can take advantage of their high level knowledge about how connections are to be used within a service in order to optimise the use of their resources. The call manager is the environment in which these call closures are loaded and executed.

An example illustrates the point: a security guard monitors video from two different rooms each with their own camera. Suppose that the rooms are adjacent and that the two cameras are connected to the same switch. We could establish two distinct connections from the cameras to the display of the security guard. However, the guard will only ever observe one camera at a time and that therefore at any given moment one of the connections is redundant. Knowing this we build a call closure which contains a connection from each camera to the display and which multi-plexes the two connections every 3 seconds. The call closure creates one connection to the display from a vci on the output port of the camera connected switch and periodically interchanges the input vci with which it is associated. Figure 6 is a schema for this example. (Ravindran, 1996) describes architectural and protocol techniques for optimised multi-cast transport, allowing many sources

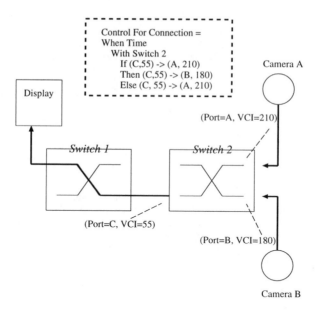

```
Control For Connection =
When Time
    With Switch 2
        If (C,55) -> (A, 210)
        Then (C,55) -> (B, 180)
        Else (C, 55) -> (A, 210)
```

**Figure 6** Example showing how call closures allows efficient resource usage in an application.

for example, to share the same distribution tree multiplexing them temporally. Call closures allow a means by which an application can define how this is to be done on *a per-application basis*.

The control defined by the application within the call request may only manipulate the connections allocated to that call, preventing calls interfering with each other; (Rooney, 1996) examines the motivation for call closures and their implementation within the Hollowman in greater detail.

In summary, the Hollowman allows application programmers to create a behaviour - defined in a dynamically loadable programming language - for the connections that make up their services and have that behaviour execute at the heart of the control architecture during the lifetime of those connections. This allows applications great flexibility over the network resources allocated to them.

## 7  HOLLOWMAN APIS

Applications at start-up, connect with their host manager and request the creation of an API instance appropriate for its needs. Three different types of API are currently present in the Hollowman:

- a BSD-socket like API;
- a service offer API;
- a call closure API.

The BSD-socket like API allows applications to do the normal: *open, listen, connect, receive, close,* type primitives on the Hollowman SAPs, no notion of trading or services is involved.

Applications which do not wish to use the notion of service are not obliged to. The second API involves importing service offers from the Hollowman trader and reifying those offers into SAPs, the reification is achieved by the use of the host manager and connection manager as described in Sections 4, 5. Thus applications can be unaware of the location of the entities they are communicating with and details such as joining to multi-cast trees are handled by the control architecture transparently, thus simplifying the application.

The third API is simply a gateway in which user defined call closures can be loaded and executed within the call manager.

## 8   IMPLEMENTATION

The Hollowman is written in C/C++ and uses Dimma (Li, 1995) for application-to-application communication. Dimma is a framework ORB, on which real-time ORBs can be constructed. Currently all of our communication is using a Dimma implementation of the standard CORBA protocol IIOP. The call closures and call manager are written in Java and interface to the rest of the control architecture using the Java-to-C API.

The test-bed in which we experiment with the control architecture contains a small set of Fore Switches attached to HP, DEC Alpha and Solaris machines, and some ATM Cameras (AVAs).

Most of the experiments have been carried out using Ariel/SNMP for communication with the switches. This is not ideal in terms of performance, but allows *any* switch running an SNMP daemon to be managed. Currently in the unoptimized version of the control architecture connection set up across a single switch requires approx. 200 milli-seconds. This breaks down to 10 % for application-to-application communication, 40 % for application processing and 50 % for communication with the switch. This is of a similar order of magnitude to that defined in (Veeraraghavan, 1995) for an implementation of B-ISUP. We confidently expect to be able to reduce this latency by an order of magnitude as more efficient implementations of *Ariel* become widely available and by the use of lighter inter-application communication mechanisms. Caching the connection effectively removes the part of the connection set-up time required for communication with the switch; lazy connection evaluation allows the partial parallelisation of the communication and processing with the different switches, thus is most useful when there are many switches.

## 9   CONCLUSION

This paper has presented the innovative Hollowman control architecture which devolves control out of the physical switches and into a distributed processing environment.

We have detailed the techniques by which:

- applications learn of the existence of services;
- physical switches are managed within the control architecture;
- connections are established between applications;
- resources associated with connections are managed.

We have showed how the flexibility of the Hollowman allows us to experiment with techniques such as connection caching and lazy end point evaluation. The concept of a *call closure* as a bundle of logically associated connections together with the control to manage those connections has been introduced.

The Hollowman demonstrates that it is possible to delegate many control functions out of ATM switches and that by doing so we permit the full exploitation of ATM. We do not however believe that in the future that there will be one, single, standard, ATM control architecture and that in consequence we are building an infra-structure in which many control architectures: Hollowman, X-Bind, TINA, Q-2931 may run simultaneously over the same network elements.

## 10   REFERENCES

Architecture Projects Management Limited  (1995) ANSAware/RT 1.0 Manual *ANSA project.*

Barr, W. J. Boyd, T. Inoue, Y.  (1993) The TINA initiative *IEEE Commun. Mag. March 1993*

Bird, R. Wadler, P.  (1988) Introduction to Functional Programming *Prentice Hall.*

Cidon, I et al (1995) The OPENET Architecture *Sun Microsystems Laboratories SMLI TR-95-37.*

Crosby, S. (1995) Performance Management in ATM Networks *Cambridge University PhD dissertation, available as technical report TR 393.*

ITU-T  (1994) Draft Recommendation Q.2931, Broadband Integrated Service Digital Network (B-ISDN) Digital Subscriber Signalling Systems No. 2, User-Network Interface layer 3 specification for basic call/connection control *ITU publication.*

Lazar A, and Lim, K.S.  (1996) Realizing a Foundation for Programmability of ATM Networks with the Binding Architecture *IEEE Journal on Selected Areas in Communication, Vol 14, Sept. 1996.*

Lee, W. Hluchyj, M. Humblet, P.  (1995) Routing Subject to Quality of Service Constraints in Integrated Communication Networks *IEEE Network July/August 1995.*

Leslie, I. et al  (1996) The Design and Implementation of an Operating System to Support Distributed Multimedia Applications *IEEE Journal on Selected Areas in Communication, Vol 14.*

Li, G. (1995) Dimma Nucleus Design *APM Technical Report, APM 1553.00.05.*

Minzer, S.  (1991) A Signaling Protocol for Complex Multimedia Services *IEEE Journal on Selected Areas in Communication, Vol 9, Dec. 1991.*

OMG  (1991) The Common Object Request Broker: Architecture and Specification *Document Number 91.12.1, revision 1.1.*

PNNI  (1994) ATM Forum contribution, Draft Specification *94-0471 R7.*

Ravindran, K.  (1996) Architectural and Protocol Frameworks for Multicast Data Transport in Multi-service Networks *ACM SIGCOMM Computer Communication Review, Jan. 1996.*

Rooney, S.  (1996) Connection Closures: Adding application defined behaviour to network connections *Submitted to Computer Communication Review, Oct 1996.*

van der Merwe, J.E. and Leslie, I.  (1996) Switchlets and Dynamic Virtual ATM Networks *Procceding's IM'97, San Diego.*

Veeraraghavan, M. La Porta, T. Lai, W.S.  (1995) An Alternative Approach to Call/Connection Control in Broadband Switching Systems *IEEE Communications Magazine, Nov. 1995.*

## 11   BIOGRAPHY

Sean Rooney received the B.Sc and M.Sc degree in Computer Science from The Queen's University Belfast in 1990 and 1991 respectively. After three years at the research center of Alcatel Alsthom at Marcoussis working in the field of network management, he started working for his PhD degree at Cambridge University.

## PART FOUR

# Multimedia Services, Applications, Policies

Chair: Liba Svobodova,
IBM Zurich Research Laboratory

# 29

# Immersive and Non-immersive Virtual Reality Techniques Applied to Telecommunication Network Management

*Mohsen Kahani, H. W. Peter Beadle*
*Department of Electrical and Computer Engineering,*
*University of Wollongong, Northfield Avenue,*
*Wollongong, NSW 2522, Australia*
*Phone: +61-42-21-3065,    FAX: +61-42-21-3236*
*E-mail: {moka,beadle}@elec.uow.edu.au*

## Abstract

In this paper, we will introduce two different three dimensional VR-based user interface for telecommunication network management. The first one is an immersive system, using HMD and other 3D input devices. The other is a WWW-based flat screen 3D collaborative user interface. The architecture of each system and our observation from it will be described. Then, a comparison will be made to show the merits and pitfalls of each.

## Keywords

Telecommunication Network Management, Virtual Reality, Virtual Reality Modeling Language (VRML), User Interface.

## 1   INTRODUCTION

The Broadband Integrated Services Digital Network (B-ISDN) based on Asynchronous Transfer Mode (ATM) technology introduces bandwidth capabilities that allow the emergence of sophisticated multimedia applications. ATM networks include the concept of logical connectivity and virtual private network (VPN) (Kositpaiboon, 1993). A virtual private network is a set of network resources, such as user-network interfaces (UNIs), and (semi) permanent virtual connections (VPC) that link the different sites of a customer together. However, this logical connectivity, although providing higher management flexibility than physical connectivity, increases the complexity of network management task.

The virtual private network concept also implies that there are some dependencies between operation of different networks, because they may share the same physical link. Consequently, some kind of collaboration among network management systems of private networks and with that of the carrier is required to effectively manage the network in real time.

Managing ATM networks requires a more decentralised approach, as well. Several organisations, from the network provider to customer site administration, may require hierarchical access to network management information. While distributing management, a centralised and integrated view of the whole system should also be provided.

The complexity of the networks and the new services that they provide has made network management more mission critical to a larger number of organisations. This has led to the development of integrated network management systems using Windows Icons Mouse Pointer (WIMP) based direct manipulation user interfaces. Despite advances in computer technology, these interfaces appear to be inadequate for visualising and manipulating the huge databases typical of modern network management systems (Lazar, 1992). One of the major drawbacks of WIMP based user interfaces is that, even for small networks, the user becomes lost among too many open windows, and in too much modeling hierarchy (see Figure 1). This means a high conceptual load for the operator and an inefficient use of human short-term memory, when the operator wants to find faulty devices, and/or observe the performance of network elements.

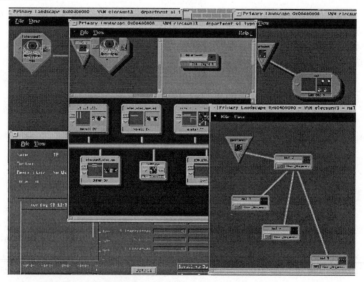

**Figure 1**- A typical view of WIMP user interfaces.

It is believed that the management of emerging networks requires greater visualisibility and interactivity than that provided by traditional user interfaces (Crutcher, 1993). The manager in these environments has to deal with tens of thousands of virtual channels, and potentially hundreds of ATM switches (Alexander 1995). To enhance the network management operating environment, we have been investigating the use of Virtual Reality (VR) user interface technology for network management applications. Two approaches have been used and some prototype systems have been implemented. In the first model (Kahani, 1995), we deployed an immersive virtual reality environment, consists of Head Mounted Display (HMD), pointing devices, joystick, and other input devices. The observation from the implementation of the system has led us to build a non-immersive, distributed, collaborative, 3D interface using WWW techniques (HTML, VRML, Java and JavaScript) (Kahani, 1996).

In this paper, firstly, the special requirements of a distributed and collaborative network management system are discussed. Then, we introduce the architecture of the immersive system and discuss our observations from it. This will be followed by the discussion of pitfalls and merits of this system that led us to implement the second prototype system. After explaining the structure of the second system, a comparison will be made between them. Finally, we conclude the paper with the discussion of the lessons learned from these implementations and explain further work.

## 2  SYSTEM REQUIREMENTS

The system proposed here is a Distributed Virtual Reality (DVR) system. DVR systems are mostly used for simulations, eg. SIMNET (Pope 1989), or computer games. However, in a network management environment, there are several issues that have to be treated differently. These issues are:

- **Bandwidth**: The amount of bandwidth used by a network management system should be as small as possible, compared to actual network traffic. Network management tasks (eg. device polling) consume a considerable amount of bandwidth by themselves, so the system should be so designed that the distributed VR user interface does not add much more traffic. Unlike some other systems in which the distributed system, itself, is the goal (eg. games or simulators), the total amount of bandwidth consumed by network management system (device polling and operators' collaborations) is considered waste, and reduces the network throughput.

- **Reliability**. Reliability is a major issue in network management. In a distributed VR game if some update messages are lost, the effect on the total system is not dramatic. In a network management environment, each individual message may carry important information, and may have a catastrophic effect on the network, if does not reach the destination. As a result, a best effort communication protocol is not suitable, and a reliable end-to-end protocol, such as TCP/IP should be considered.

- **Number of users:** Most distributed simulators have a large number of participants, spreaded over several LAN segments. As a result, the communication of update messages among users is done via either broadcasting or multicasting. In a network management environment, however, the number of participants is relatively small, and it is less likely that too many users join the system from the same LAN. So, in the absence of widely deployed point-to-multipoint and multipoint-to-multipoint services, a point-to-point unicast communication model is used.
- **Security**. For most distributed simulation systems, security is not an issue. Some of those systems have a dedicated network, which physically maintains the security, for others, such as distributed games, the data is not sensitive. None of these are true for a network management environment. As a result, special security measures must be considered.

## 3    IMMERSIVE VR USER INTERFACE

### 3.1 System architecture

The architecture of the immersive VR system is illustrated in Figure 2. This system has the basic VR elements such as 3D image rendering and 3D navigation tools. It is coupled to an existing SNMP based network management system (Cabeltron SPECTRUM). Three kinds of information are retrieved from the network management system: network configuration, topology, and performance/fault data. While the formers are nearly static and rarely need updating, the performance and fault data are quite dynamic, requiring continuous update.

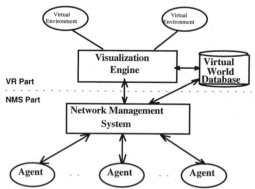

**Figure 2-** Immersive VR  system architecture.

The network configuration and connectivity information are extracted, as network topology changes, from the NMS using its Command Line Interface (CLI), and a virtual network world database is constructed, automatically. This database is used by the VR system to build the virtual environment.

To provide a real-time user interface, performance/fault data must be collected directly from the NMS. This can be achieved by establishing a direct link between NMS and VR systems, in which the VR part sends its inquiries to the NMS using CLI commands to retrieve the required data. The VR part also converts the operator's manipulation of the network to appropriate CLI commands and sends them to the NMS. The physical interface between the systems is provided by the underlying network.

After the construction of a virtual network world, the user can navigate it, by walking or flying around the network. The status of the icons in the world represents current network performance levels. For instance, the thickness of a link determines the amount of load being carried by the connection, its colour represents its operational status, and a disconnected link is represented by a broken line. Correlated alarm information is presented using speech synthesis with clues to the location of the alarm provided by spatially locating the sounds in 3D. This is in contrast to topological maps and colour based roll-up procedures used in existing WIMP based systems.

Objects in the virtual world are active so more information about their status can be obtained by walking into them for a detailed internal view. If the object is a link, walking into it will show the virtual paths within the link. If the object is a network element, walking in will show the interfaces contained in the element. If the object is a sub-network, walking in will show the layout and status of the sub-network elements. The walk in metaphor captures the hierarchically structure of the network and constrains the information presented on the screen to a comfortable level for network operators.

Navigation in the world is by using mouse, Joystick, Logitech Cyberman 3D mouse or Data Gloves. Currently, we are examining how the operator can interact with the interface in a more natural way. For example, to grab a network element, for moving, disconnecting, etc, the most natural way is to grab it with a virtual hand, using VR gloves.

The other important issue, is the representation of network element in the virtual world. Using special rendering techniques, such as texture mapping and smooth shading, the scene should be designed in such a way that it can immerse the operator, so that they can forget the interface, and act as though they are in the real world.

## 3.2 Observations

The prototype system is basic and does not incorporate texture mapping. It employs a head mounted 3D stereo display, and a Cyberman or joystick, as input device. Using this prototype system, the user can observe the hierarchy of the network and its spatial relationships. The network can freely and quickly be navigated to observe the primitive information for network elements such as faulty devices and overloaded links. We achieve this without becoming lost in a screen full of windows, the typical problem with existing WIMP based systems. A typical view of the prototype system is shown in Figure 3.

The main advantage of a VR user interface is its additional spatial dimensions, since the network's hierarchical properties become explicit (Stanger, 1992). A HMD while creating a more immersive environment, acts as an input device as well. By rotating the head, the user easily and quickly navigates into the system. A joystick or Cyberman gives more sense of moving in the virtual world than the traditional mouse. A virtual glove provides yet another powerful input device, which increases interactivity.

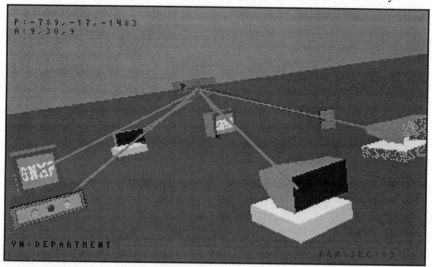

**Figure 3**- A view of immersive VR system that shows several network elements, such as router, pingable and generic SNMP devices. Colour coding has been used to show faulty devices and congested links.

The other major advantage of VR user interfaces for network management is their short learning time. As user's interaction with the system is designed to be as natural as possible, there is not much need to teach operators how to use the system. That is, if operators learn the basic principles of the interface, they can easily and quickly decide, when facing with more complicated situations, how to do the task. For instance, there is no need to teach operators how to move an object, because everybody knows how to move objects with his hands. This is in contrast to WIMP user interface, in which all actions must be taught to the operator.

The other important factor is the user's cognitive load during operation. As in WIMP user interfaces, the interaction between user and computer is not natural, the user has not only to think about 'what to do', but also 'how to do' it. For instance, if the alarms associated with an object are needed, the object has to first be selected, by clicking the mouse button on it. Then, from a menu the appropriate action must be selected. This simple task seems quite easy and straight forward. However, working with many objects in a window and with several other windows in this manner, causes confusion, because of limitations of human short term memory. While in an immersive virtual reality user interface, these kinds of tasks could be done by using a speech based interface with speech and visual acknowledgment, reducing the operator's cognitive load.

Despite these advantages, the system has some drawbacks, as well. As network management is nearly a continuous task, which takes several hours a day, the use of HMD causes some problems. Even the best available HMDs cause dizziness and eye strain if worn for a long period. Also, as it obscures the user's view, it significantly reduces the interaction and communication of the user in the real world.

The other problem is textual information. Although, a VR user interface minimises the amount of textual data by converting them to symbols in the virtual world, in a network management environment there is a significant amount of information that has to be presented to the operator as text. However, in a graphics-based user interface, proper provision of text is difficult. The situation is even worse when HMD is used.

Based on these and other limitations we decided to move toward a non-immersive approach, while maintaining the three-dimensional semantics of the view. Because of the need for a distributed and collaborative environment for effective management of forthcoming networks, a World Wide Web (WWW) based approach was chosen. The main reasons for this selection are that WWW browsers are reasonably uniform and ubiquitous, platform independent, and have low prices.

## 4   WWW-BASED USER INTERFACE

### 4.1 System architecture

The system uses a client/server architecture based on Telecommunication Network Management model (ITU-T M.3010, 1992). Each server communicates with a network management system and uses its services to get the management information. This information is sent to the clients, which are WWW browsers enhanced with Virtual Reality Modelling Language (VRML) plugin, Java and JavaScript. VRML is a three dimensional modelling language for multi-participant simulation (Bell, 1995 and Bell 1996). Java is a platform independent object-oriented language that can run in the client's environment, rather than server machine.

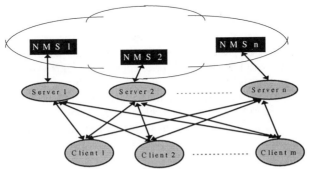

**Figure 4**- The architecture WWW-based system.

Within the virtual world, each 3D object can have a link to other objects and views that may be within the domain of another network management system (NMS). This allows an integrated view of distributed networks in which each subnet is managed by an independent NMS. Moreover, managers can collaborate with each other, in real-time, to solve the problems that involve more than one domain. Figure 4 illustrates this architecture.

Each server consists of four parts: NMS interface, Collaborative Manager (CM), object-oriented database (OODB), and HTTP server, as shown in Figure 5. NMS interface communicates with network management system via its command line interface (CLI). NMS can be any system capable of gathering information from network elements (NEs), and again in our case is Cabeltron Spectrum. The interface queries the NMS to get management information about the status of NEs, and stores them in the OODB. It also gets update information from the database and sends them to the NMS.

The collaborative manager (CM) is the core of the system. It communicates with the clients directly, or via HTTP server, through a Common Gateway Interface (CGI) script. It also coordinates the collaboration between clients, by collecting the updating information from each client, broadcasting them to the other clients, and storing them in the OODB.

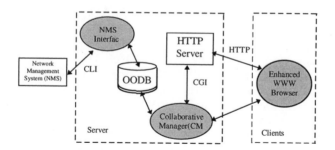

**Figure 5**-Details of client/server communication.

The scenario is as follows: The manager uses an enhanced WWW browser to connect to the HTTP server. After authentication, HTTP server asks the appropriate view from the CM via CGI protocol. The CM responds with the information in VRML format. The VRML script has several Java applets that firstly establish a TCP/IP connection between the client and the server, and secondly, control the behaviour of NEs in the client's environment. User, then, navigates into the 3D virtual world, interacts with NEs and manipulates the world scene. The position of the navigator and its manipulations' data are continually sent to the CM via the established connection. The Java applets also listen to the connection and update the world scene based on the received data.

The CM receives two kinds of data from clients. The first kind of data is only related to the virtual environment, such as notification of the changes of objects' position in the virtual world. This data is sent to all concerning participants. The other type of data concerns the real counterparts of the objects, as well, such as change of the status of a link. In this case, the NMS has to be notified of the change as well. This task is achieved through the NMS interface.

## 4.2 Observations

The system is currently being implemented. Here, we present some initial observation and results from it. The main feature of the implementation is its platform-independency. The manager can connect to the network, from any computer at any point, either remotely or locally, from his/her notebook or desktop computer, and uses the full capabilities of the system. The managers can take mobile computers with them to the fault locations, and collaborate with the managers at the central station to fix the fault quickly, and with great confidence.

As with previous system, the three-dimensional view of the network hierarchy and additional navigational facilities increase the visualisability of the network management information. The greater visualisibility means the lower probability of error and miscalculation of the manager, which directly increases the network survivability and reduces down time.

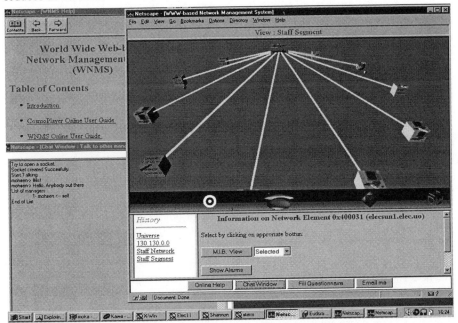

**Figure 6**- A typical view of the system.

The user initially connects to the system by requesting an HTML document from the specified HTTP server. This document asks for user ID and password, and sends them to the server, which actually calls the CM via CGI protocol. If the user is accepted, an entry in the database is created for it, and a document consisting of text and graphics frames is forwarded to it. The client then navigates into the systems, and does its management job. A typical view is also shown in Figure 6.

The impressions of people seeing the system is that, despite its preliminary implementation, its 3D view and mixture of text and graphics give the manger more flexibility than available two-dimensional commercial network management systems. Incorporating voice communication between participants and using more realistic and complex scene will improve further the system's efficiency.

Whoever has ever repaired networks acknowledges that there are some situations where one needs to be in at least two places at once. The collaboration feature of this system, addresses this problem very efficiently.

Compared to other commercial network management systems, this system is cheap, and allows the managers to use their existing computers to connect to this system for a real-time network management.


## 5   COMPARISON

In this section we will compare the systems, qualitatively. As the second system is more comprehensive and has more components, we focus our comparison mostly on the components that both systems have. In this sense, we justify why we did not build collaboration and distribution modules on top of the immersive system.

The major point in the first system is the immersion. If built properly, the users feel they are in a similar environment to the real world, with similar level of interaction. Whenever a failure occurs, the user only needs to think of 'what to do' rather than 'how to do'. This means that quicker and higher quality actions can be chosen in times of stress. However, with current technology, the level of immersion and interaction is still inadequate. Moreover, ergonomic considerations mean that existing head mounted displays are not appropriate for lengthy applications, such as network management.

The second system lacks these facilities. Instead, it benefits from some of the advantages of WIMP interfaces. Multiple frames consisting of 3D graphics and text, carry more information than a pure graphical one. Also, as mentioned earlier, in a network management environment, there is plenty of useful textual information that cannot be translated efficiently into graphical symbols. The second system can easily show them in the text frame, while proper display of text in graphical environment is more difficult, and yet to be investigated.

The level of interaction in an immersive system is much higher. Utilising VR gloves and other 3D input devices such as Cyberman and 3D mouse, navigation in the 3D world is quite easy. On the other hand, most WWW browsers only use mouse for interaction. This cause some confusion as user has to switch between different modes of movement, eg. walk vs. fly. However, it seems that forthcoming browsers will support more input devices.

Platform dependency is another important issue. Immersive systems are mostly platform dependent. Developing a system for a special platform requires a special set of libraries and utilities that usually are not available for other platforms. Even, some types of input devices, such as HMDs and gloves, are available for only few platforms. For the second system, however, the situation is different. Some WWW browser, such as

Netscape, are available for most platforms. Moreover, the browser from different platform are quite compatible, as they use the same set of standards.

The higher mobility of WWW-based approach is another advantage. The immersive system has a lot of bulky components, such as HMD, glove, and other input devices, which makes it difficult to move the system around the network. While, in the other approach, a highly mobile notebook computer can be used as the network management workstation.

The other advantage of the non-immersive system is the higher accessibility to the network management system. The prices of browsers are so low that most computers have a copy of them installed. So, the managers can virtually use any computer in the network to connect to the system, and benefit from the graphical user interface, to manage the network, remotely. With collaboration added to the system, the scope and level of management will go far beyond current systems.

With the trend of network management moving towards using HTTP and Web technology instead of or in corporation with SNMP for device polling and status notification (Wellens, 1996), the second method can be seemlessly expanded to even directly contact with the network elements. In fact, the trend towards Web-based network management is so high that some experts believe that *"the network management platform of the future may only have Web-based user interface"* (Bruins, 1996).

# 6 CONCLUSION

We are investigating the application of virtual reality user interface paradigm for managing telecommunication networks. In order to focus on user interface problem, we have used one of better network management software as the network management back-end. Then, we started our research by design and implementation of an immersive 3D virtual reality system, incorporating head mounted display and 3D input devices. The experience from this system, led us to implement a WWW-based collaborative, distributed 3D user interface, using enhanced Web browser.

In this paper, we briefly, discussed the architecture of both systems and our observation from their prototype implementations. Then we compared them in terms of their suitability for a network management environment. Each system has some advantages and drawbacks, but it seems that, for now, the performance of WWW-based system is superior to that of immersive VR system.

The prototype WWW-based system implemented here, though yet to be completed, depicts some useful features. Most commercial management systems use graphical workstations which are relatively expensive. The communication with the system using other platform is only through text-based command line interface, which is not useful for management of complex networks. On the other hand, in our implementation, the managers can connect to the system and do full network operation from any location in the network using virtually any computer. A more powerful computer can deliver a very realistic view featuring texture mapping and smooth shading, while in less powerful machines a rather primitive view, with a reasonable speed, can be shown.

The three dimensional and collaborative environment created by this system, firstly, give a greater visualisation to the system, and secondly, allow real-time communication between managers, which is necessary for management of complicated and flexible broadband networks based on ATM technology.

Finally, Although we used this system for network management purposes, the generic structure can be used in any application that require data visualisation.

### Acknowledgments

This work is supported by a postgraduate scholarship from ministry of culture and higher education of I.R. Iran granted to M. Kahani. The financial support of The Institute for Telecommunication Research (TITR) of the University of Wollongong is also hereby acknowledged.

## 7   REFERENCES

Alexander, P. and Carpenter, K. (1995) ATM net management: a status report. *Data Communication Magazine*, September issue.

Bell, G., Parisi, A. and Pesce, M. (1995) VRML 1.0 specification. *Online document http://vag.vrml.org/vrml10c.html.*

Bell, G., Marrin, C., et al. (1996) The VRML 2.0 specification. *Online document http://vrml.sgi.com/moving-worlds/.*

Bruins, B. (1996) Some experiences with emerging management technologies. *The Simple Times*, Vol. 4, No. 3, July 1996.

Crutcher, L, Lazar, A, Feiner, S and Zhou, M (1993) Management of broadband networks using a 3D virtual world. *Proc. 2nd International Symposium on High Performance Distributed Computing*, 306-15.

ITU-T Recommendation M.3010 (1992) Principle and architecture for the TMN. Geneva, 1992.

Kahani, M. and Beadle, P. (1995) Using virtual reality to manage broadband telecommunication networks. *Australian Telecommunication Networks & Applications (ATNAC'95) Conference* , Sydney, Australia, 517-22.

Kahani, M. and Beadle, P. (1996) WWW-based 3D distributed collaborative environment for telecommunication network management. *ATNAC'96 Conference*, Melborne, Australia, 483-8.

Kositpaiboon, R. and Smith, B. (1993) Customer network management for B-ISDN/ATM services. *ICC'93 Geneva Conference proceeding*, 1-7.

Lazar, A, Choe, W, Fairchild, K and Hern, Ng (1992) Exploiting virtual reality for network management. *Communication on the move-ICCS/ISITA'92*, Singapore, 979-83.

Pope, A. (1989) BBN Report NO. 7102, The SIMNET network and protocols. *technical report, BBN systems and Technologies*, Cambridge, MA.

Stanger, J. (1992) Telecommunication applications of virtual reality. *IEE Colloquium on Using Virtual World*, 6/1-3.

Wellens, C. and Auerbach, K. (1996) Towards useful management. *The Simple Times*, Vol. 4, No. 3, July 1996.

## 8 BIOGRAPHY

**Mohsen Kahani** is currently a PhD student at the University of Wollongong, Australia. He received his B.E. in 1990, from the University of Tehran, and his M.E in 1994 from University of Wollongong. His research interests includes network management, virtual reality, user interface design, and object oriented design. He is a memeber of IEEE Computer and Communication societies.

**H. W. Peter Beadle** received his B.Sc. (Hons.) and Ph.D. (Comp. Sci.) from Sydney University. He is a senior lecturer in Electrical and Computer Engineering at the University of Wollongong. His research interests include Broadband, Multimedia and Virtual Reality based telecommunications systems. His prior appointments were at the Integrated Systems Laboratory at the ETH in Zürich Switzerland and as the head of the Multimedia Applications Section in the OTC R&D laboratories in Sydney. He is a member of ACM, IEEE Computer Society and Usenix.

# 30

# Broadband Video/Audio/Data Distribution Networks — The Need for Network Management

*Alan R. Brenner*
*abrenner@gi.com*
*General Instrument*
*6262 Lusk Blvd*
*San Diego, CA 92121*

*Branislav N. Meandzija*
*bran@metacomm.com*
*Meta Communications, Inc.*
*P.O. Box 1258*
*La Jolla, CA 92038*

## Abstract

Audio-visual/data service delivery systems are based on complex network, distributed system and application architectures that enable the wide-area coordination, control, and delivery of audio, video, and data. One of the central logical components of these architectures are the transport/program/service streams used in storage and transmission of audio, video, and data. The composition of the transport/program/service streams is standardized by the Motion Picture Experts Group (MPEG) in the MPEG-2 standard.

In this paper we give an overview of the issues involved in managing MPEG-2 based audio-visual/data distribution networks and present a brief analysis of the applicability of different architectural management frameworks and standards.

## Keywords

Integrated management, broadband, MPEG-2, audio-visual service delivery.

## 1. OVERVIEW

Carriers today, Local Exchange Carriers (LECs), Interexchange Carriers (IXCs), Competitive Access Providers (CAPs), or Multiple System Operators (MSOs) are seeking new geographic and demographic markets. The nature of their services is rapidly changing from static packaged, scheduled programming to dynamic, interactive services such as: Video-On-Demand (VOD), near-VOD (NVOD), interactive shopping, interactive gaming, etc. There remains considerable debate over network architectures and distribution technologies. Clearly substantial sums of money will be spent on plant and infrastructure (in North America some estimates are in the order of $75 billion) (Tankin, 1996).

Due to the cost and nature of some high-bandwidth distribution architectures, cost recovery from subscribers is a significant issue. Consumer selection of network operator and carrier service is less of a technology issue, and more of a reflection of programming content, service availability, service quality, service flexibility and cost. Services are becoming a commodity nature with content changed frequently. Therefore the ultimate challenge for the carriers is to provide these services (telephony, video, gaming, shopping, data and telemetry) at competitive cost, minimize subscriber churn and keep service areas intact.

For the traditional LECs, IXCs, and some CAPs, this may not be much of a challenge. For the MSO community at large, the ability to control and manage network resources against acceptable levels of service availability and quality is a new dimension of the competitive contest. Promoting telephony and data services based on a network that lacks appropriate management technology and operational practices are bound to result in disaster for the MSOs. The management of broadband networks and network services is a prerequisite for success. All carriers, regardless of network management experience, are required to develop and deploy new generation management systems. This is due to the cost of distribution technology, the competitive environment, network and service complexity, and increased treatment of services as commodity.

The video, audio, and data service delivery systems deployed and operated today are an order of magnitude more complex then those deployed and operated only a few years ago. Today's systems are based on complex network, computing, and service architectures. They provide a wider variety of program delivery capabilities and services, use new technologies and vastly more complex types of equipment to enable the wide-area coordination, control, and delivery of audio, video, and data (Weiss, 1996).

The new video-audio-data-distribution networks can no longer be managed by the patch-work of management applications that have been created in an ad-hoc fashion to solve the problems of the day. Satellite Uplinks today may consist of several hundred Encoders with dozens of IP nodes each, controlled by several hundred control computers. Satellite Downlinks and Cable Headends may be on top of a pyramid of several HFC distribution networks; each with thousands of network elements serving different overlaid logical RF distribution networks; each multiplexing a variety of services ranging from digital video, to digital telephony, to Internet services. This increased complexity of managed equipment and services coupled with today's market requirements, necessitate powerful architectural solutions. These solutions should facilitate the use of off-the-shelf management technologies and tools for non-video-audio-data-distribution specific management tasks while enabling the seamless integration of newly developed video-audio-data-distribution networks.

In this paper we give an overview of the tasks and possible architectural solutions for management of video-audio-data distribution networks. First, we introduce an architectural framework based on networks deployed or planned today. Within that framework we specify common services and applications. We then outline management requirements and relate them to a combination of different off-the-shelf technologies, capable of supporting the specified common services and applications. We concentrate only on the Uplink side of the distribution network and just briefly consider the requirements of the Downlink distribution plant and Headend.

## 2. MPEG-2 BACKGROUND

The central logical components of audio-visual/data architectures are the transport/program/service streams used for storage and transmission of audio, video, and data. The composition of the transport/program/service stream is standardised by the Motion Picture Experts Group (MPEG) in MPEG-2 standard (ISO 1-4, 1994).

MPEG-2 streams are created by Encoders from other such streams and/or elementary audio/video/data/control information. Decoders recreate the elementary audio/video/data from MPEG-2 streams.

An MPEG-2 Transport Stream consists of one or more programs each containing one or more elementary streams and other streams multiplexed together. Each elementary stream consists of access units, which are the coded representations of presentation units. The presentation unit for a video elementary stream is a picture. The corresponding access unit includes all the coded data for the picture. The access unit containing the first coded picture of a group of pictures also includes any proceeding data from that group of pictures. The presentation unit for an audio elementary stream corresponds to samples from an audio frame.

Elementary stream data is carried in PES packets, which are inserted into transport packets. Transport packets are carried in Packet Identifier (PID) Streams, which correspond 1-to-1 with PES streams. The contents of the data contained in the transport packets is identified by their PID via the Program Specific Information (PSI) tables. The PSI tables are carried in the Transport Stream as additional PID streams. The PSI tables listed below contain the information required to demultiplex and present programs:

- Program Association Table
- Program Map Table
- Network Information Tables
- Conditional Access Table

While the PES layer transports the video, audio and isochronous data, additional PID streams are defined for other PSI that may include messages related to access control, text services, asynchronous data services, subtitles, etc.

Figure 1 below illustrates the structure of MPEG-2 communications represented as a OSI-style layered architecture:

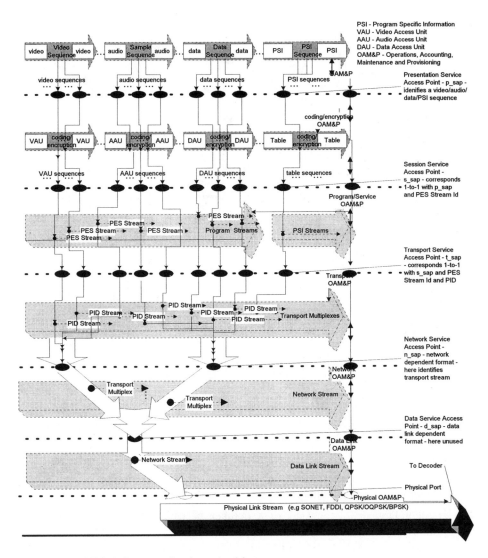

**Figure 1.** MPEG-2 Communications Architecture

## 3. PHYSICAL ARCHITECTURE

MPEG-2 streams are created by Encoders from other such streams and/or from elementary audio/video/data/control sources. Decoders recreate the elementary audio/video/data from MPEG-2 streams. Commonly, the communications system underlying MPEG-2 based audio-visual/data distribution networks can be subdivided into the following functional domains illustrated in Figure 2:

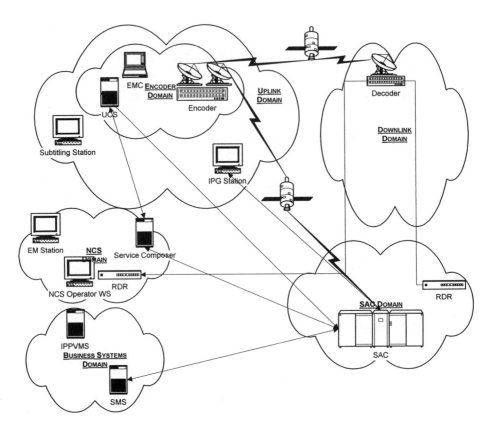

**Figure 2.** Sample Video/Audio/Data Distribution Comm. Architecture

- Network Control System (NCS) Domain — consists of program authoring stations, network management systems, entitlement management systems, etc.
- Uplink Domain — consists of a number of encoders controlled by Uplink Control Systems (UCSs) and other systems used to support the encoders such as video servers, automatic audio/video feeder networks to encoders, subtitling stations, etc. The Uplink Domain is usually subdivided into multiple:
  - ∗ Encoder Domain — consists of encoders and devices supporting encoders. Encoders are devices that generate MPEG-2 transport multiplexes from video/audio/data sources in a variety of different input formats and other transport multiplexes.
- Business Systems Domain — consist of a number of systems dealing with subscriber management and billing.
- Subscriber Authorization (SA) Domain — consists of a number of systems dealing with subscriber authorizations, key management, etc.
- Downlink Domain — can be of two types:

* * Direct Broadcast System (DBS) consisting of end-user consumer Integrated Receiver Decoders (IRDs)
  * Commercial IRDs feeding a CATV network
* Cable Headend Domain — usually overlaps with the Downlink Domain and can be of varying complexity. The Downlink Domain can include commercial IRDs, Head End Transcoders, Encoders, management systems for the Headend, and distribution plants. Optical or RF equipment is used to feed the distribution plants.
* Distribution Plant — from the Cable Headend Domain a variety of different access network technologies can be employed; these include Hybrid Fiber Coax (HFC) and Switched Digital Video (SDV).

The NCS Domain consists of Service Composers, which create programming services, schedule programs, and manage UCSs accordingly. Entitlement Management Systems (EMSs) contain the commercial IRD authorisation database, Reportback Data Receivers (RDRs) for Private Networks (using IRD diagnostic Reportback capabilities), and Network Management Stations. Commercial IRDs may be Headend receivers or a private network where all IRDs are treated as commercial units.

The main applications within the NCS domain are: defining program schedules, commercial access control, scripting using the authoring station, and management including RDR-based commercial IRD fault management.

The main components of the Encoder Domain are the Uplink Control System (UCS) and the Encoder. The UCS performs several functions (not required to be co-hosted). These functions include: Encoder control, generating required PSI, and generating messages related to access control. It consists of one or more UCSs, and one or more Encoders. The main application facilitated by the Encoder Domain is the Encoder programming and management.

Elements of the Encoder Domain interact closely with other elements of the Uplink Domain that are mostly used to create, store, and provide content for Encoders. Elements of the Encoder Domain also closely interact with elements of the Network Control System Domain. They are used to control all elements of the Uplink Domain and most significantly to provide UCSs with the program schedules and information necessary to manage Encoders. Finally, elements of the Service Access (SA) Domain interact with both UCSs and Encoders.

Figure 3 below illustrates a possible Encoder Domain configuration with the major computing and networking components:

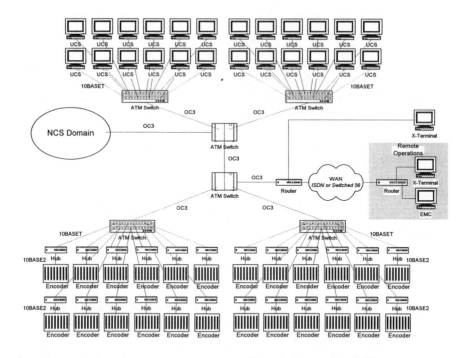

**Figure 3.** Encoder Domain  Example: Switched Virtual LANs  with ATM

The Uplink Domain includes, at a minimum, the Encoder Domain and possibly Interactive Program Guide (IPG) stations. The Uplink Domain may include different types of source equipment for video/audio, isochronous and asynchronous data, subtitles. and text services. Systems, such as the source traffic control system, automation system, and authoring station may be included as well. The main application is the generation of transport multiplexes.  The UCS-s, source equipment and auxiliary data services are managed by the NCS domain.

The Business Systems Domain consists  of Subscriber Management Systems (SMSs) and Instant Pay-Per-View Management Systems (IPPVMSs), which are used by the operator to bill IRD owners for their programming. The main application here is subscriber management. Business Systems communicate with the SA.

The SA Domain consists of Subscriber Authorization Systems (SASs), which authorize and manage all consumer IRDs, and RDRs for consumer IRDs. The main application here is subscriber management and consumer access control.

The Downlink Domain contains the population of Direct Broadcast System (DBS) IRDs, which may be consumer or commercial units. The Downlink Domain intersects with the Cable Headend Domain and the cable access plants. Currently two architectures are favored among carriers (Pugh, 1995). They are:

- Subcarrier Modulated Fiber Coax Bus (also known as Hybrid Fiber Coax) — is mostly an extension of the existing CATV fiber-coax topology.
- Baseband Modulated Fiber Bus (also known as Switched Digital Video) — closely resemble the Fiber-to-the-Curb architecture.

We consider only the HFC architecture. The Downlink Domain CATV for a large metropolitan network typically consists of a primary Central Office (CO) or Head End (HE) containing Encoders and Transcoders interconnected via multiple fiber optic links with distribution hubs. Each distribution hub is the root of a typical HFC access network to subscriber premises. The CO or HE also interfaces local digital telephony switches.

The HFC networks consist of optical fibers from the distribution hub to each fiber node, and the coaxial distribution plant from each fiber node to the subscriber premises. Two separate optical fibers between the distribution hub and the fiber node carry the downstream RF spectrum (typically, 50 to 550 or 750Mhz) or upstream RF spectrum (typically, 5 to 40 MHz). The Downlink network may be shared by different product classes such as digital telephony, switched digital service (e.g., ISDN), interactive multimedia, etc. This is accomplished by RF spectrum allocation and management over the same physical network.

Network elements of the distribution plant today do not have sufficient computing resources to utilize any of the standard network management protocols such as CMIP or SNMP. It is therefore necessary to manage the distribution plant by proxy where the actual management protocol between the proxy control unit and the network elements is most frequently RF based.

Within the HFC networks, RF Spectrum needs to be managed and coordinated with configuration management, performance management, traffic management, and fault management systems of Head End and distribution plant (Ahmed, 1996). The purpose of RF Spectrum management is to allocate the RF spectrum of the physical HFC subnetworks to multiple vendor's products supporting digital services. These products are supported using multiple logical HFC subnetworks in the same physical HFC subnetwork. The downstream and upstream RF spectra need to be coordinated.

The different domains interact for a number of different purposes. The most significant interaction types are:

- UCS/Service Composer—used mostly for communicating control information such as schedules to the UCS and reporting events to the Service Composer,

- SAS/Encoder broadcast — used mostly for communicating control information to the decoders through the encoders by means of MPEG-2 control messages.

- Encoder/Decoder broadcast — the main distribution mechanism for the MPEG-2 transport multiplexes to the consumers; all services and applications involving the Decoder rely on this link.

- Decoder/RDR — used for subscriber control and billing for services such as Impulse Pay Per View (IPPV).

- UCS/SAS — used for applications such as IPPV.

- Service Composer/SAS — used for synchronization of system wide data.
- Business Systems/SAS — used for applications such as consumer access management.

## 4.  VIDEO/AUDIO/DATA SERVICES

Any MPEG-2 service  can contain audio, video, data, text, control, and even executable components. A control stream is present for every service in the transport multiplex. That stream includes the MPEG-2 Program Map message. This channel may optionally also contain related text. The components of any service may be encrypted to strictly limit access.  The following subsections give a brief overview of Video, Near Video on Demand (NVOD), Radio, and Data services.

### 4.1  Television Service

Television service includes at least one video component, one audio or subtitling component, and a service control component.  Access to these components can be selectively controlled for each user. Additional components may include:

- Additional audio components for additional language tracks
- A data channel can supply information to augment video and audio (educational applications, or statistics related to a sporting event, for example)
- additional video components can be present to support synchronous commercial insertion

Near Video on Demand (NVOD), although a service in the sense of a consumer offering, is not handled as a transport multiplex service. By staggering the start time of multiple instances of the same service, the consumer can be transparently directed to the service that starts next.

### 4.2  Radio Service

Radio service includes at least one audio component and a control channel; typically it will also include text backdrop for display while tuned to service. Other components may include:

- A still frame compressed video component could be present for display
- Additional textual information
- A data channel component could supply information meant to augment the audio (such as title, artist, etc.), if a PC is connected to the Decoder data port
- Additional audio component could supply other language soundtracks

### 4.3  Data Service

Data service includes at least a single data component and a control channel; the control component includes the information defining the data channel to be isochronous or asynchronous. Other components may include:

- A text component could supply encrypted text, meant for display, perhaps conveying instructions or for advertising.
- A still frame video component may be present for display.
- An audio component may be present to provide instructions or advertising.

## 4.3.1 Textual Information

Two basic types of text systems are commonly supported, depending on the application:

- The main text system, which can be based on an extensible object-oriented scripting language; this system supports on-screen displays beyond the level of character-based text systems;
- Character-based text using the Latin ISO character set — used for specific message field definitions, rather than for definitions of on-screen displays with graphics.

Textual information originates at the UCS, at the SAS, or Auxiliary Data Systems that send the data to the Encoder. The Encoder acts as a gateway converting the input data to the MPEG-2 protocol for transport. Textual information commonly supported include barker messages, personal messages, or multicast messages.

## 4.3.2 IPG Service

An Interactive Programming Guide (IPG) service is similar to the text information service but can be transmitted at a low rate and stored internal to the Decoder. The IPG service may have both background trickle components and high speed demand components. Interactive Programming Guides provide the consumer with the allusion of rapid access to a large IPG database through a combination of IPG broadcast techniques.

IPG data is based on program schedule information given to the IPG provider by one or more program sources. The program source is usually a business entity different from the IPG provider, but may also be a network operator. The source program schedule will typically give at least the name and scheduled broadcast time of a program. It may contain additional information, such as program ratings or descriptive information.

## 4.3.3 Data Carousel

The data carousel service replays the same data over and over, e.g. weather reports.

## 5. ARCHITECTURAL FRAMEWORKS

A number of different standardization bodies and consortia have created architectural frameworks applicable to the audio-visual problem domain. These include DAVIC, IETF, ITU-T, ISO, Network Management Forum (NMF), the

Object Management Group (OMG), Telecommunications Information Network Architecture Consortium (TINA-C), and many others.

The Digital Audio-Visual Council[1] (DAVIC) (DAVIC1, 1995), (DAVIC2, 1995) specified a system reference model for audio-visual system (see DAVIC 1.0). According to DAVIC, the structure of the DAVIC 1.0 specification is motivated by applications, the driving factor in the audio-visual industry. It defines the essential vocabulary and provides the initial system reference model which is then used as follows:

- Physical architectural descriptions of the three major components of the audio-visual system:
  * Service Provider System (SPS)
  * Delivery System (DS)
  * Service Consumer System (SCS)
- A "toolbox" of the following:
  * High- and mid-level protocols (represented in the DAVIC system as vertical stacks). In addition, modulation, coding, and signaling techniques (pertinent to horizontal physical interfaces between entities of the DAVIC system; this is specially useful for content creators and service providers).
  * Application notes rehearse the steady state and dynamic operation of the system at each reference point. and define the protocol to be used in each case; this is especially useful for equipment vendors and system designers.

Although DAVIC 1.0 is the only widely recognized model for audio-visual systems and some parts of the DAVIC model are defined in great detail and well beyond a conceptual description, the specification is still not mature enough. Therefore, we do not relate the architectural concepts introduced here to the DAVIC Model.

It is important to understand the historical perspective of network management within broadband distribution networks. There are two fundamentally different communities providing these services. The telcos have developed sophisticated network management systems and an infrastructure to support high availability services, but have traditionally not been a major player in the general distribution of video services. The MSO community developed essentially standalone systems that are predominately reactive. Outages of service, although not desirable, were tolerated. The MSO community does not have a current network management infrastructure., and is largely unwilling to spend large sums of money for complex management features. The MSOs have been the traditional buyers of the broadband video/audio distribution equipment, and consequently the infrastructure of broadband video equipment has very little network management technology.

The challenge is thus to evolve a network management system that meets the

---

[1] DAVIC is a non-profit association based in Geneva, Switzerland. The purpose of DAVIC is to favour the success of emerging digital audio-visual applications and services, in the first instance of broadcast and interactive types, by the timely availability of internationally agreed specifications of open interfaces and protocols that maximise interoperability across countries and application services.

needs of both communities. An evolutionary approach is required that will allow the requirements of both communities to be met. As equipment from different companies must interface, paradigm shifts can be difficult to introduce. It would be unreasonable to expect the telcos to degrade their expectations of network management. Therefore, the MSOs will need to become more sophisticated and network management technology must offer the appropriate cost to benefit ratio.

In the following sections we rely on these considerations to sketch a framework and outline technologies suitable for the management of audio-video-data-distribution networks.

## 5.1 Information Model

TMN (ITU-T, 1992) is based on the $Q_3$-Interface for the Operations System to Network Element communication and for the Operations System to Operations System communication. The NMF sees the communications between the layers of TMN (i.e. business, service, network, and element management) as via the $Q_3$-Interface. The $Q_3$-Interface is based on OSI Systems Management and is not a feasible solution for embedded controllers with only EEPROM as non-volatile memory and with only 1-2 MB RAM. On the other hand, TMN is the only management standard including an object-oriented management information model with provisions for business and service management and with a wealth of standardized management functionality such as state, event, alarm, log, and test management. Thus, a feasible solution for the support of the TMN model is one of the critical requirements.

Use of the TMN management information model is the key to the successful development of a network management architecture for Video/Audio/Data Distribution networks.

While the $Q_3$-Interface may be supportable on the business, service, and network level, in most cases SNMP (Case 1-3, 1996) is the only viable solution for the network element level. Given the excellent work of XoJIDM (JIDM], 1995) which defined translators between SNMP SMI, CORBA IDL, and OSI GDMO, it is feasible and desirable to support the TMN information model by mapping between SNMP SMI, CORBA IDL, and OSI GDMO ($Q_3$-Interface basis). This paves the path for increasingly more sophisticated network management products.

## 5.2 Computing Paradigm

Object Oriented (OO) programming technology and methodology appear to be a significant break-through for software reuse and provide the foundation for the flexible software factory. Using object oriented technology, software parts could be made available for portable operations systems platforms, multimedia graphical interfaces, corporate data bases, and the domain specific code needed for the operations system.

Over the last several years the information technology industry has been rapidly developing object-oriented software-development environments (languages, library tools, databases, etc.) for the realization of object-oriented designs in computer systems. Of those environments OMG CORBA (OMG, 1995), and UNO have emerged as the standard incorporated in numerous other standard frameworks such as NMF OMNIPoint 2, TINA-C, and ISO/ITU-T ODP. The DAVIC 1.0 specification includes OMG UNO. All major computing manufacturers are currently supporting the OMG standards.

## 5.3  Element Management  Technology

Due to the popularity of the Internet technologies and the lack of comparable simple network management standards, SNMP has become the de-facto standard for network management. Almost any type of computing and networking equipment comes today with SNMP agents. SNMP management platforms are more mature and 5-10 times less expensive than comparable CMIP platforms. Embedding CMIP into resource scarce network elements may be prohibitively expensive if feasible at all. That leaves only SNMP as an open, standards based network management solution.

In many cases, the individual elements have only minimal resources for management. That type of environment is best  suited for the SNMP based Master-Agent/Subagent (e.g. AgentX - Daniele 1996, DPI -Winjen 1990, SMUX, EMANATE) for the following reasons:

- resource consumption - SNMP agents are minimalist in nature with most overhead going on ASN.1 encoding/decoding. Subagents usually don't use ASN.1 (e.g. the new IETF AgentX Draft Standard).
- Element synchronization - an SNMP set request is required to either set all variable bindings contained in the request or none. That means that if the set variables are distributed over multiple subagents, a protocol like AgentX would automatically take care of the roll-backs needed in case of errors.

The SNMP MIB design can follow the TMN management information model to allow an element manager to pass information to the next level manager using other protocols such as CORBA or $Q_3$ The use of proxy agents allows the use of a SNMP interface, even before the element design has matured to the next generation.

## 5.4  Data Base Technology

The extensive use of RDBMS engines has evolved from the historical view of MSOs. The requirements for persistent data storage have largely been program schedules, customer lists, authorization status, purchase records and various other tables.  The needs have been largely met by RDBMS systems.

RDBMS technology is stable mature technology. Most products today include support for distributed, replicated, heterogeneous data bases. This allows geographic distribution along with replication of data that allows continued operation in the prescience of network faults.  There are many tools that allow rapidly building client GUIs for the operators.  The standard SQL interface in combination with access libraries from DBMS vendors allows various business systems to interoperate. The use of this approach eliminates the need to develop protocols for the transfer of data from one machine to another, which had previously been a significant developmental effort.

Current OO design efforts are using the RDBMS for persistent storage, largely because of the maturity of the environment. While there are many constraints on storing objects in a RDBMS, the perception has been that OODBMS are not yet mature. As OODBMS technology matures, its use is expected to become more predominant. The size of the installed base of RDBMS ensures that backward compatibility of SQL would be a requirement.

As applications become more complex, the use of standard management applications will become the only feasible choice, as the cost of a custom development will become prohibitive. It is expected that persistent storage would be provided by a OODBMS with a CORBA/IDL interface, the management information model would be TMN based, and applications would be standardized, such as those under development by the NMF.

## 6. MANAGEMENT REQUIREMENTS

The management architecture must be service driven and must be able to support a variety of proprietary systems and protocols. Distributed applications and management systems within the management architecture must include:

- Service Creation and Network Operations (NCS Domain)
- SA Domain Management
- Uplink Domain Management
- Decoder access control
  - Consumer access control
  - Commercial access control
- Virtual Channel Map distribution
- Headend Management
- Distribution Plant Management
- Logical RF Access Network Management

The following subsections give a brief overview of the first five.

### 6.1 Service Creation and Network Operations (NCS Domain)

An operations center responsible for program authoring, service creation, network operations, and frequently also entitlement management. Created services are downloaded to Uplink Control Systems (UCSs) which then program the Encoders. Simultaneously, operations of the various Uplinks are monitored and controlled and are coordinated with the services supported. For that purpose the following distributed databases are maintained:

1. The database related to the programming of services carried by the transport multiplexes; the main components needing access to this database are the

Service Composer, which creates the services; the UCSs, which are instructed by Service Composers to generate the services; and SASs and Entitlement Management Systems (EMSs), which control consumer and commercial access to the services.

2. The database related to commercial Integrated Receiver Decoders and their authorizations; the main component needing access to this database are the Entitlement Management Systems (EMSs).

3. Network and system configuration information and other network and distributed system and application management related information; all components need access to at least the management information reflecting the configuration of the particular system itself.

The tasks of service creation and network operations are distributed in nature over local and wide area networks and need to be supportable by a heterogeneous computing and communications environment. The size of the data is in the order of gigabytes. The data is distributed and replicated and requires frequent updates over potentially unreliable media. The most stable and mature technology capable of satisfying this requirement is RDBMS technology. A variety of architectural scenarios are possible:

- distributed, replicated DBMS only as persistent storage with a corporate wide information model using a variety of distinct database views for the different application types; the actual applications can rely on a variety of different computing and control paradigms;
- distributed, replicated, isolated DBMSs with applications utilizing other computing and control paradigms linking the DBMSs.
- distributed, replicated DBMS with a corporate wide information model and most applications (including network management) utilizing SQL and triggers as the main means of control and asynchronous notifications (do to the lack of maturity of network management technology even the use of RDBMS clients in embedded controllers is considered as a viable monitoring and control application).

## 6.2  SA  Domain Management

The basic management requirements of the SA  Domain are:

- Management of the Decoder Database - this can be done through DBMS applications.
- Management of decoder access control, and decoders; the SAS maintains the database related to consumer Integrated Receiver Decoders and their authorizations; the main components needing access to this database are the SAS and Business Systems.
- Configuration and monitoring of communications links and service providers - that can be accomplished through SNMP.
- Management of RDRs which submit decoder unit transaction reports - this can be done with SNMP.

A feasible architecture for the SA  Domain would include a CORBA based SA

Domain Manager communicating with the RDBMS applications and SNMP Element Managers using UNO.

The information model is based on the TMN information model. This is independent from the individual realization of specific architectural components such as agents, element managers, network managers, and their communications infrastructure. The individual architectural components can be CORBA, SNMP, TMN, or other. All of these support the same information model which is TMN based.

In case of SNMP agents the NMF has specified a translation of GDMO to SNMP SMI. That translation is not regarded as a desirable basis for SNMP MIB design do to the complexity of generated SNMP MIBs. Instead, alternative approaches more in line with SNMP management paradigm are commonly used which still allow an easy mapping on element manager (or $Q_3$ Proxy) level between GDMO and SNMP SMI.

## 6.3 Uplink Management

The four main Uplink Management requirements are:
1. Encoder Programming - this is mainly accomplished by programming the UCS from the Service Composer;
2. Encoder Management - this is mainly accomplished by the UCS;
3. Uplink Network Management - the Uplink may be spread over a metropolitan area and may consist of several 1000 IP nodes, interconnected by a variety of technologies including ATM, FDDI, Ethernet, etc.
4. Source Equipment Management- various different types of source equipment (for example, video servers, video machines, etc.) are used on the Uplink side to generate the programming material. Two types of systems are used to deal with this complexity:
   a) Traffic control systems—include detailed program schedules and control the automation system; interact with Service composers
   b) Automation systems—used to automate the source equipment control; for example, to route source signals from video machines video servers, etc. to the Encoder

### 6.3.1 Encoder Programming

Encoder Programming is highly data intensive, distributed, replicated with different types of data views ranging from abstract views suitable for creating new services without regard to their implementation to concrete allocation and configuration of television service processors. Two candidate technologies are suitable for accomplishing this task.

**Data Base** - The type of data base selected in the architectural framework is a RDBMS. The RDBMS server could be located on the Service Composer with RDBMS clients on the UCS, network management stations, the SAS, etc. The advantages of the RDBMS approach include:

- data distribution - data can be easily distributed over a variety of Uplink Domain elements using mature RDBMS Client/Server technology. This enables the different Uplink Domain elements to have uniform access to the same data base.
- data replication - multiple synchronized copies of data can coexist with off-the-shelf RDBMS technology. This would enable the Service Composer to keep pieces of the programming information on the different UCSs which control the Encoders.
- multiple data views - different Uplink Domain elements need different views of the same data depending on their function.
- security - robust systems with possible C2/B1 level security
- standard interfaces - SQL.
- development cost - use of a mature technology with a large labor pool.

**MIB** - The information model selected for management information in the architectural framework is TMN. A natural solution would be to develop the TMN business and service layers for Encoder Programming using the TMN information model. Possibly concepts of TINA-C service architecture could be used. Assuming a CORBA based TMN information model implementation the advantages include:

- information model integration - the network management and the Encoder programming information model are different pieces of the same picture. The Service Composer could be realized as a CORBA/TMN manager managing UCSs which are CORBA/TMN agents via GIOP where the UCSs would be proxying Encoders.
- monitoring and control paradigm - the basic paradigm here has been created for monitoring and control with special emphasis on the temporal aspects of the processes managed. This is an especially important factor in comparison to Data Base solutions.
- open architecture - the architecture is based on international standards such as TMN and CORBA.
- data distribution, data replication, security, development cost - depend on the CORBA/TMN products selected.

## 6.3.2  Encoder Management and Uplink Network Management

The Encoder is usually realized as a complex extensible architecture with dozens of independent processors needing frequent configuration with large groups of service-specific parameters. Sometimes, Encoders are distributed over local area networks and are connected to UCSs via wide-area links. The Encoder CCAs have usually only minimal resources for management. That type of environment is best suited for SNMP based Master-Agent/Subagent (e.g. AgentX - Daniele 1996, DPI - Winjen 1990, SMUX, EMANATE) for the following reasons:

- resource consumption - SNMP agents are minimalist in nature with most overhead going on ASN.1 encoding/decoding. Subagents usually don't use ASN.1 (e.g. the new IETF AgentX Draft Standard).
- CCA synchronization - an SNMP set request is required to either set all variable bindings contained in the request or none. That means that if the set variables are

distributed over multiple subagents, a protocol like AgentX would automatically take care of the roll-backs needed in case of errors.

- MIB replication - the new Entity MIB (McCloghrie, 1996) enables an agent to have multiple MIB modules of the same type. That enables MIB replication in the CCAs and enables the manager to set the same variable to the same value in two different subagents in an atomic fashion.
- open architecture - SNMP is an IETF standard.
- development cost - off-the-shelf technology.
- management integration - almost any networking equipment in existence is SNMP manageable which would allow an integrated management of all Uplink Equipment.

### 6.3.3 Uplink and Source Equipment Management

Much of the Uplink computing equipment and network equipment such as routers, bridges, hubs, switches, etc. are commonly managed with SNMP. This integrates nicely with the SNMP Encoder management.

Both traffic control and automation systems could be easily managed by SNMP for the same reasons as indicated above.

### 6.4 RDR/Subscriber Management

RDR/Subscriber management is ideally not developed in-house but customized from one of the off-the-shelf products such as Small World which support CORBA and industry's standard management platforms such as OpenView. That would allow either application specific integration or CORBA based integration between the SAS, the Uplink , and the Subscriber Management System.

Subscriber management depends on the type of subscriber access but will likely be SNMP based.

A subscriber management application is illustrated in Figure 4:

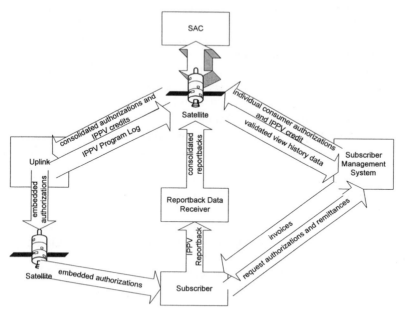

**Figure 4.** Subscriber Management and Reportback Data Receiver Application

The Subscriber Management application assumes the following sequence of actions in the case of the Impulse Pay Per View service:

1. A subscriber requests program authorization from a program packager (program packagers may, or may not, also be programmers).
2. The program packager uses a Subscriber Management System (SMS) to set up an account for the subscriber and transmit the authorization request to the SAS.
3. The SAS adds the new subscriber IRD unit address to its database, turns on the appropriate tiers, and inserts the IRD's authorization into the SAS authorization stream.
4. The Uplink sites feed the authorization stream into the Encoders, which multiplex it with the video programs. The IRD will receive its authorization as long as it is tuned to any Encoder receiving the SAS's authorization stream. It should take less than 5 seconds from the time the SMS sends the authorization request to the SAS to when the IRD receives its authorization.
5. After the subscriber purchases an IPPV program, a reportback is generated and sent to the RDR operating on a small VAX computer. The reportback can be via modem or via history cards. There can be multiple RDRs.
6. The RDRs consolidate and transmit the reportback data to the SAS.
7. The SAS regularly retrieves from each Uplink system the IPPV program log, listing the IPPV programs that have been broadcast; the SAS uses this information to validate the reportback information received from the RDRs and create View History data for the program packagers.
8. The program packagers use their SMSs to retrieve the validated View History data from the SAS.

9. Upon retrieving the View History data, the program packagers extend additional IPPV credit to their subscribers.
10. The packagers can use the SMS billing system or their own billing system to invoice the customer for their subscription and IPPV purchases.
11. The subscriber pays the program packager.

## 6.5 Decoder Access Control

Access control is frequently managed through a tiering structure, coupled with encrypted key delivery. Each Decoder is authorized for a set of services. The Decoder manages these services in the form of a bit vector called the authorization vector. The position of a bit within the vector identifies the service to which the bit corresponds. If a bit in the vector is set to zero, the Decoder is not authorized for the corresponding service.

Usually a hierarchy of secret keys is used whose distribution and maintenance creates a major synchronization issue. At the root of the hierarchy is secret information which is placed into the secure access control device in the Decoder when it is manufactured. Such information is unique to the access control device and lasts for the lifetime of the unit. The hierarchy usually consists of multiple levels.

For an IRD to be able to access a service, the two ends of the hierarchy must be linked by a complete chain. The keys for each level in the chain are generated by the Decoder using information delivered uniquely to it in a control channel message. Secret data delivered in the control channel is protected by encrypting it under a strong cipher such as the Data Encryption Standard (DES).

Access control is likely to remain a proprietary application for some time to come.

## 6.6 Virtual Channel Maps

The transport multiplex carries a large number of video, audio, text, and data service components. The concept of virtual channels offers the user a consistent view of services available, and provides a reference mechanism for various combinations of services (e.g. television with secondary audio). Each virtual channel number defines the access point for a particular service, and represents a particular combination of components. The access point is described in terms of its transport multiplex location and MPEG service number. Transport multiplex location is given as satellite and transponder (for satellite IRDs) or its carrier frequency (for cable).

The consumer set-top box upon acquisition of the transport multiplex obtains virtual channel information through the Virtual Channel Map (VCM). A VCM is a data structure that contains entries for each virtual channel in the map. Each entry has the name of the virtual channel, an identification of the RF channel and multiplex relevant to that bitstream, the service number in the multiplex, and various sub-identifiers. Set-tops usually reference a single VCM but may also reference multiple as is done for multisatellite C band configurations.

VCMs originate at the Service Composer, or at the UCS if there is no Service

Composer. The channel number space between a VCM may be partitioned between several originators, so that each originator is a designated controller of a virtual channel number. There may be different classes of VCMs such as:

- Commercial VCMs—intended for commercial broadcast users

- Consumer VCMs—intended for consumer broadcasts

- Private VCMs—intended for private network users

In addition, there may be local copy modifications to a VCM that are to be distributed only to users of the current multiplex to which they apply.

Virtual Channel Maps largely rely on   simple data distribution and synchronization which can be easily accomplished through RDBMS or CORBA based applications.

## 7.  CONCLUSIONS

An overview of the tasks and possible architectural solutions for management of video-audio-data distribution networks has been given. The  architectural framework sketched is planned for the next generation of  systems. The  two fundamentally different communities providing or planning the services outlined have historically very different approaches to network management. While the telcos have developed sophisticated network management systems and an infrastructure to support high availability services, but have traditionally not been a major player in the general distribution of video services, the MSO community developed essentially standalone systems that are predominately reactive with lower service reliability. That leads to the current status-quo in which the MSO community does not have a current network management infrastructure, and is largely unwilling to spend large sums of money for complex management features.

Considering those restrictions, it is quite difficult to completely embrace complex architectures such as TMN or TINA-C. However, we also concluded that use of the TMN management information model is  key to the successful development of a network management architecture for Video/Audio/Data Distribution networks.

In this paper we have sketched the architectural framework and  architectural requirements of a feasible solution. While the $Q_3$-Interface may be supportable on the business, service, and network level, in most cases SNMP  is the only viable solution for the network element level. Thus, the TMN Information Model has to be supportable through the SNMP information model. Given the excellent work of XoJIDM which defined translators between SNMP SMI, CORBA IDL, and OSI GDMO, it is feasible and desirable to support the TMN information model by mapping between SNMP SMI, CORBA IDL, and OSI GDMO ($Q_3$-Interface basis). This paves the path to sophisticated yet feasible and achievable management solutions.

# 8. REFERENCES

Ahmed Masuma and Vecchi Mario, Common Spectrum Management Interface MIB: Definition of Managed Objects for HFC RF Spectrum Management, twcable.com:/pub/SMA..

Case J., McCloghrie K., Rose M., Waldbusser S., Structure of Management Information for Version 2 of the Simple Network Management Protocol (SNMPv2), RFC 1902, January 1996.

Case J., McCloghrie K., Rose M., Waldbusser S., Protocol Operations for Version 2 of the Simple Network management Protocol (SNMPv2), RFC 1902, January 1996.

Case J., McCloghrie K., Rose M., Waldbusser S., Management Information Base for Version 2 of the Simple Network management Protocol (SNMPv2), RFC 1902, January 1996.

Daniele Mike and Francisco Dale, Agent Extensibility (AgentX) Protocol, Internet-Draft, June 1996.

DAVIC, DAVIC 1.0 Specifications, Description of DAVIC Functionalities, September 1995.

DAVIC, DAVIC 1.0 Specifications, System reference Models and Scenarios, September 1995.

ISO/IEC JTC1/SC29/WG11 IS 13818-1: 4 Information Technology - Generic Coding of Moving Pictures and Associated Audio: Systems, Video, Audio, Conformance, November 1994.

ITU-T, Principles of Telecommunications Management network, ITU-T Recommendation M.3010, 1992.

JIDM, Inter-Domain Management Specifications: Specification Translation, X/Open Preliminary Specification, Draft of April 17, 1995.

McCloghrie Keith and Bierman Andy, Entity MIB, Internet-Draft, May 1996.

OMG, The Common Object Request Broker Architecture Specification, Object Management Group, Revision 2.0, July 1995.

Pugh William and Boyer Gerald, Broadband Access: Comparing Alternatives, IEEE Communications Magazine, August 1995, Vol. 33 No.8.

Tankin Harry, Distilling Optimism from Hype with Practical Network Management, Communications Technology, May 1996.

Weiss Merrill, Issues in Advanced Television Technology, Focal Press, 1996.

Winjen B., Carpenter G., Curran K., Sehgal A., Waters G., Simple Network Management Protocol: Distributed Protocol Interface, Version 2.0, RFC 1592, May 1990.

# 31

# Management of an ATM based Integrated Voice and Data Network - a Pragmatic Solution

Werner Filip
IBM European Networking Center
Vangerowstr. 18
D-69115 Heidelberg
Germany
teleph. +49-6221-59-4371
fax +49-6221-59-3300
wfilip@vnet.ibm.com

Georg Zoerntlein
IBM European Networking Center
Vangerowstr. 18
D-69115 Heidelberg
Germany
teleph. +49-6221-59-4389
fax +49-6221-59-3300
gzoern@vnet.ibm.com

**Abstract**

The realization of management systems for large, heterogeneous telecommunication networks is still an open issue with hard to tackle problems. This paper reports on the architecture, design and implementation of a management for a large scale ATM based integrated voice/data network. The focus has been put on the description of a pragmatic approach to implement a standardized, TMN-based solution. The approach is being called pragmatic since various existing standardized generic building blocks have been employed for the implementation. Additionally, the functionality of already existing element management systems for the various network components has been exploited and integrated in a homogeneous user environment. Thus, it was possible to save a lot of cumbersome implementation work.

**Keywords**

ATM, Voice and Data Network, TMN, VPN, OMNIPoint, CMIP

## INTRODUCTION

New network providers and already established public network operators (PNOs) are starting to exploit advanced technologies, most prominently ATM, in a large scale. The goal is always to provide end-users with new types of services (e.g. video conferencing, enhanced voice services, corporate networks etc.), and to lower transmission costs by optimizing bandwidth utilization. A large scale exploitation of advanced network technology, means, at least currently, that it has to be integrated into an existing network infrastructure or combined with network components that implement an already well established network service like the conventional voice phone. Additionally, a powerful and extendible network management system is needed to fully bring to bear the new technology and to achieve the aspired goals.

The requirements on such a network management system are tremendous, comprising all hard to tackle issues of network management, as there are the most dominant ones:
- appropriate visualization of large networks,
- handling of huge data volumes describing the network,
- heterogenity of the network components with different management interfaces,
- diversity of the services to be managed, and
- complicated technical interrelationships of the various hardware components.

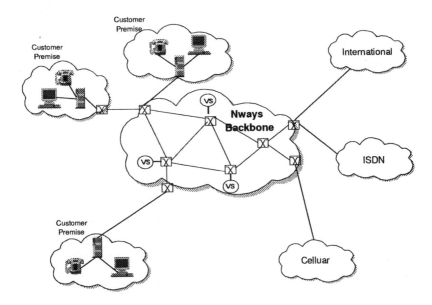

**Figure 1** Integrated Voice/Data Network

In this paper the architecture, design and realization of the network management system for a large scale, integrated voice/data network will be described. Most notably, the focus will be on the description of the pragmatic approach that has been chosen to implement the network management system. The approach is being called pragmatic since various existing generic but standardized building blocks have been exploited for implementing the network management solution. This was the only way a solution for a network of the given complexity could be provided which was cost effective and implementable in a timely fashion.

To make our explanations and reasoning clear a short outline of the network that has to be managed and its components is given. This network is being implemented country-wide by an upcoming German public network provider (PNO).

In Figure 1 the main components of the integrated voice/data network and their composition is illustrated. The basis for transferring data and voice traffic is an ATM backbone network that consists of ATM switches which are connected via E1 and/or E3 trunks. The ATM switches provide various adapters establishing different data transfer interfaces (e.g. X.25, FrameRelay etc.). For providing voice services dedicated voice switches are employed. Those voice switches implement the appropriate means for handling phone calls, maintaing subscriber numbers and do signalling between voice switches.

For both types of switching elements in the network (ATM switches and voice switches) element management systems are available providing convenient end-user interfaces (EUI) to initiate management commands. It was an important goal in the design of the integrated management solution to exploit the existing element management systems appropriately to avoid the reimplementation of already existing management functionality.

The paper has been organized in the following way: In the next section an overview on the requirements of the integrated network management system (INMS) will be given. The third section summarizes the principal architectural concepts on which the implementation of the INMS has been based on. The way, the different management layers have been structured and implemented is presented in the fourth section. The functionality of the INMS applications is sketched in the fifth section. The final section summarizes our experiences and results.

# REQUIREMENTS ON INTEGRATED NETWORK MANAGEMENT

It has been pointed out that for the two component types of the described voice/data network two element management systems already exist. Each management system allows to manage networks built up from a single component type (voice switches or ATM switches) in a convenient way. However, the management of a network that is composed of the two different network element types is not facilitated properly by either of the two management systems. Also, a disjoint use of the both management systems will not provide the proper means to manage a composed network especially a network of the size envisioned by the network provider. Thus, it has been decided to implement an integrated network management system (INM) providing an operator with a single view on the composed network and a uniform operator interface for both types of network elements.

Considering the envisioned size and complexity of the integrated voice and data network essential criteria for the integrated network management system (INMS) have been:

## *Functionality*
- **Integrated Fault Management** - All fault messages initiated from the various equipment types should be represented in a standardized format and represented homogeneously. Thus, it will be possible to interface with various tools and fault management applications in a uniform way.
- **Integrated Network Configuration Management** - The INMS has to provide the functionality to configure network topologies composed of the different network element types. Via an appropriate graphical representation of the topology the invocation of functionality for network element configuration should be uniform and simple.
- **Definition of Views** - The INMS has to provide the capability to define dynamically different views (e.g. voice view, flat view etc.) which allow for different types of network operators to concentrate on specific parts of the network. Since the network will grow over time it has to be possible to adapt the network management organization dynamically.
- **Customer Network Management Services** - The INMS has to provide appropriate means that a customer can retrieve management information for his dedicated net-

work parts (access nodes). Throughout this paper these services will also be synonymously denoted under the term VPN (virtual private network).

### Simple uniform interface

Operators should not deal with different tools for different network management functions, because this leads to high education efforts, acceptance problems, and network management operations. A uniform interface will guarantee that operators have a high productivity and a steep learning curve.

### Scalable architecture

The architecture must allow that the management system scales appropriately with the size of the network and the number of customers on the network.

### Extendability

It must be possible
- to easily integrate functions for managing new resources,
- to introduce new customer services quickly, and
- to facilitate organizational changes appropriately.

# MANAGEMENT OF TELECOMMUNICATIONS NETWORKS

The management and administration of large and complex networks, as the one outlined in the introduction, asks for a well architected system. The standards and architecture of choice for the world market is called the Telecommunications Management Network (TMN) as specified by the ITU-T standards body [1], [2], [7]. This standard provides both an operational and a technical model for network management in the telecommunications environment that is intended to allow for consistent network management of equipment and services regardless of the vendor. The operational model describes the following four layers from lowest to highest:
- **Network Element Management** - These network managers manage specific pieces of equipment such as switches, multiplexors, etc.
- **Network Management** - These managers combine the management of Network Element Managers to allow for the management of a specific network.
- **Service Management** - The Service Managers manage across networks to provide services to the customer, e.g., leased line and telephone. They also can provide a window for their customer into the service provider's company.
- **Business Management** - This is the administration layer of management and takes care of functions such as billing, receivables, etc.

The technical models are in the form of strictly defined connectivity both between the layers and within the layers and object representations or abstractions of what is being managed. This connectivity has been defined though very detailed profiles of international standard protocols that platform providers and equipment manufacturers will be required to implement.

The operational layers talk to each other in a hierarchy. For example, the Network Element Managers have manager applications that manage the Network Elements (e.g., switches). The software in the network element that corresponds with the management applications in the Network Element Manager is called an agent and, therefore, creates an agent manager relationship between the two layers.

The object models are abstractions of both physical (e.g., switches) and logical (e.g., services) entities that will be managed. These abstractions not only describe the attributes of the entity but the way the entity behaves, how it can be monitored, and how it can be controlled. These definitions are being described in a specification language (GDMO) [4] that can be utilized by application programs to implement both management and agent applications.

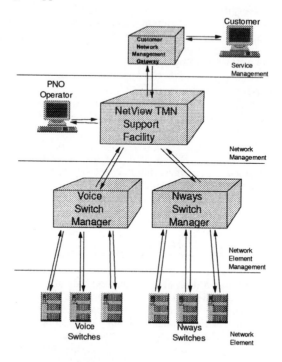

**Figure 2** INM Architecture

Following the standards the architecture depicted in Figure 2 has been chosen as the basis for realizing the INMS. IBM's NetView TMN Support Facility for AIX (TMN/6000) [8], [9], [10] with its special extensions for managing telecommunications networks has been utilized to implement the functions of the different management layers. A detailed description of the various agents that have been implemented will be given in the following chapter.

# IMPLEMENTATION OF THE MANAGEMENT LAYERS

## *Agent Structure*

The implementation concept of the INMS is primarily based on the exploitation of the X.700 Agent/Manager concept. This means that a hierarchy of Agent/Manager components are realized which finally implements the user functionality that is described in the next chapter.

An Agent/Manager component can be characterized in the following way: It acts as a manager (i.e. sending CMIP requests) to the components which are at a lower level in a Agent/Manager hierarchy and behaves as an agent (i.e. answering to CMIP request and sending event reports) to higher layer Agent/Manager components. An Agent/Manager component contains objects which are instantiations of object classes encapsulating data attributes as well as behavioural actions.

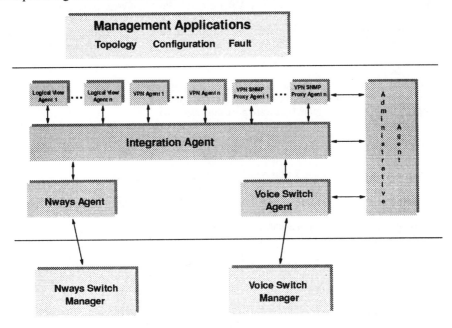

**Figure 3** Agent/Manager Structure

The TMN platform provides an environment and a framework which supports the implementation of Agent/Manager components. Compiler tools are provided that transform a GDMO based managed object (MO) specification to a large degree into executable code that linked with an agent runtime environment executes the described functionality. A specific editor is available which allows to extend the automatically generated code appropriately. Thus, hand crafted code can be integrated conveniently for those implementation aspects that could not be derived automatically from the semiformal GDMO specifciation. The object classes can optionally be generated as persistent code interfacing to an object oriented database system. Based on this environment the Agent/Manager hierarchy depicted in Figure 3 has been implemented. The different Agent/Manager components realize the following functionality:

- **Integration Agent** - The Integration Agent contains objects that model the physical topology of the interconnected voice and ATM switching elements.
- **Logical View Agent** - Logical View Agents represent alternative logical views on top of the physical topology. This allows an operator to focus on specific network parts.
- **Voice Switch Agent** - The Voice Switch Agent realizes a kind of Q-adapter functionality for the voice switch network elements, maintaining all information to allow the integrated use of the element management functionality provided by the voice switch manager (VSM).
- **Nways Agent** - The Nways Agent realizes Q-adapter functionality for the Nways network elements and maintains all information to invoke the functionality of the Nways switch manager (NSM)

- **VPN Agent** - The VPN Agents are special customer specific view agents which focus on the access nodes for a customer and their connections into the PNO network.
- **VPN SNMP Proxy Agent** - A VPN SNMP Proxy Agent provides an SNMP gateway for a VPN agent.
- **Administrative Agent** - The VPN Adminstrivative Agent contains all administrative information (addresses, entry objects etc.) for the various agents in the system.

## Object Definitions

The objects establishing a topological view on the network have been derived from the following classes of the OMNIPoint Management Information Base (MIB) [5],[6]:
- representing links (objects that transport information)
  - **circuit** - a connection between two nodes; it may consist of component circuits.
  - **facility** - refers to the physical means carrying a signal; used to carry circuits.
  - **transport connection** - established and used by two peer connection oriented transport protocol layer entities for transferring data.
- representing nodes (objects that generate/consume information)
  - **opEquipment** - represents physical components of a managed element;
  - **computerSystem** - represents the aggregate of components which as a whole is capable of performing data processing, storage, and retrieval functions.
  - **processingEntity** - represents the physical portion of a computer system;
  - **coTransportProtocolLayerEntity** - represents an instantiation of any connection-oriented transport layer protocol.
  - **clTransportProtocolLayerEntity** - represents an instantiation of any connectionless transport layer protocol.
  - **location** - refers to a place occupied by one or more objects or persons;
- representing collections of links and nodes (supporting hierarchy levels)
  - **opNetwork** - represents collections of interconnected telecommunications and management objects; an opNetwork may be nested within another opNetwork.
- **customer** - refers to a corporation, organization or individual.

# INM MANAGEMENT APPLICATIONS

In the following the functionality of the INM management applications will be characterized as they are available to the network operator. Those applications interact with the CMIP agents described in the previous chapter.

## Integrated Topology

The INM Topology will act as an integrated graphical user interface for operators to perform various network management functionality. Different topologies can be implemented via view agents and used in parallel. The topology display is based on the OMNIPoint object model. The OMNIPoint object model has been chosen, because it already supports the object class "customer", which provides the basis for the introduction of Virtual Private Networks (VPNs). The GUI will
- display different locations of network devices represented by a location symbol;
- show connections between locations and phys ical network nodes displayed as lines;
- display the status of a symbol or a line using different colors (customization is possible).
- allow to explode location symbols (double clicking), displaying their respective functional components which can be access nodes, backbone nodes, or a combination of both. If there is more than one functional unit associated with a location, the connections between those functional units are also displayed.

- show access and backbone nodes represented by different symbols (again color coding is used to indicate status information)
- provide the functionality to execute voice equipment and Nways switch symbols. That means, double-clicking on a symbol could start the appropriate management application for the real resource represented by the symbol (dependent on the element managers' supported functions). Thus, the topological views act as an integration point for the different tools managing the different systems in the network.

In the following a concrete topology as reflected in the GUI will be presented. The topology is implemented by a View Agent which is based on the Integration Agent (see Figure 4 for details).

**The PNO Icon**
An icon, representing the PNO enterprise, will be presented in the sub-map as a result of double-clicking on the OMNIPoint icon (view 0).

**Exploding the PNO Icon**
The operator can double-click on that icon and a sub-map, showing icons representing the major groups of information available through the PNO topology navigation tree, will be presented as shown in view 1. These icons are the following:
- One icon representing an instance of the [PN]:pnAgents class which contains instances representing the various agents.
- One icon representing an instance of the [PN]:pnIntegratedNetwork class containing the PNO Integrated Network Hierarchy.
- One icon representing an instance of the [PN]:pnLogicalView class containing the hierarchies of existing Logical Views.
- One icon representing an instance of the [PN]:pnVpNetwork class containing the PNO customers and its VPN Views.

All these instances are contained in the Administrative Agent (see previous section for more details) .

**Exploding the Integrated Network Icon**
When exploding this icon, a sub-map with icons representing major portions, sub-networks, of the PNO Integrated Network will be presented. Such icons represent instances of the [PN]:pnNetwork class. Any icon that represents such an instance of the [PN]:pnNetwork class can again be exploded into a sub-map showing the next level of details of the PNO Integrated Network as depicted in view 3 which shows a sub-map with the details of the 'FRANKFURT' network shown in view 2. Such submap, with the details of a network as shown in views 2, 3 and 4, can contain icons representing instances of the following classes:
- [PN]:pnNetwork
- [PN]:pnBackboneNode
- [PN]:pnAccessNode

**Exploding Links between Networks and Nodes**
As shown in views 2, 3 and 4, networks and nodes represented by such icons can be connected with a line representing the fact that at least one direct link exists connecting two equipments where one of these equipments belongs to the element represented by one of the icons and the other equipment belongs to the element represented by the other icon. Such lines, will always represent instances of the class [PN]:pnCircuit, contained in the PNO Integration Agent.

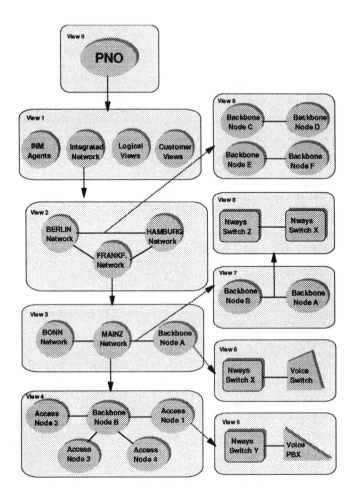

**Figure 4** Network View

**Exploding Links between Networks and Nodes**
Lines representing links between networks or between networks and nodes can be exploded into sub-maps with the actual links connecting two nodes, each one belonging to
one of the termination points of the exploded line, as shown in views 7 and 9. Note that
multiple links may exist and will be presented like in view 9.

**Exploding Links between two Nodes**
Lines representing links between two nodes, as shown in view 7, can be exploded into
sub-maps with the actual links connecting two equipment each one belonging to one of
the termination points of the exploded line. At least one pair of equipment, connected
by a line, will exist and will be presented as in view 8.
        Such lines, will always represent instances of one of the following classes:
* [PN]:pnTrunk
* [PN]:pnVoiceEquipNwaysLink
* [PN]:pnVoiceEquipVoiceEquipLink

**Exploding Nodes of a Network**
Icons representing either a backbone node or an access node can be exploded into a sub-map showing icons representing the actual equipments, like Nways Switches and Voice Equipment. Such sub-maps, with the details of backbone nodes or access nodes as shown in views 5 and 6, can contain icons representing instances of the following classes, contained in the PNO Integration Agent.

**Links between Equipment**
Icons representing Nways Switches and icons representing Voice Equipment can be connected with lines representing that a direct connection exists between them as shown in views 5 and 6. The lines connecting such icons will always represent instances of one of the following classes:
- [PN]:pnTrunk
- [PN]:pnVoiceEquipNwaysLink
- [PN]:pnVoiceEquipVoiceEquipLink

In the integrated topology of the PNO network only one line will be shown connecting two pieces of equipment even if multiple physical links exist between the two systems. It will also not be possible to explode such line into sub-maps showing details of the physical links that it represents.

## *Integrated Configuration Management*
Configuration management of the various network elements will bee provided through the Voice Switch Manager (VSM) application and the Nways Switch Manager (NSM) [11], [12], respectively. In the INM an integrated configuration management application for the network level will be provided. The INM configuration management application will support discovery functions for network elements that have been defined in VSM and NSM, and configuration management functions on the level of the OMNIPoint object model. The principle network configuration functions are listed below. The integrated configuration management will provide the functions to change dynamically the configuration of the network:
- add a new network node  - allows to introduce a new network node.
- add a new connection - introduces new connections between existing network nodes via an appropriate user interface.
- remove a network node - ensures proper removal of an existing network node (e.g. no links/connections exist for nodes being removed).
- remove a connection - guarantees topology consistency in an analogue way as the respective function for removing network nodes.
- view/change node parameters - allows convenient node parameter changes.
- view/change connection parameters - provided to handle connection parameters.
- add a new location - allows to introduce a location into the network.
- remove a location - removes a defined location from the network.
- define a new subnetwork - allows to structure a big network hierarchically.
- existing subnetwork change - allows to restructure a network hierarchy.
- remove a subnetwork - removes a defined subnetwork leaving a consistent topological view of the higher layer network.

## *Integrated Fault Management*
Voice switches report alarms to the Voice Switch Manager application, while the Nways switches report alarms to the Nways Switch Manager. The two management systems provide different interfaces and capabilities for handling alarms. In order to provide an integrated process for handling and tracking alarms a unique way for indicating alarms at one centralized point of control is required. Similarly to the integration

of the different topologies it is intended to integrate the alarm management via a CMIP agent. This allows to utilize the generic Fault Application of TMN/6000 for the integrated centralized fault management. For the VSM the VSM agent will perform the necessary mapping between the VSM message format and the event-report format as described in the standard documents [1], [2], [3].

The integrated Fault Management will provide the following functions:

- display of events/alarms in a textual way (event cards or event lists);
- highlight the network elements in the topology for which an event report has been generated;
- access the TMN browser [10] to retrieve specific information on the network element for which the event report has been generated;
- manage event report workspaces;
- manage Event Forwarding Discriminators (EFDs);

For controlling the event filters in the various agents an operator interface is provided. With this interface new filters can easily be created and maintained.

## *Customer Network Management*

As part of the INM architecture a customer network management gateway has been introduced to allow a customer to perform an integrated end-to-end management of his private network(s) connected via the PNO network. Our concept for realizing the gateway functionality is based on the X.700 agent/manager framework [3] of the TMN/6000 platform, establishing a service management layer. More precisely, via an agent interface (CMIP/SNMP) customer specific agents are introduced which provide a customer with relevant information extracted from the integrated agent on the network management layer employing manager functions. Realizing a separate agent rather than allowing a customer having direct access to parts of the integrated MIB has the following main advantages:

- enhanced security due to the separated agents;
- scalable performance since customer specific agents can be located on one or several machines.

It has to be mentioned that by providing a CMIP or SNMP agent it is the task of the customer to integrate the agent into his management environment and to look for appropriate applications to manage such an agent. However, since the applications for managing the integrated agent at the network management layer are generic, they can also be exploited for managing the customer agents. Thus, an additional service could be provided to a customer.

The MIB defined in the different customer agents will be derived from the integrated MIB at the network management layer. The chosen information model for the integrated MIB allows to define customer specific subnetworks. Those subnetworks will be extracted and stored in the different customer agents. The manager component of the X.700 agent/manager framework will take care of the synchronization between the customer and the integrated agent.

Functions for administrating the various agents at the service management layer will be provided that allows to:

- add a new customer - introduces a new customer into the PNO network;
- remove a customer - deletes a customer from the PNO network;
- configure a customer network - allows to define the customer specific network components (nodes, links, subnetworks) based on the topology;
- change a customer network - allows to change the configuration associated with a customer via the topology display.

As indicated in the overview of the management system architecture, VPNs are realized as customer specific agents. Those agents are derived from the integration agents and cooperate with them. The VPN agents will provide standardized management interfaces upon which standard application can act. To manage VPNs the following functions will be available:

- create a VPN agent - supports the creation of a VPN agent for components of the PNO network. being selected in the topology display;
- list all VPN agents - provides an overview of all created VPN agents;
- view a VPN agent - allows to visualize the configuration of selected VPN agents;
- delete a VPN agent - deletes a generated VPN agent;

## SUMMARY

The architecture, design and realization of a network management system for a large and complex telecommunication network has been presented. The main goal was to provide a well architected solution that fulfills certain requirements with respect to scalability and extendablity and which could still be implemented within a restricted time frame and with a restricted number of resources. With the conceived modular architecture based on standard building blocks and standardized communication interfaces (CMIP) the scalability and the extendability of the whole management system could be assured. The various agents and management applications can be located on different computers or colocated on the same machine. Depending on the size of the network the appropriate hardware configuration for the management system can be chosen. New management applications can be integrated utilizing standardized interfaces. New network elements can be covered extending the object model appropriately.

## REFERENCES

[1]   ITU-T Recommendations M.3000 - Overview of TMN Recommendation - 1994
[2]   ITU-T Recommendations M.3010 - Priciples for a Telecommunications Management Network - 1994
[3]   CCITT Recommendation X.700, Management Framework Definition for Open Systems Interconnection (OSI) for CCITT Applications
[4]   CCITT Recommendations X.722 (1992) / ISO/IEC 10165-4: 1992, „Guidelines for the Definition of Managed Objects"
[5]   NMF: Forum 006, Forum Library - Vol.4: OMNIPoint 1 Definitions, 1992
[6]   NMF: Forum 019, OMNIPoint 1 Object and Ensemble List, 1992
[7]   Veli Salin, „Telecommunications Management Network, Principles, Models, and Applications", Telecommunications Network Management into the 21st Century, 1993, p.72-135
[8]   GC31-8016, IBM TMN Products for AIX: General Information, 1995
[9]   GC31-8008, IBM TMN Products for AIX: Developing TMN Applications, 1995
[10]  SC31-8017, IBM NetView TMN Support Facility for AIX: User's Guide, 1995
[11]  SH11-3072, IBM Nways BroadBand Switch Manager for AIX, Getting Started, 1995
[12]  SH11-3073, IBM Nways BroadBand Switch, Distributing and Updating the Nways Switch Manager and Nways Switch Manager Control Program, 1995

# 32

# Conflict Analysis for Management Policies

*E. Lupu, M. Sloman*
*Imperial College, Department of Computing,*
*180 Queen's Gate, London  SW7 2BZ, U.K.*
*E-mail: e.c.lupu@doc.ic.ac.uk, m.sloman@doc.ic.ac.uk*

## Abstract

Policies are a means of influencing management behaviour within a distributed system, without coding the behaviour into the managers. **Authorisation** policies specify what activities a manager is permitted or forbidden to do to a set of target objects and **obligation** policies specify what activities a manager must or must not do to a set of target objects. Conflicts can arise in the set of policies. For example an obligation policy may define an activity which is forbidden by a negative authorisation policy; there may be two authorisation policies which permit and forbid an activity or two policies permitting the same manager to sign cheques and approve payments may conflict with an external principle of separation of duties. This paper reviews the policy conflicts which may arise in a large-scale distributed system and describes a conflict analysis tool which forms part of a Role Based Management framework. Management policies are specified with regard to domains of objects and conflicts potentially arise when there are overlaps between domains. It is not desirable or possible to prevent overlaps and they do not always result in conflicts. We discuss the various techniques which can be used to determine which conflicts are important and so should be indicated to the user and which potential conflicts should be ignored because of precedence relationships between the policies. This reduces the set of potential conflicts that a user would have to resolve and avoids undesired changes of the policy specification or domain membership.

## Keywords

Distributed systems management, management roles, management policies, conflict detection, conflict resolution, policy precedence.

## 1    INTRODUCTION

There has been considerable interest recently in policy based management for distributed systems (Sloman, 1994; DSOM, 1994; Magee, 1996; Koch, 1996). Separating policies from the managers which interpret them permits the modification of the policies to change the behaviour and strategy of the management system without recoding the managers. The management system can adapt to changing requirements by disabling policies or replacing old policies with new ones without shutting down the system. We are concerned with two types of policies: **authorisation** policies which specify what activities a subject is permitted or forbidden to do to a set of target objects and **obligation** policies which specify what activities a subject must or must not do to a set of target objects. The subject or target of a policy is usually expressed as a domain of objects and applies to all objects in the domain so a single policy can be specified for a group of objects. This helps to cater for large scale systems in that it is not necessary to define separate policies for individual objects in the system, but rather for groups of objects. We permit the specification of both positive and negative authorisation policies and require explicit authorisations i.e. non authorised invocations are forbidden.

In a large distributed system there will be multiple human managers specifying policies which are stored on distributed policy servers. Policy inconsistencies can arise due to omissions, errors or conflicting requirements of the managers specifying the policies. For example an obligation policy may define an activity a manager must perform but there is no authorisation policy to permit the manager to perform the activity. Conflicts can also arise between positive and negative policies applying to the same objects. In general, whenever multiple policies apply to an object there is a potential for some form of conflict but it is essential that multiple policies should apply in order to cover the diversity of management functions and of management domains. For example there may be different policies relating to security, monitoring, or configuration which apply to a set of objects reflecting different management functions which may be performed on the objects. Similarly the policies specified for the network, sub-network and workstation domains will all propagate to the network objects inside the workstation.

In this paper we describe the tools we have developed for analysing policy specifications to determine inconsistencies and conflicts. We use **roles** as the means of grouping policies related to a particular manager position and then managers can be assigned or removed from the position without changing the policies (Lupu, 1997). We also define the relationships between roles with regard to the use of shared resources or with regard to the organisational structure e.g. the manager assigned to role A has the right to assign a task to the manager assigned to role B. A large scale distributed system will have very large numbers of objects and policies distributed around the system, so the conflict detection cannot be centralised but also has to be distributed. Our use of roles and inter-role relationships provides a scope for the conflict detection and helps to limit the number of policies which have to be examined in order to determine conflicts. We assume more specific policies take precedence over less specific ones in order to automatically resolve some conflicts and so reduce the number that human administrators have to resolve. This paper focuses on techniques and tool support for off-line conflict detection and resolution, although some conflicts can be detected only at run time.

In section 2 of the paper we give more details of the domains, policies and roles which form our management framework. Section 3 discusses the types of policy conflicts we need to detect. In section 4 we explain our approach to conflict detection, policy precedence relationship and describe the tools we have implemented.

## 2 MANAGEMENT FRAMEWORK

The main components of our management framework are domains for grouping objects, a policy service to support the specification and storage of policies and roles to reflect the organisational structure, responsibilities and relationships between management positions.

## 2.1 Domains

Domains provide a flexible means of partitioning the objects in a large system according to geographical boundaries, object type, management functionality, responsibility, and authority or for the convenience of human managers (Sloman, 1994a & b). A domain groups the management interfaces of objects and may include other domains (which are called subdomains of the parent domain). An object or subdomain may be a member of multiple parent domains.

The **Domain Browser** is a tool for navigation in the domain structure. In Figure 1 the current domain /Example/Org1/Policies contains two policy objects, has one subdomain (SharedPolicies) and is a member of two parent domains (AllPolicies and Org1).

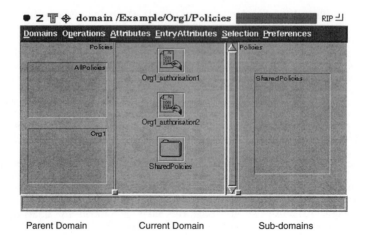

| Parent Domain | Current Domain | Sub-domains |

**Figure 1**   The domain browser.

## 2.2  Policy Service

In this section we give some examples of obligation and authorisation policies and an overview of the notation used.

**Authorisation policies** define what activities a subject (manager or agent) can perform on a set of target objects or what monitored information can be received e.g.

> **A+**  *Sregion_agents {"lu1", "lu2": enable(); disable(); reset(); off()} *Sregion
> **when** (time > 08:00)  &&  (time < 18:00)

Subjects in the Sregion_agents domain are permitted to perform enable, disable, reset or off operations on objects of type lu1 and lu2 (line units) in the Sregion domain, between hours 08:00 and 18:00.

**Obligation policies** define what activities a manager or agent must or must not perform on a set of target objects. Positive obligation policies are triggered by events. Constraints can be specified to limit the applicability of the policy based on time or attributes of the objects to which the policy refers.

> **O+**  **on** overload_event *Sregion_agent {"lu1": disable(), "lu2": enable()} *Sregion
> **when** (08:00 < time)  &&  (time < 18:00)

This positive obligation policy is triggered by an overload event and results in the agent disabling line units of type lu1 and enabling line units of type lu2.

> **O–**  x:*Sregion_agent {"lineunit": enable(); disable(); reset(); off()} *Sregion
> **when** x.state == standby

This negative obligation policy specifies that standby agents must not perform control operations on line unit objects even though they may be authorised to do so.

The general format of a policy is given below with optional attributes within brackets (the braces and semicolon are the main syntactic separators). Some attributes of a policy such as trigger, subject, action, target or constraint may be comments (e.g. */* this is a comment \*/), in which case the policy is considered high-level and not able to be directly interpreted.

identifier mode [trigger] subject '{' action '}' target [constraint] [exception] [parent] [child] [xref] ';'

The **mode** of the policy distinguishes between positive obligations (**O+**), negative obligations (**O-**), positive authorisations (**A+**) and negative authorisations (**A-**). Negative obligations should be read as "obliged not to" and can be considered as 'filters' (Moffett, 1993) to prevent the actions specified in positive obligation policies being performed under certain circumstances, which is why they cannot be triggered.

The **subject** of a policy specifies the human or automated managers and agents to which the policies apply and which interpret obligation policies. The **target** of a policy specifies the objects on which actions are to be performed. Security agents at a target's node interpret authorisation policies and manager agents in the subject domain interpret obligation policies. Both subject and target can be defined using a domain scope expression which identifies a set of objects in terms of union, difference, intersection and membership operators over sets of domains and objects. By default, policies propagate to subdomains within a domain and hence to indirect members of the parent domain, but the scope expression can limit this propagation to direct members. An advantage of specifying policy scope in terms of domains is that objects can be added and removed from domains to which policies apply without having to change the policies. The domain scope expressions are evaluated when detecting potential conflicts to determine the subject and target sets to which the policy applies. The **actions** specify what must be performed for obligations and what is permitted for authorisations. It consists of method invocations or a comment and may list different methods for different object types. Multiple actions can also be specified. The **constraint** limits the applicability of a policy, e.g. to a particular time period, or making it valid after a particular date. An **exception** mechanism is provided for positive obligations to permit the specification of alternative actions to cater for failures which may arise in any distributed system.

High level abstract policies can be refined into implementable policies. In order to record this hierarchy, policies automatically contain **references** to their parent and children policies. In addition, a manual cross reference list of policies may be kept e.g. to refer to the authorisation policy granting permission for an obligation policy's activity.

The policy service provides tool support for defining policies and disseminating polices to the relevant agents which will interpret them. It also permits policies to be enabled, disabled or removed from the agents (Marriott, 1996a & b). Policies are implemented as objects which can be members of domains (see Figure 1) so that authorisation policies can be used to control access to the policies stored in a policy server, e.g. to permit only authorised managers to define and modify policies.

## 2.3 Roles

Specifying organisational policies for human managers in terms of a **manager position** rather than the person permits the assignment of a new person to the manager position without respecifying the policies referring to the duties and authorisations of that position. The tasks and responsibilities corresponding to the position are grouped into a **role** associated with the position (which is essentially a static concept in the organisation). These definitions correspond to the concepts of classic Role Theory which postulates that individuals occupy positions inside an organisation and associated with the position are a set of activities (including the required interactions) that constitute the role of that position (Biddle, 1979).

Manager positions can be represented as domains and we can consider a **role** to be a set of management policies relating to a particular subject i.e. the Manager Position Domain (Sloman 1994a). A manager may be assigned to a role by including his **User Representation Domain (URD)** in the manager position domain, and the policies of the role will propagate to the URD and objects contained in it. The URD is a persistent representation of the manager in the system. The problem with this approach is that when a manager is assigned to multiple roles all the policies propagate to the URD, so the roles cannot be distinguished and the manager could perform a task in one role with the rights from another. An alternative way of assigning a manager to a role is by specifying a policy authorising him to create an agent in the manager position domain. The manager assigned to more than one role interacts with an adapter object in his URD which forwards the invocations to the relevant agent in the position domains. The adapter would provide a separate context for each role and thus permit the manager to keep activities pertaining to each role distinct. In this respect the adapter is similar to an X server maintaining different windows with different active shells Figure 2. Note that although logically placed in the position domains the agents may be implemented as threads of the adapter object to improve performance.

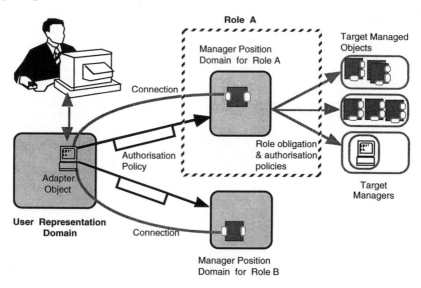

**Figure 2**   Management Roles.

There is a need for interactions between roles (e.g. delegating a task from one role to another or coordinating access to shared objects). Our management framework therefore caters for specification of role relationships using policies, interaction protocols and concurrency constraints. Figure 3 represents the extended role model as presented in (Lupu, 1997). The policies within a role or between related roles provide a scope within which to search for conflicts.

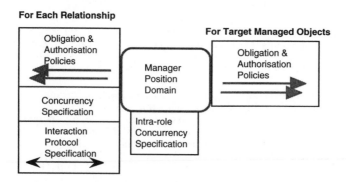

**Figure 3**   The extended role model.

# 3    CONFLICT CLASSIFICATION

**Modality conflicts** are inconsistencies in the policy specification which may arise when two or more policies with modalities of opposite sign refer to the same subjects, actions and targets. This occurs when there is a triple overlap between the sets of subjects, targets and actions as shown in Figure 4, and so can be determined by syntactic analysis of polices. There are three types of modality conflicts:

- **O+/O-** the subjects are both required and required not to perform the same actions on the target objects.
- **A+/A-** the subjects are both authorised and forbidden to perform the actions on the target objects.
- **O+/A-** the subjects are required but forbidden to perform the actions on the target objects (obligation does not imply authorisation in our case).

A second type of conflict refers to the consistency between what is contained in the policies i.e. which subjects, targets and actions are involved and external criteria such as limited resources or the overall policies of the organisation. An example of this type of conflicts arises from the principle of separation of duties (Clark, 1987) e.g. the same manager cannot authorise payments and sign the payment cheques. These conflicts are **application specific** and cannot be determined directly from the policy specifications – additional information is needed to specify the conditions which result in conflict. These can be specified as a **meta-policy** i.e. a policy about permitted policies. Several types of application specific conflicts such as: conflict of priorities for resources, conflict of duties, conflict of interests, multiple managers conflict and self-management conflict have been identified in (Moffett, 1994) and classified according to the overlaps between the subject, action and target sets.

Modality conflicts arise from overlapping domains but it is impractical to prevent these overlaps (see 4.1 a) as there is a need for multiple policies to apply to a domain to reflect partitioned responsibility and the various different management functions that can be performed on target objects e.g. different managers may be responsible for maintenance and security relating to a domain of workstations. In the following, we discuss the precedence relationships which can help to resolve modality conflicts then describe our approach to specifying meta-policies to detect application specific conflicts.

# 4    CONFLICT DETECTION

Conflict detection between management policies can be performed statically for a set of policies in a policy server or at run time. A run-time mechanism acts as a filter preventing activities that must not be performed (O-) or are not permitted (A-) (Moffett, 1993). The advantage is that all the constraints of the policies can be evaluated at run time and so all conflicts can be detected, but some conflicts may really be specification errors and should rather be detected by static analysis c.f. compile time vs. run-time error detection for programming languages. The disadvantages of static analysis are that policy constraints cannot be evaluated, as they depend on run-time state, and domain membership may change, so only potential rather than actual conflicts can be detected. Both static and run-time conflict detection are needed, but this paper concentrates on a static conflict detection tool which assists the users specifying policies, roles and relationships. In the following we discuss some principles for the detection of the modality conflicts and present an implementation of the conflict detection tool.

## 4.1  Policy Precedence Relationships

As previously mentioned, modality conflicts result from a triple overlap between the subjects, actions and targets of the policies. In a typical organisation there will be some general policies pertaining to all staff in the organisation as well as more specific policies relating to staff in a department or section. Staff may also be members of many different domains. Detecting the triple overlaps between policies with modalities of opposite signs would therefore detect many potential conflicts which do not result in actual conflicts. Using a policy precedence relationship can substantially reduce the number of conflicts indicated to the user and permit apparently inconsistent specifications. There are several principles, outlined below, for establishing this precedence. The choice between them has to be guided by which conflicts should be ignored and how easy it is for the human user to understand the decisions and selection of the conflict detection tool using this principle i.e. how intuitive the principle is.

## a)   Negative policies always have priority

It is quite common for negative authorisation policies to always override positive ones so that  a forbidden action will never be permitted. Consider the following policies:

*/\* All users are forbidden to access the system files \*/*

**P1**      A-  @/users { reboot() } @/workstations

*/\* The system administrators are authorised to reboot the workstations \*/*

**P2**      A+  @/users/sys_admin { reboot() } @/workstations

Policy P1, being negative has priority over P2 so  the system administrators are denied access to the system files, but then they cannot perform their function. To resolve this conflict it is necessary either to rewrite policy P1 or to exclude the system administrators from the /users domain. Although our access control system assumes a negative default authorisation policy and so we would not specify P1 but only P2. Database security systems often implement negative authorisation so we permit it at the specification level.

## b)   Assigning explicit priorities

A user can assign explicit priority values to policies to define a precedence ordering, but meaningful priorities are notoriously difficult for users to assign and may result in arbitrary priorities which do not really relate to the importance of the policies. Inconsistent priorities could easily arise in a distributed system with several people responsible for specifying policies and assigning priorities.

## c)   Distance between a policy and the managed objects

The concept of calculating the *distance*  between a rule (policy) and the objects it refers to has been introduced in (Larrondo-Petrie, 1990) for authorisation policies in an object oriented database. Priority is given to the policy applying to the *closer* class in the inheritance hierarchy when evaluating access to an object referenced in a query. For example the access policy applying to a foreign student is the one applying to a student and overrides the general access policy applying to a person if foreign student is a subclass of student which is a subclass of person. The distance between the policy and the objects to which it applies indicates the relevance of the policy and can be precisely evaluated from the number of levels of refinement of the organisational policies. In the general there is a compromise between the complexity and the intuitiveness of the distance to be evaluated. A distance which is intuitive may not correctly evaluate the importance of a policy in all the cases and a complex calculated distance may not be intuitive enough for the human user to understand the selection and priorities assigned to a policy during the conflict detection process e.g. the priority could be based on the product (refinement level) * (last modification date).

## d)   Specificity related to domain nesting

The principle here is that a more specific policy i.e. a policy applying to a subdomain refers to fewer objects so overrides more general policies applying to an ancestor domain. This concept has been introduced in Miró (Heydon 90) and is a particular case of the previous concept of distance. Considering the specificity of a policy with regards to the objects it applies to is an intuitive concept in a domain based system. A subdomain of objects is created for a specific management purpose – to specify a policy that differs from those applying to the objects in the parent domain. For example the system administrators are a particular group of users which have access to the system files despite the general policy denying access to all users of the system. The other policies applying to all the users apply then to the system administrators in the same way. Precedence based on domain nesting can thus be used to allow conflicting specifications by automatically resolving some conflicts.

In section 4.2 we describe how domain nesting can be used within conflict detection to reduce the number of potential conflicts. We recognise that this principle does not apply successfully to all the situations i.e. there are cases in which it is desirable that a global policy overrides more specific ones. For this purpose the conflict detection can be performed with precedence relationships optionally disabled. The following two sections examine the importance of the overlaps between domains while applying the domain nesting principle and indicates the cases where inconsistencies still remain.

## 4.2 The importance of overlaps in modality conflicts

The analysis for conflicts of a set of policies enumerates all the subject, action target tuples which have a different set of policies applying to them. This makes it easier to determine where a conflict occurs and where precedence resolves a potential conflict.

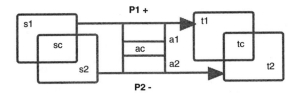

**Figure 4** Overlapping Subjects, Targets and Actions.

Consider the policies **P1** and **P2** represented in Figure 4 with P1 being positive and P2 being negative. Let us call the overlapping areas $s_c$, $a_c$ and $t_c$ for common subjects, actions and targets. The triple overlap between the policies P1 and P2 creates three tuples to which different sets of policies apply:

- P1 applies to $<s_1-s_c, a_1-a_c, t_1-t_c>$

- P2 applies to $<s_2-s_c, a_2-a_c, t_2-t_c>$

- P1 and P2 apply to $<s_c, a_c, t_c>$

Neither P1 nor P2 is more specific so a conflict is indicated to the user as their modalities are of opposite sign. Now consider a policy P3 (shown in Figure 5) defined by the tuple $<s_3, a_3, t_3>$ such that $s_3=s_c$, $a_3=a_c$ and $t_c$ is a subset of $t_3$ which is a subdomain of $t_2$.

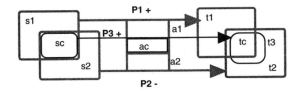

**Figure 5** Adding a policy.

We now have the following tuples and policies:

- P1 applies to <s1-sc, a1-ac, t1-tc>
- P2 applies to  <s$_2$-s$_c$, a$_2$-a$_c$, t$_2$-t$_3$>
- P1, P2 and P3  apply to <s$_c$, a$_c$, t$_c$>
- P2 and P3 apply to <sc, ac, t3-tc>

P3 is positive and is more specific than P2 so it overrides P2 in the areas where they overlap i.e. for the tuple  <s$_c$, a$_c$, t3-t$_c$> and  <s$_c$, a$_c$, t$_c$>. Since P1 and P3 have the same modality, no conflicts would be indicated.

Note that when displaying the result of a conflict detection check it is important to provide the user with the information regarding which policies conflict, where precedence overrides conflicts and to which tuples <subjects, actions, targets> these policies apply.

## 4.3  Limitations of domain nesting based precedence

The domain nesting precedence determines all policies which apply to a tuple of subjects actions and targets and gives precedence to policies which apply to a more specific set of subjects, targets or both. There are cases in which precedence cannot be established because the sets are equal, the subject sets are more specific but the target sets are less

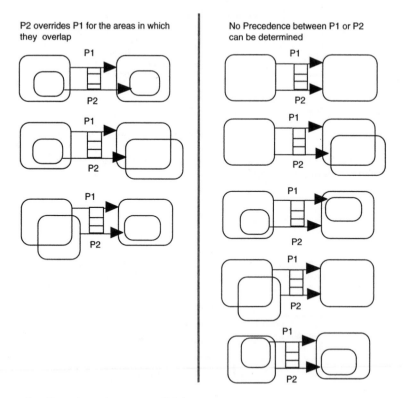

**Figure 6**    Precedence between policies.

specific or vice versa. The various situations where precedence can or cannot be established between two policies are shown in Figure 6. Note that that precedence may based on a policy's subject *or* target set so it is not an ordering relation because it is not transitive. There is no precedence relationships between obligations and authorisations since an obligation overriding an authorisation would convey the implicit assumption that the obligation implies authorisation and this is not true for our policies.

## 4.4 A conflict detection tool

### Implementation issues

The conflict detection tool detects overlaps between policies and applies domain nesting based precedence. The domains and policies are distributed among several servers so Corba invocations are used for retrieving the policies and querying domains to evaluate their sets of subjects, actions, and targets. In theory, all policies in the system need to be checked for overlaps but this is impractical. Instead, we permit the user to specify the scope of policies to be checked, for example the policies applying to particular roles or the policies of a relationship between roles. Policies or domains of policies can be dragged from the domain browser window (Figure 1) into the conflict detection window to establish the set of policies over which the conflicts are to be detected. The meta policies discussed in section 4.5 can also explicitly define the scope to which they apply.

Since there are cases in which a more specific policy should not take precedence, domain nesting precedence can be optionally disabled so all the overlaps are indicated. Finally an analysis option also permits all the tuples of subjects, actions and targets and the policies applying to them to be displayed even if there are no conflicts as it is useful to examine which policies apply to which tuples. When enabled, the precedence relationship between policies is indicated by arrows between the policy icons as shown in Figure 7.

### Example

Policies are implemented as objects in the system. This example shows the use of the conflict detection tool while specifying policies for managing other policy objects. Consider the case of three organisation domains Org1, Org2, Org3 sharing the same computer system. Each organisation has two managers Oi_m1 and Oi_m2 {i ∈ 1..3} and a set of policies which includes a domain of shared policies about the general use of the computer system (see Figure 1). Each of the organisations wants to retain control of all its policies including the shared policies. In particular each organisation has two policies stating that its managers can perform various operations on all its policies and that the managers from the other organisations are prohibited from performing the operations retract(), disable() or delete() on any of the objects contained in the organisation's policy domain. These policies have the following format (only shown for Org1):

Org1_authorisation1 **A+** @/Org1/Managers {create(); delete(); distribute(); enable();
disable(); retract()} @/Org1/Policies

Org1_authorisation2 **A-** @/Org2/Managers + @/Org3/Managers {delete(); disable();
retract()} @/Org1/Policies

Note the '@' symbol selects all non domain objects in nested domains. (With a default negative authorisation, the second policy could actually be revised to only permit create and enable to give the same effect but we will ignore that for the purposes of this example.) The managers of one organisation are subjects of three policies: one authorising all the operations on the objects of the Policies domain in their organisation and two others prohibiting some operations on the policy objects in the policy domain of the other two organisations. A potential conflict arises from the presence of the shared Policies subdomain in each organisation. Since the positive authorisation policy is more specific than the two others (it relates to the managers of Org *i* while the others relate to the managers of Org *x* + Org *y*) no conflicts are detected because the positive authorisation policies override the negative ones, as shown in the conflict detection window of Figure 7.

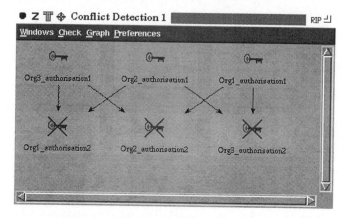

**Figure 7**    The conflict detection window. Shows positive authorisation (keys) overriding negative authorisation (crossed out keys).

If the check for conflicts is performed without the domain nesting based precedence, conflicts such as the one shown in Figure 8a are detected. The subjects, actions, targets tuple is shown in the upper part and the conflicting policies to which no precedence applies are shown in the lower part of the screen. Similarly remaining conflicts can be displayed after applying domain nesting precedence, although there are none in this example. Policy icons can be dragged from this window onto a policy editor window for viewing and revising if required.

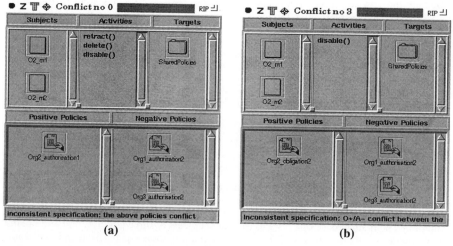

**Figure 8**    Examples of detected conflicts.

Consider a policy refined from a more abstract obligation policy specifying that the managers from **Org2** must (modality **O+**) disable the policies on the failure of the policy server.

Org2_obligation1 **O+ on** maps_failure @/Org2/Managers { disable() } @/Org2/Policies

If domain nesting precedence is used, the access should be granted since the positive authorisation takes precedence over the negative ones. With precedence disabled, the **O+/A-** conflict is also detected as shown in Figure 8b.

## 4.5 Meta-Policies

Meta-policies specify application specific consistency constraints pertaining to the contents of policies. Meta-policies constrain the set of acceptable policies in terms of their attributes. They can be expressed as logical predicates applying to the sets of policy objects determined by a domain scope expression (dse). For example a conflict for resources may arise when the number of objects in the target domain of any two policies is greater than 11.

$$\forall P1, P2 \in < dse >$$

$$fail \leftarrow P1.t\,arg\,ets + P2.t\,arg\,ets > 11$$

The solution of the following Prolog code gives all the conflicting policies.

```
checkResourcesNumber(P, Q, Res) :-
    numberOfTargets(P, N1), numberOfTargets(Q, N2), Res = N1 + N2.
checkRes(P, T) :- checkResourcesNumber(P, T, Res), P \= T, Res > 11.
check1(Bag) :- findall([P, T], checkRes(P, T), Bag).
```

We have been experimenting with meta-policies by implementing the predicate specification in Prolog for the cases presented in (Moffett, 1994). The policies contained in the Conflict Detection window are automatically translated into Prolog assertions. A Prolog process can then be started from the conflict detection tool loading the file containing the translated policies. The predicate specifying the conflict is then a query on the assertions database which gives the policies in conflict.

## 5 RELATED WORK

Our concept of domain nesting precedence is based on that of Miró (Heydon, 1990), but they only deal with authorisation policy for file system security. Sandhu (1996) presents constraints which are similar to our meta-policies, but the notation used is not described. The work presented in (Michael, 1993) relates to general policies, expressed in natural language and modelled in an Entity Relationship representation. A theorem prover is used to detect the inconsistencies. The "law governed systems" of (Minsky, 1996) implements a common global set of constraints by means of filters in every node which check that all interactions are consistent with the global law.

Another approach, used to detect feature interaction in telecommunication systems (Griffeth, 1993), considers policies as goals and applies planning techniques to resolve situations with incompatible goals. Planning techniques for conflict management are also used in Distributed Artificial Intelligence (Lander, 1994). In the case of our management policies such techniques could be used only in conjunction with the refinement of the policies. Koch (1996) uses a policy notation based on ours and establishes a semantic graph model to detect ill-behaved policy sets with unsatisfiable pre-conditions. This can also be used to perform "what-if" analysis on chains of policies prior to execution.

Deontic Logic provides the closest approximation of our management policies in the context of a logic system. A model of policies as deontic logic statements for office automation can be found in (Ong, 1993). However Standard Deontic Logic also relies on the axiom of inter-definability which defines a permission as **P = ¬ O¬ P**. No such assumption is made between our authorisation and obligation policies. However a number of new logical systems with slightly different axioms are emerging and this may be of interest for our policies.

# 6    CONCLUSIONS

This paper has presented the integration of a conflict detection tool in a more general role and policy based framework for distributed systems management. We perform off-line, static analysis of a set to policies to determine two types of conflicts: (i) modality conflicts which can be checked by analysing the syntax of the policies and (ii) application specific conflicts with external constraints which we express as meta policies. Modality conflicts arise from a triple overlap between the subjects, actions and targets of the policies, but it is not practical nor desirable to prevent these overlaps. We make use of a precedence relationship based on the specificity of the policies with respect to domain nesting to reduce the number of potential overlaps indicated to a user, as we consider this to be an effective and intuitive precedence relationship. Roles are an important management concept but also provide a scope to limit the set of policies to be analysed.

Another aspect of policy analysis relates to determining the policies applying to a particular subject or target. Our policies explicitly identify both subject and target and the domain service maintains the list of policies applying to a domain so this is comparatively easy to do, but has not yet been implemented.

We have implemented a prototype role framework which supports distributed policy and domain servers and analysis of a set of policies, indicating conflicts as well as precedence relationships. This will enable us to experiment in realistic situations and evaluate the use of the precedence relationship. Our approach is to detect as many conflicts as possible at specification time, rather than leaving them to be detected at runtime. The user can then modify the policies to remove conflicts. This has been implemented using a Corba based distributed programming environment.

Further work remains to be done on the use of dynamic run-time conflict detection within policy interpreters and what to do about conflicts which have been detected. Our meta policy specification language also need further refinement, as translating all policy specifications into Prolog assertions is a rather "heavy handed" approach.

# 7    ACKNOWLEDGEMENTS

We gratefully acknowledge financial support for the EPSRC RoleMan project (GR/K 37512) and British Telecom for the Management of Multimedia Networks project. We are grateful to Jonathan Moffett for many invaluable comments which have improved this article. We acknowledge the contribution of our colleagues to the concepts described in this paper – in particular Nicholas Yialelis and Damian Marriott.

# 8    REFERENCES

Biddle, B. and Thomas, E. Eds. (1979) *Role Theory: Concepts and Research*. New York, Robert E. Krieger Publishing Company.

Clark, D. and Wilson, D. (1987) A comparison of Commercial and Military computer security Policies. *IEEE Symposium on Security and Privacy*.

DSOM (1994) Proceedings of the IEEE/IFIP Distributed Systems Operations and Management Workshop, Toulouse (France).

Griffeth, N. and Velthuijsen, H. (1993) Reasoning about goals to resolve conflicts. *Int. Conf. on Intelligent Cooperative Information Systems*, Los Alamitos (Calif.), IEEE Computer Society Press, 197–204

Heydon, A. et al. (1990) Miró: Visual Specification of Security. *IEEE Transactions on Software Engineering*, **16**(10), 1185-1197.

Koch, T. et al. (1996). Policy Definition Language for Automated Management of Distributed System. *IEEE 2nd. Int. Workshop on Systems Management*, Toronto (Canada).

Lander, S. E. (1994). Distributed Search and Conflict Management Among Reusable Heterogeneous Agents. Ph.D. Dissertation, University of Massachusetts, Amherst, (USA).

Larrondo-Petrie, M. et al. (1990) Security Policies in Object-Oriented Databases. *IFIP Database Security, III: Status and Prospects,* Elsevier Science Publishers B.V. (North-Holland).

Lupu, E. and Sloman, M. (1997) Towards a Role Based Framework for Distributed Systems Management. *Journal of Network and Systems Management,* **5**(1) Plenum Press.

Magee J. and Moffett J. eds. (1996) Special Issue of *IEE/BCS/IOP Distributed Systems Engineering Journal* on Services for Managing Distributed Systems, **3**(2).

Marriott, D. and Sloman M. (1996a). Management Policy Service for Distributed Systems. *Proc. IEEE Third International Workshop on Services in Distributed and Networked Environments (SDNE 96)*, Macau, 2–9.

Marriott, D. and Sloman M. (1996b) Implementation of a Management Agent for Interpreting Obligation Policy. *IEEE/IFIP Distributed Systems Operations and Management (DSOM 96)*, L'Aquila (Italy).

Michael, J. (1993) A Formal Process for Testing Consistency of Composed Security Policies. Ph.D. Dissertation, George Mason University, Fairfax, Virginia.

Minsky, N. H. et al. (1996) Building Reconfiguration Primitives into the Law of a System. *IEEE Third International Conference on Configurable Distributed Systems (ICCDS 96)*, Annapolis (Maryland), 89–97.

Moffett, J. et al. (1993) The Policy Obstacle Course: A Framework for Policies Embedded within Distributed Computer Systems. Technical Report, Schema/York/93/1, Department of Computer Science, University of York (UK).

Moffett, J. and Sloman M. (1994) Policy Conflict Analysis in Distributed System Management. *Ablex Publishing Journal of Organisational Computing*, **4**(1), 1–22.

Ong, K. L. and Lee, R. M. (1993). A Logic Model for Maintaining Consistency of Bureaucratic Policies. *26th Annual Hawaii International Conference on System Sciences,* Hawaii, IEEE Computer Society Press. Vol. III, 503–512

Sandhu, R. S. et al. (1996) Role-Based Access Control Models. *IEEE Computer,* **29**(2), 38–47.

Sloman, M. (1994a). Policy Driven Management for Distributed Systems. Plenum Press *Journal of Network and Systems Management*, **2**(4), 333–360.

Sloman, M. and Twidle, K. (1994b). Domains: A Framework for Structuring Management Policy. In *Network and Distributed Systems Management.* Sloman M. ed., Addison Wesley, 433–453.

## PART FIVE

# Testing the Management Information Base

Chair: Shoichiro Nakai, NEC

# 33
# Implementation and evaluation of MIB tester for OSI management

*Keizo SUGIYAMA, Hiroki HORIUCHI, Sadao OBANA and*
*Kenji SUZUKI*
*KDD R&D Laboratories*
*2-1-15 OHARA KAMIFUKUOKA-SHI, SAITAMA 356, JAPAN*
*Tel: +81 492 78 7328*
*Fax: +81 492 78 7510*
*e-mail: sugiyama@csg.lab.kdd.co.jp*

## Abstract

This paper proposes a conformance test method for Management Information Base (MIB) and discusses an implementation and an evaluation of a MIB tester. The proposed test method is based on and an extension of the existing conformance test method for protocol entity. We show the practical solution to the scope of testing by conducting the capability test to all test cases and the behaviour test limited to the test cases actually used. In the implementation, the MIB tester generates test scenarios for capability tests automatically. We evaluate the proposed test method and demonstrate the effectiveness of the MIB tester through its application to the actual agents.

## Keywords

OSI management, conformance test, MIB, GDMO

## 1   INTRODUCTION

In line with the progress of standardisation on Telecommunications Management Network (TMN) [M3010, 1992], it is indispensable for the development of network management systems to conduct a test of OSI management.

In OSI management, a manager accesses to Management Information Base (MIB) in an agent, which is a collection of managed objects, by using Common Management Information Protocol (CMIP). The management information and the behaviour are defined by Guidelines for the Definition of Managed Objects (GDMO) templates. It is necessary for the test of OSI management to verify the behaviour of managed objects, as well as the behaviour of CMIP protocol entities. Although a conformance test method for protocol entity has been established [X290, 1992] [Serre, 1993], that for MIB has not been established so far [Ödling, 1994]. The existing conformance test method for protocol entity, however, can not be applied to that for MIB just as it is since there are differences in modelling between protocol entities and managed objects. Development of the conformance test method for MIB is strongly expected.

This paper proposes the conformance test method for MIB and describes an implementation and an evaluation of a MIB tester. First, we discuss the applicability of the existing conformance test method for protocol entity to that for MIB. We propose the conformance test method for MIB by taking account of the differences in modelling between protocol entities and managed objects. We describe the MIB tester based on the proposed test method. Finally we evaluate the proposed test method and the effectiveness of the MIB tester through the application to tests of actual agents.

## 2    OVERVIEW OF OSI MANAGEMENT AND EXISTING CONFORMANCE TEST METHOD FOR PROTOCOL ENTITY

### 2.1 OSI management

There are two roles in OSI management systems. One is a manager which has a responsibility for managing and the other is an agent which is managed by a manager. Management operations to an agent are conveyed by CMIP. It defines Protocol Data Units (PDUs) corresponding to services of Common Management Information Service (CMIS). The management information and the behaviour of a managed object are defined by nine kinds of GDMO templates: MANAGED OBJECT CLASS, PACKAGE, ATTRIBUTE, ATTRIBUTE GROUP, BEHAVIOUR, ACTION, NOTIFICATION, NAME BINDING and PARAMETER templates. These templates may refer to Abstract Syntax Notation 1 (ASN.1) modules which specify syntax of management information over CMIP.

### 2.2   Existing conformance test method for protocol entity

The existing conformance test method for protocol entity [X290, 1992] examines whether or not an implementation by vendors, so called an Implementation Under Test (IUT), conforms to the relevant protocol specification. The test is conducted by connecting with a test system. The test suite, a set of test cases, is developed from the protocol specification and the Protocol Implementation Conformance Statement (PICS) which gives information about implemented capabilities.

The test suite consists of basic interconnection tests, capability tests and behaviour tests, with the different test objective. Basic interconnection tests verify the basic capability of interconnection. Capability tests and behaviour tests verify the static and the dynamic conformance respectively.

There are two kinds of test components: the Lower Tester (LT) and the Upper Tester (UT), which control and observe the lower and upper service boundaries of the IUT respectively. The Test Coordination Procedure (TCP) is the rules for cooperation between the LT and the UT during testing. The Point of Control and Observation (PCO) is a point where the occurrence of test events is to be controlled and observed. The abstract service primitives (ASPs) are exchanged via PCOs. Four test configuration patterns exist depending on the location of the IUT and the LT, the number of PCOs and the existence of PDUs for controlling tests (i.e. TM-PDUs), as shown in Figure 1.

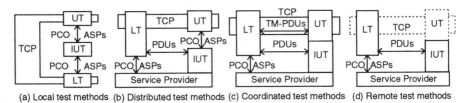

(a) Local test methods  (b) Distributed test methods  (c) Coordinated test methods  (d) Remote test methods

**Figure 1** Four test configuration patterns of conformance test for protocol entity.

## 3 PROPOSAL ON CONFORMANCE TEST METHOD FOR MIB IN OSI MANAGEMENT

### 3.1 Differences in modelling between protocol entities and managed objects

Table 1 shows the differences in modelling between protocol entities and managed objects. The behaviour of a protocol entity, modelled as a finite state machine, is determined by a current state and an input event from a peer entity. On the other hand, the behaviour of a managed object is determined not only by a set of current attribute values and a CMIP PDU from a manager but also input events from actual resources and other managed objects in MIB. The last two types of input events may result in the modification of attribute values and the emission of voluntary notifications without interacting with a manager.

Table 1     Differences in modelling between protocol entities and managed objects

|  | *Protocol entity* | *Managed object* |
|---|---|---|
| Model | ● Finite state machine | ● Abstraction of actual resource by object-oriented approach |
| Spec. description | ● State transition table/diagram<br>● Formal Description Technique (FDT) | ● GDMO definition<br>(The behaviour is described by natural language) |
| Character-ised by | ● Input events<br>● Set of states<br>● State transition<br>● Output events | ● Attributes<br>● Permitted operations<br>● Emittable notifications |
| Interaction with outside world | ● Primitives with upper and lower layers<br>● PDUs with a peer entity included in primitives | ● CMIP PDUs with a manager<br>● Actual resources<br>● Other managed objects in MIB |

## 3.2   Applicability of existing conformance test method for protocol entity to that for MIB

### 3.2.1   Test configuration

Protocol entities provide service boundaries to exchange primitives with upper and lower layers, while managed objects do not provide upper service boundary from a modelling point of view. Therefore, the local test methods and the distributed test methods, in which a PCO exists between a UT and an IUT, are not suited as a test configuration of conformance test method for MIB. Although the coordinated test methods need a UT, there are few cases that it is possible to implement an UT in MIB additionally. The remaining remote test method, which needs only a LT and an IUT, is suited as the pattern of test configuration.

### 3.2.2   Test purpose

*Basic interconnection test*
The purpose of basic interconnection tests is to establish the sufficient conformance for interconnection to be possible, without performing thorough testing. The basic interconnection test for protocol entity verifies the main features in a protocol and/or transfer syntax specification. On the other hand, as a

manager interacts with managed objects of an agent by operations and notifications, the basic interconnection test for MIB should verify the possibility of receiving and responding operations and issuing notifications for managed objects often used, by setting the CMIS parameters to typical values.

## Capability test

The purpose of capability test is to verify the existence of claimed capabilities of an IUT. The capability test for protocol entity is conducted according to a PICS. In case of managed objects, their capabilities are claimed by a Managed Object Conformance Statement (MOCS) and a Managed Relationship Conformance Statement (MRCS) for name bindings [X724, 1994]. The capability test for MIB should verify the capability of performing all permitted operations and issuing all emittable notifications for each managed object instance as follows.

- Obtain all readable attribute values by issuing M-GET.
- Modify all writable attribute values by issuing M-SET.

A 'ModifyOperator' parameter value is set according to the implementation status of each attribute.

- Invoke all actions by issuing M-ACTION.
- Create managed object instances by issuing M-CREATE.

A 'managed object instance', a 'superior object instance' and a 'reference object instance' parameter values are set according to the implementation status for 'create with automatic instance naming' and 'create with reference object'.

- Delete managed object instances by issuing M-DELETE.
- Examine all emittable notifications by receiving M-EVENT-REPORT.

## Behaviour test

The purpose of behaviour test is to verify an IUT as thoroughly as possible, over the full range of dynamic conformance requirements. The behaviour test further consists of a valid behaviour test and an invalid behaviour test.

- valid behaviour test;

While the valid behaviour test for protocol entity verifies the dynamic behaviour such as the state transition, the variation and the combination of parameter values in PDUs etc., the valid behaviour test for MIB should verify the dynamic behaviour of managed objects and that of MIB as a whole according to the BEHAVIOUR template's definition.

- invalid behaviour test;

The invalid behaviour test for protocol entity uses 1)syntactically invalid test event, 2)semantically invalid test event and 3)inopportune test event. The invalid behaviour test for MIB should use test events 1) and 2) in order to verify the invalid behaviour for a syntax/semantics error of a parameter value since management information is also conveyed as a transfer syntax of ASN.1. As to 3), this test should be conducted only if the attributes which indicate some states such as an administrative state are included in the managed object.

## 3.3  Scope of testing

As we can define all test cases of the capability test for protocol entity corresponding to items in a PICS, so we can define all test cases of the capability test for MIB corresponding to items in a MOCS/MRCS.

It is possible to define almost all test cases of the behaviour test for protocol entity since protocol entity is modelled as finite state machines, though the number of test cases becomes huge. The test can be conducted by sending possible input events for all states and observing the reactions. On the other hand, it is practically impossible to define all test cases of the behaviour test for MIB because of the following two reasons. One is that the interaction with actual resources and other managed objects should be taken into consideration. The other is that the same test case may produce different test outcomes on different occasion since the behaviour of actual resources are affected by environmental and time-dependent conditions. Consequently, we should create the test suite for the capability test to all test cases and the one for the behaviour test limited to the test cases actually used.

Hereafter we call the test suite, a *scenario*, and the test system corresponding to LT, a *MIB tester*. The MIB tester assumes that CMIP communication can be correctly performed between a manager and an agent by using the underlying protocols.

## 4    IMPLEMENTATION OF MIB TESTER

### 4.1  Fundamental design principles

- The scenario for capability test is automatically generated, and the one for behaviour test is manually created. The capability test is substituted for the basic interconnection test.
- Applicable to various agents with different GDMO definitions and naming trees, without modification of its programs.
- The program is executed under Windows NT®. The CMIP board, commercially available [Idoue, 1990], is used for protocol processing of CMIP. The C language is used for programming.

### 4.2  Realisation of MIB tester

This section describes the realisation method of each function in the MIB tester. Figure 2 shows the software module configuration required for these functions.

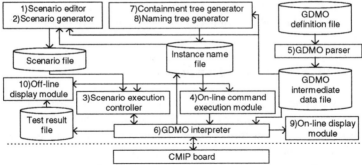

**Figure 2** Software configuration of MIB tester.

## 4.2.1 Scenario creation and execution function

This function supports the creation of a scenario, which is a sequence of CMIS operations and control commands, and the execution of the created scenario by the following modules.

1) **A scenario editor**, which inputs the parameter values for a sequence of CMIS operations and control commands manually and stores them in a scenario file.
2) **A scenario generator**, which generates operations for each managed object instance from the GDMO definitions and the naming tree, and stores them in the scenario file. The details on automatic scenario generation mechanism in a scenario generator are described in 4.3.
3) **A scenario execution controller**, which reads the scenario file and controls the execution of the scenario according to the control commands.
4) **An on-line command execution module**, which immediately issues a CMIS operation input by a test operator manually.

A test operator can specify the control of scenario execution into the scenario file. For this purpose, the scenario editor provides him with control commands of WHILE/WEND (to iterate the portion specified by a label), WAIT (to wait for the specified period), GOTO (to go to the line specified by a label), TIMER(to start at the specified time), and BELL (to sound a beep).

## 4.2.2 GDMO encoding and decoding function

This function supports the encoding and decoding of parameter values of any GDMO definitions without modification of its programs by the following modules.

5) **A GDMO parser**, which inputs the GDMO definition file and outputs the result of analysis to the GDMO intermediate data file.

6)A **GDMO interpreter**, which encodes and decodes parameter values of managed objects in CMIS primitives by referring to the GDMO intermediate data file.

### 4.2.3    Instance name management function

This function supports the management of instance names registered manually or extracted from received primitives by the following modules.

7)A **containment tree generator**, which creates a containment tree based on the NAME BINDING definition.

8)A **naming tree generator**, which creates a naming tree from received primitives or names input manually.

### 4.2.4    Test result analysing function

This function supports the analysis of the responses of operations and the notifications on a CMIS primitive level, including syntactic and semantic check of parameter values by the following modules.

9)**An on-line display module**, which displays the result of analysis for primitives and some CMIS parameter values immediately after the primitive is received.

10)**An off-line display module**, which displays the result of analysis for all parameter values of each primitive stored in a test result file.

The on-line display module displays the following items for each primitive: time for sending or receiving a primitive, InvokeID, LinkedID, Mode (i.e. confirmed/non-confirmed), ManagedObjectClass and EventType. The off-line display module displays all parameter values of a primitive sent or received as shown in Figure 3.

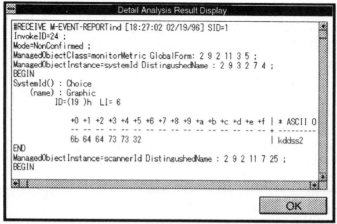

**Figure 3**    Off-line display window for M-EVENT-REPORT indication.

## 4.3 Automatic scenario generation mechanism

The following rules are applied to each managed object instance acquired by an automatic acquisition procedure, which traverses the naming tree by issuing M-GET with the scope parameter value set to 'firstLevelOnly' repeatedly. The values of CMIS primitive parameters not mentioned here such as synchronisation and filter are omitted or set to default.

### Generation of M-GET operations

The automatic acquisition procedure serves both to acquire the naming tree and to obtain attribute values in each instance, namely, to verify whether or not all attribute values are correctly read. In order to verify the effect of performed operations, an M-GET operation is also generated after an operation other than M-GET described below is generated.

### Generation of M-SET operation

An M-SET operation is generated depending on 'propertylist' in an ATTRIBUTE clause within the PACKAGE template as follows.
- 'REPLACE-WITH-DEFAULT' is included in 'propertylist'.

An M-SET with the 'modifyOperator' parameter value set to 'setToDefault' is generated.
- 'REPLACE/ADD/REMOVE' is included in 'propertylist'.

Default values for simple types of ASN.1 such as INTEGER and BOOLEAN are given by users in advance. Default values for constructed type are given as the combination of the default values for simple types. For 'REPLACE/ADD' of 'propertylist', an M-SET with the 'modifyOperator' parameter set to 'replace/addValues' and the attribute value set to the above default value is generated. For 'REMOVE' of 'propertylist', an M-SET with the 'modifyOperator' parameter set to 'removeValues' is generated for the attribute added by the previously generated M-SET or retrieved at the acquisition of the naming tree.

### Generation of M-ACTION operation

An M-ACTION is generated with the 'ActionType' parameter set to the object identifier in the ACTION template and the 'ActionInfo' parameter set to the default value as described in the above M-SET operation if the 'ActionInfo' parameter needs to be set.

### Generation of M-CREATE operation

When a 'CREATE' clause exists within the NAME BINDING template, regarding the superior managed object instance, an M-CREATE is generated according to the following rules.

- 'WITH-AUTOMATIC-INSTANCE-NAMING' exists in 'CREATE' clause.

An M-CREATE with the 'superior object instance' parameter set to the instance name of the superior managed object is generated.

- 'WITH-AUTOMATIC-INSTANCE-NAMING' does not exist in 'CREATE' clause.

An M-CREATE with the 'managed object instance' parameter set according to the ASN.1 definition of naming attribute is generated.

- 'WITH-REFERENCE-OBJECT' exist in 'CREATE' clause.

An M-CREATE with the 'reference object instance' parameter set to any managed object instance of the same class as the created one is generated.

## *Generation of M-DELETE operation*

When a 'DELETE' clause exists within the NAME BINDING template for a superior managed object, an M-DELETE operation is generated according to the following rules.

- An M-CREATE has been automatically generated.

An M-DELETE is generated for the managed object instance generated in the above M-CREATE procedure.

- An M-CREATE has not been automatically generated.

An M-DELETE is generated for any managed object instance which belongs to the superior object, such as the first one.

Figure 4 shows an example of automatic generation of CMIS operation for 'administrativeStatePackage' in the 'system' managed object class [X721, 1992]. In Figure 4, underlined parts in (a) are reflected to the underlined parameters of an M-SET operation in (b).

```
AdministrativeStatePackage PACKAGE
ATTRIBUTES administrativeState GET-REPLACE;
-----------------------------------------------
administrativeState ATTRIBUTE
WITH ATTRIBUTE SYNTAX
        AttributeASN1Module.AdministrativeState;
REGISTERED AS { 2 9 3 2 7 31 };
-----------------------------------------------
AttributeASN1Module DEFINITIONS BEGIN ::=
        AdministrativeState::=ENUMERATED{
        locked(0),unlocked(1),shuttingDown(2)}
END
```

```
M-SET request
Invokeld           1            -- set by MIB tester
BaseObjectClass  { 2 9 3 2 2 13 } -- system
BaseObjectInstance
  DN:{systemId=name:"ABC"} --already acquired
ModificationList
  {{ modifyOperator 0,           -- replace
     attributeId      { 2 9 3 2 7 31 },
     attributeValue   0 }}       -- locked
```

(a) GDMO definitions                 (b) Generated CMIS operation

**Figure 4**   Example of automatic generation of CMIS operation.

# 5   EVALUATION AND DISCUSSION

## 5.1   Proposed conformance test method for MIB

### Test configuration

In case of protocol entities, the remote test methods apply when it is possible to make use of some functions of the System Under Test (SUT) to control the IUT during testing, instead of using a specific UT. For example, the Logical Link Control protocol provides the UT functionality for Media Access Control protocol testing. As to MIB, such functions are always supported in the agent since managed objects have to be controlled by management applications of the agent. Furthermore, although some implementation may provide an explicit interface between an UT and a managed object, ASPs can not be standardised since such interface is dependent on the implementation. The remote test methods are best suited to the conformance test for MIB from these points.

### Test purpose

We can apply the framework of the conformance test for protocol entity to that for MIB since both protocol entities and managed objects have a static aspect and a dynamic aspect. The test purpose, however, can not be applied as it is.

For example, a preamble is often included in the scenario for protocol entity in order to start a test from a specified state. The preamble for MIB may be meaningless since the model of MIB is not a finite state machine and the attribute value which indicates some states may be changed without any interactions from a manager. Another example is that the human operation to actual resources should be included in test cases of the behaviour test for MIB if its behaviour is controlled by man-machine interaction. The behaviour of protocol entity is generally controllable from the LT by communicating with the protocol entity to be tested and , if necessary, using TM-PDUs.

### Scope of testing

It is difficult to conduct the thorough conformance test even for protocol entity whose models are finite state machines since test cases become huge. In case of MIB, the combination of attribute values would be almost infinite and the environmental and time-dependent conditions to actual resources would make the matter worse. As the practical solution, the conformance test for MIB should be conducted by creating the test scenario for the capability test to all test cases and the one for the behaviour test limited to the test cases actually used. For example, we can create a syntactically invalid test event with an attribute set to syntactically invalid value such as wrong tagging for ASN.1 encoding. For the semantically invalid test event, we can set the attribute to the value not permitted in 'property-list' or ASN.1 subtype definition.

## 5.2  Operability

We applied the MIB tester to the conformance test of an agent which support MIB for managed objects defined in accordance with the ISO/CCITT and Internet Management Coexistence (IIMC) [NMF26, 1993]. The MIB includes 102 managed object classes and 205 managed object instances. The MIB tester is implemented on a PC, DECpc XL560 (CPU: Pentium 60MHz) and is connected with the agents via Ethernet. The executable software size of the MIB tester is approximately 1.3 MB.

All the functions described in 4.2 was correctly performed. The scenario for the capability test was automatically generated for each instance of the MIB. For example, when an 'interface' managed object instance was specified for scenario generation and eight 'ifEntry' managed object instances were contained in it, the MIB tester generated eighteen M-GET operations, (ten for acquisition of the naming tree and eight for confirmation of M-SET operations), and eight M-SET operations to these instances, since an 'ifEntry' managed object class has one writable attribute called 'ifAdminStatus'. As a result, we were able to verify the dynamic conformance of this managed object instance.

It would be convenient to analyse the causality relationship among a sequence of CMIS primitives since an attribute value reflects the status change of an actual resource and an operation for a managed object sometimes causes future notifications. For example, a 'perceivedSeverity' attribute value varies according to the severity of an alarm and an attribute value change notification is caused by a previously issued M-SET operation. We will support this function in the future extension.

## 5.3  Automatic scenario generation

The number of instances in MIB of actual agents is huge in many cases. One example is managed objects which hold performance attributes for each termination point of the transmission equipment. Some Synchronous Digital Hierarchy (SDH) transmission equipment supports approximately sixty thousand managed object instances including performance related managed object instances. It spends so much time to create the scenario for all managed object instances manually that the automatic scenario generation function must be necessary.

The proposed scenario generation rules does not address the behaviour test. It is possible, however, to generate syntactically invalid test events to some extent by using different GDMO definitions from these for the agent. The method also does not cover the relationship among managed objects. For example, there are some cases that an action for a managed object affect the behaviour of other managed objects. We could generate such test cases from the formal description of BEHAVIOUR templates [X722/PDAM3, 1995] being now studied in ITU-T.

# 6 CONCLUSIONS

This paper proposed a conformance test method for MIB in an agent of OSI management and described an implementation of the MIB tester according to the test method. The proposed test method was based on and an extension of the existing conformance testing method for protocol entity by taking the differences in modelling between protocol entities and managed objects into account. The remote test methods were adopted as the test configuration of the conformance test for MIB. We defined the test purposes for the basic interconnection test, the capability test and the behaviour test. As to the scope of testing, we showed the practical solution by conducting the capability test to all test cases and the behaviour test limited to the test cases actually used. We have implemented the MIB tester so that test scenarios for capability tests are automatically generated. The MIB tester is flexible to GDMO definitions so as to be applicable to various kinds of agents. We evaluated the proposed test method and demonstrated the effectiveness of the implemented MIB tester through the application to the conformance test for the actual agents.

## ACKNOWLEDGEMENT

The authors wish to express sincere thanks to Prof. Yoshiyori URANO of WASEDA University for his guidance and kind suggestions.

## REFERENCES

A.Idoue, T.kato, K.Suzuki and Y.Urano (1990), Design and Implementation of OSI Communication Board for Personal Computers and Workstations, Proc. of Eleventh International Conference on Computer Communication

ITU-T Rec. M.3010 (1992), Principles for a Telecommunications Management Network

ITU-T Rec. X.290 (1992), OSI Conformance Testing Methodology and Framework for Protocol Recommendations for CCITT Applications - General Concepts

ITU-T Rec. X.721 (1992), IT - OSI - SMI: Definition of Management Information

ITU-T Rec. X.722/PDAM3 (1995), IT - OSI - SM: Guidelines for the Definition of Managed Objects - Amendment 3 (Use of Z for Managed Object Behaviour)

ITU-T Rec. X.724 (1994), IT - OSI - SMI: Requirements and guidelines for implementation conformance statement proformas associated with OSI management

J.M. Serre. , P. Lewis. and K. Rosenfeld. (1993), Implementing OSI-Based Interfaces for Network Management, IEEE Communications Magazine, Vol.31 No.5

Network Management Forum, OMNIpoint, Forum 26 (1993), Translation of Internet MIBs to ISO/CCITT GDMO MIBs

O. Ödling and S. Wallin (1994), Building MIB Applications, Proc. Of IEEE 1994 Network Operations and Management Symposium

## BIOGRAPHY

**Keizo Sugiyama** is a research engineer of Network Management System Laboratory, KDD R&D Laboratories. Since joining the Labs. in 1987, he has been engaged in the researches on implementation of OSI Protocol, EDI(Electronic Data Interchange) translator and network management systems. He received the B.E. and M.E. from Kyoto University in 1985 and 1987 respectively.

**Hiroki Horiuchi** is a senior research engineer of Network Management System Laboratory, KDD R&D Laboratories. Since joining the Labs. in 1985, he has been engaged in the researches on formal description techniques for communication protocol, implementation of OSI Protocol and network management systems. He received the B.E. and M.E. from Nagoya University in 1983 and 1985 respectively.

**Dr. Sadao Obana** is a senior manager of Network Management System Laboratory, KDD R&D Laboratories. Since joining the Labs. in 1978, he has been engaged in the researches in the field of computer communication, database and network management systems. He received the B.E., M.E. and Dr. of Eng. Degree of electrical engineering from Keio University in 1976, 1978 and 1993 respectively.

**Dr. Kenji Suzuki** is a senior project manager of R&D planning group in KDD R&D Laboratories. Since joining the Labs. in 1976, he worked in the field of computer communication. He received the B.S., M.E. and Dr. of Eng. Degree of electrical engineering from Waseda University, Tokyo, Japan in 1969, 1972 and 1976 respectively. From 1969 to 1970, he was with Philips International Inst. of Technological Studies, Eindhoven, the Netherlands as an invited student. Since 1993, he has been a Guest Professor of Graduate School of Information Systems, in the University of Electro-Communications. He is a member of IEEE.

# 34

# Design and Testing of Information Models in a Virtual Environment

*R. Eberhardt[1], S. Mazziotta[2], D. Sidou[2]*
*Swiss Telecom PTT R&D and Institut Eurécom*
*[1]CH-3000 Bern 29, +41 31 338 81 45, +41 31 338 59 59*
*eberhardt@vptt.ch*
*[2]F-06904 Sophia-Antipolis Cedex, +33 4 93 00 26 43,*
*+33 4 93 00 26 27, {mazziott\sidou}@eurecom.fr*

## Abstract

The paper introduces the TMN-based Information Model Simulator (TIMS) toolkit, a rapid-prototyping environment for TMN information models and explains how it is used in a real-life example. Based on GDMO/ASN.1, GRM and a formal behavior description, the toolkit generates TMN systems and allows both the visualization of MIBs at run-time and the access to it through CMIP. Behavior of object models is described through the use of relationships which leads to considerable simplification of the behavior and is the first step for a distributed environment. Assertion mechanisms permit the designer to verify the correct state of the model at run-time. The paper publishes the results of the tools use for the specification of a management interface for the V5 access network interface.

## Keywords

Information models, GDMO, Relationships, formal behavior, simulation

# 1     INTRODUCTION

The goal of the TIMS-project is to provide a laboratory environment for TMN-designers enabling them to prototype solutions, build mock-ups and to test them prior to standardization, procurement or network introduction. This approach, so the intention, will speed up the standardization process and improve the quality of the specifications. A mock-up can also be used to support procurement, enhance education and improve acceptance testing of the finished products (Eberhardt et al, 1996).

We believe that the gap between formal specification and actual implementation should be as small as possible. The telecom operator profits from a tighter specification while the developer can (semi-) automatically create parts of its application. GDMO is such an example. TIMS therefore focuses on formal but executable specifications of static and dynamic schema's in the TMN, i.e. the relationships between managed (and managing) objects and their behavior description.

What is required to reach this goal? Behavior specifications must be separable into a specification part and an emulation (or algorithmic) part. The specification part defines rules on the static and dynamic properties of the information model, is implementation-independent and therefore normative. The emulation part describes the algorithm of an operation within or on the information model (e.g. Ensemble-scenarios (NMF 025, 1992)) and is non-normative. Following principles stated in (Kilov, 92), the specification paradigm is declarative and advocates the use of relationships. To keep the specifications understandable and manageable, only functional behavior properties are considered. Thus quantitative or physical aspects are abstracted away (e.g. actual distribution, timing and real-time issues). Validation is based on simulation and testing of the executable specification, therefore mathematical reasoning is not used. Finally, to ensure executability and rapid-prototyping, the proposed behavior specification language is based on an existing and interpreted language: Scheme (Clinger and al, 1991).

*Plan of the Paper*
Section 2 introduces the highlights of the TIMS toolkit seen from the users perspective, namely its interfaces and the method of behavior formalisation. Section 3 shows in a practical example how TMN information model design is performed. Section 4 summarizes the results of the design of the $V5.1_{AN}$ management interface which show that the approach chosen in TIMS is viable indeed. Conclusions on the work up to now are discussed in section 5.

## 2   TIMS TOOLKIT

Clause 2.1 introduces the technical support implemented within the TIMS platform, described in detail in Mazziotta and Sidou (1996). Clause 2.2 presents the main features of the TIMS Behavior Language (BL) and its structure.

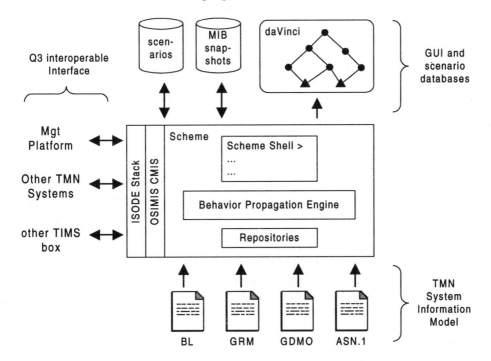

Figure 1 : TIMS and its environment

### 2.1   The Implementation

TIMS is a single toolkit running under UNIX. It is built around the OSIMIS development toolkit (Pavlou et al, 1995) and other public domain tools. TIMS consists of the simulator part called "TIMS box", support software such as the simulation manager and visualization and browsing tools. The TIMS box is built as an open system with following interfaces made available:

• Q3 interoperable interface (left hand side of Figure 1)

- the TMN information model integration (bottom of Figure 1)
- test execution environment and the GUI (top of Figure 1)

## *Q3 Interoperable Interface*
TIMS boxes may be accessed via a CMIS API and an underlying OSI protocol stack. The CMIS API and the OSI-stack used are taken from the OSIMIS library and the ISODE implementation, respectively. Providing CMIS allows for (1) several TIMS boxes to interact in different roles (agent, manager, manager/agent) and (2) to integrate with real TMN applications (e.g. commercial management platforms, real network elements, network emulators, etc.). This feature is fundamental for the reuse of TIMS specifications in procurement, as reference configurations, for education and testing of real TMN systems.

## *Input of the TMN Information Model*
TMN Information Models (IM) are composed of GDMO, ASN.1 and (recently) of GRM specifications which correspond to the static part of the specification. TIMS requires only the relationship class specification of the GRM; the relationship mapping productions are incomplete and therefore embedded in the behavior. The dynamic part consists of the behavior specification (c.f. section 2).

GDMO, ASN.1 and GRM parsers[*] provide suitable output that can be integrated in the Scheme environment. This makes all the required information about Managed Objects Classes, Relationship Classes available, e.g. to ensure the correctness of operations performed on Managed Objects. For ASN.1, a reasonable support of a value notation is needed since behaviors often include the creation and manipulation of complex ASN.1 values. Examples of such constructs are provided in section 3.

## *Graphical User Interface*
The GUI mainly corresponds to the command panel which is based on the Tk toolkit. This panel controls the Test Execution Environment and allows the visualization of the simulation. The Test Execution Environment consists of a Scenario Player, and a Snapshot Player:
- The Scenario Player enables the user to run sequences of management operations or real resource changes (e.g. emulating equipment failures). Step-wise execution allows tracing all changes in the MIB.

---

[*] The GDMO parser is based on the OSIMIS GDMO compiler front-end (Pavlou et al, 1995), ASN.1 on the ASN Free Value Tool (ASNFVT, 1992). The GRM parser was developed in TIMS.

- The Snapshot Player enables to save the current system state and to use it as starting point for scenario runs. This feature is very important in the prototyping phase when a given state is reached only after a long simulation run.

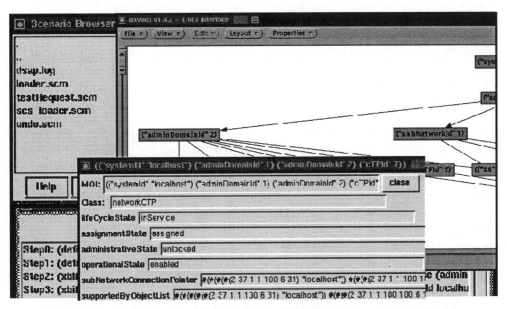

Figure 2 : Screen Shots of the TIMS Execution Environment (the graph display, the contents of a managed object and the scenario player)

TIMS presents the MIB as a graph of information objects and relationships. The visualization is done with daVinci (davinci, 1996), a generic graph visualization tool. This tool provides automatic layout generation as well as selective suppression of sub-graphs. The user, selecting a given node of the graph, can then access the managed objects' attributes and their associated values presented in form of TK widgets.

## 2.2 Behavior Specification

### TIMS Behavior Language Features
TIMS follows in essence the approach advocated by Kilov (1992) where both the use of relationship-based formalisation and asserted specifications are employed.

**Relationship-based Formalisation :** BL employs the ODP notion (X.902, 1995) of "action" to describe a behavior :   "An action is defined as something which happens, of interest for modeling purposes and associated with at least one object. [...]". Very often, an action can be associated to causal dependencies between objects. This leads to a behavior-model based on roles and relationships. In the context of TMN information models the General Relationship Model GRM (X.725, 1995) is a natural candidate.

Relationship-based Behavior Formalisation provides simpler, more readable and expressive behavior specifications because managed objects are identified through roles instead of raw attribute pointers or any other mechanism currently available to realize a relationship.

**Use of assertions :** Assertions define the pure specification aspects of the system. In TIMS, assertions are properties that are checked during the execution of the simulation. Although often considered a burden, assertions prove very valuable during incremental and component-based model development. Experience shows that the specifier can not control the whole complexity of its system and especially when there is a lot of "behavior interference". Complex behavior often emerges when relationships overlap, i.e. an object participates in several relationships in different roles. Specifying assertions is an effective method to ensure the correctness of the resulting specification at run-time (protecting the specifier from unwanted or conflicting behaviors).

## *Structure of a Behavior*

A behavior is always defined in the context of a relationship (*scope* clause) between objects fulfilling given roles. It corresponds to the execution of a piece of code (*body* clause) when it receives a message at one of its interfaces, if the guard (*when* clause) is evaluated to true (i.e. enables the execution). The body of a behavior consists of Scheme code. There is no a priori structure imposed on it. A CMIS API is provided to specify any CMIS operation within behaviors (GET, SET, ACTION, ...). In addition a GRM-like API provides access to the required relationship abstract operations, i.e. ESTABLISH, BIND, UNBIND, TERMINATE. Since usual programming features (i.e. control flow structures, variable notation...) are required, the use of an existing and simple programming language, Scheme, reveals to be a reasonable choice. The execution of the body is immediately preceded and followed by a pre-condition (*pre* clause) and a post-condition (*post* clause), respectively. Finally, each behavior is labeled by the specifier for the purpose of identification, especially during debugging.

Invariants are described by a behavior with post-condition but no body part. A relationship-invariant is a behavior with the message received being any kind of operation executing upon objects bound to the relationship. A role-invariant restricts the message target to the objects involved in the role (cf. section 3).

# 3 MODELING IN TIMS

This section provides a step-wise, example-based tutorial introduction to how modeling is typically performed in TIMS. The examples derive from the management interface model for V5 (ETS 300 376-1, 1994).

Management and domain requirements, expressed in prose or in a semi-formal representation, form the basis for identifying managed and managing resources. Placing the resources in context, i.e. building the static schema, is the logical next step. Some invariants, such as the cardinality between managed objects, will have become apparent at this point. Were we writing software, we'd sit down and begin coding - in TMN partially reflected in the Ensemble-scenarios. Coding reveals whether the choice of resources (managed objects) was correct, if the static schema is useful and also gives an indication on the ease of use of the interface. This is the point where we believe a TMN-laboratory becomes useful. The following is an example on how this is achieved in TIMS:

**Requirement**: "A field replaceable unit (FRU) is represented as a physical resource (the equipment) and the logical resource (the functionality, here called userPort). The operational state of the userPort is dependent on the physical resources it requires (card, power supply, etc.). The userPort becomes operational automatically once all physical resources have been installed."

**Give relationships a name** : Given the resources and an ER-diagram, writing GRM provides its structure, the role names, their cardinality and their relationship operations.

```
Rresource RELATIONSHIP CLASS
     BEHAVIOUR RresourceBehaviour;
     SUPPORTS ESTABLISH, TERMINATE;
     ROLE physicalRole
          COMPATIBLE-WITH equipment        [...]
     ROLE logicalRole
          COMPATIBLE-WITH userPort
          PERMITTED-ROLE-CARDINALITY-CONSTRAINT INTEGER(1)
          REQUIRED-ROLF-CARDINALITY-CONSTRAINT INTEGER(1..MAX)
          [...]
```

**Define constructors & destructors for the static schema :** "The installation of the FRU results in the automatic generation of the logical resources (userPort) and other relationships. Trigger for this behavior is the creation of the equipment through an external message (e.g. an M-CREATE)."

```
(define-behavior "create-equipment" (scope "RmanagedElement")
     (when (and (param Create?) (param moc=? "equipment")))
     (pre ...)
     (body ...
        (set! (ttpInstance (Create userPort ...)
        (Establish (operation-name) "Rresource" (rir:gen-ri-inst "Rresource")
                (("physical" equipmentInstance)
                 ("logical" (ttpInstance))))
        ...)
     (post ...))
```

**Write scenarios (management function) :** TIMS does not make a difference between code representing behavior internal to the MIB and operations between manager and agent. In the example, the operational state of userPort is set to "disabled" if it is in the "enabled" state and one of its supporting physical resources goes to "disabled".

```
(define-behavior "logical enabled->disabled "
     (scope-rel "Rresource")
     (when (and (asn=? (Get (Part (ri) "physical") "operationalState") 'disabled)
            (some (lambda (userPort) (asn=? (Get userPort "operationalState") 'enabled))
                (Part (ri) "logical"))))
     (pre ...)
     (body (for-each (lambda (userPort)
                (cond ((asn=? (Get userPort "operationalState") 'enabled)
                       (Set userPort "operationalState" 'disabled))))
                (Part (ri) "logical")))
     (post (for-each (lambda (userPort) (asn=? (Get userPort "operationalState") 'enabled)))
```

For those not familiar with Scheme, the lambda-expression allows to define unnamed functions with temporary variables (such as a function asserting the operationalState of userPort being enabled).

**Make the model water-proof :** Having written some code, the designer must now review the static and dynamic schemas and check at which points the behavior could fail due to ambiguities or bugs. Write assertions or invariants (e.g. as part of a relationship) handling these cases and run the simulation. Assertion-failures will point at problems in the code. If the simulation shows unexpected behavior this indicates weak guard statements. Depending on the run-time problems, the designer may now have to revisit all previous phases up to the design level.

**Finalise the model** : Once the model works as expected the tough work is over. The next steps involve mapping the behavior onto a system management model (called Engineering Viewpoint, in this case OSI Systems Management). The designer needs to write the relationship mapping (pointer structures, etc. ) and to implement action and notification signatures together with their parameters for behavior crossing the system management boundary. Example : "A userPort is associated to a v5Ttp using the ACTION setReciprocalPointer".

```
(define-behavior "Rv5Interface-SetReciprocalPtr"
    (scope-rel "Rv5Interface")
    (when (and (ri) (param Action?) (param Action=? "setReciprocalPointerAction")))
    (pre)
    (body (let* ((aEndObject(car (param argument)))
            (zEndObject (cadr (param argument)))
    [...]
    -- this action results in the call of an establish & bind operation between aEndObject and
    -- zEndObject.

    (post ...))
```

## 4    A CASE STUDY: V5.1 MANAGEMENT

This section provides results and metrics acquired during the implementation of the V5.1 management model for configuration management. The case study was developed in parallel to a V5 management interface specification which is to be used for procurement purposes.

The case study focused on the overall behavior of the management architecture and less on the mapping between hardware behavior and its TMN representation (e.g. state mappings of protocol engine finite state machines). The model consists of 12 managed object classes and 18 relationship classes. It covers scenarios (management functions) for the insertion, removal and configuration of ports as well as their provisioning (cross-connection). A minimal implementation counts 46 managed object instances of which 31 represent 64 kBit/s channels and can be therefore reduced while prototyping (e.g. down to 10). The average size of a simple dynamic schema (e.g. state change between managed objects) lies at 20 lines of Scheme-code. Complex schema's, such as the constructor for a v5interface MIB structure involve around 50 lines. The code is very repetitive so that cut & paste helps reduce tedious writing.

The major effort (40 %) lay in acquiring the necessary V5 domain knowledge, sometimes down to the protocol level. Implementing the V5 static schema (resource selection, GRM) was quickly done, while embedding the fragment into an overall MIB architecture turned out to be more difficult than expected (total 20%), mainly due to functional restrictions found in current GDMO libraries (M.3100). Implementing behavior itself resulted in a repeated review and refinement of the requirements, sometimes identifying new demands along the way. Behavior-design and debugging made up for another 40% of the effort. The modeling of invariants and assertions went hand in hand with the static schema (GRM constraints) and - when actually specified - during the development of each behavior.

## 4.1  Evaluation

The language Scheme proved difficult at first, as the engineers implementing the case study were more familiar with imperative languages. Once understood, however, the interpreted nature of Scheme allowed for trial-based code-development. The TMN-API's themselves were rapidly understood and easily used. The graphical interface proved useful only for educational purposes and for browsing of small MIB's. A run-time debugger for behavior traces and for the analysis of interference's between behavior executions would speed up coding.

Designing and coding the dynamic schema is very repetitive, but requires detailed knowledge of both the domain and the TMN libraries. We believe that in future it should be possible to reuse generically defined relationships and their behavior. Guards are the enabling condition for the execution of behavior. Specified incompletely, they may lead to unexpected side effects difficult to foresee by the designer. Due to GRM, many assertions and invariants came for free, saving extra coding. Not surprisingly, assertions were used only sparsely because they are often considered tedious to formulate and very large. This applies especially for post-conditions, fundamental for evaluating the consistency of the MIB.

Relationships proved to be very useful for both design and coding. Making the model accessible via CMIP requires the additional effort of implementing the relationship mapping. We believe that a relationship management service reflecting GRM-constructs would simplify management scenarios, provide for distribution transparency and aid in behavior description.

The V5 management interface specification project running in parallel benefited a lot from our design work, especially when it came to understanding the finer points of the model, its restrictions and pitfalls. The TIMS scenarios could be mapped almost 1:1 into the ensemble specification.

# 5 CONCLUSIONS

The paper gives an overview of the rationale guiding TIMS' development. It describes the TIMS TMN toolkit and its highlights. The relationship behavior formalisation is introduced and a guide on how to model TMN interfaces using TIMS is provided.

Although the initial learning curve is steep, it compares favorably with commercial toolkits. While Scheme has its advantages, it isn't widely known and not sufficiently readable to be used as specification; a specification-typesetter could be considered a solution.

The language features and the use of relationships provide everything required by ODP and align reasonably well with new methodologies suggested in ITU-T (G.851, 1996).

The greatest cost remains in acquiring the domain knowledge necessary to be able to develop a management interface. A laboratory environment helps understanding both the domain and its management design. The speed in which new TMN information models were implemented (smaller test cases with up to 8 managed objects were implemented in less than 2 weeks) indicates that it is possible to bring TMN into the laboratory.

# 5 REFERENCES

ASNFVT (1992), ASN.1 Free Value Tool, at ftp://osi.ncsl.nist.gov/pub/osikit/"

Clinger, W. and Rees, J. (1991) - Report on the Algorithmic Language, Scheme. ACM Lisp Pointers, vol. 4 (3), 1991. - Available at http://www.cs.indiana.edu/scheme-repository/doc/standards/r4rs.ps.gz.

davinci (1996) The Interactive Graph Visualization System daVinci, available at http://www.informatik.uni-bremen.de/~inform/forschung/daVinci/daVinci.html.

Eberhardt, R. and Sidou, D. et al. (1996) Executable TMN specifications with TIMS, proceedings of the NOMS'96 , IEEE. Available at http:/www.eurecom.fr/~tims

ETS 300 376-1 (1994) Q3 interface at the Access Network (AN) for configuration management of V5 interfaces and associated user ports; Part 1: Q3 interface specification, ETSI Technical Standard, December 1994

NMF 025 (1992) The Ensemble Concept and Format, Network Management Forum

G.851-01 (1996) Draft Recommendation, Managemeent of the transport network - Application of the RM-ODP framework. June 1996

X.725 (1995) ISO/IEC JTC 1/SC 21, ITU X.725 - Information Technology - Open System Interconnection - Data Management and Open Distributed Processing - Structure of Management Information - Part 7 :General Relationship Model.

X.902 (1995) Basic Reference Model of ODP - Part 2: Foundations, ISO 10746-2, ITU X.902.

Kilov, H. (1992) From OSI Systems Management to an Interoperable Object Model: Behavioural Specification of (Generic) Relationships, Proceedings 3rd Telecommunications Information Networking Architecture Workshop (TINA 92), 1992, Narita, Japan,

Mazziotta, S. and Sidou, D. (1996) - A Scheme-based Toolkit for the Fast Prototyping of TMN-systems - 1996. Seventh International Workshop on Distributed Systems : Operations & Management.

Pavlou, G. and McCarthy, K. and Bhatti, S. and Knight, G. and Walton, S. (1995) The OSIMIS Platform: Making OSI Management Simple, Integrated Network Management IV, 1995, Santa Barbara, USA

## 6    BIOGRAPHY

Rolf Eberhardt joined Swiss Telecom R&D in 1991 following his graduation in computer science from ETH Zürich, Switzerland. He has worked as TMN specification engineer, consultant and in standardization (ETSI TM2, ITU-T SG4, NMF), mainly on transmission networks. His interests focus on behavior formalisation, access networks and X-interfaces.

Sandro Mazziotta graduated in computer science in 1993 from Nice-Sophia Antipolis University, France. In 1994, he joined the Corporate Communications department of Institut Eurécom where he pursues a Ph.D. thesis. His research interests include the specification, validation and testing of the dynamic behavior of object-based distributed systems.

Dominique Sidou received his computer science degree: "Diplome d'Etudes Approfondies (DEA)", from University Paul Sabatier, Toulouse - France in 1989. His research interests focus on distributed systems: behavior modeling and validation, distributed object technologies (ODP, CORBA), and network and systems management (TMN, SNMP). He is research engineer at Institut Eurécom (Sophia-Antipolis, France).

# 35
# Formal Specification and Testing of a Management Architecture

*G. P. A. Fernandes** *and J. Derrick*
*Computing Laboratory, University of Kent, Canterbury, CT2 7NF, UK.*
*(Telephone: +44-1227-764000, Email: {gpaf,jd1}@ukc.ac.uk.)*

### Abstract
The importance of network and distributed systems management to supply and maintain services required by users has led to a demand for management facilities. Open network management is assisted by representing the system resources to be managed as objects, and providing standard services and protocols for interrogating and manipulating these objects. This paper examines the application of formal description techniques to the specification of managed objects by presenting a case study in the specification and testing of a management architecture. We describe a formal specification of a management architecture suitable for scheduling and distributing services across nodes in a distributed system. In addition, we show how formal specifications can be used to generate conformance tests for the management architecture.

**Key words:** Managed objects behaviour; Management architecture; Formal description techniques; Open Distributed Processing; Conformance.

## 1 INTRODUCTION

The importance of network and distributed systems management to supply and maintain services required by users has led to a demand for management facilities. However, fully integrated management systems which will cope with management of large-scale distributed applications and their underlying communication services are still not available. Such applications require open management to integrate their components, which may have been obtained from a number of sources; the cost of system administration will depend to a large extent on how easy it is to perform this management integration. The creation of open distributed management depends upon there being a common representation for the resources being managed. This can be achieved by the creation of a suitable family of managed object definitions.

At present the nature of the resources to be managed and the behaviour they are expected to exhibit are expressed in natural language, structured and organized using a simple specification technique set out in the Guidelines for the Definition of Managed Objects (GDMO) [ISO91]. The informal nature of this technique makes the implemen-

---

*Work supported by JNICT Program *PRAXIS XXI* (Portugal) under grant No. BD/2804/93

tation and testing of managed objects expensive, because much skilled effort is needed to interpret the specifications and construct suitable tests. Formal description techniques (FDTs) offer the promise of improved quality and cost reduction by removing errors and ambiguities from the specification and automating aspects of both implementation and testing. Indeed, interworking will depend on specification and testing and product cost will depend on the efficiency of these processes. The aim of our work is to test the applicability of FDTs to managed object specification by formally specifying a realistic and large application using an object-oriented variant of the formal technique Z.

In this paper we show how formal techniques can be used to specify a management architecture suitable for scheduling and distributing services across nodes in a distributed system. The aim of the architecture is to optimise the use of resources by distributing the load and managing the resources available in order to fulfil the requirements of application services [G. 96]. In section 2 of the paper we describe the management model we use. Section 3 discusses the management infrastructure we have been developing. Section 4 shows how we can apply an object-oriented variant of the formal language Z to the specification of interacting managed objects by formally specifying the architecture, and in section 5 we show how test generation methods can be applied to Z specifications of managed objects.

## 2   MANAGEMENT MODEL

The role of *management* is to monitor and control the system to be managed, so it fulfils the requirements both of the owners and the users of the system. The management model presented in this paper is a distributed object-oriented model based on the Open Distributed Processing (ODP) Reference Model [ITU95] and the OSI management model [ISO92a]. The Reference Model of ODP provides a framework for the standardisation of Open Distributed Processing. The OSI Systems Management provides mechanisms for the monitoring, control and coordination of those resources which allow communication to take place in the OSI environment (OSIE).

The existing approaches to management address mainly network management. Many of the ideas included in the OSI management model, a standard for network management, can be used for distributed systems. However, it must be taken into account that while network management concentrates on largely autonomous network devices, distributed systems management addresses components which are much more dependent on each other.

One of the most important ideas in OSI Systems Management is the use of object-oriented principles to define management information and interfaces. The devices in the network that are subject to management are viewed as *managed objects*. Organisational requirements require partition of OSI management into functional areas, such as security, account and fault management, or for other management purposes, such as by geographical, technological or organisational structure. To reflect this, managed objects are organised into management *domains*. Managed objects in a particular domain are subject to a common management policy, which consists of a set of rules constraining the behaviour of those objects. The ability to specify precisely management policies, independent of the implementation is an important benefit of formal specification.

## 3 MANAGEMENT ARCHITECTURE

In this section we present an architecture to support management of distributed systems, addressing in particular the issue of distributing the workload submitted to a distributed system by its users. Distributed scheduling is used in order to locate a new service on the most appropriate node, taking into account the current state of the system and the quality of service requirements of the service.

A centrally-located allocator – *Distributed System Manager* (DSM) – is responsible for taking decisions in order to determine to which node in the system each service will be allocated. To determine if a node is suitable to instantiate a service, the DSM has to compare the quality of service (QoS) requirements of the service with the resources provided by the node. Placement is based on the last known state of the system, which is stored by the DSM, and updated by the monitoring information it receives from node managers.

The foundation of any management system is a database containing information about the components being managed. This type of database is often referred to as Management Information Base (MIB). The MIB is a structured collection of managed objects which represent the components that are monitored and controlled by a management system. Each node in the system maintains a MIB that reflects the status of the managed objects at that node.

The *Node Manager* is an entity local to a node, responsible for managing the objects within that node and reporting monitoring information to the DSM. It is capable of performing management operations on managed objects on behalf of a DSM and of emitting management notifications on behalf of a managed object to inform the DSM about the occurrence of an event. The node manager monitors and controls the services instantiated on the node and collects information about the resources available. This information is stored in the local MIB.

The DSM also maintains a MIB where information about the nodes under the DSM control is stored. After initialisation, the DSM issues requests for monitoring information only when it is trying to find a suitable location for a service. The set of polled nodes will send monitoring information which will be used to update the DSM MIB.

All newly created services are instantiated by the DSM upon request by the *trader*. The (ODP) trader is an object that provides a service which accepts and stores service offers from potential providers (servers) and hands out this information on request to potential clients. The DSM selects a suitable location for the service requested and asks the local node manager to instantiate that service.

## 4 SPECIFYING MANAGED OBJECTS FORMALLY

This section illustrates how we have used an object-oriented variant of Z to specify our management architecture.

Z [Spi89] is a state based formal description technique (FDT), and Z specifications consist of informal English text interspersed with formal mathematical text. The formal part describes the abstract state of the system (including a description of the initial state of the system), together with the collection of available operations, which manipulate the state. The descriptions are given in terms of set theory and first-order predicate calculus. The *schema calculus* provides a useful (and visual) way to structure specifications, and to provide for a degree of modularity in the definition of operations. Z has proved to be one of the most enduring formal description techniques, partly because of its simplicity and readability. It has gained significant industrial usage and support over the years. Z has been shown to be a suitable vehicle for the specification of information related activities, and because of this has been considered a suitable language for use in the information viewpoint within ODP.

However, modern distributed systems are object-based, and for this reason there has been interest in extending Z to facilitate an object-oriented specification style. This allows for a proper definition of inheritance, and for encapsulation to be used to structure the specification in terms of classes and objects. Object-Z [DRS95] and ZEST [Raf94, CR92] are similar object-oriented extensions of Z. They both use the concept of a class to encapsulate the descriptions of an object's state with its related operations. In addition, they provide support for inheritance, instantiation and the description of communication between objects. In this paper we use ZEST to specify our managed objects, although a description in Object-Z would be very similar. Using an object-oriented variant of Z allows a hierarchy of classes to be developed as the Guidelines for the Definition of Managed Objects [ISO91] indicate.

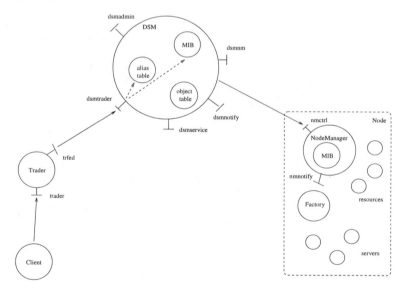

**Figure 1** The management architecture.

We specify the collection of objects shown in Figure 1. The complete specification, called *MgtSystem*, defines a number of objects (*DSM*, *Nodes*, *Trader*) with a description of how they interact. To illustrate the specification of a single object, we will consider the *DSM* object. Some familiarity with the Z language is assumed.

## 4.1 Specifying a single object - the DSM

We model a managed object class by a ZEST class specification which encapsulates a number of fixed attributes, a state schema declaring the variable attributes, and a collection of operation schemas. In a fashion similar to Z, these are enclosed by lines with the class name at the top (here *DSM*).

Inside the class are the variables and operations used in the class, some of these are for internal use only, while others form part of the interface of the class. We document this by beginning each class with a description of the interfaces. A ZEST class may have several interfaces, and each interface defines what is visible at that particular interface of the object. A name appearing in an interface corresponds to either an operation or attribute, for example, the *dsmnm* interface consists of the operations *UpdateNode* and *NoResources* (the **add** tells us these operations are included in the interface). Attributes and operations not appearing in any of the named interfaces are then internal to the class. In the formal specification, the names of the interfaces document the interface partition, and they are not used when invoking the operations.

Variable attributes are declared in a state schema, and their initial values are given by the schema *INIT* - here we specify that all variable attributes are in their own initial states (i.e. the ones given by *their INIT* schemas).

The variable attributes in the DSM are in fact instances of appropriate classes. They represent the data concerning aliases (i.e. service descriptions), objects created and the results of node monitoring (the MIB). The declaration *dsm_mib : DSM_MIB* declares *dsm_mib* to be an instance of the class *DSM_MIB*, which is specified as a ZEST class consisting of the data stored in the MIB together with operations to access and update that data. The *dsm_mib* contains information about the resources available on the nodes managed by the DSM. The DSM can manipulate the *dsm_mib* via its operations, for example, the information stored in the *dsm_mib* for a node can be updated by the node calling the DSM *UpdateNode* operation, which calls *Update_Node* in the *dsm_mib*.

Re-use is supported by the definition of *generic* classes and operations. For example, *DSMTable* is a generic class defined in terms of two generic parameters which can be instantiated with particular types (*Alias*, *Handle*, etc) in differing contexts.

---
_DSM _____

**interface** *dsmnm* **add** *UpdateNode, NoResources*
**interface** *dsmservice* **add** *InstallAlias, RemoveAlias, . . .*
**interface** *dsmtrader* **add** *LookupConstraintsFailed, LookupConstraintsSuccess, . . .*
**interface** *dsmnotify* **add** *CapsuleTerminated*

*id : DSMId*

> *alias_table : DSMTable[Alias, AliasData]*
> *object_table : DSMTable[Handle, NotificationData]*
> *dsm_mib : DSM_MIB*

> _INIT _____
> *alias_table.INIT* $\wedge$ *object_table.INIT* $\wedge$ *dsm_mib.INIT*

*UpdateNode* $\hat{=}$ *dsm_mib.Update_Node[nmId?/nodeId?, nmData?/nodeData?, TRUE/result!]*
The other ZEST operations come here

---

The behaviour of a class is described by specifying ZEST operations. Each ZEST operation describes how the output is related to the input and how the state changes as a result of invoking the operation. For example, *UpdateNode* is an operation defined in terms of an operation in the *dsm_mib* object, but with the names of the inputs and outputs changed (e.g. *nmId?* is used instead of *nodeId?* etc).

## 4.2   The Node class

The *Node* class encapsulates the node entities and operations involved in management. It includes an instance of *NM*, the node manager class, and an instance of the *Factory* class. The factory is the entity responsible for the instantiation of services.

```
┌─ Node ──────────────────────────────────────────────────────────
│  interface external add CheckRequirements, SendInfo, Instantiate, Kill
│  ┌──────────────────────────────────────────────────────────────
│  │ nm : NM
│  │ factory : Factory
│  ├──────────────────────────────────────────────────────────────
│  │ │ id : NMId
│  │ ├────────────
│  │ │ id = nm.id
│  │ CheckRequirements ≙ nm.CheckRequirements,   SendInfo ≙ nm.SendInfo
│  │ Instantiate ≙ nm.Instantiate
│  ⋮
└──────────────────────────────────────────────────────────────────
```

An *axiomatic declaration* specifies that the identity ($id : NMId$) of the *Node* class is equal to the identity of the node manager in the class ($id = nm.id$). The definition $SendInfo \triangleq nm.SendInfo$ promotes the operation *SendInfo* defined in the object *nm* to be an operation of the class *Node*.

The *Node Manager* monitors and controls the services instantiated on the node and collects information about the resources available. The node manager class is represented by *NM* below. The information concerning the resources provided by the node is stored in the *nm_mib*. This information is reported by the node manager to the *DSM* to keep the *dsm_mib* updated.

Different management domains, with different responsibilities and purposes may coexist. A *DSM* is the agent responsible for the management of the nodes which are members of one domain. The same node can be a member of different domains, being therefore under the control of more than one DSM. The node manager stores, for each DSM it is associated with, its identifier and the last time monitoring information was sent to update that DSM's *dsm_mib*. For this purpose, the *NM* state includes *updateRecord* of type *UpdateRecord* which is declared as follows (with appropriate definitions for the components):

$$UpdateEntry == DSMId \times SentTime$$
$$UpdateList == seq\ UpdateEntry$$
$$UpdateRecord == MIBUpdateTime \times UpdateList$$

*updateRecord* contains information about the last time the *nm_mib* was updated (*MIBUpdateTime*) and the last time information from the *nm_mib* was sent to each DSM.

```
┌─ NM ──────────────────────────────────────────────────────────────────
│  interface nmctrl add CheckRequirements, SendInfo, Instantiate, Kill
│  interface nmnotify add CapsuleTerminated
│
│  UpdateEntry == DSMId × SentTime
│  UpdateList == seq UpdateEntry
│  UpdateRecord == MIBUpdateTime × UpdateList
│  id : NMId
│  ┌──────────────────────────────────────────────────────────────────
│  │  nm_mib : NM_MIB
│  │  updateRecord : UpdateRecord
│  │  ┌─ CheckRequirements ──────────────────────────────────────────
│  │  │  dsmId? : DSMId
│  │  │  updated! : Bool
│  │  │  dsmId! : DSMId
│  │  │  ──────────────────────────────────────────────────────────
│  │  │  dsmId! = dsmId?
│  │  │  let mibTime == updateRecord.1 ∧ updateList == updateRecord.2
│  │  │  in (∃₁ sentTime : SentTime • (dsmId?, sentTime) ∈ ran updateList∧
│  │  │     (sentTime = mibTime ⇒ updated! = TRUE)∧
│  │  │     (sentTime ≠ mibTime ⇒ updated! = FALSE))
│  │  ┌─ SendInfo ──────────────────────────────────────────────────
│  │  │  dsmId? : DSMId
│  │  │  nmId! : NMId
│  │  │  nmData! : NM_MIB
│  │  │  dsmId! : DSMId
│  │  │  ──────────────────────────────────────────────────────────
│  │  │  nmId! = id ∧ nmData! = nm_mib ∧ dsmId! = dsmId?
│  │  Instantiate ≙ . . .
└────────────────────────────────────────────────────────────────────────
```

The operation *CheckRequirements* can be called on a node by any *DSM* that controls that node. This operation checks if the information in the *nm_mib* has not been changed since the last time it was sent to the DSM identified by *dsmId?*, in which case returns *updated! = TRUE*. If the *nm_mib* has been changed then the node manager will have to retrieve the new information from the *nm_mib*, as specified in *SendInfo*, and send it to the DSM. *Instantiate* is the operation provided by the node manager that allows a new instance of a service to be created via the factory, its definition is omitted here.

## 4.3   Specifying the interaction between objects

The complete specification contains definitions of the trader class (*Trader*) and a *Nodes* class. The interactions between objects of these classes is given by *MgtSystem*. This class contains an object of type Trader, a distributed systems manager (i.e. an object of type *DSM*) together with a set of nodes being managed (*Nodes*) on which services can be scheduled. The class *Nodes* will contain a set of objects of type *Node* together with operations to add and delete nodes etc. Operations are defined in *MgtSystem* which describe how objects in the class interact and communicate. We have omitted some of the operations and the type definitions.

---

**_MgtSystem_**

---

$trader : Trader$
$dsm : DSM$
$nodes : Nodes$
$suitableNodes : \mathbb{P}\ NMId$

---

**_INIT_**

$trader.INIT \wedge dsm.INIT \wedge nodes.INIT$

---

$PollNodes \mathrel{\widehat{=}} [suitableNodes? : \mathbb{P}\ NMId, \Delta(suitableNodes)] \bullet$
　　$\bigwedge node : suitableNodes? \bullet$
　　$((node.CheckRequirements \bullet [updated! = TRUE \wedge suitableNodes' = suitableNodes \cup \{node\}])$
　　$\vee((node.CheckRequirements \bullet [updated! = FALSE]) \mathbin{\mathring{,}} node.SendInfo \mathbin{\mathring{,}} dsm.UpdateNode))$

$InstantiateService \mathrel{\widehat{=}} \ldots$

$NewActivation \mathrel{\widehat{=}}$
　　$(DSMCreateNewActSuccess \mathbin{\mathring{,}}$
　　　　$(PollNodes \bullet [suitableNodes = \varnothing] \wedge [result! : DSMServiceStatus \mid result! = NoSuitableNode])$
　　$\vee(PollNodes \bullet [suitableNodes \neq \varnothing] \mathbin{\mathring{,}}$
　　　　$(InstantiateService \bullet [instantiateResult! = FALSE] \wedge$
　　　　$[result! : DSMServiceStatus \mid result! = FailedToCreateServer])$
　　$\vee(InstantiateService \bullet [instantiateResult! = TRUE] \mathbin{\mathring{,}}$
　　　　$DSMLookupUpdate \wedge [result! : DSMServiceStatus \mid result! = DSMServiceSuccess])))$

⋮

---

*NewActivation* specifies the behaviour corresponding to the creation of a new service instance (called an activation). When the DSM decides to create a new instance of a service, a suitable node will have to be found to allocate that instance. The DSM looks up in the *dsm_mib* for nodes that can provide the service requirements (this is specified in *DSMCreateNewActSuccess*) and polls them to check if they can still provide the same requirements.

The operation *PollNodes* specifies the sequence of operations that are performed when the suitable nodes are polled. The DSM invokes *CheckRequirements* on each node. This operation will return *updated!*, which is *TRUE* if the information in the *nm_mib* has not been changed since the last time it was sent to the DSM, and *FALSE* otherwise. In the last case the node manager will retrieve the updated information from the *nm_mib* (as specified in *SendInfo*) and send it to the *DSM* by calling its *UpdateNode* operation. *UpdateNode* will update the information in the *dsm_mib* for that node.

The $\bullet$ notation in $(node.CheckRequirements \bullet [updated! = FALSE])$ signifies enrichment in that $[updated! = FALSE]$ enriches the environment in which *node.CheckRequirements* is interpreted. Distributed conjunction, as in $\bigwedge node : suitableNodes? \bullet \ldots$ is a convenient mechanism to take the conjunction for each *node* of type *suitableNodes?* of the expression following the $\bullet$.

The operation *PollNodes* also illustrates communication in Object-Z/ZEST using the operator $\mathbin{\mathring{,}}$. The operator composes the two operations in the given order (therefore it is not commutative) with the following communication mechanism. Communication is left to right and hidden, outputs of the left operand equate to inputs of the right operand with the same basename (i.e. ignoring ! and ?) and both are hidden [DRS95]. Thus in the communication $node:SendInfo\, {}_9^o\, dsm.UpdateNode$ the outputs of the first operation are used as inputs to the second operation. Sequential chains (as in $(node.CheckRequirements \bullet [updated! = FALSE]) \, {}_9^o\, node.SendInfo \, {}_9^o\, dsm.UpdateNode)$ are interpreted left-associatively.

After *PollNodes* has been performed, the global variable *suitableNodes* will contain the set of nodes that can still provide the service requirements. *InstantiateService* specifies the behaviour corresponding to the allocation of a new service instance to a node selected from *suitableNodes*.

# 5 DERIVING TESTS FROM FORMAL SPECIFICATIONS

One potential application of formal techniques is in the automation of some or all of the process of testing. In order to support conformance testing of distributed systems, ODP defines conformance within the reference model. This interpretation includes the definition of conformance in each language. Because behaviour is represented by the Z operation schemas, a conformance statement in a Z specification corresponds to one or more operation schemas. Behaviour is said to conform if the post-condition and invariant predicates of this information manipulation are satisfied in the associated Z schemas.

Appropriate work on test generation from Z specifications includes [SC91, CW92, CS94, Ste95, Hor95], however, little of this work is specifically targeted towards distributed systems or ODP in particular. Exceptions to this include [CW92, Ste95]. [Ste95] describes important work done under the Prost project in the UK on the testability of managed object specifications. ZEST is used to specify managed objects, and an inheritance hierarchy is constructed which facilitates the construction of a sound and complete test suite. Importantly, though, the test generation aims to supply heuristics and is not automatic. The heuristics provide a collection of tests together with a residual component which makes explicit the functionality not covered by the test suite. The tests generated form an independent and orthogonal collection of tests.

Because of the inheritance hierarchy, the reuse of tests between related specifications is maximised. A prototype tool-set developed by Logica provides organisational support for the collection of test specifications as they are generated.

We illustrate the use of the method defined in [Ste95] by deriving tests for operations specified in our management architecture. The method derives a formal specification (also written in Z) of conformance tests from each managed object, by producing a collection of tests for each operation in the managed object.

The method describes three actions: partitioning, weakening and simplification, to construct a set of tests for each operation. The method is based on the idea of testing only some of the interesting inputs and only some of the consequences of the operation, a test is therefore an abstraction of the original operation. Each time an action is applied it divides an operation into several parts, each of which will either be a test or be further divided. The division must satisfy the following rules (and heuristics enforce this), where $Op$ is the operation under test and $\{T_i\}$ the collection of tests at this stage: *soundness* i.e. $\forall i \bullet Op \Rightarrow T_i$; *completeness* i.e. the collection of tests must cover the operation, so $Op = \bigwedge_i T_i$.

As an example, consider the operation *RemoveAlias* which is part of the complete DSM specification:

```
┌─ RemoveAlias ─────────────────────────────────────────────────────
│ Δ(alias_table)
│ alias? : Alias
│ status! : DSMServiceStatus
├───────────────────────────────────────────────────────────────────
│ (alias? ∉ dom alias_table.table ⇒ status! = NonExistantAlias)
│ (alias? ∈ dom alias_table.table ∧ alias_table.table(alias?).2 = TRUE ⇒ status! = AliasActive)
│ (alias? ∈ dom alias_table.table ∧ alias_table.table(alias?).2 ≠ TRUE∧
│     alias_table.table(alias?).3 = TRUE ⇒ status! = AliasPosted)
│ ((alias? ∈ dom alias_table.table ∧ alias_table.table(alias?).2 ≠ TRUE∧
│     alias_table.table(alias?).3 ≠ TRUE) ⇒
│   (status! = DSMServiceSuccess ∧ alias_table.Remove_from_table[alias?/key?][TRUE/result!]))
```

**Partitioning** the operation involves deriving a set of tests each covering a different aspect of the operations pre-condition. The partition is defined by predicates $H_i$, and for an operation $Op$ defined in terms of a declaration $D$ and a predicate $P$, we derive tests $T_i$ given by

```
┌─ Op ──────────────────────────        ┌─ T_i ─────────────────────
│ D                                      │ D
├────────────────                        ├────────────────
│ P                                      │ H_i ⇒ P
```

This will split the operation into a collection of tests such that $Op \mathrel{\widehat{=}} T_1 \wedge \ldots \wedge T_n$. For example, for the operation *RemoveAlias* we could partition as follows:

$H_1$ - $alias? \notin$ dom $alias\_table.table$
$H_2$ - $alias? \in$ dom $alias\_table.table$

Choosing the predicates $H_i$ to perform the partition is part of the testers skill, the aim is to choose a partition that simplifies what is being tested. The partitioning process generates (after simplification) tests $T_1$ and $T_2$ (where we will subdivide $T_2$ further) given by:

```
┌─ T_1 ─────────────────────────────────────────────────────────────
│ Δ(alias_table)
│ alias? : Alias
│ status! : DSMServiceStatus
├───────────────────────────────────────────────────────────────────
│ (alias? ∉ dom alias_table.table ⇒ status! = NonExistantAlias)
```

```
┌─ T_2 ─────────────────────────────────────────────────────────────
│ Δ(alias_table)
│ alias? : Alias
│ status! : DSMServiceStatus
├───────────────────────────────────────────────────────────────────
│ (alias? ∈ dom alias_table.table ∧ alias_table.table(alias?).2 = TRUE ⇒ status! = AliasActive)
│ (alias? ∈ dom alias_table.table ∧ alias_table.table(alias?).2 ≠ TRUE∧
│     alias_table.table(alias?).3 = TRUE ⇒ status! = AliasPosted)
│ ((alias? ∈ dom alias_table.table ∧ alias_table.table(alias?).2 ≠ TRUE∧
│     alias_table.table(alias?).3 ≠ TRUE) ⇒
│   (status! = DSMServiceSuccess ∧ alias_table.Remove_from_table[alias?/key?][TRUE/result!]))
```

Repeated partitioning can be applied on the input e.g. on the value of $alias\_table.table(alias?).2$ etc, to produce four tests, the last being:

$T_4$
$\Delta(alias\_table)$
$alias?: Alias$
$status!: DSMServiceStatus$

$((alias? \in \text{dom } alias\_table.table \land alias\_table.table(alias?).2 \neq TRUE \land$
$\quad alias\_table.table(alias?).3 \neq TRUE) \Rightarrow$
$\quad (status! = DSMServiceSuccess \land alias\_table.Remove\_from\_table[alias?/key?][TRUE/result!]))$

**Weakening** can now be applied, which loosens the constraints on the output (and after-state). Weakening an operation $Op$ produces a test $T_w$ and a residual part $T_r$, which documents the aspects of the operation we are not testing, with $Op \cong T_w \land T_r$. In *RemoveAlias* we are not interested in checking the output *status!*, just that the *alias_table* has been altered correctly, so we weaken the test $T_4$ to derive the weakening $T_w$. The remaining component, $T_r$, will now document the aspects of *RemoveAlias* not covered by the conformance testing.

$T_w$
$\Delta(alias\_table)$
$alias?: Alias$
$status!: DSMServiceStatus$

$((alias? \in \text{dom } alias\_table.table \land alias\_table.table(alias?).2 \neq TRUE \land$
$\quad alias\_table.table(alias?).3 \neq TRUE) \Rightarrow$
$\quad alias\_table.Remove\_from\_table[alias?/key?][TRUE/result!])$

This method works well on the individual schema level, however, most of the interesting behaviour in a managed object results from a process of combining operations using the schema calculus. Current test generation methods such as the one outlined above need to be extended to produce tests from operations defined using the schema calculus. For example, if the operations $Op_1$ and $Op_2$ produce complete and sound tests $\{T_i\}$ and $\{R_j\}$ respectively, can we derive a suitable collection of tests for the operation $Op_1 \,\sembnf\, Op_2$?

# 6   CONCLUSIONS

The use of formal description techniques is increasing within ODP, and a number of proposals to specify managed objects formally have been made [Rud91, SM92, Nor92, WJ94, ISO92b]. However, existing work in this area has concentrated on small scale case studies involving just one managed object (often the Sieve or LOG managed object). At ISINM'95 we reported on differing proposals to the formal specification of managed objects [DLT95]. Further work in the UK has produced guidelines on how to specify managed objects in Z [Zad96], and derived a method for producing tests derived from these formal specifications.

The aim of our work has been to test the applicability of these methods by specifying a larger scale case study of a new application (rather than a behaviour that is well documented). While Z is not necessarily a perfect vehicle for managed object specification, it does offer considerable benefits over current practice. For specifications where behaviour is important or subtle, GDMO clearly needs enhancement, and Z offers a wide user base and suitable facilities for abstraction. The ability to derive tests from formal specifications adds another dimension to the usefulness of the technique, although further work is needed in this area as outlined above.

# REFERENCES

[CR92] E. Cusack and G. H. B. Rafsanjani. ZEST. In S. Stepney, R. Barden, and D. Cooper, editors, *Object Orientation in Z*, Workshops in Computing, pages 113–126. Springer-Verlag, 1992.

[CS94] D. Carrington and P. Stocks. A tale of two paradigms: Formal methods and software testing. In J.P. Bowen and J.A. Hall, editors, *ZUM'94, Z User Workshop*, pages 51–68, Cambridge, United Kingdom, June 1994.

[CW92] E. Cusack and C. Wezeman. Deriving tests for objects specified in Z. In J. P. Bowen and J. E. Nicholls, editors, *Seventh Annual Z User Workshop*, pages 180–195, London, December 1992. Springer-Verlag.

[DLT95] J. Derrick, P.F. Linington, and S.J. Thompson. Formal description techniques for object management. In A. S. Sethi, Y. Raynaud, and F. Faure-Vincent, editors, *Fourth IFIP/IEEE International Symposium on Integrated Network Management (ISINM '95)*, pages 641–653. Chapman and Hall, 1995.

[DRS95] R. Duke, G. Rose, and G. Smith. Object-Z: A specification language advocated for the description of standards. *Computer Standards and Interfaces*, 17:511–533, September 1995.

[G. 96] G. P. A. Fernandes and I. A. Utting. An Object-Oriented Model for Management of Services in a Distributed System. ECOOP'96 workshop on Object Oriented Technology for Service and Network Management, 1996.

[Hor95] H-M. Horcher. Improving software tests using Z specifications. In J. P. Bowen and M. G. Hinchey, editors, *Ninth Annual Z User Workshop*, LNCS 967, pages 152–166, Limerick, September 1995. Springer-Verlag.

[ISO91] ISO/IEC JTC1/SC21/WG4 10165-4 (X.722). *Information Technology - Open Systems Interconnection - Structure of Management Information - Part 4: Guidelines for the Definition of Managed Objects*, 1991.

[ISO92a] ISO/IEC 10040. *Information Technology - Open Systems Interconnection - Systems Management Overview*, 1992.

[ISO92b] ISO/IEC JTC1/SC21/WG4 N1644. *Liaison to CCITT SG VII concerning the use of Formal Techniques for the specification of Managed Objects*, December 1992.

[ITU95] ITU Recommendation X.901-904 — ISO/IEC 10746 1-4. *Open Distributed Processing - Reference Model - Parts 1-4*, July 1995.

[Nor92] N D North. RSL specification of the log managed object. Technical report, National Physical Laboratory, UK, 1992.

[Raf94] G. H. B. Rafsanjani. ZEST - Z Extended with Structuring: A users's guide. Technical report, British Telecom, June 1994.

[Rud91] S. Rudkin. Modelling information objects in Z. In J. de Meer, V. Heymer, and R. Roth, editors, *IFIP TC6 International Workshop on Open Distributed Processing*, pages 267–280, Berlin, Germany, September 1991. North-Holland.

[SC91] P. Stocks and D. Carrington. Deriving software test cases from formal specifications. In *6th Australian Software Engineering Conference*, pages 327–340, July 1991.

[SM92] L. Simon and L. S. Marshall. Using VDM to specify OSI managed objects. In K R Parker and G A Rose, editors, *Formal Description Techniques 1991*. North Holland, 1992.

[Spi89] J. M. Spivey. *The Z notation: A reference manual*. Prentice Hall, 1989.

[Ste95] S. Stepney. Testing as Abstraction. In J. P. Bowen and M. G. Hinchey, editors, *Ninth Annual Z User Workshop*, LNCS 967, pages 137–151, Limerick, September 1995. Springer-Verlag.

[WJ94] C. Wezeman and A. J. Judge. Z for managed objects. In J. P. Bowen and J. A. Hall, editors, *Eighth Annual Z User Workshop*, pages 108–119, Cambridge, July 1994. Springer-Verlag.

[Zad96] H. B. Zadeh. Using ZEST for Specifying Managed Objects. Technical report, British Telecom, January 1996.

# TRACK III

## PART ONE

# CORBA-Based Management
## Chair: Laura Cerchio, CSELT

# 36

# Incorporating Manageability into Distributed Software

*R. Chadha and S. Wuu*
*Bellcore*
*445 South Street*
*Morristown, NJ 07960, USA*
*+1(201) 829-4869 (tel), +1(201) 829-5889 (fax)*
*chadha@bellcore.com*

## Abstract

In today's world, where software applications have evolved from mainframe-based applications to client/server-based distributed systems, the need for effective, easy to develop, and open management systems is becoming painfully evident. The goal of this work is to develop cost-effective methods and tools for managing software in a manner analogous to the way in which network elements are managed in today's networks. In order to provide a flexible, interoperable, standards-compliant solution to this problem, we have developed an infrastructure that conforms to the Telecommunications Management Network (TMN) suite of standards. The TMN is based on the OSI/CMISE systems management standards. Although telecommunications providers, both in the domestic and international markets, have been pressing equipment and operations systems suppliers to start developing TMN-compliant products, the supplier community has been reluctant to move towards OSI/CMISE management systems, largely due to the difficulty and expense of implementing OSI/CMISE interfaces to management applications.

This document gives a description of the components of an infrastructure that provides a middleware layer which shields software developers from much of the complexity of OSI/CMISE management implementation. It presents techniques that can be used to simplify the process of implementing TMN-compliant interfaces. This layer will considerably ease the process of incorporating standards-based management interfaces into software components for the purpose of management.

## Keywords

Managing distributed software, CMISE, CORBA

## 1   INTRODUCTION

In today's world, where software applications have evolved from mainframe-based applications to client/server-based distributed systems, the need for effective, easy to develop, and open management systems is becoming painfully evident. The quality of large, distributed software applications will ultimately depend upon the effectiveness of their management systems. Management systems are required to perform fault, configuration, accounting, performance, and security management. As stated in the abstract, our work focuses on the problem of managing distributed software applications. This entails a study of the whole lifecycle of an end-to-end management application, starting with the definition of an information model, or management information base (MIB); designing the interface between the management application and the managed resources (which in this case are distributed software applications); defining the functionality of the management application; and so on.

This paper describes an architecture for distributed systems management based on OSI systems management. The use of OSI management provides the inherent advantage of making available standard, open interfaces for management applications. OSI management systems have not been widely implemented and deployed, in spite of support from international standardization bodies and government organizations. Even though OSI management offers a powerful, object-oriented management model, proprietary protocols and SNMP (Case et al., 1990) are still the preferred solutions for a large section of the market. This has been attributed mainly to the difficulty of OSI management implementation. In this paper, we show how we have hidden much of the complexity of OSI implementation from developers who need to make their applications manageable. We have analyzed, from a developer's viewpoint, the steps required to make an application manageable. In order to make a software application manageable, it will be necessary for developers to instrument their applications so that an agent will be able to retrieve management information from these applications; also, these software applications must be designed to emit the appropriate event reports, as defined by the information model. As mentioned earlier, it is naturally desirable that this be accomplished with a minimum of effort, so as not to add a disproportionate burden to the developer's task. With this in mind, we examine the steps required to make an application manageable, and suggest ways to partially automate this task.

A preliminary description of this work was given in (Chadha and Wuu, 1996a), where we described an end-to-end management system that managed a set of simple CORBA software components. In this paper, we describe in greater detail the steps required to build an agent for the managed resources, and to make applications manageable. This paper is organized as follows. Section 2 outlines the contributions of this work. Section 3 describes the system architecture and information model used for our prototype development, and the various components that make up the system. In Section 4, we describe the steps required to build an agent for our system. The steps for incorporating manageability into software components appear in Section 5. Some related work is discussed in Section 6, with conclusions in Section 7.

## 2   CONTRIBUTIONS OF THIS WORK

The main contribution of this work is that it demonstrates the feasibility of applying OSI management to distributed systems management. In this paper, we show how OSI management and the OMG CORBA (OMG, 1994) framework can be easily integrated, and why this approach is a good one. We demonstrate that the task of making a distributed CORBA application manageable can be reduced to the task of implementing some automatically generated CORBA interfaces.

The most important feature of our software management infrastructure is that the developers of the distributed software application being managed do not have to have any expertise in OSI systems management. Their task is to write server code to implement management operations (such as a function to "get" an attribute value), and to make calls to remote objects in order to send them information about events that occur in the application that the management application needs to know about. Thus they only deal with CORBA interfaces and are completely shielded from details about the CMIP protocol (CCITT, 1991) and about populating ASN.1 data structures (CCITT, 1988), which can be a formidable task, and which is an integral part of OSI management.

The second advantage of our approach is that it provides a precisely defined method for going from a GDMO (CCITT, 1992b) information model (which is the starting point of any OSI management application) to an implementation. The use of CORBA interfaces which are automatically generated from a GDMO information model provides a uniform starting point for all developers who need to make their applications manageable.

Finally, due to its implementation of OSI management standards, the management infrastructure described in this paper provides an open interface for management applications. Implementing such an open interface will make the distributed software application attractive to customers who wish to implement their own management applications in order to customize the latter to fit their needs.

## 3   AN END-TO-END MANAGEMENT SYSTEM

In this section, we describe a prototype implementation of a management infrastructure for managing distributed software applications. The following sections describe the architecture of this system, the implementation platforms and the information model used.

### 3.1 The information model

We have developed an information model (using GDMO) to represent managed software applications. This model defines a new managed object class called `softwareProcess`. This managed object class inherits all the properties of the managed object class `software`, which is defined in (CCITT, 1992d). The purpose of this managed object class is to encapsulate information about network and software configuration (including distributed software dependencies), performance (including process health and status), and faults. In addition to the attributes inherited from the `software` managed object class, the `softwareProcess` managed object class contains the following attributes: `serverList`, `peerList`, `clientList`, `processId`, `processingEquipmentName`, `processName`, `processCreationTime`, `processUpTime`, `processUserId`, `processGroupId`, `processEfUserId`, `processEfGroupId`, `processArguments`, `threadsUsed`, `cpuTime`, `filesOpened`. The first three attributes listed here are pertinent for monitoring distributed processes, while the rest provide information about the process and its health. These attributes are more fully described in (Chadha and Wuu, 1996a). The `softwareProcess` managed object class can also emit a number of notifications inherited from the `software` managed object class. The complete GDMO description of this information model can be found in (Chadha and Wuu, 1996b).

## 3.2 System Architecture

Figure 1 shows the architecture of the system and the platforms chosen to implement this prototype. All applications and platforms in this prototype run on Sun SPARC-stations™ running Solaris™ 2.3. There are essentially three distinct components in this architecture: the manager, the agent, and the managed application (or managed resources). The interface between the manager and the agent is a standard CMIP interface. The manager and agent communicate using SunLink™[1] OSI 8.0, a full 7-layer OSI stack. The managed software application is implemented using the Common Object Request Broker Architecture (CORBA™[2]) (OMG, 1994). The interface between the agent and the managed application is CORBA. CORBA was developed by the Object Management Group (OMG), and defines mechanisms by which distributed objects transparently make requests and receive responses. It provides interoperability between applications on different machines in heterogeneous distributed environments. CORBA provides an Interface Definition Language (IDL) for defining the interface to an object. Section 3.3 discusses the translation of a GDMO description of an object into a CORBA IDL description. Our choice of CORBA as the distributed application platform was motivated by the emergence of CORBA as one of the prime candidates for the next generation of distributed computing platforms. The following subsections describe the manager, agent, and managed resources platforms in more detail.

---

1. Sun SPARCstation, Solaris, and SunLink are registered trademarks of Sun Microsystems, Inc.

2. CORBA is a registered trademark of Object Management Group, Inc.

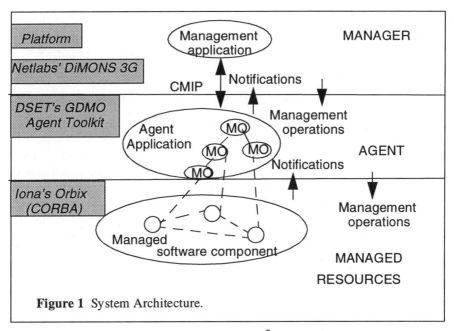

**Figure 1** System Architecture.

## 3.2.1   The manager platform: Netlabs™[3]' DiMONS 3G

In order to reduce the complexity of building an OSI manager, we used Netlabs' DiMONS 3G management platform. The DiMONS 3G platform includes GDMO and ASN.1 compilers, which are used to incorporate new GDMO object definitions into the platform. An API called the Portable Management Interface (PMI) is provided for manipulating managed object information and building management applications. Access control lists are used to govern access to management information. It should be noted that Netlabs no longer supports the DiMONS 3G product, and much of their technology has been licensed by Sun Microsystems, Inc. and is incorporated in SunSoft's Solstice™ Enterprise Manager™[4] product. We plan to migrate to Sun's platform in the near future.

## 3.2.2   The agent platform: DSET™[5]'s GDMO Agent Toolkit

The agent was built using DSET's GDMO Agent Toolkit. This toolkit provides a suite of tools for building lightweight agent applications with low memory requirements. It provides GDMO and ASN.1 compilers that generate C data structures for

---

3. Netlabs is a registered trademark of Netlabs, Inc.

4. Solstice and Enterprise Manager are registered trademarks of Sun Microsystems, Inc.

5. DSET is a trademark of DSET Corp.

access by a program. It also provides a C/C++ API called ASN.C/ASN.C++ to enable programmers to manipulate ASN.1 data structures with ease. The platform is built on top of the Distributed Systems Generator, a DSET proprietary product.

### 3.2.3    The managed resources platform: IONA's Orbix™[6]

IONA's Orbix (Iona, 1996) is a commercially available implementation of CORBA. This product provides a CORBA IDL compiler, which generates a C++ class for each IDL interface. Each operation in an IDL interface is mapped into a C++ member function. An IDL interface consists of *operation* and *attribute* specifications. Each attribute is mapped into a pair of C++ member functions: one to read (get) the value, and the other to write (set) the value. IDL attributes can also be "readonly"; these attributes map to a single function, which returns the attribute value. IDL allows one interface to inherit from another, thereby creating an inheritance hierarchy. The object-oriented nature of IDL makes it relatively easy to define a mapping between the object models of CORBA IDL and GDMO; such a mapping will be discussed in the next section.

## 3.3    GDMO to CORBA IDL translation

The OSI Systems Management suite of standards has defined a language to be used for defining managed objects, namely GDMO. However, in order to implement OSI management, it is necessary to translate this GDMO specification into a language closer to the implementation platform. Since we have chosen the CORBA platform for implementing managed applications, we need a way to translate GDMO to CORBA IDL. Fortunately, the X/Open Joint Inter-Domain Management Taskforce (XoJIDM) has been working on this problem for the past several years, and has developed a GDMO to CORBA IDL specification translation algorithm. Commercial compilers have recently become available for performing this translation. For a complete description of this translation algorithm, see (X/Open, 1994).

Figure 2 shows a fragment of an object description in GDMO, and its translation into CORBA IDL. The GDMO fragment defines an attribute (`administrativeState`) and the management operations that can be performed on it (`GET` and `REPLACE`). The corresponding CORBA IDL contains two methods, one to perform the management operation `GET` (`administrativeStateGet()`) and one to perform the management operation `REPLACE` (`administrativeStateSet()`).

The XoJIDM taskforce has also been working on a Dynamic Interaction Translation document, which provides a mapping between CMISE services and OMG services. The GDMO to CORBA IDL specification translation algorithm mentioned

---

6.  Orbix is a registered trademark of IONA Technologies Ltd.

```
ATTRIBUTES administrativeState
GET-REPLACE;

administrativeState ATTRIBUTE WITH ATTRIBUTE SYNTAX
 Attribute-ASN1Module.AdministrativeState;
MATCHES FOR EQUALITY;
```

**Figure 2 (a)** Fragment of GDMO definition of an attribute.

```
Attribute-ASN1Module::AdministrativeStateType
administrativeStateGet() raises (CMIP_ATTRIBUTE_ERRORS);

void administrativeStateSet(in
Attribute_ASN1Module::AdministrativeStateType Value)
raises (CMIP_ATTRIBUTE_ERRORS);
```

**Figure 2 (b)** Translation of GDMO into CORBA IDL.

above only addresses the syntactic translation of GDMO to CORBA IDL, and ignores the dynamic policies of object behavior defined by other GDMO templates. For example, NameBinding templates define the policies governing the lifecycle of a managed object (i.e. rules governing creation, deletion, copying, and naming of managed objects). The specification translation also does not make use of any of the OMG Common Object Services (OMG, 1995). The Dynamic Interaction Translation document is intended to deal with issues such as support for conditional packages, distribution of event reports using the OMG Event Service, and support for a Management Information Repository.

The XoJIDM Dynamic Interaction Translation provides several scenarios for CMIP-CORBA interworking. In one scenario, the managing system is CORBA-based, and the agent is in an OSI managed system, which communicates with the manager using CMIP. In such a case, a gateway function must be part of the managing system in order to translate the manager's CORBA requests into CMIP requests, to translate the agent's responses from CMIP to CORBA, and to translate CMIP event reports received from the agent into CORBA (see Figure 3). In another scenario, the situation is reversed: the managed system is CORBA-based, whereas the managing system is OSI-based. A gateway function is now required in the managed system, to translate between CORBA and CMIP (see Figure 4).

Finally, it is possible to have a scenario where both the managing and managed systems are CORBA-based, and they communicate using either CORBA, or two CMIP-CORBA gateways (one in the agent and one in the manager).

## 4   BUILDING THE AGENT: CMIP-CORBA INTERWORKING

In our architecture, the managing system is OSI-based, and the managed system belongs to the CORBA world, a situation similar to that depicted in Figure 4. This

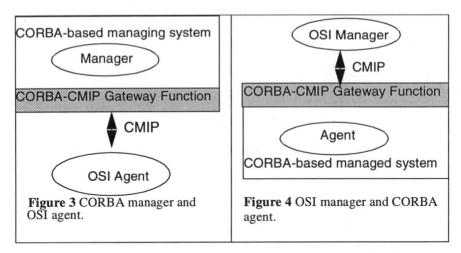

**Figure 3** CORBA manager and OSI agent.

**Figure 4** OSI manager and CORBA agent.

situation requires us to build a gateway function in the managed system. Rather than build such a system from scratch, we chose to use the infrastructure provided by the DSET GDMO Agent Toolkit, which handles much of the CMIP communications tasks, and leaves only a small portion of the gateway to be implemented by us. Thus in our system, the gateway function will be incorporated inside the agent application in the managed system.

Our agent application has to perform a number of tasks. It has to maintain a Management Information Tree (MIT) for the managed object instances; it has to translate management operations into CORBA operations to be performed on the managed software components; and it has to set up a CORBA server to receive notifications from the managed software components, which it must then translate into CMISE event reports and send to the manager. In order to be able to receive notifications sent from CORBA applications (which are the managed resources), the agent starts up a thread that implements a CORBA server. This thread waits for notifications to arrive from the managed software components, and translates them into notifications that are sent to the manager. The tasks performed by the agent are described in detail in the following subsections.

## 4.1     The agent's Management Information Tree

When an agent starts up, it can create instances of managed objects in its MIT. For our purposes, let us assume that it creates two managed object instances. The first instance created is an instance of the `system` managed object class. This instance is used as the root of the MIT. The second instance created is an instance of the `eventForwardingDiscriminator` managed object class, defined in (CCITΓ, 1992a). This is a special-purpose managed object class, whose function is defined in (CCITT, 1992c). This managed object class contains attributes which determine how incoming event reports from managed resources will be disposed of.

Two of the more important attributes of this managed object class are the `desti-nation` attribute, which specifies a list of managers to whom event reports should be forwarded, and the `discriminatorConstruct` attribute, which allows the specification of a filter that ensures that only event reports satisfying this filter are forwarded to managers whose address is specified by the `destination` attribute.

Whenever a managed software component comes up, it makes a call to the agent CORBA server (implemented as a thread within the agent application) and sends it an `objectCreation` (CCITT, 1992a) notification. This notification contains all the information needed by the agent to create an instance of the appropriate managed object class. This managed object class will typically be `softwareProcess`, or a more application-specific managed object class that inherits all properties from the `softwareProcess` managed object class. This instance is created by the agent, and an `objectCreation` notification is sent to the manager. If this software component goes down or terminates, an `objectDeletion` notification (CCITT, 1992a) is emitted.

## 4.2    Mapping CMIP requests to function calls in the agent

One of the tasks of the gateway function in Figure 4 is to take a CMIP request (received from the manager), translate it into an appropriate method call on a CORBA interface, and invoke that method. Using the DSET Toolkit, CMIP requests received from the manager are automatically mapped to function calls; however, it is the user's responsibility to specify *what* functions will be called. As an example, suppose that the manager issues an M-GET operation on a managed object with distinguished name X. This M-GET request is received by the DSET agent, and transmitted to the managed object with distinguished name X. Now, for every attribute of this managed object, the user is expected to have specified a function to be called whenever an M-GET is issued on this attribute. Thus, for every attribute of the managed object X, the corresponding user-specified function is automatically called. Figure 6 below illustrates the operation of this strategy. Note that the gateway function is actually implemented inside the agent here.

Since the user is required to specify what function is called when a CMIP request is received, let us examine how this relates to the XoJIDM translation specification. Using the example in the previous section, suppose the manager issues an M-GET request on managed object X. Suppose X has two attributes, `a1` and `a2`. According to the XoJIDM translation specification, M-GET requests for these two attributes map to the two methods

    `a1Get()` and `a2Get()`.

Therefore, we wrote a function that is called whenever a GET has to be performed on the attribute `a1` (and similarly for attribute `a2`); this function does the following:

1. Makes a call to a method called `a1Get()` implemented in the CORBA object which is represented by the managed object to which `a1` belongs

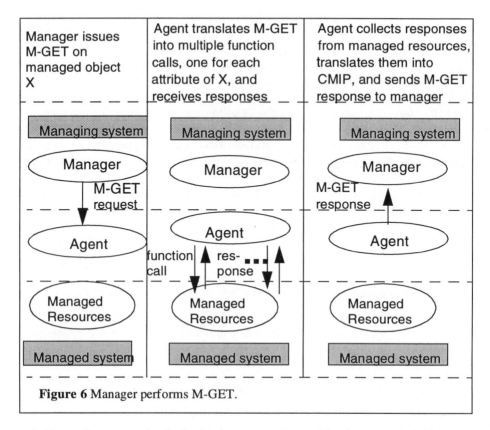

**Figure 6** Manager performs M-GET.

2. Stores the return value in the data structure allocated for that managed object in the agent.

The DSET Toolkit handles the formatting and dispatching of the M-GET response.

## 4.3    Mapping CORBA event reports to CMIP event reports

The XoJIDM translation specification maps event reports to methods. These methods can therefore be invoked by the CORBA managed resources whenever an event report is to be forwarded to the agent. Once such a method invocation is received by the agent, the following must be done:

1. The agent forwards this event report to all Event Forwarding Discriminator objects and Log objects in its MIT.

2. The Event Forwarding Discriminator objects check their discriminator construct and scheduling attributes to determine whether this event report needs

to be forwarded to a manager; if so, the event report is forwarded to all managers listed in the Event Forwarding Discriminator's `destination` attribute.

3. The Log objects check their discriminator construct and scheduling attributes to determine whether this event report needs to be logged; if so, appropriate log records are created in the agent's MIT.

The DSET Toolkit simplifies this entire process by providing an API for notifications. Thus, instead of performing steps 1 through 3 listed above, a call to an appropriate function in this API conveys all the notification information to the Toolkit engine, which takes care of steps 1 through 3.

We implemented this process in our agent by setting up a thread which is a CORBA server that waits for notifications from managed objects. Whenever a notification is received, it populates the necessary data structures and makes a call to the appropriate function in the DSET notification API.

## 4.4 How dependent is this architecture on the DSET Toolkit?

In designing this architecture, one of our primary objectives was to ensure that our design would survive platform and tool changes. The preceding sections described much of the work that we did in order to integrate the management of CORBA software components using an agent built using the DSET GDMO Agent Toolkit. However, the fact that we are using the XoJIDM specification translation makes the architecture flexible. Once implementations of gateway functions (as depicted in Figures 3 and 4) based on the XoJIDM work become available on commercial platforms, it should be a simple task to migrate our approach to any commercial platform which provides such a gateway implementation. At this stage, since no commercial gateway implementations were available, we were forced to implement our own gateway function, and were able to successfully leverage DSET's GDMO Agent Toolkit for this purpose.

## 5 MAKING SOFTWARE MANAGEABLE

The first step to making a CORBA application manageable is to select the managed object classes that are going to be used to represent the managed information. This can be done by developing new GDMO descriptions of managed object classes, or re-using existing ones. This GDMO document can then be translated automatically into CORBA IDL. The IDL thus generated consists of two parts: one contains interfaces for which the managed application is a client, and the other contains interfaces for which the managed application is a server. The managed application will act as a server when management operations are performed (see Section 5.1), and will act as a client when it needs to send notifications to the agent (see Section 5.2).

## 5.1      Instrumenting managed software components to respond to management operations

In order to implement the server part of the IDL, the developer must write code to implement the functions specified therein. An example of such a function is a function to "get" the value of an attribute. For example, using our example from Section 4.3, a function to retrieve the value of an attribute `a1` would be called `a1Get()`, and would return the current value of this attribute. Much of the task of implementing these functions can be simplified too. First, consider attributes that can be obtained from the operating system (such as, for example, the `processId`, `processUpTime`, and `cpuTime` attributes of the `softwareProcess` managed object class). Clearly, the code to obtain the values of such attributes is independent of the managed software component. Thus, these functions can be implemented once, and this code can be linked in with every managed software component. This makes the instrumentation for these attributes almost trivial. For other attributes that require explicit population by each developer (e.g. the `affectedObjectList` attribute, which specifies the object instances which can be directly affected by a change in state or deletion of a given managed object), much of the code can be written once, leaving only a few constants and parameters to be filled in by the developer. Thus the burden of instrumenting a managed software component is reduced to a minimum.

## 5.2      Emitting notifications from managed software components

The client side of the IDL is used for sending notifications; thus the semantics of the notifications (i.e. *when* does a notification need to be sent?) need to be analyzed and then implemented by populating the notification parameters and calling the notification functions wherever and whenever required. For example, in some cases, whenever the value of some attribute of interest changes, an `attributeValueChange` notification is sent to the agent. This notification contains information such as the name of the object sending the notification, the type of event, and the attribute that has changed along with its new value. Another example is that an `objectCreation` notification must be sent when the application starts up. The developer must incorporate client calls in order to send these notifications to the agent.

## 6      RELATED WORK

Although OSI management standards have been in existence for a number of years, OSI management systems have not been deployed on a large scale. This is partly due to the difficulty of implementing OSI management systems. A large number of researchers have been looking at the problem of simplifying the implementation of management systems. Some have opted for non-standards-based approaches, by defining their own tools and protocols for management. The obvious disadvantage of such approaches is the lack of standardization.

A lot of work has been done in the area of simplifying the task of implementing OSI management systems. The OSIMIS (OSI Management Information Service) platform (Pavlou et al., 1993) is an object-oriented development environment in C++ based on the OSI management model. It provides APIs for hiding the details of the underlying management services. An OSI management library is described in (Deri and Mattei, 1995). This paper also describes tools for facilitating the development of OSI management applications. Their approach for representing ASN.1 syntax using strings resembles that used by the IBM cmipWorks platform (Geiger et al., 1994). In addition to these platforms, there are a number of commercially available development tools and platforms (some of which were used in our work and mentioned in Section 3.2) that speed the process of OSI management implementation. However, none of these platforms provide any method for integrating OSI management and the OMG CORBA framework, which is one of the main features of our work.

## 7 CONCLUSION

In this paper, we described the internals of the service management infrastructure outlined in (Chadha and Wuu, 1996a). Special attention was paid to the details of building an agent, and methods for simplifying the process of making software components manageable. The construction of a CORBA-CMIP gateway allows transparent management of CORBA software components using OSI management, and hides the complexity of OSI management from the software developers.

## 8 REFERENCES

Case, G., Fedor, M., Schoffstall, M., Davin, J. (1990) "*A Simple Network Management Protocol (SNMP)*", Request for Comments 1157.

CCITT Recommendation X.208 (1988), ISO/IEC 8824, Spec. of Abstract Syntax Notation One (ASN.1).

CCITT Recommendation X.711 (1991), ISO/IEC 9596-1, Information Technology - OSI, Common Management Information Protocol (CMIP) - Part 1: Specification.

CCITT Recommendation X.721 (1992a), ISO/IEC 10165-2, Information Technology - OSI - Management Information Services - Structure of Management Information - Part 2: Definition of Management Information.

CCITT Recommendation X.722 (1992b), ISO/IEC 10165-4, Information Technology - OSI - Management Information Services - Structure of Management Information - Part 4: Guidelines for the Definition of Managed Objects.

CCITT Recommendation X.734 (1992c), ISO/IEC 10164-4, Information Technology - OSI - Systems Management: Event Report Management Function.

CCITT Recommendation M.3100 (1992d), Generic Network Information Model.

Chadha, R., Wuu, S. (1996a), *"Managing Distributed Systems using OSI Management"*, Proceedings of the Second Intl. IEEE Workshop on Systems Management, Toronto Ontario, Canada, pp. 117-126.

Chadha, R., Wuu, S. (1996b), *"Service Management Infrastructure"*, Bellcore TM-25364.

Deri, L., Mattei, E. (1995), *"An object-oriented approach to the implementation of OSI management"*, Computer Networks and ISDN Systems 27, pp. 1367-1385.

Geiger, G., Allen, W., Majtenyi, A., Reder, P. (1994), *"IBM cmipWorks: Technical paper"*, IBM.

Iona (1996), *"Welcome to Iona"*, IONA Technologies Ltd., http://www.iona.ie.

OMG (1994), *"Common Object Services Specification"*, Volume I, OMG Document Number 94-1-1.

OMG (1995), *"Common Object Request Broker Architecture 2.0 Specification"*.

Pavlou, G. Bhatti, S. N., Knight, G. (1993), *"The OSI Management Information Service: User's Manual, Version 1.0"*.

X/Open (1994), *"GDMO to OMG IDL Specification Translation Algorithm"*, X/Open Company Ltd.

## BIOGRAPHY

Ritu Chadha obtained her Ph. D. in Computer Science from the University of North Carolina at Chapel Hill in 1991. The subject of her dissertation was the mechanical generation of loop invariants for the purpose of program verification. She joined IBM in 1991, where she worked on software testing tools. In 1992, she joined Bellcore as a Research Scientist. She has worked on areas in software testing, modeling and simulation, distributed systems, and network and service management. She is currently working on issues related to Web site management. Dr. Chadha is also a part-time faculty member at Rutgers University.

Sze-Ying Wuu joined Bellcore in 1990 as a Research Scientist. She has worked on areas in distributed systems, distributed directories, multi-media conferencing systems, Internet and Intranet, network and service management, and web management. She was granted a patent in 1986 for her research work in distributed directory technology. She previously was the Engineering Manager at JvNCnet, an Internet Service Provider, where she was responsible for network management, operations, and engineering. She also worked with Bellcore as a consultant on telecommunication signaling and call processing systems from 1986 to 1989. Sze-Ying received an MS in Computer Science from the University of Wisconsin-Madison in 1985.

# 37

# Designing Scaleable Applications Using CORBA

*R.B. Whitner*
*Hewlett-Packard Company*
*Network & System Management Division*
*3404 E. Harmony Rd.*
*Fort Collins, CO 80525 USA*
*Phone: 970-229-3821 FAX: 970-229-2038*
*E-mail: rick_whitner@hp.com*

### Abstract

The Object Management Group's Common Object Request Broker Architecture (CORBA) is quickly becoming an industry-leading approach to building distributed applications. CORBA was investigated for its use to build applications that must deal with very large numbers (up to millions) of managed objects. This paper presents a set of challenges to scaleability encountered during that investigation and a set of techniques that can be used to address those challenges.

### Keywords

CORBA, scaleability, object identity, object granularity, partitioning, Entity-Warehouse Model

## 1 INTRODUCTION

CORBA is an architecture—defined by an industry consortium called the Object Management Group (OMG)—for building distributed applications. CORBA supports a computation model based on communication among objects. An object is a computational entity that supports a well-defined set of services, called its interface. Implementations of object interfaces are deployed for use by object users. Object users (clients) issue requests to objects by means of object references which encapsulate object location information. An object request broker (ORB) manages the dispatch of requests between object users and objects.

CORBA was investigated for use in a distribution infrastructure on which applications that must manage very large numbers of objects will be built. The objects in these applications typically represent real resources in the environment, such as workstations, printers, network elements, etc. These objects are referred to here as entities. The applications manage the entities in a variety of ways. For example, one application maintains a topology of relationships that exist among entities. Another renders graphical representations (maps) of various topologies. Others provide report generation, trend analysis, fault management, etc. Various protocols—of which CORBA is just one—can be used to operate on these entities.

The distribution infrastructure had to support environments scaling up to millions of entities. Additional constraints were also imposed. Constraints worth mentioning were that the infrastructure could not impose the use of proprietary extensions to CORBA, nor make the applications dependent on any particular vendor's ORB. Additionally, it was anticipated that there would be significant costs associated with transitioning legacy environments to CORBA. A model that would enable developers to easily integrate legacy applications and choose how much of CORBA to use was desired.

The investigation examined various ways of applying CORBA such that the requirements could be reasonably satisfied. The results of the investigation are presented here. Section 2 presents a list of issues that application developers will face when using CORBA for large-scale application development. Section 3 presents a model for using CORBA to build applications that will scale to support very large numbers of objects. Section 4 presents some conclusions.

## 2    CORBA SCALEABILITY ISSUES

This section describes key issues that must be addressed when building CORBA applications that have very high scaleability requirements.

### 2.1  Appropriate Level of CORBA Object Granularity

In an object-oriented environment, an object represents an abstraction of some 'real' thing that the developer defines so that a computational representation of the thing can be created. For example, a set of services that permit the management of a workstation might be organized together into a single object interface called Workstation. This interface (abstraction) can be treated programmatically as though it really were a workstation, even though physically it is not.

Object granularity is a term used to describe the relative levels of abstraction between different definitions of the same thing. A specification Y is said to be more granular (finer grained) than X when Y defines as objects the non-object characteristics of X. In CORBA terms, Y defines interfaces for parts of X that are not interfaces, as illustrated in Figure 1. In Specification X in this illustration, B is considered an object by CORBA because it is defined using the keyword *interface*; A is only a data structure. In Specification Y, both A and B are CORBA objects.

```
struct A {                              interface A {
        attribute string a;                     attribute string a;
        attribute string b;                     attribute string b;
};                                      };
interface B {                           interface B {
        attribute A a;                          attribute A a;
        attribute string b;                     attribute string b;
};                                      };

    (a) Specification X                     (b) Specification Y
```

**Figure 1** Relative levels of object granularity in a specification.

Being a CORBA object implies a certain status within a CORBA environment (first class citizenship, so to speak). Being a 'first class CORBA object' means tighter integration with other CORBA services. The CORBA object (designated using the IDL keyword interface) is the basic unit that CORBA recognizes—it is the only unit for which you can obtain object references. Common object services, such as the OMG Naming Service and Transaction Service (Object Management Group, Inc., 1996) typically are defined in terms of first class CORBA objects. Other CORBA features such as interface inheritance (with its corresponding type checking) only apply to CORBA objects.

Despite these benefits, CORBA object granularity must be considered carefully in very large environments because it directly impacts the number of CORBA objects in the system. Finer-grained approaches mean more CORBA objects, which could mean a significant increase in system resource requirements due to overhead associated with CORBA objects.

Care must also be taken to ensure the appropriate level of abstraction is being applied. For example, some of the applications referred to in this investigation use CORBA to provide a higher level of abstraction of SNMP objects. It would not be appropriate to simply create a CORBA object for each SNMP object.

## 2.2 Proprietary versus Interoperable Object References (IORs)

An object reference is a CORBA-defined data structure that clients use to make requests for object services. CORBA permits two types of object references. Interoperable Object References (IORs) (Object Management Group, Inc., 1995) enable one to reference objects across interoperating ORB domains (across different ORBs). IORs are required for inter-ORB (cross-domain) communications. ORB vendors may choose to also use the IOR for intra-ORB (single domain) communications.

Alternatively, vendors may choose to use a proprietary object reference structure for communication that takes place within their ORB's domain. Proprietary references give vendors much more opportunity to optimize the implementation of the reference. Common optimizations are to reduce resource consumption and to increase performance. For example, some ORBs support object references as small as four bytes on average using proprietary protocols. Contrast this with 100+ bytes average (or more) for an IOR in the most efficient ORB implementations.

Some ORBs exclusively use the IOR. Others use the proprietary reference by default, and require the developer to request an IOR when interoperability is needed. Some applications will virtually always need IORs. Object directory services—such as the Naming Service—are a prime example. These services know nothing about the objects registered with them; they simply hand out references on request. A client in one ORB domain who queries the directory service and receives a reference to an object in another ORB domain must receive an IOR in order to do anything with the reference. In large, enterprise environments, it is reasonable to expect that ORBs from many vendors will exist and that interoperability across those ORBs is required. One must weigh the additional overhead required for an IOR against the relative frequency that interoperability is expected.

## 2.3   Partitioning the 'Object Space'

In order to access an object, one must have a reference to it. While there are several techniques for obtaining references, the general purpose technique is to query some sort of directory service. The OMG Naming Service is probably the most commonly used directory service today for CORBA application development. This service takes a name as input and returns an object reference. Examples of other directory services are a trading service and a factory finder. (Object Management Group, Inc., 1996) These services each return references in response to some type of query.

The common CORBA technique that was observed for making objects generally available utilizes a single-tiered lookup strategy: objects are registered with the naming service, and clients query the naming service to obtain IORs. This approach does not scale well. As the number of objects increases, the burden on the directory services becomes too great. Ways of addressing this problem typically involve some kind of partitioning of the name space or a multi-tiered lookup strategy; however, there are no generally available guidelines for how to do this effectively.

Another consideration is the fact that a generic directory service has limited potential to optimize searches. The significance of this limitation increases as the number of objects grows large.

## 2.4 Object Identity

Comparisons often need to be made to determine whether two objects are in fact the same object. For example, when traversing a graph of objects, one needs to determine whether a given node (object) has already been visited. Objects need to support some notion of identity in order to permit such comparisons to be made.

There is a common misconception that CORBA object references represent object identity. Statements that lead one to conclude that object references and object identity are synonymous can be found throughout the CORBA specifications themselves. However, one important thing that an object reference does not provide is the identity of the object. Multiple references can point to the same object. CORBA supports operations on the object reference that enable one to determine if two references are equivalent, but if the references are not equivalent, the objects could still be the same.

The best way to determine the identity of an object is to ask the object itself— assuming it provides an interface for doing so. However, CORBA makes no requirement that an object support such an interface. The OMG Relationship Service (Object Management Group, Inc., 1996) defines the *IdentifiableObject* interface, but there are still two problems related to object identity when operating in very large environments.

An *IdentifiableObject* has a read-only attribute of type *ObjectIdentifier* named *constant_random_id*. The value associated with *constant_random_id* must not change during the lifetime of the object. It is not guaranteed to be unique. Its typical use is as a key in a hash table. If the two identifiers are different, one can conclude that the two objects are different. When collisions between these identifiers occur, the client can use the *is_identical* operation to ask the object whether the other object that it is being compared against is the same.

The first problem is related to the potential amount of network traffic that might be generated when having to perform a large number of comparisons. Each call to obtain the value of *constant_random_id* or to perform the *is_identical* operation could result in a remote call through the ORB. This leads one to look at caching strategies to minimize the amount of ORB traffic. This, in turn, leads to further questions about whether the identifier should be obtained in advance or obtained on demand, and whether more can be done to strengthen the notion of identity through the use of universally unique values, possibly eliminating dependence on the *is_identical* operation altogether.

The second problem is more political than technical. It has to do with the importance of object identity to the OMG. By and large, the notion of identity is very weak in CORBA. There is very little use of the *IdentifiableObject* interface in other services. This runs contrary to the strong requirement for object identity that was found in this investigation. One of the most heavily used services among the applications in this study is the topology service. That service has a very strong dependency on object identity.

## 2.5  Bulk Operations

It is often the case that one will want to perform the same operation on many objects simultaneously. Common examples are querying many objects through a single call, or dispatching an update request to many objects at the same time. The primary motivation behind bulk operations is performance and resource consumption optimizations by taking advantage of co-location of objects. The value of using bulk operations increases as the number of objects being managed increases.

Consider, for example, an application that renders a graphical map of large numbers of objects. To display a map, a certain bit of information must be obtained from each object. The straightforward approach is to iterate through the list of objects requesting each object to supply the required information. The problem is that this approach can result in a tremendous amount of network traffic. To obtain just one attribute value from 100 objects will require 100 object requests. If there is a slow distribution link between the client and the objects, performance can be unacceptable. If, however, the 100 objects were partitioned among two 'collection' objects, where the link between each collection object and the objects that it manages is fast and the link between the client and the two collection objects is slow, performance is going to be much better if two calls can be made to the collection objects asking them to obtain the information from the individual objects.

ORB implementations typically take advantage of the co-location of objects within the same server process to optimize the size of object references. For example, some object references embed location information for the server process that 'houses' an object, plus a key that uniquely identifies the object within the server. An ORB can reduce the average per-object reference size by keeping one copy of the server location—which will be the same for all objects from that server—and sharing that copy among all related references. Unfortunately, the information needed for clients to take advantage of co-location is hidden within the opaque object reference structure.

To facilitate bulk operations, a partitioning model is needed that will increase the likelihood that co-location of objects occurs. Further, the client needs a means of identifying co-located objects so that bulk operations on the right set of objects can be requested. Finally, an interface capable of supporting bulk operations on the desired objects must be provided.

## 3   SOLUTIONS TO SCALEABILITY ISSUES

The previous section identified five issues facing developers who are building applications with high scaleability requirements. This section discusses ways of addressing four of those issues. No attempt is made to address the issue of appropriate level of CORBA object granularity. Unfortunately, determining appropriate levels of abstraction is still largely an art. The appropriate level will vary from application to application.

The solutions to the issues raised in sections 2.3-2.5 are wrapped together into a model called the Entity-Warehouse Model, or Entity Model for short. In essence, the Entity Model provides structure that enables developers to design their applications in a consistent way to effectively manage the issues that arise when working with CORBA.

## 3.1 IORs to Represent Managed Entities

It was decided that if an entity was implemented as a CORBA object, the implementation would exclusively use IORs for those objects, rather than using proprietary references. The applications being developed will run on several vendor platforms, as well as on several different ORBs. These applications make heavy use of naming, trading, and topology services to gather collections of objects to operate on. Many instances of these services will be deployed throughout the enterprise. The instances will be federated together to present a single 'logical' view of the enterprise, rather than a view of disjoint islands of objects. A client is equally likely to obtain a reference to an object implemented over ORB A as from ORB B. Using IORs is the only reasonable way to ensure that the client will be able to use the object once it gets a reference to it.

## 3.2 Entities and Warehouses — A Partitioning of the Object Space

The Entity Model partitions the object space into entities and warehouses. Entities are instances of objects that generally represent the 'real' objects (physical or logical resources) that the user is interested in managing. Entities are not constrained to being CORBA objects. Warehouses are instances of objects that manage collections of entities. Warehouses may themselves be entities. Warehouses generally are CORBA objects. Warehouses may be managed by other warehouses. An installation might consist of millions of entities, managed by thousands of warehouses, residing on hundreds of management hosts throughout the enterprise.

Warehouses play a role in an entity's object lifecycle. Warehouses are often entity factories, or work in cooperation with entity factories. Warehouses are involved in the relocation and destruction of entities. Warehouses may also provide the interface for operating on the entities themselves (i.e., clients manage entities indirectly using an interface provided by the warehouse).

Warehouses also provide directory services for the entities they manage. Rather than utilizing generic directory services to locate all objects, the model encourages limiting the use of the generic services to locating the relatively small number of warehouses, and using warehouses to locate the larger number of entities.

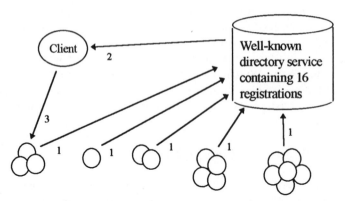

1. White circles represent entities. All entities are CORBA objects. All entities register with a well-known directory service.
2. Client obtains a reference to the desired entity from the directory service.
3. Client uses the entity's object reference to operate on the entity.

**Figure 2** Using CORBA without the Entity Model.

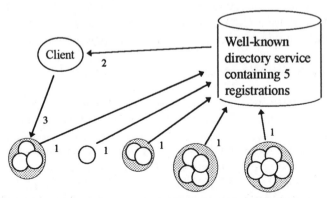

1. White circles represent entities, gray circles represent warehouses. All warehouses are CORBA objects, and register with a well-known directory service. Entities *may* be CORBA objects. Entities *may* register with a well-known directory service.
2. Client obtains a reference to the warehouse containing the desired entity from the directory service.
3. Client uses the warehouse's reference to indirectly operate on the entity in directory, or to obtain a reference to the entity itself.

**Figure 3** Using CORBA with the Entity Model.

Figures 2 and 3 illustrate CORBA with and without the Entity Model. Object partitionings that often naturally exist are formalized into warehouses. Warehouses can be located using the public object directory, but individual entities generally cannot. IDL interfaces are provided as base interfaces for both entities (*Entity*) and warehouses (*EntityWarehouse* and *EntityFactory*), as shown in Figure 4.

```
struct PartitionedIdentifier {
    string id_context;
    string id_base;
};

interface Entity {
    readonly attribute PartitionedIdentifier entity_id;
};

struct EntityKey {
    PartitionedIdentifier  id;
    Object ior;
};
typedef sequence<EntityKey> EntityKeyList;

interface EntityFactory {
    EntityKey create_entity(in string criteria);
};

interface EntityWarehouse {
    Entity get_entity(in PartitionedIdentifier entity_id);
    EntityKeyList query(in string query_string);
    void destroy_entity(in EntityKeyList key_list);
};
```

**Figure 4** IDL for the Entity Model

## 3.3  Object Identity

The Entity interface defines the read-only attribute *entity_id*, through which each entity provides a unique identifying value. *entity_id* is a struct consisting of two parts. By convention, *id_context* represents the identity of the warehouse containing the entity. *id_base*  uniquely identifies the entity within the context given by *id_context*. Individually, each part is immutable, collectively they are unique in space and time.

Because several key applications depend so heavily on object identity, a special structure called the entity key was defined. The entity key contains the entity identifier and (optionally) the IOR for the entity. Entity keys can be given to applications that depend on object identity so that those applications will not have to query the object separately to obtain the value. The entity key effectively provides a caching mechanism for the object identity value. This caching alternative was chosen over the on-demand model because of the difficulty of ensuring high cache hit rates for some of the applications. The topology service, for example, will contain very large graphs that can be queried in a variety of ways. Until usage patterns can be studied, it is difficult to ascertain if adequate hit rates can be sustained.

## 3.4 Supporting Bulk Operations

The Entity Model partitions the object space in such a way as to increase the likelihood that bulk operations can be performed with sufficient regularity to be cost effective. An examination of the *id_context* field enables one to determine what groupings exist.

A second necessary ingredient is an interface that supports bulk versions of the desired operation. It is expected that the bulk versions will be very similar to the corresponding single instance versions. The definition of these will be specific to the type of object being managed. The example in Figure 5 illustrates a single instance interface and its corresponding bulk interface.

```
interface IPHost : Host {
    attribute IPAddr ip_address;
};
typedef sequence<IPHost> IPHostList;
typedef sequence<IPAddr> IPAddrList;

interface IPHostWarehouse : HostWarehouse {
    IPAddrList get_ip_address(in IPHostList ip_host_list);
};
```

**Figure 5** Single instance interface and corresponding bulk interface

## 3.5 More on the Entity Key

There are several other aspects of the Entity Model that are not covered in this paper. One semantic of the entity key worth mentioning is the fact that the *ior* value is optional. This opens the door to a variety of implementation alternatives, such as managing non-CORBA objects. For example, applications such as the topology service accumulate entity keys, but do not examine their contents. It is not necessary for an entity to be a CORBA object for it to be manageable by such services. These entities simply must be managed by some warehouse that is a CORBA object. The entity ID can be used to identify the warehouse and the entity within the warehouse. Interfaces on the warehouse can permit operating on entities as second class objects.

# 4    CONCLUSION

With appropriate usage models, CORBA has been found to be adequate to support applications that must scale to manage very large numbers of objects. The Entity-Warehouse Model was developed to provide a usage model for dealing consistently with several challenges encountered. The Entity-Warehouse Model is an approach to using CORBA that enables developers to effectively manage complexity and the load on system resources. It requires no extensions to CORBA and is not specific to any particular ORB. Developers are not constrained to using this model, however, it is expected that in using the model, the scaleability potential of their applications will be greatly increased.

CORBA has evolved considerably since this investigation was begun. ORB implementations are maturing such that some issues originally raised in the investigation are no longer of major concern. The biggest example of this is evident in the improvement in the size of the IOR. In the first IOR implementations, the optimized size was very large (well over 100 bytes). This was clearly an issue to applications that accumulate and cache large numbers of IORs (such as the topology service). Recent modifications to the specification of the IOR, however, have enabled ORB vendors to significantly reduce that size, with expectations that sizes of 32-64 bytes are obtainable.

Other recent CORBA developments have yet to be investigated to determine the impact on scaleability. Work needs to be done to determine how to best align the Entity-Warehouse Model with specifications such as the Collections service, which provides a model for partitioning objects into collections, and the Query service, which provides a model for querying for objects. Warehouses are a collection of entities, and provide a method for locating entities via a generalized query. Additionally, the Collections and Query services must be examined to determine best practices that will ensure scaleability goals can be achieved.

# 5    REFERENCES

Object Management Group, Inc. (1996) Naming Service Specification, in *CORBAservices: Common Object Services Specification*, 3-1 - 3-18.

Object Management Group, Inc. (1996) Transaction Service Specification, in *CORBAservices: Common Object Services Specification*, 10-1 - 10-86.

Object Management Group, Inc. (1995) The Common Object Request Broker: Architecture and Specification, 7-5 - 7-6.

Object Management Group, Inc. (1996) Trading Service Specification, OMG Document *orbos/96-05-06*.

Object Management Group, Inc. (1996) Life Cycle Services Specification, in *CORBAservices: Common Object Services Specification*, 6-7 - 6-14.

Object Management Group, Inc. (1996) Relationship Service Specification, in *CORBAservices: Common Object Services Specification*, 9-19 - 9-20.

# 6     BIOGRAPHY

RICHARD B. (RICK) WHITNER is an architect for the HP OpenView program at Hewlett-Packard. Since joining HP in 1988, he has been active in the design of platforms for network and systems management, and in the investigation and application of object-oriented technologies. During the past two years he has represented HP in the Object Management Group's (OMG) Telecommunications Domain Task Force. Prior to joining HP he was a senior programmer/analyst for the Virginia Cooperative Extension Service. He holds a B.S. in Mathematics from Clinch Valley College of the University of Virginia and an M.S. in Computer Science from Virginia Tech.

# 38

# ACE: An environment for specifying, developing and generating TINA services

*P.G. Bosco, D. Lo Giudice, G. Martini, C. Moiso*
*CSELT*
*Via G. Reiss Romoli 274, 10147 - Torino - Italy ,*
*tel: +39 11 2286806, fax: +39 11 2286862,*
*{Piergiorgio.Bosco,Giovanni.Martini,Corrado.Moiso}@cselt.stet.it*

**Abstract**
The objective of the paper is to present an innovative environment (Application Construction Environment - ACE) for the specification, development and generation of Telecommunication services according to emerging standards such as TINA and CORBA. We will briefly describe the service specifications and development lifecycle requirements for distributed object applications in the telecom domain and how ACE functionalities address them.

**Keywords**
TINA, CORBA, CASE, Specifications validation, Behaviour modelling

## 1 INTRODUCTION

The adoption trend of client-server and object oriented paradigms (e.g., ISO/ODP (ISO/ODP, 1993), CORBA(OMG, 1991) and TINA (Handegard, 1995)) is now occurring in various IT domains as well as in telecom ones. Actually, no matter the domain in consideration, today no formal description techniques are available to express in a complete, compact and natural way the concepts present in the models mentioned above. This is especially true about the design phase of the development lifecycle of distributed applications.

In the IT world the available object-oriented CASE tools for distributed applications are currently more oriented to code generation than to support the whole specification process (included behavioural description).

However there is some attempt at improvement due to the rise of *defacto* standard architectures such as CORBA (OMG, 1991) and MS OLE (distributed OLE) (Brockschmidt,1995) which allow to consider architectural principles in the early stages of specification. In fact some CASE tools are able to generate applications based on these architectures. These typically generate description languages such as IDL for CORBA and OLE types for MS OLE, from the analysis models (GUI, architecture early phases, application logic and Data logic are forced to be kept separate) and finally they also generate the stubs that deal with the distribution mechanisms. This approach offers a good starting point for later design/development phases, though it represents only a small step further, since it still lacks the expression of constructs or "services" such as transactions, concurrency, typical of distributed object environments. Not to say, other requirements traditionally well considered in the telecom industry, such as more formal behavioural description, validation of specifications, performance evaluation etc. are not  usually addressed in a integrated way. Notably, only SDL-based CASE environments, approximate this level of requirement coverage.

On the contrary, standard languages familiar to the telecommunication community such as SDL or LOTOS and their latest evolutions do not have sufficiently matched the object-oriented models and behavioural concepts of the recently emerged architectures mentioned above.

In general, the lack of a complete formalism for object-oriented distributed processing models is one of the causes of the absence of CASE tools suited to support the design phases of a system according to such models. This problem can be addressed to some degree by following a pragmatic approach to apply and extend existing notations such as OMT (Rumbaugh,1991), SDL, etc. to distributed computational architectures.

## ACE Objectives

The motivations to develop ACE were to allow the specification and development of applications according to the distributed processing model of TINA and CORBA, with a particular emphasis on the aspects which nowadays are more difficult to be found in commercial tools, i.e. behavioural modelling. Also "openness" wrt the actual OO distributed platform architecture was an additional requirement, which CSELT had already started to address in previous versions (Bersia,1995).

There two main strategic reasons to the development of ACE are:
- To use ACE internally to support the specification and prototyping of services according to overall TINA architecture (Service Architecture, Computing and Information Architecture), to CORBA and OMG/COSS1 and OMG/COSS2 for the basic Object Services (OMG,1995).
- To stimulate in the industry the development of "industrial quality" tools "á la ACE": ACE today is a running prototype which acts as a "live" requirement.

Last but not least, ACE aims at providing support for the main lifecycle requirements as expressed in the TINA Service Architecture (Kobayashi,1995). It is a nut-shell for a number of tools (specification editors, performance evaluators, code generators, etc.) from which the analysts/designers/developers all together orchestrate.

## 2   REQUIREMENTS FOR TINA SERVICES DEVELOPMENT TOOLS

It is useful to identify a set of functional requirements that should be supported by a development environment during the main stages of a TINA service lifecycle.

### Browsers/Editors

The specification and the development of the components of a service should be done by means of proper editors, customised ad hoc to the formalism used for the specification. Advanced graphics, icons, links should be supported. The resulting GUI of the tool should be highly customisable and adaptive to the developer's needs.

The editors should be integrated with the other functionalities of the environment (e.g. syntactic verification could interactively be invoked when new icons are introduced in a diagram).

### Validation & Simulation

Integrated provisioning of functionalities for the validation and verification of the correctness of a service should be considered as a mandatory aspect to be supported. It could be done according to different criteria e.g. wrt the specification, wrt implementation (no programming errors), or wrt the performance and timing constraints. The confidence about the suitability of an application is gained by challenging its specification against a set of *dynamic properties*:
- Global properties: a set of constraints holds for all the system states and/or for subsets of the states (e.g. a single critical resource can be always allocated to

either zero or one resource client).

- Local properties: a set of constraints holds for the trajectories across the state space of the system and/or for the trajectories across a subset of the state space.

In both cases system state exploration (with interactive emulation or with exhaustive emulation) seems to be the only practical approach to partially achieve the goal.

### Performance evaluators

Performance of a service is one of the most important non-functional requirements. The evaluation of the performances (e.g. response time or throughput of a service) can take place at two main points in the lifecycle of application creation: the specification phase and the acceptance phase.

In both cases the system specification can be taken as the model to study, provided that the service has been implemented as specified. In the former case the objective is to estimate the upper limits in the performances of the service, as functions of the size of the resources; in the latter case the goal is to forecast the performances of the real implementation with particular resource configurations and load situations.

The distributed nature of the computing environment, adds a degree of complexity, since the communication infrastructure plays its own role in the whole performances.

### Code Generation

Given the adoption of CORBA for what regards TINA DPE (Distributed Processing Environment, computational model and services), the first main requirement of code generation is to generate CORBA IDL.

Further sophisticated code generation is also required in order to meet TINA extensions such as the support for "multiple interfaces".

Also given the requirement of better satisfying "behavioural" aspects of the Telecommunication services based on a distributed platform, generation that integrates COSS basic services such as Transactions, Events and Trading could also be foreseen. Main reference languages are the ones for which OMG has already defined a mapping (C++ ) or is going to (JAVA).

Finally we also believe that targeting to environments such as MS OLE/COM is also a requirement. In developing peripheral telecom services where MS Lans or MS Desktops are part of the scenario, one would want to take advantage of the advanced development environment present on MS platforms (e.g. for Service Creation) and of the high number of "prefabricated" SW components based on MS OCX and OLE Automation objects (Brockshmidt,1995) already available on such platforms.

Furthermore, whatever the target platform is, it is required to be able to trace the behaviour of the service components (CORBA objects, OLE objects, ...) when they are deployed. Specific selectable mechanisms for their tracing should be embedded in the generated code. The tracing tool should provide enough information in order to highlight the various interactions among service components themselves and the DPE services.

## 3    ACE FUNCTIONALITIES

This section outlines the main functionalities supported by the ACE environment satisfying the above requirements.

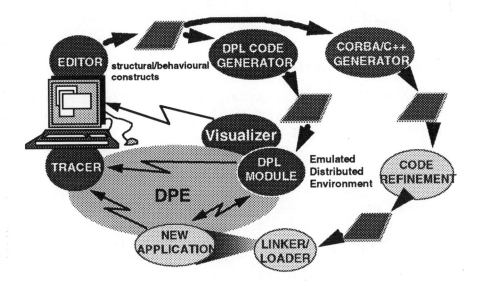

**Figure 1** Overview of the development cycle in ACE.

For the sake of clarity, in Figure 1 an overview of the main phases of the development lifecycle supported in ACE is illustrated. From the set of editors it is possible to produce code both for the simulation environment and for the real platform. The specifications can be executed in a simulated environment, for a first functional validation of the service logics, that can be traced directly on the editors in order to achieve a better comprehension of the execution of the user-level specifications. The code generated for the real platform, after a refinement process of the implementation skeletons, is ready to be installed and deployed on the DPE.

## 3.1   ACE COMPUTATIONAL MODEL CONFORMANCE

As far as conformance objectives towards TINA DPE (Handegard,1995) the following computational model (Bosco,1993) is realised in ACE:
- Creation/Deletion of object and group instances (building-blocks).
- Asynchronous/Synchronous invocations of operations: invocations can be either blocking or non-blocking.
- Accesses to attributes, to read/write their values.
- Subscription to a notification emitted by an object, indicating the notification handler, i.e. the interface to be invoked when these notifications are received.
- Emission of a notification (to all the subscribers).
- Concurrence control in multi-threaded objects: guards, semaphores and close/open nested transactions.

## 3.2   EDITING COMPONENT

The editing component consists of a suite of graphical editors. They support the specification of templates for building-blocks (BBs) and objects, including their behavioural parts, and interface templates.

Each graphical editor is associated to a (graphical) language, consisting of a set of icons and rules on their composition and interconnection through (labelled) links, as depicted in Figure 2. The icons can have attributes that are specified through dialogue-boxes: the attributes are introduced by a sequence of dialogue-boxes, where the structure of a dialogue-box depends on the attributes introduced in the previous ones. Checks for syntax and type constraints are performed during the data input.
The selection of an icon associated to a specification sheet automatically opens the corresponding editor on the associated sheet.

The BB editor (BBE) is the graphical editor to specify BB templates. Special icons represent the object templates included in the BB template, and the contract declarations (i.e. the interfaces accessible from the outside of the BB).

The object editor (OE) is the graphical editor suited to specify object templates, and relies on the behaviour editor (BE) to specify their behaviour.
The specification of an object template is structured into a hierarchy of sheets.
- The declaration part: declaration of parameters, interfaces, internal state, methods, local functions.
- The methods and functions, specified in graphical way.

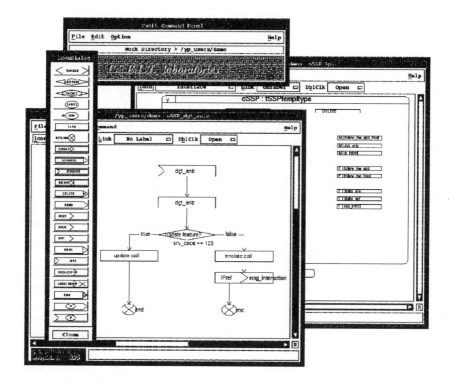

**Figure 2** Behavioural and object template editors.

The OE works on the declaration part, while the BE works on the methods/functions graphically specified (Figure 2). The language associated to OE is a set of icon types that represent the different declaration components of an object template.

Methods are specified by the BE. The language associated to this editor is a flow chart language, where the icons, derived from SDL symbols, are specialised to represent behavioural constructs of the computational model (e.g. invocation of an operation), control constructs (e.g. method start, if-then-else, return, etc) and synchronisation constructs (guards and locks).
The local functions of an object can be specified either in a graphical way through the BE or in a textual way in DPL (see next section 3.4).

The interface editor (IE) is the graphical editor to specify interfaces, and declaring in

CORBA IDL operations, attributes, data types.

*Interactive editing features*
All the graphical editors provide interactive/adaptive editing features to drive the production of specifications. For example when a new specification sheet is created the editor initialises it with the mandatory icons, or when an icon is instantiated the editor automatically opens the associated dialogue-box and highlights the mandatory editing-fields.
Syntactic controls and type checking are performed as soon as the values of a dialogue-box are confirmed.

## 3.3  SYNTACTIC AND SEMANTIC CHECKING

ACE performs a global syntax and type check of the specified templates. All the constructs and the declarations are type-checked, the syntax of the attributes of the icons is controlled and the syntax of graphical specifications is checked.
This global check can detect errors even if no errors were signalled during the editing of a template: in fact, for example, after the change of a variable declaration all the expressions where it occurs may be badly typed.
If the global analysis successfully terminates, it generates an intermediate code representation of the template, which is used as input to all the final code generators.

## 3.4  SIMULATION AND VISUALIZATION OF SPECIFICATIONS

At present, functional validation of specification is assumed to be carried on in a user-based interactive way, through the simulated execution. DPL (Distributed Processing Language) is a multiparadigm language that can be used to fast-prototype applications according to the computational model and to execute them in a (sequential) simulated environment. By exploiting Prolog features, DPL is able to model behavioural evolutions of objects dynamics, and, with the provision of an internal scheduler, to support the execution of concurrent entities within the simulation environment, by providing in this way interleaving semantics.

*Mixed mode simulation*
Though the provision of a simulated environments has proved to be sufficient for the validation of  user specified service logics, it shows a certain degree of inadequacy when developing service components making intense use of already deployed services. In this case, providing support from within ACE for their modelling does not seem very practical and feasible. This issue has been tackled devising a mixed

mode simulation mechanism, that allows simulated components to interact with applications and services executing on the real DPE (CORBA platform).

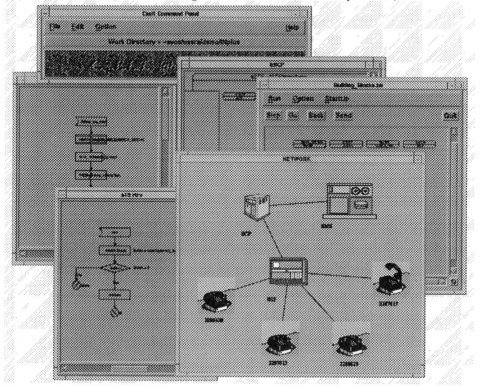

**Figure 3** A simulation session.

*Visualisation and tracing of executions*

The execution of specifications can be visualised in two different ways:

- Model oriented: the executions are traced in terms of concepts of the computational model (e.g. interactions through interfaces).
- Application oriented: the specifications are coupled with a user defined scenario that represents the agents and the artefacts involved in the application (e.g., the telephones, the switches, calls, and connections, as illustrated in Figure 3).

The model oriented visualisation is based on user selectable tracing of the behaviour diagrams and of the object states, directly on the same graphical representation of specifications. The users can interact with the execution by asynchronously invoking announcements to the objects, defining break-points or inspecting state variables.

The application-oriented visualisation is based on an interface through which the users can introduce, in an asynchronous way, inputs to the application (e.g. by means of the mouse, dialogue-boxes,...) and the application can provide outputs in a graphical way (e.g. through some animations).

The behaviour of this interface depends on the characteristics of the application and must be explicitly programmed, together with some enrichment of the specification to interact with the visualisation interface.

## 3.6 PERFORMANCE EVALUATION OF SPECIFICATIONS

The technical approach chosen for the performance evaluation of an application is the simulation of its specification, augmented with a non-functional characterisation of the significant activities, i.e.:

- Durations of activities: single actions or groups of actions can be given a duration (possibly stochastic).
- Mapping of objects and messages on computing and communication resources: each BB is "deployed" on a physical processor.

An already available process-based timed executor is being integrated into ACE ( $SPIN^+$ (Chiocchetti,1994)). A number of executions of the model are run and, out of them, estimates are calculated for the desired system parameters. The definition of the performance indexes is embedded within the system specification, as well as validation properties.

## 3.7 CODE GENERATION FOR REAL DISTRIBUTED EXECUTION

ACE provides a code generator to produce CORBA/IDL and C++ for deploying and running the applications on a real CORBA platform. At the moment, the generation is targeted for the Iona Orbix (Iona,1995) realisation, though the targeting to other platforms (Chorus,1994) is an ongoing activity. Among them, of particular relevance is the ReTINA distributed platform (Chorus,1995), a real-time CORBA platform.

The actual mapping tackles and solves the differences between the object model of the TINA computational model (objects having multiple interfaces) and the one supported by the CORBA binding with C++ language. In the process of the code generation this semantic gap is fulfilled by proper object composition and dedicated "glue code" that reduce the mismatch between the two different models.

The code generation covers the production of both the declaration part of templates and the behaviour parts which are specified through the graphical diagrams.

As far as interfaces are concerned, CORBA IDL code is generated automatically starting from the graphical specification of an interface, as well as the contrary (i.e. it is possible to import externally defined interfaces specified in CORBA IDL). Likewise, starting from the object specifications (structural and behavioural sections), skeletons C++ implementations are generated automatically from the SDL-like behavioural notation.

## 4    REFERENCE APPLICATIONS

ACE is the reference environment within CSELT for developing TINA compliant services. Experience and useful suggestions have been gained by employing the environment (though for minimal parts, and not in all its components) in the definition of the specifications and functionalities for small scale projects within our research group, regarded as proofs of concepts for the TINA Service Architecture. Among them, it has provided a useful support for the definition and observation of the specifications of the Connection Management module during the TINA World Wide Demo (Spinelli,1996) allowing to trace and analyse the service specifications. Within the ACTS ReTINA project, it is the core tool providing support functionalities for the definition, validation, and code generation of service components.

## 5    FUTURE EVOLUTIONS

The activity on ACE is still evolving, and during the next months we'll focus on 1) the extension and the upgrade of existing functionalities towards TINA and OMG standards, and 2) the inclusion of functionalities to keep pace with recent technology evolutions in the market place (e.g. support for JAVA and for the Unified Modelling Language (Booch,1996)).

Evolutions regarding 1) comprise the targeting code generation to different CORBA (2.0) platforms (ReTINA DPE) and the support for specification written in TINA ODL(Leydekkers,1995); the inclusion of other languages for expressing application functionalities (e.g. Message Sequence Charts, Use Cases) and DPE basic services abstraction. In this case, the idea is to simplify the use of low-level object services (e.g.: naming, transactions, concurrency, etc.) by providing graphical entities (Icons, symbols, etc.).

During '96 we plan also to expose the tool at major events such as: OMG Object World Trade shows, etc. In order to stimulate and divulge ACE basic philosophy and approach contacts with major vendors and tools developers various discussions are underway. A public domain version of ACE can be obtained from the authors.


## 6    CONCLUSIONS

At the present stage ACE is an advanced integrated CASE environment for distributed object oriented applications compliant to the emerging general and telecom-oriented distributed processing standards. Even if not yet exhaustively satisfied many of the requirements and gaps described in initial sections are addressed in ACE.

*Part of this work has been supported by the EEC ACTS AC048 project ReTINA.*


## 6    REFERENCES

N.Bersia, P.G.Bosco, G.Canal, R.Manione, C.Moiso, M.Spinolo (1995) A Case Environment for TINA-oriented applications, in *Proc. ISS'95*.

G.Booch, J.Rumbaugh, I.Jacobson (1996) Unified Modeling Language,Rational Rose.

P.G.Bosco, G.Giandonato, C.Moiso (1993) A distributed processing model for telecommunication services and operations software, in *Proc. TINA'93*.

K.Brockschmidt (1995) Inside OLE 2. MS Press, Redmond.

E.Chiocchetti, R.Manione, P.Renditore (1994) Specification based performance evaluation of distributed systems for telecommunications, in *Proc. Tools'94*.

Chorus Systeme (1994) COOL-ORB Programming Model. Technical Report, Paris.

Chorus Systeme (1995) , ReTINA, http://www.chorus.com/Research/retina.html

T. Handegard, N. Mercouroff (1995) Computational modelling concepts. TINA-C.

Iona (1995) Advanced Programmer Guide. Iona, Dublin.

ISO/ODP (1993) Basic Reference Model of ODP-part 3: Prescriptive Model. ISO/SC21 N8125.

H.Kobayashi, K.Moor, C.Abarca (1995) TINA Service Architecture. TINA-C.

P.Leydekkers, N.Mercouroff (1995) TINA Object Definition Language. TINA-C.

OMG (1991) The common object request broker. OMG Document Number 91.8.1.

OMG (1995) CORBA Services: Common Object Service Specifications. OMG Document Number 95.3.31.

J.Rumbaugh et al. (1991) Object-Oriented Modeling and Design, Prentice Hall.

G.Spinelli, W.Takita, G.Martini, S.Chikara (1996) TINA Connection Management Implementation over Distributed Processing Platforms, in *Proc. NOMS '96*.

# 39
# Supporting Dynamic Policy Change Using CORBA System Management Facilities

*Stephen Howard, Hanan Lutfiyya, Michael Katchabaw and Michael Bauer*
*Department of Computer Science, The University of Western Ontario*
*London, Ontario, Canada N6A 5B7*
*email:* {showard,hanan,katchab,bauer}@csd.uwo.ca

## Abstract

Automation of management tasks is an effective counter-measure to the growing complexity of distributed systems. An increasingly popular view redefines the role of the management system to include automated validation and enforcement of policy. This research proposes an architecture for a policy-driven management system which can adapt dynamically to policy change. We show how this architecture can be implemented in a CORBA distributed object computing environment on top of the recently adopted System Management Common Management Facilities.

## Keywords

Policy-driven management, CORBA management, distributed system management, distributed object computing

## 1 INTRODUCTION

New levels of automation are needed to address the growing complexity of distributed systems. We share the view that *policy-driven* management is an appropriate means to automate more of the management function. As an ideal, we believe that human involvement can be reduced to defining policy and roles and that intervention should be necessary only for critical failures. We are realistic in the sense that we recognise the infancy of the current state of research relative to that vision.

This paper proposes a basic architecture for a policy-driven management system which can support application, system and network management at a level of automation beyond that of today's management systems. We pursue the problem in a CORBA (OMG, 1995) environment, believing that CORBA is reflective of the direction in which distributed computing is being steered and thus, new challenges for management (particularly application-level management) will be found there. With its focus on object orientation, and a rich set of standard object services, CORBA presents the opportunity to explore new approaches to building management systems. A particular focus of our work is to investigate the suitability of the new OMG System

Management Common Management Facilities (X/Open, 1995) to support our architecture.

The remainder of the paper is organized as follows: In Section 2, we describe the general notion of policies and policy-driven management. Section 3 presents our architecture. In Section 4, we introduce the recently adopted OMG management services and summarize a few of the design issues we face in attempting to use these to support our architecture. Section 5 highlights some related work and Section 6 draws some conclusions and outlines our next steps.

## 2   POLICY-DRIVEN MANAGEMENT

A basic premise of this research is that the management system exists to uphold management policy. In practice, policy exists at many levels. Corporate policy makers often begin by defining high-level strategic policy, which is then refined into tactical or goal-oriented policy and then further refined into low-level activity-based policy. The policy refinement process will be difficult to automate without constraining the expression of higher-level policy in unnatural ways. Other research in this area which strives to encompass the policy refinement hierarchy within the management architecture generally allows for the storage of policies at various levels and then facilitates relationship tracking among these levels of refinement. The refinement process itself is left to human policy administrators.

Our research currently addresses only low-level policy, making the assumption that policies have been refined to the point where they can be expressed in a formal syntax. While our work requires basic policy definition services, the design of these is not our focus and therefore we have borrowed concepts from other excellent work in this area (Becker, Raabe and Twidle, 1993, Marriott, Mansouri-Samani and Sloman, 1994, Koch, Krell and Krämer, 1996). The remainder of this section outlines our view of policies; how they are specified and how they are organized.

## 2.1   Policies

In our architecture, low-level policies are treated as objects. A simple example is presented to illustrate some of the attributes included in a policy specification. The example policy defines a requirement to monitor CPU load in a distributed environment and react when an unusually high load is encountered.

```
POLICY reportHighCpuLoad {
  MODE obligation
  SUBJECT (...)/systemManager
  EVENT highCPU@(...)/*/systemMO
  ACTION $S->displayWarning
  EVENT-RECIPIENT (...)/eventLogger
  DESCRIPTION Report CPUs with extraordinarily high load
}
```

Each policy carries a name and a mode (either *obligation* or *authorization*). In the example, a management application called systemManager is the policy *subject* and is ultimately responsible for enforcing this policy. The subject is specified using a *domain expression* as described in Section 2.2.

The circumstance in which enforcement is required is defined by the *event* clause. The event expression in the example contains a single term identifying the *primitive event* highCpuLoad (discussed in greater detail in Section 2.3) and the set of managed objects from which that event may emanate. These managed objects are the *targets* of the policy and are specified using domain expressions. It is assumed that a managed object called systemMO exists on each host to collect CPU utilization (and other) information through operating system instrumentation. Note that "(...)" is not valid syntax but is used here in place of a full domain path. The "*" in the target expression is a wildard causing inclusion of any systemMO managed object below the current level in the domain hierarchy.

The enforcement actions to be taken by the subject are indicated in the *action* clause. In this example, when the systemManager receives a highCpuLoad event notification from any systemMO object, it invokes its own displayWarning method (the $S syntax indicates a method on the subject) perhaps to inform a human manager of the condition.

*Event-recipient* objects receive the same event notifications as the subject but do not carry out enforcing actions. In this example, an eventLogger object records the events in its log.

A description may be included with the policy to provide information for policy browsers and other policy administration applications.

## 2.2   Policy Domains

Policy domains (Becker, Raabe and Twidle, 1993, ISO/IEC JTC1/SC21/WG4, 1992, Sloman and Twidle, 1994) are used to group objects which share common management policy. The objects in a domain can be managed objects or other domains, allowing for the construction of domain graphs. An object can belong to multiple domains, in which case the domains are said to overlap.

Domains are used as the basis for object naming; that is, domains correspond to naming contexts and thus impose a structure and allocation scheme for the local naming of managed objects. *Domain expressions* are used in policy specifications to identify the policy's subject(s), target(s) and alternate event recipients. Domain expression resolution yields a set of object references.

## 2.3   Primitive Events

*Primitive event* specifications define significant circumstances and are bound to policies through the policy specification. In our research, we deal with three primitive event types. For *scheduled* events, the circumstance is time-based, thereby providing a mechanism for scheduling management tasks. *Initialization* events, which are associated with the instantiation of managed objects, provide a means to specify managed object initialization tasks. A great variety of other circumstances can be realized using *alarm* events which are triggered by changes in managed object state.

Like policies, primitive events are treated as objects. To illustrate some of the attributes of a primitive event, the following example defines the `highCpuLoad` event referred to in Section 2.1.

```
EVENT highCpuLoad {
  TRIGGER ALARM cpuLoad > 95
  ATTRIBUTES cpuLoad
  DESCRIPTION CPU load too high
}
```

Each primitive event has a unique name. The *trigger* identifies the event's type. For an alarm-type event, a predicate defines a condition based on managed object attributes. For a scheduled event, a time expression is given to define when the fixed, relative or repeating time when the event occurs. Initialization events require no extra information as they are always tied to object instantiation. In this example, the event is triggered if the managed object's `cpuLoad` attribute exceeds a threshold value of 95 percent. If so, the management system must generate an event report carrying the `cpuLoad` attribute value (as well as other standard information such as an identifier, event source and time stamp).

## 2.4   Compound Events

Within the `EVENT` attribute of the policy specification, it is possible to define *compound events* using pre-defined composition operators. We are currently investigating a small set of operators but have defined a mechanism which allows new operators to be added easily. Event composition semantics are complex and impose event synchronization and buffering requirements. A detailed discussion of this topic falls outside the scope of the paper. The following simple example demonstrates our intent.

```
(e1@!d1 and e2@d2) then e3@d3
```

It is here that assumed that e1, e2 and e3 are primitive events and that d1, d2 and d3 are domain expressions specifying the managed object(s) at which the events originate. The example requires that an e1 event from *all* managed objects within the d1 domain and an e2 event from *any* managed object in the d2 domain must precede an e3 event from *any* managed object in the d3 domain.

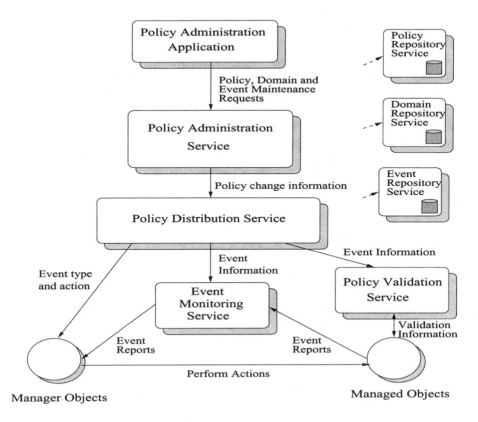

**Figure 1** Management System Architecture

## 3  MANAGEMENT SYSTEM ARCHITECTURE

The management architecture we propose is shown in Figure 1. This architecture supports policy administration, policy violation detection, management task scheduling, managed object initialization, event filtering/forwarding and reactive control. A key aspect of the architecture is that it is designed to facilitate changes in policy at runtime. The management system is able to adapt dynamically to accommodate these changes. The architectural components are described below.

### 3.1  Managed Objects

Managed objects are used to model system resources. One managed object may encapsulate a single resource or many resources to provide an appropriate management view. For application-level management, the managed object may be an instrumented process. For device-level management, the managed object is probably a "stand-in" object which interacts in a private protocol with one or more devices to represent them to management. For system level management, the managed object interacts with operating system instrumentation. The management system is unaware of any interaction between managed objects and real resources.

In the example of Section 2.1, it is intended that the `systemMO` managed objects (of which there is one per host) are simple application level objects that collect CPU load data (and probably other system-level information) from their respective operating systems.

From a developer's perspective, managed objects inherit heavily from predefined managed object classes. This inheritance will provide basic manageability as well as resource class-specific characteristics and functionality. Common to all managed objects are attributes and methods which support basic object lifecycle operations, naming, domain association, collection, and so on.

## 3.2 Manager Objects

A policy's *subject* resolves to one or more manager objects to which responsibility for the policy is assigned. Event and action information from the policy must be distributed to these objects so the manager object knows when and how to enforce the policy (that is, what methods to invoke in what circumstances). Manager objects are also managed objects, allowing them to be managed by higher-level management applications.

## 3.3 Policy Administration Applications

Authorized administrators must have a way to add and remove policies, define events, build domain hierarchies, browse policy information, and so on. A policy administration application provides a user interface to facilitate this type of activity. It is conceivable that more than one type of policy administration application may exist. Policy administration applications use the operations of the Policy Administration Service (PAS) to interact with the management system.

## 3.4 Policy Administration Service (PAS)

The Policy Administration Service controls external access to policy, event and domain information. A client views the PAS as an aggregate of the three repositories (see Section 3.5). Internally, however, there is a requirement to control the use of these "raw" services through an addional layer of functionality.

At a given instant, the management system is configured for a particular set of policies, domains and events. Because such configurations are generated and adjusted automatically, information integrity is vital. As a result, the PAS must ensure that: domain expressions are resolvable, referenced primitive events really do exist, enforcement actions specify valid objects and method signatures, new policies do not conflict with existing ones, expired policies do not linger, and so on.

## 3.5   Repository Services

Three repositories provide persistent storage for policies and associated information. Each repository is treated as a separate service, each with a well-defined service interface. Although the PAS imposes consistency requirements across the repositories at a higher level, at the repository level each individual service operates independently of the others.

The **Policy Repository Service (PRS)** allows for the storage and retrieval of policies. A policy may exist in an inactive or active state (initially, a policy is inactive). A policy can only be removed after it has been deactivated. The PRS allows clients to add or remove policies, activate or deactivate existing policies, retrieve and/or change the attributes of policies and retrieve policies matching certain criteria.

The **Domain Repository Service (DRS)** facilitates the organization of managed objects into policy domains. New domains can be defined and positioned in the domain hierarchy and existing (empty) domains can be removed. Managed object *types* can be added to, or removed from existing domains. Adding and removing managed objects themselves are operations on the managed objects, not on the DRS. Policies are not added to domains; the association between a policy and a domain is determined by the policy's target (which is a domain expression). The DRS allows clients to query a domain for the managed objects it contains, its subdomains and its parent domain. The DRS is capable of resolving a domain expression to yield a set of managed object references.

The **Event Repository Service (ERS)** stores primitive events independently of the policies which use them. This provides some opportunity for reuse of an event across multiple policies. The ERS provides a single interface which allows clients to add, remove or alter event specifications, to inquire about the attributes of a particular event, or to retrieve events matching certain criteria.

## 3.6   Policy Distribution Service (PDS)

The Policy Distribution Service analyzes policy changes and then initiates and coordinates adaptive restructuring. Manager objects must be informed of policy responsibility assignments, and about the events and enforcing actions associated with new policies. The Policy Validation Service requires event triggering conditions to update the appropriate initializer, validator, and scheduler objects. Event forwarding and filtering information must be passed to the Event Monitoring Service so it can maintain its event channel graphs.

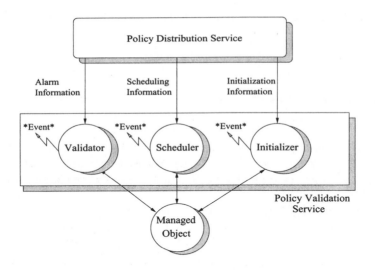

**Figure 2** Policy Validation Architecture

## 3.7   Policy Validation Service (PVS)

Each policy includes an event expression describing the situation for which enforcement is necessary. Primitive events in that expression contain triggering conditions which become the basis for validation. The three types of triggers (see Section 2.3) call for three sub-components of the PVS as shown in Figure 2. Each of these is a generic validator which accepts rules from the PDS and then validates managed object behaviour against these rules, generating event reports upon validation failure. The roles of these components are described briefly below.

1. An *initializer* generates an event report when a managed object is created. This provides an opportunity for policy subjects to carry out initialization tasks on the object.
2. A *validator* holds validation rules from alarm-type event specifications. For example, consider the `systemMO` object with attribute `cpuLoad`. When this (or any) management attribute changes, the managed object invokes a generic `validate` method on its validation object(s) (there could be more than one if the managed object belongs to multiple domains). This method validates the change against currently held rules involving this attribute (in this case `cpuLoad > 95`).
3. A *scheduler* handles schedule-type events and maintains a queue of scheduled times when event reports are to be generated. The content of these reports is dictated by the event specification.

## 3.8 Event Monitoring Service (EMS)

Conceptually, events originate at the managed object level (actually with initializer, validator and scheduler objects) and are reported to management applications. In our work, policy and event specifications must contain sufficient information to allow the system to establish event routing paths and apply necessary filters automatically. The Event Monitoring Service (EMS) accomplishes this using graphs of CORBA event channels. With information received from the PDS about new, deleted or changed event specifications, the EMS can restructure its event channel graphs accordingly. Event communication is based strictly on asynchronous, push-style event flow.

Any object in the system which can generate or recieve events must contain methods through which the EMS can control the object's registration with event channels. In the example of Section 2.1, the EMS would ask the `systemManager` object to register as a consumer to a particular event channel which has been set up to buffer and propagate `highCpuLoad` events. A similar request would be made on the managed object end (the mechanics of this are oversimplified here for the sake of brevity).

## 4 DESIGN ISSUES

Included in the OMG *Common Facilities Architecture* (OMG, 1994) are basic requirements for *System Management Common Facilities*. OMG has recently adopted an initial set of services based on a proposal from X/Open (X/Open, 1995). The major ones are outlined here (detailed coverage is beyond the scope of this paper):

- The *Managed Sets Service* is a simple collection-like service which allows managed objects to be grouped.
- The *Instance Management Service* is a specialization of the CORBA Lifecycle Service. An *Instance Manager* acts both as a factory for a single type of managed objects and as a managed set of these objects. It can report the interface definition of its supported type and can have associated with it *Initialization Policy Objects* and *Validation Policy Objects*. A *Library* object is used to collect, create and locate instance managers.
- The *Policy Driven Base Service* provides the interface for managed objects to be managed by instance managers and to be associated with *policy regions* (domains).
- The *Policy Management Service* is essentially a domain service through which policies and managed objects are bound.

The standards assume that policies are hard-coded methods in the validation policy object which are invoked from the managed object at points where state changes must be verified. In our architecture, policies are objects. Although we maintain the concept and referential relationships of the validation and initialization objects, we redefine their roles to that of generic processors that can accept rules at runtime from the Policy Distribution Service. This way, all validator objects can be instantiated from a single class (and all initializer objects from another). A generic *validate* method is invoked by the managed object (an attribute-value pair is passed to define the state change). We add equally generic scheduler objects but at the managed object level (rather than being assocatiated with the instance manager).

Managed objects in our implementation will inherit our own *managedObject* interface which will, in turn, inherit the CORBA *PolicyDrivenBase* interface. This will provide managed objects with all the management capability defined in Section 3.1.

Our Event Monitoring Service builds its event channel graphs (described in Section 3.8) using standard CORBA event channels (the implementation must support filtering rules).

Much of the functionality of our Domain Repository Service can be realized by "wrapping" the *PolicyRegionsInstanceManager* defined within the CORBA Policy Management Service.

Beyond the standard services and the "retrofits" described, we add policy administration and distribution services, repository services for policies and events, and applications for policy administration.

## 5   RELATED WORK

There is a good base of research dealing with categorization, refinement, specification and organization of management policy (Becker, Raabe and Twidle, 1993, Marriott, Mansouri-Samani and Sloman, 1994, Moffett, 1994, Wies, 1995, Sloman and Twidle, 1994).

We have found two other groups who share our pursuit of a policy-driven management system which adapts dynamically to changes in policy. Researchers at Imperial College continue to play a lead role in policy-driven management in general, and their recent work (Marriott and Sloman, 1996) extends into dynamimic policy distribution and into the CORBA environment. While we share goals, our approach differs in how policy distribution, validation and event monitoring are realized. Koch *et al.* (Koch, Krell and Krämer, 1996) also propose an architecture similar to ours. They are beginning to explore policies as CORBA objects and are investigating approaches to policy distribution. They place more emphasis on the policy hierarchy issue than we do and, again, our approach to such things as event monitoring and policy distribution are quite different.

X/Open, the consortium responsible for the CORBA System Management Common Management Facilities (X/Open, 1995), will continue to pursue a richer set of management services for CORBA. While the available services represent an important base for management application developers, the level of automation we pursue exceeds what these services deliver. Our attempt to employ these services as a foundation seems to be novel.

The Joint X/Open/NM Forum Inter-Domain Management (JIDM) Taskforce (Soukouti, 1995) is addressing the inevitible coexistence of distributed computing environments and the desire for management systems to extend beyond their native domain. Their work includes algorithms for translating specifications among CORBA IDL, SNMP MIB and OSI GDMO. They have also proposed gateway solutions to inter-domain scenarios such as using OSI management for CORBA objects. While the value of this work is readily apparent, our approach is based on a "purist" view in which the management model is more closely coupled with the OMG architecture.

## 6 CONCLUSIONS AND FUTURE WORK

In this paper we have proposed a general architecture for a policy-driven management system in a distributed object computing environment. A key requirement for the management system is that it be able to adapt dynamically to changing policy. The CORBA environment has been chosen as the target for an implemetation of our architecture. Of particular interest in the implementation is whether the recently adopted system management services can be used. In this paper we have mapped out a strategy for doing this.

The implementation is ongoing. Because the CORBA management facilities are new, and no vendor products currently implement them, we must build prototype versions of these in addition to our own management services. We are currently constructing our prototype management system on OS/2 using IBM's implementation of CORBA (SOMobjects 3.0). We anticipate several rounds of architecture and prototype refinement before we can begin to address more complex (and higher-level) management issues.

## REFERENCES

[1] Becker, K., Raabe, U., Sloman, M. and Twidle, K. (editors) (1993) *Domain and Policy Service Specification: IDSM Deliverable D6, SysMan Deliverable MA2V2.*

[2] ISO/IEC JTC1/SC21/WG4 (1992) *Management Domains Architecture* (Working Draft).

[3] Koch, T., and Krell, C., and Krämer, B. (1996) Policy Definition Language for Automated Management of Distributed Systems. *Proceedings of the IEEE Second International Workshop on Systems Management,* Toronto, Canada.

[4] Marriott, D., Mansouri-Samani, M. and Sloman, M. (1995) Specification of Management Policies. *Proceedings of the Fifth IFIP/IEEE International Workshop on Distributed Systems: Operations and Management (DSOM '94),* Toulouse, France.

[5] Marriott, D. and Sloman, M. (1996) Implementation of a Management Agent for Interpreting Obligation Policy. *Proceedings of the Seventh IFIP/IEEE International Workshop on Distributed Systems: Operations and Management (DSOM '96),* L'Aquila, Italy.

[6] Moffett, J. (1994) Specification of Management Policies and Discretionary Access Control, in *Network and Distributed Systems Management* (ed. M.S. Sloman). Addison-Wesley.

[7] OMG (1994) *Common Facilities Architecture* (OMG Document Number 94.11.9).

[8] OMG (1995) *The Common Object Request Broker: Architecture and Specification (Revision 2.0)* (OMG Document Number PTC/96.03.04).

[9] Sloman, M. and Twidle, K. (1994) Domains: A Framework for Structuring Management Policy, in *Network and Distributed Systems Management* (ed. M.S. Sloman). Addison-Wesley.

[10] Soukouti, N. (1995) *Toward Managing CORBA Objects via OSI Network Management Mechanisms* (Draft Paper).

[11] Wies, R. (1995) Using a Classification of Management Policies for Policy Specification and Policy Transformation. *Proceedings of the Fourth International Symposium on Integrated Network Management (ISINM '95)*. Santa Barbara, California.

[12] X/Open (1995) *Systems Management: Common Management Facilities, Volume 1* (Preliminary Specification).

## ACKNOWLEDGEMENTS

This research is supported by the IBM Centre for Advanced Studies and the Natural Sciences and Engineering Research Council of Canada.

## ABOUT THE AUTHORS

**Stephen L. Howard** is a Ph.D. student in the Department of Computer Science at the University of Western Ontario.

**Hanan L. Lutfiyya** is an Assistant Professor in the Department of Computer Science at the University of Western Ontario. Dr. Lutfiyya received a Ph.D. in computer science from the University of Missouri at Rolla.

**Michael J. Katchabaw** is a Ph.D. student in the Department of Computer Science at the University of Western Ontario.

**Michael A. Bauer** is a Professor in the Department of Computer Science and Senior Director of Information Technology Services at the University of Western Ontario. Dr. Bauer received a Ph.D. in computer science from the University of Toronto.

## PART TWO

# Network Monitoring Policies
## Chair: James W. Hong, Pohang University

# 40
# Deriving Variable Polling Frequency Policies for Pro-active Management in Networks and Distributed Systems

*P. Dini, R. Boutaba*
*Computer Science Research Institute of Montreal*
*1801, McGill College street, # 800, Montreal, Qc, H3A 2N4, Canada*
*e-mails: {dini, rboutaba}@crim.ca*

### Abstract
Management activities are based on the state of distributed system components, relations of these components, and their behaviour. Since management policies are applied across an abstraction of distributed systems, the quality of decisions is dependent on the representation fidelity of the system real state. Obviously, the data collection process updating the abstract representation of real distributed systems has a major significance to carry out relevant management decisions. Existing approaches promote a fixed polling frequency for all components. We present shortcomings of these issues and propose adapted polling formulae considering the behavioural parameters of managed components. Models for the flexible polling frequency and for the frequency adaptor are presented. Based on these models, we propose adapted formulae considering both the availability features and dynamics features in a linear and exponential approach. Policies for applying these formulae are proposed with respect to the component grouping and frequency classification. Implementation aspects are discussed.

### Keywords
*Variable polling frequency, pro-active management, distributed systems*

## 1 INTRODUCTION

Two paradigms are currently used for the system management namely, *the platform-centred management* (commonly used by proprietary architectures) and *the distributed management* (recommended by OSI and ODP standards). Distribution is viewed either as *heterogeneous management entities* localized on specific sites and cooperating for a common problem, or *homogeneous management entity units*, distributed on many sites (so called SMAEs in OSI

This work was partially funded by the Ministry of Industry, Commerce, Science and Technology, Quebec, under IGLOO project organized by the Computer Science Research Institute of Montreal, and by a grant from the Canadian Institute for Telecommunication Research (CITR) under the Networks of Centres of Excellence Program of the Canadian Government.

standards). Management activities are based on the state of DS (Distributed System) components, relations of these components, and their behaviour. Since *management policies* are applied across an abstraction of distributed systems, the quality of decisions is dependent on the representation fidelity of the system real state. Obviously, the data *collection process* updating the abstract representation of real DSs has a major significance to carry out relevant *management decisions*.

## 1.1 Architectural issues

Updated data are collected into two phases namely, at *physical level* (from real resources to standardized records) and at *management level* (from management agents to their correspondent managers). As shown in Figure 1, these data are collected from the real resources either by the source initiative as change events (*physical notifications, management notifications*), or by specialized software agents (*physical actions, management actions*). At the *management level,* received data are recorded within managed object MIBs (Management Information Bases) each time an update action or notification occurs.

In the *platform-centred management* approach, a manager receives from or interrogates its own agents with respect to the state of MIB components, via *management notifications (event approach),* and respectively *management actions (commands approach). Centralized management* policies are consequently applied on MIBs across appropriate agents. A similar aspect raises within a *distributed management* platform. However, in this approach, decision policies are derived by knowledge exchanges between managers.

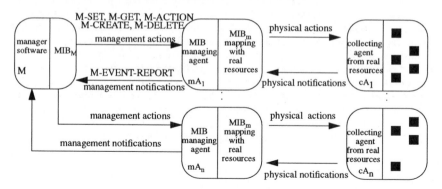

**Figure 1** Overall Architecture of the OSI Management System.

As previously presented, managers must accurately know the real state of managed components. Within the *event approach*, a real DS component automatically sends notifications (at the physical level) as state change events, which are conveyed by specialized agents to MIBs. The MIB agent sends in turn this notification (at the management level) to its own manager. The more the number of changes is high, the more the amount of traffic increases, and the system performance implicitly decreases. Even further, many times, the notification data may not be relevant for the system management.

In the *commands approach*, supervisors (agents for the real resources, or managers for agents) send actions to collect data (eventually updated). Collecting commands could be issued periodically with a variable or fixed frequency. The process of collecting information at regular intervals is known as *polling*. The result of a polling operation is either new data (as *information message*), or no new changes (*control message* with no reports).

There are three *important features* of this process namely, the *polling interval* defined as the amount of time between two consecutive polling operations within which the polled component has not transmitted new data, the *walk time* representing the amount of polling interval consumed by polling messages, and the *response time* referring to the amount of time between a command and the appropriate response. Evaluation of these features depends on several management aspects.

Among the two most polling relevant aspects are the *standardized management communication and recorded information*, and the *adaptation of the polling frequency* with respect to significant evolving system parameters. For the former, the problem is solved only at the *management level*, where the polled resource must accept the management messages as used by its callee, for example SNMP (Snmp 1990), or CMIP (Cmip 1991) protocols (Figure 1). Data on real resources are recorded in special management data bases called MIB (Mib 1990) as managed objects, and accessed by management agents. At the *physical level* there are only private solutions. Several attempts for a generic agent (Claudé 1990) must be developed. Adaptation of the polling frequency could conform with either general conditions (traffic, specific topology), or component behavioural aspects (state, notifications).

In DSs, two types of *polling mechanisms* are currently used: the *roll-call polling* and the *hub polling* (Stalling 1993). The former implies that each subordinate resource is in turn interrogated, while its manager is waiting to the *response message*. The latter supposes that interrogated subordinates are tightly coupled within a loop. The manager sends a *polling operation* to a loop head-component which is propagated up to the loop tail-component. Head- and tail-components receive and respectively send the start and the stop polling cycle flags from/to their callee. Further, all loop components, except the loop tail, pass the polling operation request to its neighbour. All called components send a message to the callee (*information* or *control*). Roll-call polling mechanism is used by SNMP across Internet (Stalling 1993). Hub polling is frequently combined with the roll-call polling and used within intelligent hardware resources at the CAN (Chip Area Network) or BAN (Board Area Network) level, since the hub polling is strongly dependent on the topology of resources (known as *daisy chaining* mechanism (Liu and Gibson 1986).

## 1.2 Shortcomings of existing approaches

Managing by the *event approach* is expensive in time costs. Even in the *command approach* the *polling frequency* must continuously be adapted in order to avoid *control messages* as responses. Several existing approaches propose a statistical computation of the polling interval according to the network traffic (Schwartz 1977), buffer occupancy (Konheim and Meister 1974), or particular parameters of network topologies (Ahuja 1982). This computation is rather static because mainly the average values are considered. Callison claims that values of the real-system parameters must be *re-evaluated* at a fixed temporal intervals (based on *newer statistics* on the previous parameters) in order to ensure the data validity (Callison 1994). Then, the polling frequency is globally established considering statistical combinations of many network parameters. Its update is valid for all polled components. In the following we presents several undesired results due to these approaches.

Let us take a simple example based on Figure 1, where a manager $M$ polls $n$ agents called $mA_i$ ($i = 1..n$). Conforming to the existing approaches all agents are equally treated within a polling cycle, that is, $M$ sends a message to $mA_1$ and then, it waits for the $mA_1$'s response. $mA_1$ will be revisited after the polling of all $mA_i$ ($i = 2...n$). If all these n-1 pollings return no new changes, the network has been unnecessary *overloaded*. Further, during this time, $mA_1$ could have many changes to transmit, that will be *delayed* in the next polling cycle. In related works, the polling frequency is the same for all components, determining either a high management traffic which diminishes system performances in the case of no new changes, or a loss of possible relevant changes, for those components which are less stable. We consider that the polled components must be classified upon several behavioural or managerial criteria. For each group, a distinct polling frequency could be used.

In *de facto* systems, the roll-call polling mechanism is used. A distinct polling frequency is useful even if the components are polled in parallel. Here, the problem is how to classify polled components in groups having the same polling frequency. Polling frequency can be updated with respect to either many *global parameters* (as in previous works), or *specific parameters*, such as *component performances*. Component performances refer to *qualitative parameters* of component's services. Our thesis is that the polling frequency must be adapted with respect to the *component's behaviour*. Several current behavioural parameters concerning the *availability features* (Dini, Bochmann, Boutaba 1996) and *dynamics features* (Dini 1996) have been identified. The proposed approach in this paper takes a *basic polling frequency*, computed by one of previous statistical methods, and *adapts it* to be *more accurate* with respect to the component behaviour.

This paper focus on adapted formulae of polling frequency in DSs, based on behavioural component parameters which are currently measured or computed. In Section 2 we present several related works treating *statistical polling formula* and *behavioural parameters* with respect to the *component availability* and *dynamics features*. Section 3 focuses on our proposals, reporting on polling model, frequency adaptor model, and adapting formulae. Two kinds of corrections are proposed namely, linear and exponential. In Section 4 we present results of the implementation of these proposals. Conclusion and future work presented in Section 5 conclude the paper.

## 2 RELATED WORK

An analysis of the *polling interval* concerning the *roll-call mechanism* is presented in (Agbaw 1994). Agbaw discusses the *polling interval* of the *roll-call mechanism* used by SNMP and presents factors which determine the elapsed time between successive polling operations of the same agent. The calculus implies a number of N agents subordinated to a manager, the average time ($\Delta$) required to perform a single poll, and the desired polling interval for a manager engaged full-time in polling.

Other statistical analyses of the performance in order to establish a polling interval are presented in the literature. Schwartz identifies the *polling interval* and the *response time* as two significant network performance facets (Schwartz 1987). However, the response time is considered more directly related to user needs. All equations presented in by Schwartz are related to the relationship between the *walk time* (amount of polling interval consumed by polling messages), *polling interval*, and *response time*. In this approach, the polling interval is a random variable, depending on the *network traffic* in both polling directions. Schwartz's model is *traffic-based*. The polling interval is not updated with respect to the behaviour of the polled component.

Konheim's and Meister's (1974) statistics model is *buffer occupancy-based*. Computing formulas focus on the communication invariants at each component (packet-arrival rate, frame-length, average of the walk time). The calculus is oriented to get the polling process time, rather than to capture the influence of the system component behaviour. Consequently, this is a complementary approach to our proposal.

Ahuja's model is based on a particular case, where polled components respond only by *control messages* (no information). In this particular approach, the time distance between adjacent components is equal and invariant. A computing formula for the polling interval is obtained (Ahuja 1982). The model could approximate the minimum polling interval for a single level of a hierarchy manager-agents.

The foreseen works prescribe formulas to compute the polling interval by only taking into account the propagation time issues referring to either the component topology, traffic parameters, or communication aspects related to the polling process itself. Callison claims that values of the real-system parameters must be *re-evaluated* at a fixed temporal intervals (based on *newer statistics* on the previous parameters) in order to ensure the data validity (Callison 1994). Otherwise, the correctness of the collected information does not necessarily imply correct management decisions, since real-time systems involve active components (Shin and Parameswaran 1994). However, we consider that a flexible correlation of the polling frequency with *behavioural parameters* of a DS components can substantially reduce the *management traffic*. These behavioural parameters are presented in details by Dini, Bochmann and Boutaba (1996) and Dini (1996), and briefly described in Dini, Bochmann, Koch, and Krämer (1997).

## 3 PROPOSAL

Our proposal refers to three following aspects namely, the *model* for a variable polling frequency, *adapted polling frequency formula*, and *criteria for component classification* in groups having the same polling frequency. We are considering here only the polling at the *management level*.

### 3.1 Models for flexible polling frequency

In Figure 2, we present the current model versus our proposed model. In the *existing model*, the same polling frequency is initially established at a fixed value for all polled components. This value is adapted with respect to the network traffic, the network topology, and other global network parameters. We propose a new polling model, where polled components are classified in groups. A distinct polling frequency, e.g. $f_{01}$ or $f_{02}$ in Figure 2, is used for each group, which represents the basic frequency computed for the existing system component, but corrected with a *component type depending factor* ($\zeta$).

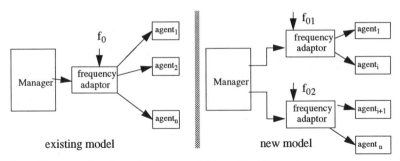

**Figure 2** Polling models using a fixed or a variable polling frequency.

In our solution, we consider that the basic frequency $f_0$ is deduced from formulae of existing approaches which apply various computing criteria, as presented further. If one considers only the number of polled components (n) and the average time required to perform a single poll ($\Delta$), the polling frequency should guarantee that $f \leq 1/n\Delta$ (Stalling 1993). Consequently, in this case, $f_0 = 1/n\Delta$.

In the traffic-based approach, the average scan time ($\bar{t}_c$) is calculated assuming a fixed walk time for all stations, considering the arrival rate of packets at each station, and including the total traffic intensity on the common channel (Schwartz 1987). For those DS components polled upon Schwarz's criteria, the basic polling frequency is $f''_0 = 1/\bar{t}_c$. In the topology-based approach, the poll scan time T is computed following Ahuja's criteria, which assumes that the distances between adjacent stations are equal. The basic polling frequency for these systems is $f'''_0 = 1/T$. Since all these frequencies ignores behavioral aspects related to DS components, the basic frequency represents the minimum polling frequency established with respect to global conditions. This basic frequency will be modified by using the correction coefficient, whose computing formula will be proposed later. Consequently, the new polling frequency must dynamically be adapted.

## 3.2 Frequency adaptor model

Since different subsystems could correspond to different previous criteria, the proposed frequency adaptor model allows to compute the variable polling frequency using $f_0$, $f'_0$, $f''_0$, or any other basic frequency. For each component or group of components, the adaptor must compute the appropriate coefficient ($\zeta$) based on behavioural parameters. The proposed model has four distinct blocks, as shown in Figure 3.

- *the component health parameters block* represents the current availability and dynamics features, as discussed in Section 2.2 (usually, a specific MIB-like database);
- *the coefficient computing block* receives these current values (6) and computes the appropriate $\zeta$ coefficient (5). We will propose $\zeta$ computing formula in Section 3.3.
- *the frequency adaptor block* receives several input data having distinct meanings. From the discriminator block, there is the *activate* (4) command which starts the frequency adaptor activity. The entry (1) ensures different basic frequencies (it can always be modified), whereas the entry (2) is reserved for an imposed basic frequency due to management goals. The variable frequency $f_{0i}$ is the new polling frequency for the component *i*.

- *the discriminator block* has three functions. First, it automatically performs the correction based on entries called (1), (2), and (4), giving a variable polling frequency noticed $f_{0i}$. Second, based on the entry (5), it captures certain abnormal values, and commands the highest polling frequency.

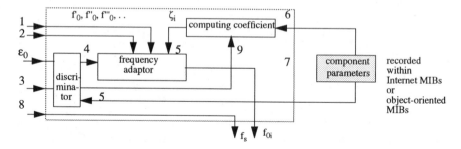

**Figure 3**  Model of the adaptor for a variable frequency.

Concrete conditions will be discussed later. Third, according to the manager policy, the discriminator imposes the computing formula to the coefficient computing block (9). This issue refers to either components classified in the same group, or to a single component which has changed its own behaviour. The entry (8) ignores all previous variants and imposes a polling frequency as requested by the manager ($f_s$). The activity performed within this model has three options: either the manager requests the polling with a well-defined frequency (8), the manager imposes a basic frequency $f_0$ (1 and 2) which will be adapted by the coefficient $\zeta$, or the discriminator detects an anomaly of h(t) value ranges. Since the first case is manager-dependent, we are concerned in the following with the computing polling frequency of the last two cases.

## 3.3 Adapting formulae

Various adapted formulae have been proposed to be used by the coefficient computing block.

Conforming to the existing approaches, we assume that a polling frequency $f_0$ is previously computed with respect to network global parameters. Two approaches are possible to adapt the polling frequency:

- *using behavioral parameters:* Behavioural parameters summarized in Dini, Bochmann, Koch and Krämer (1997) may be used independently, or in combination to compute new polling frequency values. These parameters are related to operational aspects of managed objects. From a generic formula $[f_\zeta, (1 + \zeta)f_0$, where $\zeta$ is a correction coefficient, four different adapted polling frequency formulae have been deduced. An agent-based management using these formulae is presented in the paper cited above at this conference.

- *using new state change models:* In order to predict the behavior of a managed object and apply pro-active management policies, a refinement of state change model have been proposed in Dini, Bochmann, and Boutaba (1996). We consider aspects related to the usage state, current availability, and costs. This paper presents several policies related to this approach.

In both cases, the proposed formulae try to adapt the polling frequency to current changes of behavioural parameters or states. If the discriminator receives (input 5) behavioural parameter values below several thresholds guaranteed by the component, it disallows the calculus of the coefficient computing block and allows the manager to apply other appropriate policies, i.e. a well specified polling frequency $f_s$. If the current availability tendency is stable, the discriminator stops the coefficient computing and $f_0$ is applied.

## 3.4 Polling algorithms

The following scenarios could be used by a manager to apply a self-adapting managing policy based on the state of managed objects.

### Simple examples

Let us consider a simple system of a server with many clients. Various situations can be identified, as presented in the following examples.

- *example 1.* The manager uses a polling frequency policy to *get state* values of managed objects with a fixed polling frequency $f_0$. The managed object representing the server can send traps or notifications concerning its usage state as follows: if the number of clients exceeds 10, the server send *active1* event, if the number exceeds 15, the server send *active2* event, whereas the *active3* event is sent when the number of clients is 19. 20 represents the internal event of the server managed object causing the state change notification of type *usageState* from *active* to *busy*.

- *example 2.* The manager uses a polling frequency policy to get state values of managed objects with a fixed polling frequency $f_0$. The health of the managed object representing the server varies according to variations of its operational state (Dini, Bochmann, Boutaba 1996). The managed object representing the server can send *alarm1*, *alarm2*, and *alarm3* traps or notifications appearing at well-defined current availability thresholds. For example, alarm1 corresponds to *currentAvailability = 0.9*, alarm2 to *currentAvailability = 0.85*, and alarm3 to *currentAvailability = 0.75*.

- *example 3.* The manager uses a polling frequency policy to *get state* values of managed objects with a fixed polling frequency $f_0$. If within a large period of time, there are no notifications from the managed object representing the server, or the *get state* does not capture a state change, the traffic has been usefulness charged. A function adapting the polling frequency with respect to costs incurred by polling without changes versus changes occurring and not captured in time has been proposed by Agbaw (1994). This solution is valid only in polling-driven systems, such as Internet.

## Flexible polling frequency algorithms

Since CMIP-based systems allow the polling approach by the primitive M-GET, while the SNMP-based systems acts as event-driven across the primitive Trap, different implementation solutions are specific to event-driven (CMIP-based) or polling-driven (SNMP-based). Consequently, regardless of which approach is considered, an harmonization of these two ways must be performed by an adequate *flexible polling frequency policy*. Since the current availability value represents the confidence that the managing system has on the real resource availability, it seems naturally to accommodate the polling frequency with the current availability. A higher polling frequency is used to poll managed objects displaying a low current availability, following an adaptation function. Even in the event-driven approach, the polling frequency policy could detect the de-activation of notification services. However, if a managed object has a stable usage state or operational state, the management traffic is usefulness increased for a class of managed objects, while a uniform polling frequency could lead to loss useful information regarding other managed objects. According to these cases, we develop three flexible polling frequency policies concerning either i) the usage state, ii) the current availability, or iii) the costs incurred by usefulness polling versus changes lost. Hereafter, E represents the component type and F represents the component identifier.

*i) flexiblePollingFrequency1 policy.* This policy adapts the basic polling frequency $f_0$ to $f_{adapted} = (1 + th_i) \times f_0$, where $th_i$ corresponds to the notification *active$_i$*. These $th_i$ are specific to different component types. For example, a CPU decreases its QoS according to the number of clients. When this number increases, the probability to cause failures because of lack of capacity becomes greater.

```
. . . . . . . . . .
flexiblePollingFrequency1
. . . . . . . . . .
functional behavior
 input:    E:type, F: name, active_i, f_0, th_1, th_2, th_3
 output: f_adapted
. . . . . . . . . .
 begin
    compute f_adapted = (1 + th_i) x f_0
 end begin
. . . . . . . . . .
```

*ii) flexiblePollingFrequency2 policy.* This policy adapts the basic polling frequency $f_0$ to $f_{adapted} = f_0 / h_i$, where $h_i$ thresholds correspond to the notifications *alarm$_i$*. These $h_i$ are specific to different component types. Since the current availability value represents the confidence the managing system has on the real resource availability, it seems naturally to accommodate the polling frequency with the current availability. A higher polling frequency is used to poll managed objects displaying a low current availability.

```
. . . . . . . . . .
flexiblePollingFrequency2
. . . . . . . . . .
functional behavior
 input_1: E:type, F: name, alarm_i, f_0, h_1, h_2, h_3
 input_2: E:type, F: name, f_0, h_1, h_2, h_3
 output: f_adapted
. . . . . . . . . .
 begin1
    compute f_adapted = (1 + h_i) x f_0
 end begin1
. . . . . . . . . .
 begin2
    invoke currentAvailabilityFunction (Dini 1996)
    if currentAvailability > h_1 then f_adapted = f_0
    if currentAvailability > h_2 then f_adapted = f_0/h_1
```

```
             if currentAvailability > h₃ then f_adapted = f₀/h₂
             else f_adapted = f₀/h₃
          end begin2
. . . . . . . . . .
```

*ii) flexiblePollingFrequency3 policy.* This policy is based on costs of the usefulness polling $C_p$ versus costs incurred by changes lost $C_l$. According to these costs, Agbaw proposed a function to adapt the polling frequency (Agbaw 1994). Let us suppose that we have these costs for each type of system component.

```
. . . . . . . . . . . . .
flexiblePollingFrequency3
. . . . . . . . . .
functional behavior
 input: E:type, F: name, Cp, Cl, f₀
 output: f_adapted
. . . . . . . . . .
 begin
    invoke f_Agba (Agbaw 1994)
    f_adapted = f_Agba

 end begin
. . . . . . . . . .
```

## Combined polling policy

Polling policy combines these three *flexible polling policies* with the get state management action to self-adapt the polling to current state changes, threshold notifications, or costs.

```
. . . . . . . . . .
polling policy
. . . . . . . . . .
f_polling = f₀
functional behavior
 input: E:type, F: name, th₁, th₂, th₃, h₁, h₂, h₃
         [alarm1 | alarm2 | alarm3] | [active1 | active2 |
         active3], C_p, C_l

 output: E:type, F: name, f_polling
. . . . . . . . . .
begin
    if alarm_i then flexiblePollingFrequency2(alarm_i)
                    and
                    f_polling = f_adapted
    fi
    if active_i then flexiblePollingFrequency1(active_i)
                    and
                    f_polling = f_adapted
    fi
    else get state (E, F)
        if currentAvailability < h1
            then flexiblePollingFrequency2(h)
            and
            f_polling = f_adapted
        else flexiblePollingFrequency3
            and
            f_polling = f_adapted
        fi
    fi
end begin
. . . . . . . . . .
```

According to the behavior of the *combined polling policy*, the frequency of polling is adapted to various constraints, coming from the unexpected behavior of system components.

## 4 IMPLEMENTATIONS ISSUES

Several aspects arise applying these formulae. Commonly, even for a single component, the correction coefficient continuously varies. It is a heavy task to adapt the value of the polling frequency each time a change occurs. Then, a clustering of these values is useful. Second, having a distinct polling frequency for each component makes heavy the management activity. Frequently, the managers want to cover a area of components under the same umbrella. Then, a clustering of components must be adopted. Finally, the increasing complexity of computing formulae needs a manager policy to correctly select and link together the formulae's expressions.

### 4.1 Clustering polling frequency values

For one component or for a group of components, there is a single basic formula given by $f_0$. Usually, when a polling request is issued, returned data are received by the same manager component. If the manager has $n$ distinct possibilities to process polled data, we propose a clustering of the largest interval $[f_0, (1 + \zeta)f_0]$ in intervals having the length equal to the ($\zeta$ x $f_0$) / n. For the exponential adaptation (Dini, Bochmann, Koch and Krämer 1997), the interval is $[f_0, (1 + e^\zeta)f_0]$ and the ratio is $e^\zeta$ x $f_0$.

### 4.2 Component classification policies

Components behave a large spectrum of current parameter values and states. However, the polling frequency can not follow each variation. Consequently, the domain range of parameter values must be divided into many subranges. For each subrange, a unique polling frequency value is established. It will be useful to group components having the same subranges into classes in order to compute a single frequency. Sometimes, other grouping criteria are used (geographical, administrative, etc.). In this case, the manager must decide a policy to choose the reference values to represent a group and select the group's $\zeta$.

*Current availability-based classification policy*
The policy we propose to classify parameter values considers several distinct cases:
   • $C_1$ represents special components having a fixed polling frequency, regardless their current parameter values. This class has two opposite subclasses namely,
   • $C_{11}$ denotes those components which are not critical, and then, their polling is done by using the basic frequency;
       $C_{11i}$ (i = 1..n) component uses the basic frequency $f_{0i}$;
   • $C_{12}$ refers to critical components requiring a fixed high polling frequency;
   • $C_2$ contains the remainder which is commonly classified as following,
   • $C_{2i}$ (i = 1..5) represents components having the observed availability value within the interval $[(4+i)$x $0.1, (5+i)$x $0.1]$
   • $C_{26}$ represents those components with a low current availability (h < 0.5 or other constraints) for what the manager does not poll.

*Groups polling policy*
Components belonging to the same class ($C_{ij}$) could be grouped in order to diminish the computing task. However, if for different reasons, such as administrative or functional dependencies, components of distinct classes must form a polled group having the same polling frequency, the question is how to determine the reference component.
   We propose a selection policy with the most restrictive condition. Among group's compo-

nents, the manager must poll only the representative component of the group. The election is first performed by following the previous classification. Inside a class, the polling frequency is determined by that component having the maximum polling frequency.

## 4.3 Using a flexible polling frequency policy

Currently, we are experimenting these ideas within the context of automatic reconfiguration management. Several implementations issues belonging to the used platform are presented below. First, these computing formulae considering the behavior parameters inherit timing aspects presented in (Dini Bochmann Boutaba 1996) (Dini 1996), i.e., the availability and dynamics parameters are clock-time-dependent. However, the basic polling frequency currently used in SNMP-base management systems is fixed (10 seconds) between managers and proxy-agents, whereas the polling frequency between proxy-agents and the SNMP-agents could be modified, as presented in (Agbaw 1994). Referring to Figure 2, the proxy agents appear between an OSIMIS manager M and an SNMP-agent $mA_i$, in order to correctly map specific features of managed objects versus management protocols. Agbaw's formulae include also probabilistical evaluations of costs with respect to the information loss, or control message (unnecessary polling).

Other management aspects can benefit of the present proposal. For example, existing management tools can implement these algorithms to facilitate the automatic management of networks. Additionally, in self-testing systems, where some testing components are charged to automatically test critical functional components, a more accurate information can be obtained if the polling frequency is adapted according to the dynamic results of tests (Dini 1995). For example, when some tests are applied to a system component, the polling frequency may be increased, in order to accurately capture the real state of the concerned component into a small time vicinity.

Current network management stations products, such as *NetView/6000* (IBM), *SunNet Manager* (Microsystems), and *Open View* (Hewlett-Packard) help human operators to manage networks. They consider a graphic facilities to scroll and read MIB information, and a window to choose the polling interval, e.g. MIB Data Collection Dialog Box of NetView/6000. The polling interval value is established by the human operator as results of lecturing MIBs. The operator can also stop and restart the activity of data collection. Decisions of polling frequency and of stopping or restarting the collection are *human-based*. In open large systems this task becomes difficult and with not relevant results. Therefore, existing tools can implement policies to adapt the polling frequency.

## 5 CONCLUSION AND FUTURE WORK

In order to enhance the management policies in distributed systems we have presented in this paper a variable polling frequency based on the dynamic features of a managed object. Since existing approaches promote a fixed polling formulae for all components, there are no implemented mechanisms to automatically vary the period of polling. The management tools envisaging to easily allow to a human operator to change the polling frequency have not this feature. Our proposal are based on the current evaluation of component properties, that are explicitly described by its availability and dynamics features. The model proposed for the adaptor of a variable frequency permits to compute the correction coefficients and to discriminate between different management polling policies with respect to discriminator criteria. The proposed formulae can adopt a linear or exponential approach to correct the basic polling frequency established by previous studies. In order to better select the polling frequency we propose component classification rules based on the component properties. Consequently, several distinct classes are obtained. Finally, a criterion for component group is proposed. We have equally presented implementation issues referring to our experiment platform using the OSI-

MIS/ISODE distributed environment and SNMP-agents.

The ongoing studies are presently done on formulae including also probabilistical evaluations of costs with respect to the information loss, or control message (unnecessary polling). Additionally, a combination of health criteria and management constraint criteria imposed by costs could be used to build automatic management policies.

## 6 REFERENCES

Agbaw, C. (1994) Management Data Collection and Gateways, *M.Sc. Thesis*, McGill University, 1994.

Ahuja, V. (1982) Design and Analysis of Computer Communication Networks, McGraw-Hill, Computer Communication Networks, 1982.

Callison, H. R (1994) A Periodic Object Model for Real-Time Systems, *Research Paper, Department of Computer Science and Engineering*, FR-95, University of Washington, Seattle, WA, IEEE 1994.

Cana (1991) ISO/IEC 9596-1:1991, Information Technology - Open System Interconnection - Common Management Information Protocol - Part 1: Specification, CAN/CSA-Z243.142-91.

Claudé, M. (1990) *Unification de Contextes Hétérogènes par un Agent Générique OO*, Thèse de Doctorat de l'Université Pierre et Marie Curie, février 1990.

Dini, P., Bochmann, v. G., Koch, T., Krämer, B. (1997) Agent Based Management of Distributed Systems with Variable Polling Frequency Policies, *Proceedings of The IM'97 Symposium*, San Diego, 1997.

Dini, P., Bochmann, v. G., and Boutaba, R. (1996) Performance Evaluation for Distributed System Components, *The Second IEEE Systems Management Workshop*, Toronto, Ontario, Canada, June 19-21, 1996.

Dini, P. (1996) New Aspects for Run-time QoS Evaluation in Networks and Distributed Systems, Tutorial, *European Simulation Multiconference 1996*, Budapest, Hungary, June 2-6, 1996, pp. 3-11.

Dini, P. (1995) Management Policies of Active and Passive Tests in Distributed Systems, *Technical Report, IGLOO Project*, CRIM/University of Montreal, May 1995.

Konheim, A.G. and Meister, B. (1974) Waiting Lines and Times in a System with Polling, *Journal of the ACM*, 21, no. 3, July 1974, pp. 470-490.

Liu, L. and Gibson, G. (1986) *Microcomputer Systems: The 8086/8088 Family*, Prentice-Hall, 1986

Mib (1990) RFC 1213, *Management Information Base for Network Management of TCP/IP-based internets: MIB II*, eds: K. McCloghrie, M. Rose.

Schwartz, M. (1977) *Telecommunication Networks: Protocols Modeling and Analysis*, Addison-Wesley Publishing Company, 1987.

Schwartz (1987), *Computer-Communication Networks Design and Analysis*, Prentice Hall, Inc., 1987.

Shin G.K. and Parameswaran, R. (1994) Real-time Computing: A New Discipline of Computer Science and Engineering, *Proceedings of the IEEE*, vol. 82, no. 1, January 1994.

Snmp (1990) RFC 1157, *A Simple Network Management Protocol*, M. Schoffstall, M. Fedor, J. Davin, J. Case (05/10/90).

Stalling, W. (1993) *SNMP, SNMPv2, and CMIP: The practical Guide to Network Management Standards*, Addison-Wesley Publishing Company, 1993.

# 41

# Agent based Management of Distributed Systems with Variable Polling Frequency Policies

*P. Dini, G. v. Bochmann*
*University of Montreal*
*CP 6128, Succursale CENTRE-VILLE, Montreal, H3C 3J7, Canada.*
*email:* `dini,bochmann@crim.ca`

*T. Koch, B. Krämer**
*FernUniversität Hagen*
*Datenverarbeitungstechnik, 58084 Hagen, Germany.*
*email:* `thomas.koch,bernd.kraemer@fernuni-hagen.de`

## Abstract

Management activities are based on the state of distributed system components, relations of these components, and their behaviour. Since management policies are applied across an abstraction of distributed systems, the quality of decisions is dependent on the representation fidelity of the real system state. Obviously, the data collection process updating the abstract representation of real distributed system components has a major impact on the quality of management decisions. Gathering most topical management data improves the quality of management decisions, but requires a high degree of monitoring activity. This is contradictory to the request for low impact management systems, where the amount of system resources used for management purpose should be as small as possible.

In this paper we present a twofold approach to this problem: First a high level management architecture is described where monitoring is performed by distributed agents with generic functionality for filtering and event creation. The distribution of active management agents reduces the amount of management related traffic and avoids a potential bottleneck on a centralized management station. Second an adaptive polling frequency approach is presented which enables the monitoring agents to adapt their polling frequency automatically to different behavioral parameters of managed components. The automatic adaptation reduces the performance impact of the agents significantly while at the same time a high accuracy of management relevant information about critical components is ensured. Implementation aspects of the introduced management architecture in a CORBA environment are also discussed.

## Keywords

Agent based monitoring, Adaptive polling frequency, Distributed systems management

---

*Supported by the BMBF, Project No. INF 26

# 1    INTRODUCTION

Two paradigms are currently used for system management: the platform-centered management (commonly used by proprietary architectures) and the distributed management (recommended by OSI and ODP standards). Distribution is viewed either as heterogeneous management entities localized on specific sites and cooperating for a common problem solution, or homogeneous management units, distributed on many sites (so called *System Management Application Entity (SMAE)* in OSI standards). Management activities are based on the state of DS (Distributed System) components, relationships of these components, and their behavior.

Managers must accurately know the state of managed components. Within the notification approach, a DS component automatically sends notifications upon state changes to specialized agents. The agent forwards this notification to its own manager. The amount of traffic increases with the number of changes and the system performance decreases. Beside that, many times the notification data may not be relevant for the system management.

In the polling approach, supervisors (agents for the real resources, or managers for agents) send actions to collect data (eventually updated). Collecting commands could be issued periodically with a variable or fixed frequency. The process of collecting information at regular intervals is known as polling. The result of a polling operation is either new data (as information message), or no new changes (control message without reports).

The polling procedure is characterized by three important features: the *polling interval* defined as the amount of time between two consecutive polling operations within which the polled component has not transmitted new data; the *walk time* representing the amount of polling interval consumed by polling messages; and the *response time* referring to the amount of time between a command and the appropriate response. Evaluation of these features depends on several management aspects.

Gathering most topical management data usually improves the quality of management decisions, but requires a high degree of monitoring activity. This is contradictory to the request for low impact management systems, where the amount of system resources used for management purpose should be as small as possible. In this paper we present a twofold approach to this problem: In Section 2 a high level management architecture is described where monitoring is performed by distributed agents with generic functionality for filtering and event creation. Section 3 shortly defines behavioral parameters which are used in Section 4 as input for an adaptive polling frequency approach which enables the monitoring agents to adapt their polling frequency automatically with respect to different behavioral parameters of managed components. Implementation aspects of the introduced management architecture in a CORBA environment are discussed in Section 5.

# 2    ARCHITECTURAL ISSUES

A comprehensive management architecture requires the provision of several generic management services. Figure 1 provides an overview of the architectural concept. Different management applications share a set of generic management services. These services are focused on management issues only, distribution transparencies are provided by appropriate middleware components. The integration of different management applications into a common architecture allows simplified cooperation and coordination of activities from

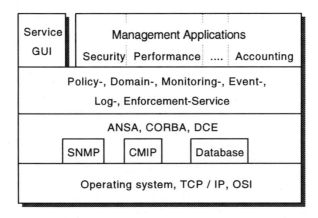

**Figure 1** Overview of the management architecture

different management areas. This integrity is required to avoid unexpected side effects where activities in one management application has a major impact on the result of activities in another area. The performance management of multimedia applications, for example, depends usually on the underlying network, which is managed by a network management application. Coordination of both management activities allows us to reach an optimized result for both areas.

A common graphical user interface (GUI) provides access to the management services. This interface allows the human administrator editing of policies and domains, activation and deactivation of services and monitoring of management activity. It is important to note that the management services in this architecture are considered to be a distributed application running on an appropriate middleware platform. State of the art management protocols, like the *Simple Network Management Protocol (SNMP)* or the *Common Management Information Protocol (CMIP)*, and an object oriented database are enclosed within the middleware and therefore transparent to the management services and applications. Elementary services, like communication or data-storage, are provided by standard operating systems.

Figure 2 provides an overview of the runtime environment. Updated data is collected in two phases namely, at a physical level (from real resources to standardized records) and at a management level (from management agents to their correspondent managers). As shown in Figure 2, these data is collected from the real resources either by the source initiative as change events (physical notifications, management notifications), or by specialized software agents (physical actions, management actions).

At runtime most management activities are triggered by events created from monitoring results. Monitoring provides actual values about the state of managed objects either by requesting the values at a regular interval or by receiving notifications from the managed objects. Evaluation of monitored values may result in an event if a predefined threshold

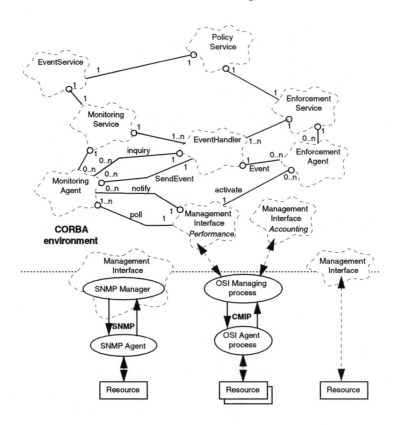

**Figure 2** Agents and Interfaces in a CORBA environment

or a combination of critical values is reached. The functionality of the monitoring service comprises operations for:

- Instantiation and removal of monitoring agents based on informations provided by the event service.
- Management of agents including behavior adaption to different polling intervals, thresholds, etc.
- Provision of monitored values on request from the management service.

Monitoring is performed by generic monitoring agents (Koch and Krämer 1996, Koch 1995) which are installed by the monitoring service and located close to the managed object.

The management architecture is realized as a distributed CORBA application, therefore the interfaces are described in CORBA IDL. Figure 2 provides an overview of several possible management interfaces. On the left side an interface to the Internet management environment is displayed. Access to the SNMP manager is performed with method invocations on a CORBA interface. Number and complexity of interface functions depend on the capability of the SNMP manager. The management interface will be implemented as CORBA server and the capabilities are fixed with the code. Therefore the functionality of the SNMP manager is not a subject of flexible policy definitions. Any change to the behavior requires recoding of the CORBA server. It is an important design decision to provide most general but still powerful capabilities. The minimum capabilities are prescribed by the functionality of the SNMP protocol. This simple approach is obviously unsufficient for comprehensive management tasks. Enhanced capabilities depend strongly on the type of resource that will be managed. The use of the CORBA interface inheritance mechanism could provide a solution. An elementary SNMP management interface to the CORBA environment provides the SNMP functionality only and serves as parent class for more specialized interfaces providing enhanced functionality for certain types of resources. This approach additionally ensures openness to new resources that may provide an enhanced SNMP MIB.

A similar problem occurs with the inclusion of the OSI management environment. A different partitioning of interface functionality is used for representation of an OSI managing process in the middle of Figure 2. Obviously the same style could be used for the representation of the SNMP manager or conversely. The problem of assigning appropriate capabilities to the interfaces could also be solved in the same way as proposed for the SNMP interface.

Figure 2 also illustrates the major advantage of a management environment that is implemented as a CORBA application. The architecture is open to any management protocol as long as an appropriate interface is defined. This allows the inclusion of applications as well as system and network components into a comprehensive management environment. The CORBA standard additionally ensures interoperability between different management service objects, running on different CORBA implementations.

## 3 BEHAVIORAL PARAMETERS

Unexpected operational state changes of system components imply that the service availability of components can not be guaranteed in general. State changes could either refer to the operational availability of a component or to the quality of the provided service. In the sequel, we shortly present these parameters. The interested reader can find additional details on the health behavioral parameters in the appendix, and complete information in cited papers.

### 3.1 Availability parameters

Several run-time measurable features for the operational availability have been proposed and defined by Dini, v.Bochmann and Boutaba (1996) as follows. *Current availability* $h(t)$ is a continuous function of time, defined as a quotient between the amount of time in which the resource has been in the enabled state, commonly called operational time, and the observation period, i.e. the period between the time $t_0$ of the event start and the time

$t$ of the end of the observation period. The notation $t_i$ represents time stamps attached to each event occurring in a system.

## 3.2   Dynamics parameters

Dini (1996) characterized dynamic aspects of the component behavior by several computable features. These features refer to either stability or instability of the component. The stability order $k$ evaluates how long an operational state holds, with respect to the preceding change. There are three instability measures: the *instability order*, the *repeatability order*, and the *multiplicity order*. The instability order $p$ gives a measure of how long in time an operational state holds. The repeatability order $r$ defines a local instability within a vicinity of a change, whereas the multiplicity order $m$ refers to a long time instability. Each order is dynamically computed as an integer number and the appropriate time stamp is attached to it. Time stamp lists corresponding to $p, r$, and $m$ orders form the time behavioral history of a system component.

## 4   ADAPTIVE POLLING FREQUENCY

In this section we propose a model for an optimized monitoring agent that adapts its polling frequency automatically according to predefined criteria. The monitoring agent is initialized with a basic frequency computed for the monitored system component. We assume that the basic frequency $f_0$ is deduced from formulae of existing approaches presented by Stallings (1993) and Schwartz (1987). Since all these approaches ignore behavioral aspects related to the monitored components, the basic frequency $f_0$ represents the minimum polling frequency established with respect to global conditions. This basic frequency will be modified by use of a correction coefficient. The proposed model allows two different adaptation styles: The polling frequency is either adapted with respect to the behavior history of the monitored component, or with respect to the current value of the monitored attribute.

## 4.1   History based adaptation

We assume that a basic polling frequency $f_0$ is computed with respect to global system parameters and known by the agent. This basic frequency will be adapted by the correction factor $\zeta_i$ according to the following equation:

$$f_{0_i} = f_0 \left( 1 + \zeta_i \right) \qquad \text{for } i = 1, \dots, 4 \tag{I}$$

Consequently, the basic value of $f_0$ can be calculated independently according to global statistical parameters, whereas the right polling frequency is increased by the fraction $\zeta_i f_0$. This term signifies the correction we apply by taking into account the health behavioral parameters presented in Section 3. Next we will present two sets of formulae which linearly or exponentially correct the basic polling formula.

### *Linear adaptation*
Four different formulae are proposed for the calculation of the correction factor $\zeta_i$.

*Formula 1*    The *current availability* computed at the last time stamp $t$ is represented by $h(t)$. The unavailability at $t$ is $1 - h(t)$. Then, a first correction factor is defined with respect to these two related parameters. The quotient between the availability and unavailability is a relevant evaluation of the operational state. The first formula that we propose is to adapt the basic polling frequency by:

$$\zeta_1 = (1/h(t) - 1) \tag{1}$$

*Formula 2*    If the dynamics parameters are critical features for a component, we can combine them with the availability features. For the moment, we consider only the stability orders (called $k$ and $k'$ for the enabled and disabled operational state, respectively). Intuitively, with increasing values of $k$ or $k'$ the polling interval should increase, i.e. the polling frequency decreases. Therefore the polling frequency is conversely proportional to the stability orders. Since $k$ and $k'$ are finite integers, for $k \neq 0$ and $k' \neq 0$, we propose:

$$\zeta_2 = \zeta_1 + max(1/k, 1/k') \tag{2}$$

*Formula 3*    Commonly, an unavailability described by an instability order $p$ with $(t_i - t_{i-1} = 10^{-p})$ is considered to be in the range $[p_{min}, p_{max}]$, where $p_{mim}$ is the lowest measured value and $p_{max}$ is the highest value accepted for a component type. Consequently, $1/(p_{max} - p_{min})$ is the ratio for each $p$'s unity which can be added to the previous equation (2).

$$\zeta_3 = \zeta_2 + p_0/(p_{max} - p_{min}), \text{where } p_0 = max\{p_i | t < t_i\} \tag{3}$$

*Formula 4*    For some network components, the repeatability and multiplicity orders are relevant management features. Since $m = 1$ and $r = 1$ are the lowest limits, we assume that $m_{max}$ and $r_{max}$ are extreme limits accepted for an order $p$. Naturally, when a component behaves with $r > 1$ and/or $m > 1$, the polling frequency must be adapted in a certain manner. If we consider $m_0$ and $r_0$ similarly to $p_0$, we could improve the coefficient expression:

$$\zeta_4 = \zeta_3 + m_0/(m_{max} - 1) + r_0/(r_{max} - 1) \tag{4}$$

*Remarks:*    With respect to the availability, Formula 1 multiplies the basic polling frequency by a coefficient in the range $[0, 1]$, if we admit a minimum availability of 0.5. Each of the subsequent formula increments the maximum value of the coefficient by one, with a maximum increase of $f_{0i} = 6 \times f_0$ if Formula 4 is used.

## Exponential adaptation

Another way to weight the current behavioral parameter values is to consider the exponential function, which better emphasizes new values with respect to their magnitude. Let us consider a large range of the interval $[a, b]$ described by $\exp([a, b])$. If $z \in [a, b]$, the relation $z \leq (e^z - 1)$ holds for all $z$. Consequently, we have a larger spectrum to distribute the polling frequency within the same range of $\zeta_i$. Each of the previous formulae becomes $\zeta_i' = \exp(\zeta_i)$ and the maximum polling frequency is given by $f_{max} = f_0 \times 6 \times (e - 1)$, where $e = 2.73\ldots$

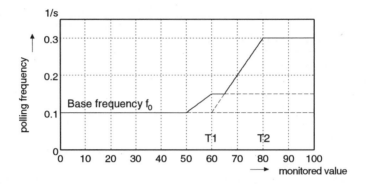

**Figure 3** Threshold based frequency adaptation

## 4.2   Threshold based adaptation

The second adaptation style is based on the idea that in general the polling frequency should be increased if the measured value gets close to a predefined threshold. The specification of the threshold based frequency adaptation is very simple. The system administrator specifies a base frequency $f_0$ and up to $n$ thresholds with individual characteristics. The concept is illustrated in Figure 3 with a base frequency $f_0 = 0.1\ ^1/_s$. Here the monitored value is in the range $[0, 100]$, which is typically for some kind of load measured in %. The monitoring agent will create a notification at the threshold values 60 ($T_1$) and 80 ($T_2$), respectively. A more detailed description of the agent behavior and the optional characteristics for each threshold is given by Koch, Krell and Krämer (1996), and in Koch and Krämer (1996). If a frequency adaptation is desired, the administrator may optionally specify three additional values for each threshold:

**Factor:** The factor $\alpha_j$ describes the desired polling frequency at the corresponding threshold $T_j$. The desired frequency is calculated by multiplication of the base frequency with $\alpha_j$.

$$f_{T_j} = f_0 \times \alpha_j$$

The example in Figure 3 uses the factors $\alpha_1 = 1.5$ for $T_1$ and $\alpha_2 = 3$ for $T_2$.

**Low:** The low value $\beta_{lj}$ describes the frequency adaptation for measured values *below* the corresponding threshold $T_j$. For simplified definition the agent assumes a linear increase from the base frequency to the increased frequency defined with $\alpha_j$. The frequency on a slope is therefore described by

$$f_{lj}(v_k) = a_{lj} \times v_k + b_{lj} \tag{II}$$

where $v_k$ is the last monitored value. Based on this definition $\beta_{lj}$ defines the value where the increased slope will meet the base frequency. The following equations are used to determine the parameters $a_{lj}$ and $b_{lj}$:

$$a_{lj} = \frac{\alpha_j - 1}{T_j - \beta_{lj}} f_0 \qquad \text{(III)}$$

$$b_{lj} = f_0 \left(1 - \frac{\alpha_j - 1}{T_j - \beta_{lj}} \beta_{lj}\right) \qquad \text{(IV)}$$

In the example the values $\beta_{l1} = 50$ and $\beta_{l2} = 60$ are used.

**High:** The high value $\beta_{hj}$ mirrors the functionality of $\beta_{lj}$ for the degrading slope *above* the threshold $T_j$. The equations (II), (III) and (IV) are used with an index $h$ instead of $l$ for definition of the slope. In the example no values for $\beta_{hj}$ are defined, therefore the agent uses by default a vertical slope for all values above the threshold.

The polling frequency will be adjusted in every measurement cycle. The agent always uses the highest value if several slopes are applicable. The resulting curve is printed as a bold line in Figure 3.

The threshold based approach could be used in combination with the history based adaptation as presented in Section 4.1. In the combined approach, the base frequency $f_0$ will be adapted by (I) according to the history of the managed component.

# 5 IMPLEMENTATION ASPECTS

As explained in Section 2 our architecture is implemented in a CORBA environment, we use the Orbix implementation (IONA 1995). The monitoring agent is therefore realized as a CORBA object consisting of several modular components. The modularity allows an easy exchange or enhancement of the internal components to create a more specialized agent (Perrow, Hong, Lutfiyya and Bauer 1995).

Figure 4 gives an overview of the generic agent. The agent provides two interfaces: The service interface is used to retrieve monitoring information from the agent, like the current value $w_k$ or the content of the buffer. The management interface provides functionality for controlling the agents behavior.

**Organizer.** This component coordinates the internal activity of an agent and provides the functionality for both service and management interface. At startup time the organizer reads configuration information from an initialization file and passes the appropriate parameters to the components (indicated with dashed arrows in Figure 4). Monitoring is performed automatically with an adaptive measurement frequency. To ensure data integrity, invocations on any of the interfaces are blocked until a complete measurement and evaluation cycle is finished.

The **Filter.** transforms an input stream of values (here $v_k$) into an output stream ($w_k$) according to a number of predefined filter functions. They include identity, median with window size, and medium with window size.

The **Trigger** function performs a call to the Event Handler whenever a predefined

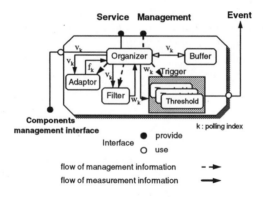

**Figure 4** Monitoring agent with frequency adaptation

condition occurs on the filtered data stream. An arbitrary set of threshold values with different characteristics for each threshold can be programmed.

The **Adaptor** implements the algorithms for the calculation of an adequate polling frequency as described in Section 4. The adaptor learns about every measured value $v_k$ and provides a new frequency $f_k$ for the organizer.

**Buffer.** This component stores up to $N$ previous measurements, where the buffer size $N$ is adjustable through the management interface. The buffer is organized as a ring buffer, that is, as soon as the buffer is filled for the first time, it starts to override old values according to a FIFO strategy.

## 6   CONCLUSION

We have presented a distributed management architecture in which monitoring is performed by distributed agents with generic functionality for filtering, polling frequency adaptation and event creation. The distribution of active management agents reduces the amount of management related traffic and avoids a potential bottleneck on a centralized management station. The automatic adaptation reduces the performance impact of the agents significantly while at the same time a high accuracy of management relevant information about critical components is ensured. Several adaptation strategies allow a flexible configuration of the agent according to the individual requirements of the managed component.

## REFERENCES

Dini, P. (1996) New Aspects for Run-time QoS Evaluation in Networks and Distributed Systems. In *European Simulation Multiconference*, pages 3–11, Budapest, Hungary,

June 1996. Tutorial.

Dini, P., v.Bochmann, G. and Boutaba, R. (1996) Performance Evaluation for Distributed Systems Components. In *2nd Int. IEEE Workshop on Systems Management*, pages 20–29, Toronto, Canada, June 1996. IEEE. ISBN 0-8186-7442-3.

Koch, T. and Krämer, B.J. (1996) Rules and agents for automated management of distributed systems. *Distributed Systems Engineering Journal* **3**, pages 104–114.

Koch, T., Krell, C. and Krämer, B.J. (1996) Policy Definition Language for Automated Management of Distributed Systems. In *2nd Int. IEEE Workshop on Systems Management*, pages 55–64, Toronto, Canada, June 1996. IEEE. ISBN 0-8186-7442-3.

Koch, T. (1995) Rule Based Management Architecture with Smart Agents. In M. Sloman and T. Usländer, editors, *Proc. of the Int. Workshop on Services for Managing Distributed Systems*, Karlsruhe, September 1995. Fraunhofer IITB.

IONA Technologies Ltd., Dublin, Ireland. *Orbix Programmer's Guide*, 2.0, November 1995.

Perrow, G.S., Hong, J.W., Lutfiyya, H.L. and Bauer, M.A. (1995) The Abstraction and Modelling of Management Agents. In Sethi, Raynaud and Faure-Vincent, editors, *Integrated Network Management, IV*. IFIP, Chapman & Hall, pages 466–478.

Schwartz, M. (1987) *Computer-Communications Networks Design and Analysis*. Prentice-Hall.

Stallings, W. (1993) *SNMP, SNMPv2, and CMIP: The Practical Guide to Network Management Standards*. Addison-Wesley.

# APPENDIX 1   AVAILABILITY AND DYNAMICS HEALTH PARAMETERS

## Availability parameters

*Current availability.*   *Observed availability* $h(t)$ is a feature of a component representing the availability of the component's services up to a given time. Its value defines how long this component has been in operational state with respect to the observation period $T$.

$$h : [t_0, T] \to [0, 1] h(t) = t_{op}/(t - t_0)$$ where $t_{op}$ is computed within $[t - t_0]$

Within a period $[t_0, T]$, the availability $A(T)$ is considered as equal to $h(T)$. The availability of a component type is commonly defined as the average of the $h(T)$ of all components of this type across a period $[t_0, T]$. Since the availability is a global evaluation, $A$ is calculated across many periods $T_1, T_2, T_3, \ldots$

*Minimum current availability.*   Across an observation period, the minimum value of $h(t)$ is called *minimum observed availability* and is defined as:

$$h_{min} : [t_0, T] \to [0, 1]$$
$$h_{min}(t) = min\{h(t)|t_0 \le t \le T\}$$

Since we have concluded that the local extreme values of $h(t)$ occur when a state change event arrives, $h_{min}(t)$ is among these peak point values.

*Weighted current availability.*　　As defined above $\underline{h}$ is a global evaluation without distinction between later or recent $h(t)$ values. Since the health is the current observed availability, it seems necessary to emphasize the recent health values more with respect to the earlier ones. We introduce the weighted average of observed availability which exponentially weights the later $h(t)$ values. Consequently, the $h(t)$ values will be taken into account by an exponential power. This power is dependent on the time and the change intervals, as follows:

$$\overline{h}(t_i) = \frac{h(t_{i-1}) + l \times h(t_i) \times \exp(t_i - t_{i-1})}{1 + \exp(t_i - t_{i-1})} \quad \text{for } (i > 0)$$

where $l = 0$ if the operational state within $[t_{i-1}, t_i]$ is disabled, and $l = 1$ if the operational state within $[t_{i-1}, t_i]$ is enabled.

## Dynamics parameters

*Stability order.*　　The *stability order* $k$ accurately qualifies a stationary state with respect to change frequencies. We say that a state has the order $k$ of stability if for $t_i - t_{i-1} \approx k \times t_{i-1}$ no state change event occurs.

*Instability order.*　　The *instability order* $p$ refers to state change events only (no commands) and consequently depends only on the $t_i$ time stamps. We define the instability order of magnitude $p$ if $\delta_{i-1,i} \approx 10^{-p} \times t_{i-1}$. Explicitly, the calculus formula is obtained by a logarithmic function $p = \lceil -\log(t_i - t_{i-1})/t_{i-1} \rceil$, where $\lceil a \rceil$ is the greatest integer less than or equal to $a$. The instability order is therefore an integer number.

Within a given time period $t_T$, a component could have different instability orders $\{p_1, p_2, \ldots p_s\}$. The instability of its operational state is first described by $[p_{min}, p_{max}]$, where $p_{min} = min\{p_i | 1 \le i \le s\}$ and $p_{max} = max\{p_i | 1 \le i \le s\}$, and second, by the complete set of order values.

*Repeatability order.*　　If the instability occurs consecutively, say $r$ times, we call that the *instability of order* $p$ has a *repeatability of order* $r$.

Thus, if $\delta_{i+j-2,i+j-1} \approx 10^{-p} \times t_{i+j-2}$, for $j = 1, 2, \ldots, r$, the managing objects must consider not only the health value, but equally, the $(p,r)$-*instability* of order $p$ and *repeatability* of order $r$.

An instability order $p_i$ could have, in turn, several repeatability orders $\{r_{i0}, r_{i1}, \ldots r_{iw}\}$. As for the instability order, each $p_i$ order is characterized by $r_{i/min} = min\{r_{ij} | 0 \le j \le w, 1 \le i \le s\}$ and $r_{i/max} = max\{r_{ij} | 0 \le j \le w, 1 \le i \le s\}$. Semantically this is equivalent to the minimum, and respectively maximum number of consecutive change intervals of the same range $p$, within an approximation given by the error of the function $f(x) = \lceil x \rceil$. The repeatability order is also an integer number.

*Multiplicity order.*　　The *multiplicity order* $m$ represents the number of times a real resource has been involved into an instability of order $p_i$ (with or without a repeatability order) during a given period. Consequently, at any pair $(p_i, r_{ij})$ one could attach the multiplicity order $m_{(pi,rij,T)}$ which captures the number of times the instability of order $p_i$ and repeatability $r_{ij}$ have been raised within the $t_T$ time period. The multiplicity order is an integer number.

# 42

# A nonblocking mechanism for regulating the transmission of network management polls

*A. B. Bondi*
*AT&T Labs*
*PO Box 3030*
*Crawfords Corner Road*
*Holmdel NJ 07733-3030*
*U.S.A.*
*Tel. +1 908 949 1921, Fax. +1 908 949 1720.*
*E-mail: andre.bondi@att.com*

### Abstract

We propose and analyse the performance of a new mechanism that allows an arbitrary number of network management polling requests to be outstanding at any instant, while preventing the flooding of the network with polling messages. An advantage of the mechanism over one previously described in the open literature is that it continues to function smoothly, without freezing for prolonged periods, even if many failed nodes are polled in succession.

### Keywords

Network management polling, status polling, SNMP.

## 1. INTRODUCTION

In IP-based and SNMP-based network management systems, the status of managed nodes is ascertained by sending polling messages at scheduled times or on demand according to need (Rose, 1991). Care must be taken to ensure that bursts of polling messages do not interfere unduly with user traffic. On the other hand, because timely dispatch of polling messages is essential to network management, a polling dispatch mechanism should not prevent polls from taking place for any significant length of time.

One way to reduce the risk that the user's network is flooded with polling messages is to restrict the number that may be unacknowledged at any instant (Sturm, 1995). This is done in such systems as HP OpenView Node Manager and implementations of OpenView on other systems, e.g., AT&T OneVision.[1] We shall refer to this method as the constrained outstanding poll method of degree $N$, or COP-N.[2] In the implementations just mentioned, $N$ is set to three. Under COP-N, the rate at which polls are emitted is inversely proportional to the time taken to resolve a poll, because there is no explicit control on the polling rate. Therefore, the polling rate may be higher than desirable if the acknowledgment time is short and all polled nodes are responding. This is particularly noticeable when nodes are being discovered. As the network management system "learns" about nodes about which it has no record in its database, it must verify their status and existence by polling them, too. Polling is done by means of ICMP (internet control message protocol) *ping* messages. Polls may occur in bursts, especially when a network management system is being brought up for the first time (AT&T/HP, 1994).

Suppose that the network management system allows unacknowledged polls to be outstanding to at most three nodes any instant. If the first attempted *ping* (ICMP message) to a node fails to elicit a response after a timeout interval of ten seconds, at most three subsequent attempts follow. The timeout interval associated with each attempt is set to twice that of the previous attempt. The sequence of attempts terminates if an acknowledgment to the current *ping* is received before a timeout expires, or after the timeout has expired on the fourth attempt. If no response has been received by the network management system after all four attempts, the node in question is declared to be inoperative, and the colour of the corresponding icon on the network management display is changed accordingly (Sturm, 1995). We call this sequence of attempts a polling cycle. The maximum total time between the beginning of the first attempt and the end of the fourth attempt is 150 seconds.

We have already seen that COP-N does not protect the network from bursts of polls when nodes are responsive. It has also been reported that COP-N may prevent polling activity when nodes are unresponsive (Sturm, 1995). Consider a scenario in which three failed or unreachable nodes are polled in rapid succession. Since no more than three nodes may have unacknowledged polls concurrently, it will be impossible to determine the status of any other node for the next 2.5 minutes. Similarly, five minutes will elapse before any polling can be done if six unresponsive nodes are polled in succession, 7.5 minutes if nine unresponsive nodes are encountered, etc (Sturm, 1995). If many nodes are unreachable, e.g.,

1. All information in this paper about the HP OpenView and AT&T OneVision network management systems is contained in published sources. OpenView is a trademark of Hewlett Packard. OneVision is a trademark of AT&T Corp. and of NCR Corp. (formerly known as AT&T Global Information Solutions).

2. This term has not been used previously.

because of a failed router, the time from status or configuration change to operator notification may be excessive.

One might wish to overcome the freezing of polling activity by increasing the number of outstanding polling messages. This would be ill advised. Apart from failing to prevent freezing if more than $N$ unresponsive nodes are polled in sequence, the increase in $N$ would allow bigger bursts of polls to be released, to the extent that acknowledgments might not be processed fast enough by the network management station (Sturm, 1995).

In this paper, we propose and model the performance of an alternate mechanism for allowing an arbitrary number of nodes to be polled concurrently without receiving acknowledgments. The mechanism prevents flooding with *ping* messages by ensuring that polling messages are transmitted at a rate that does not exceed a prescribed level. It might also be tuned to ensure that the kernel of the host operating system, which performs protocol handling in operating systems such as UNIX$^{TM}$, is not overtaxed by returning polling messages. It also ensures that polls of unresponsive nodes do not impede the monitoring of other nodes. We shall call this method of regulating the emission of polling messages the regulated poll emission method (RPE).

## 2. DESCRIPTION OF THE RPE METHOD

Status polling requests may be generated in bursts, depending on how they are scheduled. To prevent *ping*s from being transmitted at an excessive rate, they should be scheduled for transmission in rapid succession at a controlled rate specified by the network administrator. One method of doing so is a modification of the leaky bucket algorithm similar to that proposed to regulate the transmission of ATM cells (Eckberg, Luan, & Lucantoni, 1990). In this algorithm, a set number of *ping*s could be transmitted within a specified time frame. *Ping*s in excess of this number would be queued for later dispatch.[3] Alternatively, *ping*s would be queued for transmission a minimum time apart, thus reducing the burstiness of *ping* transmissions and capping the bandwidth they use. Under the RPE method, polling messages would be dispatched at intervals of length not less than some quantity $\tau$. Let us suppose that all polling messages have constant length, and that the time to emit each one is $C$, with $C < \tau$. Then the maximum poll transmission rate is simply $1/\tau$, regardless of the states of nodes polled or the sequence in which they are polled.

To speed the processing of acknowledgments, records of unacknowledged *ping*s would be registered in an ordered data structure indexed by target IP address.

---

3. *Ping* messages would never be discarded as might be the case with ATM cells, selectively or otherwise, because this could lead to a false diagnosis of node failure.

To speed the management of timeouts, records of unacknowledged *pings* would be kept in another ordered data structure, indexed by the time at which a timeout is scheduled. Each record in one data structure would contain a pointer to its mate in the other, to facilitate rapid removal upon timeout or upon receipt of an acknowledgment, whichever occurred first.

The RPE method has the following advantages:

1.  The method allows an arbitrary number of status polls to be outstanding simultaneously.

2.  The method permits rapid update of status changes on operators' network maps.

3.  The method prevents the release of bursts of status polling messages onto the network.

However, the method does have limitations also. While the RPE method would certainly regulate the transmission of outbound ICMP messages, there would not be any equivalent rate control for the reply (acknowledgment) messages. Under stable conditions, the inbound message arrival sequence would not be unlike that of the outbound sequence. However, we do not know if the echos sent in response to *pings* would arrive in a bursty manner, especially if they were collectively delayed at a (possibly) transient bottleneck on their return path.

In the next section, we shall model the duration of a status polling cycle. Then, we shall use this model to compare the throughput and delay characteristics of the COP-N and RPE methods.

## 3. MODELING THE DURATION OF A STATUS POLL CYCLE

The duration of a poll is determined by the number of polling attempts needed to determine if the polled node is reachable and responsive or either unreachable or unresponsive. In the proposed method, this duration is the amount of time during which a pending poll request must be stored. In the method described in (Sturm, 1995), this is the amount of time during which a dispatch slot (or queue in the terminology of (Sturm, 1995)), cannot be used to perform another poll.

Figure 1 shows the components of the duration of a polling cycle consisting of $K$ attempts. The duration of the $k$th attempt ($1 \leq k \leq K$), consists of the polling request emission time, $C$, and the lesser of the timeout interval $T_k$ and the time to receive an acknowledgment, $A_k$. The probability that the $k$th polling attempt is unacknowledged is $p_k$, and that it was received before the timeout, $1 - p_k$. In OpenView/OneVision, there are $K = 4$ attempts, with $T_1 = 10$ seconds, $T_2 = 20$ seconds, $T_3 = 40$ seconds, and $T_4 = 80$ seconds. Since $\sum T_i = 150$ seconds, 2.5 minutes may elapse from the time a node is polled to the time at which it is declared unreachable, or down.

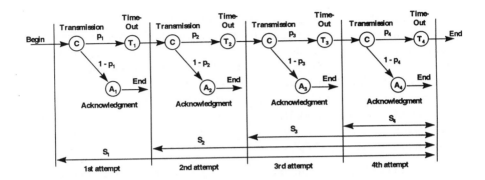

**Figure 1** Components of a polling cycle with 4 attempts.

A polling attempt is unacknowledged if any of the following events occur:

1. An ICMP polling message is lost on its outward journey, say with probability $\gamma_1$.

2. A node might be down, or it might simply not be reachable because all paths to it are blocked for some reason, say with probability $\gamma_2$. For example, the node might only be reachable via a single bridge or router, which itself could be inoperative.

3. The corresponding acknowledgment packet might be lost on its return journey owing to network congestion, say with probability $\gamma_3$.

4. The acknowledgment does not return prior to the expiration of the current timeout, say with probability $Pr(A_k > T_k)$.

It follows that $p_k$ is given by

$$p_k = 1 - (1-\gamma_1)(1-\gamma_2)(1-\gamma_3)Pr(A_k < T_k) \tag{1}$$

provided all four events occur independently.

The acknowledgment return time $A_k$ is also influenced by the three factors affecting the $\gamma_i$s. Because the length of $A_k$ depends on network topology as well as on congestion levels, determining the form of its distribution may be complex. For the purpose of illustration, we shall assume that $A_k$ is exponentially distributed with a mean equal to a round trip time commensurate with the network being managed. In a local area network, $E[A_k]$ might be less than 10 msec. For nodes that are further away, the round trip time may be much larger, e.g., 100-200 msec. Under severe congestion at the destination or intermediate nodes, $E[A_k]$ may even be of the order of a few seconds.

Notice that if all nodes are unreachable or unresponsive, $\gamma_2 = 1$ and hence $p_k = 1$. Similarly, regardless of the state of the nodes, if all ICMP packets are lost in either one or both directions, $\gamma_1 = 1$ or $\gamma_3 = 1$, and again $p_k = 1$, as expected. Also, since the timeout interval $T_k$ doubles for successive values of $k$, the probability that the acknowledgment fails to arrive before timeout expiry diminishes, and the failure of the $k$th polling attempt tends to be due more to either or both of the risks of packet loss or the unresponsiveness of the polled node. Finally, we verify the consistency of the equation by observing that if the timeout goes off immediately, i.e., if $T_k = 0$, $Pr(A_k < T_k) = 0$ and $p_k = 1$ as expected.

Equation (1) contains the implicit simplifying assumption that $p_k$ only depends on $k$ through the increasing timeout interval $T_k$. This is very likely to be true for the factors accounting for polling message loss, since congestion will fluctuate quickly relative to the lengths of the timeout intervals. If we had information about the likelihood of a node being unresponsive on one polling attempt given that no response was received in previous attempts, $\gamma_2$ could be made dependent upon $k$. However, responses might not be received for a variety of reasons. For example, the node might be switched off, in the course of being rebooted, or otherwise on the verge of restoration.

The value of the $\gamma_i$s may vary with the sequence of nodes either currently scheduled for polling or for which polls have been initiated but not yet resolved. For instance, if a failed router lies on the unique path to a large number of managed workstations that are scheduled for polling within a short amount of time, the value of $\gamma_2$ will tend to one. On the other hand, if the router and all paths to it are fully operative, $\gamma_2$ will tend to zero.

To obtain upper bounds on the throughputs of either polling method, it is necessary to determine the mean duration of a polling cycle. We may model the duration of an entire polling cycle consisting of $K$ attempts as follows. Let $C$ denote the time to emit an ICMP polling message. Let $S_k$ denote the remaining time to resolve the poll at the beginning of the $k$th attempt, i.e., the combined holding times of the $k$th and subsequent attempts, if any. Then, $S_K$ is the time to resolve the final polling attempt of the cycle, while $S_1$ is the entire time to make the determination that the polled node is reachable or not. For attempts other than the $K$th, a subsequent attempt is necessary with probability $p_k$ and is unnecessary with probability $1 - p_k$. If a $(k+1)$th attempt is not needed, the duration of the $k$th attempt will be the time from emission to the return of the acknowledgment. If it is needed, the remainder of the polling cycle includes the $k$th timeout and the duration of the $k + 1$ attempt, $S_{k+1}$. Therefore,

$$S_k = C + (1 - p_k)A_k + p_k(T_k + S_{k+1})   k = 1, 2, ..., K - 1. \tag{2}$$

At the $K$th attempt, the remainder of the polling cycle lasts $A_K$ if the acknowledgment returns before the $K$th timeout expires and $T_K$ otherwise. Hence,

$$S_K = C + (1-p_K)A_K + p_K T_K.$$ (3)

By inspection, we see that the components of the average polling cycle time are given as follows:

1. The mean duration of the final stage of the polling cycle is given by

$$E[S_K] = E[C] + (1-p_K)E[A_K] + p_K E[T_K].$$ (4)

2. The mean duration of the remaining time of the polling cycle at the beginning of the $k$th attempt is

$$E[S_k] = E[C] + (1-p_k)E[A_k]$$
$$+ p_k(E[S_{k+1}] + E[T_k]) \quad k=K-1, K-2,...,2,1.$$ (5)

The average duration of a polling cycle, $E[S_1]$, may be obtained by backward recursion beginning with $E[S_K]$.

Finally, we derive an approximate expression for the expected number of polling attempts in a cycle. We need this expression to determine the maximum expected bandwidth used for polling with either scheme. The expression is approximate because our formulation of it assumes that the events on successive polling attempts within a cycle are independent. They clearly are not, because if a node is inoperative or unreachable on one attempt it will almost certainly be unreachable on subsequent attempts. Under the assumption of independence, only one attempt is required in a cycle with probability $1-p_1$. Only two attempts are required with probability $p_1(1-p_2)$. $K$ attempts are required with probability $\prod_{i=1}^{K-1} p_i$. If $N_P$ denotes the number of polling attempts required within a cycle, under the assumption of independence,

$$E[N_P] = \sum_{k=1}^{K-1} k(1-p_k) \prod_{j=1}^{k-1} p_j + K \prod_{k=1}^{K-1} p_k.$$ (6)

If $s_p$ denotes the size of a polling message, and $X$ denotes the maximum throughput under whichever polling dispatch mechanism is used, the maximum polling bandwidth demand is bounded above by $s_p X E[N_P]$.

## 4. POLLING CYCLE THROUGHPUTS AND FREEZING

### 4.1 COP-N method

For the COP-N method described in (Sturm, 1995), the maximum rate at which polling cycles may be initiated if at most $N$ polls are outstanding is

$$X_N = N/E[S_1].$$ (7)

Thus, the throughput will be driven down if the probability that an ICMP poll is acknowledged is small for whatever reason, i.e. if $p_k$ is close to 1. The system will freeze for a period of at most $\sum_{i=1}^{K} T_i$ seconds, the sum of the timeout intervals, if $N$ unreachable nodes are polled in rapid succession.

If the network is functioning perfectly, the maximum throughput is given by

$$X_N^* = N/(a + C)$$ (8)

where $a$ is the mean acknowledgment time in the absence of failures or packet losses and $C$ is the time to emit an ICMP poll. If $a$ is small, bursts of polls may swamp the network at rates high enough to degrade application performance.

## 4.2  RPE method

Let $\tau$ be the minimum time between ICMP poll emissions. Then $1/\tau$ is the maximum rate at which the network mangement engineer chooses to issue polling requests. We must have $\tau > C$ because of hardware and software limitations. However, we must also have $\tau < E[S_1]/N$ for this method to have a higher throughput than the COP-N method. Let $M$ be the maximum allowed number of nodes for which polling requests may be outstanding. The value of $M$ is equal to the maximum number of entries on a list of unacknowledged polls, and would be determined by memory constraints and/or by performance considerations. Typically, $M$ will be much larger than $N$.

The average number of outstanding polling requests resident in memory is equal to the product of the the rate at which polls are emitted and the average duration of a polling cycle, i.e., $XE[S_1]$. This number must be less than the maximum number of unresolved polling cycles, $M$. Hence, we must have $XE[S_1] < M$. But $X < 1/\tau$, so we finally have the following bounds on the node polling rate:

$$X < \min \left[ 1/\tau, M/E[S_1] \right]$$ (9)

Of these, $\tau$ is a tunable parameter whose value would be determined by network and operating system constraints, $C$ is determined by the polling message size and the bandwidth of the subnetwork closest to the network management workstation, while $M$ is a tunable parameter whose value need only be restricted if memory is scarce. Under normal operating conditions, $\tau$ would be the parameter controlling the polling rate. The polling would only be constrained by $M$ if $E[S_1]$, the average duration of a polling cycle, were to become very large for any reason. $M$ should be made large enough to allow a polling rate of $1/\tau$ under normal conditions. Any of the following factors could limit the (desired) polling rate:

1. the need to keep the polling rate down to avoid saturating the operating system in the network management station, thus necessitating an increase in $\tau$,

2. the need to keep the polling rate down to avoid saturating the managed network with polling traffic, again necessitating an increase in $\tau$,

3. very long propagation delays or network congestion, resulting in long acknowledgment return times, and hence a large value of $E[S_1]$,

4. a high incidence of node failure, again resulting in a large value of $E[S_1]$.

If the first two constraints were loose, $M$ would have to be made suitably large to sustain throughput. Conversely, if either of the first two constraints were tight, the value of $M$ would not be critical. On the other hand, a large value of $M$ would ensure that polling would continue without interruption even if a large number of failed nodes were encountered in succession. Polling would never be frozen unless $M$ unreachable nodes were polled in rapid succession. In the limiting case, $M$ would never be an effective constraint if it were greater than or equal to the number of nodes under management or the number of nodes that could be discovered. If freezing did occur under this method, it could be indicative of a very serious failure in the network, e.g., it could indicate the failure of a router or bridge that is the sole means of accessing the nodes in question. By contrast, in the COP-N method with $N \ll M$ polls outstanding, the polling process would freeze as soon as $N$ failed or unresponsive nodes were polled in succession. On the other hand, if both nodes and network are reliable, COP-N may (perhaps intermittently) swamp the network with polling messages, while RPE could be tuned to prevent this from happening by choosing an appropriate value for $\tau$.

## 5. DELAY ANALYSIS

A polling dispatch mechanism should be able to emit polls in a timely manner once they have been scheduled, whether or not the nodes to be polled are reachable. In this section, we compare the delay characteristics of the proposed rate control mechanism with those of the limited slot mechanism. Because polls may sometimes occur in bursts or else infrequently (Sturm, 1995; AT&T/HP, 1994), we shall not attempt to perform steady-state queueing analyses of the two polling mechanisms. Instead, we shall perform an analysis of the behaviour of the two mechanisms when a fixed number of polls is scheduled for dispatch, of which a given number of unreachable or failed nodes will be polled consecutively.

Suppose that $n > N$ nodes are scheduled for dispatch at time $t$, and that of these, the first $N$ nodes are either down or otherwise unreachable, while the states of the other $n - N$ nodes are unknown. Suppose further that $M$, the maximum permissible number of initiated unresolved polling cycles under RPE, is much greater than the average number of nodes with initiated unresolved polling cycles, $E[S_1]/\tau$.

*Case 1: COP-N Method.* The polling cycles for the $N$ failed nodes last $\sum_{k=1}^{K} T_k$ seconds. During this period, no other nodes can be polled. Then, we may expect that the status poll of the last node in the queue will be initiated at time $t + (n - N - 1)E[S_1] + \sum_{k=1}^{K} T_k$ seconds, for $n \geq N + 1$.

*Case 2: RPE Method.* Suppose that there is no effective limit on the number of unresolved polling requests, i.e., that $M$ is very large. If $n$ ($< M$) nodes are scheduled for polling at time $t$, and the minimum time between polls is set to be $\tau$, the $n$th polling cycle will be initiated at time $t + (n - 1)\tau$ seconds, whether or not the first $N$ nodes to be polled are unreachable. Moreover, the $n$th polling cycle will be initiated at time $t + (n - 1)\tau$ regardless of which or how many of the $n - 1$ nodes polled before it are unreachable. This shows the robustness of the method under adverse conditions.

For the specified scenario, the delay performance will be better with the rate controlled mechanism if $\tau$ is set so that

$$(n - 1)\tau \leq (n - N - 1)E[S_1] + \sum_{k=1}^{K} T_k \tag{10}$$

In particular, for $n = N + 1$, i.e., for the node queued just behind the first $N$ at time $t$, the $n$th polling cycle will be initiated earlier with the rate controlled mechanism provided that

$$\tau \leq \frac{\sum_{k=1}^{K} T_k}{N} \tag{11}$$

Since $\tau$ will typically be on the order of seconds or a fraction of a second, while the sum of the timeout intervals is 150 seconds, the advantage of the rate controlled mechanism is clear, especially if it is quite likely that nodes will be unresponsive for any reason.

## 6. NUMERICAL ILLUSTRATIONS

### 6.1 Analysis of numerical results

To illustrate our performance model, we shall compare throughput bounds and sample delays for the COP-N method and the RPE method under different scenarios. We make the following assumptions:

1. The acknowledgment return time $A_k$ is exponentially distributed with a mean that is independent of the polling attempt in the cycle. The mean depends on the network topology and traffic conditions in a manner that we shall not attempt to capture. For a network with remote nodes, this mean might be

100 milliseconds. For a local area network, it might be as low as 2 milliseconds.

2. The default size of a polling message (*ping*) is 64 bytes.

3. The bandwidth of the local area network to which the network management station is attached is 10 Mbps.

4. At most four polling attempts are made within a polling cycle. The length of the $k$th timeout is $10 \times 2^{k-1}$ seconds, for $k = 1,2,3,4$.

5. Unless otherwise stated, under the RPE method, the maximum number of allowed outstanding polls is effectively infinite.

The effects of different parameter values may be easily determined by coding the formulae in a spreadsheet, as illustrated in Table 1.

As shown in Figure 2, the mean number of polls per polling cycle (marked on the left vertical axis) ranges from an inherent minimum of one to the programmed maximum of four as the probability of polling an unresponsive node increases. This quantity is only mildly dependent on the *ping* return time. The mean polling cycle duration (marked on the right vertical axis) ranges from a small level to the sum of the timeout intervals as the probability of polling an unresponsive node increases from zero to one. The time is shorter when the average *ping* return time is shorter, as one might expect. However, this effect disappears when nodes are unresponsive, since the large timeout intervals become the dominant factor. Consequently, under COP-N, the rate at which nodes may be polled diminishes as the probability of polling an unresponsive node increases, while the number of polling attempts per cycle rises from one to four as one would expect. Because COP-N is almost frozen when nodes are unresponsive, the bandwidth due to polling under COP-N diminishes, too.

In contrast, as illustrated in Figure 3, under RPE, provided enough memory is allocated for outstanding polls, the maximum number of nodes polled per second is constant. The coincidence of the dashed and solid lines for RPE shows that the node polling rate is independent of the *ping* round trip time. To avoid reducing the node polling rate when the probability of polling many unresponsive nodes in sequence is close to one, one must store a number of pending polls equal to the sum of the timeout intervals multiplied by the maximum polling rate, $1/\tau$. In Figure 3, this is reflected in the slight drop in the node polling rate when the maximum permitted number of outstanding polls $M$ is 1000 rather than 1500.

As shown in Figure 4, the bandwidth required for polling increases somewhat under RPE as the probability of polling an unresponsive node increases, because the expected number of polls per node increases. When the probability of encountering an unresponsive node is low and the *ping* return time is low, COP-N may have a much higher node polling rate and consequent demand for bandwidth than is desired. Also, the node polling rate of RPE is insensitive to the presence of unresponsive nodes. It is worth noting that the range of node polling rates and

**Table 1:** Illustration of the polling cycle analysis

| Mean ping return time | 0.01 | sec |
|---|---|---|

| Ping emission time | 5.12E-05 sec |
|---|---|

| Polling message loss probabilities (gamma1, gamma3) | 0.0001 |
|---|---|

| Poll message size | 64 bytes | Min time between emissions tau | 0.1 sec | Network Mgt Station LAN bandwidth (Mbps) | 10 Mbps |
|---|---|---|---|---|---|

Timeouts (s)

| | | Pr(Ak<Tk) | | | |
|---|---|---|---|---|---|
| T1 | 10 | 1.0000 | | COP Number of outstanding polls | 3 |
| T2 | 20 | 1.0000 | | RPE max outstanding polls M | 1000 |
| T3 | 40 | 1.0000 | | | |
| T4 | 80 | 1.0000 | | | |

| Prob unresponsive node encountered (gamma2) | p1 Prob first polling attempt fails | p2 Prob 2nd polling attempt fails | p3 Prob 3rd polling attempt fails | p4 Prob 4th polling attempt fails | Mean polling cycle duration | COP-N Throughput (nodes/sec) | Polls/cycle | COP-N b/width (kbps) | RPE max polling rate (nodes/sec) with no limit on no. of outstanding polls M | RPE max b/width (kbps) with no limit on no. of outstanding polls M | RPE method max avg polls outstanding | RPE Method delay of 10th poll when first 3 nodes down (sec) | COP-N Method delay of 10th poll when first 3 nodes down (sec) | RPE max polling rate with limited M (nodes/sec) | RPE max polling b/width with limited M (kbps) |
|---|---|---|---|---|---|---|---|---|---|---|---|---|---|---|---|
| 0.00 | 0.00020 | 0.00020 | 0.00020 | 0.00020 | 0.01 | 248.92 | 1.00 | 127.47 | 10.00 | 5.12 | 0.12 | 0.9 | 150.0 | 10.00 | 5.12 |
| 0.01 | 0.01020 | 0.01020 | 0.01020 | 0.01020 | 0.11 | 26.28 | 1.01 | 13.59 | 10.00 | 5.17 | 1.14 | 0.9 | 150.2 | 10.00 | 5.17 |
| 0.05 | 0.05019 | 0.05019 | 0.05019 | 0.05019 | 0.57 | 5.28 | 1.05 | 2.85 | 10.00 | 5.39 | 5.68 | 0.9 | 150.9 | 10.00 | 5.39 |
| 0.10 | 0.10018 | 0.10018 | 0.10018 | 0.10018 | 1.26 | 2.38 | 1.11 | 1.35 | 10.00 | 5.69 | 12.61 | 0.9 | 152.1 | 10.00 | 5.69 |
| 0.20 | 0.20016 | 0.20016 | 0.20016 | 0.20016 | 3.26 | 0.92 | 1.25 | 0.59 | 10.00 | 6.39 | 32.62 | 0.9 | 155.4 | 10.00 | 6.39 |
| 0.30 | 0.30014 | 0.30014 | 0.30014 | 0.30014 | 6.54 | 0.46 | 1.42 | 0.33 | 10.00 | 7.26 | 65.44 | 0.9 | 160.9 | 10.00 | 7.26 |
| 0.40 | 0.40012 | 0.40012 | 0.40012 | 0.40012 | 11.83 | 0.25 | 1.62 | 0.21 | 10.00 | 8.32 | 118.26 | 0.9 | 169.7 | 10.00 | 8.32 |
| 0.50 | 0.50010 | 0.50010 | 0.50010 | 0.50010 | 20.02 | 0.15 | 1.88 | 0.14 | 10.00 | 9.60 | 200.19 | 0.9 | 183.4 | 10.00 | 9.60 |
| 0.60 | 0.60008 | 0.60008 | 0.60008 | 0.60008 | 32.23 | 0.09 | 2.18 | 0.10 | 10.00 | 11.14 | 322.29 | 0.9 | 203.7 | 10.00 | 11.14 |
| 0.70 | 0.70006 | 0.70006 | 0.70006 | 0.70006 | 49.75 | 0.06 | 2.53 | 0.08 | 10.00 | 12.97 | 497.48 | 0.9 | 232.9 | 10.00 | 12.97 |
| 0.80 | 0.80004 | 0.80004 | 0.80004 | 0.80004 | 74.07 | 0.04 | 2.95 | 0.06 | 10.00 | 15.12 | 740.65 | 0.9 | 273.4 | 10.00 | 15.12 |
| 0.90 | 0.90002 | 0.90002 | 0.90002 | 0.90002 | 106.86 | 0.03 | 3.44 | 0.05 | 10.00 | 17.61 | 1068.59 | 0.9 | 328.1 | 9.36 | 16.48 |
| 1.00 | 1.00000 | 1.00000 | 1.00000 | 1.00000 | 150.00 | 0.02 | 4.00 | 0.04 | 10.00 | 20.48 | 1500.00 | 0.9 | 400.0 | 6.67 | 13.65 |

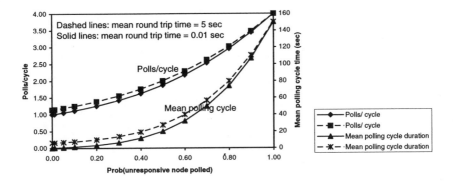

**Figure 2** Polls per polling cycle (left axis), mean polling cycle duration (right axis).

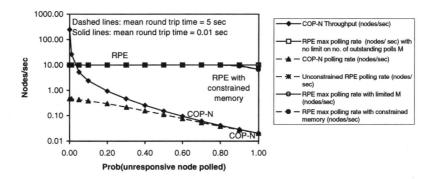

**Figure 3** Nodes polled per second.

bandwidth usage is much lower under RPE than under COP-N. This means that RPE enables one to control the bandwidth required by polling much more tightly than can be done with COP-N. This is achieved by tuning $\tau$ until the required tradeoff between node polling rate and bandwidth usage is achieved.

Delay is illustrated in Figure 5. Under RPE, the delay to the first poll of the 10*th* node with the first three nodes down is small and constant. Under COP-3, the delay to the first poll of the 10*th* node with the first three nodes down increases from a level slightly greater than the sum of the timeout intervals to a level that is considerably larger.

Finally, let us analyse the memory requirement to store unresolved outstanding polls under the RPE method. The average number of outstanding initiated polling cycles is equal to the mean polling cycle time multiplied by the maximum throughput. The average number of outstanding polls increases dramatically with the probability of encountering an inoperative node, as shown in Figure 6. To keep the node polling rate constant, the maximum number of nodes for which polls are

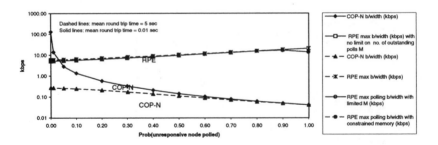

**Figure 4** Bandwidth used by polling (kbps).

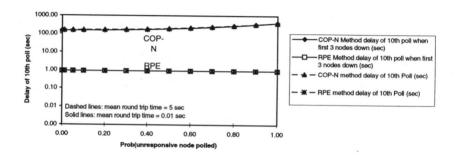

**Figure 5** Delay of 10th poll (seconds).

unresolved, *M*, must be suitably large. However, if the probability of polling an unresponsive node, $\gamma_2$, is 0.5 or less, this number is not exorbitant, although it is sensitive to the round trip time. As $\gamma_2$ approaches one, the timeouts become the dominant factor in determining the average polling cycle time, and hence the average number of outstanding polls at maximum throughput. In our illustration, storage would be required for 1500 outstanding polling records. In practice, a high value of $\gamma_2$ would be undesirable, and perhaps indicative of a major problem in the managed network.

## 6.2 Discussion

Our analysis corroborates previously reported observations (Sturm, 1995) about the behaviour of COP-N. Let us first consider polling rates and polling bandwidth usage. COP-N cannot prevent surges of polling activity when *ping* return times (mean round trip times) are short. At the same time, polling activity all but ceases if the likelihood of polling an unresponsive node is high. By contrast, the proposed

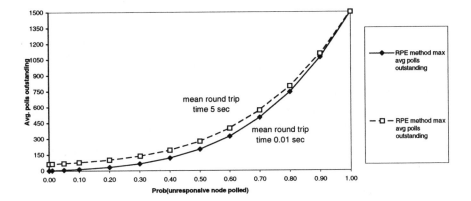

**Figure 6** Average number of unresolved polls under RPE.

RPE method limits network polling bandwidth when *ping* round trip times are short and sustains it at a constant level when unresponsive nodes are encountered, provided that the network management system allows records for a sufficient number of outstanding polls to be stored. Thus, the RPE method may be tuned to limit polling bandwidth or sustain polling activity according to need. These points are highlighted in Figure 3 and Figure 4. With regard to polling delays, under COP-N, polling is effectively suspended until all timeouts associated with polls of unresponsive nodes have expired. On the other hand, RPE limits delays by ensuring that polls may be transmitted at regular intervals if needed, regardless of the state of the network.

## 7. CONCLUSION

We have presented RPE, a method for regulating the transmission of polls in a network management system that will sustain polling at a specified rate when multiple unresponsive nodes are encountered. Beneficial features of this method include (i) that it may be tuned to ensure that polls do not swamp the network with management traffic and (ii) that the node polling delay is insensitive to the state of the network. Hence, RPE overcomes some of the network management difficulties inherent in a previously implemented method, as described in the open literature.

### Acknowledgments

This work was begun while the author was on assignment to AT&T Global Information Solutions (AT&T GIS) from AT&T Bell Laboratories. He has benefited from discussions with Dennis Zavila and Steve Leonard of NCR (formerly AT&T GIS), Gagan L. Choudhury and Kathleen Meier-Hellstern of AT&T Laboratories (formerly AT&T Bell Laboratories), and Alan Shepherd and Glen Shirey of Hewlett Packard.

## 8. REFERENCES

AT&T/HP (November 1994). *AT&T OneVision Node Manager Fundamentals for Network Operators* (CW 9118-9000-1847). AT&T/Hewlett Packard.

Eckberg, A. E. Jr., Luan, D. T., & Lucantoni, D. M. (1990). An Approach to Controlling Congestion in ATM Networks. *International Journal of Digital and Analog Communication Systems*, *3*, 199-209.

Rose, M. T. (1991). *The Simple Book: An Introduction to Management of TCP/IP-based Internets*. Englewood Cliffs: Prentice Hall.

Sturm, R. (March 1995). Q&A. *OpenView Advisor*, *1*(2), 11-13.

Sturm, R. (April 1995). Q&A. *OpenView Advisor*, *1*(3), 9-12.

## Biography

André Bondi is a Senior Technical Staff Member in the Teletraffic and Performance Analysis Department at AT&T Labs (formerly AT&T Bell Laboratories). He has worked on a large variety of performance aspects of telecommunications, including network management, operator services, computer performance, ATM, call routing, and intelligent networking. Prior to joining AT&T, he was an assistant professor of computer science at the University of California, Santa Barbara. Dr. Bondi holds a Ph.D. in computer science from Purdue University, an M.Sc. in statistics from University College London, and a B.Sc. in mathematics from the University of Exeter.

## PART THREE

# Fault Management I
Chair: Heinz-Gerd Hegering,
University of Munich

# 43

# Fault Isolation and Event Correlation for Integrated Fault Management

*S. Kätker, M. Paterok*
*IBM European Networking Center*
*Vangerowstr. 18, D-69115 Heidelberg, Germany*
*Tel.: +49 6221 59 4404 / 4312*
*e-mail:* {kaetker,paterok}@vnet.ibm.com

### Abstract

An algorithm for fault isolation and event correlation in integrated networks is presented. It reconstructs fault propagation during run-time by exploring relationships between managed objects and provides improved focus and efficiency compared to similar algorithms. The functionality of the algorithm is generic, straightforward, efficient, and applicable for different management architectures such as SNMP and TMN. Clearly defined interfaces allow parallel execution of problem investigation, integration of different management architectures and systems, and incorporation of other correlation techniques.

Prototype and production level systems have been implemented for TMN- and SNMP-based management systems to validate the concept and for use in real networks. This development was done after an extensive study of requirements and existing techniques, the results of which are also presented.

### Keywords

Fault Management, Service Management, TMN, Fault Isolation, Event Correlation

## 1  INTRODUCTION

Aside from the growth in size and speed, telecommunication networks have become more and more complex in their layered structures and services. Upcoming integrated backbones using Asynchronous Transfer Mode (ATM) and Synchronous Digital Hierarchy (SDH) simultaneously carry voice, data, and video traffic, thereby dramatically increasing the number of protocols in use. A typical scenario comprises applications running over TCP/IP which is using ATM which in turn runs on SDH over dark fiber lines. As a result, a broken link in the underlying backbone or any other problem in the base transmission services causes a large number of higher layer services to fail. These services may not even be directly connected to the failing component, i.e. a failing SDH link would disturb operation of the TCP/IP networks and connections including applications in possibly very large numbers.

This type of problem causes hardship to the network management systems: floods of events are generated by the management systems of the various networks. If there is no integrated management system the operators need to correlate the problems on the various network management consoles manually using their experience. This is a daunting task which usually takes too long in the face of high downtime costs, if it is possible at all. An integrated management system, if present, is flooded with events from all the different subnetwork management systems.

A problem which is largely unresolved in practice (and in theory) is that of automated fault isolation and event correlation for integrated networks. *Event correlation* means that all problem indicators (i.e. events) whose generation have been caused by the same underlying problem are grouped together. *Fault isolation* stands for the automated detection of the root problem which is the cause of the problem. In this paper, an algorithm is presented which allows to determine the root cause of network problems and correlate the events associated with the problem. The algorithm requires a limited integration of the management system. It is based on model traversing and allows incorporation of different correlation techniques. The algorithm is currently being implemented and will be used for the management of real-life telecommunication networks. It was patented by IBM (Kätker and Paterok, 1996).

The requirements which an event correlation solution for integrated telecommunications networks should meet are listed in Section 2. The different approaches and techniques known so far are also discussed. The algorithm is presented and explained in Section 3 and aspects of its implementation are described in the following section. A real-life correlation example is shown in Section 5 before the paper ends with the conclusion.

# 2   REQUIREMENTS AND TECHNIQUES

This sections lists the critical requirements for effective fault localization and event correlation. The existing literature is discussed in light of these requirements. The suitability of these techniques may differ depending on the management layer where it is applied.

## 2.1   General Requirements

In order to be effective, the technique has to cope with the following general requirements:

- **Integration with Network Management Infrastructure**
  Each correlation algorithm requires the underlying network management sys-

tem to perform specific functions, e.g. information requests. These functions must be easily embeddable into existing network management systems in practical use.

- **Flexibility**
  Networks are dynamic with respect to their structures and components. It must be possible to adapt the correlation system to changes in the network topology, the component types and versions, and the services offered. Implementation of the changes must be straightforward and controllable to keep failures as a result of changes to a minimum level.

- **Performance and Parallelism**
  Algorithms requiring a central component to be contacted upon each correlation step will suffer from performance problems when large numbers of events flow in. Parallelism in this case covers external parallelism where the algorithm can be distributed on several physical machines and internal parallelism which stands for the use of threads on the same machine. Requests for network information usually take much more time than the correlation itself, in particular in faulty networks where information requests are frequently timed out. It is therefore important to perform these requests in parallel by use of threads.

- **Functional Distribution and Combination of Techniques**
  Levels of integration are limited in most network management systems in use now and in the near future. The algorithm must therefore lend itself to distribution over different network management systems with possibly different architectures. These different systems may perform correlation autonomously using different techniques, as long as they comply to clearly defined and straightforward interfaces.

- **Robustness Against Incomplete Data**
  The technique must be able to perform its task in the face of incomplete data, i.e. missing events and timed-out information requests caused by faulty networks.

## 2.2   Fault Isolation and Event Correlation Techniques

Most known techniques can be classified by one of four different approaches. Subsequently, these approaches are sketched, the key references are mentioned and the advantages and drawbacks are discussed in light of the requirements listed above.

### Model Based Reasoning Tools

Many published concepts and implementations use artificial intelligence methods for fault management. To overcome the lack of structure in the rule bases, the second generation of expert systems, called model based reasoning systems (Kehl and Hopfmüller, 1993, Jakobson and Weissmann, 1993), use different techniques to represent structural knowledge (using object-oriented models) and heuristical knowledge (using rules).

A maximum of flexibility and the ease of customization of basic correlation rules are the major advantages of model based reasoning systems. Major disadvantages include limited performance and lack of parallelism. Customization and maintenance of large rule bases is a critical task. Also, rule based systems frequently suffer from lack of communication with their environment, e.g. for exchange of information. Model based reasoning systems play a very important role in the area of device level fault management and for executing escalation rules or fault management policies, e.g. to determine which trouble tickets have to be created under which circumstances. In this role the expert system serves as an integrator between the different fault management techniques.

## Fault Propagation Models

These models describe which symptoms will be observed if a specific fault occurs. Kliger et al. (1995) use the concept of causality graphs to decode the set of observed symptoms. Bouloutas et al. (1992) use intersections of sets of possible failures that may cause a given symptom.

Fault model based approaches require an a priori specification of the fault-causes-symptom relationship for any fault that might happen in the system. This is hard to obtain in practice in particular when the network is subject to change. If the fault-causes-symptom knowledge is easy to determine for a given system, fault propagation model based tools can provide a robust and efficient event correlation solution.

## Model Traversing Techniques

These algorithms reconstruct fault propagation during run-time by using relationships between managed objects. Jordaan and Paterok (1993) present an algorithm for event correlation based on object model traversing in a management system which is based on the Telecommunications Management Network (TMN, (Aidarous and Plevyak, 1994)). Starting with the managed object that emitted the trigger event related managed objects are discovered stepwise and their events are correlated. The algorithm presented in this paper is related to the algorithm from (Jordaan and Paterok, 1993) but has clear advantages in the direction and efficiency of fault isolation. The technique described by Jordaan and Paterok is targeted to a specific information model, it provides correlation, but no directed search or fault isolation. Based on an object-oriented network model an event correlation system model for the physical layer of large telecommunication networks is presented by Houck et al. (1995). The service dependency graph introduced in (Kätker, 1996) models a client server application and the network it uses as a set of abstract services and the dependencies between the services. A generic algorithm is presented that traverses the dependency graph to extend fault isolation and event correlation to the application level.

Model traversing techniques work best when managed object relationships are graph-like and easy to obtain. Their strengths are high performance, potential parallelism, and robustness against network change. Their disadvantage is a lack of flexibility, in particular if fault propagation is complex and not well-structured.

## Case Based Reasoning Tools

The goal of these tools is to suggest a solution for a given problem by citing a solution that solved a similar problem in the past. Therefore, case based reasoning tools are usually not used for fault isolation, but for the following fault diagnosis task (see Dreo and Lewis (1995) for an example).

This technique analyses the phenomenons of a given problem, i.e. no knowledge about the fault propagation is required. However, a number of similar problems have to be resolved before the user can start to compare past and present data. Incomplete or wrong results may occur if the network configuration is changing. Also, algorithms for finding similar, not necessary equal, problems by comparing symptoms is non-trivial ('fuzzy query').

## 2.3 Layer-Specific Requirements

The requirements and suitable techniques for fault isolation and event correlation vary significantly depending on the management layer and function where it is applied (see Stallings (1993) for the functions of the various layers).

### Network Element Layer

This layer is involved with a variety of different hardware devices which often do not support standard management protocols or object models. Event correlation must therefore deal with incomplete information and support proprietary interaction with devices. Since component behavior upon failure heavily varies depending on device type and manufacturer, the corresponding fault isolation technique should provide a maximum of flexibility, e.g. if a new equipment type or version is installed. Event rates are mostly low to moderate.

These requirements make expert systems a suitable tool for network element layer event correlation, provided that the expert system is tightly integrated with the management system.

### Network Layer

In general, any communication network can be treated as a layered graph. More specifically, some graph-based standard models exist (e.g. the ITU M.3100 generic network model (ITU, 1995), which is widely used in the TMN and TINA (Richter et al., 1995) arena) to model various kinds of telecommunication networks. As these network models represent the layered physical and logical structure of a telecommunications network event storms occur specifically in this layer. Since the overall network frequently changes because nodes are added and/or removed, network layer event correlation should explore the network structure during runtime.

That suggests the use of model traversing techniques for network layer event correlation.

### Service Layer

Currently no standard models have been defined for the service layer. They would be characterized by a complex, hierarchical structure of objects representing services, applications, subcomponents, and the dependencies between them, resulting is some sort of dependency graph (see TINA (Dupuy et al., 1995) for approaches). Because high event rates have to be expected for service layer management, a dependency graph walking algorithm is a suitable technique.

## 2.4   Combination of Techniques

The above survey shows that there is no single technique dominating for all aspects of real-life network environments. To provide comprehensive fault isolation and event correlation, it is therefore recommended to combine different techniques in a distributed and partly autonomous fashion. In particular, expert systems and model traversing algorithms offer complementary capabilities, the first being good at complex, flexible, unstructured, non-distributed problems, while the latter are generic, high performing, simple, and can be distributed. The current development described in Section 4 is capable of combining these two techniques. A model traversing algorithm identifies the failing network node and correlates possible events from other nodes. An expert system analyses the devices themselves and executes fault management policies by, e.g., setting trouble ticket parameters, issuing rerouting requests, or paging technical support staff.

The remainder of this paper concentrates on the description of a generic model traversing technique for network layer fault isolation.

## 3   ALGORITHM

The algorithm uses relationships between network hardware and software components and their managed objects to search for the root cause of a problem. This search is carried out using three types of information request functions. To enable the algorithm for a specific network management system only these functions need to be implemented. The algorithm can therefore easily be adapted to different networks and management systems.

Assume that a connection A→Z from node $A$ to node $Z$ fails and an event is generated as a result. This event triggers the algorithm which performs the following steps:

### Horizontal Search
Starting from node $A$, the next component (node) along the path of the connection A→Z is identified and the current operational state of this node and the connecting link is determined. This is repeated for every hop along the path until one such component is non-operational.

If all components are operational, then no problem could be detected for the trigger event and the event is declined or passed on for manual treatment. If a problem is identified, it replaces the current problem as the root cause of the problem. If the root cause has already been detected as a result of a problem search triggered by another event, event correlation takes place, i.e. the second event detecting the problem is appended to the first one and further problem search is performed for the first problem only. Horizontal search is carried out by continued application of the following two functions for information request:

getNextHop(A → Z, B)  get the node representing the next instance downstream from *B* that is used for the communication between *A* and *Z*.

getOperationalState(A) get the operational state of *A*.

## Vertical Search

Once horizontal search has detected a component as being faulty it is determined whether this component is an elementary object (e.g. a repeater) or whether it consists of multiple components (e.g. a link using a path or connection in an underlying network). If the component is composed the problem search is carried one level lower to be continued at the underlying network level. This is supported by the information request getLowerLayerSAP, which determines the underlying path's start and end points. Horizontal search is then performed for the path between these two points.

getLowerLayerSAP(A) get the node representing the lower layer Service Access Point (SAP) that is used by *A*.

## Termination of Search

A search terminates when one of the following applies:
- A non-operational component was identified by the horizontal search algorithm. If this component is elementary, i.e. the search cannot proceed in an underlying network, the problem is reported to the operator.
- If the horizontal search algorithm cannot detect a non-operational component in the path the problem is declined and either purged or reported to the operator for manual treatment.
- The root cause of the problem was identified and is identical to the root cause of another open problem. The problem is appended to the other problem, search stops for the appended problem and may be continued for the other problem.

The overall outline of the algorithm is as follows: horizontal search is applied until a faulty component is detected. If this component is composed, vertical search is applied for the start and end points of the failing component to carry the search to the next lower level. Then horizontal search starts for the lower level. All this is repeated until a termination rule applies, e.g. the failing component is elementary.

Figure 1 explains how the fault isolation algorithm works for an example network by subsequently applying the three information request functions. The scenario is an TCP/IP network running over a multi-segment LAN. A LAN segment fails and causes a TCP connection to abort. The corresponding TCP event triggers the start of the algorithm. Horizontal search using getNextHop and getOperationalState verifies that connectivity at the TCP layer is lost. To further explore the problem in the underlying IP network vertical search takes place by applying the function getLowerLayerSAP to the two end points of the TCP connection. During the following horizontal search, getNextHop uses the IP routing table to first find out that

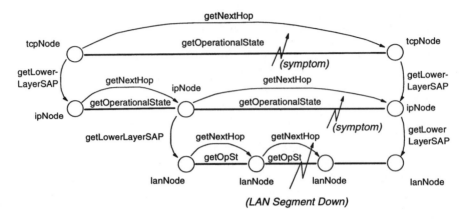

**Figure 1** Illustration of the Algorithm for an IP Network Running on a LAN.

the first hop is operational and then detect that the second hop is not. Vertical search carries the search to the next lower layer, the LAN level. Horizontal search then detects that the LAN segment between the two middle lanNodes (which normally represent bridges) is down. The result of `getLowerLayerSAP` reveals that this is an elementary object, i.e. there is no underlying network from a management perspective. Search stops and the result is communicated to the operators.

The algorithm will be further illustrated by the implementation aspects described in Section 4 and in the example given in Section 5 which contains event correlation.

## 4   IMPLEMENTATION ASPECTS

IBM is developing prototypes and production level systems that implement the algorithm while meeting as many of the general requirements from Section 2 as possible. To support integration with different network management infrastructures the kernel algorithm is separated from the actual information access modules (see Figure 2). The Information Access module implements the three elementary functions `getNextHop`, `getLowerLayerSAP`, and `getOperationalState` for different management systems and techniques without changing the kernel part of the algorithm. Experience from prototypes and product development shows that the three elementary functions can be provided for a variety of standard and proprietary infrastructures. Prior to the implementation described in this paper, prototypes have been developed for TCP/IP networks (based on SNMP, MIB-II) and for Token Ring LANs (based on a proprietary information model). See Stallings (1993) for the Simple Network Management Protocol (SNMP) and MIB-II.

The kernel part drives the fault isolation process, administers the problems including correlation and handles the threading. Internal parallelism is implemented using DCE threads. Each thread executes the algorithm for a given symptom in

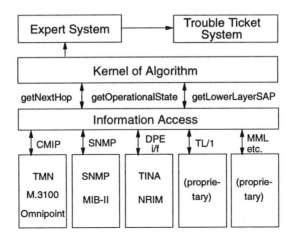

**Figure 2** Architecture of the Fault Isolation and Event Correlation System.

parallel resulting either in correlation and termination or in continued search. External parallelism is implemented by migration of threads from one fault isolation and event correlation agent to another. Each agent is responsible for a given management domain or layer. If a thread detects that further diagnosis is needed for an object that is located in another agent, the problem reference and the identification of the object is transferred to the neighboring agent. Since the migration interface is well defined, an interface definition language like CORBA's IDL (OMG, 1995) can be used in order to extend the external distribution over technology and administrative boundaries. Threads may migrate between public network operators or to and from a customer site as well.

Maximum performance is ensured by placing the correlation and isolation process as close to the monitored networks and devices as possible. The distributed algorithm was therefore implemented in the network level agents whenever the architecture allows, e.g. for the TMN-based version using the M.3100 model. If this is not possible, e.g. for the SNMP-based version, the algorithm was implemented in the network manager components using internal parallelism. However, caching of relevant object information can be used to improve performance if needed.

The information access interface was implemented for both the TMN-based M.3100 information model (ITU, 1995) and the SNMP-based MIB-II. In the following, the TMN-based approach is discussed in more detail. In terms of OSI management the information access interface acts as a manager application to the M.3100 network level agent that maintains the managed objects for a given subnetwork.

The managed object classes `trail` and `connection` provide the information for the algorithm as shown in Figure 3. A `trail` represents an entity that is responsible for the integrity of transfer of characteristic information while a `connection` represents an entity that is responsible for the transparent transport of information. `trail` and `connection` contain attributes that point to the `trail`/`connection` termination points (`a-TPInstance` and `z-TPInstance`) in the corresponding network

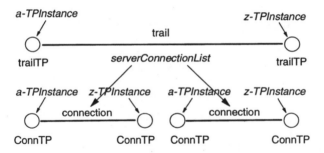

**Figure 3** Relationships Provided by the M.3100 Standard Object Model.

elements that are logically or physically connected by the `trail` or `connection`. A `trail` may contain a `server connection list` attribute that points to the connections that serve the `trail`, i.e. are used to build the `trail`. `connections` may contain a `server trail list` attribute that points to the the `trails` that serve this `connection`, i.e. the `trails` that are used to build the `connection`.

The 3 basic functions provided by the information access interface are implemented as follows:

- `getOperationalState(A)`
  get the `operationalState` attribute of the `trail` or `connection` given by A.
- `getLowerLayerSAP(A)`
  get the first `connectivity` object in the `serverConnectionList` or `serverTrailList` of the `trail` or `connection` given by A.
- `getNextHop(A → Z, B)`
  get the next `connectivity` object in the `serverTrailList` or `serverConnectionList` of the `connection` or `trail` given by A → Z after the `connectivity` object given by B. If B is undefined return the first `connectivity` object. If B is the last `connectivity` object in the list return undefined.

Implementation of these three functions enabled the generic algorithm to perform on TMN systems using M.3100 for network layer modelling.

# 5  EXAMPLE

This section illustrates the algorithm and its distributed implementation by a simple example. Consider a client server application running over a TCP Wide Area Network (WAN) connection. The WAN part of the TCP connection is provided by an ATM virtual path as sketched in Figure 4.

The ATM network uses SDH for the physical links between the switches. There are two management domains in this network. The customer running the client server application operates an SNMP management domain responsible for the management of his IP network. The customer is using a virtual private network (VPN)

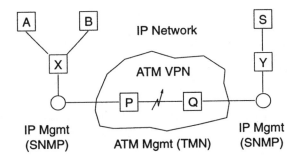

**Figure 4** IP over ATM Example Scenario.

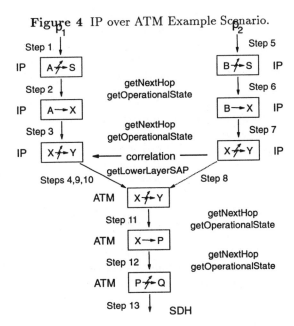

**Figure 5** Steps Performed in the Example Scenario.

from a network provider. The ATM VPN part is managed by the network provider using the TMN management framework and an M.3100 based network model for the ATM network. Each domain is controlled by a separate intelligent agent that is capable of executing the fault isolation and event correlation algorithm independently.

The steps which the algorithm performs are shown in Figure 5. Assume a physical link failure in the ATM VPN between switches $P$ and $Q$. As a consequence, no communication is possible from clients $A$ and $B$ with the server $S$. This problem is reported by the client $A$ to the IP manager (Step 1). The network layer fault isolation creates a problem $p_1$ and starts the search algorithm for this problem in a separate thread. After exploring that the connectivity between $A$ and $X$ is operational (Step 2), the thread detects the breakdown of the connection between the IP routers $X$ and $Y$ as the cause for the TCP connection problem (Step 3). The root cause of the current problem is updated to reflect this fact (Step 4) and the search will be continued because this problem has not yet been identified by another search thread.

Client $B$ reports a communication problem on the TCP connection between $B$ and $S$ to the SNMP management as well. A new problem $p_2$ is created and a thread starts to diagnose the TCP connection (Step 5). This thread executes in parallel to the thread that is working on the TCP connection problem $p_1$ between $A$ and $S$. The thread working on problem $p_2$ determines the cause for problem $p_2$ in the IP connectivity problem between $X$ and $Y$ (Steps 6,7). Since there is already an open problem registered for the managed object representing this connection in the agent, the thread appends (correlates) problem $p_2$ with problem $p_1$ and terminates (Step 8).

The thread for problem $p_1$ now migrates the problem to the intelligent agent responsible for the ATM VPN because the IP connection between $X$ and $Y$ is served by a path through the ATM WAN (Step 9; application of `getLowerLayerSAP`). It displays to the operator that a problem in the link between $X$ and $Y$ was detected and reported to the network operator running the link. A new thread is created in the ATM agent (Step 10) and the thread in the IP agent terminates or waits for a response from the ATM agent. This thread determines the `trail` object representing the ATM path that serves the IP connection between $X$ and $Y$, checks its operational state, and all its server connections. It detects that there is connectivity between $X$ and $P$ (Step 11) but not between the switches $P$ and $Q$ (Step 12). This is the root cause of the problem and is notified to the fault application of all management domains that have dealt with symptoms of the problem (in this case, the fault application for the ATM VPN management domain and the fault application for the IP management domain at the client site). If the underlying SDH network has a management domain as well, the problem could be passed to this system and the search is continued there (Step 13).

# 6  CONCLUSION

A solution for fault isolation and event correlation in integrated networks is presented. It is based on the model traversing approach, i.e. it reconstructs fault propagation during run-time by investigating relationships between managed objects. The proposed algorithm has specific advantages compared to other fault isolation and event correlation solutions.

The algorithm is simple, i.e. its dynamic behavior is straightforward and can be reconstructed. Since a single symptom event is sufficient to trigger the fault localization procedure, the algorithm is robust against loss of events and is able to perform its task based on incomplete data. The algorithm can operate at very high speed because the kernel algorithm can run very efficiently in a distributed and parallel fashion. Experience has shown that the system is able to cope with event storms of several hundreds of events per second. The technique was designed to incorporate other correlation techniques. This flexibility is of utmost importance for real-life environments with its integration of different networks with different requirements. The algorithm is generic, i.e. it is applicable for different network types and management architectures. The simple, well defined interface between the kernel algorithm and the Information Access modules easily supports different management paradigms and network protocols. The system has been implemented by IBM for SNMP-managed IP networks using MIB-II as a prototype and is being realized for TMN based systems using the M.3100 information model in a production level system. The experiences made during operation validated the approach.

Further studies on the applicability of the technique to other integrated networks are currently ongoing. Particular focus is put on the integration with expert systems and on dependency-graph based techniques for fault isolation on the application layer.

## ACKNOWLEDGEMENTS

The authors would like to thank Prof. Kurt Geihs from the University of Frankfurt and the members of the IBM ENC Systems and Network Management Department for their helpful suggestions.

## REFERENCES

Aidarus, S., Plevyak, T. (1994) *Telecommunications Network Management into the 21st Century*. IEEE Press, New York.

Bouloutas, A., Calo, S.B., Finkel, A. (1992) Alarm Correlation and Fault Identification in Communication Networks. *IBM Technical Report*, TR-17967.

Dreo, G. and Valta, R. (1995) Using Master Tickets as a Storage for Problem Solving Expertise. *Proc. 4th IFIP/IEEE International Symposium on Integrated Network Management*, 328-40.

Dupuy, F., Nilson, C., and Inoue,Y. (1995) The TINA Consortium: towards networking telecommunications information services. *IEEE Communication Magazine*, **33**, No. 11, 78-83.

Houck, K., Calo, S.B., Finkel, A. (1995) Towards a Practical Alarm Correlation System. *Proc. 4th IFIP/IEEE International Symposium on Integrated Network Management*, 519-30.

ITU (1995) Recommendation M.3100 Generic Network Information Model.

Jakobson, G. and Weissmann, M.D. (1993) Alarm Correlation, *IEEE Network*

Jordaan, J.F. and Paterok, M. (1993) Event Correlation in Heterogeneous Networks Using the OSI Management Framework. *Proc. ISINM'93, IFIP TC6/WG 6.6 International Symposium on Integrated Network Management*, 683-95.

Kätker, S. and Paterok, M. (1996) *System zur Überprüfung eines Datenübertragungsnetzwerks*. German Patent No DE 44 28 132 C 2, US Patent No GE 994 021 N.

Kätker, S. (1996) A Modelling Framework for Integrated Distributed Systems Fault Management. *Proc. IFIP/IEEE International Conference on Distributed Platforms*, 186-98.

Katzela, I. and Calo, S.B. (1995) Centralized vs. Distributed Fault Localization. *Proc. 4th IFIP/IEEE International Symposium on Integrated Network Management*, 251-61.

Kehl, W. and Hopfmüller, H. (1993) Model-Based Reasoning for the Management of Telecommunication Networks. *Proc. ICC'93, IEEE International Conference on Communications*, 13-17.

Kliger, S., Yemini, S., Yemini, Y., Ohsie, D., Stolfo, S. (1995) A Coding Approach to Event Correlation. *Proc. 4th IFIP/IEEE International Symposium on Integrated Network Management*, 266-77.

Lewis, L. (1993) A Case-Based Reasoning Approach to the Resolution of Faults in Communication Networks. *Proc. ISINM'93, IFIP TC6/WG 6.6 International Symposium on Integrated Network Management*, 671-82.

Nygate, Y.A. and Sterling, L. (1993) ASPEN – Designing Complex Knowledge Based Systems. *Proc. 10th Israeli Symposium on Artificial Intelligence Computing Vision, and Neural Networks*, 51-60.

Nygate, Y.A. (1995) Event Correlation using Rule and Object Based Techniques. *Proc. 4th IFIP/IEEE International Symposium on Integrated Network Management*, 279-89.

OMG (1995) The Common Object Request Broker: Architecture and Specification, Rev. 2.0.

Richter, L., Wakano, M., and Oshigiri, H. (1995) *Telecommunications Information Networking Architecture - Network Resource Information Model.* TINA-Consortium Document No TB_C2.LSR.001_1.0_93.

Stallings, W. (1993) *SNMP, SNMPv2, and CMIP; The Practical Guide to Network Management Standards.* Addison Wesley.

# BIOGRAPHY

**Stefan Kätker** received his Diploma in Computer Science from the University of Erlangen-Nürnberg, Germany, in 1992. Since 1992 he is working in the Systems- and Network Management Department at the IBM European Networking Center in Heidelberg. Currently he is architect and scientific consultant for TMN fault management products and solutions. Research interests include SNMP- and TMN-based fault, application, and distributed systems management.

**Martin Paterok** received his Diploma and Dr.-Ing. in Computer Science from the University of Erlangen-Nürnberg, in 1985 resp. 1990. After working at IBM's T.J. Watson Research Center in Yorktown Heights he joined the IBM European Networking Center Heidelberg in 1991. He was staff member and manager of the Systems and Network Management Department. He is now manager of the Department for Mobile Data Communication. Research interests include integrated systems management and wireless data communication, in particular telematics and intelligent highway.

# 44

# Non-Broadcast Network Fault-Monitoring Based on System-Level Diagnosis

*Elias Procópio Duarte Jr.*\*  
*Takashi Nanya* †  
*Tokyo Institute of Technology*  
*Dept. Computer Science*  
*Ookayama 2-12-1 Meguro-Ku*  
*Tokyo 152 Japan*  
*{elias,nanya}@cs.titech.ac.jp*

*Glenn Mansfield*  
*Shoichi Noguchi*  
*Sendai Foundation for Applied*  
*Information Sciences*  
*Tsutsujigaoka 5-12-55*  
*Miyagino-Ku Sendai 983 Japan*  
*{glenn,noguchi}@tia.ad.jp*

## Abstract

Network fault management systems are mission-critical, for they are most needed during periods when part of the network is faulty. Distributed system-level diagnosis offers a practical and theoretically sound solution for fault-tolerant fault monitoring. It guarantees that faults don't impair the fault management process. Recently, results from the application of distributed system-level diagnosis applied for SNMP-based LAN fault management have been reported [1, 2]. In this paper we expand those results by presenting a new algorithm for diagnosis of non-broadcast networks, applied to point-to-point network fault management. In the algorithm, nodes test links periodically, and disseminate link time-out information to all its fault-free neighbors in parallel. Upon receiving link time-out information a node computes which portion of the network has become unreachable. This approach is closer to reality than previous algorithms, for it is impossible to distinguish a faulty node from a node to which all routes are faulty. The diagnosis latency of the algorithm is optimal, as nodes report events in parallel, and latency is proportional to the diameter of the network. The dissemination step includes mechanisms to reduce the number of redundant messages introduced by the parallel strategy. We present a MIB for the algorithm, and a SNMP-based implementation. The evaluation of algorithm's impact on network performance, shows that the amount of bandwidth required is less than 0.1% for popular link capacities. We conclude demonstrating the integration of LAN and WAN fault diagnosis into a unified framework.

## Keywords

Distributed System-Level Diagnosis, Network Fault Management, SNMP

---

\* also with Federal University of Paraná, Informatics Dept., C.P. 19081 Curitiba PR, CEP 81531-990, Brasil. The author has a scholarship from the Brazilian research council, CNPq.

†also with University of Tokyo, Research Center for Advanced Science & Technology, Komaba 4-6-1 Meguro-ku, Tokyo 153, Japan

# 1   INTRODUCTION

Fault management is the set of activities required to guarantee network availability, even in the presence of network faults and performance degradation. Fault management must thus be fault tolerant, for network faults should not impair the system that is meant to solve them. Fault management can be broadly subdivided into monitoring and control. Monitoring is the process employed for obtaining information required about the components of a network, in order to make management decisions and subsequently control their behavior. In this paper we present a fault-tolerant approach for non-broadcast network fault-monitoring based on the long standing theory and practice of distributed system-level diagnosis.

Current SNMP-based fault-management systems are based on the manager-agent model, in which a fixed manager station queries a set of agents for management information. This centralized scheme is inherently unreliable, for if the manager becomes faulty, network management stops on the entire network.

Hierarchical schemes are also popular, in which machines that participate in management form a tree, where the root is the main manager, leaves are agents and internal nodes run both the agent server and a monitoring application. An internal node of the tree monitors nodes in the subtree of which it is the root, and reports to its parent in the tree. Hierarchical schemes are also inherently unreliable, for whenever an internal node becomes faulty, monitoring stops on part of the network.

System-level diagnosis offers a theoretically sound and practical framework for fault-tolerant network monitoring: even if any part of the network becomes faulty, fault-free nodes are able to diagnose the system. Recently, results from the application of distributed system-level diagnosis applied for SNMP-based LAN fault management have been reported. In [1] the Adaptive Distributed System-level Diagnosis (ADSD) algorithm was implemented using SNMP. In this algorithm, nodes are assumed to be fully connected, and the testing topology is a ring, such that the number of tests is as low as one per node per testing round. For a network of $N$ nodes, the diagnosis latency of the algorithm is $N$ testing rounds, i.e. proportional to the number of nodes in the network.

In [2] another algorithm for LAN fault-diagnosis, Hi-ADSD (Hierarchical ADSD), was introduced and implemented using SNMP. In this algorithm, the number of tests is the same as in ADSD, but the testing topology is initially a hypercube, and diagnosis is reduced to $\log^2 N$ testing rounds. Those algorithms employ a distributed strategy for fault management, in which a collection of network nodes perform network diagnosis, and the human manager may attach an interface to any of these nodes to receive diagnostic information.

In this paper we expand those results by introducing an algorithm for diagnosis in non-broadcast networks, applied to point-to-point network fault management. In the algorithm, a node tests links periodically, and disseminates link time-out information to all its fault-free neighbors in parallel. Upon receiving link time-out information a node computes which portion of the system has become unreachable. This new approach to diagnosis, based on *link time-out* and *node unreachability* is closer to reality than previous approaches. There are two reasons for this improvement: (1) it is impossible to distinguish a node fault from the fault on all the paths to that node; (2) in previous algorithms, two fault-free nodes in disconnected components keep the old status for each other, which may not correspond to reality.

A node joining the algorithm disseminates information about itself, and collects diagnostic information from its neighbors. The diagnosis latency of the algorithm is optimal, as nodes report events in parallel, and latency is proportional to the diameter of the network. The dissemination step includes mechanism to reduce the number of redundant messages introduced by the parallel strategy. We present a MIB for the algorithm, and a SNMP-based implementation. The evaluation of algorithm's impact on network performance shows that the amount of bandwidth required is less than 0.1% for popular link capacities. We conclude demonstrating the integration of LAN and WAN fault diagnosis into a unified framework.

The rest of the paper is organized as follows. Section 2 reviews system-level diagnosis, including algorithms for LAN fault management. Section 3 reviews algorithms for diagnosis on networks of general topology, and includes the specification of the new algorithm for non-broadcast networks. In section 4 we present a MIB and a SNMP-based implementation of the algorithm. In section 5 we evaluate its impact on network performance. Section 6 concludes the paper, showing the integration of LAN, and WAN fault diagnosis into a unified framework.

## 2 SYSTEM-LEVEL DIAGNOSIS

Consider a system consisting of $N$ units, which can be faulty or fault-free. The goal of system-level diagnosis is to determine the state of those units. For almost 30 years researchers have worked on this problem, and the first model of diagnosable systems was introduced by Preparata, Metze, and Chien, the *PMC Model* [3]. In the PMC model units are assigned a subset of the other units to test, and fault-free units are able to accurately assess the state of the units they test. The PMC model assumes the existence of a *central observer* that, based on the syndrome, can diagnose the state of all the units. In this paper, we will use alternatively the word node for unit, and network for system.

Early system-level diagnosis algorithms assumed that all the tests had to be decided in advance. An alternative approach, which requires fewer tests, is to assume that each unit is capable of testing any other, and to issue the tests adaptively, i.e., the choice of the next tests depends on the results of previous tests, and not on a fixed pattern. Hakimi and Nakajima called this approach *adaptive* [5].

Early system-level diagnosis algorithms assumed the existence of the previously mentioned central observer. This situation was changed by Kuhl and Reddy [6, 7], who introduced *distributed* system-level diagnosis, in which fault-free nodes reliably receive test results through their neighbors, and each node independently performs consistent diagnosis. Important distributed system-level diagnosis algorithms include [8] and [9]

System-level diagnosis algorithms proceed in testing rounds, i.e., the period of time in which each unit has executed the tests it was assigned. To evaluate those algorithms, two measures are normally used: the total number of tests required per testing round and the diagnosis *latency*, or delay, i.e., the number of testing rounds required to determine the state of the units.

The Adaptive Distributed System-level Diagnosis algorithm, *Adaptive DSD*, was introduced by Bianchini and Buskens [10, 11]. Adaptive DSD is at the same time distributed and adaptive. Each fault-free node is required to perform the minimal number of tests per testing interval, i.e., one test, to achieve consistent diagnosis in at most $N$ testing rounds. There is no limit on the number of faulty nodes for fault-free nodes to diagnose the system. Later, the Hierarchical Adaptive Distributed System-Level Diagnosis (Hi-ADSD) algorithm [2] was introduced and implemented with SNMP. By using a testing assignment that is initially a hypercube, Hi-ADSD has diagnosis latency of $\log^2 N$ in the worst case.

## 2.1    Algorithms for Diagnosis in Networks of General Topology

Up to this point, we reviewed system-level diagnosis algorithms for SNMP-based LAN diagnosis, in which the network is assumed to be fully connected, e.g., an Ethernet or a network based on switches. From this point on we review algorithms for diagnosis in non-broadcast networks, which can be applied for point-to-point network fault management. We introduce our new algorithm in the next section.

In [12] Bagchi and Hakimi introduced an algorithm for system-level diagnosis in networks of general topology. Initially each fault-free node knows only about its own state, and of its physical neighbors. Fault-free processors form a tree-based testing graph. Diagnostic messages are sent along the tree. The number of messages required by this algorithm to achieve diagnosis is shown to be optimum. Unfortunately the algorithm is not executed *on-line*, i.e., no processor can become faulty or be repaired during the execution of the algorithm. This characteristic rules out the application of the algorithm for WAN fault diagnosis.

In [13, 14] Bianchini *et.al.* introduced and evaluated through simulation the Adapt algorithm. The Adapt algorithm can be executed on-line: when a given node becomes faulty, a new phase begins in which other nodes reconnect the testing graph. The underlying testing assignment of Adapt is a minimally strongly connected digraph over the physical network. To build the testing graph, Adapt employs a distributed procedure that requires massive amounts of large diagnostic messages to be exchanged among the nodes.

Recently Rangarajan *et.al.* [15] introduced another algorithm for system-level diagnosis for networks of arbitrary topology that can be executed on-line. The algorithm, which we call here *RDZ*, for the author's initials, builds a testing graph that guarantees the optimal number of tests, i.e., each node has one tester. Furthermore it presents the best possible diagnosis latency by using a parallel dissemination strategy. Whenever a node detects an event, it sends diagnostic information to all its neighbors, which in turn send it to all its neighbors, and so on.

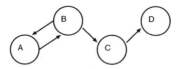

**Figure 1** A *jellyfish* fault configuration.

Although the RDZ algorithm presents the best possible diagnosis latency, and the best possible number of testers per node, it does not diagnose link faults and also a node fault configuration which the authors call *jellyfish faulty node configuration*. In this fault configuration, between two connected components there is a set of nodes such that part of those nodes test each other in a cyclic fashion, and other tests emanate from the cycle. If all nodes in the jellyfish become faulty simultaneously, nodes in the connected components won't diagnose that situation. It should be noted that a jellyfish may involve from one to an arbitrary number of nodes.

Consider figure 1. All nodes form a jellyfish, in which there is a cycle (node A and node B) and tests emanating from the cycle (from node B to node C to node D). If both nodes A and B become faulty, nodes C and D won't be able to diagnose the fault. The same is true if nodes A, B, and C become faulty, i.e., node D doesn't detect the event. The RDZ algorithm cannot be applied for network fault monitoring, for it is unacceptable to have an arbitrarily large portion of the network to become faulty in an undetected fashion.

# 3 A NEW ALGORITHM FOR DIAGNOSIS ON NON-BROADCAST NETWORKS

In this section, we introduce a new algorithm that diagnoses link time-outs, and node reachability, using the minimum number of tests, i.e. one per link, and also presenting the optimal latency. Before introducing the algorithm, consider figure 2. In fault situation A the node is fault-free, but all links leading to that node are faulty, in fault situation B, the node itself is faulty. From test results it would be impossible for any other node in the system to determine which is the actual situation. Our algorithm is based on this fact: a link may *time-out* to a test, and if all links to a given node have timed-out, then the node is *unreachable*. Thus links may be in one of two states *fault-free, timed-out* and nodes may be *fault-free* or *unreachable*. This approach to fault diagnosis on wide-area networks is closer to reality, for links are usually made up of not only wires but may also involve a number of network devices, hubs and gateways.

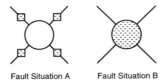

Fault Situation A    Fault Situation B

**Figure 2** Ambiguous fault configurations.

To keep the number of tests minimum, there is one tester per link. As a link always connects two nodes, and nodes have unique identifiers, the node with the highest identifier tests the link at each testing interval. If the link *times-out*, i.e., the neighbor doesn't reply to the test, and in the past testing interval it did, then there is a new *fault event*. Analogously, if the link has timed-out in the past testing interval, and it does carry a reply this time, then there is a *repair event*.

The algorithm employs a *two-way test*. This guarantees that the *jellyfish* fault configuration is always detected, even keeping the minimum number of tests. When node A is testing the link to node B, node A gets the local time at B, and stores that result on B. In this way, not only node A knows about the state of B, but also node B can monitor the tester activity. If a threshold is decided for the maximum interval between link tests, then a node can time-out the tester whenever the threshold is exceeded. When a node detects

a link time-out or a tester fault, it starts or continues testing the link until it ceases to
time out, such that, when the link recovers again, only the node of highest identifier tests
the link.

Each node keeps a state counter for each link in the system, which is initially zero, and is
incremented at each new event information received for that link. This permits a node to
identify redundant messages. After a new event is discovered, each node propagates event
information to all neighbors. This parallel dissemination strategy is the same employed by
the RDZ algorithm. Besides the nodes identifier, and the status counter, each diagnostic
message carry information about which nodes have processed the message. In this way,
the number of redundant messages is reduced, and messages do not cycle in the network.
After receiving a message, each node appends its own identifier to the list of nodes that
has processed the message. Furthermore it appends the identifiers of the neighbors to
which the message was already sent. For a full discussion and evaluation of this approach
please refer to [15]. It should be clear that, as messages are short, the impact of this
strategy on network performance is small. In section 5 we evaluate the percentage of link
bandwidth required to run the algorithm.

After a node receives information about a link event, it runs an algorithm (like the
breadth-first tree) to compute the system connectivity, thus discovering which portions
have become reachable or unreachable.

The data structures of the algorithm are thus:

- A Link table indexed by link identifier, containing a status counter for the link, and the
  last time the link was tested. The counter is initially zero, and an even value indicates
  a fault-free state; The last-test-time is updated only on nodes connected to the link
  and such that the node doesn't test the link, but its neighbor;
- A Link-Events table, containing at each entry the link identifier, the state counter of
  the link, and a list of nodes that have already processed the message as seen by the
  sending node.

The algorithm in pseudo-code is:

```
BEGIN
 /* at node i */
 DO FOREVER
   FOR each link i-j, that connects node i to node j
     IF (i > j) OR (node j is faulty)
     THEN test link i-j; /* get local time at node j */
          IF link i-j is fault-free
          THEN set last-time-tested on j;
          IF there is a new event
          THEN add event to new-event table;
     ELSE /* check link tested by neighbors */
     IF last-time-tested > testing interval threshold
     THEN add event to Link-Events table;

   FOR each entry in Link-Events table
        IF entry carries new information
        THEN update link counters;
```

```
        FOR each neighbor k of node i
            IF k has not received the message
                THEN set event information to k's new-event table;
    compute node reachability;
    SLEEP(testing interval)
END;
```

## 3.1   An Example Execution

Consider the example system in figure 3. Initially all links and nodes are fault-free. Each node starts testing links as depicted by arrows, and exchange test information with neighbors. Eventually each node receives information about all links.

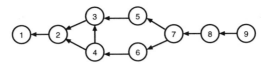

**Figure 3** The testing assignment on an example non-broadcast network.

Now consider the first event depicted in figure 4, in which link 3-5 is faulty and times-out. This time-out will be detected by node 5, and immediately disseminated to node 7. This in turn will disseminate to node 8 (and from there to node 9), and node 6. Node 6 disseminates information to node 4, and from there to node 3, node 2 and node 1. Node 2 disseminates the information to node 3. Now, if node 3 timed-out out the tester (link 3-5) before the information arrives from node 2, then node 3 will also disseminate information on the time-out. If a node, say node 2, receives two diagnostic messages about the same event it will only disseminate the first of them, because the second is recognized as old information. Thus, the highest number of messages per event per link is two. After all nodes receive and process diagnostic messages, they run an algorithm to compute system connectivity, and conclude that all nodes are still connected.

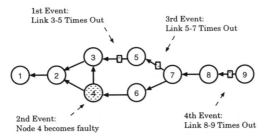

**Figure 4** A series of events occur in the network.

In the second event depicted in figure 4, node 4 became faulty. Node 6 detects a time-out on link 6-4, and after the testing threshold expires, node 2 and node 3 detect time-outs on links 4-2 and 4-3 respectively. The system now is divided into two connected components, one consisting of node 1, node 2 and node 3, the other consisting of node 5, node 6, node

7, node 8 and node 9. As on each component a node detects and disseminates the event, diagnostic information will eventually reach every fault-free node in the system.

Now consider the third event, link 7-5 becomes faulty and times-out. The resulting system has 3 connected components, the first consisting of node 1, node 2, and node 3; the second of node 5 alone; and the third of node 6, node 7, node 8, and node 9. In the first component not one node detects the event, because it is already disconnected from the rest of the system. In the second component, node 5 eventually times out on the test of link 7-5 and realizes it is disconnected from the system, i.e., every other node is unreachable. At the third component, node 7 initially detects link 7-5 time-out and the event is disseminated to the other nodes in the component.

If still another link, 9-8, becomes faulty and times-out, node 9 detects the event and recognizes it is disconnected from the system. Node 8 times-out on the testing threshold of link 9-8, and disseminates event information to node 7 and node 6. The other nodes in the network are already in disconnected components.

After these events, when faults are repaired, nodes testing corresponding links will detect the events, and disseminate the information to other nodes in their connected components. Eventually the whole system becomes a unique connected component, and every node receive diagnostic information about all links.

## *Correctness*

Here we give an informal discussion of the correctness of the algorithm. Consider a connected component of the system, made up of fault-free nodes and such that between any pair of those nodes there is a fault-free path. The neighborhood of the component is defined as the set of links that timed-out in the previous testing-round. Clearly, any new event in the component or in its neighborhood is detected by nodes in the component. This is assured by the testing strategy, in which there is a two-way test on each link from any node of the component. Now consider that one event has occurred. If a fault-free node or link has become faulty, then one node in the component will detect the fault, and forward it to other neighbors. As each node forwards new information to all neighbors, information will eventually reach all nodes in the component. If the fault breaks the component in two, then nodes on both components will detect a link time-out, and disseminate the information on their respective components. Now consider a repair event: if a test succeeds on a link that had been timing-out, the two nodes (tested and tester) exchange diagnostic information, and disseminate this information to their neighbors. Thus event information is always disseminated to every fault-free node in the component.

Event counters guarantee that old information is recognized as such. Furthermore, those links that have odd event-counters are timed-out links and those that have even-counters are fault-free links. This is guaranteed because a counter is only incremented when a new event happens, from timed-out to fault-free or opposite. As the counter is initially zero, for a fault-free initial status, and when it times-out it is increased to 1 and so on, an even value will always indicate a fault-free state, and an odd value a faulty state.

## 4   SNMP-BASED IMPLEMENTATION

In this section we present an implementation of the algorithm for non-broadcast network fault management based on SNMP [17, 18, 19]. Each node running the algorithm keeps two tables. The first table keeps information about each link in the network: its identifier, the state counter, and the time it was tested. The time field is only used by nodes that test a link to implement the two-way testing strategy. We give below the corresponding ASN.1 table.

```
LinkState OBJECT-TYPE
    SYNTAX  SEQUENCE OF LinkStateEntry
    ACCESS  not-accessible
    STATUS  mandatory
    DESCRIPTION
        "This is an array that contains link status information."
    ::= { diagnosis 1 }

linkStateEntry OBJECT-TYPE
    SYNTAX  LinkStateEntry
    ACCESS  not-accessible
    STATUS  mandatory
    DESCRIPTION
            "Each entry of linkState shows if a link is timing-out
             or fault-free, according to the status counter"
    INDEX   { linkID }
    ::= { LinkState 1 }

LinkStateEntry ::=
    SEQUENCE {
        linkID          DisplayString,
        StatusCounter   Counter,
        TestedTime      TimeTicks  }
```

The second table, LinkEvents, is a dynamic table, in which event information is added by the local agent and its neighbors. After each testing interval, all entries in the table are processed. Each entry contains the identifier of the link that suffered the event, the timestamp for that event, and a string containing the identifiers of all the nodes that have already processed the message. The ASN.1 table is given below.

```
LinkEvents OBJECT-TYPE
    SYNTAX  SEQUENCE OF linkEventsEntry
    ACCESS  not-accessible
    STATUS  mandatory
    DESCRIPTION
        "This is a dynamic table to which information
         about new link events are added."
    ::= { diagnosis 2 }

linkEventsEntry OBJECT-TYPE
    SYNTAX  LinkEventsEntry
    ACCESS  not-accessible
    STATUS  mandatory
    DESCRIPTION
```

```
            "Each entry of linkEvents carries the link identifier,
             the status counter for the new event, and a sequence
             of identifiers of nodes that have processed the event"
   INDEX    { linkID }
   ::= { LinkEvents 1 }


LinkEventsEntry ::=
     SEQUENCE {
          LinkID           DisplayString,
          StatusCounter    Counter,
          Path             DisplayString  }
```

In our implementation nodes set neighbors tables, and thus security measures must be taken, specifically assignment of restricted access permission. It should be clear that from the LinkState table that the complete network configuration is available to each node, which can calculate the system connectivity at any time. Works on generating network configuration information automatically have been reported [16], and can be employed to build the LinkState table.

## 5   IMPACT ON NETWORK PERFORMANCE

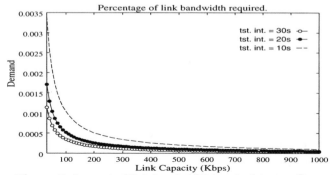

**Figure 5** Amount of link bandwidth required to run diagnosis.

At each testing interval, each link carries one message from the tester to the neighbor. Furthermore, for any new event in the network, each link will carry usually one, and at most *two* messages about the event. The reason is that after updating the state counter, a node does not forward any other message that contains known information. The link will carry two messages only if both nodes send information at the same time. Thus, the total number of messages per event required by the algorithm is at most $2 * L$, where $L$ is the number of links.

The graph in figure 5 shows the impact of the algorithm on network performance, by showing the percentage of link bandwidth required by diagnostic messages. The graph shows links of different capacities, and results are shown for different testing intervals, of

10 seconds, 20 seconds, and 30 seconds. We consider a fault rate $\lambda$ of 0.001. The size of SNMP messages assumed is 128 bytes. Results show the percentage of bandwidth required is always less than 0.1%, on links from 28.8Kbps to 1Mbps.

## 6 CONCLUSION

In this paper we presented a new distributed algorithm for system-level diagnosis on non-broadcast networks. The purpose of the algorithm is to allow each node to independently detect which portions of the network are faulty or unreachable. We show that in some cases it is impossible to distinguish between the two cases.

A node running the algorithm executes link tests at a testing interval. The algorithm employs the minimum number of tests, i.e., one per link. Of the two nodes connected by a link, the one with highest identifier is the link tester. We assume nodes have local memory, and tests are built in such a way that both ends of a link detect a link time-out in case of link or one node failure.

Upon detecting a new event, diagnostic information is disseminated in parallel, and the algorithm has the minimum diagnosis latency, i.e., proportional to the diameter of the network. Mechanisms are included to reduce the amount of redundant messages. As each message is small, containing information about one event, and any link carries at most two messages, the impact of the algorithm on network performance is small. A MIB and SNMP implementation were presented.

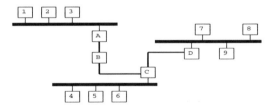

**Figure 6** A small internet.

As future work we discuss here an integrated approach for internet fault monitoring. This approach can be achieved by running specific algorithms for diagnosis on broadcast networks (LAN's), like Hi-ADSD, together with the algorithm introduced in this paper. Consider the small internet in figure 6. Nodes with identifiers from 1 to 9 are connected to broadcast networks. Node A, node B, and node D have a link to a broadcast network, and to a non-broadcast network. Node B is takes part only in the non-broadcast network. For the two algorithms to run cooperatively, it is sufficient that nodes only on a broadcast network run an algorithm for diagnosis on the LAN to which it belongs; nodes not on a non-broadcast network run the algorithm for diagnosis on that network; nodes that are on a broadcast network, but also have a link to another network must run both a LAN diagnosis algorithm, and a WAN diagnosis algorithm. This means these nodes execute tests according to the two algorithms, and also carry the necessary data structures to hold information about the entire system. In this way, a truly fault-tolerant network fault management system can be deployed, in which any fault-free node can diagnose the whole system.

# REFERENCES

[1] E.P. Duarte Jr., and T. Nanya, "An SNMP-based Implementation of The Adaptive DSD Algorithm for LAN Fault Management," *Proc. IEEE/IFIP NOMS'96*, pp. 530-539, Kyoto, April 1996.

[2] E.P. Duarte Jr., and T. Nanya, "Hierarchical Distributed System-Level Diagnosis Applied for SNMP-based Network Fault Management", *Proc. IEEE 16th Symp. Reliable Distributed Systems*, Niagara, September 1996.

[3] F. Preparata, G. Metze, and R.T. Chien, "On The Connection Assignment Problem of Diagnosable Systems," *IEEE Transactions on Electronic Computers*, Vol. 16, pp. 848-854, 1968.

[4] S.L. Hakimi, and A.T. Amin, "Characterization of Connection Assignments of Diagnosable Systems," *IEEE Transactions on Computers*, Vol. 23, pp. 86-88, 1974.

[5] S.L. Hakimi, and K. Nakajima, "On Adaptive System Diagnosis" *IEEE Transactions on Computers*, Vol. 33, pp. 234-240, 1984.

[6] J.G. Kuhl, and S.M. Reddy, "Distributed Fault-Tolerance for Large Multiprocessor Systems," *Proc. 7th Annual Symp. Computer Architecture*, pp. 23-30, 1980.

[7] J.G. Kuhl, and S.M. Reddy, "Fault-Diagnosis in Fully Distributed Systems," *Proc. 11th Fault Tolerant Computing Symp*, pp. 100-105, 1981.

[8] S.H. Hosseini, J.G. Kuhl, and S.M. Reddy, "A Diagnosis Algorithm for Distributed Computing Systems with Failure and Repair," *IEEE Transactions on Computers*, Vol. 33, pp. 223-233, 1984.

[9] R.P. Bianchini, K. Goodwin, and D.S. Nydick, "Practical Application and Implementation of System-Level Diagnosis Theory," *Proc. 20th Fault Tolerant Computing Symp*, pp. 332-339, 1990.

[10] R.P. Bianchini, and R. Buskens, "An Adaptive Distributed System-Level Diagnosis Algorithm and Its Implementation," *Proc. 21st Fault Tolerant Computing Symp*, pp. 222-229 , 1991.

[11] R.P. Bianchini, and R. Buskens, "Implementation of On-Line Distributed System-Level Diagnosis Theory," IEEE Transactions on Computers, Vol. 41, pp. 616-626, 1992.

[12] A. Bagchi, and S.L. Hakimi, "An Optimal Algorithm for Distributed System-Level Diagnosis," *Proc. 21st Fault Tolerant Computing Symp.*, June, 1991.

[13] M. Stahl, R. Buskens, and R. Bianchini, "On-Line Diagnosis on General Topology Networks," *Proc. Workshop Fault-Tolerant Parallel and Distributed Systems*, July 1992.

[14] M. Stahl, R. Buskens, and R. Bianchini, "Simulation of the Adapt On-Line Diagnosis Algorithm for General Topology Networks," *Proc. IEEE 11th Symp. Reliable Distributed Systems*, October 1992.

[15] S.Rangarajan, A.T. Dahbura, and E.A. Ziegler, "A Distributed System-Level Diagnosis Algorithm for Arbitrary Network Topologies," *IEEE Transactions on Computers*, Vol.44, pp. 312-333, 1995.

[16] G.Mansfield, M.Ouchi, K.Jayanthi, Y.Kimura, K.Ohta, Y.Nemoto, "Techniques for automated Network Map Generation using SNMP", Proc. of INFOCOM'96, pp.473-480, March 1996.

[17] M. Rose, and K. McCloghrie, "Structure and Identification of Management Information for TCP/IP-based Internets," *RFC 1155*, 1990.

[18] J.D. Case, M.S. Fedor, M.L. Schoffstall, and J.R. Davin, "A Simple Network Management Protocol," *RFC 1157*, 1990.

[19] K. McCloghtie and M.T. Rose, "Management Information Base for Network Management of TCP/IP-based Internets," *RFC 1213*, 1991.

# BIOGRAPHY

**Elias Procópio Duarte, Jr.** is a PhD student in Computer Science at Tokyo Institute of Technology, Tokyo, Japan. He is also an Assistant Professor at the Department of Informatics of Federal University of Paraná, Brazil. He has an MSc degree in Telecommunications from the Polytechnical University of Madrid, Spain, 1991. He also has an MSc degree in Computer Science from Federal University of Minas Gerais where he received his BS in Computer Science in 1988. Main research interests include distributed systems and computer networks, their dependability, management, performance evaluation, and algorithms. He is a student member of the IEEE and the ACM.

**Glenn Mansfield** obtained his Master's degree in 1977 from Indian Institute of Technology, Kharagpore, India in the field of Nuclear and Particle Physics followed by his Masters in Physical Engineering in 1979 from Indian Institute of Science, Bangalore, India. After working with Tata Consultancy Services, Bombay, India as a senior systems analyst for five years, he obtained his Ph.D. specializing in Logic programming, from Tohoku University, Japan. He has worked as a research associate in the computer center of Tohoku University for a period of 3 years and is currently chief scientist at Sendai Foundation for Applied Information Sciences, Japan. His areas of interest include expert systems, logic programming, computer networks and their management, use of the Internet for education. He is a member of the Internet Society, the ACM, the IEEE and the IEEE Communications Society.

**Takashi Nanya** is a professor in the Research Center for Advanced Science & Technology at the University of Tokyo, and also a professor in the Departmnent of Electrical Engineering at Tokyo Institute of Technology, Tokyo, Japan. His research interests include fault-tolerant computing, computer architecture, design automation and asynchronous computing. He was a visiting research fellow at Oakland University, Michigan, in 1982, and at Stanford University, California, in 1986-87. He received his B.E. and M.E. degrees in mathematical engineering and infomation physics from the University of Tokyo in 1969 and 1971, respectively, and his Dr.Eng. degree in electrical engineering from Tokyo Institute of Technology in 1978. He is a member of the IEEE, ACM, IEICE, and the Information Processing Society of Japan.

**Shoichi Noguchi** is the director of the Sendai Foundation for Applied Information Sciences, and also a professor at Nihon University, Japan. He received his B.E., M.E. and D.E. degrees in Electrical Communication Engineering from Tohoku University in 1954, 1956 and 1960 respectively. He was a professor at Tohoku University until 1993. He is active in the fields of information science theory, computer network fundamentals, parallel processing, computer network architectures and knowledge engineering fundamentals. He is the president of the Information Processing Society of Japan since 1995.

# 45

# Proactive Management of Computer Networks using Artificial Intelligence Agents and Techniques

*Marco Antonio da Rocha*
Federal University of Rio Grande do Sul (UFRGS)
National Supercomputing Center and Computer Science Institute (CESUP)
91520-130, Porto Alegre, RS, Brazil
Fone/FAX: +55.51.339-46-99, E-mail: rock2@cesup.ufrgs.br

*Carlos Becker Westphall*
Federal University of Santa Catarina (UFSC)
Network and Management Laboratory (LRG)
Caixa Postal 476 - Campus Trindade
88040-970, Florianópolis, SC, Brazil
Fone: +55.48.231-97-39, FAX: +55.48.231-97-70, E-mail: westphal@lrg.ufsc.br

## Abstract

This work was developed in the area of Computer Network Management. The work is intended to establish a strategy for the implementation of proactive management in the available management environment, i.e. the National Supercomputing Center, for the management of networks associated with the use of agents. The work was motivated by a need to explore the use of agents to identify symptoms of proactive management problems which might occur in networks, and especially to recognize a problem using artificial intelligence techniques and take reactive measures to solve it, configuring a proactive management application for the prevention of problems in computer networks.

## Keywords

Proactive management, baseline, artificial intelligence, rules of production, management networks

## 1    Introduction

Due to the constant interaction between different types of computer network users, the need is increasingly felt for better organizational techniques in order to manage the resources offered. There no longer exists a single user profile, standard solutions are inadequate, and it is now necessary to divide efforts and apply diversified solutions which attend to the interests of each work group which makes up the network. Network management is a distributed application which involves the exchange of information among the management processes, with the goal of monitoring and controlling diverse network resources. The processes involved with a specific area assume two important roles: Manager and/or Agent. The Manager is part of a distributed application which generates operations and receives reports. The Agent is part of a distributed application which generates objects associated with it (responding to requests for operations from the Manager and giving reports which reflect the operation of the objects). Admitting that the management tools don't encompass the wide range of network problems and that they aren't always applied by network operators, it becomes necessary to apply other management mechanisms in order to overcome the most apparent shortcomings.

Using this as a starting point, a need was felt to cover this particular subset of computer networks that are not attended to by currently available management tools and which address the individual differences of each organization which separate it from the others, as well as making it easier to administrate. Thus, it becomes possible to have two possible operations in a computer network: Reactive Operation, in which the problems are reported to the manager for action, and Proactive Operation, in which the management should be capable of detecting problems, and avoiding them, before they occur.

Therefore, this paper seeks to present an adopted strategy for the implementation of proactive management in computer networks. The work that motivated the elaboration of this report was driven toward the implementation of an agent and the joint utilization of network monitoring and commands of the UNIX operating system for the management of computer networks, where the equipment makes up the network of the Institute of Computer Sciences and of the National Supercomputing Center, including a SunNet Manager platform, Sun workstations, Sun OS 4.1 operating system compatible with UNIX 4.3 BSD, Sun OS C compiler, Open Windows 2.1 and PCs in a network configuration with the National Supercomputing Center were used.

This paper is organized in the following manner: in Section 2, the relative aspects of network management in general are addressed, a proactive management scheme is presented, as well as observations on how artificial intelligence techniques can be applied in this paradigm; in Section 3, a detailed treatment is presented of a strategy for implementation of a minimum proactive management prototype, as well as a rule set for use in detecting network problem symptoms; in Section 4, a commentary is made as to the results obtained; finally, in Section 5, the final conclusions are presented.

## 2    Network Management and Proactive Management

The flow of information in a network should be reliable and rapid, which implies that the data undergo constant monitoring, so as to filter or even detect problems which can result in losses. A network can exist without management mechanisms, although its users might encounter difficulties with congestion, security, routing, etc. Management is oriented towards controlling activities and monitoring network resource use. Simplified, the basic duties of network management are: obtain information from the network, treat these data for possible diagnosis and execute solutions to the problems. To reach these objectives, the <u>management functions</u> should be contained in diverse components of the network, allowing for the discovery of, prevention of and reaction to problems [WES 91, WES96].

To solve the problems associated with network management, the ISO, by OSI/NM, proposes three models: the Organizational Model, which establishes a hierarchy among management systems within one management environment, dividing the environment to be managed into various domains; the Informational Model, which defines the objects of management, their interrelationships and the operations made upon these objects. A MIB is needed to store managed objects; the Functional Model, which describes the functions - error, configuration, performance, account and security - of management. In this way, the concept of network management provides the administrator with sufficient means by which to distribute the network's resources to its users, while, due to the quantity of information available, allowing for proactive management.

## 2.1    Proactive Management

As previously stated, the concept of proactive management involves the anticipation of possible problems which may occur in a computer network and their detection before they occur, and not merely reporting their existence. It is fundamental that abnormal network operation be observed, that symptoms be collected and that larger problems which may come to occur be diagnosed correctly, or that anomalies be registered when it is not possible to collect enough evidence which associates an event with a known problem. It is also necessary to maintain constant observation of the network, so that based on this knowledge enough data can be collected in order to identify what might be a symptom and relate it to a known problem. In an ideal situation, the LAN management tools establish a framework within which devices like smart hubs can monitor network activity and place information at distributed management platforms which automatically will generate trouble ticket, operational costs and usage reports [JAN 93].

These same platforms should provide services through the application of management with the capacity of configuration and planning. In Proactive Management, it is necessary to have a comparison of many management tools so that, when data is computed, a report can be created which indicates the cause to be inspected. The majority of these tools work with thresholds which establish the limits where the reporting of events like number of errors, specific types of packets and other parameters regarding the selection of intervals is to begin. It is also important to add related databases which makes for easier access and more versatile use of management data, making the integration of other tools possible.

Through the measurement of normal activity within a given period of time and the identification of performance based on statistical calculations, it is possible to establish a body of data with normal function parameters which we call the baseline. This baseline can be used in Proactive Management by a set of functions to establish a valid statistic which characterizes the normal operation of the network during a new period of time for a specific interval, assessing the levels of traffic at different times on different days [JAN 93].

## 2.2    Artificial Intelligence and Proactive Management

As we have seen earlier, network management is a complex job, in which we find support for other areas of applications, among which Artificial Intelligence and expert systems are especially prominent. The principal management products already use these to facilitate management functions. Artificial Intelligence can be used to anticipate problems that would leave the network inoperative. In this way, adopting Artificial Intelligence can contribute to Proactive Management. System monitoring can be used to project the network's performance, comparing data taken with a baseline which takes into account the proper choice of corrective action. The job of interpreting and diagnosing a network malfunction is a strong point in expert systems, which can rely on inferences regarding the collected data. Design and planning of new installations can be performed with a good knowledge based installed.

The use of expert systems has recently increased, principally in those areas where complex functions are performed but where few specialists exist. Computer networks fit this context due to their wide dispersion and growing use. Services offered by a network become indispensable; many people depend on them so much that an investment in their management becomes viable. A quick cost-benefit survey shows the following advantages: better quality of service: with the dissemination of the specialist throughout all segments of the network. The administrator's job is facilitated, resulting in better performance; greater agility, lower costs and greater productivity in the execution of services permitted by automation; higher reliability, with decreased decision-making time; training support for improved human resources preparation.

A expert system is divided into four distinct phases: acquisition of knowledge; knowledge base; inference machine; explanatory interface. The acquisition of knowledge is a phase involving extraction and formulation of knowledge from an expert for use in a expert system. In this process, work is performed with "knowledge engineers", technicians specialized in the job of helping experts put their knowledge into the expert system using practical rules and knowledge structuring. As the expert puts forth his or her knowledge, the knowledge engineer represents it as a set of heuristic rules which, when coded, drive the process by a mass of information, making the process more efficient. Thus, obtaining these rules is an important step in the acquisition of knowledge.

The knowledge base stores the knowledge of the expert and differentiates from a conventional database in that it is active in nature, permitting updates conforming to the context. The structure of the knowledge base will depend on the type of knowledge represented. To have deductive knowledge, the base will usually be composed of rules. To have modeling of physical structures, causal links or interrelationship between models, the ideal structure may be a semantic network.

The inference machine selects and applies the appropriate rule during each step in the expert system, manipulating the knowledge base. The inference machine can base itself on premises or elementary bits of information, and tries to achieve its objective through a combination of the two. In this case, it is said that it finds its way forward. The machine can also base itself on an objective and verify the needed premises using the facts involved and arrive at a conclusion. In this case, it is said that it works backwards. Inference machines that use a mixture of these approaches are those which are most successful, since, in most cases, the choices made in the inference process are reproductions of the processes a human would be likely to employ.

In a expert system, the knowledge of the problem's domain is organized separately from the other system knowledge, such as the procedures or steps for problem solving, or interaction with the user represented by the explanatory interface, which defines how to present the knowledge. This division is intentional because these systems divide themselves according to knowledge base (the store of specialized knowledge) and inference machine (which unites the procedures for fixing problems, or steps for solving them). The combination forms what is called a knowledge-based system. The base contains facts and rules, and the inference machine decides how to apply these rules and in which order so as to obtain new knowledge. Once the specialized knowledge is separated, it becomes easier for the designer to manipulate procedures [ROC 94a].

## 3 Implementing a Minimum Proactive Prototype

In the following section, the implementation of a minimum proactive prototype is presented for the purposes of making a practical validation of the work. Considering what has been stated until now, the objective of proactive management is to develop a proactive approach in one of the functions of the functional management model, with performance management being studied as a consequence of the desire to foresee drops in network performance and at the same time anticipate the events which may come to clog the network or result in greater damage. It should not only advise of the occurrence, but also take measure to avoid letting such problems expand and become critical.

To reach this goal, a method was idealized where the process remains active, periodically monitoring remote systems and analyzing the results along with other information taken from the baseline, where the manager receives traps from agents, and which plays a role in the diagnosis of network problems, taking action according to threshold levels or merely communicating a fact which has occurred. In short, proactive operation. [ROC 94a]

### 3.1 Proactive Management Applied to a Performance Analysis of a Network Gateway Machine

During this article, individual details of the presently achieved workflow will be presented, showing the adopted strategy for its elaboration, but abstracting specific details such as the SunNet Manager interface and the method of developing an agent for the management environment [SUN 89]. The task of monitoring and observing the network is best achieved by agents with resident action in the machines to be managed. In view of this, Agent 6 was created with the intent of measuring quality and operation in the network communication services by monitoring the volume of traffic and verifying the number of errors. These characteristics will later be used as a database for establishing a baseline.

Having these data, the following workflow was determined: run agent 6, Hostmem, Hostif and Hostperf in some sub-network nodes, measuring congestion and other statistics in different hours; based on the results, establish a baseline or base with known points as a measure of standard deviation for measurements in normal situations; implant the diagnostic model, so as to activate rules each time the measurement taken is beyond the standard parameters contained in the baseline, stipulated as: for each measurement taken from the monitoring agents that arrives within reach of the timed measurement in which it occurred, a diagnostic module will be triggered to verify whether of not problems exist according to the module's rules; if the diagnostic module verifies the a problem exists which could result in a performance drop or congestion, rules will be used to determine the motives for the event; in a third moment, after verifying the previous, measures are taken to avoid the problem, reporting the anomalies found to the network administrator and suggesting corrective measures.

## 3.2    Environment of the Experiment and Strategies Used

In this experiment, the local network at CESUP (the National Center for Supercomputing) was used, which represents a highly heterogeneous network. At CESUP, IBM-PC compatibles, Apple Macintosh computers, Silicon Graphics workstations, Suns and a Cray YMP2E supercomputer are employed, and more specifically, the SUN workstation sub-network.

The Macintosh computers communicate to each other using the AppleTalk protocol. The physical connection is achieved via a hub, which is the star configuration in this sub-network. The AppleTalk sub-network connection with the CESUP local network is by a gateway.

The 143.54.22 sub-network is an Ethernet segment, and there are found the Silicon Graphics and Sun stations. Another Ethernet segment is sub-network 143.54.1, which is the backbone of UFRGS, where the Tchepoa router (port of entry to the Tche network), Vortex (name server of the university network) and Routcc (router for the Downtown Campus). Remote access to CESUP, for users from other institutions, is done via Tchepoa and Vortex. Access by UFRGS users is via Routcc.

Uniting these two sub-networks, we find the Darwin station, which plays a key role in the CESUP local network. This machine is the only access node to the Gauss supercomputer (CRAY YMP2E) for both local and remote users. Connection to Darwin and the Gauss is via proprietary CRAY protocol. Darwin also serves the local network archives, e-mail, names (DNS) and yellow pages (NIS).

## 3.3    Agent 6

To obtain a practical validation of this experiment, the problems of performance and congestion were also chosen, since the results in those areas would serve as feedback for other studies. The objective was to find a mid-region for performance drop at a given workstation and when levels near this region were reached, send a message to the network administrator alerting the beginning of a drop in performance and congestion. The term congestion is defined here as the point where many packets are present in a branch of the sub-network which cause a degradation in performance, i.e. when the number of packages sent does not correspond with the number of packets distributed. Congestion differs from traffic volume primarily in the distribution of transmissions. If the volume is high, but is being managed by just one machine, then the network congestion, from the point of view of that machine, is not considered to be high.

As cited in [ROC 94], the average used for measuring network congestion is based on a comparison of the number of attempts made at transmission by the interface of one host of an Ethernet channel with the number of collisions which occurred in this transmission. If the percentage of collisions is high, this indicates that the network is congested, for there is little free time without someone transmitting. This measurement obviously depends on the machine in question.

If the machine in which the agent is running is generating high traffic and the rest of the machines on the network are generating almost nothing, then the number of collisions will be low, not reflecting the true volume of transmission achieved. The true volume of traffic can only be measured by counting all of the packets that pass through the network. Meanwhile, this verified difference will not adversely affect the average desired congestion, since if the station at which the agent is running is transmitting too much, it does not perceive congestion. And for the rest of the stations, there will only be congestion when they try to transmit more, which will also be reflected at the agent's station by an increase in the number of collisions. To obtain a percentage of collisions, the relation between the number of collisions and the number of attempts at transmission which occurred during the interval between the present time and the most recent consultation [ROC 94].

Agent 6 was installed in the Darwin workstation, which is the server for the local network and the gateway for CESUP, being monitored the ie0-bus interfaces of the UFRGS network and the ie1-bus of the CESUP local network. Agent 6 operated in two basic modes supported by the SunNet Manager (SNM) platform: data monitoring and event monitoring.

For data monitoring, the agent remained in continuous execution, making a consultation at each time interval. At the end of each consultation, the data recovered was sent to the SNM manager. This periodic sending of data permits the generation of graphs (see figure 1) by the SNM platform, as well as the storage of recovered data from previous consultation by the same agent, which can then generate statistics about the status of the communication system. Since the SNM platform supports the automatic generation of events like "send mail" and permits the specification of various types of thresholds, the implementation of this kind of operation by Agent 6 brought no burden to its development. To the contrary, this facility was used in sending alerts to the network administrator.

Simple monitoring of data at each small interval in time does not permit easy conclusions as to whether or not the network is seriously congested or not, since flurries of transmission happen normally. For example, if it were possible to obtain the proposed average at each attempt at transmission, the data recovered would always indicate 0 or 100% congestion, which would be difficult for one to interpret. This situation is aggravated in the event monitoring mode, since it does not interest the network administrator to receive an event informing that the rate of collisions was, for example, 30% during the last second. On the other hand, if a measure of congestion is obtained during the entire life of the system, the average would be quite low and would not adequately reflect the problems which had occurred during determined periods.

In order to allow for more intelligent control of a congestion situation on the network, it is necessary for the agent to give an average reading over the last N time intervals. The choice of this number N should be based on modeling the system's situation adequately in relation to real time and according to the administrative needs. From this average, we can then have a wider vantage point and truly manage useful events, informing that network congestion is on the rise and is surpassing the limits imposed for correct operation.

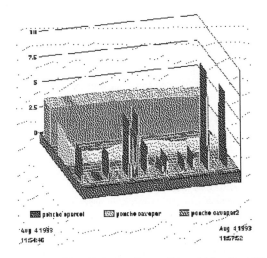

Figure 1: Graphic with the Attributes of Agent 6 in a Real Situation

The implemented agent possesses unique table called "output". In this table, all the parameters corresponding to the table of the like-named etherif and three more attributes specific to Agent 6 are recovered. These attributes are "opercol" -- percent of collisions at each basic interval; "oaveper" -- average percent over the last 60 intervals; and "oaveper2" -- average percent over the last 900 intervals.

The calculation of the averages is updated at each interval. This calculation does not consume much CPU time because the algorithm used maintains the sum of the last 900 and 60 intervals in memory, allowing it to merely subtract the oldest average, add the newest, and make two divisions (60 and 900) to obtain the latest averages. The calculation of the average at each interval (each consultation) is performed by dividing the difference in collisions by the number of attempts at transmission in a period. This number of attempts at transmission is calculated as the sum of the difference of attempts at retransmission and the difference of packets transmitted successfully. The average calculated over 60 basic intervals exhibits successive elevated peaks which appear as reasonably large increases in transmission congestion, indicating that the situation is critical in that period. The average calculated over 900 basic intervals exhibits low sensitivity to alterations caused by any application requiring large transfers. To the contrary, its high stability better reflects the state of utilization of the network in general, for all users and during a longer period of time (above 15 minutes).

## 3.4      Building a Baseline

This station was continuously monitored by Agent 6 for 24 hour periods, taking measurements at 10 and 15 minute intervals, seeking the best interval of measurements per hour, with a total of 4 to 6 measurements per hour for a daily total of 96 to 144 measurements. After making the measurements, evaluations were made regarding the progress and profile of the traffic measured during this period and the objects which best contributed to the establishment of reference levels of operation were selected, respectively, for Hostperf, Hostmem, Hostif and Agent 6. These are: Hostperf      => cpu - percent of CPU utilization; ipkts - number of incoming packets; opkts - number of outgoing packets; ierrs - number of incoming errors; oerrs - number of outgoing errors; colls - number of collisions; Hostmem => mbuf - percent of buffers used; memused - bytes used by the network; memfree - bytes free to the network; mem - percent of bytes used by the network; Hostif =>  ipkts, opkts, ierrs, oerrs and colls - same meaning for ie0 and ie1 interface; Agent 6 => oaveper, opercol, oaveper2 - mentioned in the previous section; odrops - cumulative number of packets in the output queue; obuff - current number of buffers transmitting;

Each agent collects data from the gateway machine and generates ASCII files which contain the values encountered in each parameter that they must measure. The objects selected are extracted from these archives via a "parser" program which is specific for each agent, written in C language, which collects them in an intermediate file, calculating the average and the standard deviation of the values of the object for each hour. This intermediate file is passed by another general parser program, which also calculates the average and the standard deviation of the values of each object, uniting each monitoring day within its respective weekday, thus obtaining the average standard deviation for each hour of each day throughout the period monitored. With this baseline, we have an average standard operation for the network for each hour of each day of the week.

## 3.5      The Diagnostic Module

The diagnostic module was elaborated with all the characteristics of a expert system, utilizing the four phases of this type of system which are presented here. Methodologies proposed by [LEA 95] for the construction of knowledge trees were employed in the process of knowledge acquisition. To complement this, a bibliographic review of available material was performed and soon after the first attempts at modeling the system (represented using a Generic Semantic Model, or GSM), were made. In the figure 2, the final modeling of the system is presented.

The knowledge tree construction methodology, also proposed by [LEA 95], associates diagnostics with their determining factors, which are proposed according to importance. By this process, a set of trees was built which represents the knowledge base for the system. From these trees the rules composing the rule base of the diagnostic module were derived, and it was identified that the diagnostics are realized on four level. In Figure 3, three of them can be seen: the lowest is the parameter level, where the state of each parameter is identified, being that it is evaluated as a function of its value as well as of the average and the corresponding deviation at present days and hours; the partial diagnostic level contains diagnostics which are made through parameter analysis, though they are not important for the user, and; the upper level of final diagnostics, which are presented to the user. Also, at the level of the tree appearing in the diagram, the suggestion level was applied, where, as a function of the final diagnostics, suggestions are made to the network administrator.

Knowledge is represented by facts and rules. The facts are generated from the monitoring and baseline files. The implementation of this system was achieved in Prolog. The files that arrive via the system have their format converted to the Prolog fact form. The facts have the following format: agent (name-attribute, value); previous (name-attribute, value); agent (interface, name-attribute, value); previous (interface, name-attribute, value).

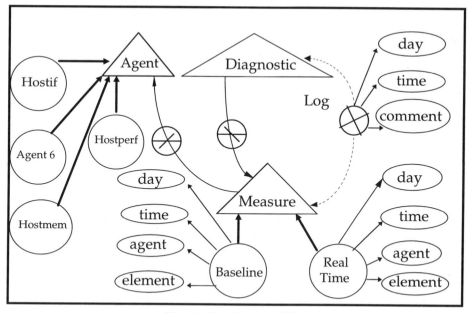

Figure 2: Modeling using GSM

The hostif agents and Agent 6 depend on the interfaces of each machine, and for this reason they also save this value. The previous value of each parameter should be saved in the case of cumulative measurements, and only the difference between the previous and present ones are important.

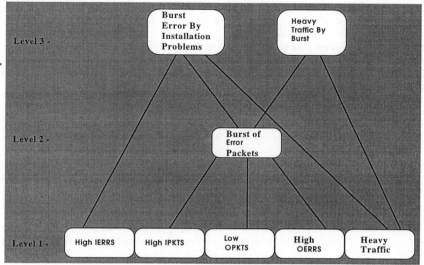

Figure 3: Partial Representation of the Knowledge Tree

As the prototype was implemented in Prolog, operators were used for defining a syntax for rules which represent the knowledge of domain. Having defined these operators, the rules can be written in a more accessible and adequate syntax by the specialist, and included in the interpreter to facilitate the implementation of the inference machine: theory "Teoria" (Theory); rule "N" (N); if "ListaDeCondições" (ListOfConditions); then "Consequência" (Consequence); because "Explanação (Explanation). Where "Teoria" - identify the ruleset. The inference machine can be triggered for just one set of rules, instead of surveying the entire rule base; "ListaDeCondições" - a conjunctive list, in Prolog format, of conditions; "Consequência"- routine Prolog call; "Explanação" - translation of the rule in natural language, used as an option for explanation.

```
teory final
rule 6 :
if
      [parametros('CPU high'),
      parametros('OPERCOL high'),
      parametros('OAVEPER high'),
      parametros('OAVEPER2 high'),
      parametros('IPKTS high'),
      parametros('OERRS low')]
then
      goal(final('Low performance'))
because
'Was detect use of CPU over normal and parameters OAVEPER e OAVEPER2 greater than
zero, with more packets in and out of machine grather than normal. This machine is
loaded because have high number of colisions and rate of error'.
```

Figure 4: Low Performance Rule

Figure 4 illustrates an example of a rule used in the system, with the above-mentioned syntax: The first set of rules serves as a parameter which identifies situations which are abnormal with respect to the parameters given by the agents. Therefore, the value obtained in the monitoring file generated by the agents is compared to the average baseline value, taking into account that the baseline represents a "normal" network standard, per hour and day of the week. The second set is called partial and detects alert situations that should be investigated by the system, but not presented to the user.

The final set takes into account the parameter and final diagnostics to identify situations which should be presented to the network administrator. The rules in the suggestion set take into account the final diagnostics and supply suggestions to the administrator for avoidance or elimination of the problem. Mapping between knowledge tree levels and the rule sets is direct. The organization of the rules base in "rulesets" makes it easier to implement the inference machine, as well as making the search more efficient. The inference machine was implemented upon the resolution mechanism of Prolog itself, upon which the prototype was implemented. Given a set of rules, the mechanism takes each rule of that set and tests its premises, backtracking. The Prolog algorithm is showed in figure 5.

The algorithm treats a rule, with or without an explanation (because), tests the conditions set represented by a list, executes the rule action, marks the region triggered [goal(marca-trace(Teoria, N)]. The register of rules that have been triggered is later used by the explanation mechanism. After a rule is tested, a failure is forced so that, by backtracking, the rest of the regions can be tested. In the prototype, this inference mechanism is invoked for each rule set, sequentially. This highly procedural strategy makes inference easier to implement.

The explanatory interface seeks the diagnostics at the top of the tree that have been detected, sets up the path that led to each diagnosis and a text using the "because" sentences from each region triggered on the path. The following is a representation of this algorithm: assume T as the theory set in the rule base; "t" is the top theory in the theory hierarchy T; identify the set R of rules which pertain to the theory t which have been triggered; for each rule r of R; 1. set up the path C of rules triggered that led to trigger of r; 2. follow the path C, from top to base of knowledge tree, treating "because" sentences.

```
diag(Teoria):-
      (teory Teoria rule N : if LCond then Cons;
      teory Teoria rule N : if LCond then Cons because _),
      testa_cond(LCond),
      call(Cons),
      goal(marca_trace(Teoria,N)),
      fail.
diag(_).

testa_cond([]).
testa_cond([C|R]):-
      call(C),
      testa_cond(R).
```

Figure 5: Prolog Algorithm for Inference Machine

The explanation mechanism sets up the entire explanation of all paths at once, responding because it was given a suggestion by the proactive management system, because each final diagnosis was given, and so on. Thus, the model has conversion blocks; knowledge base; inference engine; explanation; interface.

The conversion block is responsible for receiving the monitoring and baseline files and converting them to the Prolog fact format. These facts will form the knowledge base. Based on the knowledge base, the inference engine is used for analysis of the values of the parameters and for inferring a diagnosis. This diagnosis is supported by an explanation which indicates the motive(s) for the problem, as well as suggesting possible resolutions for it. From the interface, the network administrator receives information from the system as well as the suggestions and has the possibility of expressing an opinion, agreeing or not with the diagnosis given, Beyond this, it should, in whatever situation, describe which of the approaches were followed in trying to solve the problem.


## 4    Results

For security reasons, the present prototype runs on an IBM/PC compatible microcomputer connected to the network, and Arity Prolog was implemented. For testing, the monitoring files generated by the Darwin machine agents had to be copied directly to the prototype directory, being that, during this test period, only one type of problem was verified.

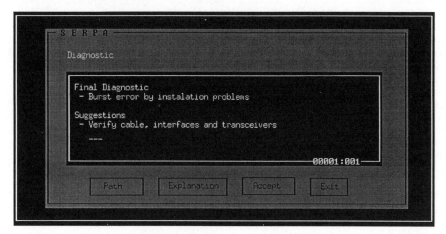

Figure 6: Diagnostic Demonstration Screen

Once executed, the prototype automatically converted the monitoring data and consulted the baseline statistics, using the hour and date of the system as a base, it informed the user of the final diagnosis and of suggested actions to be taken by the network administrator, as shoed in figure 6.

From here, it was possible to request an explanation the system interface is comprised of: Central Window, which is used to show diagnostics; Path Option, a simplified explanation, showing which rules were triggered, but without assembling them; Explanation Option, explains the rules used by the system; Accept Option, registers the file in the log; Exit Option, to leave the system.

If the explanation has a number of lines above that available on the monitor, exceeding the window's size, as happened in the example, it can be scrolled down.

The Accept Option registers those suggestions which have been accepted by the administrator in a log file; if he or she leaves the system without accepting any, this fact is also registered in the log file. The purpose of the log is to validate the system itself, making it possible to later investigate the situations in which the administrator did not accept the suggestions, so as to create a record which can be used by the inference machine to construct diagnostics. The automation of this process is very important, because the daily work routine of the network administrator normally does not have time to manually procure a log file, principally due to the size of these files.

The prototype proved its utility, since the symptom "flurry of error packets" enabled the administrator to take rapid measures, due to the fact that notification of the event was received before the event was completed.

## 5    Conclusion

With the objective of validating the use of proactive management in computer networks, this study proposed a strategy for the implementation of proactive management using artificial intelligence, more specifically, systems based on knowledge. The area of network management chosen for this experiment was that of operation management, analyzing the problem of performance drops in a "gateway" machine (network server), observing how its operation influenced the network.

Theoretical and practical activities were performed to verify that agents were indeed viable for network monitoring and, with this data in hand, it was possible to construct a database containing the normal state of the network. Thus, it was possible to verify the operation of the "gateway" machine (network server). To this end, the following steps were performed: Agent 6 was developed to be responsible for monitoring the network and analyzing the performance of the Ethernet channel; "parser" programs were developed in C language to place monitoring files in the baseline file component format; a baseline was assembled with the Agent 6 results plus the three proprietary Hostmem, Hostperf and Hostif agents; a diagnostic module was elaborated with characteristics of a expert system to inquire as to the existence of performance problems, and if confirmed, establish the motives which caused the drop in performance. Once executed, the prototype automatically converted monitoring data and consulted the statistical information of the baseline, taking the system date and time as a base, informed the user of the final diagnosis and offered suggested actions to be taken by the network administrator.

The prototype proved its usefulness, in that the "flurry of error packets" diagnosis, given its subtle characteristics, would otherwise demand much time to detect, while using the tool enabled the administrator take rapid measures, due to the fact that notification of the event was received before the event was completed. It would be interesting to focus on this work and more fully research ways to amplify the diagnostic module and perfect this prototype. In addition, we can cite the following items to be improved: to implement automatic interconnection of the diagnostic module and the monitoring files; to improve explanation interface to emit suggestions to users by e-mail automatically, and; to implement other areas of network management, such as configurations and failures.

In the way of new tendencies, we can cite the following items which we believe to be a natural sequence of this work: to extend the work to other technologies like FDDI, ATM, etc.; to develop the diagnostic module with a coexisting system; to use the diagnostic module on natural networks; to develop proactive management utilizing a combination of artificial intelligence with simulation.

The problem of network management is far more complex and requires careful treatment in order not to create new problems when trying to solve others. This holds even more true for performance management, traditionally treated separately. In proactive management, every care should be taken to anticipate problem occurrences and to not induce administrators to seek solutions for virtual situations. This article, besides presenting an implementing strategy, showed results from the intended model which served to demonstrate that the operation can be attained. The experiment demonstrated that CESUP possesses a well-constructed network, with stable operation and performance, which did not allow us a greater number of problems to be inspected, though the problem which did occur was detected and reported to the administrator, confirming proactive management's capabilities. Its operation and principals can be used as a base for new works to be performed on the proposed model.

## REFERENCES

[COM 91]   COMER, D. E.  Internetworking with TCP/IP: Principles, Protocols, and Architeture. Volume 1. Seg. Edição. Prentice-Hall, Englewood Cliffs, NJ, 1991.

[DEF 96]   DE FRANCESCHI, A.S.M.; KORMANN, L.F.; WESTPHALL, C.B. Performance Evaluation for Proactive Network Management. In: Proceedings of  IEEE/ICC´96 International Conference on Communications, Dallas, Texas, USA, Jun. 23-27, 1996.

[HUL 87]   HULL, R., KING, Roger.  Semantic Database Modeling: Survey Applications and Research Issues. ACM Computing Surveys. v.19, n.3 Setp., 1987.

[JAN 93]   JANDER, Mary.  Proactive LAN Management: Tools that Look for Trouble to Keep LANs Out of Danger. Data Communications. Mar., 1993.

[LEA 95]   LEÃO, Beatriz. Class notes from COMPO2 course CPGCC/UFRGS. Porto Alegre, 1995.

[POS 81]   POSTEL, J.B.  Internet Protocol. Request for Comments 791, DDN Network Information Center, SRI International, Setembro, 1981, 45pp.

[POS 81a]  POSTEL, J.B.  "Internet Control Message Protocol - DARPA Internet Program Specification". Request for Comments 792, 1981.

[ROC 94a]  ROCHA, M.A. A Strategy for Implementation of Proactive Management in Computer Networks. Research Report, Porto Alegre, UFRGS-CPGCC, Mar. 05, 1994.

[ROC 94]   ROCHA, M.A. Computer Network Management through New Agents. Proceedings of the XII Brazilian Symposium on Computer Networks, p.113-133, Curitiba, PR, Brasil 1994.

[STE 90]   STEVENS, W.R.  UNIX Network Programming.Prentice-Hall Inc. Englewood Cliffs, NJ, 1990.

[SUN 89]   Sun Microsystem Inc. SunNet Manager Tutorial - How Write an Agent, 1989.

[SUN 89a]  Sun Microsystem Inc. Network Programming Guide, 1989.

[SUN 90]   Sun Microsystems Inc. SunOS Reference Manual. Vol I, 1990.

[WES 91]   WESTPHALL, C. B.  Conception et développement de l'architecture d'administration d'un réseau métropolitain. Thèse de Doctorat nouveau régime. Université Paul Sabatier. Toulouse, 16 Juillet 1991.

[WES 96]   WESTPHALL, C. B. & KORMANN, L.F.  Usage of the TMN Concepts for Configuration Management of  ATM Network. International Symposium on Advanced  Imaging and Network Technologies. Germany, Berlin Oct. 7-11, 1996.

## ABOUT THE AUTHORS

**Marco Antonio Rocha**

Obtained his B.Sc. degree in Informatics at PUC - Pontifícia Universidade Católica - Rio Grande do Sul, Brazil, in 1989 and a M.Sc. degree in Computer Science at Federal University of Rio Grande do Sul in 1996. At present, he is lecturer at the University of ULBRA in the south of Brazil and also works at CRT which is the Brazilian Telecom carrier placed in Rio Grande do Sul.


**Carlos Becker Westphall**

Obtained a degree in Eletrical Engineering in 1985 and a M.Sc degree in 1988, both at the Federal University of Rio Grande do Sul, Brazil, and a  Dr. degree in Informatics - Network Management - at the Université Paul Sebatier, France, in 1991. Presently, he is a Professor in the Department of Computer Science at the Federal University of  Santa Catarina - Brazil, where he acts as the leader of the Network and Management Laboratory and also coordinates the multi-institutional PLAGERE project (Platforms for Network Management) funded by the Brazilian National Research Council (CNPq).

## PART FOUR

# Information Models

Chair: George Pavlou
University College London

# 46

# Event Modeling with the MODEL Language[*]

D. Ohsie[†]
Columbia University
Department of Computer Science
New York City, NY 10027
ohsie@cs.columbia.edu

A. Mayer[‡], S. Kliger, S. Yemini
System Management Arts (SMARTS),
14 Mamaroneck Ave,
White Plains, NY 10601
alain, kliger, yemini@smarts.com

## Abstract

Event modeling is an essential component of event correlation systems; this paper introduces the MODEL language, which comprises the event modeling component of SMARTS' InCharge™ event correlation system. We demonstrate the features of the MODEL language through examples from the multimedia Quality of Service (QoS) domain. In addition, we provide a comparison of MODEL with the event modeling capabilities of other event correlation systems; we demonstrate that MODEL generalizes the capabilities of other systems and is more flexible.

## Keywords

Event correlation, event modeling, multimedia.

## 1    INTRODUCTION

Network management consists mainly of monitoring, interpreting, and handling of events or exceptional condition in the operation of the network. Event correlation is the process of automatically grouping related events based on their underlying common cause, thereby compressing the event stream and identifying potentially hidden problems. NetFACT (Houck et al. 1995), SINERGIA (Brugnoni et al. 1993), IMPACT (Jakobson and Weissman 1995), ECXpert (Nygate 1995) and the authors' own InCharge™ (formerly DECS) (Yemini et al. 1996) are all examples of such systems. An event correlation system consists of two basic components: an

---

[*]The process for event correlation and problem reporting described in this paper is covered by U.S. Patent No. 5,528,516.

This research was supported in part by Air Force Contract No. F30602-95-C-0262.

[†]This author's research was supported in part by NSF grant IRI-94-13847.

[‡]This authors current affiliation is Bell Labs / Lucent Technologies, 600 Mountain Ave, Murray Hill, NJ 07974, alain@research.bell-labs.com

*event definition and propagation model* (or simply *event model)*, and a *reasoning algorithm.* The event model describes the underlying system, while the reasoning algorithm processes incoming events and correlates based on the knowledge contained in the event propagation model. The event model in turn consist of a *class-level* model, and a run-time *object topology.* The class-level model describes the general rules for propagating events from objects of one class to another, while the object topology describes a particular instantiation of the run time model which reflects the current state of the actual system.

As an example of event modeling, consider the scenario in Figure 1 from the Multimedia Quality of Service (QoS) domain. Here, a video sender, an electronic classroom located on the local area network LAN2, wishes to transmit some live video to a receiver located on LAN1 using the video tool `vic` (McCanne and Jacobson 1995a) which utilizes the UDP transport protocol. The UDP connection transports IP packets through routers D, C, B, and A which connect the LAN domains through a router backbone domain. The router backbone domain uses physical-layer wide-area network (WAN) domains. Similarly, an audio sender, Internet phone, on LAN4 wishes communicate with a receiver on LAN3, using the audio-tool `vat` (McCanne and Jacobson 1995b). Its IP packets are routed via F, C, B, and E. These transmissions, plus other, unrelated traffic cause the rate of packets

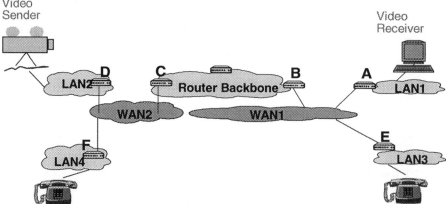

**Figure 1** Multimedia over a multi-domain network.

arriving at C to be too high. Consequently the buffer at C overflows, causing the multimedia transmissions to lose packets. Note that congestion of this nature is the most common cause of packet loss on the Internet. The packet losses at router C will propagate to all UDP connections which router C is a part of. Since UDP does not retransmit lost packets, these losses will in turn propagate to the multimedia transmissions and hence the quality at the receiver may become unacceptable.

The *class-level event model* for the scenario described above consists of the following: a definition of the "poor video quality" event, a rule describing the propagation of router congestion to packet loss and then to poor video quality, and optionally a "high packet loss" event at the router level. The *object* topology consists of the individual routers and multimedia applications and their relationship in the underlying network. A *reasoning algorithm* would infer the presence of the congestion problem based on the poor video and audio quality and the event model illustrated.

In previous work (Kliger et al. 1995), we described the coding approach to event correlation which is the reasoning algorithm of our InCharge™ event correlation system (Yemini et al. 1996). There, we showed how the symptoms of each problem in a modeled system could be treated as a code for that problem, and that elementary techniques from coding theory could be profitably applied to event correlation. That work presupposes that there is a causality graph which maps each problem to its immediate (possibly unobservable) causal effects and in turn to others, until an observable symptom event is caused. Thus, computing the code for a problem involves computing a closure over the causal relationships emanating from the problem.

In this work, we describe the MODEL language for the definition of object and event models. MODEL supplies an object-oriented data model complete with inheritance and overloading. It also provides instrumentation capabilities to automatically tie attributes in the model to SNMP MIB variables. More importantly, MODEL supplies two feature which are essential to event correlation. First, it provides a declarative specification of events in the form of boolean expressions over attributes in the object model. This allows the definition of events to be integrated into the model of the objects in which the event occurs. Second, MODEL allows the user to specify *local* event propagation rules in which we show how to construct the causality graph from the combination of the class-level event propagation model and the current object topology. Often, event propagation patterns depend heavily on the way in which objects are currently interconnected; changing the topology of the modeled objects will drastically alter the observed symptoms of a problem. We will show that MODEL's approach to defining event propagation is superior to the event modeling capabilities of existing systems, because it can handle topology dependent event propagation through the use of event overloading. In addition, MODEL is correlation algorithm independent, so it can actually be substituted for the event modeling subsystems of existing correlation systems to improve their generality and ease of use.

The rest of this paper is organized as follows. In Section 2, we will use scenarios from the multimedia QoS domain among others to provide a "description by example" of the MODEL language. In Section 3, we will outline the process of developing reusable event libraries in MODEL. Section 4 will provide a critical comparison of the MODEL language with the event modeling capabilities of other event correlation system.

## 2    THE    MODEL    LANGUAGE    AND    QOS MANAGEMENT

In this section, we give an in-depth "introduction by example" to the MODEL language. We begin with an example from the multimedia management domain. Consider again the example configuration of Figure 1 and the following scenario: Due to high traffic volume, router C experiences congestion. As a consequence, its buffers overflow and incoming IP packets get lost. Since the video and the audio receiver are endpoints of a UDP connection which is layered over router C, they both experience the same type of QoS violation: an average transmission rate that is drastically below tolerance. A correlator, using the knowledge provided by the corresponding model, should report a high probability that the problem causing these violations is located in the domain of the router backbone.

We will begin by considering a simple example of a causal relationship, that of congestion causing lost packets. First we must define what we mean by "high packet discards". Let us assume that the router implements the IP protocol and is instrumented via SNMP. We can then measure the total number of discarded packets by querying the SNMP MIB-II variables ipOutDiscards and ipInDiscards:

```
interface IPRouter: IP
{
    instrumented attribute long ipInDiscards;
    instrumented attribute long ipOutDiscards;
    attribute long discardsThreshold;

    event PacketDiscardsHigh "The level of discarded packets is high" =
    (delta    ipInDiscards    +    delta    ipOutDiscards)    /    delta    _time    >
    discardsThreshold;

    instrument SNMP;
}
```

In this example, the `attribute` statements defines measurable properties of the IP protocol entity. The `event` statement defines the circumstance under which the event can be said to have occurred. In this case, the event `PacketDiscardsHigh` will be deemed to have occurred whenever the sum of the changes `ipInDiscards` and `ipOutDiscards` per time exceeds a threshold. The `delta` keyword indicates that the difference between the new and old values of the attribute are desired. The `_time` keyword refers to the time at which samples are taken. Thus this event is triggered when the discard rate reaches the threshold.

Here we digress for a moment to reflect on the relationship between MODEL and SNMP. The ipInDiscards and ipOutDiscards attributes are automatically instrumented via SNMP; no additional programming is required to keep these attributes updated with current values. In addition, a utility program called *mib2model* can be used to parse SMI MIB definitions and generate the corresponding MODEL classes automatically. Thus all features of the MODEL language essentially extend the functionality of the underlying SNMP MIBS. This approach meshes well with the SNMP philosophy; the underlying device must implement only the simple SNMP protocol and can thus concentrate its resources on its task (in our example, routing packets). The event management system provides higher level services using dedicated management resources. MODEL enables the event modeler to ignore this distinction and concentrate on simply modeling the events without regard to who supplies the information. In our example, we have effectively extended the power of the standard SNMP MIB to include our newly defined event.

Now, let us return to modeling the congestion problem at the router. We want to express the fact that there is a causal relationship between the congestion problem and the high packet discard event (with probability 1.0):

```
problem Congestion "High congestion" = PacketDiscardsHigh 1.0;
```

This line would be added to the MODEL class definition above. Note that this is a semantic declaration in the form of a rule; however, it does not have any specific algorithmic or operational meaning. It simply expresses the fact that there is a causal relationship between these two events. The inclusion of the problem and symptom in the scope of a single class obviates the need to write the rule as follows:

```
Congestion(IPRouter(X)) -> PacketDiscardsHigh(IPRouter(X));
```

We have modeled a local symptom which indicates the problem of Congestion. However, we would also like to relate the problem to the other observed symptoms at the multimedia application level. In this way, anomalies observed at the multimedia level can be correlated with the problem detected at the lower level.

Problems in one object propagate to related objects via *relationships*. In our example, the Congestion problem would propagate to higher level connections which are layered over the congested IP node. Thus we would add the following statement to indicate the relationship between IP nodes and connections:

```
relationshipset Underlying, TransportConn, LayeredOver;
```

The keyword `relationshipset` indicates that many connections may be layered over a single IP node. Now, we would like to express the fact that the congestion problem causes both the local symptom PacketDiscardsHigh, and propagates those discards as losses in the higher level connection:

```
problem Congestion "High congestion" =
        PacketDiscardsHigh 1.0, ConnectionPacketLossHigh 0.8;

propagate symptom ConnectionPacketLossHigh =
        TransportConn, Underlying, PacketLossHigh;
```

Note that we have added the symptom ConnectionPacketLossHigh to Congestion problem and that we have used a causal probability of 0.8, where a value of 1.0 indicates complete certainty. This indicates that congestion at the IP node may not cause packet losses on all connections above it, depending on the circumstance surrounding the congestion; we would not want to rule out congestion simply because a single connection which is layered over the node is not experiencing problems.

The `propagate symptom` statement says that the symptom ConnectionPacketLossHigh refers to an event in a related object, namely the event PacketLossHigh in any TransportConn which layered over this IP node. Now, we will continue the example by presenting the MODEL code which further propagates the problem to its observable symptom in the multimedia layer:

```
interface TransportConn
{
    propagate symptom PacketLossHigh =
               Port, ConnectedTo, PacketLossHigh;
}

interface UDPPort: Port
{
    propagate symptom PacketLossHigh =
               Appl, Underlying, PacketLossHigh;
}

interface MM_InPort: Appl
{
  instrumented attribute long MinRate;
  instrumented attribute long MaxRate;
  instrumented attribute long MsgCounter;
  instrumented attribute long ActTime;

  computed attribute ActualRate = (MsgCounter)/(_time - ActTime);

  event BadRate = (MinRate > ActualRate) || (ActualRate > MaxRate);

  problem PacketLossHigh = BadRate 1.0;
}
```

Note that a TransportConn simply propagates the packet losses to the Ports to which it is connected; a UDPPort (which, being a subclass of Port, inherits from Ports) in turn propagates the packet losses to Applications which are LayeredOver the port. For simplicity, the relationships which are utilized for this propagation, ConnectedTo and Underlying, are not defined here. Typically they would be inherited from generic link and node classes in the Netmate hierarchy, which is described in Section 3.

The multimedia receive port, MM_InPort, is a subclass of Appl. Therefore, it receives, via inheritance, the PacketLossHigh symptom from the UDP_Port which it is LayeredOver. The PacketLossHigh event in the MM_InPort has a single locally defined symptom, thus we again utilize the *problem* statement to define its symptom. In this case, PacketLossHigh causes the observable symptom BadRate, which indicates the reception rate is out of tolerance. Since this symptom is observable, it is defined using the *event* statement and an expression to detect the symptom. This example also demonstrates the use of expressions to define attributes as shown in the definition of the attribute ActualRate.

The combination of the propagate symptom statement and one-to-many relationships allow the MODEL language to express complex problem-symptom relationships in a compact form. For example, suppose that there were many multimedia connections layered over the same congested router (possible causing the congestion). In this case, there will be many UDP connections (subclass of TransportConn) layered over the single IP object. The congestion problem may cause symptoms in any or all of the connections which are layered over the IP object.

Now consider trying to write a single rule to express the relationship between the Congestion problem and its symptoms. First, we would have to include complex conditions to identify which multimedia receivers were related to which IP nodes. The MODEL approach of expressing propagation over existing relationships of the object model provides the proper level of abstraction by separating the causal knowledge from the knowledge of the network topology. In addition, by chaining objects together, MODEL can express propagation paths of arbitrary length with ease, while a single rule would require increasing complexity as the propagation paths lengthened.

In addition, the rule language would have to provide some type of *for all* construct, or else there would have to be multiple versions of the rule, one for each possible configuration of multimedia connections over the IP nodes. By breaking the propagation knowledge into discrete units of propagation from a single object to a related object, different topologies at run-time can be handled with a single model. Note however that the main advantage of the rule based paradigm is retained; causal knowledge is expressed in a declarative fashion, independent of the inference engine which uses it.

Up to now, we have focused on multimedia modeling; however, we have been careful to use classes which are not multimedia-specific wherever possible (e.g., IPRouter, TransportConn). This enables us to reuse the invested modeling effort for other applications. To illustrate how MODEL provides for such *modularity of modeling*, we show how to extend our model to a database client domain. This domain will exhibit an entirely different set of symptoms as a result of the congestion at the router (which is a problem *common* to both domains). MODEL allows us to utilize the existing model, and to extend it by adding subclasses and overloading the event propagation in these subclasses to match the behavior of the newly modeled objects.

Database applications typically utilize TCP connections to access database servers. Since TCP connections are reliable, they must retransmit packets which are discarded by underlying IP nodes. Thus the symptom propagation pattern for TCP clients differs somewhat from that of UDP clients. We will use the event overloading capabilities of MODEL to express this difference:

```
interface TCPPort: Port
{
    problem PacketLossHigh =
                ApplicationDelay 1.0, TCPRetransmissionsHigh 1.0;

    propagate symptom ApplicationDelay = Appl, LayeredOver, Delay;
    propagate symptom TCPRetransmissionsHigh =
                TCPConn, PartOf, RetransmissionsHigh;
}

interface TCPConn: TransportConn
{
    readonly intrumented attribute long tcpRetransSegs
                "The total number of segments retransmitted - that \n"
                "is, the number of TCP segments transmitted \n"
                "containing one or more previously transmitted \n"
                "octets.";

    event RetransmissionsHigh = tcpRetransSegs > Threshold;
}

interface DBClient: Appl
{
    problem Delay = TransactionTimeout 1.0, ServerLongLockHolding 1.0;
    propagate symptom ServerLongLockHolding =
                DBServer, ServedBy, LongLockHolding;

    event TransactionTimeout imported;
}
```

Note that TCPPort is derived from Port, but has a different definition for PacketLossHigh than UDPPort, reflecting the different effect packet loss has on a TCP connection. Specifically, the lost packet symptom eventually propagates to the TCP protocol entity which experiences a high rate of retransmission, while the application layered over the node experiences delays; in contrast, the UDP port propagates the lost packet symptom to the application, since it doesn't perform retransmission.

In the case of database clients, the application delay event is further specialized to cause transaction time-outs and long lock holding periods on the server. Note that the event TransactionTimeout is defined as *imported*. This indicates that the event cannot be detected by querying attributes of the data model. Instead an outside entity is responsible for notifying the event correlator of the occurrence of this event. This give maximum flexibility to the modeler to include events which might otherwise be difficult or impossible to monitor.

The event overloading capability of MODEL allows for the creation of very abstract and powerful models, because at each stage of the propagation, the modeler must only concern himself with the immediate effects of a problem on the higher layer. The details of how this effect manifests itself can then be altered by simply deriving a new subclass and refining the definition of events in the subclass. Thus we can express the general notion that congestion at a node causes losses on connections which are layered over the node without having to specify the exact effects of these losses. Subtyping and refinement allow the modeler to specify these effects differently for TCP and UDP connections.

The MODEL language contains many other features which are beyond the scope of this paper. The interested reader is referred to (System Management Arts 1996b).

## 3    CLASS LIBRARIES IN MODEL

As we have shown, MODEL provides an object oriented modeling framework with inheritance. This makes it ideal for developing extensible class libraries for event modeling. In the examples above, we simply added relationships, attributes and events to the model when needed. In actual MODEL development, we have found that a three stage modeling process works best.

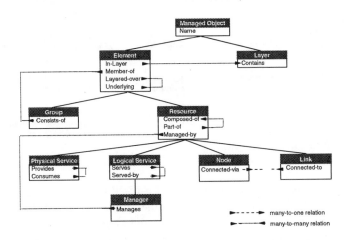

**Figure 2** Netmate class hierarchy

In the first stage, a generic library of networking classes is used to define the basic relationships between objects in any modeled system. This set of classes is called the *Netmate hierarchy* is detailed in (Dupuy et al. 1991) and depicted in Figure 2.

The next stage consists of data modeling. Data modeling involves deriving domain specific classes from the Netmate classes and adding the appropriate attribute and instrumentation statements to produce an accurate data model of the domain. In this stage, the *mib2model* translator described above is used to generate class definitions to represent those objects which are instrumented via SNMP MIB's.

The third stage involves adding the actual event propagation information to the model, either directly into the second stage data model, or into subclasses of this model.    At this stage, it may be necessary to add additional relationships and attributes to the data model, if it is seen that event propagation occurs over relationships that were not contemplated in the Netmate model, or that important events cannot be monitored in the original data model.

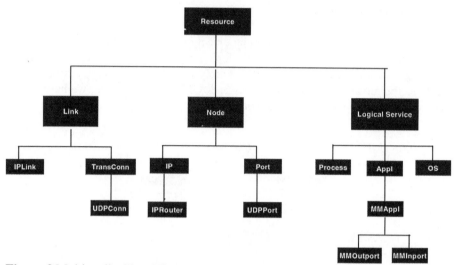

**Figure 3** Multimedia Class Hierarchy

Using this methodology,  we have developed a Multimedia QoS management library.  Figure 3 illustrates the class hierarchy of the Multimedia library.  Note that the "root" node is actually the *resource* class of the Netmate class library. The attributes of classes in the library are instrumented via the QoSMIB (Florissi 1996). QoSMIB provides quality of service metrics which are important to diagnosing problems in the multimedia domain.  Since QosMIB has an SMI specification and can be accessed via SNMP, we utilized the *mib2model* translator to build a number of the classes in our multimedia library.  Consider, for example, the *MM_InPort* class (introduced in section 2) which represents a multimedia receiver.   The *MinRate* attribute of this class represents the minimal transfer rate necessary to support the receiving application; this attribute is retrieved automatically from the QoSMIB.

Examples of other domains for which libraries have been developed include problems in the T1  and T3 connections of telecommunications service providers, TCP/IP data networks and  low earth orbiting (LEO) satellite networks.  These examples illustrate that MODEL provides a basis for developing event libraries for a wide variety of problem domains.

## 4     COMPARISON TO EXISTING SYSTEMS

In this section, we perform a comparison of  MODEL with the event modeling methods of other event correlation systems in the literature.  The NetFACT (Houck et. al. 1995) event model has three classes of object: paths, nodes, and shared resources.  There are three relationships via which events propagate: Nodes and shared resources have "dependencies" on shared resources; nodes and paths are "connected" to one another; and paths are "composed of" underlying nodes and paths.  The NetFACT event model is thus ideally suited for expression in the MODEL language.  We have captured the NetFACT event model in  about forty lines of MODEL code; space limitations preclude its inclusion in this section.

The NetFACT correlation algorithm involves a voting scheme whereby each symptom event counts as a vote for any problem which may have caused it.  This algorithm can be applied to the output of any MODEL language model by simply tracing the propagation backward from symptom to problem.   In addition, a MODEL back-end could generate code to automatically tally up votes for each problem via a method generated for each symptom event.   Thus, MODEL completely generalizes the NetFACT event model, while giving the users flexibility to add their own new classes, relationships and event propagation rules.

SINERGIA  (Brugnoni et. al. 1993) expresses its event model via forward chaining rules which match a particular network topology and use the status of each object in the topology to generate a fault hypothesis for that portion of the network. The generated hypotheses are then fed to a search algorithm which searches for the most likely combination of fault hypotheses.  The MODEL event model differs from that of SINERGIA in that instead of specifying particular network topologies and writing rules for each one, the **propagate** statement is used to express the way in which events propagate generally.  The expected events for a particular topology can then be generated automatically based on the actual  instantiated objects.

SINERGIA's rules closely match the "data sheets" which specify the domain knowledge which is input to the system.  Thus, generating MODEL code for SINERGIA would require an additional level of abstraction to be performed. However, if this conversion can be achieved properly, then the resulting MODEL code is more general than the original SINERGIA rules and could be used to generate fault hypotheses for arbitrary topologies.  In fact, the SPRINTER event simulator (Manione and Montanari 1995) uses a MODEL-like event propagation model to discover missing and improper rules in the SINERGIA rule base.  In addition, writing rules for problems where the events are propagated a very long distance from the problem would seem to be difficult in the SINERGIA methodology, as the size of a SINERGIA rule increases exponentially as the number of components involved.  IMPACT (Jakobson and Weissman 1995) also uses a rule-based approach to define when a correlation rule matches the network topology; thus it stands in the same relation to MODEL as SINERGIA.

ECXpert (Nygate 1995) uses rules to define when an incoming event can be correlated with an event or set of events which were previously received.  Thus, an

ECXpert rule is similar to a MODEL **propagate** statement, in that it specifies the relationships between events, rather an entire topology of events in a single rule. However ECXpert rules are not as well integrated into the object model as MODEL; thus, ECXpert rules involve string matching to determine event type and database lookup to verify that events have been received from related objects. In addition, ECXpert rules are not completely declarative; the user must specify the rules in terms of an incoming "new event" and the existing "old event" in the context of a particular correlation group to support the correlation algorithm, rather simply providing a relationship between events. In addition, all relationships are defined between alarms; there is no way to specify a problem which itself cannot be observed. Finally, ECXpert requires numbering the events with a precedence indicating which level in the correlation tree the event is expected to occur. This requires one to view the correlation tree as a whole instead of simply providing local propagation rules which expand into a correlation tree based on the current network topology.

## 5   CONCLUSION

In this paper, we have introduced the MODEL language and showed its application to event modeling. We have shown that MODEL provides a flexible framework for declaratively expressing event propagation which compares favorably to the modeling capabilities of existing systems. Finally, we have also shown how MODEL can be applied develop reusable event libraries and have outlined such a library for multimedia QoS management.

## 6   REFERENCES

Bolot, J.C., Turletti T., and Wakeman I. (1994) Scalable Feedback Control for Multicast Video Distribution in the Internet. *ACM SIGCOMM 1994.*

Brugnoni S., Bruno G., Manione R., Montariolo E., Paschettra E., and Sisto, L. (1993). An Expert System for Real Time Fault Diagnosis of the Italian Telecommunications Network. In: Hegering, H.-G. and Yemini, Y. (editors). *Third International Symposium on Integrated Network Management,* San Francisco 18-23 April 1993. The Netherlands, North Holland, 617-628.

Busse I., Deffner B., and Schulzrinne, H. (1995) Dynamic QoS Control of Multimedia Applications Based on RTP. *International workshop on high-speed networks and open distributed platforms 1995.*

Dupuy, A., Sengupta, S., Wolfson, O., and Yemini Y. (1991) NetMate: A Network Management Environment. *IEEE Network Magazine.*

Florissi P. (1996). QoSME: QoS Management Environment. Ph.D. Thesis, Columbia University, 1996.

Fry M., Ray, P., Seneviratne, A., and Witana, V. (1996). Multimedia Service Delivery with Guaranteed Quality of Service. *IEEE Network Operations and Management Symposium 1996.*

Houk K., Calo, S., Finkel, A. (1995). Towards a Practical Alarm Correlation System. In: Sethi, A., Raynaud, Y., Faure-Vincent, F. (editors). *Fourth International Symposium on Integrated Network Management,* San Francisco, 1995. London, Chapman & Hall, 226-238.

Jakobson, G. and Weissman, M. (1995). Real-time Telecommunication Network Management: Extending Event Correlation with Temporal Constraints. In: Sethi, A., Raynaud, Y., Faure-Vincent, F. (editors). *Fourth International Symposium on Integrated Network Management,* San Francisco, 1995. London, Chapman & Hall, 290-302.

Kliger, S., Yemini, S., Yemini, Y., Ohsie, D., S. Stolfo (1995) A Coding Approach to Event Correlation. In: Sethi, A., Raynaud, Y., Faure-Vincent, F. (editors). *Fourth International Symposium on Integrated Network Management,* San Francisco, 1995. London, Chapman & Hall, 266-277.

Kumar V. (1996). *MBone, Interactive Multimedia on the Internet.* New Riders.

McCanne, S. and Jacobson V. (1995a). *vic: A Flexible Framework for Packet Video.* ACM Multimedia.

McCanne, S. and Jacobson V. (1995b). *vat: A Visual Audio Tool.* LBL.

Manione, R. and Montanari, F. (1995). Validation and Extension of Fault Management Applications through Environment Simulation. In: Sethi, A., Raynaud, Y., Faure-Vincent, F. (editors). *Fourth International Symposium on Integrated Network Management,* San Francisco, 1995. London, Chapman & Hall, 238-249.

Nygate, Y. (1995). Event correlation using rule and object based techniques. In: Sethi, A., Raynaud, Y., Faure-Vincent, F. (editors). *Fourth International Symposium on Integrated Network Management,* San Francisco, 1995. London, Chapman & Hall, 278-289.

Paxson, V. (1996). End-to-End Routing Behavior in the Internet. In: *ACM SIGCOMM '96.*

Seneviratne, A., Fry, M., Withana, V., Horlait, E. (1994). Quality of Service Management for Distributed Multimedia Applications. In: *IEEE Conference on Computation and Communication 1994.*

System Management Arts. (1996a). *MODEL Language Reference Manual,* White Plains, New York, 1996.

System Management Arts. (1996b). *MODEL Developer's Guide,* White Plains, New York, 1996.

Turletti, T. and Bolot, J.-C. (1994). Issues with multicast distribution in heterogenous packet networks. In: *6th International Workshop on Packet Video.*

Yemini, S., Kliger, S., Mozes, E., Yemini, Y., and Ohsie, D. (1996). High Speed and Robust Event Correlation. *IEEE Communications Magazine,* May 1996.

# 47

# Network management services using a temporal information model

*T. K. Apostolopoulos, V. C. Daskalou*
*Dept. of Informatics, Athens Univ. of Economics and Business*
*76, Patission Street, 104 34 Athens, Greece*
*Tel: +30 1 8203173, Fax:+30 1 8226204,*
*E-mail:{thodoros, dxv}@aueb.gr*

## Abstract

In this paper we address the issue of time as an attribute of the network management information. More precisely, we incorporate the temporal dimension in the management information model proposed by the Internet Engineering Task Force (IETF), as it is described by SMI. The core of the proposed network management information model is the *Temporal Management Information Base (TMIB)*, a conceptual representation of the diachronic behaviour of network resources. We define the architecture as well as the services supporting our view. Finally, we give the exact definition of one key service using ASN.1 formulation.

## Keywords

temporal management information base, temporal network management services, modelling and interpretation of management information

## 1 INTRODUCTION

The today's picture of networking is composed by a very heterogeneous environment supporting multivendor applications upon a variety of underlying switching systems and transmission facilities. The need to control these complex and heterogeneous networks has introduced the concept of network management. A lot of research effort has been given in order to solve problems arising in this area and to establish standards that could be used across a broad spectrum of product types (e.g. hosts, routers, bridges, switches, telecommunication equipment) in a multivendor environment.

The general architecture of a network management system is based on a client-server architecture, where the server is called agent, while the client is the manager. Each network component has an agent which maintains a local Management Information Base (MIB). The MIB is a conceptual representation of the network resources that provides the network manager with the ability to observe and control the current behaviour of network elements. The manager and the agents can communicate through a network management protocol such as the SNMP (Case, 1990) or CMIP (ISO 9596) protocols. The interaction between the manager and the agents admits the retrieval and/or update of the MIB information in a way enabling the implementation of various network management functions.

Time is an attribute of most real world phenomena, and in this paper we address the issue of time as an attribute of the network management information. More precisely, we incorporate the temporal dimension in the management information model proposed by the Internet Engineering Task Force (IETF), as it is described by SMI (Rose, 1990). A similar approach is presented elsewhere (Shvartsman, 1993). The core of the proposed network management information model is the *Temporal Management Information Base (TMIB)*, a conceptual representation of the diachronic (past and current) behaviour of network resources. The temporal network management model and the TMIB design are presented in detail in section 2. The architecture that implements the proposed information model as well as the general services are described in section 3. In section 4 we present an analytical view of one significant temporal management service. In section 5 we discuss the main issues of our approach.

## 2 TEMPORAL NETWORK MANAGEMENT INFORMATION MODEL

Two different information models were defined in order to represent the network management information: one from ISO (ISO 10165) and another one from IETF (Rose, 1990). Both models are centered around the so-called managed objects, which represent an abstraction of network resources and form the Management Information Base (MIB). The IETF's model classifies the managed objects in two types: *scalar objects* and *table objects*. The table objects are two dimensional arrays of scalar objects and at a given time they consist of multiple row entries. The scalar objects are simple MIB variables which can have at most one instance at a given time. The management information contained in the MIB can be classified according to the various criteria as it is described in (Apostolopoulos, 1996) and (Harista, 1993). In particular, taking into account the frequency that management information variables change their values, we conclude to two broad classes of objects (Apostolopoulos, 1996):

- *Quasistatic objects*, which describe the current network configuration (e.g. the number of host interfaces, the routing table, etc.) and their values do not change very often. For example, in MIB-II (McCloghrie, 1991) we can characterise as quasistatic objects the information contained in the `system` group, the `ifTable` (information about the interfaces on a host), the `ipRouteTable` (routing table), etc.
- *Dynamic objects*, which are related to network events (e.g. the transmission of packets) and their values change very often during time. This information involves mainly the objects that are of type Counter and Gauge.

In this paper we propose a model that provides a diachronic representation of the network management information. The model uses a Temporal Management

Information Base, TMIB in order to represent the past and current behaviour of network resources. This new information model is based on the IETF's model and it extends it in order to include the temporal nature of the management information. The incorporation of the temporal dimension in the IETF's model admits the adoption of the TRDM temporal database model. The TRDM, that is presented in (Snodgrass, 1987), extends the standard relational data model by timestamping each row.

According to the new temporal management information model proposed in this paper, the TMIB is a collection of historical table objects that represent a diachronic view of the management information. The TMIB tables represent the past and current network state. The TMIB consists of two types of historical tables: *interval tables* and *event tables*. Interval tables consist of a set of explicit columnar management information objects and of two implicit time objects. These time columnar objects, *validFrom* and *validTo*, represent the time interval [validFrom, validTo) during which the state of the management information is valid. Event tables consist of a set of explicit columnar management information objects and of one implicit time object *validAt*. This time object refers to the instant that an event described by the explicit columnar objects took place. In order to extend the IETF's management information model we should transform the scalar and table objects that constitute the IETF's MIBs into TMIB tables. For this purpose we use the following rules:

1. The MIB table objects are mapped onto corresponding TMIB tables. The explicit non-temporal columnar objects of the derived TMIB table represent the columnar objects that constitute the MIB table object.

2. The scalar objects of MIB groups are mapped onto non-temporal columnar objects of TMIB tables.

| nodeID | ifIndex | ifType | ifAdminStatus | ifOperStatus | ifLastChange | Valid time | |
|--------|---------|--------|---------------|--------------|--------------|------------|---------|
| | | | | | | (from) | (to) |
| pegasus | 1 | 6 | 1 | 1 | 500 | 570 | 1100 |
| pegasus | 2 | 15 | 1 | 2 | 800 | 850 | ∞ |
| pegasus | 3 | 6 | 1 | 2 | 900 | 950 | 1200 |
| pegasus | 1 | 6 | 1 | 2 | 1050 | 1100 | 1800 |
| pegasus | 3 | 6 | 1 | 1 | 1180 | 1200 | ∞ |
| pegasus | 1 | 6 | 1 | 1 | 1700 | 1800 | ∞ |

**Figure 1** Part of the quasistatic `ifTable` TMIB table.

Taking under consideration the criterion that classifies the network management information into quasistatic and dynamic managed objects, the TMIB tables (interval and event tables), according to the type of managed objects they represent, are characterised as:

1. *Quasistatic tables:* Tables that represent **only** quasistatic management information. The non-temporal columnar objects of a quasistatic TMIB table represents **only** the quasistatic part of the corresponding MIB table or quasistatic scalar objects of MIB groups.

2. *Dynamic tables:* Tables that represent only dynamic management information. The non-temporal columnar objects of a dynamic TMIB table represents **only** the dynamic part of the corresponding MIB table or dynamic scalar objects of MIB

groups. The explicit columnar objects of a dynamic TMIB table represent delta values of the corresponding MIB objects.

The reason for the characterisation of a TMIB table as quasistatic or dynamic is based on the type of operations that can be performed upon each type of table. The information included into quasistatic TMIB tables can be the argument in a retrieve, append, delete or replace database operation, while the information included in dynamic TMIB tables can be an argument only to a retrieve operation.

The TMIB tables (interval and event tables) are historical tables that can be classified in two categories: 1) *simple tables* and 2) *capitalised tables*.

| nodeID | ifIndex | ifInOctets | ifOutOctets | Valid time (from) | (to) |
|--------|---------|------------|-------------|------|------|
| pegasus | 1 | 2000 | 1200 | 100 | 130 |
| pegasus | 2 | 4250 | 3000 | 100 | 130 |
| pegasus | 3 | 550 | 630 | 100 | 130 |
| pegasus | 1 | 2050 | 1320 | 130 | 160 |
| pegasus | 2 | 4700 | 3200 | 130 | 160 |
| pegasus | 3 | 400 | 850 | 130 | 160 |
| ...... | ..... | ... | .... | ... | ... |

**Figure 2** Part of the dynamic `ifTable` TMIB table.

**Simple tables** are constructed and maintained by using the temporal database model augment by the appropriate network monitor and network executor procedures. Our model provides monitor procedures with user-definable polling interval for quasistatic and dynamic TMIB tables. Examples of simple TMIB tables are illustrated in Figures 1 and 2. Figure 1 illustrates a part of the quasistatic `ifTable` TMIB table. It is an interval table that represents the quasistatic part of the MIB-II `ifTable` managed object. In this table we can see the history (the past and the present) of the values taken by each columnar managed object. For example, we can see that during the time interval [570, 1100) the interface with `ifIndex` = 1 was up, during the interval [1100, 1800) it was down and since time 1800 until now, e.g. during the interval [1800, ∞), is again operational. Each row lists the values of the columnar objects during the time interval [*validFrom, validTo*). The current information is represented with rows having the *validTo* time object equal to ∞. Figure 2 illustrates a part of the dynamic `ifTable` TMIB. It is an interval table representing a dynamic part of the MIB-II `ifTable` for node "`pegasus`". As this is a dynamic table, each row represents delta values of the columnar managed objects during the interval [*validFrom, validTo*). For example, the first row in this table says that during the interval [100, 130) the node "`pegasus`" at the interface with `ifIndex` =1 received 2000 octets and sent 1200 octets.

**Capitalised tables** serves the need for capitalisation of the network management information because of the vast amount of data included in simple tables (especially in dynamic tables). This type of tables is constructed by applying the appropriate aggregate operations on simple TMIB tables. Based on the simple historical table of Figure 2, an example capitalised TMIB table representing the history of the total number of `ifInOctets` received by all the interfaces of node "`pegasus`" is illustrated in Figure 3. In order to produce this capitalised table we determine the periods of time for which no new rows were inserted in the table presented in Figure 2. For each such period we calculate the sum of ifInOctets from the rows that where valid during that period. For example, during the interval [100, 130) the total number of ifInOctets is 6800 because it is calculated only from the rows {pegasus, 1, 2050, 1320, 100, 130} and {pegasus, 1, 4700, 3200, 175, ∞} that where valid during that interval.

| nodeID | sumTotalIfInOctets | Valid time | |
|--------|--------------------|------------|------|
|        |                    | (from)     | (to) |
| pegasus | 6800 | 100 | 130 |
| pegasus | 7150 | 130 | 160 |
| .... | ...... | ... | ... |

**Figure 3** Part of a dynamic capitalised TMIB table.

## 3    ARCHITECTURAL MODEL

In this section we will describe the architectural model and the services that incorporate directly the proposed network management perspective. In order to present this architectural model we define two new objects: the *temporal network management system (TNMS)* and the *temporal agent (t-agent)*. We define as TNMS an object that manages a *Temporal Management Information Base (TMIB)* which represents diachronically the global network state. A t-agent is an object physically located in a specific network node which has the responsibility of collecting network management information of all the network elements belonging to its *area of responsibility* and storing this information in a number of TMIB tables. The proposed network management architecture consists of one TNMS and a number of t-agents as illustrated in Figure 4.

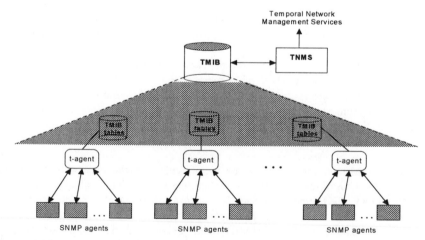

**Figure 4** The proposed network management architecture.

The TMIB design follows the distributed model because it consists of the total number of TMIB tables that reside in the t-agents constituting the entire managed network. The communication between the TNMS and the t-agents does not need a specific network management protocol. This can be accomplished directly because the TNMS can consider that the TMIB tables residing on every t-agent are tables constituting a distributed database. All the well known techniques and protocols borrowed from the distributed database era may be used in order to achieve a consistent global state of the overall network.

The architectural model admits a user-definable set of t-agents that should be established in the network management environment. More precisely, the network administrator should establish each t-agent implementation in a specific network node and define its area of responsibility. In order to define the area of responsibility of each t-agent, the network manager should divide all the network components and/or the networks that he wants to manage in a number of non-intersecting sets of SNMP agents. Each t-agent maintains a number of TMIB tables. These tables describe the diachronic network state that reside in the SNMP MIBs of all the network elements in each own area of responsibility (see Figure 4).

Each t-agent is associated with a unique object identifier that characterises the agent in the entire network. Each t-agent, in order to collect the network management information stored in the TMIB table, uses polling procedures, as well as event-driven mechanisms, that are based on the SNMP protocol.

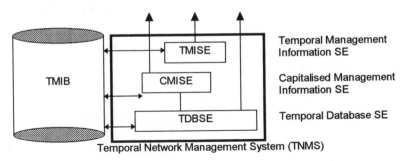

**Figure 5** The structure of the Temporal Network Management System.

One important parameter of the TNMS design is to have the network administrator interact solely with the TMIB, so that from the user's point of view the TMIB embodies diachronically the network. This can be done by offering to the user a set of temporal network management services for the monitoring and control of the past and current behaviour of network resources. The TNMS, in order to provide these services, is constructed by the following service elements, as illustrated in Figure 5:

1. *Temporal Management Information Service Element (TMISE)*: It provides services for the definition of objects needed by the TNMS operation. It provides also services for the TMIB tables creation and manipulation.

3. *Capitalised Management Information Service Element (CMISE)*: It provides services for the capitalisation and aggregation of the historical network management information.

4. *Temporal Database Service Element (TDBSE):* It provides services for the monitoring and control of the historical behaviour of network resources which is stored in the appropriate quasistatic and dynamic simple TMIB table objects.

In the sequel, we will describe the proposed services. The TMISE provides the following services:

*Definition services:*

- *TM-DEFINE-TABLE:* Service used to define an object that will describe the structure of a TMIB table. More precisely, it is used to define the explicit columnar objects and the implicit time objects that will constitute a TMIB table used by the TDBSE or the CMISE.
- *TM-DEFINE-EVENT:* Service used to define an object that will describe an event and the causes that can trigger this event.
- *TM-DEFINE-FILTER:* Service used to define an object that represents the constrains that can be applied on the values of the explicit columnar objects of a TMIB table (selection constrains) and the constrains that can be applied on the values of the implicit time objects (temporal constrains).

*Table creation and manipulation services:*

- *TM-CREATE:* Service used to create a TMIB table object associated with the TDBSE or the CMISE respectively. The structure of the TMIB table should have been already defined by the user with the *TM-DEFINE-TABLE* service.
- *TM-COPY:* Service used to create a copy of a TMIB table object as a whole, or to create a copy of some of its columnar objects, or to create a copy of some of its rows.
- *TM-REMOVE:* Service used to drop a TMIB table object as a whole, to delete a number of its explicit columnar objects or a number of table rows.

The CMISE provides the following services:

- *CM-RETRIEVE:* Service used to retrieve the capitalised information stored into TMIB tables objects related to the CMISE.
- *CM-EXECUTE:* Service used to control the procedure that fills an already created CMISE TMIB table with capitalised information coming from the aggregation of past or current information included in TDBSE TMIB tables.

The *TDBSE* provides the following services:

- *TD-MONITOR:* Service used to control the polling mechanism of network monitors that store the historical network management information in TDBSE TMIB tables.
- *TD-RETRIEVE:* Service used to retrieve the historical information stored into TMIB tables objects. This service allows the user to select past and current views of the network behaviour.
- *TD-APPEND, TD-DELETE, TD-REPLACE:* Services used to change the current state of network resources by creating, deleting and updating rows to the quasistatic TMIB tables.
- *TD-EVENT-REPORT:* Service used to inform the user that an event has occurred. This event can be an already user-defined event concerning thresholds upon values of TMIB table objects or a SNMP trap event.

- *TD-ACTION*: Service used to perform simultaneously a set of operations to different quasistatic TMIB tables. An action is atomic that is, either all or none of the operations have to be done as a unit in order to maintain network configuration integrity.

# 4   THE TEMPORAL NETWORK MANAGEMENT SERVICES

In this section, due to lack of space, we will present only the TM-DEFINE-TABLE service of the TMISE. This service can be used to define the structure of a TMIB table associated with one of the TDBSE or the CMISE. This service can create the structure for two types of tables:
1. *simple tables:* these TMIB tables can contain simple (not capitalised) past or current network management information and can be associated with the TDBSE.
2. *capitalised tables:* these TMIB tables can contain capitalised past or current network management information and can be associated with the CMISE.

The TM-DEFINE-TABLE operation takes as argument a TMIBStructure which can be of type TMIBSimpleStructure, if the user wants to define the structure of a simple table, or of type TMIBCapitalizedStructure, if the user wants to define the structure of a capitalised table. The result of the TM-DEFINE-TABLE operation is the identifier of the object that defines the table structure. The ASN.1 definition of this service is the following:

```
tm-define-table  OPERATION
                 ARGUMENT TMIBStructure
                 RESULT   TMIBStructureID
                 ERRORS   {invalidStructure, noSuchTableType}
        ::= 1
invalidStructure ERROR
    PARAMETER    TMIBStructure
    ::= 1
noSuchTableType  ERROR
    PARAMETER    TableType
    ::= 2
TMIBStructure ::= CHOICE {
    simple       [0] TMIBSimpleStructure,
    capitalised  [1] TMIBCapitalisedStructure }
TMIBStructureID ::= ObjectDescriptor
TMIBSimpleStructure ::= SEQUENCE {
    tableType              TableType,
    tableColumnarObjects SEQUENCE OF MIBObject }
MIBObject ::= SEQUENCE {
    mibObjectID            [0] OBJECT IDENTIFIER,
    mibColumnarObjects     [1] SEQUENCE OF OBJECT IDENTIFIER
                               OPTIONAL }
TableType ::= ENUMERATED {quasistatic (0), dynamic (1)}
TMIBCapitalisedStructure ::= SEQUENCE {
    tableType              TableType,
    tableColumnarObjects SEQUENCE OF TMIBColumnarObject}
```

```
TMIBColumnarObject ::= CHOICE {
    simple          ObjectID,
    aggregate       AggregateObject }
TMIBTableObjectID ::= ObjectDescriptor
ObjectID ::=SEQUENCE {
    tmibTable       TMIBTableObjectID,
    tmibColumn      ObjectDesctiptor }
AggregateObject ::= CHOICE {
    count    [0]  AggregateTerm,
    countU   [1]  AggregateTerm,
    sum      [2]  AggregateTerm,
    sumU     [3]  AggregateTerm,
    avg      [4]  AggregateTerm,
    avgU     [5]  AggregateTerm,
    stdev    [6]  AggregateTerm,
    stdevU   [7]  AggregateTerm,
    any      [8]  AggregateTerm,
    min      [9]  AggregateTerm,
    max      [10] AggregateTerm,
    first    [11] AggregateTerm,
    last     [12] AggregateTerm,
    var      [13] SEQUENCE {
                  aggregateAttribute   EventExpression,
                  parameters           AggregateParameters},
    rate     [14] SEQUENCE {
                  aggregateAttribute   EventExpression,
                  parameters           AggregateParameters,
                  per          Span} }
AggregateTerm ::= SEQUENCE {
    aggregateAttribute    TMIBColumnarObject,
    parameters            AggregateParameters }
AggregateParameters ::= SEQUENCE {
    partition        [0] SET OF TMIBColumnarObject OPTIONAL,
    window           [1] Span OPTIONAL,
    aggregateFilter  [2] FilterObjectID OPTIONAL }
```

The `TMIBSimpleStructure` type has two components: the `tableType`, that specifies the type of the TMIB table as quasistatic or dynamic and the `tableColumnarObjects` that specifies the TMIB table columnar objects. As mentioned in the previous section, the columnar objects of a simple historical TMIB table represent some MIB objects. So, for the `tableColumnarObjects` component we use the `MIBObject` type, which has the `mibObjectID` and `mibColumnarObjects` components. With the component `mibObjectID` we can specify the object identifier of the SNMP MIB table or group that its objects will be depicted on the TMIB table columnar objects. Moreover, with the `mibColumnarObjects` component, which is a sequence of the object identifiers, we can define explicitly which of the MIB objects included in the group or table with object identifier the `mibObjectID` will be depicted on the TMIB table columnar objects. For example, if we want to define a the TMIB table in Figure 2 we should write:

```
tmibTableStructure TMIBSimpleStructure ::=
{
tableType    dynamic,
tableColumnarObjects {
    {mibObjectID    1.3.6.1.2.1.2.2 ,
     mibColumnarObjects
        {1.3.6.1.2.1.2.2.1.1, 1.3.6.1.2.2.2.1.10,
        1.3.6.1.2.2.2.1.16}}}
}
```

The `TMIBCapitalisedStructure` type has two components. The first component is the `tableType` that specifies if the table is quasistatic or dynamic. The second component, the `tableColumnarObjects`, specifies the explicit columnar objects of the capitalised table. The `tableColumnarObjects` component is a set of objects of `TMIBColumnarObject` type. The `TMIBColumnarObject` type is a choice of two types of components. This means that a capitalised TMIB object can have as columnar objects the columnar objects of a simple TMIB table or aggregate objects that will contain aggregate information of simple TMIB tables. The `AggregateObject` type is a choice of fifteen different types of aggregate operators as specified in (Snodgrass, 1993). As we can see from the `AggregateObject` type definition, the service user can specify any of the following aggregates:

1. *Simple aggregates:* `count`, `sum`, `avg`, `any`, `min`, `max` and `stdev`. The operators `countU`, `sumU`, `avgU` and `stdevU` are used when we want unique aggregation, that is the aggregation is performed over the set of strictly different values in an columnar object.

2. *Temporal aggregates:*

- `first`: returns, at each point in time, the oldest value of the given attribute, that is, the one associated with the first valid row
- `last`: analogous to `first`. It returns at each point in time, the newest value of the given attribute, that is the one associated with the row with the latest *from* time.
- `rate`: computes the average growth or decrease experienced by values of an attribute over time, and is applicable only to numeric attributes in event relations. The return value indicates growth or decrease per time unit, e.g. Kbits/sec, packets/sec.
- `var`, VARiability of time spacing: computes the degree of inequality of the time spacing within a given set of events

For the definition of the `count`, `countU`, `sum`, `sumU`, `avg`, `avgU`, `stdev`, `stdevU`, `any`, `min`, `max`, `first` and `last` aggregates we use the `AggregateTerm` type. This type has two components: first the `aggregateAttribute` component that represents the columnar object on which the aggregate should apply, and second, the `parameters` component that represents the parameters that control an aggregate operation. The `parameters` component is of type `AggregateParameters` that includes three components:

1. `partition`: component used for the specification of the management objects whose values would be used for the partition of the rows that participate in the aggregate operation.

2. `window`: this component represents the moving window in cumulative aggregates. An aggregate may or may not take into account rows that are no longer valid. By

taking this criterion under consideration, we can divide the aggregates into two types: *cumulative aggregates*, whose values for each point $t$ in time are computed from all rows that have been valid in the past, as well as those valid at $t$ and *instantaneous aggregates*, whose values for each point $t$ in time are computed only from the row valid at time $t$. For cumulative aggregates, we must specify how far in the past to include rows used to compute a value at time $t$. For instantaneous aggregates the `window` component has the value `"instant"`. If all previous rows are to participate, the value `"ever"` is used. Intermediate cases, such as using only those rows valid during some interval in previous time, are specified using values like `"second"`, `"hour"`, `"day"`, etc. Such aggregates are termed *moving-window aggregates*.

3. `aggregateFilter`: this component is the identifier of a filter object that the service user should have already defined by using the TM-DEFINE-FILTER service. It is used in order to apply specific constrains to the values of the explicit columnar objects and to the values of the implicit time objects that participate in the aggregate operation.

# 5   DISCUSSION

In this paper, we have proposed a temporal network management information model. Within this framework we have presented the primitive services provided by the model, as well as the architecture that supports them. The main aspects of our approach are:

- The direct incorporation of the time parameter as an attribute of the network management information, in a way that maintains the already existing semantics as far as the information management model is concerned.
- The adoption of the temporal database model for storing the historical network management information as well as the development of well-defined services for doing network management related operations.
- The adoption of a distributed database architecture, a fact that facilitates the easy implementation of the proposed approach. This can be done using well-known from the database era protocols, directly in the implementation of our approach.
- The formal definition of primitive network management services using ASN.1 formulation. This can result in a concrete and compact implementation of our view.
- The possibility for developing easily, complex network management applications based on the primitive services that have been defined.

Our view has been supported by the implementation of a prototype that partially implements the proposed model. In this simplified version we have focused on SNMP MIB-II. The experimental use of this prototype exhibits very positive results as far as the advantages of our approach is concerned. The description of the first version of the prototype is presented in (Apostolopoulos, 1996).

# 6    REFERENCES

International Organization for Standardization, Open System Interconnection, *Common Management Information Protocol (CMIP)*, International Standard Number 9596.

International Organization for Standardization, Open System Interconnection, *Management Information Model*, International Standard Number 10165.

Apostolopoulos, T.K. and Daskalou, V.C. (1996) Temporal Network Management Model: Concepts and Implementation Issues. *Computer Communications* (in press).

Case, J.D. Fedor, M.S. Schoffstall, M.L. and Davin J.R. (1990) *A Simple Network Management Protocol*. RFC 1157, DDN Network Information Center, SRI International.

Haritsa, J. Ball, M. Roussoloulos, N. Baras, J. and Data, A. (1993) Design of the MANDATE MIB in Integrated Network Management III (ed. H-G. Hegering and Y. Yemini) IFIP Working Group 6.6., San Francisco.

McCloghrie, K. and Rose, M.T. (1991) *Management Information Base Network Management of TCP/IP based internets: MIB-II*. RFC 1213, DDN Network Information Center, SRI International.

Rose, M.T. and McCloghrie, K. (1990) *Structure of Management Information for TCP/IP based internets*. RFC 1155, DDN Network Information Center, SRI International.

Shvartsman, A.A. (1993) Dealing with History and Time in a Distributed Enterprise Manager. *IEEE Network*.

Snodgrass , R.T. (1987) The Temporal Query Language Tquel. *ACM Transactions on Database Systems*, Vol. 12, No.2, 247-298.

Snodgrass, R.T. Gomez, S. and McKenzie E. (1993) Aggregates in the Temporal Query Language TQuel. *IEEE Transactions on Knowledge and Data Engineering*, Vol.5, No.5, 826-842.

# 7    BIOGRAPHIES

**Theodore K. Apostolopoulos** received his diploma in electrical engineering (1979) and his Ph.D. in Informatics (1983) from National Technical University of Athens. Since 1993 he is an Associate Professor in the Dept. of Informatics of Athens Univ. of Economics and Business. His research interests include telecommunication and computer networks, distributed systems and databases. He has more than 30 publications covering the above scientific areas.

**Victoria C. Daskalou** received her bachelor degree in Informatics from the Dept. of Informatics of Athens Univ.of Economics and Business in 1992, where currenlty she is a Ph.D. student since 1993. Her research interests include network management, databases and intelligent networks.

# 48
# Meta Managed Objects

*J. Seitz*
*Institute of Telematics*
*University of Karlsruhe*
*Germany*
*Phone +49 721 608 4020*
*Fax +49 721 38 80 97*
*E-Mail seitz@telematik.informatik.uni-karlsruhe.de*

## Abstract

Network management has become a crucial affair in today's business. This task needs computer assistance due to the complexity and dimension of the networks. But this computer assistance has to cope with a problem, that has been running through the history of computer communication: due to the lack of one standard in network management heterogeneous management protocols and management models have emerged. In order to manage a large computer network several management standards have to be considered.

To avoid dealing with different management application that can only manage a certain part of the whole network a way of integrating the different management worlds has to be achieved. This paper explains a way to integrate different network models by introducing so called Meta Managed Objects. The described work is part of a project carried through at the University of Karlsruhe [Sei94a].

## Keywords

Network management model, integration of management architectures

## 1 MOTIVATION

Today's business rely heavily on communication and cooperation services provided by computer networks. Failures in these networks can lead to fatal consequences. Thus, network planning is a crucial task and when running the network bottlenecks have to be identified and failures have to be located before they lead to unbearable problems. Of course, all these tasks of network planning and network management must be computer-aided, since the complexity and the dimensions of today's networks make the network operator reach his limits soon when trying to cope with the problems by himself.

In order to carefully plan and to comprehensively manage computer networks, certain standards have evolved dealing with the way to define information needed for these tasks and to have access to this information. Besides proprietary management solutions like IBM's NetView for their SNA networks, two different management worlds have emerged: on the one hand there is a management framework defined by the ISO for their Open Systems Interconnection (OSI), on the other hand the Internet

Society has developed the Simple Network Management Protocol SNMP with its own management model. Although SNMP was thought as an transitional solution it has proved to be practical and has therefore been extended to overcome its weaknesses due to its simplicity.

Nevertheless, there is no hint on one and only future management standard. All the standards will coexist and there are different markets for them. There is no substitute for SNMP and its successors in the area of managing and planning Local Area Networks, while the ISO management framework is mainly applied by the carriers of public and wide areas networks, especially in Europe. But with the interworking between local and public networks there will always be the problem to deal with both standards at the same time. Let us have a look at the ATM management model as defined by the ATM Forum [AC95] (cf. Figure 1). There are five management interfaces: M1 and M2 define the interface between the customer's management system and the ATM end-station. M3 is the Customer Network Management Interface. With M4 the interface between private and public networking technologies is defined. Finally, M5 is the interface between different Carrier Network Management Systems.

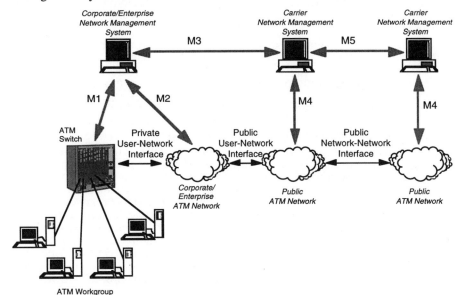

**Figure 1** The ATM Management Model.

Obviously, at the public user-network interface the two management worlds meet. But that means, that the (usually SNMP-based) corporate/enterprise network management system tries to get information out of components based on the ISO management framework. For the carrier network management system the problem is vice versa. Thus, the management application has to make a difference between the ways to have access to the management information according to the supported management architecture. But this would lead to very complex and voluminous implementations. To overcome this situation different approaches have been worked out that will be introduced in section 2. Section 3 details the concepts of so called Meta Managed Objects, that integrate different management models. The practicability of this approach is described in section 4 by a simple but comprehensible example. Finally, section 5 concludes this paper with some aspects that are deeply associated with the concepts of Meta Managed Objects but could not be described in detail due to the spatial limitations of this paper.

## 2    DIFFERENT APPROACHES TO INTEGRATED NETWORK MANAGEMENT

Let us first consider the two management worlds. Although there is the opinion, that »any API that tries to be management protocol independent will capture the worst of both worlds« [Ros93], many approaches have been made in order to make these worlds meet. There are different ways to approach the goal of integrating the management worlds:

- First, management agents have been developed that are independent of the used management protocols, i.e., they support different management protocols [MBL92]. Thus, any management application may retrieve information from this agent independently of the used management protocol. However, these agents are very complex and afford plenty of processor and memory capacity. Thus, they are not practicable for network elements that are limited in memory and computing performance.

- Second, there is the solution of implementing a management gateway between the different worlds. These gateways are also called management proxies [Cha93], [Gre93] and work usually only one-way, i.e., they support management applications based on one management protocol and agents based on another. Besides, these proxies usually support a MIB especially designed for the management information found in the hidden management agents. Thus, they are not very flexible and the management application must differentiate, whether it addresses an agent based on the same management protocol or a heterogeneous agent via the proxy.

- Third, management platforms have evolved dealing with more than one management architecture. E.g., the Distributed Management Environment (DME) as defined by the Open Software Foundation (OSF) [OSF92] supplies the traditional management framework based on OSI and internet management and has added an object-oriented framework using the common object request broker architecture (CORBA). Nevertheless, this architecture has proven to be very difficult to implement [Mar94], that indeed no implementation is available covering all the architecture's aspects. Another example for an integrating management platform is Hewlett-Packard's OpenView AdminCenter [HP94] using SNMP, RPC and CORBA to have access to management information. There they use special model languages to integrate different management models based on SMALLTALK.

Although these solutions may all help to integrate the different management worlds, they still lack in some points. The next section will discuss their disadvantages and show another way of integrating different management protocols, namely the concept of Meta Managed Objects.

## 3    THE CONCEPT OF META MANAGED OBJECTS

The described approaches have different disadvantages:

- Whenever a new management architecture comes up, all the transforming entities, i.e., the management gateways or the protocol independent agents, have to be re-implemented. The same occurs whenever there are new managed objects to deal with.

- On the other hand, there is new overhead produced: with the management gateways you have some machines doing the transformation of management protocols and information and with the protocol independent agents the agent's code is so voluminous that it might not work on network elements with little memory such as bridges or hubs.

So the best would be a flexible interpretable approach as it is realized by the concept of Meta Managed Objects. Therefore, a Meta Managed Object (cf. Code Segment 1) consists of

- the object's name in order to instantiate the object;
- the object's access rights;
- a list of the object's attributes to which we will come later on;
- a list of possible operations that can be performed on this object;
- a list of reports the object can send out whenever some critical or exceptional situation has occurred.

```
META MANAGED OBJECT <ObjectName>
  ACCESSRIGHTS <AccessRights>
  ATTRIBUTES  {<AttributeList>};
  OPERATIONS  {<OperationList>};
  REPORTS     {<ReportList>}.
```

**Code Segment 1** Definition of a Meta Managed Objects.

The integration of different management models within these Meta Managed Objects is achieved by correlating items of these models in the lists of attributes, operations or reports as mentioned above. Thus, let us have a closer look at the way an object's attribute is described (cf. Code Segment 2). This attribute is definitely identified by a name (after the key word ATTRIBUTE), its type is determined by the key word TYPE and its access rights are defined by the key word ACCESS. Then there is a correlation list containing the following information:

```
ATTRIBUTE <AttributeName>
 TYPE <AttributeType>;
 ACCESS <AccessRight>;
{PROTOCOL <Protocol1>;
    RETRIEVE <Protocol1.Service>; ATTRIBUTE <Protocol1.Informationmodel>;
    MODIFY <Protocol1.Service>; ATTRIBUTE <Protocol1.Informationmodel>}
{PROTOCOL <Protocol2>;
    RETRIEVE <Protocol2.Service>; ATTRIBUTE <Protocol2.Informationmodel>;
    MODIFY <Protocol2.Service>; ATTRIBUTE <Protocol2.Informationmodel>}
{PROTOCOL <...>;
    ... }
```

**Code Segment 2** Definition of an attribute within a Meta Managed Object.

- With the key word PROTOCOL the management protocol to which the following information relate is defined.
- The statement with the key word RETRIEVE describes the service of the management protocol to use in order to have access to the management information

corresponding to this attribute.

- With the key word ATTRIBUTE is defined where to find the corresponding information in the management model when using the determined management protocol. Since there might not be only a 1:1 relation between attributes it is possible to define a computation of the attribute's value by giving a numerical expression consisting of arithmetical operations and values defined in the management model. E.g., the attribute's value corresponds to the sum of x and y divided by 2, whereby x and y are managed objects or attributes of managed objects in the management model of the management protocol defined with the key word PROTOCOL.

- In the next line the key word MODIFY leads to the management protocol service you have to use in order to modify the management information.

- Again the key word ATTRIBUTE gives the hint to the corresponding information in the management model on which the determined management protocol is based on. It is necessary to differentiate between reading and modifying the information, because if there is not only one management item concerned (cf. point 2) you must define how a modification reflects on the values of the different management items.

The operations of a Meta Managed Object can be found in a list of operations. Each operation is defined in terms of its name (unique within the Meta Managed Object) and the different ways to perform this operation according to the different management protocols and models. Therefore, for each management protocol the operation and the concerned attributes are listed (cf. Code Segment 3).

```
OPERATION <OperationName>
{PROTOCOL <Protocol1>;
    OPERATION <Protocol1.Oper>; ARGUMENTS <Protocol1.Informationmodel>}};
{PROTOCOL <Protocol2>;
    OPERATION <Protocol2.Oper>; ARGUMENTS <Protocol2.Informationmodel>}};
{PROTOCOL <...>;
 ...};
```

**Code Segment 3** Definition of an operation within a Meta Managed Object.

Finally, the Meta Managed Object contains a list of reports, too, that might be received whenever an exceptional situation occurs. Since management agents based on different management protocols use different means to report such a situation (e.g. via traps in SNMP or via event reports in CMIS/P), the received report and the report that the user expects have to be mapped. This mapping is done in the report definition, where the report is identified by a name, where the report's output values are specified, and where the report is correlated to the possible reports in other management models (cf. Code Segment 4).

```
REPORT < ReportName>
REPORTVALUES <ReportValuesTypes>
{PROTOCOL <Protocol1>
    REPORT <Protocol1.Report>; RESULT <Protocol1.Informationmodel>}};
{PROTOCOL <Protocol2>
    REPORT <Protocol2.Report>; RESULT <Protocol2.Informationmodel>}};
{PROTOCOL <...>
 ...}
```

**Code Segment 4** Definition of a report within a Meta Managed Object.

Of course, you cannot presuppose that there is always a mapping between different management models. There will always be discrepancies between the models leading to some lines in the definition of a Meta Managed Object not to be completed. Therefore, the value Null can be specified to point out that this information cannot be found in the concerned management model. If an attribute of a Meta Managed Object might result from a sum of two attributes found in a specific management model, it is likely that a modification of the Meta Managed Object's attribute cannot be performed within the given management model.

Anyway, a correlation between different management models has been achieved. Nevertheless, this is only a *model correlation*, that must be instantiated according to the managed object classes being instantiated in managed objects. There are two cases to be detailed:

If there is only one instance per managed element, then the instantiation is simple. In order to identify the instance the Meta Managed Object's name an the agent's address are concatenated. Then there is a list of pairs mapping the Meta Managed Object's attributes with attributes in the agent's MIB, the Meta Managed Object's operations with the operations supported by the agent and the Meta Managed Object's reports with the event reports or the traps the agent is able to send out.

The second case is somewhat more complex. The Meta Managed Object might be instantiated several times per managed element. Thus the identifiers for the instances must be extended by a number that has to be incremented whenever a new instance is created or has been encountered. Then again, the Meta Managed Object's attributes, operations, and reports have to be mapped within lists. Let us have a closer look on how the instances might be created: the application might ask for all the instances of a Meta Managed Object. Then, the addressed agent is accessed (by using the supported management protocol) and all the values of the objects or attributes, operations and reports in the agent's MIB that appear in the Meta Managed Object's description are requested. The results are then ordered (e.g. combining the attributes of an ISO/OSI managed object or gathering the attributes of a row in an SNMP-table) to form the instances of the Meta Managed Object giving them unique identifiers as described above.

Thus, there is not only a mapping of the different management models, but also a correlation of the managed objects to be supported by an agent. The following section will detail this procedure by giving an example. This might appear rather simple but in order to concentrate on the core of the described approach it can explain its functioning.

## 4   EXAMPLE OF A META MANAGED OBJECT

In order to illustrate the concept of Meta Managed Objects let us have a look at an example concerning systems management. In a local area network the system administrator wants to generate a login procedure that chooses the machine to log on depending on the momentary number of users on that machine. Thus, he tries to equally load the machines he is responsible for. To make the problem more complex, the network consists of heterogeneous machines that support different management models. To retrieve the desired information, i.e., the number of users logged on, the login procedure has to take care of the supported management models what makes the implementation very complex and inflexible.

The solution would be the introduction of Meta Managed Objects. Therefore, let use assume the following three management models supplying the information about the momentary number of users:
1. In the „Host Resources MIB" [GW93] there is the Managed Object hrSystemNumUsers, that supplies the desired information:

```
hrSystemNumUsers OBJECT-TYPE
    SYNTAX Gauge
    ACCESS read-only
    STATUS mandatory
    DESCRIPTION
            "The number of user sessions for which this host is
            storing state information.  A session is a
            collection of processes requiring a single act of
            user authentication and possibly subject to
            collective job control."
    ::= { hrSystem 5 }
```
**Code Segment 5** SNMP managed object hrSystemNumUsers.

2.  If the information is supplied by an ISO/OSI management agent, it might implement the Management Information Base according to the OSIMIS (OSI Management Information Service) [UCL92] based on the ISO-Development Environment ISODE (to simplify the definition of this object we have concentrated on the GDMO-definition of a package for this object, that is a collection of attributes and notifications contained in the object) :

```
uxObj1Package PACKAGE
    BEHAVIOUR uxObj1Behaviour;
    ATTRIBUTES
        uxObj1Id                GET,
        sysTime                 GET,
        wiseSaying              GET-REPLACE,
        nUsers                  GET,
        nUsersThld              GET-REPLACE;;;
    NOTIFICATIONS
        nUsersThldExceeded;;;
    REGISTERED AS       { uclPackage 50 };
```
**Code Segment 6** ISO/OSI managed object package uxObj1Package.

3.  Finally, if the machine does not supply any management agent at all, there is still the possibility to get the desired information via a UNIX system call (who) and counting the number of entries:

```
rsh <node> `who | wc -l'
```
**Code Segment 7** UNIX system call.

These three possibilities to access and to model the information might be integrated into one Meta Managed Object:

```
MANAGED OBJECT UnixSystem
ACCESS ReadOnly
ATTRIBUTE NumberOfUsers
 TYPE INTEGER;
 ACCESS ReadOnly;
{PROTOCOL SNMP;
    RETRIEVE Get-Request; ATTRIBUTE 1.3.6.1.2.1.25.1.5;
    MODIFY NULL; ATTRIBUTE NULL}
{PROTOCOL CMIP;
    RETRIEVE M-Get-Request; ATTRIBUTE UxObj1.nUsers;
    MODIFY NULL; ATTRIBUTE NULL}
{PROTOCOL rsh;
    RETRIEVE 'who | wc -l'; ATTRIBUTE NULL;
    MODIFY NULL; ATTRIBUTE NULL}
```

**Code Segment 8** Meta Managed Object `UnixSystem`.

Of course, one can add the information needed to report the machine being overloaded by users. In Code Segment 6 there is the notification `nUsersThldExceeded` defined, that is sent out, whenever the number of users exceeds a given threshold. The same effect could have an SNMP-trap, if it is defined in the SNMP-MIB. Thus, adding a report definition to the Meta Managed Object would extend its functionality.

Obviously, this approach is very flexible. Whenever a new management protocol or even a new way of accessing the desired information comes up, the Meta Managed Object has to be extended by only the statements of the new protocol and the relevant services and attributes to retrieve and to modify the information.

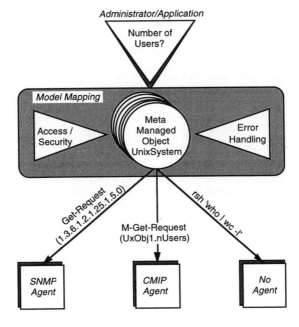

**Figure 2** Illustration of the exemplary Meta Managed Object.

Figure 2 describes the example. The different ways to get to the information are transparently hidden by the concept of Meta Managed Objects. The user respectively the application is simply asking for the number of users from a specified node and gets an INTEGER as the answer.

# 5   CONCLUSION

The described approach of Meta Managed Objects is a way of integrating different management models. It is very flexible since it is interpreted and dynamically adopted to the instances found in the addressed management agents. Nevertheless, the Meta Managed Objects are only one aspect of an integrating management architecture. Further items to bear in mind are:

1. *the integration of management protocols:*
   Since not only the management models but also the management protocols differ, a way to integrate these protocols has to be worked out. Together with the Meta Managed Objects a Generic Management Protocol GMP has been defined [Sei94b], that maps the services of any management protocol onto the GMP-services and vice versa. Thus, e.g., one can request an SNMP-service, that is then transformed into a CMIP-request by using GMP as an intermediate protocol. The answer is then re-transformed into an SNMP-response.

2. *the interpretation of error messages:*
   Just like management models and management protocols differ, they also send out different error massages. Thus, these error messages must be mapped as well in order to avoid misunderstandings. For example, if a CMIP-M-Get-Request has resulted in an error message „noSuchAttribute", a SNMP-based application would expect to receive an error message „noSuchName".

3. *the integration of security mechanisms:*
   Network management is a very critical tasks. With wrong operations you will not only not remove the reported network problem but cause a total net breakdown. Thus, only authorized users and/or applications may change management information. To make sure that authentication and authorization work, security mechanisms have to be provided. Nevertheless, there are different security levels in different management architectures. An integration of these security mechanisms provided by different management standards has to be achieved so that a management application can choose the level of security it needs to carry through its management task.

The co-working of the described mapping mechanisms is presented in [Sei94a]. The architecture has been implemented as a prototype and has proven its practicability in an AI-based network management tool, that has been developed since 1991 [Sei91], [Röt92], [SSS96]. Anyway, the architecture has not been tuned in items of speed or throughput, but it worked fine with a non-time-crucial expert system.

The future work is focused on the management of agent-based distributed systems using the described architecture [SS95]. Furtheron, the architecture is adopted to the task of tariffing broadband services. As already mentioned in section 1, especially with ATM the different management worlds will meet needing integrated management solutions.

# 6 REFERENCES

[AC95]  P. Alexander and K. Carpenter. *ATM Net Management: A Status Report.* Data Communications International, Vol. 24, No. 12, September 1995, pp. 110–116.

[Cha93]  A. Chang: *ISO/CCITT to Internet Management Proxy.* Internet Draft, March 1993.

[Gre93]  U. Gremmelmaier: *The Internet Proxy Agent — A Link between OSI and TCP/IP Management.* In Conference Proceedings of the Third Joint Workshop on High Speed Networks, Paris, France, March 1993.

[GW93]  P. Grillo and S. Waldbusser: *Host Resource MIB.* Request for Comments 1514. Network Working Group, September 1993.

[HP94]  Hewlett-Packard GmbH, Network & System Management Division: *HP OpenView Admin Center.* External Specification, Böblingen, 1994.

[Mar94]  J.S. Marcus: *Icaros, Alice and the OSF DME.* In Proceedings of the Fifth IFIP/IEEE International Workshop on Distributed Systems: Operations & Management (DSOM'94), October 10–12, 1994, Toulouse, France, pp. 4.1– 4.21.

[MBL92]  S. Mazumdar, S. Brady, and D.W. Levine: *Design of Protocol Independent Management Agent to Support SNMP and CMIP Queries.* Research Report RC 18246 (80062), IBM Research Division, T.J. Watson Research Center, Yorktown Heights, NY, USA, August 13, 1992.

[OSF92]  The Open Software Foundation: *OSF Distributed Management Environment (DME) Architecture.* May 1992.

[Röt92]  J. Röthig: *Considerations on the Choice of Bandwidth Balancing Moduli in DQDB Metropolitan Area Networks.* In Proceedings of European Optical Communications and Networks — Tenth Annual EFOC/LAN '92 Conference, Papers on Networks, Paris, France, June 24–26, 1992, pp. 235–240.

[Ros93]  M.T. Rose: *Network Management: Status and Challenges.* Connexions — The Interoperability Report, Vol. 7, No. 6, June 1993, pp. 11–17.

[Sei91]  J. Seitz: *An Architecture for an Expert System Aiding in Network Planning and Network Management.* In K. Majithia (ed.): Silicon Valley Networking Conference SVNC, Proceedings, Santa Clara, CA, USA, April 1991, Maple Press pp. 197–202.

[Sei94a]  J. Seitz: *Integration heterogener Netzwerkmanagementarchitekturen.* No. 289, Series 10: Computer Science/Communications Technology, VDI-Verlag, Düsseldorf, 1994 (German).

[Sei94b]  J. Seitz: *Towards Integrated Network Management: A Generic Management Protocol.* In Proceedings of the Fifth IFIP/IEEE International Workshop on Distributed Systems: Operations & Management (DSOM'94), October 10–12, 1994, Toulouse, France, pp. 6.1– 6.18.

[SS95]  G. Schäfer and J. Seitz: Managing Agent-Systems: A Key Issue for Dealing with Complexity. In Proceedings of the Intelligent Agents Workshop, Oxford Brookes University, Oxford, England, November 23, 1995, pp. 85–86.

[SSS96]    G. Schäfer, J. Schiller, and J. Seitz: *A Toolkit for Rapid Prototyping of Expert Systems for Integrated Network Management.* In M.H. Hamza (ed.): Proceedings of the Fourteenth IASTED International Conference „Applied Informatics", Innsbruck, Austria, February 20–22, 1996, pp. 65–67.

[UCL92]   *A Guide to Implementing Managed Objects Using the GMS.* Universal College London. Version 2.99.1/Draft 1, October 1992.

## 7   BIOGRAPHY

Jochen Seitz has studied Computer Science at the University of Karlsruhe in Germany. Then, he joined the University's Institute of Telematics in 1989, where he finished his Ph.D. thesis in 1993. He is still with the institute, preparing courses on telematics and working in the area of application, network and systems management. He is mainly interested in the areas of security and accounting management.

# 49

# RelMan : a GRM-based Relationship Manager

*Emmanuel Nataf, Olivier Festor, Laurent Andrey*
*INRIA Lorraine*
*Technopôle de Nancy-Brabois - Campus scientifique. 615, rue du*
*Jardin Botanique - B.P. 101 54600 Villers Lès Nancy Cedex France.*
*Tel: (+33) 83.59.20.48, Fax: (+33) 83.27.83.19 email:*
*{nataf,festor,andrey@loria.fr}*

## Abstract

The modeling of physical and logical network resources provides the main interface to network management software applications. The object-oriented information model defined in the OSI managment framework provides notations to specify such interfaces. This framework has been recently extended with a relationship model. Relationship modeling adds informations upon isolated object specifications and offers new services for the management of these relationships. In this paper, we address some practical issues on network management architectures. To implement this new model, we start from relationship specifications up to the provision of a relationship service within the manager role of a managment system. Our study outlines some lacks in the standardized relationship model and extends the notations with new statements to allow the development of a generic relationship provider. We also present a simple prototype that works on our Managed Object Development Environment. This prototype is a first implementation towards a complete tool for relationship management.

## Keywords

Management, Applications, GRM, Information Model, OSI, Relationships, TMN, Architecture.

## 1 INTRODUCTION

The need for relationships in the design phase of Information Models has been recognized in various places and several proposals have been made in this direction [Kilov 95, Clemm 93b]. Some proposals have been precursors for the General Relationship Model such as the work done by A. Clemm [Clemm 94, Clemm 93a]. Other ones have proposed their own notation such as S. Bapat in [Bapat 94].

Despite the standardisation of the General Relationship Model (GRM) , which is in fact recent [CCITT.X.725 95], very few Information Models currently include such specifications and many persons seem to be reserved to use it for three mains reasons: lack of experience, unavailabilty of applications and poorly identified application domains.

Thus, one challenge for its acceptance is the provision of valuable results showing that the GRM is useful along different phases of the development of Information Models. Concerning the use of these relationship specifications, several results have already been obtained in combining GRM and behavior specifications for the formal description of information models. The results are reported in the following studies: [Sidou 95] [Marshall 93] [Festor 94] [Keller 95].

From our point of view, the GRM may also be very useful in the operating phase of a management application. A first approach to this issue was proposed in [Clemm 93a] where relationships have been implemented in an agent as Managed Objects. The current standard GRM allows more representations of relationships in agents, i.e. operations, name-bindings, attributes. Unfortunately, the existing approaches do not address all these representations.

In this paper we present a novel way of exploiting GRM specifications in the operating phase of a management system. This approach is based on defining a service level at the manager side called RelMan which allows both standard CMIS operations on MOs and relationship management operations based on the standardised GRM notation. RelMan supports all types of relationship representations. This element provides a full relationship management interface to any applications and is generic in the way that it can act on any loaded information model containing GRM and GDMO* specifications.

The remainder of the paper is organized as follows. Section 2 provides an overview of the GRM standard and illustrates some of its limits. Section 3 contains a presentation of our RelMan application and how this application exploits the GRM-based relationship specifications found within an Information Model. Finally, section 4 concludes the paper and outlines some research issues we are currently working on.

# 2  THE GENERAL RELATIONSHIP MODEL

Some information specification documents based on the OSI network management framework [ISO-10040 92] contain a mix of Managed Object Class (MOC) definitions in GDMO and entity-association (E/A) diagrams, as in [ITU-T.M.3100 95, NMF 94]. The use of the Chen model [Chen 76] or derived ones, allows the specification of relationships between MOCs. Afterwards these relationships must be represented in the MOCs themselves with attributes or extra MOCs. Specifiers and implementers must deal with several formal methods and most important, they must share the same semantics. This is not a trivial assumption [Kilov 92]. The GRM provides a mean of expressing relationships in way similar to GDMO. Moreover, it enables a uniform specification of relationship representations in this

---

*Guidelines for the Definition of Managed Object [ISO-10165.4 92]

Management Information Model (MIM) [ISO-10165.1 92]. In this section, we will summarize the main features of this model and give a simple example of its use.

## 2.1 Basic Characteristics

The final text of the GRM [ISO-10165.7 95] provides a conceptual model and notational tools for designing Managed Relationship Classes (MRCs) and their representation in terms of MIM features. The MRC specifications permit the description of behavior dependencies between Managed Objects (MOs, that are instances of MOCs). The specifications also enable the management of these MOs as a whole through a Managed Relationship (MR, that is an instance of a MRC). All described dependencies are maintained inside the MOs through the representation means. MRC specifications are defined in terms of roles, operations, attributes and behaviors.

- Each MO in a MR fulfils at least one of the relationship role. One role is defined by a compatible MOC that constraints the effective MOC characteristics of the MO fulfilling the role. A role is also constrained by the number of MOs that can fulfil it and by the number of MRs (of the same MRC) in which a given MO fulfils the same role.
- Operations are generic. They are: ESTABLISH for creating a new instance of a MRC, TERMINATE for deleting a MR, QUERY for information retrieval purpose, BIND and UNBIND for adding or removing one or more MOs from a MR, NOTIFY as a notification facility from MRs, and USER-DEFINED for anything else. These operations can have an extra label, e.g. to distinguish two ESTABLISH operations within one MR.
- Specification of relationship attribute emphasize perticular MO attribute. Indeed each relationship attribute must belong to exactly one of the participant MO and so their read/write capabilities are thoses defined by the MOC.
- Behavior definitions are, as usual in this context, expressed in natural language.

## 2.2 A simple example (from M.3100)

In this section, an example of using the GRM is extracted from the ITU-T M.3100 recommendation [ITU-T.M.3100 95]. On this information model, we perform reverse engineering to obtain the corresponding GRM specification. Figure 1 shows an E/A reprensentation of the cross connection relationship that exists between MOs of the `terminationPoint` MOCs (we will only say `terminationPoint` MOs), or group of `terminationPoint` MOs (GTP MOs) and `crossConnection` MOs inside a managed element (as a switch). The diagram lacks of precision and the behavior in the `crossConnection` MOC adds extra constraints. The `terminationPoint` MOC is an abstract class and there is constraints for the choice of the subclasses. By example, when the cardinality is n = 2, there should be one `connectionTerminationPointSource` MO and one `connectionTerminationPointSink` referenced by attributes in the `crossConnection` MOC. Moreover, the E/A does not specify that only `GTP` MOs must be cross connected with `GTP` MOs (and `terminationPoint` derived MOCs together).

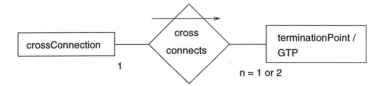

**Figure 1** Entity/Association diagram from the recommendation M3100

Other behavior disseminations between the E/A diagram and MOC specifications exist in the model. Here we focus only on the case above and write the MRC specification in GRM accordingly. The resulting specification is illustrated in figure 2. There are three declared generic operations (lines 3-5), and two attributes in 6 and 7 respectively. We define three roles, the role *from* and *to* (lines 8-11 and 16-19) for the `terminationPoint` MOC (lines 9 and 17) and one role *connector* (12-15) for the `crossConnection` MOC. There must be exactly one MO fulfilling each one (lines 10 and 18) and each of theses MOs can only be in at most one MR of this class in the same role (lines 11, 16 and 19, the maximum is at 1), but can exist without participating at this relationship (the minimum is at null). This specification is more detailed than the simple diagram of figure 1; it shows the structure of the relationship and explicits the operations that can be applied on it. We do not detail here in which other MOCs configuration this MRC can be reused by derivation. We do this with interconnection between managed elements and point to multi point cross connections in a managed element. The behavior (line 2) is not given for conciseness. The actual one is constructed with the behavior specifications of the `crossConnection` and `terminationPoint` MOCs from which statements about the cross connection have been selected.

The second step of the modelling process aims at representing MRCs in the information model with the use of the RELATIONSHIP MAPPING template [Sibilla 93, Kilov 95]. Figure 3 shows one example of mapping for the above MRC. For one MRC, there can be several mappings. Thus the mapping template must indicate which MRC it maps (line 2). For each role of the referenced MRC, one should specify the MOCs (that must be compatible with the MRC role), as in lines 4 and 6. One can also specify the representation means, i.e. what are the MIM features used to save the MRs composition inside the MOs fulfilling the various roles. This is done in line 5 where we use an attribute to represent MOs in the *from* role. The relationship attributes are mapped onto some of the MOC in the *connector* role (line 7). The next statement concerns the mapping of relationship operations into system operations (lines 8-13). As an example, the ESTABLISH operation is mapped onto a CREATE CMIS service on the *connector* role, i.e. a CREATE operation of the *crossConnection* MOC. We also show the mapping of the QUERY operation where one relationship operation is mapped on two management system operations (lines 11-13). At execution time, these operations are all encapsulated in CMIS requests. The behavior specification (not detailed here) contains state-

(1)  *pointToPointCrossConnection* RELATIONSHIP CLASS
(2)    BEHAVIOUR *pointToPointCrossConnectionBehavior*;
(3)    SUPPORTS ESTABLISH,
(4)     TERMINATE,
(5)     QUERY;
(6)    QUALIFIED BY *signalType*,
(7)     *directionality*;
(8)    ROLE *from*
(9)     COMPATIBLE WITH *terminationPoint*
(10)     PERMITTED-ROLE-CARDINALITY-CONSTRAINT *1..1*
(11)     PERMITTED-RELATIONSHIP-CARDINALITY-CONSTRAINT *0..1*
(12)    ROLE *connector*
(13)     COMPATIBLE WITH *crossConnection*
(14)     PERMITTED-ROLE-CARDINALITY-CONSTRAINT *1..1*
(15)     PERMITTED-RELATIONSHIP-CARDINALITY-CONSTRAINT *0..1*
(16)    ROLE *to*
(17)     COMPATIBLE WITH *terminationPoint*
(18)     PERMITTED-ROLE-CARDINALITY-CONSTRAINT *1..1*
(19)     PERMITTED-RELATIONSHIP-CARDINALITY-CONSTRAINT *0..1*;
(20) REGISTERED AS {*m3100.Relationship.pointToPoint*};

**Figure 2** A relationship class for the cross connection

ments about the representation, explaining that the attribute *fromTermination* is one of the `crossConnection` MOC, or giving some indications to assign value to the encapsulated CMIS service parameters.

## 2.3 Current limits

The GRM helps MIB modellers to specify dependencies between MOs, but the MRCs are not integrated enough in the MIM. Currently, by the means of GDMO, MIB specifications and CMIS/CMIP definitions, a network management application can only be written in terms of MO manipulations that will be effectively sent over the network to distant MOs (i.e. by CMIS operations or actions). In this context, we have identified the following limits within the GRM:

- the MRs do not explicitly exist in MIBs;
- the MRC template does not enable the description of relationship operation parameters;
- a relationship attribute could not be computed from severals MOs attributes (see [Bapat 93] for this issue);
- the specification of the behavior is not formally expressed, this is one of our research topic and results are emerging [Festor 94, Sidou 95, Nataf 96], recent works as [Keller 95] are studied.

```
(1)    circuitPointToPointConnectionMapping RELATIONSHIP MAPPING
(2)          RELATIONSHIP CLASS pointToPointCrossConnection;
(3)          BEHAVIOUR circuitPointToPointConnectionMappingBehavior;
(4)          ROLE from RELATED-CLASSES circuitTerminationPointSource
(5)            REPRESENTED-BY ATTRIBUTE fromTermination,
(6)          ROLE connector RELATED-CLASSES crossConnection
(7)            QUALIFY signalType directionality,
                 . . .
(8)             OPERATIONS MAPPING
(9)              ESTABLISH MAPS-TO-OPERATION
(10)               CREATE OF connector,
                 . . .
(11)             QUERY MAPS-TO-OPERATION
(12)               GET fromTermination OF connector
(13)               GET toTermination OF connector;
(14)    REGISTERED AS {m3100.RelationshipMapping.CircuitPointToPoint};
```

**Figure 3** A relationship mapping for the cross connection

As MOC, MRC implementation is a local matter outside the scope of standardization. We think we can extend the GRM towards a better integration of both the MIM and the network management applications. In the next section, we show how a logical uniform repository for MRs can be automatically generated from GRM specifications and how a set of relationship information services can be provided to either the network engineer or any application in the Manager role of a management interaction.

## 3  THE RELMAN RELATIONSHIP MANAGER

RelMan is a software tool that takes place as an application service element (ASE) at the Manager side of a Manager/multi-Agent architecture. RelMan is strongly coupled with CMIS. It has an internal framework designed to receive GRM and GDMO specifications. These specifications must be already known by agents on the managed network.

### 3.1  Goals

RelMan adds two new concepts in the standardized framework.

- A Relationship Management Information Base (RMIB) that is a logical repository of MRs (we do not impose a particular database scheme here; relationship database does not mean relational database).
- A Common Relationship Information Service (CRIS) that is a CMIS equivalent for the relationship management. It is more detailed than the generic definitions

of the GRM concerning the relationship management operations and notifications (part 8.1 of the GRM standard).

## 3.2 Architecture

We propose a relationship management model that extends the system management model of [ISO-10040 92] from which figure 4 is partially extracted. The "Managing Open System" box contains our extension, the "Communicating" large arrow and the "Managed Open System" box are those of the standard.

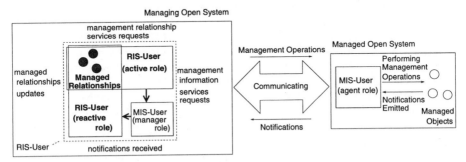

**Figure 4** Relationship management model

The Relationship Information Service User (RIS-User) manages MRs by relationship operation invocations and notification receptions. We define a **relationship management interaction** as a single relationship operation, a single notification or an identified set of logically related relationship management operations and notifications, during which the manager role does not change. In an interaction, one relationship operation or notification can be mapped on several system operations or notifications respectively. Thus, one **relationship management interaction** can imply several **system management interactions**[†]. Furthermore a MR can bind MOs in several MIBs and the **systems management interactions** generated by a **relationship management interaction** may not occur between all the same MIS-User in the agent role (but we assume with the same RIS-User in the relationship manager role).

Figure 4 details some parts of the RIS-User. The RIS-User in the active role requests management relationship services (establish, terminate, query, ... ) to MRs. These requests will be translated in management information services requests in accordance to the OPERATIONS MAPPING specifications and will be sent to the MIS-User in the manager role (the one of the standard). The later receives notifications (we will see later which ones in the next section) from managed systems and forwards them to the RIS-User in the reactive role. This RIS-User has two mains functionalities. On one hand it must translate system notifications to relationship ones (with the help of the OPERATIONS MAPPING specifications). On the other it

---

[†]defined in [ISO-10040 92]

must look at the state of MOs that participate in MRs in order to guarantee their consistency and update the MRs accordingly.

## 3.3    The Relationship Management Information Base

The Relationship Management Information Base (RMIB) is part of the RIS-User where each MR is modelled with sufficient information in order to enable relationship operations requests to be performed.

A MR can be modelled in the RMIB by a tuple of the global identifier of each MO it binds. We propose an optimization of the tuple size based on the use of the REPRESENTED-BY specification of a role. In a MRC mapping, roles may be interrelated in several manners (ATTRIBUTE, NAMING, RELATIONSHIP OBJECT and OPERATION). As an example, the mapping given in figure 3 shows that MOs in the *from* and the *to* role are represented by attributes of the MO in the *connector* role. In this case, one MR could be modelled in the RMIB only by the identifier of the MO in the *connector* role. We thus define the *connector* role as **representative** of the MRC and its mapping. The MOCs that are **representative** can be found in a determinist manner from MRCs mapping specifications. All the required information can be obtained from **representative** MOs via CMIS requests.

Optimizing the tuple size of representative MOs leads to some performance issues. By example, to find MOs participating in a MR, one should first query (CMIS GET) the **representative** MOs to retrieve other MO identifiers, and then act on these MOs. If this relationship is often used, the MRC will be better modelled by the whole MOs tuple. On the other hand, the use of the representative MOs prevents the growth of the RMIB and avoids major consistency problems of the repository since information is distributed across agents.

Without regarding the relationship operations defined in the MRCs, the RMIB can provide a general query service that enables the discovery of all the MRs of a given MRC. Other provided services are dependent of relationship operations.

## 3.4    Relationship services

The relationship services are provided to manage MRs. These relationship operations may be linked to a set of standard CMIS operations. A full description of relationships services and their mapping onto CMIS services is given in [Clemm 94]. Conversaly to this work, we do not suppose that each MRC is represented by a specific MOC but by an arbitrary relationship mapping. Our work is in progress to formally define the behavior of these services. FDTs as [Schneider 92, CCITT-Z-105 95] combined with ASN.1, will be probably used.

## 3.5    RIS-User in the reactive role

A MR's state depends on the participating MOs states and MOs can be accessed by other MIS-User (or RIS-User). The only means for a manager to know about MOs states is to poll them frequently or to wait for notifications emitted from the

MOs (or a mix). In our architecture, we construct the RIS-User in the reactive role (figure 4) in order to receive notifications and update in the RMIB the MRs from which sender MOs belong too. Two notifications types are taken into account:

- notifications that are the system operations mapped from a NOTIFY relationship operation;
- notifications that are side effects of MOs manipulations by a manager (i.e. directly performed CMIS operations by a manager).

By example, lines 9 and 10 of the figure 3 leads the RIS-User to wait for *objectCreation* (defined in [ISO-10164.1 92, ISO-10165.2 92]) from the *connector* role. A good integration of these notification scrutations requires the creation of `eventForwardingDiscriminator` MOs within the agents [ISO-10164.5 92] [ISO-10165.5 92].

## 3.6 Early results

We have implemented the RelMan layer over the OSIMIS [Pavlou 95] platform combined with our MODERES‡ [Festor 95] toolkit for Information Model processing. Within this application we currently provide to a human manager the possibility to request and act upon multiple agents through either a standard CMIS interface or a relationship interface which is the union of all relationship operations defined within the known Information Model(s).

Early results are very encouraging. First, the use of a relationship-oriented interface is very smart and easy to use since one can combine in one relationship operation several standard CMIS calls. Second, the extensions to the GRM defined within RelMan are also useful in earlier stages of the development, i.e. at the specification stage and especially if one wants to define formally the Information Model before any implementation.

Thus such a relationship-based service provides a more flexible and powerful interface to applications and network managers. Moreover this service requires little overhead.

One issue we have not addressed so far, is the problem of inter-agent relationship consistency and update. Update over multiple agents is possible in RelMan but no support for Distributed Transaction Processing to ensure consistency of updates is currently provided.

Last but not least, GRM specifications appear to be easily readable. GRM provides just two templates which appears to be a reasonnable size and complexity extension to GDMO. But beware of this apparent simplicity: writing GRM specifications is a very difficult task that requires a lot of efforts.

---

‡Managed Object Development Environment by RESEDAS

# 4  CONCLUSION AN FUTURE WORK

In this paper, we have presented a new application domain for GRM specifications within Information Models. With the development of the RelMan prototype we have shown the feasibity of automated exploitation of GRM specifications as a service at the manager side. In this approach we have also identified some limits of the current GRM notation to allow a complete automation of its exploitation. To overcome this, we have slightly extended the notation and integrated these extensions in the RelMan application. By developing the GRM support at the manager side, we allow relationships to be represented within or between agents by all representations defined within the standardized GRM mapping model, i.e. name-bindings, attributes, operations, relationship objects.

Our current work focusses on extending the service provided by the RelMan in providing a standardized representation of the relationship instances as a MIB within the manager. This manager will then act as an intermediate manager. To upper level managers it provides both CMIS and CRIS relationship services. Theses services are then mapped to standardized CMIS operations sent to the lower level agents hosting managed objects. This is done as part of our JASOM intermediate manager project. Another domain we are currently working on is the link of a fault management application to RelMan in order to allow this application to collect parts of its configuration information needs from the relationships among MOs. These relationships are there not hard-coded in the fault management application but defined like MOs in the Information Model and expressed in GRM.

Future work covers research on the design process of the relationships among MOs and its integration in the design process of management information models. In fact, it is not clear in the MIM design process when and where relationships have to be defined. Early work on this subject by our group tends to distinguish between two relationship types: application specific and generic ones whereas the first ones may cover some of the later ones. Another research domain covers the knowledge add-on provided by explicit relationship specifications in the Information Model. In this case the GRM should embed a knowledge description language to provide more description power to Information modellers. This will allow applications to be more flexible and exploit better the Information Model. In this context, the JASOM application will probably feed us with a set of primary requirements.

# REFERENCES

[Bapat 93] S. Bapat. *Richer Modeling Semantics for Management Information*. H.-G. Hegering et Y.Yemini, éditeurs, *Integrated Network Management III*, pages 15–28. IFIP Transaction C-12, Elsevier Science Publisher B.V. (North Holland), April 18-23 1993.

[Bapat 94] Subodh Bapat. *Object-Oriented Networks : Models for architecture, operations and management*. Prentice Hall, 1994.

[CCITT-Z-105 95] Comité Consultatif International Télégraphique et Téléphonique (CCITT), *SDL Combined with ASN.1 (SDL/ASN.1)*, Norme Internationale,

CCITT-Z-105, Janvier 1995, SG.10 Q.6 Proposed New Recommendation.

[CCITT.X.725 95] Comité Consultatif International Télégraphique et Téléphonique (CCITT), *Information Technology - Open Systems Interconnection - Structure of Management Information - Part 7: General Relationship Model*, International Standard, CCITT.X.725, June 1995.

[Chen 76] P.PS. Chen. The entity-relationship model - toward a unified view of data. *ACM Transaction on Database System*, 1(1):9–36, March 1976.

[Clemm 93a] A. Clemm. *Incorporating Relationship into Management Information. 2nd Network Management and Control Workshop*. IEEE, 1993.

[Clemm 93b] A. Clemm et O. Festor. *Behavior, Documentation and Knowledge: An Approach for the Treatment of OSI-Behavior. Fourth International Workshop on Distributed Systems: Operations and Management*, 1993. October 5-6, 1993, Long Branch, New-Jersey, USA.

[Clemm 94] Alexander Clemm. *Modellierung und Hanhabung von Beziehungen zwischen Managementobjekten im OSI-Netzmanagement.* Thèse de Doctorat, Ludwig-Maximilians-Universität München, 1994.

[Festor 94] O. Festor. *OSI Managed Objects Development with LOBSTERS.* 1994. Fifth International Workshop on Distributed Systems: Operations and Management, 12-16 Septembre 1994, Toulouse, France.

[Festor 95] O. Festor. *MODE: a Development Environment for Managed Objects.* pages 616–628, 1995. in [ISINM'95 95].

[ISINM'95 95] ISINM'95. *Integrated Network Management, IV.* Chapman & Hall, 1995. Proc. IFIP IEEE 4th Int. Symp. on Integrated Network Management, Santa Barbara, CA, 1-5 May, 1995.

[ISO-10040 92] International Organization for Standardization (ISO), *System Management Overview*, Interim Final Text, ISO-10040, January 1992.

[ISO-10164.1 92] International Organization for Standardization (ISO), *System Management - Part 1: Object Management Function*, International Standard, ISO-10164.1, January 1992.

[ISO-10164.5 92] International Organization for Standardization (ISO), *System Management - Part 5: Event Report Management Function*, International Standard, ISO-10164.5, January 1992.

[ISO-10165.1 92] International Organization for Standardization (ISO), *Structure of Management Information - Part 1: Management Information Model*, International Standard, ISO-10165.1, January 1992.

[ISO-10165.2 92] International Organization for Standardization (ISO), *Structure of Management Information - Part 2: Definition of Management Information*, International Standard, ISO-10165.2, January 1992.

[ISO-10165.4 92] International Organization for Standardization (ISO), *Structure of Management Information - Part 4: Guidelines for the Definition of Managed Objects*, International Standard, ISO-10165.4, January 1992.

[ISO-10165.5 92] International Organization for Standardization (ISO), *Structure of Management Information - Part 5: Generic Management Information*, International Standard, ISO-10165.5, January 1992.

[ISO-10165.7 95] International Organization for Standardization (ISO), *Structure of Management Information - Part 7: General Relationship Model*, International

Standard, ISO-10165.7, June 1995.

[ITU-T.M.3100 95] International Telecommunication Union - Sector Telecommunication (ITU-T), *Generic Network Information Model*, International Standard, ITU-T.M.3100, November 1995.

[Keller 95] J. Keller. *An Extension of GDMO for Formalizing Managed Object Behaviour*. 1995. FORTE'95.

[Kilov 92] Haim Kilov. *Understand → Specify → Reuse : precise specification of behaviour and relationships. DSOM*, 1992.

[Kilov 95] H. Kilov. *Understanding the semantics of collective behavior: the ISO General Relationship Model*. 1995. ECOOP95 one day workshop: Use of Object-Oriented technology for Network Design and Management. Aarhus, Denmark, August 7-11, 1995.

[Marshall 93] L S. Marshall. *Using VDM to Specify Managed Object Relationship. FORTE*, 1993.

[Nataf 96] E. Nataf, O. Festor, A. Schaff et L. Andrey. *Validation d'un modèle de gestion d'interconnexion de commutateurs à l'aide de système de transitions étiquetées. CFIP'96*, 1996.

[NMF 94] Network Management Forum, *Switch Interconnection Management: Configuration Management Ensemble*, NMF, November 1994.

[Pavlou 95] G. Pavlou et all. *The OSIMIS Platform: Making OSI Management Simple*. Hall Chapman, éditeur, *Integrated Network Management IV*, pages 480,493, 1995.

[Schneider 92] J.M. Schneider. *Protocol Engineering: A Rule-based Approach*. Vieweg, 1992.

[Sibilla 93] M. Sibilla et all. *Management Information Base : Guidelines Leading From Generic Specification of Managed Object Relationship to Consistent Implementation. Distributed Systems : Operation and Management*, October 5-6 1993.

[Sidou 95] D. Sidou. *TIMS : a TNM-based Information Model Simulator, Principles and Application to a Simple Case Study. Distributed System Operation and Management*, October 19-22 1995.

## PART FIVE

# Fault Management II
Chair: Gabi Dreo Rodosek,
Leibniz Supercomputing Center

# 50
# Divide and Conquer Technique for Network Fault Management

*Kohei OHTA   Takumi MORI   Nei KATO   Hideaki SONE*
*Glenn MANSFIELD   Yoshiaki NEMOTO*
*Graduate School of Information Science, Tohoku university*
*Aramaki Aza Aoba, Aoba-ku, Sendai, JAPAN*
*e-mail: kohei@nemoto.ecei.tohoku.ac.jp*

### Abstract

From the perspective of fault management, traffic characteristics contain symptoms of faults in the network. Symptoms of faults aggregate and are manifested in the aggregate traffic characteristics generally observed by a traffic monitor. It is very difficult for a manager or an NMS to isolate the symptoms manifested in the aggregate traffic characteristics. Symptoms get obscured by other symptoms. At times there are too many symptoms clouding the symptom space, making the task of symptom isolation practically impossible. In this work we present a powerful technique, the divide and conquer technique, wherein symptoms are iteratively isolated from the aggregate observable. This provides a tractable mechanism for symptom isolation, fault detection and analysis. The symptom isolation technique makes it possible to use a simple thresholding mechanism for detecting abnormalities. We have implemented the system using the popular SNMP-based RMON technology. Using dynamically constructed filters to suppress already detected symptoms in the observed aggregate, fresh symptoms are isolated. Experimental results show a significant improvement in the fault management capability and accuracy.

### Keywords

fault management, fault detection, RMON, traffic monitoring

## 1  INTRODUCTION

In the fault management framework alarms are received by an NMS which has the responsibility of correlating the alarms and locating the faults. A fault may cause several alarms in the network, and several faults may coexist at the same time leading to a cascade of alarms. Mapping alarms to faults is a challenging problem. Several approaches have been suggested e.g. use of coding techniques[KLI95], network configuration information[HOU95][GLN96] etc.. Yet the basic requirements of network fault management[DUP89][STA93] are far from being realized.

Alarms are generated by entities in the network when they sense an abnormality. For example, an agent may be configured to generate an alarm when it sees too many

ICMP[RFC792]-destination unreachable packets. An alarm may be generated when a router *drops* the default route, or when the operational status of an interface goes *down*.

Detecting an abnormality is a challenging problem. It involves knowing what is *normal* [LAB91]. By comparing an observed value with the normal value an entity may decide whether there is an indication of an abnormality. What is normal for one network may not be so for another network. Dauber[DAU91] suggests a procedure called *baselining* to know what is normal for a network.

In general an NMS collects information about the network. From the perspective of fault management, this information will be analyzed for *symptoms* or indications of abnormal health. The NMS may employ some thresholding mechanism to decide whether the status of a Managed Object (MO) or, the combined status of a set of MOs is indicative of abnormal health or otherwise[GLN92][KOH95]. In the above example, if the number of ICMP destination unreachable packets exceeds a threshold, say *alpha*, an *alarm* or an *event* may be triggered indicating an abnormality in the network and calling for action/intervention. More often than not, this event is the consolidated effect of several symptoms. Each ICMP destination unreachable packet indicates that some destination is unreachable and, as such, is a symptom. It is the task of the NMS to detect the presence of symptoms, isolate and identify each symptom and, diagnose and control the fault indicated by the symptoms.

Symptoms have their own respective characteristics. The *amplitude* of a symptom is a measure of the observable which will be thresholded to decide whether the observable represents an abnormality. The *persistence* of a symptom is a measure of the duration of the symptom and the *frequency* of a symptom is the number of times the symptom has been observed in a given period.

The existence of multiple faults in a network complicates the analysis of symptoms. Some faults and the corresponding symptoms are persistent. If there are persistent symptoms manifested in an aggregated MO, (e.g. ICMP destination unreachable packets), fresh symptoms are likely to get obscured. It is likely that new events are not triggered as the event is already *ON*. Dynamically adjusting the threshold to the "normal" state of the network is a non-trivial problem. Thus some or maybe most of the symptoms occurring in the presence of a persistent symptom are likely to remain hidden, unnoticed. In such situations, the simple thresholding mechanism is a failure at detecting fresh abnormalities and setting new alarms.

Most networks have their own specific eccentricities - which manifest themselves as symptoms. These are *known problems* that are probably under examination. From a managers point of view it would be beneficial to be able to detect fresh problems, i.e., to view the health of a network minus the known problems.

In this paper, we present a powerful technique, the divide and conquer technique, wherein symptoms are iteratively isolated from the aggregated observable. This provides a tractable mechanism for fault detection and analysis. In section 2 we discuss the traffic monitoring approach to fault management, in section 3 we present the divide and conquer technique for fault isolation and management, in section 4 we discuss an SNMP-based practical implementation using widely available RMON-Agent technology, in section 5 we discuss the performance of the proposed technique based on results of experiments on an operational network, followed by conclusion in section 6.

## 2  TRAFFIC MONITORING FOR NETWORK FAULT MANAGEMENT

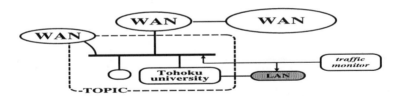

**Figure 1** The transit network TOPIC and a stub LAN

### 2.1  Traffic monitoring

Network management primarily involves monitoring aggregate characteristics. Number of packets, number of collisions, number of broken packets, number of ICMP packets etc. are examples of aggregate characteristics. Aggregate characteristics provide only a macro-view of the network. For example by looking at the number of ICMP destination-unreachable-packets(ICMP-DUR) the NMS can infer that one or more destinations cannot be reached. However for actual fault detection and subsequent diagnosis it is important to know which destination(s) was unreachable and from which source. This micro-view can be obtained by examining the relevant packets in the network.

Special purpose agents, *network monitors*[TCPDUMP][NNSTAT][RFC1757], generally on network entities dedicated to management, provide information about the network traffic as seen by the network entity or monitor. So, a monitor on an Ethernet segment would provide information on all the packets transiting that Ethernet segment. Apart from the macro-view these special agents may also be, in some cases dynamically, configured to provide the micro-view from the network entity i.e. to examine the packets in the network.

This mode of fault management by traffic monitoring has some strong points. It consumes less bandwidth, is more effective and may detect faults from network traffic characteristics, causes of which may lie inside or even outside the management domain.

### 2.2  Traffic characteristics in a typical network

Traffic characteristics vary from network to network and more so between different types of networks. Characteristics of a transit network are much different from that of a LAN. Fig.1 shows a typical medium-scale network comprising of several WANs and LANs. The area enclosed within the dotted line is the Ethernet backbone of TOPIC(Tohoku Open Internet Community, AS2503)[TOPIC], a network which connects several universities, colleges, museums, and other academic organizations.

Tab.1 shows some statistics of the traffic on the TOPIC network and on another stub LAN connected to TOPIC. The data is obtained by analyzing the packets collected by a traffic monitor over a 24-hour period. The figures of packets dropped are only rough estimates.

**Table 1** Traffic characteristics

| target network | TOPIC backbone | LAN |
|---|---|---|
| number of sender hosts | 19028 | 570 |
| number of receiver hosts | 19317 | 538 |
| number of host pairs | 109660 | 1479 |
| number of ICMP dest unreachable packets | 17176 | 82 |
| number of ICMP dest unreachable sender hosts | 482 | 18 |
| number of ICMP dest unreachable receiver hosts | 565 | 12 |
| number of ICMP dest unreachable host pairs | 1194 | 19 |
| number of lost packets | 167975 | 9501 |
| total packets | 38064708 | 6638313 |

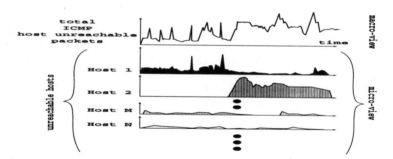

**Figure 2** Macro-view & Micro-view

The data shows that for the small scale LAN the traffic is small, the number of sources and destinations are small, so are the number of errors packets e.g. ICMP Destination Un-Reachable packets (ICMP-DUR). In this case, it may be possible to have a fault detection system do an exhaustive examination of each packet in the network.

But the data for the medium-size transit network TOPIC shows very different characteristics. The sample data shows that there were 17176 ICMP-DUR packets during the period of observation. This figure is abnormal. Yet it is only a macro-view and requires the relevant packets to be analyzed. Analysis of the relevant packets showed that 482 destinations were unreachable from 565 sources. In other words, there were 482 symptoms of probable faults in the network during the period of observation.

To examine the time pattern of the symptoms consolidated in the macro-view of ICMP-

DUR packet count we filtered out all the ICMP-DUR packets from the data sample. Then we sorted and separated the different *Destination-IP addresses* in the filtered sample. Each unreachable destination represents a symptom. The various symptoms are shown in fig.2. It is evident from the analysis that the aggregated macro-view represented by the total number of ICMP-DUR packets is a consolidation of several symptoms - each representing the unreachability of a different host, shown in the graphs of fig.2.

The above analysis was done offline. The point to be noted is that, it is impractical to examine the internals of each packet for online-management purposes even with dedicated machines[KIM93][STA93].

In Fault Management it is necessary to use the macro-view to select a potential problem spot, an event. And, then focus on the micro-view of that event.

# 3 FAULT DETECTION SYSTEM WITH DIVIDE AND CONQUER TECHNIQUE

## 3.1 Fault detection model

In our model (fig.3) of fault detection there are essentially three parts. An event, E, is detected in the Event-detection phase. This event is essentially a macro-view that indicates the presence of one or more symptoms of faults. The NMS focuses on the set of symptoms $S'$, $S' \subseteq S$, that are likely to have triggered the event, and finds $S''$ the set of symptoms that did trigger E. The NMS then proceeds with the diagnosis of the symptoms and corrective procedures for the corresponding faults. Further, having isolated the symptoms, it filters out these symptoms from subsequent observations. Detailed explanation of each phase follows.

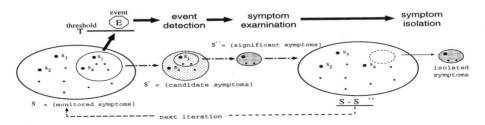

**Figure 3** Fault detection model

### Event detection
Ideally an event E is triggered when, the amplitude, $a(s_i)$, of a symptom $s_i$ exceeds some threshold $t_i$ for $s_i$.

$$a(s_i) > t_i \implies E = true$$

However, for practical purposes, instead of applying the threshold to individual symptoms $s_i$, it is more effective to to apply the threshold to an aggregated set of symptoms $S = \{s_i\}$.

$$\sum_i a(s_i) > T, s_i \in S \implies E = true$$

It is clear that two or more symptoms may cooperate to trigger an event. Thus the NMS has to carry out the process of identifying the symptoms that did trigger the event. Further, the threshold $T$ is in general set for a single symptom. This threshold when applied to the aggregated amplitude of several symptoms is small and, is almost always exceeded.

### Symptom examination

Though several symptoms $S'$, $S' \subseteq S$, cooperate to trigger an event, all symptoms are not significant. The NMS will use some mechanism to isolate the significant symptoms $S''$, $S'' \subseteq S'$.

$$S'' = \{s_i | s_i \text{ is significant}, s_i \in S'\}$$

The significance of a symptom may be gauged by one or more characteristics of the symptom e.g amplitude, frequency, persistence or by specific knowledge about the symptom. The symptoms are then used for diagnosis.

### Symptom isolation

To avoid the preponderance of already detected, persistent symptoms , the set of significant symptoms $S_t''$ from the Symptom examination phase in the $t^{th}$ cycle are suppressed in the subsequent event detection phase. Thus the symptom space $S_{t+1}$ in the $t + 1^{th}$ cycle is given by

$$S_{t+1} = S_t - S_t''$$

## 3.2   The Divide and Conquer technique

Fig.4 shows the basic components for our proposed event-driven, dynamic symptom isolation system. First, an event is detected by monitoring aggregated characteristics. Next, symptoms that triggered the detected event are analyzed. Significant symptom(s) are isolated from the aggregated characteristics of the subsequent cycle, in the last component. This cyclic process isolates the symptoms iteratively.

## 4  IMPLEMENTATION OF THE DIVIDE AND CONQUER TECHNIQUE

The authors implemented and experimented with the Divide And Conquer Technique. The implementation was on the SNMP platform and used the RMON-MIB[RFC1757] for traffic analysis. The RMON-MIB is a powerful tool for traffic analysis and provides

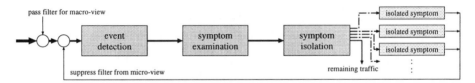

pass filter for macro-view

event detection

symptom examination

symptom isolation

isolated symptom

isolated symptom

isolated symptom

suppress filter from micro-view

remaining traffic

**Figure 4** Basic components of iterative isolation

mechanisms for applying filters for analyzing traffic, and to configure alarms and events, by using SNMP-set and get constructs.

An NMS in general needs to watch out for several types of events. The data for some of these events are readily available from counter values of predefined MOs. For the remaining events the RMON Agent needs to be configured to generate the data by analyzing the traffic. This generally involves applying filters and counting packets that pass the filter. Though theoretically an infinite number of filters may be applied and a watchout may be maintained for all possible events, there is a practical limit to the number of filters that can be used and hence on the number of events that can be watched out for simultaneously. To get around this problem we used sampling techniques. It essentially involves time-division multiplexing - in each time slot a particular event is checked for by applying the appropriate the set of *pass* filters.

Though the NMS in general scanned for several events, in the following we will concentrate on events due to the ICMP-Host-Unreachable(ICMP-DUR) packets. Since there is no ICMP-DUR counter readily available, we had to configure the RMON-Agent to generate the count. This was done in the **Event Detection** phase by sampling the traffic and filtering for ICMP-DUR packets. In this phase *suppress* filters are applied to suppress known symptoms (detected in the **Symptom Examination** phase), if any. The sampling was necessitated for overall performance purposes. An event would be triggered when the number ICMP-DUR packets exceeded the threshold.

On being notified of an event (by a trap) the NMS would begin the **Symptom Examination** phase. The NMS starts off a packet-capture process. All ICMP-DUR packets are captured and analyzed by the NMS. The analysis is carried out by sorting the packets by the IP-address of the unreachable destination (in the IP-header that is sent as data in the ICMP packet). Each unreachable destination is a symptom. The symptoms are then arranged in frequency of occurance. In the **Symptom Isolation** phase, the Top N symptoms greater than the threshold are selected as significant symptoms indicating real faults, needing diagnosis. Filters are made corresponding to these symptoms and are used as suppression filters in the **Event Detection** phase.

## 4.1 Structure of the system

An outline of the implementation is shown in fig.5. In the figure, rounded rectangles represent the RMON-MIB MOs, shaded areas represent the components of the concept, and shaded rectangles represent sub-processes. Arrows on continuous lines represent packet flows, arrows on broken lines represent instructions from the NMS.

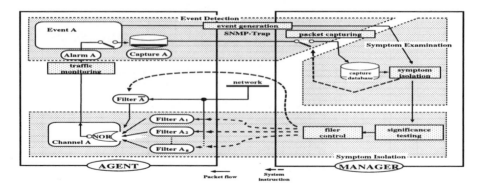

**Figure 5** Implementation using RMON technology

## 4.2   Event detection

In this phase traffic monitoring and event generation takes place. The manager sets the filter and channel MOs for a particular macro-view or event, Filter A and Channel A in fig.5. The agent monitors the target traffic which passes through the pass-filter for the macro-view covering the set of symptoms $S$. An alarm $A$ is generated when the monitored statistic is over the predefined threshold $T$. The alarm $A$ in turn triggers an event $E$ which starts the packet capture process in the capture MO and sends a notification about the event to the NMS. (Via a trap).

## 4.3   Symptom examination

The Symptom examination process comprises of two sub-processes, packet capturing and symptom analysis. Once the NMS receives notification of the event E, it will want to examine the symptoms that are manifested in the traffic characteristics. For this purpose, the NMS needs a trace of the traffic. The NMS obtains the trace from the RMON-agent by polling the corresponding capture MO, Capture A in fig.5. From the trace the NMS finds the symptoms $S'$ that cooperated to trigger the event.

## 4.4   Symptom isolation

In this phase the NMS picks the significant symptoms $S''$ from $S'$, using some criteria, for diagnosis. The NMS decides whether a symptom is significant or not by analyzing its characteristics. A simple decision can be made based on frequency. The symptom that is occurring frequently, probably needs attention and is significant. On the other hand there may be a database of symptoms which may be looked up to ascertain the severity of an event. For example, if a DNS server has become unreachable - it is certainly a severe fault that may affect the network users. In the pilot implementation, we used the Top-N method to determine whether a symptom is significant. The symptoms are sorted by frequency, and the top $N$ symptoms are considered to be significant. The figure N is implementation specific and in general depends on the number of available filters, etc.

Next, the manager sets the filters corresponding to the isolated symptoms to suppress these already-detected symptoms in the subsequent traffic. The reduced set of symptoms are

$$S_{t+1} = S_t \cap \overline{S''}$$

In the RMON architecture the above translates to

$$S_{t+1} = \overline{\overline{S_t} \cup S''}$$

In fig.5, Filter $\overline{A}$ corresponds to $\overline{S_t}$ and Filters $A_1 \sim A_N$ correspond to the isolated symptoms in $S''$. Channel A corresponds to the RMON-MIB channel object where the Filters A, $A_1 \sim A_N$ are combined. When the threshold for a particular macro-view is crossed an alarm is triggered. This alarm is configured in the RMON-MIB alarm MO, Alarm A in fig.5. The alarm A in turn is configured to trigger an event MO, Event A, which sets off a trap to the NMS notifying the NMS of the event and, starts the packet capture in the MO Capture A.

## 5  PERFORMANCE EVALUATION

### 5.1  Fault detection by monitoring ICMP packet flow

To evaluate the proposed system, we experimented on fault detection by monitoring the ICMP traffic in the network. ICMP belongs to the TCP/IP protocol suit and is primarily used between network entities to notify problems about network reachability, congestion, packet loss, etc.

### 5.2  System configuration for evaluation

The experimentation and system evaluation was carried out on the TOPIC backbone network which is an Ethernet(fig.1). A dedicated RMON-Agent was connected to the Ethernet. We concentrated on the ICMP destination unreachable packets.

The traffic was sampled every 5 minutes for a duration of 1 minute. For traffic analysis the sampled traffic of the most recent 180 minutes was used. Furthermore, since the RMON device we used offered a maximum number of 10 filters - we employed 8 for symptom (suppression) filters and 2 for the event (pass) filters.

### 5.3  Results and Observations

*Order from Chaos: tractability of event monitoring*
Fig.6-a shows the total number of ICMP-DUR packets. A very large number of ICMP-DUR packets are observed and there is a wild fluctuation in the number. It is difficult to

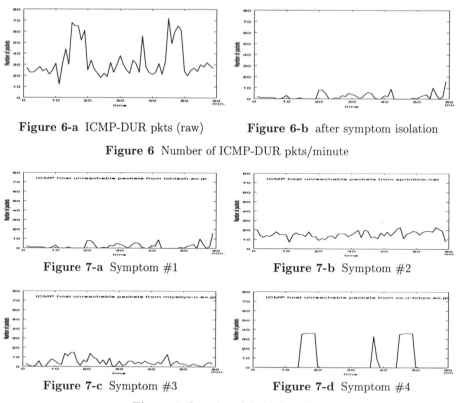

**Figure 6-a** ICMP-DUR pkts (raw)          **Figure 6-b** after symptom isolation

**Figure 6** Number of ICMP-DUR pkts/minute

**Figure 7-a** Symptom #1          **Figure 7-b** Symptom #2

**Figure 7-c** Symptom #3          **Figure 7-d** Symptom #4

**Figure 7** Samples of divided traffic

make much sense from the graph apart from forming the general idea that there are "too many" ICMP-DUR packets. Also, setting an event threshold at any sensible value seems to be impossible ! On the other hand fig.6-b shows the count of ICMP-DUR packets after the symptoms have been isolated. This graph shows that the problem-space (number of unresolved or unknown symptoms of faults present in the traffic) is well within control. There are no wild fluctuations. Occasional peaks do get resolved i.e. the corresponding symptoms do get identified and isolated quickly.

In fig.7 we show the characteristics of some of the isolated symptoms corresponding to the one hour period in fig.6. (The order of the graphs is insignificant). From the graphs it is evident that for detection purposes symptoms may be individually subjected to a uniform threshold. This is a very significant result corroborating our claim of the usability of a simple thresholding mechanism for fault detection and management.

## Symptom detection capability

To measure the impact on symptom detection capability we compared the performance of our system which uses the the divide and conquer technique with a system that doesn't.

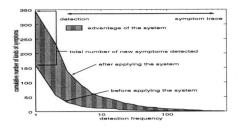
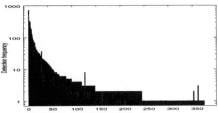

**Figure 8** Overall evaluation          **Figure 9** Detection accuracy

Both systems used identical RMON-Agents to set appropriate filters for alarms, to obtain event information and to carry out packet capture. Both systems analyzed the captured traffic to identify symptoms and employed similar thresholding mechanisms.

The results are shown in fig.8. The ordinate represents the number of symptoms, and the abscissa represents the minimum number of times the corresponding symptoms were detected. We can see that the number of symptoms which were detected more than 10 times using the divide and conquer technique, is 50 and, 12 otherwise. Of particular interest is the number of symptoms detected (at least once). It is evident that using the divide and conquer technique the symptom detection capability is significantly ( > 200% ) improved.

### Symptom detection accuracy

Fig.9 compares the accuracy with which the two systems detect symptoms. The figure shows the number of times each symptom is detected by the system using the divide and conquer mechanism and the one that doesn't. The dark and the gray bars represent the number of times the corresponding symptom has been detected by the divide and conquer system and the traditional system, respectively. The symptoms are ordered by frequency of occurance (as seen by the system that employs the divide and conquer technique). It is clear that the divide and conquer mechanism does increase the accuracy of symptom detection significantly. In the case of some symptoms it does appear that divide and conquer mechanism is less accurate. Particularly towards the tail end of the spectrum. It may be noted that there is an inherent latency in the symptom detection mechanism. It takes a finite amount of time after the event detection for the system to start the capture process, and then do the symptom examination. Symptoms which have very short persistence, disappear in this time and thus elude detection. This syndrome is present in both systems. Yet, since the two systems are not synchronized in the strict sense, some symptoms elude one system and are caught in the other. The concentration of symptoms that eluded the divide and conquer technique based system at the tail of the spectrum, is attributable to the ordering of the symptoms in the figure.

## 6   CONCLUSION

In this paper, we have focussed on the issue of network fault management by traffic monitoring. We have proposed a simple and powerful technique, the divide and conquer technique which makes the problem of detecting symptoms of faults in the network more tractable. By applying this technique, one can use simple thresholding mechanisms for setting alarms and detecting events in the network. We have implemented the system using standard network management protocols and technology. We have compared its performance with conventional systems. The divide and conquer mechanism does enhance the performance significantly- the spectrum of symptoms detected are broadened and the accuracy with which symptoms are detected is increased.

# REFERENCES

[KLI95] S. Kliger et.al "A coding approach to Event Correlation ", Proceedings , Fourth International Symposium on Integrated Network Management, 1995.

[HOU95] K. Houck et.al "Towards a Practical Alarm Correlation System ", Proceedings , Fourth International Symposium on Integrated Network Management, 1995.

[GLN96] G.Mansfield, M.Ouchi, K.Jayanthi, Y.Kimura, K.Ohta, Y.Nemoto, "Techniques for automated Network Map Generation using SNMP", Proc. of INFOCOM'96, pp.473-480, March 1996.

[DUP89] Dupuy.A, et.al " Network Fault Management :A User's View", Proceedings , First International Symposium on Integrated Network Management(May), 1989.

[STA93] William Stallings: "SNMP, SNMPv2, and CMIP - the Practical Guide to Network Management Standards -", Addison-Wesley Publishing Company 1993.

[RFC792] SJ. Postel "Internet Control Message Protocol ",RFC 0792 09/01/1981

[LAB91] Lee LaBarre "Management by Exception: OSI Event Generation, Reporting and Logging ", Proceedings , Second International Symposium on Integrated Network Management, 1991.

[DAU91] Steven M. DAUBER, "FINDING FAULT", BYTE Magazine, March 1991.

[GLN92] G.Mansfield, et.al "An SNMP-based Expert Network Management System",IEICE Trans. COMMUN.,Vol.E75-B,NO.8, pp.701-708, August 1992.

[KOH95] K.Ohta, et.al "Configuring a Network Management System for Efficient Operation", International journal of network management, Vol.6, No.2 March-April 1996.

[TCPDUMP] V. Jacobson, C.Leres, and S.McCanne, "tcpdump", available via a-FTP to ftp.ee.lbl.gov, June, 1989.

[NNSTAT] R.R. Braden and A. Deschon. NNStat, "Internet statistics collection package. Introduction and User Guide", Technical Report RR-88-206, ISI, USC, 1988. Available for a-ftp from isi.edu.

[RFC1757] S. Waldbusser, "Remote Network Monitoring Management Information Base", RFC 1757 02/10/1995.

[TOPIC] Description about TOPIC at: http://www.topic.ad.jp/

[KIM93] Kimbery C Claffy and George C. Polyzos "Application of Sampling Methodologies to Network Traffic Characterization", IEEE SIGCOM'93 1993.

[RFC1157] M. Schoffstall et al.: "Simple Network Management Protocol SNMP", RFC1157,1990.

# 7 BIOGRAPHY

**Kohei Ohta** received his B.E and M.S degrees in information engineering from Tohoku University in 1993 and 1995, respectively. Now he is a doctoral student at Graduate School of Information Sciences of Tohoku University. He has been engaged in research on network management.

**Takumi Mori** is a senior of information engineering of Tohoku University. He has been engaged in research on network management.

**Nei Kato** received his M.S. and the Dr.Eng. degrees in information engineering from Tohoku University in 1988 and 1991, respectively. Now he is an associate professor at Graduate School of Information Sciences, Tohoku University. He has been engaged in research on pattern recognition, neural network and computer networking. Dr. Kato is a member of IEEE and the Information Processing Society of Japan.

**Hideaki Sone** received his B.E. in Electrical Engineering and M.E. and Doctoral Degree in Electrical Communications from Tohoku University, Sendai, Japan. He is an Associate Professor in Research Institute of Electrical Communication, and is working with Computer Center, Tohoku University. His main research interest lies in the fields of communication systems and instrumentation electronics. Dr. Sone is a member of the IEICE, the IEEJ, the SICE, the IPSJ, and the IEEE.

**Glenn Mansfield** obtained his Master's degree in 1977 from Indian Institute of Technology, Kharagpore, India in the field of Nuclear and Particle Physics followed by his Masters in Physical Engineering in 1979 from Indian Institute of Science, Bangalore, India. He obtained his Ph.D. specializing in Logic programming, from Tohoku University, Japan. He is currently chief scientist at Sendai Foundation for Applied Information Sciences, Japan. His areas of interest include expert systems, logic programming, computer networks and their management, use of the Internet for education. He is a member of the Internet Society, the ACM, the IEEE and the IEEE Communications Society.

**Yoshiaki Nemoto** received his B.E. and M.E. and Ph.D. degrees from Tohoku University, Sendai, Japan, in 1968, 1970 and 1973 respectively. He is a professor with Graduate School of Information Sciences, Tohoku University. He has been engaged in research work on microwave networks, communication system, computer network system, image processing and handwritten character recognition. Dr. Nemoto is a member of the IEEE and the Information Processing Society of Japan.

# 51
# Automated Proactive Anomaly Detection

*C. S. Hood and C. Ji*[†]
*Department of Computer Science and Applied Mathematics*
*Illinois Institute of Technology, Chicago, IL USA 60616*
*chood@charlie.cns.iit.edu*

[†]*Department of Electrical, Computer and Systems Engineering*
*Rensselaer Polytechnic Institute, Troy, NY USA 12180*
*chuanyi@ecse.rpi.edu*

## Abstract

To address the increasing complexities of fault management, we propose an automated, proactive monitoring system using adaptive statistical techniques. Requiring only a minimal amount of network specific information a priori, the system continually collects data, uses the data to learn the normal behavior of the network, and detects deviations from the norm. The proposed system is thereby able to detect unknown or unseen faults. Experimental results on real network data demonstrate that the proposed system can detect abnormal behavior before a fault actually occurs.

## Keywords

Fault detection, proactive network monitoring, statistical learning methods, MIB, SNMP

## 1 INTRODUCTION

Fault management is the part of network management responsible for detecting and identifying faults in the network. Interest in fault management has increased over the past decade due to the growing number of networks that have become a critical component of the infrastructure of many organizations, making faults and downtime very costly. In addition, as computer networks evolve from providing only "best effort" service to providing a range of service guarantees to accommodate real-time applications (e.g. video), higher levels of reliability are required. By preserving network reliability, fault management lays the foundation for the stringent Quality of Service (QoS) requirements placed on networks by real-time applications.

As the fault management problem becomes more important, it has also become more difficult. This can be traced primarily to the dynamic nature and heterogeneity of current networks. Fundamental changes to the network occur much more frequently due to the growing demands on the network and the availability of new, improved components and applications. With network components and applications developed in an open environment, a network can be configured by mixing and matching several vendors' hardware and software. While this allows the network to utilize the latest technologies and be customized to the needs of the users, it also increases the risk of faults or problems [18].

Previous research in fault management has covered approaches such as expert systems [7], Finite State Machines (FSMs) [13], advanced database techniques [17], and probabilistic approaches [3]. A review of communication network fault detection and identification can be found in [8]. The approaches mentioned above require specification of the faults to be detected. This limits the performance of these approaches since it is not feasible to specify all possible faults. In addition, changes in network configuration, applications and traffic can change the types and nature of faults that may occur, making modeling faults more difficult and in many cases impractical. Research using learning machines to detect anomalies [11] addresses the issue of fault modeling, but does not provide a method for correlating the information collected in space or time.

The problem we will tackle is automated fault detection without specific models of faults. We propose an adaptive learning system for network monitoring. The system learns the normal behavior of each measurement variable. Deviations from the norm are detected and the information gathered is combined in the probabilistic framework of a Bayesian network. Benefits from this approach include the ability to detect unknown faults, the ability to correlate information in space and time, and the ability to detect subtle changes occurring before the actual failure. This allows faults to be detected when they are developing so the network manager has time to take corrective action to prevent outages or downtime. In addition, this approach requires minimal amounts of network specific information, so it can be generalized across network nodes and types of networks. Our approach is tested on a computer network. We monitor the Management Information Base (MIB) variables collected within the Simple Network Management Protocol (SNMP) framework. No specialized hardware is required for monitoring.

The paper is organized as follows. Section 2 provides background material on Bayesian networks. Our intelligent monitoring approach is described in Section 3. Detailed information about the data we collected is given in Section 4, and Section 5 contains results and comparisons. Conclusions and areas for further investigation are discussed in Section 6.

## 2   BAYESIAN NETWORK BACKGROUND

A Bayesian network, also called a belief network or a causal network, is a graphical representation of relationships within a problem domain. More formally, a Bayesian network is a directed acyclic graph (DAG), where certain conditional independence assumptions hold [7]. The nodes of the DAG represent random variables. The conditional independence assumptions are as follows:   Given a DAG $G = (N,E)$, where $n \in N$ is a node in the network and $e \in E$ is a directed arc.  For each $n \in N$, let $p(n) \subseteq N$ be the set of all parents of $n$, and $d(n) \subseteq N$ be the set of all descendents of $n$.  For every subset $W \subseteq N - (d(n) \cup \{n\})$, $W$ and $n$ are conditionally independent given $p(n)$.  In other words, for any node in the DAG, given that node's parents, that node is independent of any other node that is not its descendent.  Figure 1 illustrates the independence assumptions for a Bayesian network similar to the one we will use in our monitoring system.

These assumptions allow us to estimate the conditional probabilities of any of the nodes (or random variables) in the Bayesian network given the observed information or evidence. The strength of Bayesian networks is that they provide a theoretical framework for combining statistical data with prior knowledge about the problem domain.  Therefore, they are particularly useful in practical applications.

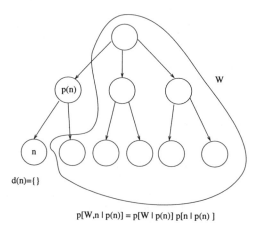

$$p[W,n \mid p(n)] = p[W \mid p(n)] \, p[n \mid p(n) \,]$$

**Figure 1**  Example of Bayesian network independence assumptions.

Bayesian networks have been widely used for medical diagnosis [15] [4], troubleshooting [5], and in the communication network field, they have been proposed to diagnose faults in Linear Lightwave Networks [6].  In [6] other methods have been used for detection and the Bayesian networks are used for diagnosis only.  In this work, we propose using a Bayesian network as a mechanism to combine information from different variables for the purpose of detecting anomalies.

## 3   MONITORING SYSTEM

We propose an automated monitoring system that is able to detect anomalies without specific models of the behavior to be detected.  The premise of this approach is that anomalous or unusual network behavior is an indication of a fault within the network. The logical flow of information through the monitoring system components are shown in Figure 2.

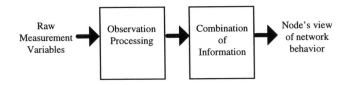

**Figure 2** Logical flow of information through the monitoring system.

The system resides locally, allowing each node in the network to compose a picture of the network's health. To get this picture, measurement information must be combined with prior knowledge about the network. To accomplish this, the monitoring system has two main components; observation processing and combination of information. The raw measurement variables are processed to estimate the probability of each measured variable at a given time. The probabilities are then combined using a Bayesian network to provide a broader picture of the network's behavior. By doing this locally, we can correlate this information in time and space. This allows the central network manager to receive a more complete, less noisy picture of each node's view of network health. This can ease the alarm correlation problem. It also allows the node to take corrective actions if necessary.

## 3.1 Observation processing

The goal of the observation processing part of the monitoring system is to take the raw measurement variables and transform them into a set of measures indicating the behavior of each variable. Each measurement variable is a time series. Many of the measurement variables are representative of network traffic. To date, the characterization of network traffic signals is an active research area [9]. Therefore, the signals (i.e. measurement variables) to be processed are not considered to be well understood and as such there is not an optimal, or even standard method to characterize the behavior of these signals.

In processing the information we use a change detection methodology. Since the behavior of the network is dynamic, the behavior of the measurement variables change frequently. As most changes are related to network traffic, simply detecting that a change has occurred is not enough. The goal is to try to recognize the changes that are important in terms of fault detection. We do this by characterizing the behavior of the measurement variables.

*Segmentation*
One of the challenges presented by the network dynamics is the non-stationarity of the observations. Since our goal is to extract pertinent information, we need to group the time series data in some way so that features can be calculated. To do this we segment the data into variable length pieces. Each piece contains a portion of the time series that is statistically similar.

There are two primary benefits realized from segmentation: (1) the statistics calculated from each segment are more representative of the signal, and (2) signal processing techniques requiring a stationary signal can be used within each segment. In terms of monitoring, the segmentation provides the benefit of temporally correlating the observations. Since many of the network signals are bursty, the temporal correlation can help distinguish between a burst and a change in the nature of the signal. The sequential segmentation algorithm described in [1] is used. Once the observations have been grouped into segments, the pertinent information must be captured from each segment.

## Feature extraction

Before our approach to feature extraction is discussed, we first need to examine the shortcomings of commonly used methods. Thresholds are the primary method currently used in both practice [10] and research [3] for detecting abnormal behavior. The feature is not the value of the threshold itself, but the information on whether or not the threshold has been exceeded by a particular measurement variable. One of the difficulties with thresholds is properly setting the threshold level, since they are highly dependent on the traffic level.

While properly set thresholds do a good job of detecting large rises and falls in a measurement variable, more subtle behavior changes are missed. For example, a change in the variance of a signal, or a subtle change in the mean will not be detected using thresholds. These types of changes may be symptoms of something problematic in the network - this behavior is unusual for the variable. Detection of the more subtle signs of problems may allow corrective action to be taken to avoid a bigger problem. Identification of the problem also becomes easier with a more complete description of the symptoms rather than just the extreme cases.

Our goal is to extract information that will help determine whether the behavior of the measurement variable is normal or abnormal. Ideally we want to capture information in the signal that will change or become abnormal only when a problem is occurring. To do this we would need a feature that is invariant to the network traffic patterns and other influences that cause the non-stationarities. This is an open problem, so we choose a feature that changes along with the network and continually adapt the model of normal behavior.

To detect the more subtle changes in the nature of the measured variables, we use the parameters of an AR(2) or second-order autoregressive process as features. The AR(2) process is defined as

$$y(t) = a_1 y(t-1) + a_2 y(t-2) + \varepsilon(t),$$

where $y(t)$ is the value of the signal at time $t$, $a_1$ and $a_2$ are the AR parameters that we use as features, and $\varepsilon(t)$ is white noise. Additional features have been investigated in [6].

## Learning the behavior

The features provide the ability to detect subtle changes, but they must first be used to establish a description of normal behavior. The description of the behavior is in the form of a probability distribution. Changes can then be detected relative to the distribution.

Ideally, for each sample we would like to estimate its likelihood given that, (1) the network is operating normally, and (2) there is a fault. To do this, we need to know exactly when the network as a whole, and each of its functions is operating normally. This type of information is not typically available. We are able to get some information about the health of the network from the tools currently used to report problems (i.e., log files). The information available contains reports on certain types of serious network problems that have occurred. From a learning perspective, we can use these reports as labels.

One way of using the labels is to learn the probability distribution of each measurement variable, given that its related network function is abnormal. This sounds promising, but in fact is extremely difficult due to the sparse nature of abnormal data. With so few examples of problems, we do not have the variety (i.e., many examples of different problems) or the depth (i.e., several examples of the same type of problem) to effectively learn the distribution [2].

Instead we learn only the likelihood of the sample given the network function is normal. We define normal behavior to be the behavior of the variable during the time period when the distribution is learned. This time period will be referred to as the learning window. Since what we are learning is the usual behavior of the variable during the learning window, there is no guarantee that this always corresponds to the network or network functions being healthy. If there is a problem in the network that impacts the behavior of the network for a significant portion of the learning window, the problematic behavior will be learned as the usual behavior. We assume that this is rarely the case.

## 3.2 Combination of Information

The goal of this part of the monitoring system is to combine the processed measurement variable information into higher-level measures of network behavior. These higher-level measures provide the monitored node's view of the network behavior. These measures can be used to trigger local control actions or a message to a centralized network manager.

Each measurement variable measure is combined in the probabilistic framework of a Bayesian network. The Bayesian network is used because it provides a method for estimating probabilities and it allows observed information to be combined with prior knowledge.

We begin by defining the random variables or nodes in the Bayesian network. There are two types of variables, those that are observed and those that are not observed and thus need to be estimated. We will call the variables that are not observed the internal variables. The observed variables directly correspond to the MIB variables. The internal variables are defined to be Network, Interface (IF), IP, and UDP. The IF, IP, and UDP variables correspond to the MIB groups. Logically they represent different types of network functionality. The MIB variables within a group are the measurement variables for that network function (nf).

The Network variable is defined to correspond to all of the network functionality. In this work, we assume that the network is comprised only of the IF, IP, and UDP functions since these were the functions being used at the router where we were monitoring. Other network functions can easily be added. The structure of the Bayesian network is shown in Figure 3. The arrows between the nodes in Figure 3 go from cause to effect.

In our model, the health of the network is the most general information estimated and can be considered to be an underlying influence on the rest of the nodes in the Bayesian network.

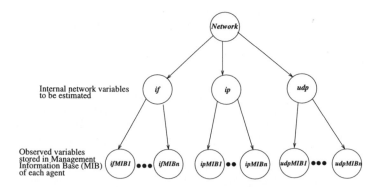

**Figure 3** Bayesian network for fault detection.

The overall health of the network directly influences the health of the three functions of the network (i.e., IF, IP, and UDP). This is indicated in our model by the arrows from the network random variable to the IF, IP, and UDP random variables.

The model has been designed based on the intuitive relationships from the structure of the MIB. The Bayesian network model requires the conditional independence assumptions described in Section 2. These assumptions are reasonable since each of the network functions represent independent functional components of the network. These components may fail independently, although there is a relationship between the functions, and serious problems in one component can eventually impact the other components. Since propagation of a fault through the functional components depends on the type and location of the fault (i.e., faults may propagate from low-level functions to high-level functions, or vice versa) [16], it is difficult to incorporate a propagation structure into the model that will accommodate all types of propagation.

In addition, the relationships between the nodes of the Bayesian network (network functions and MIB variables) are complex and not well understood. Therefore, as a starting point, we have proposed a simple model where no a priori relationship between the network functions is assumed, given the overall health of the network is known. Therefore we have not assumed a fault propagation structure in our model. Alternative Bayesian network structures that assume fault propagation structures have been investigated in [6].

Since we are monitoring locally, all of the evidence or probabilities estimated from the observed MIB variables is available to the system. This enables the system to calculate the desired posterior probabilities using a complete and current set of observations.

## 4   DATA COLLECTION

Data for this work was collected from the RPI Computer Science Department network. The network as shown in Figure 4 is comprised of 7 subnetworks, or subnets, and two routers. The individual nodes (e.g., workstations, printers, etc.) on the subnets are not shown. Router 1 is the gateway between this network and the campus network, with all the traffic to and from the campus and the outside world flowing through this router. Router 2 mainly routes the local traffic flowing between the subnetworks. A large portion of this traffic is access from workstations to the fileservers. Data was collected from Router 2, the internal router.

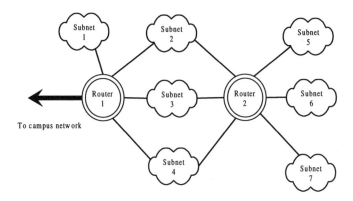

**Figure 4** Configuration of the monitored network.

The data was collected by polling the router using SNMP queries. The router was polled every 15 seconds. All of the available router MIB variables were collected. Of these variables, we studied the 14 that were active. The other variables changed infrequently and during the times of recorded faults, they rarely provided information that was not provided already by the active variables.

We monitored the router for approximately seven months. The monitoring was continuous for most of the time. During this time the log files generated by the syslog function were also saved to label the faults within the data. As syslog is not targeted specifically for network errors, only a subset of all network problems will be reported by syslog. The network related problems reported by syslog are usually severe. In fact, the type of network problem we found to be most commonly reported by syslog was server not responding. This type of message indicates a severe problem that may be occurring for a number of reasons (e.g. server down, path unavailable). The syslog messages do not provide any information about the cause of the problem.

## 5 EXPERIMENTAL RESULTS

The system proposed for anomaly detection was tested on a set of 10 faults observed on the network in Figure 4 between October 1995 and March 1996. Most of the faults (9 out of 10) have been recorded as server not responding. The remaining fault is a report of excess Ethernet collisions on one of the subnets. Due to the mechanism currently used to log faults on this network (syslog), we could only observe types of faults where a service provided by the network is not operational. This mechanism provided accurate reports of severe faults, but no account of less severe network faults.

Although nine of the faults studied were the same, we did not observe the same types of changes in the data from fault to fault. This can possibly be traced to the fact that the faults were caused by different sets of circumstances or root causes. Even if the root causes were similar, the problems could have manifested themselves differently due to the network environment at the time they occurred. Specific root cause information may not be available without knowledge of the implementation details of the nodes comprising the network.

The results along with comparisons with threshold methods are shown for one of the faults. The results for the remaining faults are summarized.

## 5.1 Our Results

One of the faults the system was tested on was a fault that was reported as server not responding between 6:33 am and 6:36 am on December 23, 1996. The fileserver that was not responding was on Subnet 2. A total of 13 machines reported this problem ( 7 on Subnet 2, 4 on Subnet 3, and 2 on Subnet 4). The results are shown in Figure 5.

The asterisks denote the fileserver downtime period. Abnormal behavior is detected in the Network approximately 12 minutes before the server is reported unreachable. Anomalies are present before the problem in all three network functions, but only IP detects an anomaly during the crash. Similar results were obtained when a 4 hour learning window was used.

It is important to keep in mind the structure of the network in Figure 4. The router that we are monitoring continues to route all other traffic normally. The fileserver being down is not problematic to the router, but still we are able to detect the problem by monitoring the router. The router is able to detect that there is an anomaly in the network through changes to its MIB variables. Therefore, the results that we have are from the routers view of the network.

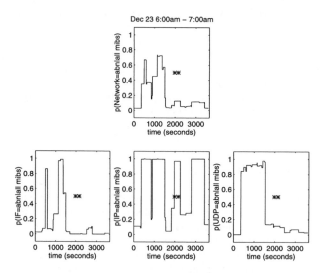

**Figure 5** Results using 1 hour learning window.

## 5.2 Composite results

More generally, 7 out of the 10 faults studied were detected using a 1 hour learning window and 5 out of 10 were detected using a 4 hour learning window. We considered a fault to be detected if the posterior probability that Network is abnormal is greater than 0.5.

These results need to be put into perspective. Ideally, we could calculate the number of times that a fault is detected when there is no fault (false alarms). Since we only have labels from the log file for the severe faults, it is not clear where other faults should or should not be detected. Therefore, to get some measure of the sensitivity we calculated the percentage of time that the posterior probability of the Network, IF, IP, and UDP is abnormal. The following figures indicate the percentage of abnormal time.

Network abnormal
- 6.29% of time using 1 hr learning window
- 3.85% of time using 4 hr learning window

IF abnormal
- 35.97% of time using 1 hr learning window
- 24.85% of time using 4 hr learning window

IP abnormal
- 42.03% of time using 1 hr learning window
- 33.32% of time using 4 hr learning window

UDP abnormal
- 42.24% of time using 1 hr learning window
- 30.29% of time using 4 hr learning window.

## 5.3 Comparisons

Since thresholds are commonly used to detect faults, we compared our results to those obtained using a feature set of an upper and lower threshold. The feature is whether the particular variable is within the thresholds or not. To combine the information from each MIB variable, we counted the total number of variables exceeding their thresholds at each time instance. The thresholds were calculated using learning windows of 1 hour, 4 hours and 1 week. The first two correspond to the windows used by the monitoring system. The third corresponds to the common practice of determining threshold levels using a large amount of data.

The results from all of these methods are shown in Figure 6. The asterisks denote the period where the fileserver was down. The results for 1 and 4 hour learning windows are identical. Both have small peaks where thresholds have been exceeded by 3 of the 14 MIB variables. The results using a 1 week learning window are essentially the same for the entire hour, thereby providing no useful information.

Abnormalities at the highest level of our monitoring system (Network) are being detected a small percentage of the time. Therefore the detection of the test faults is significant. On the other hand, the faults that were not detected may not have had symptoms present at the router, or the features we used may not have captured adequate information to detect symptoms that were present.

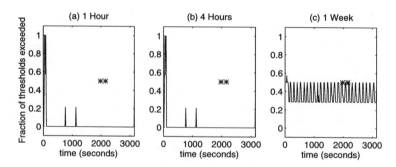

**Figure 6** Results using thresholds.

## 6    CONCLUSION

We have shown that it is possible to use an adaptive learning machine to detect network faults without using models of specific faults. The Bayesian network provided a theoretical framework within which we were able to use prior knowledge to determine a structure and learn the normal behavior of the measurement variables. The system was tested on real data involving a fileserver crash on a computer network. It was able to successfully detect something abnormal approximately 12 minutes before the fileserver crashed. The detection was done from a router on the network which was operating properly. The router detected that something was wrong in the network from changes to its measurement variables.

With early detection, the network manager can be warned of impending failure, take corrective action and avoid the failure and costly downtime. Our approach is able to accomplish early detection by recognizing deviations from normal behavior in each of the measurement variables, correlating this information in time and then combining the information in a probabilistic framework.

This work is viewed as a first step in the direction of an automated fault management system that can generalize from network to network with minimal network specific information required a priori. The fault management problem is very complex and the nature of the problem evolves as networks evolve. Future work involves expanding the scope of the experiment, as well as further investigation of the methods.

The Bayesian formulation of this detection problem can be extended to incorporate more types of information or MIB groups. It can also be extended to combine the observations of several nodes in the network at the central network manager. With this extension, the Bayesian network's ability to estimate probabilities with incomplete information could be utilized. In addition, the use of learning to identify other possible structures for the Bayesian network is also an area for further investigation. Another area to investigate is better usage of the fault information that is available. This information can be used to help detect and diagnose known faults and to improve the feature extraction.

# 7 REFERENCES

[1] U. Appel, A.V. Brandt, "Adaptive Sequential Segmentation of Piecewise Stationary Time Series," Information Sciences, vol. 29, 1983, pp. 27-56.

[2] C. Cortes, L.D. Jackel, W-P Chiang, "Predicting Failures of Telecommunication paths: Limits on Learning Machine Accuracy Imposed by Data Quality," Proceeding of the International Workshop on Applications of Neural Networks to Telecommunications 2, Stockholm, 1995.

[3] R.H. Deng, A.A. Lazar, W. Wang, "A Probabilistic Approach to Fault Diagnosis in Linear Lightwave Networks," IEEE JSAC, vol. 11, no. 9, Dec. 1993, pp. 1438-1448.

[4] D. Heckerman, "A tractable algorithm for diagnosing multiple diseases," Proceedings of the Fifth Workshop on Uncertainty in Artificial Intelligence, Windsor, ON, 1989, pp. 174-181.

[5] D. Heckerman, J.S. Breese, K. Rommelse, "Decision-Theoretic Troubleshooting," Communications of the ACM, vol. 38, March 1995, pp. 49-57.

[6] C. Hood, "Intelligent Detection For Fault Management of Communication Networks," Ph.D. Dissertation, Rensselaer Polytechnic Institute, 1997.

[7] G. Jakobson, M.D. Weissman, "Alarm Correlation," IEEE Network, Nov. 1993, pp. 52-59.

[8] A.A. Lazar, W. Wang, R. Deng, "Models and Algorithms for Network Fault Detection and Identification: A Review," ICC, Singapore, Nov. 1992.

[9] W. Leland, M. Taqqu, W. Willinger and D. Wilson, "On The Self Similar Nature of Ethernet Traffic (extended version)," IEEE/ACM Trans. Networking, vol. 2, pp. 1-15, Feb., 1994.

[10] E.L. Madruga, L.M.R. Tarouco, "Fault Management tools for a Cooperative and Decentralized Network Operations Environment," IEEE JSAC, vol. 12, no. 6, Aug. 1994, pp. 1121-1130.

[11] R. Maxion, F. Feather, "A Case Study of Ethernet anomalies in a Distributed Computing Environment," IEEE Trans. on Reliability, vol. 39, Oct. 1990, pp. 433-443.

[12] J. Pearl, Probabilistic Reasoning in Intelligent systems: Networks of Plausible Inference. San Mateo, CA: Morgan Kaufman, 1988.

[13] I. Rouvellou, "Graph Identification Techniques Applied to Network Management Problems," Ph.D dissertation, Columbia University, 1993.

[14] P. Smyth, "Markov Monitoring with Unknown States," IEEE JSAC, vol. 12, 1994, pp. 1600-1612.

[15] D.J. Spiegelhalter, A.P. Dawid, S.L. Lauritzen, R.G. Cowell, "Bayesian Analysis in Expert Systems," Statistical Science, vol. 8, no. 3, 1993, pp. 219-288.

[16] Z. Wang, "Model of network faults," Integrated Network Management I, B. Meandzija and J. Westcott (eds.), New York, NY, Elsevier Science Publishing Company, 1989.

[17] O. Wolfson, S. Sengupta, Y. Yemini, "Managing Communication Networks by Monitoring Databases," IEEE Transactions on Software Engineering, vol. 17, no. 9, 1991.

[18] Y. Yemini, "A Critical Survey of Network Management Protocol Standards," Telecommunications Network Management Into the 21$^{st}$ Century, S. Aidarous and T. Plevyak (eds.), New York, NY, IEEE press, 1994.

# 52
# Generating Diagnostic Tools for Network Fault Management

*Mihaela Sabin, Robert D. Russell, and Eugene C. Freuder*
*Department of Computer Science*
*University of New Hampshire*
*Durham, NH 03824*
*mcs,rdr,ecf@cs.unh.edu*

### Abstract

Today's network management applications mainly collect and display information, while providing limited information processing and problem-solving capabilities. A number of different knowledge-based approaches have been proposed to correct this deficiency, evolving from rule-based systems through case-based systems, to more recent model-based systems. Part of this evolution has been the recognition of the importance of *constraints* in a management context. This makes possible the assimilation into network management of a mature, theoretically developed technology from artificial intelligence, namely, the *constraint satisfaction problem* (CSP). In this paper we investigate the role of constraints in manipulating management data, and give an example of the use of the constraint satisfaction framework in diagnosing problems arising with Internet domain name service configurations. We also present ADNET, a system for automatically constructing C++ diagnostic programs from a model written in a simple modeling language.

### Keywords

Network fault management, configuration fault management, model-based diagnosis, constraint satisfaction, diagnostic tools.

## 1 INTRODUCTION

Today's network management applications mainly collect and display information, while providing limited information processing and problem-solving capabilities. The technological stages in managing network data classify network management applications into three distinct categories, suggestively characterized in (Rose 1993) as "browsers, mappers, and (very few) thinkers". Two essential factors explain the lack of the "thinkers": (1) the heterogeneity of managed objects which themselves are not problem solvers, and (2) the missing formalism and, consequently, technology for relating MIB modules within a configured, operational network. Thus, beyond the technical aspects of accessing formally defined managed objects through management protocols, an indispensable task is the semantic interpretation of this extremely heterogeneous data (Meyer *et al.* 1995): "Since the semantic heterogeneity of managed data has grown explosively in recent years, the task of developing meaningful applications has grown more onerous". To address the problem of effective management applications, a framework for management applications should be centered around a global network model which explains the flow of data or the occurrence of events in the network with regard to some functions performed by the network.

In recent years, more knowledge-based systems developed in the network management domain have adopted the *model-based* approach for representing and reasoning about management information. In this approach, the building blocks of the network model are the elementary structural and behavioral descriptions which define the network components: devices, links, services. The network model, however, describes not only its constituents, but also how they relate to each other. The model-based reasoning paradigm has been successfully applied in the diagnosis task (Hamscher *et al.* 1992).

In this paper we propose a *constraint-based modeling and problem-solving* approach to network fault management. The approach is based on the concept of *constraints* which capture the structure and behavior of the network to be diagnosed, and which represent relations among network components. The network is found faulty if some constraints cannot be satisfied, in which case the violated constraints are precise indicators of the cause of the fault. The main contributions of the paper are:

- to show how an existing technology for reasoning with constraints can be applied to network management,
- to introduce a declarative modeling language,
- to present a system for automating the synthesis of special purpose diagnosticians, and
- to illustrate how these can be applied in a sample model of a high-level network service.

This paper is organized as follows. In the next section we define the constraint-based approach to the diagnosis problem, and the algorithmic solution to it. Section 3 describes the general structure of a diagnostician implementing this approach. Section 4 presents the architecture of ADNET, a system that builds these diagnosticians. Based on the example described in Section 5, Section 6 illustrates the ADNET modeling language. In Section 7 we briefly review existing knowledge-based approaches to network management, and in the last section we summarize our results.

## 2  DIAGNOSIS AS CONSTRAINT SATISFACTION

The constraint satisfaction problem (CSP) paradigm has proved its applicability in various areas of artificial intelligence, such as design, configuration, simulation, scheduling, and diagnosis. Its success has been assured by both the simplicity of formulating the problem, and the diversity of continually improved algorithms to solve it. In a constraint satisfaction model of a diagnosis problem, the *constraints* capture the structure and behavior of the system to be modeled, and represent relations among the attributes of the system components. A component attribute is modeled as a CSP *variable*, characterized by possible *values* to which the variable can be instantiated, according to either the design specifications or observation measurements on the modeled attribute. The diagnosed system is found faulty if some constraints cannot be satisfied, in which case the violated constraints are precise indicators of the cause of the fault.

We formulate the diagnosis problem as a partial CSP (PCSP) (Freuder and Wallace 1992), where the partial solutions stand for the minimal sets, under set inclusion, of violated constraints, also called minimal diagnoses (Sabin *et al.* 1995). For synthesis tasks such as configuration, the constraint problem is of a more dynamic nature. (Mittal and Falkenhainer 1990) defines the dynamic CSP (DCSP) formalism to take into account conditional activation of those parts of the CSP (variables and constraints) which are relevant to the current configuration decisions. However, we have adapted Mittal's approach in two aspects, in order to deal with the task of diagnosing configuration problems, rather than configuration itself (Sabin *et al.* 1995). First, the domains of values are not restricted to predefined sets of values, instead they can be acquired at search time by observing the network. Second, the DCSP which models the configuration of some network service is solved as a partial CSP, where the partial solutions leave minimal sets of constraints unsatisfied. These partial solutions capture the inconsistencies between actual observations and model predictions, and hence pinpoint the faults in the diagnosed system.

Combinations of branch-and-bound and CSP techniques have been used in algorithms that search for a solution that leaves *minimum-cardinality* sets of constraints unsatisfied (Freuder and Wallace 1992). We have adapted one of these algorithms to search for solutions with *minimal sets* of unsatisfied constraints, in order to provide a more comprehensive explanation of the possible faults. The algorithm keeps track of the best solutions found during the search, in the sense that any proper superset of these solutions is discarded, and any solution is replaced by its proper subset when such a subset is found. The algorithm serves as the inference engine of the diagnostic tool described next.

## 3  THE GENERATED DIAGNOSTIC TOOL

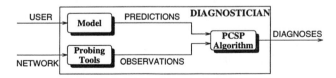

**Figure 1** Architectural description of a diagnostician for network services

Automatic diagnosis employs a diagnostic engine which outputs the expected diagnoses based on the *predictions* of the model of the system to be diagnosed, and the *observations*, or measurements, performed during the system operation (Figure 1). Measuring the actual system behavior can be fully automated by incorporating probing or monitoring tools in this diagnosis scheme. The modeling descriptions are expressed in a form congenial to the human network manager, and embed only declarative knowledge, as found for example in system specifications manuals. The user of the diagnostician is shielded from the details of how the information is to be used, or what algorithms process this information. The model of correct (and possibly faulty) behavior of the network service is expressed in CSP terms, as explained in Section 6. The PCSP algorithm for computing the minimal diagnoses is used to check the consistency between the model predictions and system observations.

# 4  A SYSTEM FOR GENERATING DIAGNOSTIC TOOLS

The diagnosis scheme implemented in a diagnostician can be further automated if specialized diagnosticians are automatically generated to handle different categories of problems. For example, the model component in Figure 1 can describe network service problems in one of the following categories:

- user interaction problems, when a network service is improperly used,
- protocol operation problems caused by incompatibilities between the end-systems on which the protocol operates, and
- configuration problems, when the network service configuration contains missing or conflicting information.

Each model description is compiled into a C++ diagnostic program that handles the types of problems described in the model. The CSP formulation for each of these groups of problems has been detailed in (Sabin *et al.* 1995). In this paper we focus on the prototype system that automates the construction of constraint-based diagnosticians.

The automatic diagnosis system for network services (ADNET) constructs constraint-based diagnostician programs in C++ from a library of network probing tools, a library of constraint-based diagnosis tools, and a model. Figure 2 outlines the ADNET architecture. The model is written in the ADNET constraint-based modeling language. The resulting diagnostician is compiled and linked with a library of network probing tools and a library of constraint-based reasoning tools, to form an operational diagnostician. We describe each of the ADNET components next.

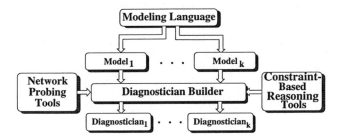

**Figure 2** The ADNET architecture for the automatic construction of diagnosticians

The *models* are written in the ADNET modeling language and formulate a standard or dynamic CSP corresponding to those aspects of a network service which are to be diagnosed. The ADNET modeling language has the advantages of constraint programming in general, namely, the language is:

- declarative: stating the constraints does not require the user to envision how the information is to be used,
- natural: constraints are expressed in a form congenial to the user,
- efficient: heuristics and inference methods can mitigate the problems of combinatorial search.

The *constraint-based reasoning tools* form a C++ library that comprises different CSP algorithms and heuristics used to solve the constraint-based diagnosis problem. In the current implementation of ADNET, the diagnostician builder is programmed to use only the branch and bound algorithm for solving dynamic partial CSPs, discussed in Section 2. Maintaining a repository of CSP techniques has the advantage of enabling another interesting developing direction, namely, of tailoring the generated diagnosticians to the problem at hand to meet better efficiency requirements.

The *network probing tools* perform two functions in the ADNET architecture. First, they can provide the model with complete domains of values, which have been previously stored in the network at network configuration time. Tools that dynamically probe the network configuration parameters save the user from a preliminary phase of manually gathering this modeling data, and, more importantly, guarantee that the model always accurately reflects the current configuration values being used in the real network. Second, the probing tools can collect observational information for those CSP variables whose values depend on the actual network operation. Both features are available in the ADNET modeling language through the variable declaration construct. A variable declaration includes the function call that probes the network and gives the expected values (predefined as configuration data, or currently observed in the running network). Unlike modeling, probing the network is a device-dependent task. Some probing is general enough to be applied to a wide variety of networks, such as the Internet **ping** program to test the reachability of another site on the network. Other probing programs may be more specific to the type of the network resource to be examined. Thus, the ADNET system provides for both user-defined and system predefined probing functions.

## 5   AN EXAMPLE: DIAGNOSING DNS CONFIGURATIONS

The ADNET diagnostician for DNS configurations can diagnose configuration problems with DNS. The common DNS error message: unknown host may have different causes related to the various underlying configurations of the DNS. For each of these configurations, the diagnostician program constructed by ADNET figures out the problem and provides a more meaningful message. Before we give such an example, we briefly present the DNS configuration as it is required by BIND (the DNS implementation written for Berkeley's 4.3BSD UNIX), running on DEC OSF/1:

- The file /etc/svc.conf is consulted to see what services are available and in what order they are to be used, as indicated by the hosts statement in this file. The possibilities are local and bind. *Local resolution* is used for small networks configured by a single administrator, with no traffic to outside world, whereas *bind resolution* becomes a "must" if the local network is connected to a larger network.
- The *local resolution* is provided by consulting the file /etc/hosts which contains a table of known names and IP addresses.
- The *bind resolution* is provided by contacting a server daemon, called **named**. The client that accesses the name server is called a *resolver*, and its configuration is defined in the /etc/resolv.conf file. A resolver creates the query, sends it across a network to a name server, interprets the response, and returns it back to the program that requested it. If the /etc/resolv.conf file is present, it is consulted to find an ordered

list of the IP addresses of server daemons to be contacted. Resolution fails if none of the servers responds, or if the first one that does respond is not able to resolve the name. If the /etc/resolv.conf file is not present, an attempt is made to contact the *local server daemon*, running on the local host.

- Each name server has its own configuration, as a *primary server*, *secondary server*, or *forwarder*. While the resolver configuration requires, at most, one configuration file, several files are used to configure **named**. For example, the boot file defines the type of the server and the location of other configuration files, such as the database files, the loopback address file, and the cache data file.

Although the above specification of the DNS configuration is not complete, it can be used to completely model the local resolution service. The diagnosis shown in Figure 3 explains the simple problem caused by incompletely specifying the /etc/hosts file, when only the local option is specified in the /etc/svc.conf file. The repair procedure for this problem is straightforward once the cause of the problem is clearly explained. More subtle

---

*Command:*          telnet alpha
*DNS Error Message:* unknown host: alpha
*Configuration:* **Local resolution only is indicated in** /etc/svc.conf. alpha **name is not in** /etc/hosts.
*Diagnosis:*       **\*\*\* Local resolution failed. No** alpha **in** /etc/hosts.

**Figure 3** DNS configuration diagnosis: incompletely specified host table

problems, such as forgetting to increment the serial number of the primary's zone files, or even syntax errors in the boot and database files, can be diagnosed by a diagnostician compiled with ADNET, if a complete DCSP model of the DNS configuration is provided.

## 6 ADNET MODELING LANGUAGE

Based on studies of FTP and DNS network services, and models built for them using the CSP formalism, we have designed a simple language for describing models in CSP terms. A model specified in the ADNET modeling language consists of four sections. The first two sections define (1) the set of *variables* and their corresponding *domains of values*, and (2) the set of *constraints*, also called *compatibility constraints*. The next two sections specify the *activation* of those variables and constraints which are relevant to the observed configuration decisions. Figure 4 presents that part of the local resolution model for the DNS example which describes the variables that may play a role in localizing the faults with the configuration of local resolution, and the constraints that restrict the values these variables can take.

The variables are defined with VAR statements. Each variable has a *name* and a *domain of possible values*. The modeling language offers two built-in mechanisms for specifying the domain of values for a variable. The DEF slot of the VAR statement specifies the domain of values known at modeling time, while the ASK slot represents the mechanism for supplying values at diagnosis time, either from the running network directly or from the user who observes the network. Referring to Figure 4, the remote-host variable has the value returned by the ASK function prompt-user, while the variable ping-path has a single predefined value ''/sbin/ping'', specified in the DEF slot. ASK functions invoke system calls for internally managed information, such as currently running processes, or prompt the user. The ADNET system offers some built-in network probing tools, such as prompt_user and ping, and provides the means to add new functions, customized for specific measurements. User-defined ASK functions, such as resolve-services and resolve-host shown in Figure 4, are written as *generator functions*, allowing the user to investigate network resources and get possible values for the current variables, one value at a time. The function is called repeatedly to return all values in the domain, until the values are exhausted or an error condition is encountered. In the latter case, the error message reports the violation of the unary constraint implicitly associated with ASK-ed variables, and is added to the diagnostic messages of the current violated constraints.

```
VAR remote-host        ASK prompt-user(''Remote host name:'')
VAR ping-path          DEF ''/sbin/ping''
VAR ping-response      ASK ping($remote-host)
VAR services-file      DEF ''/etc/svc.conf''
VAR resolution-type    ASK resolve-services($services-file, ''hosts'')
VAR hosts-file         DEF ''/etc/hosts''
VAR hosts              ASK resolve-host($hosts-file)
CON $remote-host IN $hosts
    ''*** Local resolution failed. No $remote-host in $hosts-file.''
```

**Figure 4** The VAR and CON declarations of the DNS local resolution model

The constraints are defined with CON statements. A constraint is defined on a *set of variables* and can be specified either extensionally, as the *set of tuples* of values allowed by the constraint, or intensionally, as *predicates*. The ADNET modeling language provides the standard logical, set, and relational operators. The operands can be constants (strings, numbers), current values of instantiated variables selected with $variable-name, or values returned by user-defined function calls. In addition, the user can supply his own predicates, taking as arguments any of the above. Part of the constraint declaration is a required *diagnostic message* issued in case the constraint is violated.

In general, the configuration task consists of assembling parts into a whole. Since the parts are taken from a larger (but fixed) set, some parts will never be used in the configured system. Moreover, the process of configuration reflects dependency relations as to how some parts of the configurable system require/exclude other parts to/from being present, unconditionally, or under specific conditions. The process of configuration starts with some key parts, always required in the system. To model these new features, the DCSP extensions to the standard CSP are:

- among all the CSP variables, some of them become *active variables*, as they are relevant to the configuration decisions checked at some point during the search,
- variable activity is controlled with the *activity constraints*,
- there is an initial set of active variables, called *start variables*.

A program written in the ADNET modeling language for diagnosing a configuration problem adds to the VAR and CON sections in Figure 4 two more sections:

- the START section, which defines the set of initially active variables, and
- a section that defines the set of activity constraints (*always require variable* and *require variable* constraints, introduced by the ARV and RV statements, respectively).

Figure 5 completes the DNS configuration model described in Figure 4 with the DCSP information, namely, the declaration of the START variables, and the activity constraints.

```
START  remote-host
ARV    remote-host ⇒ (ping-path ping-response)
RV     $ping-response = ''unknown'' ⇒ (services-file resolution-type)
RV     $resolution-type = ''local'' ⇒ (hosts-file hosts)
```

**Figure 5** The START variable and activity constraints of the DNS local resolution model

The START variable `remote-host` has to be instantiated no matter what configuration decisions are followed in the model. Once the value of this variable is known, two other variables become part of the diagnosis process: `ping-path` and `ping-response`. These variables are always required by the `remote-host` variable, regardless of its value, as the ARV constraint shows in Figure 5. The remaining activity constraints in the example are *require variable* constraints. These constraints activate certain variables based on the values assigned to the already active variables. For example, if the `ping-response` variable has the value "unknown", then the information about the service file `/etc/svc.conf` becomes part of the search space. Similarly, when the value "local" is observed for the `resolution-type` variable, only the host table information is further explored.

Assembling the CSP specification in Figures 4 and 5, we obtain the complete model for diagnosing the DNS local resolution. The declarative nature of the CSP formulation permits us to easily extend this model for diagnosing BIND resolution as well. To provide a basic insight on how further information about DNS configuration can be added to the local resolution model, we make use of a graphical representation of the ADNET modeling language constructs. Figure 6 illustrates the DNS local resolution model and outlines how the BIND resolution model can be built from the DNS problem description given in Section 5. The conventions in this graphical representation are simple:

- the circles, labeled CON, are the compatibility constraints, whose violations provide the diagnostic message associated with them.
- the arrows are the active variables, and are labeled with their names. These variables change dynamically in response to decisions made during the course of problem solving. An outward arrow shows the required activation of the variable, while an inward arrow shows the active variable requiring the activation.
- the variable activity is controlled by the activity constraints, drawn as boxes. They are either ARV constraints or RV constraints. In Figure 6, the ARV constraint in the Local Resolution Model activates the variables `ping-path` and `ping-response` once the variable `remote-host` is active. The RV constraints activate other variables if some already active variables satisfy some condition. Since the condition of the RV constraints

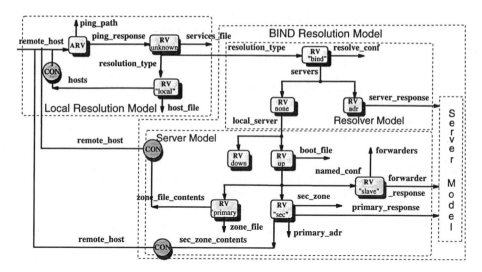

**Figure 6** DNS configuration model formulated as a DCSP

in the example in Figure 6 requires some already active variable to have a specific value, the box representing the activity constraint is labeled with that particular value.

Although the BIND Resolution Model shown in Figure 6 is not completely specified, the description given is still a working model, in the sense that the diagnosis covers the configuration problems produced by the failure of any of the compatibility constraints specified in the model, and the failure of the consistency check of any implicit unary constraint associated with the access to the domain values of a variable.

## 7   RELATED WORK

Knowledge-based technologies are characterized by utilizing domain knowledge represented in a declarative form, and human expertise expressed as rules of inference applied to the domain knowledge for performing a specific task. Knowledge-based systems for network fault management evolved from rule-based reasoning (RBR) systems to case-based reasoning (CBR) systems, and, more recently, to model-based reasoning (MBR) systems. In the following, we briefly illustrate this line of evolution, and outline the limitations each approach has encountered, as well as the solutions that have been proposed either within each approach or, more radically, by another one.

The well-known problems inherent to the *RBR systems* are the brittleness problem, or the impossibility of coping with unforeseen situations, and the knowledge acquisition problem, which arises when knowledge base growth endangers the manageability and consistency of the knowledge base itself. These problems are not critical in small, homogeneous, relatively static networks, but cannot be ignored in today's telecommunications networks, with their high rate of technological change. Thus, RBR systems become unmaintainable and unpredictable as more ad hoc rules are added to the knowledge base, with the imminent effect of proliferating unintended rule interactions and conflicts. However, different strategies of structuring the knowledge base hierarchically and providing higher-order relationships among constituent modules help diminish these problems, as is shown in the expert systems presented in (Frontini *et al.* 1991), (Schröder and Schödl 1991), (Lor 1993).

A *CBR system* addresses the brittleness and knowledge acquisition problems by exhibiting learning and adaptability capabilities. Past experience is accumulated and retrieved whenever identical or possibly similar situations are encountered. Unanticipated cases are solved by adapting existing ones. Once solved, the cases are "learned" by being added to the case repository. The prototype system for network traffic management described in (Goyal 1991) is one example of how a case-based reasoner can recognize, treat, and monitor traffic routing problems. Another example is presented in (Lewis 1993), where the diagnosis functionality is built on top of a trouble ticket system. The critical factors in CBR, however, are the similarity metrics based on which the retrieval of cases similar to the current one is possible, and the adaptation methods that ensure the transformation of the current case into one for which the solution is already known.

The *MBR paradigm* has emerged from the need to overcome the long-term, ever-growing dependency of the system on the experience gained with the system itself. The expert knowledge, which forms the empirical associations collected into rules in a RBR system, or the similarity metrics and adaptation functions in a CBR system, is now formalized into the *model* of the system to be managed. (Jordaan and Paterok 1993) describes a prototype event correlation application which needs "very little or no preconfigured knowledge", compared to RBR systems. The underlying idea is that in practice almost all objects modeling the network are related in some fashion, but just a few relationships prove to describe fault propagation effectively, and cover the vast majority of, otherwise ad-hoc, heuristic associations. Keeping modeling simple is the underlying idea in (Crawford *et al.* 1995), where the approach outlined in (Jordaan and Paterok 1993) is further formalized.

A distinct modeling technique that has recently emerged in the field of network management is the utilization of constraints. However, the constraint-based technique has not been explicitly related to the existing work on *constraint satisfaction problems* (CSPs) as they are well-known in the artificial intelligence community. For example, (Goli *et al.* 1995) proposes a constraint-based solution to the problem of checking MIB update validity, and describes the design of a network constraint management system to implement this approach. Another application of constraints is the modeling of temporal relations for the event correlation task, as presented in (Jakobson and Weissman 1995). The model captures temporal constraints and, thus, reasons about time. Constraints are also used to describe connectivity and containment relationships for the network configuration model. The description language presented in (Pell *et al.* 1995) is centered around the constraint concept. It supports network fault management through the means of checking all the constraints that define the characteristics of a particular network resource.

# 8 CONCLUSION

In this paper we proposed a constraint-based modeling and problem-solving approach to the diagnosis of network services. This approach is based on the concept of constraints which capture the structure and behavior of the system to be diagnosed, and which represent relations among system components. The automatic diagnosis employs a diagnostic engine which outputs the expected diagnoses, based on the model of the system and the observations performed during system operation. We further automated this diagnostic scheme along two directions: (1) a modeling language is provided to describe, in a declarative way, the structure and behavior of the network service, and (2) observational data can be dynamically requested from the running network by incorporating general-purpose and user-defined probing tools in the diagnostic system. We showed how these features have been incorporated in ADNET, a prototype system which automates the construction of specialized C++ diagnostic tools. We also described the ADNET modeling language and illustrated its use in a model for diagnosing configuration problems with DNS. The ADNET architecture makes possible the integration of diagnostic tools within existing network management platforms, so that the collection and display of managed data can be complemented with the problem-solving capabilities of ADNET diagnosticians.

## ACKNOWLEDGMENTS

This material is based on work supported by the National Science Foundation under Grant No. IRI-9504316, and by Digital Equipment Corporation, for which we would like to especially acknowledge the contributions of Neil Pundit and Ed Valcarce.

## REFERENCES

Crawford, J., Dvorack, D.L., Litman, D., Mishra, A.K. and Patel-Schneider, P.F. (1995) Device representation and reasoning with affective relations. In *Proceedings of the 14th International Joint Conference on Artificial Intelligence*, 1814-1820.

Freuder, E.C. and Wallace , R.W. (1992) Partial constraint satisfaction. *Artificial Intelligence*, **58**, 21-71.

Frontini, M., Griffin, J. and Towers,S. (1991) A knowledge-based system for fault localization in wide area networks. In I. Krishnan and W. Zimmer, editors, *Integrated Network Management, II*, 519-530. Elsevier Science Publishers B.V., North-Holland.

Goli, S.K., Haritsa, J. and Roussopoulos, N. (1995) Icon: A system for implementing constraints in object-based networks. In A.S. Sethi and Y. Raynaud, editors, *Integrated Network Management, IV*, 537-549. Chapman & Hall, London,.

Goyal, S.K. (1991) Knowledge technologies for evolving networks. In I. Krishnan and W. Zimmer, editors, *Integrated Network Management, II*, 439-461. Elsevier Science Publishers, B.V., North-Holland.

Hamscher, W., Consosle, L. and de Kleer, J., editors (1992) *Readings in Model-Based Diagnosis*. Morgan Kaufmann Publishers, San Mateo, CA.

Jakobson, G. and Weissman, M. (1995) Real-time telecommunication network management: extending event correlation with temporal constraints. In A. S. Sethi and Y. Raynaud, editors, *Integrated Network Management, IV*, 291-301. Chapman & Hall, London.

Jordaan, J.F. and Paterok, M.E. (1993) Event correlation in heterogeneous networks using the OSI management framework. In H.-G. Hegering and Y. Yemini, editors, *Integrated Network Management, III*, 683-695. Elsevier Science Publishers B.V., North-Holland.

Lewis, L. (1993) A case-based reasoning approach to the resolution of faults in communication networks. In *Integrated Network Management, III*, 671-682. Elsevier Science Publishers B.V., Amsterdam.

Lor, K.-W. E. (1993) A network diagnostic expert system for Acculink multiplexers based on a general diagnostic scheme. In H.-G. Hegering and Y. Yemini, editors, *Integrated Network Management*, 659-669. Elsevier Science Publishers, B.V., North-Holland.

Meyer, K., Erlinger, M., Betser, J., Sunshine, C., Goldszmidt, G. and Y. Yemini (1995) Decentralizing control and intelligence in network management. In A.S. Sethi and Y. Raynaud, editors, *Integrated Network Management, IV*, 5-15. Chapman & Hall, London.

Mittal, S. and Falkenhainer, B. (1990) Dynamic constraint satisfaction problems. In *Proceedings of the 8th National Conference on Artificial Intelligence*, 25-32.

Pell, A.R., Eshgi, K., Moreau, J.-J. and Towers, S.T. (1995) Managing in a distributed world. In A.S. Sethi and Y. Raynaud, editors, *Integrated Network Management, IV*, 95-105. Chapman & Hall, London.

Rose, M.T. (1993) Challenges in network management. *IEEE Network*, 7(6), 16-19.

Sabin, D., Sabin, M., Russell, R.D. and Freuder, E.C. (1995) A constraint-based approach to diagnosing software problems in computer networks. In *Proceedings of the 1st International Conference of Principles and Practice on Constraint Programming,*.

Schröder, J. and Schödl, W. (1991) A modular knowledge base for local area network diagnosis. In I. Krishnan and W. Zimmer, editors, *Integrated Network Management, II*, 493-503. Elsevier Science Publishers, B.V., North-Holland.

# BIOGRAPHIES

**Mihaela Sabin** received her MS in Computer Science from the Polytechnic Institute of Bucharest, Romania, in 1984. Currently, she is working towards her PhD in Computer Science at the University of New Hampshire. Her research interests include constraint satisfaction, diagnosis, modeling, and network management. She is a student member of AAAI and IEEE. Her home page address is http://www.cs.unh.edu/~mcs.

**Robert D. Russell** is an associate professor in the University of New Hampshire Department of Computer Science. His research interests include network protocol development, LAN-based parallel programming, ATM Quality of Service specification, and network management. He is a member of IEEE and ACM.

**Eugene C. Freuder** is a professor in the University of New Hampshire Department of Computer Science and Director of its Constraint Computation Center. He is a Fellow of the American Association for Artificial Intelligence. He is the founding editor of *Constraints*, An International Journal (Kluwer Academic Publishers) and a member of the Organizing Committee of the International Conference on Principles and Practice of Constraint Programming. His home page address is: http://www.cs.unh.edu/ecf.html.

## PART SIX

# Intelligent Agents
## Chair: Morris Sloman, Imperial College London

# 53

# An Agent-based Approach to Service Management - Towards Service Independent Network Architecture

*Gísli Hjálmtýsson*
*AT&T Research*
*600 Mountain Avenue, Murray Hill, NJ 07974, (908)582-5495,*
*gisli@research.att.com*

*A. Jain*
*AT&T Laboratories*
*101 Crawfords Corner Rd, Holmdel, NJ 07733-3030,*
*(908)949-5856, akj@hostare.att.com*

## Abstract

With deregulation of the telecom industry the intense competition for customers is driving service providers to offer new and sophisticated services at an increasing rate. Simultaneously, the Internet is attracting vendor creativity and putting a fatal pressure on the traditional telecom pricing structure. Whereas the telecom industry still leads in service quality, there is growing need for a cost effective architecture for network and service management comparable in responsiveness and flexibility to the Internet, yet capable of maintaining high service quality for increasingly complex services.

Although autonomous agents have been proposed for networks and distributed systems, they have largely been considered as reasoning entities exhibiting some form of intelligent behavior. Viewed more as autonomous objects, however, agents provide a powerful abstractions even when the agents task is more mundane in nature. In particular, the ability to move from one location to another goes beyond the strong level of modularity provided by object orientation, by disassociating each autonomous agent from a particular location or environment

In this paper we propose a new agent-based architecture for service management and provisioning. We describe an agent-based service environment and argue how such an environment supports rapid service creation and enables transparent services across authority domains.

## Keywords

Service management, agents, active networking, network architecture, signaling

# 1   MOTIVATION

As more players rush to become providers of communication services, the intense competition for customers is driving service providers to offer increasingly sophisticated services at accelerating rate. Although partially caused by deregulation of the telecom industry, the explosive growth of the Internet is fundamentally changing the landscape of communication, creating a new class of service providers that are putting fatal pressure on the traditional telecom pricing structure. The uniform service model of the Internet and localized control provides flexibility and responsiveness for service creation that is attracting vendor creativity. A key factor in this architecture is separation of responsibility and limited integration of service semantics into the underlying network. The network is responsible for delivering packets, the end systems are responsible for providing semantics to the packets delivered.

In contrast, the telecom infrastructure is a tightly knit web of hardware and software, where service logic is interwoven with more primitive capabilities at all levels of the network. For example, the most successful enhanced service in telephony, the 1-800 service, has connotation at all levels of abstraction. At the lowest level it represents indirect addressing, whereas at service level it means name resolution, load balancing and time-of-day sensitive routing. At the highest level it implies reverse charging. In addition to the blurring of concepts, the implementation is even more tightly coupled, with network element recording sensitive to reverse charging, and service level performing the indirect address resolutions. Whereas integration promotes performance, introducing new services becomes complex and costly, particularly since introducing a new service incurs cost for all existing services. Moreover, this rigid architecture causes long delays when introducing new services. Although some of this can be attributed to legacy, without decisive departure from current service architectures, newer transport networks will inherit this legacy (ATM Forum, 1995).

However, when compared to the Internet, the telecom industry still holds a significant lead in service quality. For example, whereas Internet telephony has become available, its quality is comparable to what the phone network provided for international calls a decade or more ago. A similar difference holds for other multimedia applications. The availability of the phone network is unsurpassed (99.99999%). In essence, whereas the Internet excels in recovering from failures, the phone network hardly ever has them[*]. Moreover, although the Internet never completely blocks a connection, service is frequently too poor to be of value. In contrast the predictable high quality service of the phone network is offered with practically no call blocking. Whereas demands for high standards of quality are in part responsible for the tight integration, there is a growing need for cost effective architecture for service and network management comparable in responsiveness and flexibility to the Internet, yet capable of maintaining the high service quality of the telecom industry for increasingly complex services.

---

[*] Although network elements may fail, there is a tremendous amount of hot sparing in the telecom world, thus hiding element failures from customers. Even at micro level every quantum of a conversation is treated reliably.

We therefore seek an architecture that uniformly separates network management from service management, while offering customized support to dynamically created services. Uniformity and separation of responsibility provides flexibility and responsiveness. Since all services are treated the same, no new facilities are needed to add or enhance services. Separation of responsibility localizes the scope of modifications, physically, at abstract level, and in terms of component modification. Whereas a uniform service model simplifies service management, removing service semantics from network elements simplifies network management. Maintaining high service quality of increasingly complex services requires customized support, including the ability to intervene or control an ongoing conversation. Agents provide a powerful abstraction that provides customized support, while operating in a uniform environment. Separation of responsibility is achieved by the agent environment providing an opaque interface - the service semantics are encapsulated in the respective service agents; the network facilities support primitive and generic abstractions but are oblivious to services and their semantics.

This paper is a paradigm paper centered on two themes. Uniformity and separation of responsibility bring flexibility and responsiveness. Customization and control provide service quality. Current network architectures trade one for the other but do not support both. We propose a new agent-based architecture for service management and provisioning, which is uniform and strongly separates service management from the low level network, yet provides sufficient customized control to maintain quality service. We show how this architecture, supports rapid service creation and real time provisioning while hiding the underlying network specifics. We furthermore argue how the agent-based architecture simplifies service management and enables transparent services across domains of authority.

After discussing related work, we analyze in Section 3 how these themes translate into architectural requirements. In Section 4 we describe the agent-based architecture. A key component of this architecture is a uniform agent environment supported by a limited set of basic network primitives, that together act as an opaque interface separating the network proper from the service level. Services are implemented by autonomous service agents that encapsulate service semantics. We show how this architecture and the agent abstraction simplifies service management. Several examples of service agents are given:

- elementary examples in Section 4.1,
- an example of mobile services in Section 4.2.2,
- an example of a call screening agent providing transparent service across domains in Section 4.2.3, and
- a customer agent in Section 4.3 illustrating rapid service creation and its enhancement in Section 4.4.

The paper concludes with a summary and discussion in Section 5.

## 2 RELATED WORK

Although autonomous agents have been proposed for networks and distributed systems, they have largely been considered as reasoning entities exhibiting some form of intelligent behavior (Etzioni and Weld, 1994), (Genesereth and Ketchpel, 1994), (Lashkari, Metral, and Maes, 1994), (Magendanz, Rothermel, and Krause, 1996). However, viewed more as autonomous objects, agents provide a powerful abstraction for distributed computing and networking, even when the agents task is more mundane in nature. In particular, the ability to move from one location to another (Gray, 1996), (Gray, Kotz, Nog, Rus, and Cybenko, 1996), goes beyond the strong level of modularity provided by object orientation, by disassociating each autonomous agent from a particular location or domain.

The work herein is related to (Tennenhouse and Wetherall, 1996), proposing the use of mobile agents to dynamically change network functionality. In comparison, our work is more aimed at complex services and service management and emphasizes separation of services from the underlying network. (Magendanz, Rothermel, and Krause, 1996), provides a taxonomy of intelligent agents for network management and discusses the potential of agent-based solutions for network and service management. Providing some of the functionality of agents, several projects are experimenting with Java (Arnold and Gosling, 1996) for service creation and management. In a related project we are working on mechanisms, implemented in a C++ library, to implement and support agents (Hjálmtýsson and Gray, 1996).

Other management frameworks that share some of the objectives of our work, include CORBA (Spec., 1995), (Vinoski, 1993) and the TINA-C (Chapman, 1995), (de la Fuente, 1994), (Berndt and Minerva, 1995) effort. Whereas the CORBA effort focuses on enabling heterogeneous systems to inter-operate, by defining interfaces, services, and information exchange mechanisms, our proposal is an environment and a methodology to create and manage the services themselves. While the TINA consortium is addressing most of the issues we cover our approach differs substantially. In particular, this work draws on some of the work on open signaling (Lazar, 1994), (Lazar and Stadler, 1993), (Wu, 1996), (Hjálmtýsson, 1997), on streamlining of signaling and management.

## 3   MOVING SERVICE SEMANTICS OUT OF THE NETWORK

In addition to the above arguments for removing service semantics from the network, the benefits of integration and service specific network optimizations are diminishing. The current phone network is engineered to support telephone conversations. Accordingly, engineering and controls make assumptions about call duration, bandwidth, and quality requirements to optimize interactive voice conversations. However, large proportion of the network capacity is currently used for other services, including data modems, facsimile and very short request-reply exchanges (e.g., credit card authorizations). Moreover, the telephone infrastructure transports most data traffic, including Internet traffic, on leased virtual private networks. Currently growing at much faster rate than telephony, the volume of these traffic classes already invalidates many of the phone-call-engineering assumptions.

Furthermore, as resource capacity increases, radically changing assumptions about signaling bandwidth, database performance and processing capabilities, the need for tight integration is further reduced. Current signaling in the telecom world, conducted on a physically separate network, is static in topology and capacity is limiting both in terms of performance and flexibility. Furthermore large proportion of signaling is service related, thus representing service semantics within the network. Carrying service level signaling as an ordinary (guaranteed service) connection, strengthens the separation between the network and the service level, and enables service level signaling to scale up with transport capacity. Exploiting modern processing power and high speed networking, combined with algorithms to hide latency, out-of-network service management can perform competitively with existing signaling systems.

A streamlined transport infrastructure, free of service semantics, with network signaling and control limited to element communication and primitive information exchange, amounts to a RISC like networking architecture. Each network component, and the network as a whole performs very basic and streamlined operations, accessed by higher levels through a very limited set of basic primitives (reduced instruction set). The service agents, playing the role of a smart compiler, translate service logic into series of primitive instructions. Such a network architecture is inherently flexible and promotes the use of commodity parts for processing and transmission, both of which steadily provide increased performance per unit cost. In contrast, customized facilities are becoming increasingly costly.

## 3.1 Network infrastructure that contains no service semantics

The challenge in building a network supportive of the service quality of the telecom world, still having a service model as flexible and responsive as the Internet's, requires a uniform service environment maintaining strong separation of network and service level concerns, while enabling customized service control of network resources. Figure 1 depicts the two layers of such an architecture, the **network level** separated from the **service level** by a thin interface. This interface consists primarily of basic primitives to network capabilities. The service environment, in which services are created, controlled and performed, consists of an execution environment and virtual network defined by the network interface. Uniformity is achieved by making this environment uniform throughout the network infrastructure.

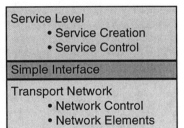

**Figure 1** Moving Service Semantics out of the Network

Separation of network concerns from service concerns is achieved by limiting the interaction between the layers to the primitives defined by the interface. As a result, since the service level accesses network resources only through the interface primitives, the physical resources, the hardware and the software implementation, is hidden and of limited consequence to the service level. Conversely, since all services are treated the same by the network new services can be introduced or existing services enhanced without any new facilities or support from the network. This way, service and network management becomes independent, significantly reducing their respective complexities.

Whereas service logic is moved **up** from the transport network infrastructure, the service level is supported throughout the network. Therefore, service logic exists throughout the network, customizing network control (still through the interface's primitives) for each service, to ensure highest service quality. In contrast, the traditional Internet service model relegates service semantics to the edges of the network, disabling service dependent inner network behavior.

## 4    AGENTS PROVIDE NETWORK INDEPENDENCE

To construct a uniform service environment enabling customized service control of network resources while maintaining strong separation of network and service level concerns, we propose an agent-based service level, with network resources hidden beneath an agent environment. Services are implemented in **service agents**, which encapsulate service semantics, including service specific data and the logic to interpret the data. The service agents operate in an agent environment which provides generic agent support - e.g., creation, destruction and execution of agents - plus access to the network primitives. Together, the network primitives and the agent environment provide an abstract network interface, separating services from the underlying network. Assuming that the generic agent support, and the limited set of network primitives is the same across the network, service agents see a uniform service environment.

The agents themselves represent autonomous objects that are mobile within the environment. Beside the ability to move, the agents are not required or assumed to exhibit intelligent behavior. The service agents themselves can, however, be of arbitrary complexity, notwithstanding being composed of other components, or cooperating with objects or agents that exist outside the agent environment. Access to the network is limited to the primitive interface provided through the agent environment. In many ways therefore these agents can be viewed as a generalized processes, and the agent environment as an operating system, controlling access to the underlying network hardware by limiting it to a small set of network primitives.

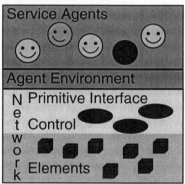

**Figure 2** Separating the Network and the Service Infrastructure.

In this framework, all services are implemented by a service agent. Moreover, all service semantics is contained and encapsulated within the agent. We distinguish a subset of service agents as **customer agents**, whose responsibility is primarily towards the service user. Whereas the goal of all service agents is to make a service available, the goal of a customer agent is to customize it for individual users. In particular, all user specific information - data, rules, or user supplied logic - is encapsulated in a corresponding customer agent. From the networks point of view all agents look the same. Conversely, seen through a service agent all networks look the same. In general, **network interface agents**, a generic low capability service agents, encapsulate network specifics; more elaborate service agents define a service, with customer agents providing customized user support. A service agent can execute at personal devices, customer premise or within the network, in any combination, provided that they provide an agent execution environment. This way the agent abstraction helps separating the service function from concerns about service location.

## 4.1 Some Elementary Agents

To make the discussion more concrete consider some examples of service agents, the most basic being a regular phone agent, a POTS[†] agent. Given a phone number (or in general a name) it establishes a phone connection, performing an indirect address resolution if needed. In particular, the agent will translate the name into an Internet address, or resolving a 1-800 name to a network address. Upon connection the agent identifies the type of receiving device (fax, voice mail, busy, no answer), thereby enabling enhanced service agents to react accordingly. Furthermore, the agent issues an indication upon call completion. While simple, this agent is already an enhancement of current POTS service. More importantly, the agent unifies telephony service, making it independent on transport technology (e.g., current phone network, Internet phone, etc.), while inviting other agents to build on it to provide an enhanced service.

For a more elaborate example, consider a call forwarding service where incoming calls are forwarded to a set of destination depending on "caller" identification. Assume the service is implemented by a Forward-Agent, which uses a network primitive to forward a connection. When subscribing the customer supplies a list of names (e.g., phone numbers) and rules to determine whereto each name is to be forwarded. The customer specific information and logic is encapsulated in a "customer" agent, and stored in the network. Upon an incoming connection request, the customer agent is invoked, who then screens the incoming name, applies the customer rules and forwards the call. Assuming an agent environment and that a network primitive to forward a connection exist, this service can be provided without any service specific modification to the network or the agent environment. The service semantics is fully encapsulated within the Forward-Agent and the customer specific "helper" agent.

In contrast, whereas the Internet service model would put such a service completely out of the network, the traditional telecom approach would involve modifications at all levels of the network. Unlike the Internet however, for performance, cost and customer service reasons it may be advantageous to offer the service from within the domain of the service provider. However, introducing new database(s) to store the list of numbers and their associated forward number, plus modifying the infrastructure to implement the service logic, affects existing services and results in growing complexity for network and service management. Instead, an agent approach encapsulates the service semantics - the data and the logic to interpret the data - in an agent that operates in a general purpose agent infrastructure accessing network facilities only through an interface to very basic primitives.

---

[†] POTS - Plain Old Telephony Service

## 4.2 Mobile Agents Enhance Service Management

In addition service management benefits from enhanced modularity and separation from network management, the mobility of service agents support service management, by conceptually separating the service abstraction from location of execution, by simplifying resource allocation, and by enabling transparent services across service domains. Various "interface" approaches achieve the separation of service definition from service implementation, resulting in a growing set of interface primitives (e.g., an API). However, a major conceptual problem arises when a part of this interface is to be implemented in one location, e.g., inside the network, and the rest in another, say customer premise. For example, consider a service API provided at the network interface. Suppose a subset of services is moved locally to the customer premise (e.g., become available at a new local PBX). The conceptual problem is: what does this mean for the interface? One option is to say that the full interface is available locally, partially implemented locally and partially relayed to the network interface. An alternative is to say that there is a locally available interface for some services, but not for others, the locally available interface overriding a portion of the full interface residing at the network. Neither one is very elegant. Conceptually either the local interface is relaying requests to the network interface, or the two implementations are conceptually different and thus so are the interfaces.

On more practical terms, after decades of debate about whether the network intelligence belongs inside, outside, or at the boundary of the network, the reality is that it appears at all of these locations. Whereas a home-office corporation may employ an answering machine and its only workstation, a large corporation may implement a rich set of services in their local PBX. Yet some other services may only be available from a service provider. Of course the users are indifferent to such details, assuming that the service is consistently delivered. Service agents provide a metaphor that captures this rather simply: offering a service at a different location conceptually corresponds to moving the agent. More generally service agents allow definition, and implementation of services independent of their domain of execution. Indeed, there is an increasing market for various services offered only within a local domain.

### 4.2.1 Agents Enhance Resource Management

Whereas transport capacity is largely managed below the agent environment, the agent abstraction enhances resource management through mobility, uniformity and encapsulation. In particular mobile agents facilitate load balancing by migrating agents. While objectives in a communication system may not require strict balancing of load, overloaded processing nodes recover by migrating agents to lightly loaded nodes in the network. In fact, to economize further on network resources an agent might be ejected to customer premise for processing. Service encapsulation within the service agent gives a conceptually convenient unit of relocation. Uniformity of the agent environment, reduces migration complexity, as simple cost measures are sufficient to select a destination of migration. In fact, a new service management dimension opens up with agent mobility as the service environment and service domains can be reconfigured rapidly.

Agent mobility furthermore enhances service quality, as services are moved to where needed on demand. This improves service availability, and response seen by users as most of the time services are provided locally, still available globally. Moreover, migrating service to a local domain improves resource utilization without sacrificing service sophistication. In contrast, consider for example mobile communication. If a user from New Jersey, currently visiting California receives a call from across the street (in California of course), it would be desirable to most of the call processing locally. In particular minimizing transport resources implies routing the conversation instate. However, except for the most elementary services, where migrating the dumb state and leaving it to the Californian network provider to interpret the state is insufficient. Mobile service agents solve this problem as the agent encapsulates the state and the logic to interpret the state.

The uniformity of the agent environment simplifies resource allocation and provisioning, as resources become interchangeable. Since databases become agent-bases, storing encapsulated agents, they are used as generic data stores, and thus equally suited for any agent. As mentioned above, same applies to processing units, all of which implement the generic agent environment. Apart from the current residency of agents at a particular node, every node appears equal to a service agent. Therefore, per service resource forecasting is not needed (or can be estimated more crudely) significantly reducing resource provisioning, and enhancing the scalability of the infrastructure.

The service agent abstraction is valuable in maintaining the service itself, as service support updating service logic and the data that defines the service without interruption of service. Service agents are transportable autonomous program entities, inherently encompassing the concept of dynamically introducing and removing a code fragment into/from an executing program. Exploiting the encapsulation properties of agents, this is achieved while preserving program level type safety and maintaining service level abstractions. Honoring the agents conceptual integrity thus helps in resolving version problems resulting from and update in the code implementing a service. As a consequence, the agent abstraction supports service evolution on reasonably small component granularity, and helps solve the difficult problem of service maintenance.

## 4.2.2 Transparent Service Across Domains

Consider a new service that is offered only in a limited geographical area. Such limitations might be transient for new services, or could arise from market segmentation or operational reasons. For example, while the infrastructure for wireless services is being built, the intra-domain service support is physically limited to the region covered by the already deployed equipment. To provide maximum coverage, the service provider may therefore sub-contract (and resell) services from other network providers. Similarly, marketing strategies in foreign markets (or otherwise geographically segregated markets) may differ significantly, thus resulting in different services being supported in different parts of the infrastructure. Although owned by the same provider, each service is therefore supported only in some sub-domain. In either case, however, customers do not want to be inconvenienced, and expect the service they subscribe to be available even when they cross the domain boundaries.

In spite of this desire, with current network architectures transparent service across segmented domains is not really an issue - it simply cannot be done. One reason for this is that current network architectures are not designed for heterogeneous support for sophisticated services. Currently, either relatively simple services are negotiated and standardized, and then offered throughout the network (and across multiple carriers); otherwise services are only available within a particular domain, and simply not available elsewhere. This remains true even when older switches/routers - supporting only a fraction of the latest standard - are inter-operable with newer switches, since typically the newer services are unavailable on paths using an old switch.

Instead, consider the uniform agent environment. Rather than encompassing service semantics for all supported services, the agents ability to move and perform throughout the environment, provides for transparent services across domains. A service normally offered in one domain, is in fact universally available, by having the corresponding service agent migrate on demand. Heterogeneity arises as service agents reside primarily where their services are offered (or deemed likely to be needed by service management) but are not universally distributed throughout the network.

### 4.2.3  An Example - Advanced Call Screening

Consider an advanced call screening service, which beside offering call blocking allows the customer to have incoming calls be processed based on caller identification. Capabilities could include, redirecting to other numbers (e.g. secretary), redirecting to voice mail, or pass the call through all based on customer preference. While not inconceivable in the current network two difficult problems arise. The more general is the problem of added network complexity as databases and code must be changed within the network and potentially some additional (special) equipment added. Still, the more problematic is how to this type of service on more global scale. Provided that a service provider operates globally, it is only natural for a customer subscribing to a service in one region, say the US, to demand the same abroad, e.g., Europe while within the service region of the provider. However, for various reasons a service offered in on region may not be offered in others.

Implementing this service in an agent environment, a call screening agent supports transparent service by migrating to where the service is requested. Unlike services like mobile, where state information (data) is migrated to where the mobile is registering, the agent brings not only data but also the knowledge of how to interpret the data. This removes the need for advance service provisioning, as the migration of the agent constitutes service provisioning on demand. Furthermore, since the agent only utilizes general purpose facilities, the marginal impact is negligible, and thus service specific resource provisioning is not required.

## 4.3 Agents Support Rapid Service Creation

The agent environment supports rapid service creation, as a new service does not require any new network support, nor coordination with other service or network providers. In particular a new service does not imply provisioning of service specific facilities - databases, or code - other than encapsulated within the respective service agent. Furthermore, a service provider could unilaterally introduce a service without any negotiations or announcements to other service providers. Still the new service would receive support comparable to any existing service. Indeed, a new player, potentially without any network infrastructure, could become a service provider simply by creating a new service agent and introducing it into the environment (of course financial settlements would need to be negotiated).

To illustrate rapid service creation consider the following example service, which we will call *"the panic button."* Subscribers identify a list of phone numbers, in some order of preference, and can specify what constitutes a successful *panic*-resolution. As an example parents could leave a *panic-button* capable device (could be a cellular phone, smart card etc.) with their children. The ordered phone number list would contain, each parents work phone number, home number, number of friends and relatives. Upon a *panic-call*, the list is processed in order, either until successful or the list is exhausted.

A traditional telecom solution would be to introduce a database to store the list of phone-numbers, and the processing logic for a panic-call. Panic-call processing then would involve identifying the customer (which could for example be the device identifier), fetching from the database the panic-list of phone numbers, and finally processing each of the numbers. Since, this new service may interfere with other services already in the network, careful system integration and service provisioning (and perhaps re-provisioning of some of the existing services) would have to be performed.

An alternative would be to push all of the intelligence out of the network, and have a smart terminal, for example a PC, implement the service completely out of the network. Call processing in this framework is simply seen from the network as a series of call requests coming from the same source, the smart terminal providing all the call processing logic. While possible, it requires substantial processing capabilities of the terminal device and requires control information (e.g. about call completion status) to propagate out of the network. Whereas the latter requires standardized information exchange, the former translates into more costly equipment.

Using a customer agent - a *panic-agent* on the panic-button device is used to implement this service. We assume that an POTS service agent exists, capable of processing regular phone calls, returning either a successful connection, or a completion indication (busy/no answer/voice mail). Upon a panic-call the panic-agent is invoked from the panic-button agent pool with the list of numbers. The agent then cooperates with the POTS service agent, processing each number on the list by issuing requests to the POTS-agent, and receiving back call completion information.

The key advantages of the agent approach are:

1.  No service specific database or any other support needs to be provisioned before the panic-button service can be offered. Still, the service is able to take advantage of being executed within the network. In particular, internal network information is available to the customer agent.
2.  Given that a network supports the POTS service agent, the panic-button service can be offered by a third party network services reseller who has no control of the physical network or network resources.

## 4.4 Out of Network Service Enhancement

Although the above panic-button example could potentially be offered out-of-network, while non-traditional, the above example is still as an in-network service offering. This means that although storage and other resources are not customized for the service, they are committed and owned by the service provider and reside within the network. In this context, out-of-network service offering, therefore means that the service is offered without any network resources commitment or control except during execution. In particular the user data and the code to interpret it - i.e., the customer agent - is not retained within the network or the service providers domain, but is instead injected into the network from customer premises upon service request. Out-of-network services for example facilitate third party service provider owning no network infrastructure.

To illustrate, out-of-network service enhancement, consider the following enhancement of the above panic-button service. Suppose now, that beside a POTS-service agent, an email service agent - an agent who accepts messages and delivers them to an Internet email address - becomes available in networks. Suppose an independent third party service provider decides to offer an enhanced "panic-button" service, by allowing the list of addresses to contain email addresses as well as phone numbers. Furthermore, since the service provider is not a network provider, the service will be offered as a small panic-device similar to a beeper. The service is implemented by an enhanced customer agent, which in addition to recognizing phone numbers recognizes email addresses.

Upon a panic-call, the device establishes a connection to the network, and injects the customer agent, carrying the panic-list, into the network. Once within the network, the agent processing the panic list as, issuing requests to the POTS agent, or issuing a panic message to the designated email recipient with the help of the email service agent. After the agent completes the list, it is destroyed by the agent environment. Observe that as before, no service specific support or provisioning is needed. Moreover, the network maintains no persistent knowledge of this enhanced service, no data nor logic. Indeed, after the panic-agent is destroyed the network acts as if nothing had happened.

In contrast a traditional network centric solution would require redesign of the supporting databases, and other provisioning, in addition to code modification. As for any new service, introducing the enhancement into the network could also cause potential interference with other existing services.

## 5 SUMMARY AND DISCUSSION

We have described an agent-based paradigm for service management, that exhibits the flexibility of the Internet while supporting customized high quality of the telecom world. We have proposed an architecture for this paradigm and shown how agents provide simple solutions to many complex service management tasks. The architecture factors services out of the network into a separate service layer, separated from the network by an opaque interface. Services are implemented by service agents, completely encapsulating service semantics, the data and the logic to interpret the data. In particular, the underlying network is free of service specific support. Instead, the service agents operate in a agent environment, accessing the network only through a limited set of basic primitives. The environment and the set of network primitives is the same across the network, resulting in a uniform view from the service agents.

We have shown how the agent abstraction, the uniformity of the agent environment and the agents ability to move helps in reducing service management complexity. Specifically, the agent-based service environment supports rapid service creation and real time provisioning. Moreover, agent-based service management enables transparent services across domains of authority. The agent abstraction conceptually clarifies some common problem in service management, such as service migration to a local domain, and service maintenance. More generally, the ability of agents to move from one location to another goes beyond the strong level of modularity provided by object orientation, by disassociating each autonomous agent from a particular location or environment. The uniformity of the agent environment simplifies resource management, thereby reducing provisioning cost and signaling requirements, as network resources become interchangeable.

This work is part of a larger project on lightweight networking aiming to streamline signaling, simplify network and service management, and exploit high level modeling abstractions at low levels of networking. In (Hjálmtýsson, 1997) we report on a lightweight call setup primitive and protocol, supporting both connection and connectionless service, while encompassing enough generality to facilitate arbitrary parameters on call setup. In particular, the light weight call setup supports the agent ideas contained herein. Although agents have been proposed before for networks and distributed systems, the low performance of agent-based systems have relegated them to tasks like user interfaces or web-searching, where performance is not a primary issue. We have developed a C++ library implementing agents, the agents themselves written in C++, compiled and run natively. This paper is a strand of a general theme investigating flexibility not only as a design goal but as a performance metric, with the aim to develop methods and metrics to quantify some of the guidelines of software engineering that postulate loose coupling and encapsulation. With the deregulation of the telecom industry and the omnipresence of the Internet, we believe that the ultimate performance of any communication architecture will hinge on its ability to handle service diversity and volatility.

## 5.1 Acknowledgements

Special thanks to Albert Greenberg for his help on the presentation of this work.

## 6   REFERENCES

K. Arnold and J. Gosling (1996) *The Java Programming Language.* Addison-Wesley, Reading, MA.

ATM Forum 95-0221R2, (1995) *Draft PNNI Signaling.*

H. Berndt and R. Minerva editors (1995) Definition of Service Architecture. *TINA-C Baseline Document*, Version 2.0.

M. Chapman editor (1995) Overall Concepts and Principles of TINA. *TINA-C Baseline Document*, Version 1.0.

The Common Object Request Broker : Architecture and Specification, Rev. 2.0 (1995)

O. Etzioni and D. Weld (1994) A Softbot-Based Interface to the Internet. *Communications of the ACM*, **37-7**, 72-6.

L. A. de la Fuente editor (1994) Management Architecture. *TINA-C Baseline Document*, Version 2.0.

M. R. Genesereth and S. P. Ketchpel (1994) Software Agents. *Communications of the ACM*, **37-7**, 49-53

R. S. Gray (1996) Agent Tcl: A flexible and secure mobile-agent system. *Proceedings of the Fourth Annual Tcl/Tk Workshop (TCL 96)*, Monterey, California

R. Gray, D. Kotz, S. Nog, D. Rus and G. Cybenko (1996) Mobile agents for mobile computing. *Department of Computer Science, Dartmouth College.*

G. Hjálmtýsson and R. Gray (1996) Dynamic Classes Enhance Maintainability of Critical Applications. *AT&T Research Manuscript.*

G. Hjálmtýsson (1997) Lightweight Call setup - Supporting connection and connectionless services. To appear at *the International Teletraffic Congress ITC'97.*

Y. Lashkari and M. Metral and P. Maes (1994) Collaborative Interface Agents. *Proceedings of AAAI'94*

A.A. Lazar and R. Stadler, (1993) On Reducing the Complexity of Management and Control of Future Broadband Networks. *Proceedings of the Workshop on Distributed Systems: Operations and Management*, Long Branch, NJ.

A.A. Lazar (1994) A Research Agenda for Multimedia Networking. *The Workshop on Fundamentals and Perspectives on Multimedia Systems, International Conference Center for Computer Science*, Dagstuhl Castle, Germany.

T. Magedanz, K. Rothermel, and S. Krause (1996) Intelligent Agents: An Emerging Technology for Next Generation Telecommunications? (1996) *in Proceedings of INFOCOM'96*, San Fransico, California, 464-472

D. L. Tennenhouse and D. J. Wetherall (1996) Towards an Active Network Architecture. *Computer Communication Review.*

S. Vinoski (1993) Distributed Object Computing with CORBA. *C++ Report.*

D. Wu (1996) An Efficient Signaling Structure for ATM Networks. *Proceeding of INFOCOM'96*, San Francisco, 844-854

## 7 BIOGRAPHY

Gísli Hjálmtýsson received a B.S degree in Applied Mathematics and Computer Science (1987) from University of Rochester, and M.S. (1992) and Ph.D. (1995) in Computer Science from University of California, Santa Barbara.

Dr. Hjálmtýsson joined AT&T Bell Laboratories, Murray Hill, in 1995, and is currently at AT&T Labs - Research. In 1993, he was a visiting scientist at Telecom Australia's, Telecom Research Laboratories in Melbourne.

His current research is in performance evaluation of networks and information systems, lightweight signaling, active networking, and new opportunities resulting from the convergence of computer networking and traditional telecommunication.

# 54
# Distributed Network Management with Dynamic Rule-Based Managing Agents

*Markus Trommer*
*Lehrstuhl für Datenverarbeitung, Technische Universität München*
*Arcisstr.21, D-80333 München, Germany*
*Email: Trommer@e-technik.tu-muenchen.de*

*Robert Konopka*
*santix software GmbH*
*Max Planck Str. 7, D-85716 Unterschleißheim, Germany*
*Email: Kono@santix.de*

## Abstract

The classical SNMP based network management architecture pursues a centralized paradigm. Its elements form two layers: centralized managers and distributed agents.

In this paper, the architecture is expanded by an intermediate layer introducing Managing Agents for Information Control (MAgIC) acting as dual-role entities. MAgICs gain, process and modify data from the layer below and can be analyzed and configured from the layer above using a separate MIB. A MAgIC offers frequently used network management functionality with a standardized interface to ease the workload of management applications and to reduce the traffic caused by management activities.

The steady progress in technology leads to the requirement of an extensibility of the functions provided by a MAgIC. A rule-based approach has been chosen to realize even complex management tasks. Rules to process management information can be defined and compiled. A Rule-MIB is available on the MAgIC, in order to store the rule definitions in table structures. SNMP is used to transfer the compiled rule definitions to the MAgIC. This approach is shown in a simple example.

## Keywords

distributed network management, dynamic rules, intelligent agents, knowledge-based preprocessing, management architecture, managing agents, SNMP

## 1. Introduction

Today, a network management architecture has to fulfil many requirements. All kinds of management tasks should be supported by the architecture. Due to rapid changes in network technology, management should be flexible in accomodating these changes. A common interface to other management applications will increase user acceptance.

Due to the geographical increase in networks, distributed network management becomes more and more important in the design of management architectures. Netload and workload of the computing environment is reduced by doing as much management as possible in the subnets. During the search for a management architecture which meets these requirements we decided to use the SNMP-framework as the base. The modifications of this architecture, that are necessary to support distributed management, are covered in Chapter 2.

In Chapter 3, managing agents are introduced, which can change their functionality due to dynamical loadable rules. An example for the way a managing agent can be controlled by dynamic rules stored in a MIB is shown in Chapter 4. Conclusions and remarks about what has to be done in the future can be found in Chaper 5.

## 2. The multilayer network management architecture

Management systems differ in the degree of logical flexibility and geographical order of the components in the network. The difference in the flexibility regarding the assignment of management objects to the areas of responsibility is a criterion for classifying the management models [HeAb94].

### 2.1 Common architecture of the SNMP management framework

The basic principle is a client-server-architecture, as shown in Figure 1. Data relevant for management is stored in several Management Information Bases (MIBs) which are administrated by agents.

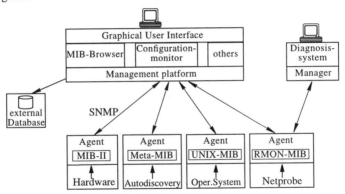

Figure 1 : Common architecture

Managers can be built as stand-alone solutions for small problems. Several managers can be run independently in a single system. In most cases, they are vendor-specific solutions with hardly any capability to communicate with other managers on the same or on other sections of the network.

In practice, developers of management applications tried to integrate the stand-alone solutions [HeAb94]. Therefore, platforms were created to offer basic management functions. For example, all management applications can use a common communication module for SNMP-access to MIBs or a common graphical user interface. Standardized service access points have to be defined and implemented for every module to be integrated.

Apart from the advantages of the platform approach compared with stand-alone solutions, there are some disadvantages. All management tasks are concentrated in a single package of programs. Because of this, the platform becomes not only very large and hard to maintain, but also imposes extremely high demands on the host, that the management software is running on. Another disadvantage is the traffic on the local area network caused by the management of the network. The traffic in a computer network should not increase too much by the management of this network. Using a centralized approach, the demand would be difficult to accomodate.

To compensate for these disadvantages, there are activities to distribute network management tasks functionally and geographically in the network. This can be supported by using management policies for the definition of abstract management goals [MaSlo96][Wies95]. Our approach uses a symmetric solution for the distribution of management tasks by extending the classical SNMP architecture [SNMP2].

## 2.2   Principle of the multilayer architecture

In the multilayer model, a new layer is inserted between the agent processes and the manager processes (see Figure 2). The managers can be either stand-alone solutions or platforms. This intermediate layer runs processes which preprocess data from the MIBs in the layer below and make it available to managers; This layer also hands over commands from the manager to the agents. To implement this feature, intermediate layer processes have a dual functionality. They are agents as well as managers.

Depending on the direction of the information flow, intermediate layer processes can be seen as either preprocessors or command distributors and interpreters.

Preprocessors are called Managing Agents for Information Control (MAgIC). They can access the contents of any MIB via SNMP. From the view of the accessed agent, managing agents behave like managers, which send queries and receive the answers. The special characteristic of the intermediate layer is, that the results of evaluations are stored in MIBs as well. This permits surveying the MIBs via SNMP. From the view of a manager, which reads information out of the managing agent MIB, the managing agent looks like a standard agent.

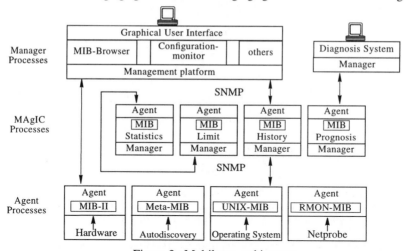

Figure 2 : Multilayer architecture

Information coming from an agent is handed over to the manager in a concentrated form. The quantity of data is reduced in this way, while the quality increases.

Commands sent from the manager to the intermediate layer are interpreted by the MAgIC, which then generates commands to the underlying agent. This is done by simply copying the commands to a determined set of agents or by transforming a complex command to several simpler commands.

A similar approach called 'dual-role-entities' is reported, but not worked out in the manager-to-manager MIB [RFC1451].

## 2.3 Features of the multilayer architecture

With the multilayer architecture, new ways of processing management data are possible. The capability to build hierarchies and to cascade the managing agents faciliates the introduction of distributed network management.

- Reduction of Netload
  In traditional architecture, common tasks are in the manager station, because complex algorithms are to be executed. With multilayer architecture, standard functions can be realized by managing agents. The managing agent is configured by the manager; it receives the order when (period of time) it should read out which data (object id) and from which MIB and how it should process the data. These parameters are sent to the managing agent via SNMP.

- Building hierarchies
  Because managing agents behave like agents from the view of a manager and behave like a manager communicating with agents, it is possible to connect several managing agents in series to build hierarchies. When a managing agent is used repeatedly, a hierarchical system of MIBs can be built, which reflects the structure of the network. This may be the physical structure or any logical structure of the observed network.

- Cascading of managing agents
  Different managing agents may also be cascaded. In this way, complex tasks can be reduced to elementary tasks. The result is a toolkit of managing agents, which can be assembled depending on the needs of the system manger.

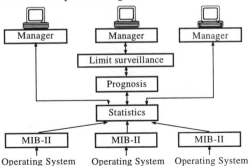

Figure 3 : Cascading managing agents

An example is shown in Figure 3. The agents of the MIB-II, a statistics preprocessor, a prognosis preprocessor, a limit preprocessor, and some management processes are connected. Of course, other managers can access the results in the statistics MIB and in the Prognosis-MIB. Therefore, these managers are able to draw different kinds of conclusions for their management tasks.

The system saves time and netload. The division into areas of responsibility (domains), which is common to the Internet, is supported by this architecture [Scha95].

## 2.4 Flexible managing agents

Owing to the raising complexity of computer networks and systems, it is difficult to provide all the functionality needed for a distributed management from scratch. Steady progress in connectivity and technology requires a dynamic adaptation and extension of management functions. A conventional management platform would deal with these requirements by simply updating parts or even the complete software. As mentioned in chapter 2.1, this is a vendor specific solution, which only works for monolithic management environments using one central platform.

A dynamic concept for managing agents, as proposed in this paper, will solve these problems in a standardized way, providing site-specific functions in a heterogenous system. There are at least three possible ways to realize a dynamic behavior of managing agents:

## *Parameterized functions*

This solution consists of a fixed set of rules. Only definite parameters can be configured in order to modify the behavior of a managing agent. This leads to simple implementations, because the basic structures of the controlling program are not subject to dynamic changes. On the other hand, changes of functionality force a partial or complete redesign.

An example of parameter-controlled rulesets for statistics, limit surveillance, history and prognosis can be found in the so called SLHP-MIB-agent defined at the Chair for Data Processing [KoTr95]. The objects to be processed are the parameters, which can be configured without constraints. There are several types of processing available (e.g. statistical calculations, limit surveillance, etc.), but these alternatives cannot be changed or extended.

## *Distributed applications*

The area of distributed systems and applications plays a major role in today's research activities. There are several proposals for the distribution and remote execution of code sequences to realize management functions. One example is a preprocessing agent that uses a stack architecture as programming model, controlled by sending a kind of assembler directive [SiTr95]. There are also attempts to use high level languages, like Java, in order to transport new functionality to managers and agents. Although Java can be precompiled to a byte code that is suitable to run on a well-defined virtual machine, an interpreter is still needed to run the process.

All these approaches have the advantage of a maximum flexibility, but there are also some unresolved problems. Apart from the lack of standarization, the security issues seem to be a major problem when using "mobile code". In addition, a new communication path has to be defined for the transmission of code.

## *Dynamic rules*

In the area of network and systems management, a knowledge-based approach is already used because of its:

- automatic evaluation of the system conditions
- recognition of complex impact chains
- ability to offer proposals for possible solutions

Apart from research activities ([KoKrRo95], [HoTr96]) knowledge-based systems are used for problem solving in trouble ticket systems and tools for security analysis (e.g. COPS [FaSp90]). One method to implement knowledge-based systems is using rules executed by an inference engine [GiRi89]. This machine also has to do some interpretations, but rules follow a strict systematic. So they can be managed by using common MIB structures. In consequence, this approach avoids the problems mentioned, because the interface is standardized and even security mechanisms, as proposed in [RFC1909] and [RFC1910], are available.

## 3.   Rule-based managing agents

### 3.1   Dynamic rules and rulesets

Rule-based systems are one of the most popular type of expert systems. They allow the modelling of knowledge in a natural way. Rules can be easily encapsulated and expanded. Furthermore, they faciliate the reasoning of conclusions made by the system. Due to these advantages, the principle structure of rules, provided by rule-based expert systems, is chosen to control the behavior of managing agents.

The proposed structure of rules consists of two major parts: conditions and actions. Usually an action is addressed as "conclusion" or "consequent". The term "action" is used in this context to emphasize the active character of this part.

In the *conditional part*, one or more comparative operations can be connected by logical operators (e.g. AND, OR). A comparative operation evaluates the relation between two operands using one of the following operators:

$$> \quad < \quad \geq \quad \leq \quad \neq \quad =$$

For a SNMP-based managing agent, the operands of a comparison can be a single constant, a reference to the attribute of a managed object or some of them combined to a mathematical expression.

`(sysServices(7) & 1) = 1 AND sysDescr(1) = "Flintstone Router"`

In the example, the condition is met if the value of the object sysServices(7) of the MIB-II has the least significant bit set (logical AND with bit mask 1) and the string of the system-description has the correct content. In this case the rule is said to be "activated".

The *action part* of a rule may consist of a list of actions, too. These are performed, if conditions in the conditional part of the rule are met. There are several possible types of actions. Setting of attributes of managed objects is one of the fundamental activities, which are necessary to do a proactive network management. Apart from that, messages to other management entities can be initiated in order to propagate processed conclusions. This could reduce the amount of management traffic needed for polling the state of a MAgIC. For the chaining of rules an additional action is defined, which allows the activation of rules or rulesets.

A ruleset is a collection of correlated rules, which can be activated as a whole. This leads to a reduction of consumed resources, because the MAgIC does not have to process all rules at the same time. In addition, rulesets with different tasks may be distributed to several MAgICs in order to keep the workload small.

It is well worth noting, that, in this definition, rules have **no** third part (ELSE) containing a list of actions to execute when the condition fails.

## 3.2    The rule-based MAgIC

Basically, a MAgIC is a dual-role-entity, which contains two SNMP-interfaces, a manager and an agent as described in chapter 2.2. Apart from those, the major element of a rule-controlled managed agent is the rule processing unit, as shown in Figure 4.

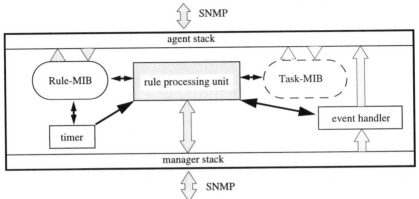

Figure 4 : The basic structure of a rule-based managing agent

It can be implemented as a simple state machine periodically processing the rules of a ruleset in sequential or even random order. This approach works for small rulesets including no more than about 100 rules.

To perform complex management tasks, preference should be given to an inference engine. The inference engine decides which rules are to be executed according to a system of priorities. This reduces the workload of the MAgIC, because only rules with the highest priority will be executed. For an experimental integration of SNMP, the inference machine provided by a commercial expert system shell is used [Stei95].

Rules are structured and stored in a so called Rule-MIB. So, the rules can be defined, read and modified by a management application. The detailed structure of this MIB will be shown in chapter 3.3. Furthermore, the actual management tasks defined by the rulesets in the Rule-MIB need a management information base to represent conclusions drawn and to offer a parameterized configuration interface. In contrast to the Rule-MIB, the so called Task-MIB does not have to be completely defined in advance. Managed objects can be instantiated at the time of creating or changing rulesets or rules. In addition, even rules are able to create new elements in the Task-MIB by simply referencing new objects.

Aside from the Task-MIB, also an event handler is integrated into the MAgIC. It allows the sending of SNMP traps or inform-requests to superior managers or MAgICs. On the other side, it can collect messages from the inferior layer and makes them available to the expressions in the conditional part of the rules.

In order to reduce the workload, there is a timer to control the activity of the rulesets. It allows the definition of time-slices, when certain rulesets are processed.

All interactions between the timer, the event handler and the rule processing unit are based on the principle of object identifiers. Rule-MIB is the common interface for configuration and data interchange. Therefore, even the distribution of basic functionality of the MAgIC, e.g. event handling, to another system is possible.

## 3.3    The Rule-MIB

The Rule-MIB is divided into five sections, each of it represented by a node in the MIB-tree, as shown in Figure 5.

iso(1).org(3).dod(6).internet(1).private(4).enterprises(1).wilma(860)

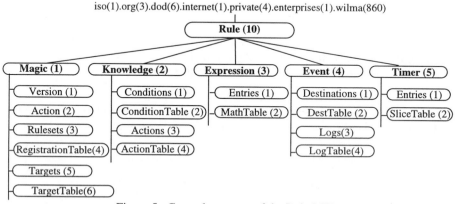

Figure 5 : General structure of the Rule-MIB

Configuration information for the managing agent itself is defined in the first node. Aside from basic functions for restarting and identifying the version of MAgIC, a registration table is defined. All rulesets stored in the MIB get an entry in this table. This makes it easier to create, activate, deactivate, or delete rulesets. A reference to the management application, which has registered the ruleset in the Rule-MIB, is also stored to coordinate the access to specific table entries. Furthermore, there is a need to define groups of hosts, called targets, in the target table. A rule is applicable to a set of targets referenced in the registration table.

Note that the actual rules are located in the Knowledge(2)-subtree. Because of the two parts of a rule, as shown in chapter 3.1, there are two tables in this part of the MIB: the condition table and the action table.

In the empty condition table, as shown in Figure 6, there are three columns defined for a single comparison: LeftOperand(2), RightOperand(4) and Operator(3). The LogicalOperator(5) is used to combine multiple lines to one conditional part of a rule. It also signals the last entry for a specific rule.

**ConditionTable (2)**

| Index (1) | LeftOperand (2) | Operator (3) | RightOperand(4) | LogicalOperator (5) |
|-----------|-----------------|--------------|-----------------|---------------------|
| 1.1.1<br>1.1.2<br>1.2.1 | | | | |

**ActionTable (4)**

| Index (1) | Type (2) | LeftValue (3) | RightValue (4) |
|-----------|----------|---------------|----------------|
| 1.1.1<br>1.2.1<br>1.2.2 | | | |

Figure 6 : General structure of the condition table and the action table

The structure of the action table is similar to the structure of the condition table. With the Type(2)-column, several actions are defined. LeftValue(3)- and RightValue(4)-columns store none, one or two parameters an action (e.g., the assignment of a value to an object instance) needs.

Both tables have an identical indexing scheme. An index consists of three numbers:

- ruleset number
- rule number
- number of the condition or action

Using this scheme, any number of conditions can be combined with any number of actions to form a rule, which is uniquely identified.

In order to permit the calculation of arithmetical and logical expressions in comparisons and actions, an Expression(3)-subtree is defined. Bivalent operations can be grouped in a recursive way to complex expressions. This is done by using the MathTable(2) with a indexing scheme similar to the tables in Figure 6. Expressions, therefore, are addressed by the object identifier describing the relevant entry in the MathTable.

The Event(4)-subtree of Rule-MIB contains tables for the logging and filtering of received messages. Furthermore, destination lists for messages generated as a result of an action of the MAgIC itself are defined.

Finally, the Timer(5)-subtree controls the administration of time slices when rulesets have to be activated.

## 3.4    Uniform object identifiers

A managing agent has to collect and modify the attributes (=instances) of managed objects located on various components. In order to do so, it is necessary to address each managed object of any agent in the system. The addressing of managed objects by object identifiers, as defined in the SNMP framework, is not sufficient to uniquely identify an attribute. The object identifier specifies only the position of the object in the Internet registration tree. For a complete description, the transport address of the agent is needed. In classical SNMP, the User Datagram Protocol is used for the transportation of messages. The missing information therefore consists of the IP address of the agent and the UDP port number, which normally has the value 161. Nevertheless, other port numbers or even transport mechanisms, e.g. an OSI protocol, can be used [RFC 1906].

To solve the problem of ambiguous object instance names, we introduce Uniform Object Identifiers. The notation for these descriptors is similar to those of Uniform Resource Locators (URLs) in the World Wide Web [RFC1738]. Because of its widespread use, this notation is well-known. A Uniform Object Identifier consists of the following parts:

- A *protocol name*. This is `snmp` in the standard case, but there are also other names defined for the use of SNMP over different transport protocols, for example `snmp_tcp`.

- The *network address* of the agent. In principle, the address can be specified in a numerical or a symbolic form using domain and host names. Again, it is not necessary to use the IP protocol, although that will be the common choice.

- An optional *port number*: These numbers can be provided for specifying a specific communication port differing from the standard.

- The *object identifier*. This is the identifier derived from the position in the Internet naming tree in numerical form.

The following shows an example for the sysDescr object instance from the MIB-II located on the host with the IP-address 10.2.3.4. The agent listens on the unusual UDP port 1166:

```
snmp://10.2.3.4:1166/1.3.6.1.2.1.1.1.0
```

The wildcard '*' is also allowed for parts of an object identifier:

```
snmp://*/1.3.6.1.2.1.1.*.0
```

This defines all instances in the system-group (1.3.6.1.2.1.1) of the MIB-II. The wildcard instead of the host address has a special meaning. It includes all hosts defined by a specific view, as e.g. the targets in the Rule-MIB.

As a result of the introduced format, the determining of references to managed objects has to be handled in a different manner than before. When specifying an object identifier in a MIB, the ASN.1 type DisplayString is used for practical reasons instead of creating three separate managed objects with the appropriate type definitions. This offers remarkable advantages. The complete identifier can be transferred using one binding. Identifiers can be read and set by standard applications like MIB browsers. Keeping the single parts of the expression together leads to atomic get- and set-operations, where no undefined or misleading constellation can occur.

## 4.   Scenario

The usage of rule-based managing agents will be explained in a small example. An important aspect for distributing management tasks to the network is the preprocessing in respect to specific systems. Actions have to be taken depending on hardware, operating system or application software versions. A possible task for the management could be the control of the available mass storage for the operating system. Hosts running out of system space can cause a lot of errors, which are difficult to diagnose. The amount of free space, which is necessary for a reliable operation can vary among the systems. In consequence different rules for every version of the operating system are necessary.

Information about the operating system and the filesystem is given by a UNIX-MIB-agent running on every host to be managed. UNIX-MIB was defined at the Chair for Data Processing in order to permit systems management operations using SNMP, similar to the Host Resources MIB [RFC1514], but more specific.

In Figure 7, an exemplary networking environment is given. Several standard Ethernet segments are connected with bridges to a backbone. The network can be seen as part of a greater private network, which is formed by WAN connections to a central department. As shown in the picture, there are different types of hardware platforms.

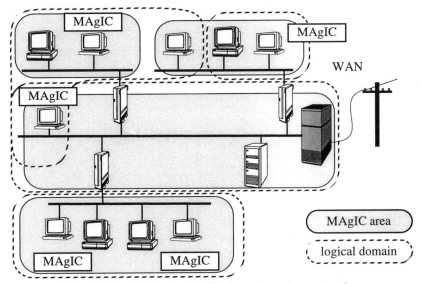

Figure 7 : Distribution of MAgICs in an exemplary network

To observe the system, at least one MAgIC is placed in every physical subnet. As shown in the picture, the area handled by a MAgIC may be different to the logical domains of the network.

In this example only three of the objects in the UNIX-MIB are used:

- *kernelSystem* which describes the UNIX versions that is running on the observed system
- *fsMountPath* describes the directory of the UNIX-System the partition is mounted to
- *fsMountFreeSpace* shows the size of the free disk space on the partition in KBytes

To execute the given task the MAgIC needs the rules, which control its behavior, and a list of targets which must be kept under surveillance. The targets are stored in the target table of the MAgIC.

Here is an example for a rule:

```
IF
cond ( */kernelSystem = "HP-UX" ) AND
cond ( */fsMountPath = "/" ) AND
cond ( */fsMountFreeSpace < 5000 )

THEN
act ( sendMessage; Admin1; "Low root disk space on HP" )
act ( setObject; localhost/TASK-MIB.Status; 3 )
```

The rule checks wether a system running HP-UX has less than 5 MB of free disk space on the root partition of the harddrive. When these conditions are true, a message is sent to the manager Admin1. In addition, a status object in the TASK-MIB of the MAgIC itself is set to the value 3, which means e.g. "resource problem".

This rule is defined in the management application. After the rules are compiled to a MIB-structure they are transfered to one or more MAgICs via SNMP. In Figure 8 the structure and the resulting contents of the tables representing the rule are shown.

**ConditionTable (2)**

| Index (1) | LeftOperand (2) | Operator (3) | RightOperand(4) | LogicalOperator (5) |
|-----------|-----------------|--------------|-----------------|---------------------|
| ⋮ | | | | |
| 3.1.1 | s://*/1.3.6.1.4.1.860.2.2.1.1 | = | "HP-UX" | AND |
| 3.1.2 | s://*/1.3.6.1.4.1.860.2.4.4.24.1.13 | = | "/" | AND |
| 3.1.3 | s://*/1.3.6.1.4.1.860.2.4.4.24.1.6 | < | 5000 | ——— |
| ⋮ | | | | |

**ActionTable (4)**

| Index (1) | Type (2) | LeftValue (3) | RightValue (4) |
|-----------|----------|---------------|----------------|
| ⋮ | | | |
| 3.1.1 | Notify | 172.16.1.17 | "Low root disk space on HP" |
| 3.1.2 | Set | s://127.0.0.1/1.3.6.1.4.1.860... | 3 |
| ⋮ | | | |

Figure 8 : Example of entries in the Rule-MIB

# 5.    Conclusions and future work

A multilayer network management architecture is introduced in this paper. The new layer consists of preprocessors called MAgICs. It is shown that the architecture meets the requirements of modern network and systems management. Distributed network management is supported by the concept of building hierarchies and cascading MAgICs. Netload is reduced by transfering parts of the management tasks into the subnets. Flexibility is reached by using preprocessors controlled by dynamic rules. Because SNMP is the only interface used, MAgICs fit in almost every existing management environment.

Until now the rule engine used to interpret the rules is very simple. The future research will concentrate on building an inference engine to enforce knowledge-based preprocessing.

Parts of the implemented software can be downloaded for noncommercial use by anonymous FTP from the Chair for Data Processing server.

URL: `ftp://ftp.ldv.e-technik.tu-muenchen.de/dist/WILMA/`

# Acknowledgements

The authors whish to thank all the members of the WILMA group at the Chair for Data Processing and Prof. Joachim Swoboda for their feedback and their support. The WILMA-Group is a team of Ph.D. and diploma students from Technical University Munich, Germany. WILMA is a German acronym for 'knowledge-based LAN management' (Wissensbasiertes LAN-Management).

# References

[FaSp90]    D. Farmer, E. Spafford: The COPS security checker system, in *USENIX Conference Proceedings*, Anaheim, CA, Summer 1990

[GiRi89]    J.C. Giarratano, G. Riley: Expert Systems: Principles and Programming, PWS-KENT Publishing Company, Boston, 1989

[HeAb94]    H.-G. Hegering, S. Abeck: Integrated network management and systems, Addison-Wesley UK, 1994

[HoTr96]    M. Horak, M. Trommer: Architektur für ein dezentrales, hierarchisches Sicherheits-management; *Proceedings der Fachtagung SIS '96*, VDF Hochschulverlag, Zürich, 1996

[KoKrRo95]    T. Koch, B. Krämer, G. Rohde: On a Rule Based Management Architecture, in *Proceedings of the second international workshop on services in distributed and networked environments*, June 1995

[KoTr95]    R. Konopka, M. Trommer: A Multilayer-Architecture for SNMP-Based, Distributed and Hierarchical Management of Local Area Networks, in *Proceedings of the Fourth International Conference on Computer Communications and Networks*, Las Vegas, September 1995

[MaSlo96]    D. Marriott, M. Sloman: Management Policy Service for Distributed Systems, in Proceedings of the third international workshop on services in distributed and networked environments, Macau, June 1996

[RFC1451]    J. Case, K. McCloghrie, M. Rose, & S. Waldbusser: Manager-to-Manager Management Information Base, April 1993

[RFC1514]    P. Grillo & S. Waldbusser: Host Resources MIB, September 1993

[RFC1738]    T. Berners-Lee, L. Masinter & M. McCahill: Uniform Resource Locators (URL), December 1994

[RFC1909]    K. McCloghrie: An Administrative Infrastructure for SNMPv2, February 1996

[RFC1910]    G. Waters: User-based Security Model for SNMPv2, February 1996

[Scha95]    H.N. Schaller, A concept for a hierarchical, decentralized management of the physical configuration in the Internet, in: K. Franke, U. Hübner, W. Kalfa, *Kommunikation in Verteilten Systemen*, GI/ITG-Fachtagung, Springer-Verlag, Berlin, 1995

[SiTr95]    M. R. Siegl, G. Trausmuth: HIERARCHICAL NETWORK MANAGEMENT: A Concept and its Prototype in SNMPv2, in *Proceedings of the Joint European Networking Conference (JENC)*, 1995

[SNMP2]    J. Case, K. McCloghrie, M. Rose & S. Waldbusser: Version 2 of the Simple Network Management Protocol (SNMPv2), SNMPv2 Working Group, RFCs 1902 to 1908, January 1996.

[Wies95]    R.Wies: Using a classification of management policies for policy specification and policy transformation, in *Proceedings of the fourth international symposium on integrated network management* (ed. A.S.Sethi et al.), May 1995

# Biographies

Markus Trommer received a diploma in electrical engineering and information technology from the Technical University of Munich, Germany. Currently he is a Ph.D. student at the Chair for Data Processing and since 1996 head of the WILMA group. His research activities focus on intelligent agents and knowledge-based, distributed network and systems management. He is a member of IEEE and GI.

Robert Konopka received a diploma in electrical engineering and information technology from the Technical University of Munich, Germany, in 1990. He spent five years at the Chair for Data Processing and the WILMA group researching on expert systems for network and systems management. Since 1996 he works as a consultant for systems management at the Santix Software GmbH, Munich, Germany.

# 55
# Delegation Agents: Design and Implementation

*Motohiro SUZUKI, Yoshiaki KIRIHA, and Shoichiro NAKAI*
*C&C Research Labs., NEC Corp.*
*1-1 Miyazaki 4-Chome, Miyamae-ku, Kawasaki, Kanagawa, 216,*
*Japan. Tel: +81-44-856-2314. Fax: +81-44-856-2229.*
*E-mail: motohiro@nwk.cl.nec.co.jp, kiriha@nwk.cl.nec.co.jp,*
*nakai@nwk.cl.nec.co.jp.*

## Abstract

We have developed a management agent that adapts the delegation concept to achieve efficient distributed network management. In conventional delegation, a network management operator details management operations in an operation-script that describes management operation flow and provides such network management functions as event management and path tracing, in the form of function objects. The operator sends this script to agents to execute. In our approach, the operator sends only a script skeleton describing management operation flow alone; function objects are built into the agents themselves. This helps keep management traffic low. Each function object contains three operational objects: enhanced, primitive, and communication. Each enhanced operational object (EOO) provides a script skeleton with a network management function. A primitive operational object (POO) provides an EOO with managed object (MO) access functions. A communication operational object (COO) provides an EOO with a mechanism for accessing the functions of other remote EOOs. This design increases the efficiency of the delegation approach as applied to network management systems. We have tested our design by applying it to path tracing and found that it works well enough to suggest the feasibility of applying it to such other distributed network management functions as connection establishment and release, fault isolation, and service provisioning.

## Keywords

distributed network management, delegation agent, scripting language

## 1 INTRODUCTION

With the explosive growth in the size of networks linking small and high-performance computers, it has become increasingly difficult for a single centralized manager to manage entire networks. In current standardized management protocols, such as the common management information protocol (CMIP) and the simple network management protocol (SNMP), an operator must manage huge quantities of information and must control network elements (NEs) with such primitive functions as get, set, create, delete, and action. This results in an excessive processing load and extremely high manager-agent communication overheads. This weakness is particularly pronounced in the execution of network management functions that require distributed processing spread over several agents, e.g. the path tracing function proposed by the Network Management Forum (NMF) (NMF, Forum014). The function of this path tracing is to search for the termination point managed object (MO) instances which make up a specified network connection, and this search will be spread over several agents. Consequently, tracing a virtual path (VP) in an ATM network, which requires the large frequency of accessing MOs in the individual agents, causes very high manager-agent communication overheads.

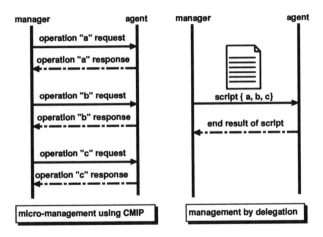

**Figure 1** Micro-management and management by delegation.

To cope with this problem, one recently reported concept of distributed management, "management by delegation" (Yechiam, 1991), appears to be especially promising. In this concept, functions delegated to agents are specified in an operation-script that describes a management operation scenario in a system or scripting language. As Figure 1 illustrates, this drastically reduces the need for communication between manager and agent. Additionally, much of the information processing traditionally performed by the manager can now be handled by the agent.

In the global analysis of the concept reported in (Maria-Athina, 1996), delegation is categorized into two types: "static" and "dynamic." In static delegation, agent functions are predefined and cannot be changed or added to during its running time. In DEAL (Simon, 1996), for example, agents simply have the function of translating structured query language (SQL) operations into such SNMP operations as get, set and get-next. While static delegation may be suitable for this type of task, there are other types for which it is not. In monitoring remote networks, such as RMON MIB (RFC1271, 1991), for example, each time a new MO definition is added to the environment, the operator must re-program and re-compile the application to deal with the new MO definition. That is, since static delegation agents cannot dynamically extend their functions, they are unable to cope with the new requirements or unexpected situations that may arise during their running time.

Dynamic delegation is much more flexible. With it, an operator can provide agents with an operation-script that describes the implementation of any new functions to be added. In this way, the operator can respond to new requirements and resolve unexpected situations. In the application monitoring of remote networks described above, for example, an operator needs only to send to the agent a new operation-script that describes the new functions needed to deal with the new MO definition.

While much discussion has been reported on the concept of applying dynamic delegation to network management, however, little actual design and implementation has been conducted, particularly in comparison to the great amount that has been conducted with respect to static delegation. In our study, we have devoted the bulk of our efforts to these two needed areas: actual design and implementation.

In Section 2 of this paper, we describe the special requirements of "network management by delegation." In Section 3, we propose a new management agent architecture for satisfying these requirements. In Section 4, we describe the implementation of our design, and in Section 5, we summarize the study.

# 2   DESIGN REQUIREMENTS

In this section, we discuss the requirements for applying dynamic delegation to network management.

**Format independence**: Operation-scripts in conventional delegation agents are written in a system language, such as C or C++, and must be at least partially written to suit the specific format in which each individual agent is implemented. Consequently, in order to use an operation-script that has been written for one format in a different format, a network management operator must appropriately modify the format-dependent part of it, i.e., that part which depends on the management application programming interface (API) implemented in the agent, such as OSIMIS (George, 1995) and XMP (X/Open). In order to avoid all the problems that this creates, operation-scripts should be format independent.

**Agent-agent communication**: When a network management function requires distributed processing spread over several agents, these agents must be able to communicate with one another without adding significantly to the operator's processing load or to manager-agent communication overheads.

**Secure execution**: When a single operation-script is sent to a number of different agents, it may happen that operations permissible for one agent are not permissible for another, and a control mechanism is needed to prevent agents from executing operations that are, for them, illegal.

**Common process sharing**: When the same single process is described in a number of different operation-scripts that are to be executed simultaneously, a considerable amount of waste will be incurred. Furthermore, the operator will have had to write the processes in each of those individual scripts. Efficiency requires the elimination of this overlap.

## 3   DELEGATION AGENT ARCHITECTURE

In this section, we describe the agent architecture we have developed for use with dynamic delegation. In the architecture, we propose a new operation-script format and execution mechanism that helps satisfy the requirements described in Section 2.

## 3.1   The script skeleton and function objects

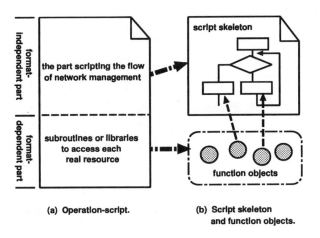

(a) **Operation-script.**

(b) **Script skeleton and function objects.**

**Figure 2** Operation-script, script skeleton, and function objects.

We propose here a new operation-script format that is independent of individual agent implementation formats. The operation-script now sent to an agent contains only a "script skeleton," which describes format-independent operational flow alone; function objects, which provide such network management functions as event management, path tracing, etc., are built into the agents themselves (Figure 2(b)). Additionally, to execute network management functions that require distributed processing efficiently, a function object has a mechanism for agent-agent communication. In the mechanism, to invoke functions provided by EOOs existing remote agents, a function object sends script skeletons to the agents.

A network management operator needs only to write a single script skeleton to be sent to all agents, regardless of their individual implementation formats, which helps keep network traffic from the manager to the agents low.

## 3.2    Operation-script execution mechanism

Figure 3 illustrates the operation-script execution mechanism. When the script processor receives a script skeleton from a manager, its execution control module checks whether all required function objects are allowed to bind with the script skeleton. This check is performed on the basis of an access control list contained in the delegation information repository (DIR). If all the objects are allowed to be bound to the skeleton, the script processor performs the binding, executes the skeleton, and sends execution results to the manager. Since the execution control mechanism checks the permissibility of any script skeleton, there is no worry of an agent's performing illegal operations.

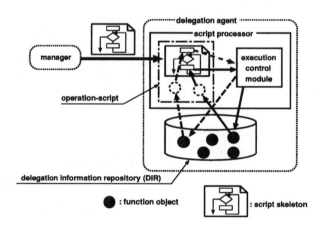

**Figure 3**  Operation-script execution mechanism.

The use of function objects itself helps greatly in sharing common processes. That is, sharing function objects among script skeletons reduces a considerable amount of waste in executing the skeletons. Moreover, to improve the efficiency of the script skeleton execution, the same mechanism contained in function objects is shared among the objects. In a typical example of the sharing, function objects contain the same mechanism for agent-agent communication.

# 4 IMPLEMENTATION

In this section, we discuss how the manner in which function objects are implemented helps to apply our architecture to dynamic delegation efficiently. Regarding the manner, to achieve concurrent MO access and provision of a simple and powerful API, we employ the concept of component-ware, in which function objects consist of three types of operational objects: "enhanced," "primitive," and "communication."

## 4.1 Operational objects

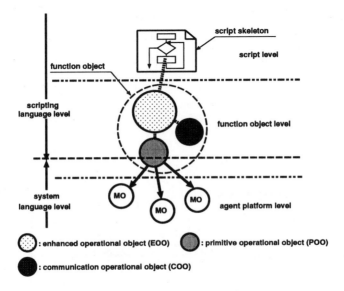

**Figure 4** Function object design.

Enhanced operational objects (EOOs) contain an API for interfacing with script skeletons. Primitive operational objects (POOs) access MOs for EOOs. Communication operational objects (COOs) provide EOOs with a mechanism to achieve agent-agent communication. Figure 4 illustrates the relationships among these objects. Each function object here is composed of an EOO, a POO, and a COO. The script skeleton specifies a function provided by the EOO, and the EOO orders its POO to access all relevant MOs. The POO does this and returns their attribute values to the EOO for processing. The EOO then returns the results of this processing to the skeleton. In the following, we discuss individual objects in more detail.

**Enhanced operational objects (EOOs):** An EOO is capable of complex processing of MO attribute values, and it contains an API for interfacing with the script skeleton. To help locate MOs that have been specified in a script skeleton,

EOOs contain a function for translating abstract names into the data to which they correspond, which is written in a scripting language that specifies data types (`string`, `integer`, etc.), as well as into attribute labels and the distinguished names (DNs) of MOs. Abstract names are simple and easily understood by an operator. An EOO communicates with EOOs existing in the remote agents to execute network management functions that require distributed processing. To communicate with remote EOOs, the EOO utilizes a mechanism provided by COOs.

**Primitive operational objects (POOs):** POOs access one or more MOs according to requirements of EOOs. To facilitate easy handling of MO attributes in an EOO, a POO translates a data written in a scripting language into the corresponding data written in a system language, and vice versa. For example, a POO translates a character string that represents the DN of an MO instance into the DN-structure in C++. POOs utilize a thread mechanism to access multiple MOs concurrently in order to achieve real-time processing.

**Communication operational objects (COOs):** COOs provide EOOs with a mechanism for agent-agent communication. This mechanism is particularly useful in implementing network management functions that require distributed processing at the agent level. The COO first locates a remote EOO providing a required function and then sends a script skeleton invoking the required function to the remote EOO for execution.

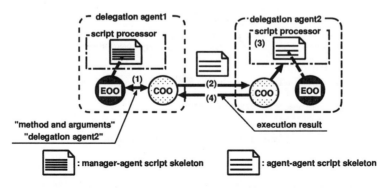

**Figure 5** Agent-to-agent communication.

Figure 5 shows this process in which agent-agent communication is achieved via COOs between two delegation agents called "delegation agent1" and "delegation agent2." An EOO of "delegation agent1" first gives its COO the names of the required function and of an agent that contains it (in this case "delegation agent2"), as well as the arguments used to invoke the function (Figure 5(1)). This COO sends a script skeleton containing these arguments to the COO of "delegation agent2" (Figure 5(2)), which forwards the skeleton to the agent's script processor. The script processor locates the EOO containing the required function, executes the skeleton with it (Figure 5(3)), and gives the results of the execution to the COO to be sent back to the COO of "delegation agent1" (Figure 5(4)).

## 4.2 Case study: path tracing

To test the feasibility of our delegation agent architecture, we have applied it to the task of tracing a VP trail in ATM networks. Figure 6 illustrates an example of tracing a VP trail called "vpTrail#1" between two ATM NEs called "node_A" and "node_B." A manager sends a script skeleton invoking the function for tracing "vpTrail#1" to "delegation agent1" managing "node_A." In the skeleton, an operator describes the abstract name specifying the required VP trail (in this case "vpTrail#1"). In "delegation agent1," "vpTrail#1" is translated into the DN of the VP trail termination MO instance (e.g. the *vpTTPBidirectional*-MO instance (Alex, 1996)) to enable MO access. If tracing the VP trail requires invoking the path tracing function provided by the other delegation agent (in this case "delegation agent2" managing "node_B"), "delegation agent1" sends a script skeleton for tracing "vpTrail#1" to "delegation agent2." In "delegation agent1," the skeleton is automatically created with the DN of an MO instance for initiating MO access in "delegation agent2." Then, "delegation agent2" sends the execution result of the skeleton to "delegation agent1," and "delegation agent1" sends the tracing result back to the manager.

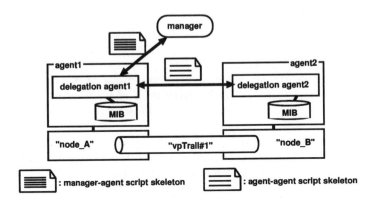

**Figure 6** Tracing a VP trail.

Currently, we have implemented the operational objects and script skeletons in Java (Ken, 1996) except the MO access portion of the POO, which is written in C++. We used TCP/IP socket libraries to transfer script skeletons. The results of the implementation show that sending only a script skeleton to an agent enables the management traffic between a manager and an agent to be kept low, and it works well enough to suggest the feasibility of applying our design to other distributed network management functions as well as path tracing.

# 5   CONCLUSION

This paper has proposed a new architecture for management agents that is designed to achieve efficient distributed network management. In this architecture, which employs the dynamic delegation concept, a network management operator describes management operations in an operation-script, and the script is sent to an agent to execute. Adopting the delegation concept helps operators cope dynamically with any situation in network management.

To implement the dynamic delegation concept efficiently in network management systems, we combine the use of a script skeleton with that of function objects. A script skeleton is independent of any agent implementation format and describes management operation flow. Function objects provide network management functions and are stored in the agent. To help keep management traffic low, only a script skeleton is sent to an agent, where it is dynamically bound with operational objects in order to execute a complete operation-script. In designing function objects, we have employed the concept of component-ware, in which function objects consist of multiple operational objects. This design supports efficient distributed network management: it keeps management traffic low and extends agent functions dynamically.

# REFERENCES

Network Management Forum. Application Services: Path Tracing Function (Forum 014).

Yechiam, Y., German, G. and Shaula, Y. (1991) Network Management by Delegation. 2nd International Symposium on Integrated Network Management.

Maria-Athina, M. and Gabi, D. (1996) Delegation of functionality: aspects and requirements on management architectures. 7th Distributed Systems: Operations & Management.

Simon, Z., Michel, L. and Jean-Pierre, H. (1996) DEAL: delegated agent language for developing network management functions. 1st International Conf. and Exhibition on the Practical Application of Intelligent Agents and Multi-Agent Technology.

RFC 1271 (1991). Remote Network Monitoring Management Information Base. IAB.

George, P., Kevin, M., Saleem, B. and Graham, K (1995). The OSIMIS platform: making OSI management simple. 4th International Symposium of Integrated Network Management.

X/Open Preliminary Specification. System Management: Management Protocols API. X/Open Company, Ltd.

Alex, G. (1996). Access Network Management Modeling. IEEE Communications Magazine, March, 62-72.

Ken, A. and James, G. (1996). The Java Programming Language, Addison-Wesley Company.

**Motohiro SUZUKI** received his B.E. and M.E. degrees in information systems engineering from Osaka University in 1992 and 1994, respectively. He joined NEC Corporation in 1994 and is now a member of the Network Research Laboratory, C&C Research Laboratories. He has been engaged in the research and development of network management systems.

**Yoshiaki KIRIHA** received his B.E. and M.E. degrees in electronic communication engineering from Waseda University in 1985 and 1987, respectively. He joined NEC Corporation in 1987 and is now a member of the Network Research Laboratory, C&C Research Laboratories. He has been engaged in the research and development of network management systems and distributed artificial intelligence systems.

**Shoichiro NAKAI** received his B.E. and M.E. degrees from Keio University in 1981 and 1983, respectively. He joined NEC Corporation in 1983 and has been engaged in the research and development of local area networks, distributed systems, and network management systems.

# 56
# A Spreadsheet-Based Scripting Environment for SNMP*

Pramod Kalyanasundaram, Adarshpal S. Sethi, Christopher M. Sherwin, and Dong Zhu
(email: {kalyanas, sethi, sherwin, dzhu}@cis.udel.edu)

Department of Computer and Information Sciences
University of Delaware, Newark, DE 19716

## Abstract
The existing SNMP management framework does not effectively support a hierarchical management strategy. Further, existing MIBs have a static structure and do not permit dynamic user organization of management information. This paper presents a spreadsheet paradigm that allows users to dynamically configure management information and set up control at an intermediate manager. This paradigm augments the basic SNMP framework by providing value added functionality at a proxy node so that it can function as an intermediate manager. The design of a proxy MIB, a scripting language, and event model that form an integral part of the paradigm are presented with future research directions.

**Keywords:** Spreadsheet Paradigm, Scripting Language, Event Model, SNMP, Proxy Agent, Intermediate Manager

## 1 INTRODUCTION

A hierarchical management strategy is an effective means of managing the large and complex internetworks that are in use today. However, the most popular management framework, the SNMP framework (which includes both the SNMP and the SNMPv2 protocols) [RM90, CFSD90, RM91, CMRW96], is largely used in a flat model with a single manager communicating with a large number of agents. The SNMP framework defines the concept of a *proxy agent* as an agent that acts on behalf of other agents. But traditionally, SNMP has used proxy agents in a *pass-through* role, wherein a proxy might facilitate the implementation of administrative or security policies but otherwise passes the manager requests and agent responses through in an essentially transparent mode. The fact that a proxy can be used as an intermediate manager for hierarchical management is recognized by SNMPv2, and the protocol even includes an Inform Request PDU intended for manager-to-manager communication. However, it cannot be effectively used because the framework lacks essential support for hierarchical management; it provides no means for managers to delegate tasks to intermediate managers or to communicate with the intermediate managers during the execution of these tasks.

Management by delegation is a well-known strategy [YGY91, Gol95, GY95] for implementing hierarchical management, but so far the SNMP community has been unable to take advantage of it because the delegation primitives have not been integrated with the SNMP framework. In this paper, we present a new paradigm – which we call the *spreadsheet paradigm* [SK94, KSS96] – that incorporates management by delegation concepts into the SNMP framework to facilitate hierarchical management.

*This work was supported in part by the U.S. Department of the Army, Army Research Laboratory under Cooperative Agreement DAAL01-96-2-0002 Federated Laboratory ATIRP Consortium.

The main objectives of the spreadsheet paradigm are: 1) to introduce a powerful intermediate manager that enhances (but preserves) the existing SNMP framework, provides value added functions, and supports delegation. 2) to provide an environment that supports user configurability of management information independent of the underlying MIB structure. 3) to support basic primitives, events and operations via a scripting MIB and language that allow a user to build fairly complex network management tasks. 4) to present the user with an abstraction and interface that is easy to comprehend and use.

The two essential features supported by this paradigm are: 1) specification of dynamic relationships between objects *across MIBs* and 2) flexible, hierarchical event building. Relationship specification allows two or more objects belonging to different MIBs to be related by a logical or temporal conditions. Hierarchical event building allows simple events (in cells) to be used to build more complex events.

Existing models [CMRW96] allow only event definitions and object relationships that are predefined in the MIBs supported by an agent. RMON [Wal95] allows users to set up monitoring information including event and threshold specification. However, the events that can be specified are fairly restrictive. Moreover, the user cannot set up operations to be performed on management information across nodes. Also, the user cannot set up control functions that can be executed. To allow managers to perform more sophisticated control functions, there is a need for a paradigm that permits *dynamic* specification of object relationships and event definitions. The spreadsheet paradigm facilitates dynamic configuration of information and control by providing the ability to selectively structure management information into views [†]. Customizable structuring of management information is a key feature needed by network managers that is missing in the SNMP framework.

This research uses the concept of management by delegation (MBD)[YGY91] but differs from the conventional MBD in that the delegation aspects are incorporated into the existing framework and hence will conform to the security aspects defined by the framework. Research has been done to provide the user with different views of MIBs [AY95]. These extensions make use of additions to the structure of management information (SMI) and special compilers to compile the MIBs. However, these extensions provide the user with just table operations as would be permitted by a database engine and the user cannot build custom views that are specific to a user's environment. Studies of temporal and event models have been done by several researchers, [Has95, CJS93] etc. Our work uses the concepts identified by these research directions in temporal and event models, and adapts these results to suit the spreadsheet paradigm.

This paper is organized as follows: Section 2 introduces the spreadsheet paradigm; Section 3 presents the Spreadsheet Management Information Base (MIB) design ; Section 4 covers the Spreadsheet Scripting Language (SSL) features; Section 5 describes the event model supported by the spreadsheet paradigm; Section 6 includes an example and Section 7 summarizes the conclusions and outlines future directions.

---

[†]The term *view* has a different meaning in SNMP terminology. We use this term to indicate user structured management information set up in the spreadsheet.

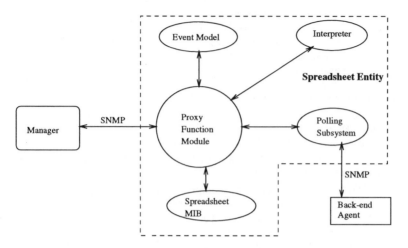

**Figure 1** Spreadsheet Entity

## 2   THE SPREADSHEET PARADIGM

An essential component of the spreadsheet paradigm is an abstraction of a spreadsheet. A spreadsheet is composed of *cells* arranged in a two-dimensional matrix consisting of *rows* and *columns*. A cell is the fundamental unit of operation in the spreadsheet paradigm. A cell contains a *control information part* and a *data part*. The control part dictates the rules for collection of information or relationships between objects. The data part contains the data collected as a result of executing the control information specification. For instance, the control part may specify that the cell should contain the result of summing two or more counters in different nodes (or different MIBs). The data part contains the result of such a summation. More details on the operation of the spreadsheet paradigm and its various scenarios are available in [SK94, KSS96].

Figure 1 shows the basic components contained in an implementation of the spreadsheet entity located at a proxy (or an intermediate manager). The proxy accepts SNMP requests from the manager and enters them into the spreadsheet that is maintained at the proxy. The control information entered into the spreadsheet contains scripts written in the Spreadsheet Language (SSL), a specially designed scripting language for use with the spreadsheet paradigm. The scripts are interpreted locally at the proxy which may result in SNMP requests to be forwarded to one or more agents. The agents' responses are processed as specified by the scripts in the spreadsheet cells. As a result of this processing, updates may occur in the data contained in one or more cells.

The spreadsheet MIB (*ssmib*) implements the spreadsheet abstraction using SNMP tables. User operations on cells map to operations on tables that are part of this MIB. The proxy function module coordinates the activities of the various components at the proxy node. When the proxy receives an SNMP request from the manager, the proxy function module performs the necessary operations on the spreadsheet MIB to implement the cell abstraction. Once the request has been carried out, the proxy function module responds to the manager that requested the operation. If a control value is entered into a cell, the objects that need to be polled are forwarded to the polling subsystem. The proxy module interacts with the event model to perform event based processing of the spreadsheet. When the script in a cell needs to be executed, the proxy function module interacts with the interpreter to process the cell control information.

When a user sets up information in a cell, there is a need to constantly update the value(s) contained in the cell. This can be achieved by polling the managed objects referenced in the cell. Thus, the spreadsheets cells at the proxy reflect the current values of the managed objects in the cells (within a certain time granularity). The polling subsystem supports cell-based polling entry creation, deletion and retrieval.

The polling subsystem optimizes the number of polls issued to the back-end agents by grouping variables based on: 1) time intervals and 2) hosts. The polling subsystem collects the variables that need to be polled at a single agent and issues the minimum number of poll requests to satisfy the polling specification for the variables. Also, if a single variable at a given agent is to be polled at different frequencies, the polling subsystem computes that minimum frequency of polling that will satisfy all the polling requests for the variable.

The spreadsheet paradigm supports a scripting language that can be used by a user to set up sheets of control. This language is described in Section 4. The spreadsheet language is interpreted by an interpreter and scripts that are set up in the various cells can be executed under the control of this interpreter. The interpreter performs the functions of syntax checking, run time error checking, detection and reporting.

The spreadsheet supports both the request/response (synchronous) and event (asynchronous) modes of operation. In the synchronous mode, the manager requests some operation to be performed using one of the standard SNMP protocol operations and the proxy responds after processing the request. In the asynchronous mode, the user sets up events to be watched, and actions associated with such events. On occurrence of any of the watched events, the proxy carries out the associated actions which may include notifying the manager. This mode of operation allows the manager to successfully delegate some of its routine tasks to the proxy.

## 3 · MIB DESIGN

The proxy based *ssmib* captures the control and data part of a cell using two tables: 1) control table and 2) data table. The structure of the MIB is shown in Figure 2. The control table (*controlTable*) is made up of a sequence of *controlEntry* elements. The *controlEntry* contains the column variables that constitute a row in the *controlTable*. The variables that form the control table index are: 1) the primary internet address of the manager who is the owner of the row 2) the spreadsheet number 3) the column id of the cell and 4) the row id of the cell. The *controlEntry* also contains a *controlstring* field that holds the control information of the cell.

Similarly, the *dataTable* is made up of a sequence of *dataEntry* elements. The *dataEntry* contains the column variables that form a row in the *dataTable*. The *dataTable* index variables are the same as those of the *controlTable* with one difference – there is an additional index variable *dataValueIndex*. This additional index variable helps to uniquely identify a particular data value in a cell that contains multiple data values.

The relationship of the control and data tables with a cell is shown in Figure 3. The

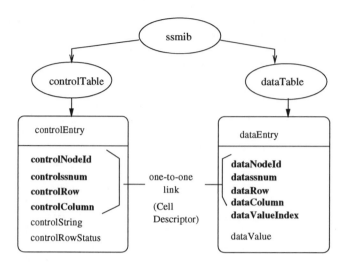

**Figure 2** Spreadsheet MIB

common variables (NodeId, ssnum, Row and Column) establish a one-to-one correspondence between the control and data tables and will be referred to as the *cell descriptor*. The cell descriptor variables uniquely identify a cell at a proxy node.

In order to create a cell, the manager creates a row in the control table. To set the control portion of the cell, the manager sets the *controlstring* variable in the *controlEntry*. Depending upon the number of values defined by the control portion of the cell, an appropriate number of rows are created in the data table with the common cell descriptor values. The cell abstraction is thus represented as a row in the *controlTable* (the control part) and one or more rows in the *dataTable* (the data part) of the cell. The manager can delete a row in the control table, and based on the control to data table association, the appropriate rows in the data table are removed. To modify the control part of a cell, the manager performs a **Set** on the specified cell which translates to a **Set** on the appropriate row in the control table. If the control portion needs to be retrieved, this operation translates to a retrieval of a row from the control table. The same operations and translations apply to the data portion of the cell. The only difference arises when there are multiple values in the data portion of the cell. In such a case, the MIB design permits the retrieval of one or more values using the standard **GetBulk** operator. This is achievable since all the values in a cell have the same cell descriptor as a prefix.

Although multiple managers can set up one or more spreadsheets at the proxy, each spreadsheet is controlled by one and only one manager. This may cause some cells to be duplicated. However, duplication of cells does not affect the polling since the polling subsystem computes an optimized polling stream.

Consider the following example. A manager (node id $m$) wants to set up a cell (with row id $i$ and column id $j$ in a spreadsheet $s$ such that it contains the same managed object on $n$ different nodes. This results in the manager forwarding a set request to create a row in the control table. The OID of the cell will be the OID of the control table suffixed by

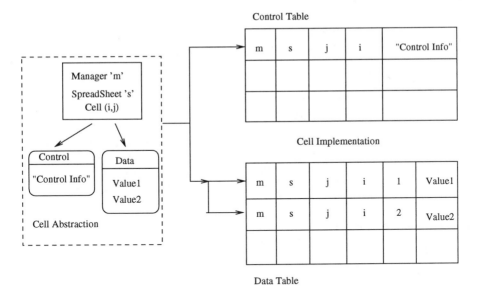

Control Table

| m | s | j | i | "Control Info" |
|---|---|---|---|---|
|   |   |   |   |   |
|   |   |   |   |   |

Cell Implementation

Data Table

**Figure 3** Cell abstraction and implementation

$m.s.j.i.$ The cell column id precedes the cell row id since this will be consistent with the way the SNMP **GetNext** operator works. If a *GetNext* is performed repeatedly, starting at a given cell, then all the cells in the first column of the spreadsheet will be returned before the cells in the second column of the spreadsheet are returned. Once the row is created, the manager can issue a set request to set up the control portion of the cell by setting the *controlstring* variable of the row with index $m.s.j.i.$. As part of the control information set up, for each variable encountered in the control variable, a unique row is created in the dataTable with the index $m.s.j.i.v$ where $v$ takes values 1 through $n$ (since there are $n$ managed objects).

Deleting a cell results in the $n$ data rows ($m.s.j.i.1$ through $m.s.j.i.n$) being deleted, followed by the control row with index $m.s.j.i$ being deleted. The semantics for setting the control part of the cell can be defined as a delete followed by a create or deleting the data rows associated with the old control information, changing the control information and creating new data rows that correspond to the new control information. In order to retrieve the control portion of the cell, the manager issues a get request to the proxy with the OID of the control table suffixed with index $m.s.j.i$. In order to retrieve the data portion of the cell, the manager can perform one of two operations:

1) perform a sequence of $n$ **GetNext** operations that will return the $n$ values contained in the cell, or 2) perform a **GetBulk** with the OID of the *dataTable* suffixed with the cell descriptor $m.s.j.i$ and specify a value of $n$ for the repeaters.

A key aspect of the MIB design has been demonstrated in the above example. Although each of the $n$ variables contained in the cell belongs to different nodes, the user (or manager) can retrieve the values in an order that is required by the user and which has been set up by the user. The user can thus configure the management information selectively and view it in an order that is different from that of the underlying MIB. Such reordering of management information will allow users to set up summaries and different views of the management information. Thus, a user can view only the information that is needed and in an order that conforms to the user's current needs. This dynamic configuration of the MIB does not exist in the current SNMP management framework.

# 4   SPREADSHEET LANGUAGE (SSL)

A language that targets a network management environment must be able to support features that facilitate the specification of network management tasks coupled with user flexibility and expressive power [Hol89]. This section describes the main features of the spreadsheet language (SSL) that forms an integral part of the spreadsheet paradigm and supports the development of network management scripts.

SSL supports features that can be broadly classified as: 1) standard procedural language features and 2) spreadsheet paradigm specific features. The standard procedural language features include: operators (arithmetic, logical and relational), control flow constructs, expression evaluation and assignment, and local variable support within a cell. The paradigm-specific features include: cell access, managed object and polling specification, multiple values in cells (or customizable views), and event specification.

SSL supports standard arithmetic operators like '+', '-', '/', '*' and relational operators like greater than ('>'), less than ('<'), and not equal to ('<>'). Cells and numbers can be used in arithmetic and relational operations. SSL supports logical-and (&&), logical-or (||) and logical-not (!) operators. All these operators allow the user to specify relationships between any set of arbitrary objects within the management domain.

Control flow constructs like **if...else...endif** and **while** allow a user to set up conditional and iterative scripts in cells. The semantics of these constructs are similar to that of a standard procedural language. However, iteration has an implicit, user-configurable, maximum loop count that ensures that a single script will not run forever.

The SSL permits a user to specify a fully qualified managed object in both symbolic and dotted decimal format. A fully qualified managed object is a combination of both the OID of the managed object and the host on which it resides. For instance, if a user wanted to monitor *tcpActiveOpens* on host *stimpy*, the user could do this by specifying *tcpActiveOpens@stimpy*. This feature allows the user to identify and access any managed object in the management domain. A user can optionally specify a polling interval for a managed object. This option allows the variable to be polled at the user-specified frequency instead of a default frequency.

The language is tightly coupled to the spreadsheet paradigm and offers facilities to access and manipulate cells. A user can specify cells using a [spreadsheet:row:col] specification (assuming a default manager). Thus, [1:2:3] would refer to a cell that occupies the second row and third column in the first spreadsheet. The language also supports assignments to cells. This allows cells to be cleared and copied. In addition, the SSL allows cells to be named (labeled) and the symbolic names to be used in scripts. Each cell supports a set of special local variables for temporary storage during script processing.

An important feature supported by the SSL is that it allows a user to set up and access multiple managed objects in a cell. A user can specify a list of objects to indicate that multiple values must be stored in the cell. Each individual value can be accessed using [s:r:c].n where $n$ represents the *nth* value in the cell. For example, a user can set up instances of a single counter (e.g. an error counter) from different nodes in a management domain within a single cell. When the manager retrieves the contents of this cell, a summary of the error counters on the nodes of interest can be obtained.

The SSL allows dynamic configuration of a spreadsheet using the *activate* and *deactivate* statements. Using these statements, a user can enable or disable the script contained in a cell. For example, if a set of counters need to be collected for a specified period of time until a fault occurs, a user could down-load the script that collects the counters but can deactivate it on occurrence of the fault. The *activate* statement allows the user to activate the collection script later. The script contained in a cell that is deactivated cannot be executed unless it is activated again. A cell cannot be activated or deactivated when the script in the cell is executing.

The SSL combines standard procedural language constructs with event specification constructs to provide a simple yet powerful platform for developing scripts that perform network management tasks. The event model is described in detail in Section 5. The following simple example demonstrates some useful features of the spreadsheet language and an important concept of the information model of the spreadsheet paradigm, which is the MIB view. The following example cell is set up to poll 8 counters from two different MIB-II groups on two different hosts and sum the information.

The polled values and summation information are all stored in a single cell as multiple values. The lexicographical order of these values are $$.1, $$.2, upto $$.12, where $$ is a special variable that represents the cell's value. In this way the cell presents to the manager a different MIB view which includes non-contiguous managed objects from different MIB groups, and even across different hosts. With cells properly set up to utilize this feature, significant power, flexiblity and efficiency could be achieved by the management application.

```
Cell [1,1]:
    label: cellMibViewExample
    action:
    {
        // Poll counters at sol
        $$.1 = tcpActiveOpens@sol.cis.udel.edu;      // TCP group counters
        $$.2 = tcpPassiveOpens@sol.cis.udel.edu;
        $$.3 = ipInReceives@sol.cis.udel.edu;        // IP group counters
        $$.4 = ipInHdrErrors@sol.cis.udel.edu;

        // Poll counters at tweety
        $$.5 = tcpActiveOpens@tweety.cis.udel.edu;   // TCP group counters
        $$.6 = tcpPassiveOpens@tweety.cis.udel.edu;
        $$.7 = ipInReceives@tweety.cis.udel.edu;     // IP group counters
        $$.8 = ipInHdrErrors@tweety.cis.udel.edu;

        // Summing:
        $$.9  = $$.1 + $$.2;  // Total TCP connections at sol
        $$.10 = $$.5 + $$.6;  // Total TCP connections at tweety
        $$.11 = $$.3 - $$.4;  // Total IP datagrams 0 with
                              // no header errors at sol
```

```
      $$.12 = $$.7 - $$.8;   // Total IP datagrams recveived with
                             // no header errors at tweety
}
```

## 5   OVERVIEW OF THE EVENT MODEL

The SNMP framework is predominantly synchronous. The primary source of asynchronous processing is the use of traps from the agent to the manager. This section describes an event model that provides asynchronous processing support for the spreadsheet paradigm and enhances the value-added capabilities provided by the proxy.

Events form the basis for the event model. An *event* is an occurrence that causes: 1) a change in the control or data part of one or more cells 2) a system related change (e.g., a timer tick, SNMP PDU receive or send) or 3) the execution of one or more cells in a spreadsheet.

Events can be either *basic* or *user-defined*. *Basic events* are intrinsic to the event model and are either SNMP or system related. These events form the basic building blocks for an event hierarchy. *User-defined (or derived)* events are those events that are built using a combination of basic and other user-defined events. The basic events supported by the event model are: *mgrget, mgrset, eventget, timer, poll, activate and deactivate*. Of these, all events except *activate* and *deactivate* are system events and cannot be generated by the user. When an event occurs, the event id and event specific details are made available to the receiving cell.

An *mgrget* event is generated when a manager issued **Get Request** is received by the proxy. The cell whose OID is part of the request is the recipient of this event. An *mgrset* event is similar to the *mgrget* event except that it is generated when a manager issued **Set Request** is received by the proxy. An *eventget* event is generated when, as part of event processing, a request is generated for executing the script contained in a cell. The *mgrget, mgrset* and *eventget* events are implicitly enabled for all cells in the spreadsheet. A user of the spreadsheet will have to explicitly disable these events, if necessary.

A *timer* event is generated on every clock tick. A user can optionally specify a time value as part of the timer event specification. A non-zero value indicates that the timer event specification is not true on every time tick, but is only true on those ticks that align with the time interval specified. For example, if a timer event specification contains *timer(5)*, it implies that the event condition will be triggered every 5 seconds . This feature is useful to perform the periodic, repetitive tasks that are typical in network management applications. A *poll* event is generated when a poll response is received by the proxy for an OID that is contained in a cell. The poll event contains the new value of the variable.

A cell in a spreadsheet is capable of generating the following events: *value change, event occurred, invalid value* and *error event*. A *value change* event is generated when a cell's value changes. This event is useful in detecting value based event specifications. An *event occurred* event is generated when the event condition specified in the cell has occurred. This event is useful to trigger dependent cells in an event cell dependency hierarchy. An *invalid value* event is generated when one or more values contained in the cell become invalid due to the dynamic nature of the variables contained in the cell. An *error* event is generated when an error is encountered during script processing.

To support the event model, all cells in a spreadsheet are modeled as *event cells* (i.e., they support event generation and receipt). *Event cells* are divided into two categories: *executable cells* and *event-based cells*. The control part of a cell is composed of two parts: an implicit or explicit *event-specification part* (or event expression) and an *action part*. The event-specification part acts as a filter that looks for one or more events to have occurred before executing the action part of the cell. An event is specified using the **on** SSL keyword. The action part of the cell is a set of SSL statements (also referred to as a *script*) that can be executed to perform some management function or subset thereof.

The general structure of a cell that is based on an event condition is:

```
on: (event_expression) ';'
action: SSL_Statement_block
```

where *SSL_Statement_block* is a set of valid SSL statements. The event expression could be a basic event or a derived event and is similar to a boolean expression. Basic and derived events can be combined using the boolean operators defined in SSL (i.e., ||, &&, !). When used in conjunction with the **on** keyword, the expression is treated as an event.

Executable cells allow a manager to request the execution of a script and return the value of the cell that results from the execution of the script. This is a synchronous operation and corresponds to the traditional SNMP framework manager-agent interaction except that a down-loaded script is executed on the proxy before a value is returned. In an executable cell, the following basic events are automatically enabled: 1) manager get request and 2) event processing. Thus, a cell containing executable SSL statements could be executed both by manager request and as part of event processing. A variation on this is the *timed-execution cell* that permits a user to set up a periodic execution of a cell based on some time criteria.

An event-based cell is a refinement of the executable cell. The refinement is in the event specification. A user can specify either basic or derived events as triggering criteria for the action part to execute. This allows a user to use previously defined events in the spreadsheet as triggering criteria. The user can thus build a hierarchy of events, all of which are based on the set of basic events supported by the event model.

The event model operation is shown in Figure 4. When a basic event occurs, the *event dispatcher* forwards the basic event to an *event scheduler*. The event scheduler schedules the cell execution based on priority and other criteria that may be specified for event selection and execution. The event scheduler identifies the target cell for the scheduled event and forwards the event to the appropriate cell or cells. This event, when received by the target cell, causes that cell to process its event condition. If the event condition matches, the action part of the event cell is executed. As part of the execution, other cells may be scheduled for execution and new events may be generated. These events are again forwarded to the event scheduler for further processing. Thus, when a basic event occurs, the cells that are dependent upon the basic event or on events generated as part of event processing are scheduled for execution. This event processing propagates until all the cells that are dependent upon the basic event or on events generated as part of the event processing have completed their execution.

Events can be combined using logical operators to form new events. However, it is important to define the semantics of such event expressions. The parsing of event expressions has to be different since events are dynamic and can occur in any order. The processing of events is contingent upon the occurrence of the event. For instance, if an event expression

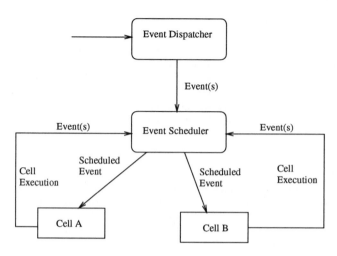

**Figure 4** Event Model Support for the Spreadsheet Paradigm

such as *event1* || *event2* is specified, and further let us assume without loss of generality that *event1* occurs at time *t1* and *event2* occurs at time *t2*, and *t1* < *t2*  (i.e., *t1* chronologically precedes *t2*), then the derived event *event1* || *event2* is signaled at time *t1* when *event1* occurs. However, the event expression *(event1 && event2)* is true only at the time instant when both *event1* and *event2* have occurred.

## 6   AN EXAMPLE

The following simple example helps to illustrate the essential features of the SSL vis-a-vis the event model, more specifically timed execution, event hierarchy and event processing. In the example, the number of connections established to certain TCP ports of a host computer is calculated. When a threshold value is crossed, an alarm is triggered and the manager is notified.

```
Cell [1,1]: Periodically download tcpConnTable at host Stimpy.
    label: tcpConnTableStimpy;
    on: timer(5 min); // Download this table every 5 min.
    action: $$ = table tcpConnTable@stimpy.eecis.udel.edu;

Cell [10,1]: TCP port to which the number of connections is calculated.
    label: localPort;

Cell [10,2]: Using the downloaded tcpConnTable of host Stimpy, calculate
    connections to the local port specified in cell localPort.
    label: countPort;
    action:
```

```
{
  $$ = 0;  // Init count
  foreach $1 in tcpConnTableStimpy do
      if ($1.tcpConnState == 5) // Established(5)
      && ($1.tcpConnLocalPort == localPort) then
  $$ = $$ + 1; // ++ count
      endif;
  done;
}
```

Cell [2,1]: Periodically calculate connections established to FTP port.
```
  label: countFTPConn;
  on: timer(10 min); // Execute this script every 10 min.
  action:
   localPort = 21; // FTP port is 21
   exec countPort; // Execute script of cell countPort
   $$ = countPort; // Get result of execution
```

Cell [3,1]: Periodically calculate connections established to HTTP port.
```
  label: countHTTPConn;
  on: timer(15 min); // Execute this script every 15 min.
  action:
   localPort = 80; // HTTP port is 80
   exec countPort; // Execute script of cell countPort
   $$ = countPort; // Get result of execution
```

Cell [3,2]: Alarm cell - when there are more than 50 connections established
to HTTP port of host Stimpy, inform the management station.
```
  label: countHTTPConnAlarm;
  on: countHTTPConn > 50; // When number of connections > 50 do action
  action: inform mgr1@sol.cis.udel.edu HTTPConnAlarm;
// HTTPConnAlarm is the event ID assigned for this event
```

Cell [3,3]: Enable the above alarm.
```
  label: activateHTTPAlarm;
  action: activate countHTTPConnAlarm;
```

Cell [3,4]: Disable the above alarm.
```
  label: deactivateHTTPAlarm;
  action: deactivate countHTTPConnAlarm;
```

The timed-execution cell tcpConnTableStimpy (cell [1,1]) periodically downloads the tcpConnTable of host Stimpy. Based on this, two other timed-execution cells (cell [2,1] and [3,1]) periodically calculate the number of FTP and HTTP connections using the executable countPort cell (cell [10, 2]). Cell localPort (cell [10,1]) is used by countPort to know which port to calculate on. If the countHTTPConnAlarm cell (cell [3,2]) is in the activated state, whenever the number of HTTP connections is greater than 50, the manager is informed.

Cells [3,3] and [3,4] illustrate the dynamic configuration aspects of the spreadsheet paradigm. A manager can enable or disable countHTTPAlarm and consequently allow or disallow the corresponding event to be reported.

# 7  CONCLUSIONS

In this paper, a spreadsheet paradigm for network management that uses a proxy architecture along with the motivation for choosing such a paradigm is presented. The components of the spreadsheet paradigm (including a proxy MIB (ssmib), spreadsheet language (SSL) and an event model) are described. The design of a proxy MIB to support the spreadsheet paradigm is covered. The important features of the language (SSL) are outlined followed by the architectural aspects of the event model. Examples that illustrate the essential features of the language have been provided.

A prototype implementation of the spreadsheet paradigm based in a proxy agent is currently in progress. The implementation includes a graphical user interface that can be used by a manager to construct, load and execute scripts in the proxy. Real world applications that can be implemented using the spreadsheet language are also being explored. Future research efforts will include a detailed investigation of the performance and security aspects of the paradigm. Currently, the event model does not support temporal operators and logical ordering of events. Such temporal aspects will improve the power of the event model. Although the history mechanism can be built using the polling and cell related features provided, the actual semantics of such a history mechanism need to be finalized.

In developing this paradigm, the following major goals have been realized: 1) the paradigm facilitates user customization of management information irrespective of the underlying information structure; 2) a distributed network management environment is created that allows the manager to successfully delegate routine management tasks.

The paradigm presented in this paper does not modify the existing management framework, but augments it. The authors hope that the spreadsheet paradigm will deliver a powerful, flexible platform for script based network management in an SNMP framework.

# REFERENCES

[AY95] K. Arai and Y. Yemini. MIB view language (MVL) for SNMP. In A.S. Sethi, Y. Raynaud, and F. Faure-Vincent, editors, *Integrated Network Management IV*, pages 454–465. Chapman and Hall, London, 1995.

[CFSD90] J. D. Case, M. S. Fedor, M. L. Schoffstall, and C. Davin. *Simple Network Management Protocol (RFC 1157)*, May 1990.

[CJS93] T. Clifford, G. Jajodia, and S. Snodgrass. *Temporal Databases: Theory, Design and Implementation*. Benjamin Cummings, Redwood City, CA, 1993.

[CMRW96] J. Case, K. McCloghrie, M. Rose, and S. Waldbusser. *Protocol Operations for Version 2 of the Simple Network Management Protocol (SNMPv2) (RFC 1905)*, January 1996.

[Gol95] Germán Goldszmidt. *Distributed Management by Delegation.* PhD thesis, Columbia University, New York, NY, 1995.

[GY95] Germán Goldszmidt and Yechiam Yemini. Distributed Management by Delegation. In *The 15th International Conference on Distributed Computing Systems*, Vancouver, Canada, June 1995.

[Has95] M. Hasan. An Active Temporal Model for Network Management Databases. In A.S. Sethi, Y. Raynaud, and F. Faure-Vincent, editors, *Integrated Network Management IV*, pages 524–535. Chapman and Hall, London, 1995.

[Hol89] D. Holden. Predictive Languages for Management. In *Integrated Network Management I*, pages 585–596. North Holland, Amsterdam, May 1989.

[KSS96] P. Kalyanasundaram, A.S. Sethi, and C. Sherwin. Design of A Spreadsheet Paradigm for Network Management. In *Proceedings of the 7th IFIP/IEEE Workshop on Distributed Systems: Operations and Management*, L'Aquila, Italy, October 1996.

[RM90] M. T. Rose and K. McCloghrie. *Structure and Identification of Management Information for TCP/IP based internets (RFC 1155)*, May 1990.

[RM91] M. T. Rose and K. McCloghrie. *Management Information Base for Network Management of TCP/IP based internets: MIB-II (RFC 1213)*, March 1991.

[SK94] A. S. Sethi and P. Kalyanasundaram. A Spreadsheet Paradigm for Network Management. In *Proceedings of the 5th IFIP/IEEE Workshop on Distributed Systems: Operations and Management*, Toulouse, France, October 1994.

[Wal95] S. Waldbusser. *Remote Network Monitoring Management Information Base (RFC 1757)*, February 1995.

[YGY91] Y. Yemini, G. Goldszmidt, and S. Yemini. Network Management by Delegation. In I. Krishnan and W. Zimmer, editors, *Integrated Network Management II*, pages 95–107. North Holland, Amsterdam, 1991.

**Pramod Kalyanasundaram** received his B.E. degree from the University of Madras, India in 1985 and his M.S. degree from the University of Delaware in 1992. He is currently working towards his Ph.D degree in Computer Science at the University of Delaware. His research interests include network management, operating systems, distributed systems and computer network protocols. He is also a member of IEEE.

**Adarshpal S. Sethi** received his MS and PhD degrees from the Indian Institute of Technology, Kanpur, India. He is currently Associate Professor in the Department of Computer & Information Sciences at the University of Delaware where he has been since 1983. In the past he has been a Visiting Scientist at IBM Research Laboratories, Zurich, Switzerland. His research interests are in network management protocols and architectures, congestion control and resource management, and protocol testing. Dr. Sethi was Co-Chair of the Program Committee for ISINM '95. He is a member of ACM and the IEEE Computer and Communication Societies.

**Chris Sherwin** is currently working toward his B.S. degree in Computer Science at the University of Delaware. His research interests include networks, operating systems and distributed systems.

**Dong Zhu** received his B.S. in Computer Science from Chongqing University, P.R.China in 1988, and M.S. in Computer Science from University of Delaware in 1996. He is currently working towards his PhD degree in Computer Science at the University of Delaware. His research interests include network management, network protocols, and formal specification of protocols. He is a member of ACM.

# KEYNOTES
# AND PANEL SESSIONS

## PART ONE

# Keynote Addresses

# 57

# The first anniversary of the Tivoli/IBM merger: the present and future of TME 10

*Franklin H. MOSS,* President/CEO of Tivoli
and General Manager of IBM's
Systems Management Division, U.S.A.

March of 1997 marks the first anniversary of one of the most unusual mergers in the history of the software industry. One year ago, tiny Tivoli Systems assumed responsibility for IBM?s world-wide systems management business - a billion dollar operation - under their own banner and architecture, TME. As a result of the merger, Tivoli is now at the center of an industry movement to provide IT organizations with open integrated solutions for managing complex network computing environments standing from the desktop to the mainframe. Today, the company's TME10 is used by thousands of companies worldwide to reduce the total cost of ownership and speed new applications into deployment - and keep them there.

Frank Moss will talk about what the merger has meant for customers as well as the management industry. He will also talk about Tivoli?s focus on the important arena of applications management, which is literally seen as the future of management in a network world.

# 58
# When societies are built from bits

*William H. DAVIDOW,* General Partner, Mohr,
Davidow Ventures, U.S.A.

Information age societies will look very different from our current ones. In the past what mattered most was geography, natural resources, and physical assets. In the future intangible assets will be king. The ability to process and move information will displace much of the physical infrastructure of society. Fewer office buildings, hotels, roads, and shopping malls will be needed. Businesses will be disintermediated, distribution channels will vanish, and the middle manstock brokers, retailers, distributors, travel agents, etc.will become increasingly less important. Of course the form and role of government will change as well. For example governments have gone to war to protect physical assets such as land that other countries wanted. In the future people will steal intellectual assets and governments will have no one to attack. Microsoft software, for example, is stolen by the citizens of China, not the government.

No one can predict the form that society will take in the future but it is possible to comprehend the forces that will be at work and the directions in which they will push society. Bill Davidow will describe some of the forces that will be at work and give some examples of how these forces might reshape societys social and physical infrastructure.

# 59

# The future of integrated network and systems management

*Yogesh GUPTA,* Senior Vice President, Computer Associates, U.S.A.

The issue of managing distributed systems and networks has recently received enormous attention. End-to-end management of the enterprise computing infrastucture has become the mantra of IT managers. With the growth of the Internet, management from anywhere is becoming a reality. However, the ultimate goal is to achieve self-managing and self-tuning networks and systems.

This talk will discuss the current state of the industry with respect to integrated network and systems management and what the future holds. Topics covered include the issues and challenges faced by network and systems managers, the evolution and impact of relevant technologies and standards, the state of the current crop of commercial products and direction in which they are evolving.

# 60
# Why bad things happen to good systems

*Lawrence BERNSTEIN,* Technologist, Price Waterhouse,
Chief Technologist, Network Programs, Inc.,
Executive Director, Bell Labs (ret.), U.S.A.

Network Management Systems are generally in the class of high performance on-line systems. With the TMN standard finally being taken seriously they are becoming part of the network elements themselves and becoming real-time systems. But they are still software systems.

Software systems suffer from a reputation of being buggy. Sometimes they are. In other cases, the customer buys a system and immediately insists on having it changed. This can and usually does introduce bugs.

Before customers buy a system, their interest is in features, price and schedule. Once they get it, their attention turns to reliability, response time and throughput. It is the job of the supplier to make sure that all six areas are considered at the time of purchase and managed throughout the entire system life cycle.

In one real life situation, a customer bought an 'off-the-shelf,' working product and refused to buy a critical module. After an initial test at the suppliers shop, the customer asked for 33 enhancements. One of these required a 20% change of the system. The developers were too optimistic about their ability to make the requested changes, and so the 'law of unintended consequences' frustrated the users and kept the developers working late into the night.

In this case, the customer lost patience, tossed out the network management system and all parties had their day in court - too many days. With good project and program management, common goals, patience and realistic expectations, this story could have had a happy ending.

## PART TWO

# Panel Sessions

# 61

# Distinguished experts panel: Integrated management in a virtual world

**Chair:**      *James HERMAN,* Vice President of Northeast Consulting Resources, Inc., U.S.A.

**Panelists:**    *Keith WILLETTS,* TCSI Corporation, U.S.A.
*Brent BILGER,* Cisco Systems, U.S.A.
*Timothy HINDS,* U.S. Environmental Protection Agency, U.S.A.
*Yogesh GUPTA,* Computer Associates, U.S.A.

The panel discussion addresses IM '97's theme, "Integrated Management in a Virtual World," reflecting our increasing interest in overall management solutions applied across all types of networks, enterprise connection systems, distributed computing systems and applications. It continues to move our attention beyond the purely technical to focus on real and virtual, comprehensive management solutions in a world increasingly filled with virtual corporations, virtual LANs, inter-enterprise networking, real and virtual service management, outsourcing, and electronic commerce.

Each of the panelists will address different aspects of the complexity of management in an increasingly virtual environment.

# 62
# Web based management

**Organizer:** *Subrata MAZUMDAR,* Bell Laboratories, U.S.A.

**Chair:**  *Bruce MURRIL,* Network Management Forum, U.S.A.

**Panelists:**  *Steve BOURNE,* Cisco Systems, Inc., U.S.A.
*Randy PRESUHN,* BMC Software, Inc., U.S.A.
*Ray WILLIAMS,* IBM, U.S.A.
*Subodh BAPAT,* Sun Microsystems, Inc., U.S.A.

With the introduction of WEB browser and java related technologies, network management platforms, applications and agent development are being re-engineered to fit this new framework. Network equipment and software vendors are rushing to adopt WEB related technologies for competitive advantage.

In this panel, network management experts will introduce and comment on proposed industry initiatives for web based manages: Java management API (JMAPI), and Web based enterprise management (WBEM). The panelist will discuss merits/demerits of various approaches for web based management.

# 63
# Challenges in managing intranets

**Chair:**     *Sigmund HANDELMAN,* IBM T.J. Watson
Research Center, U.S.A.

**Panelists:**  *Alan GEBELE,* Bellcore, U.S.A.
*Greg BOWMAN,* Tivoli Systems, Inc., U.S.A.
*Chad KEESLING,* Chrysler Corporation, U.S.A.
*Glenn GIANINO,* Computer Associates, U.S.A.

The Internet has won over in the corporate networks and intranets are becoming a vital support to their business. New paradigms of mangagement are emerging to extend to cover business processes and activities. Large multi-site corporations will need to manage clusters of intranets bridged through public networks, giving rise to new management challenges.

This panel will discuss management issues in this new environment and will combine points of view from Research, Commercial Applications and actual Industrial experience.

# 64
# Managing mobility

**Chair:**      *Nikos ANEROUSIS,* AT&T Research, U.S.A.

**Panelists:**  *Umesh AMIN,* AT&T Wireless, U.S.A.
                *Geoff MOSS,* Motorola, U.S.A.
                *Malathi VEERARAGHAVAN,* Lucent Technologies, U.S.A.

Mobile networks are expanding today at an outstanding rate both in terms ofsize (number of customers) and scope (areas of coverage). There are many types of mobile networks currently in service: cellular voice, satellite messaging, wireless personal communications, etc. In addition, a number of diverse technologies are used to provide these services, and more are to appear in the future. Managing mobile networks poses many interesting challenges primarily related to tracking user mobility and integration with the existing network management systems of wireline networks.

This panel attempts to highlight the network management issues facing mobile networks. First, we take a close look on how wireless networks can be managed, by examining the cellular voice network and the Iridium wireless personal communication service. We highlight the commonalities between managing different wireless services and the main differences with managing wireline networks, and examine how current areas of research in network management can help meet the management challenges of the wireless world. The discussion will attempt to highlight research areas that need to be explored further and identify the technologies (either standards-based or proprietary) that will play a key role in managing the wireless environment.

# 65
# New technologies bridging the gap between computer and telecommunications

**Chair:**    *Roberta S. COHEN,* Paradyne Corporation, U.S.A.

**Panelists:**    *William J. BARR,* Bellcore, U.S.A.
*Piergiorgio BOSCO,* CSELT, Italy
*TBD,* Hewlett Packard, U.K.
*TBD,* Sun Microsystems, U.S.A.

Computing and communications have long been the rivalrous siblings of the second half of this century. Each has promoted and cared for the other while quietly positioning itself for dominance. The ever increasing speed with which bits travel over the communications networks is constantly being challenged by the speed with which processors perform their instructions. Gigabit networks now service multi-megabit computers priced for individual, even recreational use.

Network and systems management has benefitted from this rivalry with Web-based network and systems management solutions the latest evolutionary twist. Devices and components were once the objects of interest for management systems; now the software and services operating on those piece parts are taking center stage.

This panel of distinguished experts addresses the effects that the convergence of computing and communications, or the continued competition of the two, have had on integrated management of networks and systems. The panel is drawn from the international telecommunications industry (CSELT - Italy and Bellcore) and from the international computing industry (Hewlett Packard - U.K. and Sun).

# 66
# Open network control

**Chair:**     *Aurel A. LAZAR,* Columbia University, U.S.A.

**Panelists:**     *Jit BISWAS,* Institute of Systems Science, Singapore
*Ian LESLIE,* University of Cambridge, UK
*Dimitrios PENDARAKIS,* IBM T.J. Watson Research
Center, U.S.A.
*Rolf STADLER*, Columbia University, U.S.A.

The vertical integration of ATM switches has stymied the deployment of innovative signaling and service technologies on ATM networks. This panel will discuss current proposals for realizing open signaling and service creation platforms that are inspired by established models of network management.

# POSTERS

# 67

# Implementation of Duplicate MD (Proxy-Agent) with distributed functions and high reliability

*YUN Su-Hun, Yoshiaki Shoji, Yoritaka Ohta*
*Second transmission division, NEC Corporation*
*1753 Shimonumabe Nakahara-ku Kawasaki 211 JAPAN*
*Tel: +81-44-435-5534,   Fax: +81-44-435-5661*
*E-mail: yun@std.trd.tmg.nec.co.jp, shoji@std.trd.tmg.nec.co.jp,*
*yohta@std.trd.tmg.nec.co.jp*

## Abstract

Implementation of MD which provides the mediation function of an interface is a key to TMN-based management of the non-Q3 interface NE. We have implemented MD which has the mediation function of the Q3 interface in a conventional agent to Proxy-Agent. We have also duplicated Proxy-Agent to improve the reliability of the network management system and have succeeded the performance improvement by distributing the functions. This paper, first of all, posits the management of agent and MIB by defining them as the managed objects, which is a part of the system management function and is at the same level management as configuration management function for network and NE. Secondly, this paper presents the classification of the Proxy-Agent functions while considering relationships between a CMIP operation from Manager and access patterns of Proxy-Agent to NE and MIB. Thirdly, this paper proposes the performance improvement of the agent in the case of mixed CMIP operations in the real operational case by breaking down the agent functions to "concentrated functions type" and "distributed functions type," which is defined in relation to the access patterns of the agent, in a duplicate Proxy-Agent system. Finally, in this paper, to keep synchronization of MIB, we propose updating the duplicate MIB by two-step CMIP operations.

## Key Words

CMIP agent, Proxy-Agent, MD (Mediation Device), duplicate system, distributed functions, fault tolerance, high reliability and performance

# An Open Distributed VPN Management System for a Multi-Domain Management World

*H. Khayat, L. H. Bjerring, R. S. Lund*
*L. M. Ericsson A/S, Sluseholmen 8, DK-1790 Copenhagen,*
*Denmark*

### Abstract
Virtual Private Network (VPN) services aim at providing users a single management interface to a heterogeneous global network infrastructure. Various interoperability problems have to be overcome in order to implement this service in a way that it appears to be simple and easy to use. These interoperability problems occur because the global network infrastructure has many owners (expressed as administrative heterogeneity), is constructed by many network technologies (expressed as technological heterogeneity), and because different portions of the network support different management services and modelling representations (expressed as service heterogeneity).

In the ACTS project Prospect a VPN architecture has been defined for managing such a network infrastructure. The base principles underlying the architecture are: (i) to keep things simple, and (ii) to clearly separate between service core and service adaptation functions. This leads to a layered management architecture where the upper layer is responsible for implementing the service logic (mainly concerned with domain level routing), the lowest layer is concerned with adapting heterogeneous management technologies (SNMP, TMN, CORBA) to the technology chosen for implementing the VPN (CORBA), and the middle layer is responsible for adapting the management models associated with each network technology domain to a common, technology-independent abstraction useful for the VPN service logic. The abstraction follows principle (i) of being simple, just as the model applied at the user interface is simple.

A VPN service has been implemented to provide end-to-end connection management support over a heterogeneous pan-European network, providing support for an integrated tele-eduaction service developed by Prospect. The network consists of a set of customer premises networks (ethernet and ATM LAN technologies) managed by SNMP-based management systems, interconnected by a public ATM network managed by a TMN-based management system.

### Keywords
Virtual Private Network, Inter-Domain Management, End-to-End Management

# 69
# SNMP-based Network Security Management

*P. C. Hyland*
*TASC, Inc.*
*12100 Sunset Hills Road, Reston, VA 22090, USA, (703) 834-5000,*
*FAX: (703) 318-7900, Email: pchyland@tasc.com*

*R. Sandhu*
*George Mason University*
*4400 University Drive, Fairfax, VA 22030, USA, (703) 993-1659,*
*FAX: (703) 993-1638, Email: sandhu@gmu.edu*

## Abstract
Security policy and security techniques have long been major research topics, but little work has been reported on management of distributed security applications other than Intrusion Detection Systems (IDS). We believe management of secure applications will see dramatic growth as adequate tools become available. This paper reviews security management issues and suggests management objects for use by Simple Network Management Protocol (SNMP) tools to collect, log, and display events and alert operators to possible trouble. We focus on the issues of adapting the necessary status and control mechanisms (management infrastructure and agents) to accommodate security management needs. To make a complete security Management Information Base (MIB), additional MIB modules must be defined for firewalls (packet filtering and proxy servers), security audit trails and IDS management parameters. Security firewalls and IDS applications are assessed for management via SNMP and are proposed as future case studies. The Packet Filter Information Protocol (PFIP) is presented as a method to allow propagation of packet filter information among compatible firewall hosts and routers in an IP-based network. We use a scenario of corporate firewalls to assess concepts of correctness, sufficiency and completeness. The properties of SNMP, version 2 that support secure management operations are reviewed and we propose further work toward an SNMP-based Security Management prototype.

## Keywords
Firewalls, Management Information Base (MIB), Network Management, SNMP, Packet Filter Information Protocol (PFIP), Security Application Management, Security Management.

# Managing personal communication systems in a multi-domain environment

*A. Richter, M. Tschichholz*
*GMD FOKUS, Hardenbergplatz 2, 10623 Berlin, Germany*
*Phone: +49-30-25499-200 • Fax: +49-30-25499-202*
*E-Mail: {richter, tschichholz}@fokus.gmd.de*

**Abstract**

The ACTS project PROSPECT is investigating the integrated management of a variety of services in a multi-domain tele-education environment. In the project's second trial the focus is on support for mobile students and teachers, enabling them to access and use advanced multimedia services from locations other than their home base. A Personal Communication System (PCS) is being defined that can support mobile user requirements in such a tele-education environment.

Support for mobility has several implications for management. The management systems of multiple organisations need to co-operate to support the provision of tele-educational services to mobile students and teachers with the particular features required by the users. Specific PCS management issues are concerned with user profile management and management of user access to their customised working environment. The question of how to account and charge for resource usage when services can be used in domains other than the user's home domain must be considered, as must the requirements for security when crossing multiple domains. The PCS-specific management functions fall into three different categories. First, providing locality-independent access to a service, second, customising access and use of services from different locations for individual users, and third dealing with charges caused by service usage from external domains. Many of the current architectural concepts available today are not sufficient for modelling the management of personal communication systems in a multi-domain environment. The PCS is therefore enhancing these concepts to provide for the management of student and teacher mobility in the PROSPECT context.

# A Layered Architecture for Capacity Planning in Hybrid Networks

*U. Datta and L. Lewis*
*Cabletron Systems, Inc.*
*486 Amherst Street, Nashua, NH 03063 USA*
*Phone 603-337-5737, Fax 603-337-5615, Email udatta@ctron.com*

**Abstract**

An important challenge of network intelligence is capacity evaluation and planning (CEP). CEP is defined typically as follows:

1. monitoring current network characteristics;
2. understanding environmental constraints/considerations;
3. forecasting future needs and technology;
4. evaluating technical opportunity;
5. creating most appropriate, consistent, and coordinated plans on a long, medium, and short term basis;
6. modifying plans based on results of actual implementations in order to provide ongoing cost effective and timely communication services to the users.

Here we propose a framework for intelligent CEP. We cast the CEP task into a multi-layer architecture, where each layer subsumes the prior layers. In this respect we borrow from the *subsumption architecture* in the robotics community. Importantly, if a Layer N task fails there is no interference with the lower-layer tasks. Our layers correspond roughly to (i) current capacity evaluation, (ii) supervised what-if scenarios and capacity evaluation thereof, (iii) unsupervised what-if scenarios, and (iv) automated capacity control.

A goal of our work is to accommodate the evolution from packet-based, to hybrid, to cell-based ATM networks. In addition, our approach is link-centered as opposed to device-centered, where an evaluation of link capacity is relative to a link's work compared to its contribution to traffic throughput. An evaluation of overall network capacity is a function of evaluations of all link capacities.

**Keywords**

Capacity Planning, Performance Management, Change Management

# Network management agents supported by a Java environment

G.W. R. Luderer, H.Ku, B.Subbiah and A.Narayanan
*Arizona State University, Tempe, AZ, USA.*
*{ E-mail: luderer@asu.edu }*

In this paper we investigate how a network management architecture might take advantage of the Java environment. Network management is by definition a distributed application. It should therefore be no surprise that a new language environment like Java that is purposely developed for the network environment should find a natural match in this application. Problems such as portability across platforms or independence from underlying hardware and software, security aspects such as authentication, enforcement of access restrictions, and preservation of the integrity of the communication process can be addressed easily in a Java environment. The navigator/browser architecture offers an ideal framework for the network management system applications.

A Java-based network management system contains management applications based on the browser model and agents in the network elements running in respective local Java environments as agent processes. The manager/browsers monitor and control the network elements. The agent process at a network element performs the network management agent functions, including interaction with the manager. The communication between the manager and the agent is at present carried out by Java classes transported across a network in bytecode format. This approach avoids the current use of ASN.1 Basic Encoding Rules in the SNMP environment. The Management Information Base-II (MIB) standard is supported at the agent. The security features inherent in the Java architecture provide an additional layer of security in heterogeneous environments.

The manager browser uses a Graphical User Interface, which contains different panels for various management functions. The user selects an agent and corresponding MIB variables for monitoring from the list of agents discovered by the manager browser. The manager encodes this information into a Java class and transmits it to the agent. The result of the query is displayed upon receiving the response from the agent. The main function of an agent is to respond to queries from a manager, parse the MIB tree and report an event to the manager by a trap. The agent is executed as a background process with a main thread listening to incoming requests and creating respective worker threads. The agent interacts with the MIB using the existing ASN.1 standard.

A system for experimentation has been created to demonstrate feasibility and carry out performance measurements. The performance of a Java agent and an SNMP agent based on a system from CMU have been compared, specifically the response times of the SNMP *get* command for *system* and *ip* groups. This response time depends primarily on the hardware platform on which the agent is running. Measurements were performed for a manager running on a Sun Microsystems SparcStation 5 under Solaris and Java agents and SNMP agents. As is to be expected, compared to the "plain" SNMP implementation, the Java-based implementations show increases in response time of 132% (Solaris agent) and 241% (Windows95 agent) when doing *system group get* commands; for *ip group* commands the respective values are 102% (Solaris) and 164% (Windows95). The current implementation shows feasibility, but for performance reasons a mechanism such as remote method invocation or transfer of object states is needed.

In conclusion, use of the Java language and environment for network management. is feasible and has inherent advantages. However, further evolution of the Java environment is necesssary before these advantages make this approach practical.

# Integrating SNMP and CMIP alarm processing in a TMN management environment

*F. Munoz-Mansilla*
*Professor at the University of Alcala de Henares (Spain)*
*J.Sanchez.University of La Coruna (Spain).Nowadays at UCLA.*
*V. Carneiro. Professor at the University of La Coruna (Spain).*
*J. Coego. University of La Coruna (Spain).*
*C/ Almirante Francisco Moreno, 5 bajo. 28040 Madrid. Spain.*
*Ph.:+34 1 4590002 Email:{fmm,jagrelo,victor,javier}@cesat.es*

## Abstract

In a TMN management environment where SNMP and CMIP alarms coexist, CMIP notifications comply with TMN standards and provide complete information, whereas SNMP traps provide much less information, deviating from TMN standards. In order to handle this situation and make an homogeneous processing of alarms we propose an adaptor in charge of receiving any type of alarm and forwarding it in the same format (recommendation X.721 for the alarm types defined in X.733) independently of its original protocol. Besides this purpose, alarm contents are enhanced using additional information maintained in a repository.

The adaptor allows to model the required additional information. When a CMIP alarm is received the adaptor completes the information it carries. In the case of a SNMP alarm, the adaptor would integrate the trap contents in the TMN architecture (mediation device function, MDF). We include other TMN processes such as historic reporting of management data or alarm filtering and correlation.

An experimental prototype has been implemented using a management platform which allows shared access to information and distributed alarm management.

## Keywords

Network management, alarm processing, SNMP, CMIP, TMN.

# 74

# Distributed Service and Network Management Using Intelligent Filters

C. Cicalese, (703) 883-1914, (703) 883-5862 (fax), cindy@mitre.org
J. DeCarlo, (703) 883-7116, (703) 883-5862 (fax), jdecarlo@mitre.org
M. Kahn, (703) 883-7356, (703) 883-5862 (fax), mkahn@mitre.org
E. D. Zeisler, (703) 883-5768, (703) 883-5862 (fax), ezeisler@mitre.org
The MITRE Corporation
1820 Dolley Madison Blvd., Mail Stop W658
McLean, VA 22102-3481 USA

H. Folts, (301) 614-5229, (301) 614-0267 (fax), folts@eos.nasa.gov
NASA Goddard Space Flight Center
Building 32, Room N230B
Greenbelt, MD 20771 USA

## Abstract

This work addresses two capabilities important to the management of distributed services. First, in order to be able to determine the root cause of a service-affecting fault, network management information must be correlated with service management information. Second, since distributed services often span many management domains, it must be possible to communicate trouble information securely from one management domain to another.

A prototype was developed to demonstrate how these ideas could be applied to managing NASA's Earth Observing System Data and Information System (EOSDIS). The prototype integrated commercial off-the-shelf products such as Hewlett-Packard's OpenView and Remedy's Action Request System with the Open Group's Distributed Computing Environment (DCE). The event correlation component was constructed using the C Language Integrated Production System (CLIPS). The service manager collected management information from the simulated service and network elements by a combination of polling and sending. Upon determination of the root cause of a fault, a trouble ticket could be automatically generated and, if necessary, could be forwarded securely to the appropriate management domain using an authenticated DCE remote procedure call.

# 75

# TINA Service Management Principle

## - In Search of A Paradigm Beyond TMN -

*H. Kamata*          *T. Hamada*

*OKI Electric*          *Fujitsu Ltd.*

*TINA-C c/o Bellcore, NVC-1C234, 331 Newman Springs Rd., Red Bank, NJ 07701, U.S.A., ph: +1 908 758 2715, fax: +1 908 758 2865, e-mail: hamada@tinac.com*

*Keywords:* TMN, Management Context, Service Transaction

TMN [1][2] takes a resource-oriented, bottom-up approach, upon which the service and the business aspects come last. TINA service management [3] takes the opposite direction towards TMN. TINA service management is a mapping of FCAPS management requirements onto its resource layer.

The TINA service layer passes its management requirements in the form of *Management Contexts (MgmtCtxt),* which are carried by a construct called *Service Transaction.* MgmtCtxts are subsequently interpreted, and the results are then mapped onto the TINA resource layer. Although the TINA resource layer can be built on a monolithic object-oriented framework known as DPE, it is still fully interoperable with TMN, allowing itself to interact with Managed Objects (MO) in various layers of TMN. TINA service is session-oriented, and a set of FCAPS MgmtCtxts are bound to each service session. With the service transaction carrying MgmtCtxts, a consistent service management throughout a distributed, heterogenous environment is guaranteed. Execution of a service-transaction (ex. accounting management) consists of the following three phases.

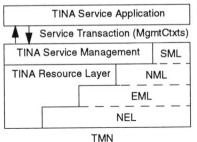

TMN

1. *Set-up Phase*: MgmtCtxts are interpreted, and necessary resources such as accounting-log record, event-channels are reserved
2. *Execution Phase*: Being allowed to access authorized data/information sources, the TINA service session is executed. Following the Accounting MgmtCtxt, accounting events may be logged or reported.
3. *Wrap-up Phase*: Reports from performance management or fault management can be summarized. If the QoS level was satisfactory, the service transaction concludes. Otherwise recovery actions such as charging compensation may be taken.

## References

[1] Principles for a Telecommunication Management Network, ITU-T Recommendation M.3010, 1992.

[2] Generic Requirements for Operations Based on the Telecommunications Management Network (TMN) Architecture, Bellcore, GR-2869-CORE, Oct. 1995.

[3] Service Architecture Ver.4.0, TINA-C, Dec. 1996.

# INDEX OF CONTRIBUTORS

# KEYWORD INDEX